DATE DUE

MAY - 2 2007

Control Theory for Partial Differential Equations:
Continuous and Approximation Theories

This is the second volume of a comprehensive and up-to-date three-volume treatment of quadratic optimal control theory for partial differential equations over a finite or infinite time horizon and related differential (integral) and algebraic Riccati equations. Both continuous theory and numerical approximation theory are included. An abstract space, operator theoretic treatment is provided, which is based on semigroup methods, and which is unifying across a few basic classes of evolution. A key feature of this treatise is the wealth of concrete multi-dimensional PDE illustrations, which fit naturally into the abstract theory, with no artificial assumptions imposed, at both the continuous and numerical level.

Throughout these volumes, emphasis is placed on unbounded control operators or on unbounded observation operators as they arise in the context of various abstract frameworks that are motivated by partial differential equations with boundary/point control. Relevant classes of PDEs include: parabolic or parabolic-like equations, hyperbolic and Petrowski-type equations (such as plate equations and the Schrödinger equation), and hybrid systems of coupled PDEs of the type that arise in modern thermo-elastic and smart material applications. Purely PDE dynamical properties are critical in motivating the various abstract settings and in applying the corresponding theories to concrete PDEs arising in mathematical physics and in other recent technological applications.

Volume II, after an introductory chapter that collects relevant abstract settings and properties of hyperbolic-like dynamics, is focused on the optimal control problem over a finite time interval for such dynamical systems. A few abstract models are considered, each motivated by a particular canonical hyperbolic dynamics. Virtually all the regularity theory needed in the illustrations is provided in detail, including second-order hyperbolic equations with Dirichlet boundary controls, plate equations (hyperbolic and not) and the Schrödinger equation under a variety of boundary controls or point controls, and structural acoustic models that couple two hyperbolic equations.

Volume I covers the abstract parabolic theory for both the finite and infinite horizon optimal control problems, as well as the corresponding min–max theory, with PDE illustrations. Recently discovered, critical dynamical properties are provided in detail, many of which appear here in print for the first time.

Volume III is in preparation.

Irena Lasiecka is Professor of Mathematics at the University of Virginia, Charlottesville. She has held positions at the Control Theory Institute of the Polish Academy of Sciences, the University of California, Los Angeles, and the University of Florida, Gainesville. She has authored or coauthored over 150 research papers and one other book in the area of linear and nonlinear PDEs. She serves on the editorial boards of *Applied Mathematics and Optimization*, *Journal of Mathematical Analysis and Applications*, and the *IEEE Transactions on Automatic Control*, among others, and she holds, or has held, numerous offices in the professional societies SIAM, IFIP, and the AMS.

Roberto Triggiani is Professor of Mathematics at the University of Virginia at Charlottesville. He has held regular and visiting academic positions at several institutions in the United States and Europe, including Iowa State University, Ames, and the University of Florida, Gainesville. He has authored or coauthored over 140 research papers and one other book in PDEs and their control theoretic properties. He currently serves on the editorial boards of *Applied Mathematics and Optimization*, *Abstract and Applied Analysis, Systems and Control Letters*.

EDITED BY G.-C. ROTA
Editorial Board
B. Doran, M. Ismail, T.-Y. Lam, E. Lutwak
Volume 75
Control Theory for Partial Differential Equations II

To Maria Ugenti, Janina Krzeminska, Antoni Lech,
and Giuseppe Triggiani

ENCYCLOPEDIA OF MATHEMATICS AND ITS APPLICATIONS

Control Theory for Partial Differential Equations: Continuous and Approximation Theories

II: Abstract Hyperbolic-like Systems over a Finite Time Horizon

IRENA LASIECKA   ROBERTO TRIGGIANI

CAMBRIDGE UNIVERSITY PRESS

PUBLISHED BY THE PRESS SYNDICATE OF THE UNIVERSITY OF CAMBRIDGE
The Pitt Building, Trumpington Street, Cambridge, United Kingdom

CAMBRIDGE UNIVERSITY PRESS
The Edinburgh Building, Cambridge CB2 2RU, UK   http:// www.cup.cam.ac.uk
40 West 20th Street, New York, NY 10011-4211, USA   http:// www.cup.org
10 Stamford Road, Oakleigh, Melbourne 3166, Australia
Ruiz de Alarcón 13, 28014 Madrid, Spain

First published 2000

Printed in the United States of America

Typeface Times Roman 10/13   *System* LATEX 2_ε [TB]

A catalog record for this book is available from the British Library.

Library of Congress Cataloging-in-Publication Data
Lasiecka, I. (Irena), 1948–
Control theory for partial differential equations : continuous and
approximation theories / Irena Lasiecka, Roberto Triggiani.
p.   cm. – (Encyclopedia of mathematics and its applications : v. 74–75)
Contents: I. Abstract parabolic systems – II. Abstract hyperbolic-like
systems over a finite time horizon
ISBN 0-521-43408-4 (v. 1). – ISBN 0-521-58401-9 (v. 2)
1. Differential equations, Partial.   2. Control theory.
I. Triggiani, R. (Roberto), 1942– .   II. Title.   III. Series.
QA377.L37   1999   99-11617
515′.353 – dc21   CIP

ISBN 0 521 58401 9 hardback

Contents

Contents of Volume I

Preface

This three-volume treatise presents, in a unified framework, a comprehensive, in-depth, and up-to-date treatment of quadratic optimal control theory for (linear) partial differential equations (PDEs) over a finite or infinite time horizon and related differential (integral) and algebraic Riccati equations. Both continuous theory and numerical approximation theory are included. An abstract space, operator theoretic treatment is provided, which is based on semigroup methods, and which is unifying across a few basic classes of evolution.

While addressing all three volumes regarding the basic, broad-range theme covered and the philosophy of approach followed, this preface focuses mostly on Volumes I and II for specific details. Indeed, driven also by recent, new PDE models such as they arise in modern technological applications, the treatment of this work has grown far beyond the original intentions and the anticipated plan. As a result, two volumes now appear in print, with a third one in preparation. A justification for the criteria that have dictated the selection of a natural subdivision of the entire work into three volumes is given below.

This treatise is a much expanded outgrowth, at least in the ratio 1 to 10, of the authors' Springer-Verlag Lectures Notes in Control and Information Sciences, Volume 164, entitled: *Differential and Algebraic Riccati Equations with Applications to Boundary-Point Control Problems: Continuous and Approximation Theory.* TheseLecture Notes, published in 1991, contained a comprehensive account of the theories that were available at that time, along with an array of numerous illustrative PDE applications with boundary/point control. However, most technical proofs were referred to the literature. A completion of these Lectures Notes was therefore called for, which inevitably stimulated an extension of their range of coverage with the addition of both new theoretical topics, as well as new PDE models and applications of modern technological origin. These, in turn, required further still theoretical analysis.

The basic dynamics is an abstract equation $\dot{y} = Ay + Bu$, where A (free dynamic operator) is the generator of a strongly continuous (s.c.) semigroup on the Hilbert (state) space Y, and where B (control operator) is an unbounded operator with a degree of unboundedness up to the degree of unboundedness of A. Moreover, u is the control function, which runs over the class of L_2-functions in time, with values in a Hilbert space U. All the boundary/point control problems for PDEs can be modeled by this abstract equation, for specific choices of the operators A, B and of the spaces U, Y. The dynamics is further penalized by a (quadratic) functional cost, containing an observation operator R, to be minimized over a preassigned finite or infinite time horizon. The theory of this problem culminates with the analysis of the corresponding differential or algebraic Riccati (operator) equations, which arise in the (pointwise) feedback synthesis of the optimal solution pair $\{u^0, y^0\}$. This problem, which originated in the late 1950s in the context of ordinary differential equations (with A, B, R matrices of appropriate size) has long been considered a truly central issue – a "battle-horse" – in deterministic optimal control theory, and related stochastic filtering theory, of dynamical systems. In the finite dimensional context, the solution in pointwise feedback form, via Riccati equations, of both the deterministic and the stochastic versions of this problem, has been known since the 1960s, through the work of Kalman and Kalman-Bucy, respectively.

These volumes present the far-reaching, technical extension of the deterministic problem, aimed at accommodating and encompassing multidimensional PDEs with boundary/point control and/or observation, in a natural way. Thus, throughout this work, emphasis is placed on unbounded control operators and/or, possibly, on unbounded observation operators as well, as they arise in the context of various abstract frameworks that are motivated by, and ultimately directed to, PDEs with boundary/point control and observation. A key feature of the entire treatise is then a wealth throughout of concrete, multidimensional PDE illustrations, which naturally fit into the abstract theory, with no artificial assumptions imposed, at both the continuous and numerical level. Justification of the abstract models adopted rests, unequivocally, with their intrinsic ability of capturing the characterizing dynamical properties of specific, relevant classes of PDEs, which motivate them in the first place. Regarding abstract modeling, the flow runs unmistakenly from an understanding of the concrete into the proper abstract.

Naturally, to extract best possible results and tune the technical tools to the problem at hand, it is necessary to distinguish at the outset between different types of PDE classes: primarily, parabolic-like dynamics versus hyperbolic-like dynamics, with further subdistinctions in the latter class. This is due to well-known, intrinsically different dynamical properties between these two classes. As a consequence, they lead to two drastically different basic abstract models, whose defining, characterizing features set them apart. Accordingly, these two abstract models need, therefore, to be investigated by correspondingly different technical strategies and tools. As a consequence, different types of distinctive results are achieved to characterize the two classes. All this dictates that the abstract theory needs to bifurcate at the very outset into a parabolic-like model and hyperbolic-like basic models; moreover, in the latter class, a further distinction into finite and infinite time horizon is called for, to account for different, critical properties between these two cases.

Thus, Volume I contains the optimal control theory for the parabolic-like class, over both the finite and the infinite time horizon, where the s.c. semigroup generated by A is, moreover, analytic; while Volumes II and III refer to the optimal control theory for the hyperbolic-like class over a finite or, respectively, infinite time horizon. This includes hyperbolic dynamics as well as Petrowski-type PDEs such as platelike models, Schrödinger equation, etc.

As already emphasized, purely PDE dynamical properties are critical in motivating the various abstract settings, as well as in applying the corresponding theories to concrete PDEs arising in mathematical physics and in other technological endeavors. This is particularly true in the case of hyperbolic-like dynamics. Unlike the parabolic-like class, which offers a certain degree of flexibility in the choice of the abstract space setting (subject to established parabolic regularity theory), by contrast, the framework in the case of hyperbolic-like dynamics is far more rigid. It requires a preliminary knowledge of the space of optimal regularity theory – apurely PDEs problem – and thus leaves no choice. Moreover, regarding the infinite time optimal control problem, the most complete theory is achieved in the cases (which occur most often, but by no means always) where the space of optimal regularity of the solution under L_2-control coincides with the space of exact controllability (or of uniform stabilization) – in other words, where the map from the class of admissible L_2-controls to the state space is surjective at some finite time. In short: In the hyperbolic-like case, optimal regularity theory is an intrinsic, critical, essential prerequisite factor in the analysis of the corresponding optimal control problem, which rigidly depends upon it (while a margin of latitude exists in the parabolic-like case, once parabolicity has been established). Accordingly, optimal regularity theory of many hyperbolic-like dynamical equations considered in the illustrations is an intrinsic part of the present volumes. A more detailed description is given below in the synopsis of Volume II. The inclusion, on the one hand, of this massive regularity theory and, on the other hand, of new PDE dynamics such as thermo-elastic plate equations and various models of coupled PDEs arising in structural acoustics, helps explain the explosion of this subject matter into three volumes.

Throughout this work, special emphasis is paid to the following topics:

(i) Abstract operator models for boundary/point control and observation problems for PDEs.

(ii) Identification of the space of optimal regularity of the solutions, typically under L_2-controls in time, and particularly for the class of hyperbolic and Petrowski-type systems or coupled PDEs problems; it is with respect to the norm of this space that the solution is then penalized in the cost functional.

(iii) Identification of the regularity properties of the optimal pair of the optimal control problem, particularly, in the parabolic-like case over a finite or infinite interval, and in the hyperbolic-like case over a finite interval. In the hyperbolic-like case over an infinite time horizon, the optimal pair need not be better than the original L_2 regularity in time, inherited from the optimization problem.

(iv) Verification of what we call the "finite cost condition" (F.C.C.) in the infinite time horizon problem and related algebraic Riccati equations, which guarantees the existence of at least one admissible control yielding a finite cost functional. In the case of parabolic-like dynamics, the F.C.C. is most readily verified via uniform feedback stabilization, as the unstable space of the dynamics is, at most, finite dimensional. By contrast, in the case of hyperbolic-like dynamics, the F.C.C. is verified via a study of the related exact controllability problem, or of the related (generally, more challenging) uniform stabilization problem, by means of an explicit, dissipative, boundary, velocity feedback operator. Exact controllability/uniform stabilization of hyperbolic-like dynamics is a topic in its own right, intimately connected with, yet distinct from, the main thrust of the optimal control problem of the present volumes. A vast literature exists, including treatments in book form. We shall return to these topics in Volume III.

(v) Constructive variational approach to the issue of existence of a solution (Riccati operator), and possibly uniqueness, of a corresponding differential or algebraic Riccati operator equation.

(vi) Development of numerical algorithms that reproduce numerically the key properties of the continuous problems. This can be done directly in the parabolic-like case. By contrast, the hyperbolic-like (conservative) case requires that a regularization procedure be performed first, before passing to the approximation analysis.

A brief description of the contents of the first two volumes follows.

Volume I focuses on abstract parabolic systems (continuous and approximation theory), where the s.c. semigroup of the free dynamics is, moreover, analytic. Save perhaps for some possible refinements, the overall theory in this chapter and companion notes is essentially optimal. This includes both the finite (Chapter 1) and infinite horizon (Chapter 2) optimal control problems, as well as the corresponding min–max theory with nondefinite quadratic cost (Chapter 6). Here, both control operator and disturbance operator are of the same "maximal" degree of unboundedness allowed with respect to the free dynamics operator. A lengthy Chapter 3 presents many multidimensional PDE illustrations with boundary/point control and observation. They include not only traditional, classical parabolic equations such as the heat equation with Dirichlet- or Neumann-boundary control, or point control, but also second-order equations with "structural" or "high" damping, as well as thermo-elastic plate equations with no rotational inertia term. For the latter two classes, recently discovered, critical dynamical properties are proved in details. These include "parabolicity" (analyticity of the corresponding semigroup) and uniform stability. Various appendices in Chapter 3, taken cumulatively, provide a self-contained subvolume focused on thermo-elastic parabolic plate equations, whose theory has become available only over the past year or so. Chapter 4 provides a detailed numerical approximation treatment, with appropriate convergence properties (possibly, with rates of convergence) of all the quantities of interest: optimal control, optimal solution, Riccati operator, gain operator, optimal cost, etc. Finally, Chapter 5 provides detailed PDE illustrations of numerical schemes that fit into the theory of Chapter 4. Regarding the theoretical treatment, the analysis in Volume I is almost exclusively operator-theoretic and is based on singular integrals as they arise in the description of the control-solution (state) map, by virtue of the key property of analyticity of the free dynamics semi-group (generated by the operator A). As it turns out, analyticity of the free dynamics compensates, in this case, for the unboundedness of the control operator or of the disturbance operator. Indeed, such analyticity yields a controlled smoothing of the control-solution map and of its adjoint. Once applied to the optimality conditions characterizing the optimal pair, such double smoothing snowballs into a bootstrap argument, which eventually leads to higher regularity of the optimal pair (over the initial regularity inherited from the optimization problem) and – finally – to a smoothing property of the Riccati operator. As a consequence, the gain operator is bounded from the state to the control space, a distinctive, critical property of the parabolic-like class. In applications to concrete PDEs, elliptic theory and identification of domains of appropriate fractional powers with Sobolev spaces play a critical role.

Volume II considers the optimal control theory for hyperbolic or Petrowski-type PDEs over a finite time horizon. It begins with an introductory chapter (Chapter 7) that collects relevant abstract settings and abstract properties of these dynamics that are to be used in subsequent chapters. It then considers three different abstract frameworks. The abstract model of Chapter 8 is motivated by the optimal control problem for second-order hyperbolic equations with Neumann-boundary control and Dirichlet-trace observation. The abstract model of Chapter 9 is motivated by wave and Kirchoff elastic plate equations, under the action of point control. It also includes two models of coupled PDE systems, such as they arise in noise reduction problems in structural acoustics. Both systems are subject to point control, which models the action of smart material technology. One couples the wave equation for the pressure in the acoustic chamber with a Kirchoff equation for the elastic displacement of the moving wall. It is an example of hyperbolic/hyperbolic coupling. Instead, in the second system, the elastic wall is modeled by an Euler–Bernoulli equation with structural damping, thus giving rise to a hyperbolic/parabolic coupling. Finally, the abstract model of Chapter 10, which further builds on that of Chapter 9, looks at first artificial and complicated. Actually, it is a natural framework, which simply extracts the correct settings for problems such as second-order hyperbolic equations with Dirichlet-boundary control, numerous other plate equations with a variety of boundary control, as well as the Schrödinger equation with Dirichlet-boundary control. All the relevant regularity theory, some of which is new, of these dynamical PDEs is provided in detail, subject to the exclusions noted below. Indeed, in contrast with parabolic theory, the regularity theory of hyperbolic and Petrowski-type equations (such as plate equations and Schrödinger equations) demand a broader array of purely PDE techniques to obtain sharp/optimal interior- and trace-regularity properties. They include energy methods, or multipliers methods, at the differential level or pseudo-differential/microlocal analysis level, which were discovered much more recently than parabolic techniques. This contrast between the two basic classes of dynamical systems – parabolic-like versus hyperbolic- or Petrowski-type equations – was already emphasized in the preface to the authors' Lectures Notes. Accordingly, Volume II contains in detail most of the needed regularity theory (both interior and trace regularity) of the many hyperbolic-like PDE systems here considered. Exceptions include the more recent regularity theory of first-order hyperbolic systems and of second-order hyperbolic equations with Neumann boundary datum, which require a treatment based on the technical apparatus of pseudo-differential operators and microlocal analysis. For these, appropriate references to the recent literature are given.

As already noted, Volume III (in preparation) will cover optimal control problems for hyperbolic-like dynamics (both continuous and numerical approximation theory) and for coupled PDE systems, over an infinite time horizon.

Further information on this treatise in the context of available books is contained in the introductory section of Chapter 0.

Acknowledgments for the First Two Volumes

Though almost all of the results presented in the first two volumes are taken from the authors' original research work in both optimal control theory and PDE theory, there are a numbers of friends whom we wish to thank very warmly, and to whom we wish to express our gratitude. In chronological order, first A.V. Balakrishnan, who, with his original work of the mid-1970s, provided the initial spur into abstract modeling for parabolic equations with Dirichlet boundary control, and who soon thereafter graciously introduced us to it. Next, we wish to thank some of our coauthors, from whom and with whom we have learnt about this fascinating subject, for joint work reported in these first two volumes: G. Da Prato on the abstract differential Riccati equation in the hyperbolic case, via a direct approach, and J. L. Lions on the regularity of second-order hyperbolic equations with Dirichlet-boundary datum. Our frequent visits over the years to the mathematically rewarding environment of the Scuola Normale Superiore di Pisa, Italy, and the stimulating exchange of correspondence with J. L. Lions on a priori inequalities of wave equations are also greatly appreciated. Special thanks for reading portions of the manuscript and offering their comments are due to P. Acquistapace (Chapter 1), A. Ichikawa (Chapter 6), S. K. Chang (Chapter 10) and, particularly, to L. Pandolfi, for his insight on the entire first draft.

In addition, we wish to express our deep appreciation and gratitude to Ms. J. Riddleberger for her superb, accurate, fast, and dedicated work in typing the various drafts of the chapters of these volumes; for keeping track of them; for safeguarding the resulting disks, etc.; and, beyond all that, for typing many of our papers that have entered these volumes. Since joining the University of Virginia, our research output would have been much hindered without her highly professional services.

    It is a pleasure to acknowledge the understanding and assistance in this project of Ms. Lauren Cowles, Editor of Mathematics and Computer Science at Cambridge University Press.

    During the many years in which these volumes were written, along with the papers on which they are based, our work has been continuously sponsored by the National Science Foundation, Division of Mathematical Sciences, whose support is greatly appreciated. In the early years, our work was also supported by the Air Force Office of Scientific Research. Over the past few years, we were fortunate to receive much appreciated support by the Army Research Office, in the program managed by Dr. L. Bushnell. Occasional sponsorship over the years by the Italian Consiglio Nazionale delle Ricerche and by the Scuola Normale Superiore di Pisa is acknowledged with gratitude.

7

Some Auxiliary Results on Abstract Equations

In the present chapter we collect, in Sections 7.1 through 7.5, a number of results that will be invoked repeatedly throughout the present volume as well as Volume III. They concern (1) regularity results of the input $\to$ solution map, and its adjoint map, over both a finite or an infinite time interval and (2) generation and abstract trace regularity under unbounded perturbation. In addition, in Section 7.6, we provide an abstract regularity result for damped second-order equations of interest in itself. Illustrations thereof are given in Section 7.7 and in Chapter 9, Section 9.10.4.

7.1 Mathematical Setting and Standing Assumptions

Throughout this chapter X and U are reflexive Banach spaces and X^* and U^* are their dual spaces. For a given $0 < T < \infty$ fixed, we shall study the operator L,

$$(Lu)(t) = \int_0^t e^{A(t-\tau)} Bu(\tau)\, d\tau, \quad 0 \le t \le T, \tag{7.1.1}$$

corresponding to the mild solution

$$x(t) = e^{At} x_0 + (Lu)(t) \tag{7.1.2}$$

of the abstract equation

$$\dot{x} = Ax + Bu \in [\mathcal{D}(A^*)]', \quad x(0) = x_0, \tag{7.1.3}$$

subject to the following standing assumptions:

(H.1) $A : X \supset \mathcal{D}(A) \to X$ is a linear operator, which is the infinitesimal generator of a strongly continuous (s.c.) semigroup e^{At} on X.

(H.2) B is a linear, continuous operator $U \to [\mathcal{D}(A^*)]'$, where A^* is the X-adjoint of A, and $[\mathcal{D}(A^*)]'$ is the dual of $\mathcal{D}(A^*)$ with respect to the pivot space X, or equivalently,

$$A^{-1} B \in \mathcal{L}(U; X). \tag{7.1.4a}$$

645

For considerations of L, over a finite time interval, we may assume, without loss of generality, that $A^{-1} \in \mathcal{L}(X)$, for otherwise we replace (7.1.4a) with

$$(\lambda_0 I - A)^{-1} B \in \mathcal{L}(U; X), \quad \lambda_0 \in \text{resolvent set } \rho(A). \tag{7.1.4b}$$

(H.3) (Abstract trace regularity) Given $0 < T < \infty$ and $1 \le q \le \infty$, there exists a constant $C_T > 0$ [which depends on T and on q, but dependence on q is omitted], such that

$$\begin{cases} \int_0^T \|B^* e^{A^* t} x^*\|_{U^*}^q \, dt \le C_T \|x^*\|_{X^*}^q, & x^* \in \mathcal{D}(A^*), \ 1 \le q < \infty; \\ \|B^* e^{A^* t} x^*\|_{L_\infty(0,T;U^*)} \le C_T \|x^*\|_{X^*}, & x^* \in \mathcal{D}(A^*), \ q = \infty, \end{cases} \tag{7.1.5a}$$

so that the closable (cf. Remark 7.1.1 below) operator $B^* e^{A^* t}$ admits a continuous extension (which may then be denoted by the same symbol) satisfying

$$B^* e^{A^* t} : \text{continuous } X^* \to L_q(0, T; U^*), \quad 1 \le q \le \infty, \tag{7.1.5b}$$

that is,

$$\begin{cases} \int_0^T \|B^* e^{A^* t} x^*\|_{U^*}^q \, dt \le C_T \|x^*\|_{X^*}^q, & x^* \in X^*, \ 1 \le q < \infty; \\ \|B^* e^{A^* t} x^*\|_{L_\infty(0,T;U)} \le C_T \|x^*\|_{X^*}, & x^* \in X^*, \ q = \infty. \end{cases} \tag{7.1.5c}$$

Here B^*, the dual of B, satisfies $B^* \in \mathcal{L}(\mathcal{D}(A^*); U^*)$, after identifying $[\mathcal{D}(A^*)]''$ with $\mathcal{D}(A^*)$. Moreover, $e^{A^* t}$ is a s.c. semigroup on X^*.

Remark 7.1.1   By a change of variable and use of the semigroup property, if (7.1.5) holds for one fixed $0 < T < \infty$, then (7.1.5) holds for $2T, 3T, \dots$, hence for *any* $0 < T < \infty$.

Remark 7.1.2   The original s.c. semigroup e^{At} on X of assumption (H.1) can always be extended as a s.c. semigroup on the extrapolation space $[\mathcal{D}(A^*)]'$. We shall continue to use the notation e^{At} for such an extension. Moreover, solely under (H.1) and (H.2), it is plainly always the case that

$$L : \text{continuous } L_p(0, T; U) \to C([0, T]; [\mathcal{D}(A^*)]'), \quad 1 \le p \le \infty. \tag{7.1.6}$$

Remark 7.1.3   Under (H.1) and (H.2), if $u \in H^1(0, T; U)$, then formula (7.1.1) for L yields, after integration by parts,

$$\begin{aligned} (Lu)(t) &= -\int_0^t \frac{de^{A(t-\tau)}}{d\tau} A^{-1} B u(\tau) \, d\tau \\ &= e^{At} A^{-1} B u(0) - A^{-1} B u(t) + \int_0^t e^{A(t-\tau)} A^{-1} B \dot{u}(\tau) \, d\tau \\ &\in C([0, T]; X). \end{aligned} \tag{7.1.7}$$

The above computations are justified on the extrapolation space $[\mathcal{D}(A^*)]'$, via Remark 7.1.2. However, the final result lies in X, at least for $u \in H^1(0, T; U)$, as noted in (7.1.7). Thus, (7.1.7) says, in particular, that the operator

$$L_T u = \int_0^T e^{A(T-t)} Bu(t) \, dt = A \int_0^T e^{A(T-t)} A^{-1} Bu(t) \, dt \qquad (7.1.8)$$

is densely defined as an operator $L_p(0, T; U) \supset \mathcal{D}(L_T) \to X$, $1 \le p \le \infty$. Moreover, L_T – being in (7.1.8) (right) the composition of a closed boundedly invertible operator A with a bounded operator – is closed [Kato, 1966, p. 164]. An adjoint of L_T is:

$$(Sx^*)(t) = B^* e^{A^* t} x^*, \quad x^* \in \mathcal{D}(S), \ 0 \le t \le T, \text{ a.e.,} \qquad (7.1.9a)$$

which is closable [Kato, 1966, p. 168] as an operator $X^* \supset \mathcal{D}(S) \to L_q(0, T; U^*)$, $1 \le q \le \infty$, as noted above (7.1.5b), where $1/p + 1/q = 1$. The unique maximal extension of the operator in (7.1.9a), which is mentioned in assumption (H.3) above, is the adjoint L_T^* of L_T. Since L_T in (7.1.8) is densely defined and closed, then $(L_T^*)^* = L_T$. Thus, hypothesis (H.3) = (7.1.5b) is now paraphrased by saying

$$(L_T^* x^*)(t) = B^* e^{A^* t} x^* : \text{ continuous } X^* \to L_q(0, T; U^*), \qquad (7.1.9b)$$

or equivalently for $1 \le q < \infty$, and a fortiori for $q = \infty$:

(H.3*)

$$L_T : \text{continuous } L_p(0, T; U) \to X, \quad 1 \le p \le \infty, \quad \frac{1}{p} + \frac{1}{q} = 1. \quad (7.1.10)$$

The point is that, although the extrapolation space $[\mathcal{D}(A^*)]'$ acts as an all-encompassing, backup space for the regularity of L, and for performing computations, solely under (H.1) and (H.2), according to (7.1.7), L may be more regular. Indeed, more precisely, we have that

$$L : \text{continuous } L_p(0, T; U) \to C([0, T]; X), \qquad (7.1.11)$$

with $1 \le p < \infty$, if and only if (H.3) holds true with $1 < q \le \infty$, or, with $p = \infty$, provided that (H.3) holds true with $q = 1$. This is the content of Theorem 7.2.1 below.

Remark 7.1.4 We note explicitly that, if there exists a point $T, 0 < T < \infty$, such that

$$L_T u = \int_0^T e^{A(T-\tau)} Bu(\tau) \, d\tau \in X, \quad \forall u \in L_2(0, T; U), \qquad (7.1.12)$$

then, for all $0 < t_1 < T$, we likewise have

$$L_{t_1} u = \int_0^{t_1} e^{A(t_1-\tau)} Bu(\tau) \, d\tau \in X, \quad \forall u \in L_2(0, t_1; U). \qquad (7.1.13)$$

Indeed, choosing at first u smooth, say $u \in C([0, T]; U)$, we write for any $0 < t < T$:

$$L_T u = \int_0^t e^{A(T-\tau)} Bu(\tau) \, d\tau + \int_t^T e^{A(T-\tau)} Bu(\tau) \, d\tau \qquad (7.1.14)$$

$$= \int_0^T e^{A(T-\tau)} Bu_{\text{ext}}(\tau) \, d\tau + \int_0^{T-t} e^{A(T-t-\sigma)} Bu_t(\sigma) \, d\sigma, \quad (7.1.15)$$

where $u_{\text{ext}}(\tau)$ extends by zero $u(\tau)$ for $t < \tau \leq T$ in the first integral, whereas in the second integral we set $u_t = u(t + \sigma)$, $\tau - t = \sigma$. Extending u to all $u \in L_2(0, T; U)$ and setting $t_1 = T - t$, we obtain, by assumption (1.12),

$$L_{t_1} u_t = L_T u - L_T u_{\text{ext}} \in Y, \quad \forall \, u \in L_2(0, T_1; U), \qquad (7.1.16)$$

as desired, and (7.1.13) is established.

7.2 Regularity of L and L^* on $[0, T]$

As anticipated at the end of Remark 7.1.3, the "trace" regularity (H.3) $= (7.1.5)$ of L_T^*, equivalently the final state "interior" regularity (H.3*) $= (7.1.10)$ of L_T, is equivalent to the following "interior" regularity for L.

Theorem 7.2.1   *Assume (H.1) and (H.2).*

(i) For $1 < q \leq \infty$ [respectively, $q = 1$], hypothesis (H.3) is equivalent to [respectively, implies] the following regularity property of the operator L in (7.1.1):

$$L : \text{ continuous } L_p(0, T; U) \to C([0, T]; X), \quad 1 \leq p \leq \infty, \qquad (7.2.1a)$$

that is, there exists $k_T > 0$ such that

$$\|Lu\|_{C([0,T];X)} \leq k_T \|u\|_{L_p(0,T;U)}. \qquad (7.2.1b)$$

(ii) For $1 \leq q \leq \infty$, hypothesis (H.3) implies that the adjoint operator L^* satisfies

$$(L^* v)(t) = \int_t^T B^* e^{A^*(\tau-t)} v(\tau) \, d\tau \qquad (7.2.2a)$$

$$: \text{ continuous } L_1(0, T; X^*) \to L_q(0, T; U^*). \qquad (7.2.2b)$$

The operator L^* is the adjoint of L in the sense that, with $1/p + 1/q = 1$,

$$(Lu, v)_{p,X;q,X^*} = (u, L^* v)_{p,U;q,U^*}, \qquad (7.2.3)$$

where the notation on the left and on the right of (7.2.3) denotes, respectively, the duality pairing between $L_p(0, T; X)$ and $L_q(0, T; X^*)$, and between $L_p(0, T; U)$ and $L_q(0, T; U^*)$.

Proof.

(i) Step 1 We first prove that, under (H.1) and (H.2), assumption (H.3) = (7.1.5) for $1 \le q \le \infty$ implies that

$$L : \text{ continuous } L_p(0, T; U) \to L_\infty(0, T; X). \tag{7.2.4}$$

To this end, with $v \in L_1(0, T; X^*)$ and $u \in L_p(0, T; U)$, we compute from (7.2.3) and (7.1.1), with $1 < p < \infty$: To this end, with $v \in L_1(0, T; X^*)$ and $u \in L_p(0, T; U)$, we compute from (7.2.3) and (7.1.1), with $1 < p < \infty$:

$$|(Lu, v)_{\infty, X; 1, X^*}| = \left| \int_0^T ((Lu)(t), v(t)) \, dt \right|$$

$$= \left| \int_0^T \int_0^t \left(u(\tau), B^* e^{A^*(t-\tau)} v(t) \right) d\tau \, dt \right|$$

$$(t - \tau = \sigma) \quad \le \int_0^T \left\{ \int_0^t \|u(\tau)\|_U^p \, d\tau \right\}^{\frac{1}{p}} \left\{ \int_0^t \|B^* e^{A^* \sigma} v(t)\|_{U^*}^q \, d\sigma \right\}^{\frac{1}{q}} dt$$

(replacing t with T in both integral signs, and using (H.3) = (7.1.5c))

$$\le \|u\|_{L_p(0,T;U)} \int_0^T C_T \|v(t)\|_{X^*} \, dt$$

$$= C_T \|u\|_{L_p(0,T;U)} \|v\|_{L_1(0,T;X^*)}. \tag{7.2.5}$$

[In the above estimates, either one considers $v(t) \in X^*$ a.e. or one takes $v(t) \in C([0, T]; X^*)$ and extends estimate (7.2.5) to all of $v \in L_1(0, T; X^*)$.] An obvious variation leads likewise to (7.2.5) also for $p = \infty$, or $q = \infty$. Then (7.2.4) is established for $1 \le q \le \infty$.

Step 2 Let $u \in L_p(0, T; U)$. By taking now u_n smooth, say $u_n \in C^1([0, T]; U)$, with $u_n \to u$ in $L_p(0, T; U)$, and integrating $(Lu_n)(t)$ by parts as in (7.1.7), we then see that $(Lu_n)(t) \in C([0, T]; X)$ by (7.1.7), whereas $Lu_n \to Lu$ in $L_\infty(0, T; X)$ by (7.2.4), and thus $Lu \in C([0, T]; X)$. The continuity of L in (7.2.4) is then improved to the continuity of L in (7.2.1), as desired.

Step 3 Conversely, the continuity of L in (7.2.1) implies the continuity of L_T in (7.1.10), and hence, equivalently for $1 \le p < \infty$, the continuity of L_T^* defined by (7.1.9) given by (H.3) = (7.1.5b).
(ii) Statement (7.2.2b) follows from (7.2.1) by duality. $\square$

Remark 7.2.1 In application to partial differential equations in this volume we shall use Theorem 7.2.1 in the Hilbert space setting, with $X = X^*$ and $U = U^*$ Hilbert spaces, and $p = q = 2$.

Remark 7.2.2   The conclusion of Theorem 7.2.1 for L applies also to the operator

$$u \to \int_t^T e^{A(\tau-t)} Bu(\tau) \, d\tau, \tag{7.2.6}$$

as we shall need, for example, in Chapters 8 and 9. Similarly, at times, as in the forthcoming Chapter 8, Section 8.2.1, we shall start with the assumption

$$Re^{At} B : \text{continuous } U \to L_q(0, T; Z), \tag{7.2.7}$$

for A and B as in (H.1) and (H.2) = (7.1.4), and with R a suitable operator $R \in \mathcal{L}(Y; Z)$ and U and Z Hilbert spaces [the case $q = 1$ will be relevant]. Then, a variation of the proof of Theorem 7.2.1 yields: For $1 < q \le \infty$ [respectively, for $q = 1$], property (7.2.7) is equivalent to [respectively, implies] the following property:

$$(L^* R^* f)(t) = \int_t^T B^* e^{A^*(\tau-t)} R^* f(\tau) \, d\tau \tag{7.2.8a}$$

$$: \text{continuous } L_p(0, T; Z) \to C([0, T]; U), \tag{7.2.8b}$$

where $1/p + 1/q = 1$. Indeed, the counterpart of Step 1 in (7.2.5) is now, by (7.2.8a), for $f \in L_p(0, T; Z)$, $1 \le p < \infty$, $g \in L_1(0, T; U)$,

$$|(L^* R^* f, g)_{\infty, U; 1, U}| = \left| \int_0^T \int_t^T \left(f(\tau), Re^{A(\tau-t)} Bg(t) \right)_Z d\tau \, dt \right|$$

$$\le \int_0^T \left\{ \int_t^T \|f(\tau)\|_Z^p \, d\tau \right\}^{\frac{1}{p}} \left\{ \int_t^T \left\| Re^{A(\tau-t)} Bg(t) \right\|_Z^q \, d\tau \right\}^{\frac{1}{q}} dt$$

$$\text{(by (7.2.7))} \qquad \le \|f\|_{L_p(0,T;Z)} \int_0^T C_T \|g(t)\|_U \, dt$$

$$= C_T \|f\|_{L_p(0,T;Z)} \|g\|_{L_1(0,T;U)}. \tag{7.2.9}$$

An obvious variation leads likewise to (7.2.9) also for $p = \infty$. Thus, (7.2.9) says that

$$L^* R^* : \text{ continuous } L_p(0, T; Z) \to L_\infty(0, T; U). \tag{7.2.10}$$

Next, the $L_\infty(0, T; U)$-regularity in (7.2.10) is lifted up to the $C([0, T]; U)$-regularity in (7.2.8b), by an approximation argument, as in Step 2, above.

First, if $f \in H^1(0, T; Z)$, then integration by parts on (7.2.8a) yields only under (H.1) and (H.2):

$$(L^* R^* f)(t) = \int_t^T B^* A^{*-1} \frac{de^{A^*(\tau-t)}}{d\tau} R^* f(\tau) \, d\tau \tag{7.2.11}$$

$$= B^* A^{*-1} e^{A^*(T-t)} R^* f(T) - B^* A^{*-1} R^* f(t)$$

$$- \int_t^T B^* A^{*-1} e^{A^*(\tau-t)} R^* \dot{f}(\tau) \, d\tau \in C([0, T]; U). \tag{7.2.12}$$

Next, let $f \in L_p(0, T; Z)$. By taking now f_n smooth, say $f_n \in C^1([0, T]; Z)$, with $f_n \to f$ in $L_p(0, T; Z)$, we then see that $L^* R^* f_n \in C([0, T]; U)$ by (7.2.12), whereas $L^* R^* f_n \to L^* R^* f$ in $L_\infty(0, T; Z)$ by (7.2.10), and thus $L^* R^* f \in C([0, T]; Z)$, as desired.

7.3 A Lifting Regularity Property When e^{At} Is a Group

Under assumptions (H.1) and (H.2) and, moreover, with e^{At} a group, the next result lifts the time regularity from L_p to C, while preserving the same space regularity. For the classes of PDEs, we have in mind, and for which the original hypothesis (7.3.1) applies, the assumption that e^{At} is a group is automatically satisfied. The result is false for general semigroups; see Remark 7.3.1.

Theorem 7.3.1   *Assume (H.1) and (H.2), and, moreover, that $G(t) = e^{At}$ is a s.c. group. Furthermore, suppose that the operator L in (7.1.1) satisfies*

$$L : continuous\ L_p(0, T; U) \to L_p(0, T; X), \quad 1 \le p < \infty. \qquad (7.3.1)$$

Then, in fact, property (H.3) = (7.1.5) holds true, and then

$$L : continuous\ L_p(0, T; U) \to C([0, T]; X). \qquad (7.3.2)$$

Proof.   By duality on (7.3.1) we obtain that the operator L^* in (7.2.2a) satisfies, with $1/p + 1/q = 1$:

$$L^* : continuous\ L_q(0, T; X^*) \to L_q(0, T; U^*). \qquad (7.3.3)$$

We now use the group assumption on e^{At} and apply L^* to a *special* function, given by the dual free ($B \equiv 0$) dynamics of (7.1.3), backward in time, that is, to

$$\hat{f}(\tau) = e^{A^*(-\tau)} x^* \in L_q(0, T; X^*), \quad x^* \in X^*. \qquad (7.3.4)$$

Then, from (7.2.2), we obtain

$$(L^* \hat{f})(t) = \int_t^T B^* e^{A^*(\tau - t)} e^{A^*(-\tau)} x^* \, d\tau$$

$$= (T - t) B^* e^{A^*(-t)} x^* \in L_q(0, T; U^*), \qquad (7.3.5)$$

recalling (7.3.3), and hence $B^* e^{A^*(-t)} x^* \in L_q(0, T - \epsilon; U^*)$, for any $0 < \epsilon < T$. Since this conclusion holds true for any finite T and any $\epsilon > 0$ small (recall Remark 7.1.1), we then obtain that

$$B^* e^{A^*(-t)} x^* \in L_q(0, T; U^*), \quad x^* \in X^*, \qquad (7.3.6)$$

continuously; that is, writing $x^* = e^{A^* T} y^*$, or $y^* = e^{A^*(-T)} x^* \in X^*$, we have from (7.3.6)

$$B^* e^{A^*(T - t)} : continuous\ X^* \to L_q(0, T; U^*), \qquad (7.3.7)$$

which is (H.3) = (7.1.5c), as desired. The equivalence between (H.3) and (7.3.3) noted in Theorem 7.2.1(i) completes the proof.   □

Remark 7.3.1   The lifting regularity result of Theorem 7.3.1 can be applied to mixed problems for second-order hyperbolic equations, for Euler–Bernoulli equations, for Kirchoff equations, etc.; see Lasiecka and Triggiani [1983; 1991]. A few applications will be made in Chapter 9, Section 9.8 in the study of the regularity of wave and Kirchoff equations with interior point control.

Remark 7.3.2   The assumption that e^{At} be a s.c. *group* is crucial in the above theorem, in the sense that if e^{At} is only a s.c. semigroup even if holomorphic, the conclusion of the theorem is false. As an example illustrating this, let $y(t, x)$ be the solution of a corresponding *parabolic* equation with, say, zero initial condition and with forcing term u in the Dirichlet boundary conditions,

$$\begin{cases} y_t = \Delta y, & \text{in } (0, T] \times \Omega = Q_T; \\ y(0, \cdot) = 0, & \text{in } \Omega; \\ y|_{\Sigma_T} = u, & \text{in } (0, T] \times \Gamma = \Sigma_T, \end{cases} \tag{7.3.8}$$

as in Chapter 3. Then the corresponding free solution ($u \equiv 0$) is described by a s.c., holomorphic semigroup on $L_2(\Omega)$ (which therefore is not a group). We have that the map $u \to y$ is continuous from $L_2(\Sigma_T) \to L_2(Q_T)$ (even $L_2(\Sigma_T) \to L_2(0, T; H^{\frac{1}{2}}(\Omega))$), see Chapter 3, Section 3.1, yet the map $u \to y(T)$ from $L_2(\Sigma_T)$ to $L_2(\Omega)$ is *not* continuous. In fact, for a preassigned $0 < T < \infty$, one may construct $u \in L_2(\Sigma_T)$ whose corresponding solution y satisfies $y(T) \notin L_2(\Omega)$ [Lions, 1971, p. 202], even in the one-dimensional case.

Remark 7.3.3   The above proof extends almost verbatim to other settings as well. For instance, one may replace assumption (7.3.1) by

$$L : \text{continuous } H^1(0, T; U) \to L_2(0, T; X), \tag{7.3.9}$$

$U = U^*$, $X = X^*$ (Hilbert spaces), and then the conclusion (7.3.2) becomes

$$L : \text{continuous } H^1(0, T; U) \to C([0, T]; X). \tag{7.3.10}$$

We shall not need these settings, however.

Proof of implication. (7.3.9) $\Rightarrow$ (7.3.10).   Under assumption (7.3.9), the counterparts of (7.3.6) and (7.3.7) are now

$$B^* e^{A^*(-t)} x \in [H^1(0, T; U)]' \tag{7.3.11}$$

continuously in $x \in X$, and

$$B^* e^{A^*(T-t)} : \text{continuous } X \to [H^1(0, T; U)]', \tag{7.3.12}$$

respectively, where $[H^1(0, T; U)]'$ is the dual of $H^1(0, T; U)$ with respect to

$L_2(0, T; U)$ as a pivot space. From here, it follows that

$$L_T u = \int_0^T e^{A(T-t)} B u(t)\, dt : \text{continuous } H^1(0, T; U) \to X. \qquad (7.3.13)$$

Finally, with $v \in L_1(0, T; X)$ and $u \in H^1(0, T; U)$, one obtains

$$\int_0^T (Lu)(t), v(t))_X\, dt = \int_0^T \left(\int_0^t e^{A(t-\tau)} B u(\tau)\, d\tau, v(t) \right) dt$$

$$= \int_0^T \int_0^t (u(\tau), B^* e^{A^*(t-\tau)} v(t))_U\, d\tau dt \qquad (7.3.14)$$

$$\leq \int_0^T \|u\|_{H^1(0,t;U)} \|B^* e^{A^*(t-\cdot)} v(t)\|_{[H^1(0,t;U)]'}\, dt \qquad (7.3.15)$$

$$(\text{by } (7.3.12)) \qquad \leq \int_0^T \|u\|_{H^1(0,T;U)} c_t \|v(t)\|_X\, dt \qquad (7.3.16)$$

$$\leq c_T \|u\|_{H^1(0,T;U)} \int_0^T \|v(t)\|_X\, dt, \qquad (7.3.17)$$

as desired, which is the counterpart of (7.2.5). In going from (7.3.15) to (7.3.16), we have recalled (7.3.12) and used that $c_t \leq c_T$ for all $t \leq T$, by duality on the original assumption (7.3.10). Then (7.3.17) yields conclusion (7.3.10) as desired, by an approximation argument, as in Step 2 in the proof of Theorem 7.2.1.

7.4 Extension of Regularity of L and L^* on $[0, \infty]$ When e^{At} Is Uniformly Stable

In this section, in addition to (H.1), (H.2), and (H.3), we assume the following "uniform exponential stability" property of e^{At}, that is, stability in the uniform operator topology:

(H.4) There exist constants $M \geq 1$, $\omega > 0$, such that

$$\|e^{At}\|_{\mathcal{L}(X)} \leq M e^{-\omega t}, \quad t \geq 0. \qquad (7.4.1)$$

As a consequence of (H.4), we extend the continuity of the operator L in (7.1.1) from $T < \infty$ to $T = \infty$. We shall provide a direct statement and a direct proof in Section 7.4.1 and a dual statement and a dual proof in Section 7.4.2.

7.4.1 Direct Statement; Direct Proof

Theorem 7.4.1.1   *Assume (H.1), (H.2), (H.3) = (7.1.5), and (H.4) = (7.4.1). Let $1 \leq p < \infty$. Let $\epsilon > 0$ be such that*

$$-\omega + \epsilon < 0, \qquad (7.4.1.1)$$

with ω the constant in (7.4.1). Then hypothesis (H.3), in its version given by (7.2.1), that is, L: continuous $L_p(0, T; U) \rightarrow C([0, T]; X)$, $1 \leq p < \infty$, can be improved to the following statements:

$$e^{\epsilon t} L \; : \; \text{continuous } L_p(0, \infty; U) \rightarrow L_p(0, \infty; X) \qquad (7.4.1.2)$$

$$: \; \text{continuous } L_p(0, \infty; U) \rightarrow C([0, \infty]; X), \qquad (7.4.1.3)$$

where the latter is the space of X-valued continuous functions bounded on $[0, \infty]$, that is, bounded under the sup norm.

Proof. Proof of (7.4.1.2). We first prove (7.4.1.2) for $\epsilon = 0$. We let $f(t) = \|(Lu)(t)\|_X^p$ for $u \in L_p(0, \infty; U)$ and obtain

$$\int_0^\infty \|(Lu)(t)\|_X^p \, dt = \int_0^\infty f(t) \, dt = \sum_{n=0}^\infty \int_{nT}^{(n+1)T} f(t) \, dt$$

$$= \sum_{n=0}^\infty \int_0^T f(nT + t) \, dt. \qquad (7.4.1.4)$$

By splitting the interval $[0, nT + t]$ in $[0, T]$, $[T, 2T]$, etc., and with a change of variable, we compute for $n = 1, 2, \ldots$, via (7.1.1) and the semigroup property

$$f(nT + t) = \|(Lu)(nT + t)\|_X^p = \left\| \int_0^{nT+t} e^{A(nT+t-\tau)} Bu(\tau) \, d\tau \right\|_X^p$$

$$= \left\| \sum_{j=1}^n e^{A((n-j)T+t)} \int_0^T e^{A(T-\tau)} Bu((j-1)T + \tau) \, d\tau \right.$$

$$\left. + \int_0^t e^{A(t-\tau)} Bu(nT + \tau) \, d\tau \right\|_X^p. \qquad (7.4.1.5)$$

Using the assumed stability $(7.4.1) = $ (H.4) and the continuity (7.2.1b), which is equivalent to (H.3), we get

$$f(nT + t) \leq k_T^p \left\{ M e^{-\omega t} \sum_{j=1}^n e^{-\omega(n-j)T} \left(\int_0^T \|u((j-1)T + \tau)\|_U^p \, d\tau \right)^{\frac{1}{p}} \right.$$

$$\left. + \left(\int_0^T \|u(nT + \tau)\|^p \, d\tau \right)^{\frac{1}{p}} \right\}^p$$

$$\leq k_T^p c_p \left\{ M^p e^{-\omega p t} \left(\sum_{j=1}^n e^{-\omega(n-j)T} \|u\|_{L_p((j-1)T, jT; U)} \right)^p \right.$$

$$\left. + \|u\|_{L_p(nT, (n+1)T; U)}^p \right\}, \qquad (7.4.1.6)$$

where in the last step we have used $(x + y)^p \leq c_p(x^p + y^p)$ for $0 \leq x, \ y < \infty$ with $c_p = 2^p$. [Dunford, Schwartz, 1957, p. 120; Martin, 1976, p. 161]. Thus, setting

$$v_j = \left\{ \int_{(j-1)T}^{jT} \|u(\tau)\|^p \, d\tau \right\}^{\frac{1}{p}} = \|u\|_{L_p((j-1)T, jT; U)}$$

$$\text{so that } \sum_{j=1}^{\infty} v_j^p = \|u\|_{L_p(0, \infty; U)}^p, \tag{7.4.1.7}$$

we rewrite (7.4.1.6) via (7.4.1.7) as

$$f(nT + t) \leq k_T^p c_p \left\{ M^p e^{-\omega p t} \left(\sum_{j=1}^{n} e^{-\omega(n-j)T} v_j \right)^p + v_{n+1}^p \right\}. \tag{7.4.1.8}$$

By Hölder's inequality with $1/p + 1/q = 1$, we have

$$\sum_{j=1}^{n} e^{-\omega(n-j)T} v_j = \sum_{j=1}^{n} e^{-\omega(n-j)T/q} \, e^{-\omega(n-j)T/p} \, v_j$$

$$\leq \left\{ \sum_{j=1}^{n} e^{-\omega(n-j)T} \right\}^{\frac{1}{q}} \left\{ \sum_{j=1}^{n} e^{-\omega(n-j)T} v_j^p \right\}^{\frac{1}{p}}$$

$$\leq \left(\frac{1}{1 - e^{-\omega T}} \right)^{\frac{1}{q}} \left\{ \sum_{j=1}^{n} e^{-\omega(n-j)T} v_j^p \right\}^{\frac{1}{p}}. \tag{7.4.1.9}$$

Raising (7.4.1.9) to the power p and substituting into the right-hand side of (7.4.1.8) yields, for $n = 1, 2, \ldots,$

$$f(nT + t) \leq C_{p, \omega, T} \left\{ e^{-\omega p t} \sum_{j=1}^{n} e^{-\omega(n-j)T} v_j^p + v_{n+1}^p \right\}, \tag{7.4.1.10}$$

where $C_{p, \omega, T} = (2k_T)^p [M/(1 - e^{-\omega T})]^p$. Then (7.4.1.9) implies

$$\sum_{n=1}^{\infty} \int_0^T f(nT + t) \, dt \leq \text{Const}_{p, \omega, T} \left\{ \sum_{n=1}^{\infty} \sum_{j=1}^{n} e^{-\omega(n-j)T} v_j^p + \sum_{n=1}^{\infty} v_{n+1}^p \right\}. \tag{7.4.1.11}$$

But one readily sees that

$$\sum_{n=1}^{\infty} \sum_{j=1}^{n} e^{-\omega(n-j)T} v_j^p = \left(\sum_{k=0}^{\infty} e^{-k\omega T} \right) \sum_{j=1}^{\infty} v_j^p = \frac{1}{1 - e^{-\omega T}} \|u\|_{L_p(0, \infty; U)}^p \tag{7.4.1.12}$$

by rearranging the (positive) terms of the double sum, which form an infinite lower triangular matrix, and summing up along the diagonals of it; in the last step in

(7.4.1.12), we have recalled (7.4.1.7) (right). Thus, (7.4.1.12), inserted in (7.4.1.11), yields, also by virtue of (7.4.1.7) (right),

$$\sum_{n=1}^{\infty} \int_{0}^{T} f(nT + t) \, dt \leq C_{p,\omega,T} \|u\|_{L_p(0,\infty;U)}^{p}. \tag{7.4.1.13}$$

As to the term $n = 0$, we have

$$\int_{0}^{T} f(t) \, dt = \int_{0}^{T} \|(Lu)(t)\|_{X}^{p} \, dt \leq T k_{T}^{p} \|u\|_{L_p(0,T;U)}^{p} \tag{7.4.1.14}$$

by (7.2.1b). Combining (7.4.1.14) with (7.4.1.13) and recalling (7.4.1.4) we finally obtain

$$\int_{0}^{\infty} \|(Lu)(t)\|_{X}^{p} \, dt \leq \text{const} \|u\|_{L_p(0,\infty;U)}^{p}, \tag{7.4.1.15}$$

as desired, where const depends on k_T, T, p, and ω. Thus, (7.4.1.15) proves (7.4.1.2) at least for $\epsilon = 0$.

The case $\epsilon > 0$ subject to (7.4.1.1) can be proved exactly in the same way. Thus, the regularity (7.4.1.2) of L is proved.

Proof of (7.4.1.3).   We first prove (7.4.1.3) for $\epsilon = 0$, indeed, we first prove that

$$\sup_{0 \leq t \leq \infty} \|(Lu)(t)\|_{X} \leq \text{const} \|u\|_{L_p(0,\infty;U)}. \tag{7.4.1.16}$$

It suffices to use a subset of the preceding argument for (7.4.1.2). Recalling (7.4.1.5) we write

$$\|(Lu)(nT + t)\|_{X} = [f(nT + t)]^{\frac{1}{p}}$$

$$= \left\| \sum_{j=1}^{n} e^{A((n-j)T+t)} \int_{0}^{T} e^{A(T-t)} Bu((j-1)T + \tau) \, d\tau \right.$$

$$\left. + \int_{0}^{t} e^{A(t-\tau)} Bu(nT + \tau) \, d\tau \right\|_{X}$$

$$\leq M e^{-\omega t} \sum_{j=1}^{n} e^{-\omega(n-j)T} \left\| \int_{0}^{T} e^{A(T-\tau)} Bu((j-1)T + \tau) \, d\tau \right\|_{X}$$

$$+ \left\| \int_{0}^{t} e^{A(t-\tau)} Bu(nT + \tau) \, d\tau \right\|_{X}. \tag{7.4.1.17}$$

Thus, by (7.2.1b), we obtain from (7.4.1.17)

$$\|(Lu)(nT + t)\|_{X} \leq M k_T e^{-\omega t} \sum_{j=1}^{n} e^{-\omega(n-j)T} \|u\|_{L_p(0,\infty;U)} + k_T \|u\|_{L_p(0,\infty;U)}$$

$$\leq \left[M k_T \left(\frac{1}{1 - e^{-\omega T}} \right) + k_T \right] \|u\|_{L_p(0,\infty;U)}, \tag{7.4.1.18}$$

and (7.4.1.16) is proved. Since, by Remark 7.1.2, $Lu \in C([0, T]; X)$ for *any* $T > 0$ with $u \in L_p(0, \infty; U)$, then (7.4.1.16) proves (7.4.1.3) at least for $\epsilon = 0$.

The case $\epsilon > 0$ subject to (7.4.1.1) can be proved exactly in the same way. Thus, the regularity (7.4.1.3) of L is proved.   □

7.4.2 Dual Statement; Dual Proof

In this subsection we essentially reprove Theorem 7.4.1.1 by using the dual hypothesis (H.3) = (7.1.5) rather than its direct equivalent version (7.2.1). In the general form given in this section, this result will be repeatedly invoked in Volume III, by specializing the operator F below to various cases.

Theorem 7.4.2.1 *Let A be the generator of a s.c. semigroup e^{At} on the Banach space X (assumption (H.1)), which, moreover, is uniformly stable, that is, satisfies (H.4) = (7.4.1).*

Let $F : X \supset \mathcal{D}(F) \supset \mathcal{D}(A) \to Z$, where Z is another Banach space, be a linear operator, so that $Fe^{At}x$ is well defined for $x \in \mathcal{D}(A)$ for all t. Assume, further, that Fe^{At} can be extended in the following sense:

$$Fe^{At} : \text{continuous } X \to L_p(0, T; Z), \quad 1 \le p < \infty; \qquad (7.4.2.1a)$$

that is,

$$\int_0^T \|Fe^{At}x\|_Z^p \, dt \le c_T \|x\|_X^p, \quad 1 \le p < \infty, \ x \in X, \qquad (7.4.2.1b)$$

for some finite $T > 0$. Then, for all $\epsilon > 0$ such that $-\omega + \epsilon < 0$, that is, satisfying (7.4.1.1), we have

$$Fe^{(A+\epsilon I)t} : \text{continuous } X \to L_p(0, \infty; Z); \qquad (7.4.2.2a)$$

that is,

$$\int_0^\infty \|Fe^{(A+\epsilon I)t}x\|_Z^p \, dt \le C_{\omega,\epsilon,T,p} \|x\|_X^p, \quad x \in X, \qquad (7.4.2.2b)$$

where $C_{\omega,\epsilon,T,p}$ is obtained explicitly in (7.4.2.4) below.

Proof. We compute for $n = 1, 2, \ldots$ and $t - (n-1)T = \tau$:

$$\int_{(n-1)T}^{nT} \|Fe^{(A+\epsilon I)t}x\|_Z^p \, dt = \int_{(n-1)T}^{nT} \|Fe^{(A+\epsilon I)(t-(n-1)T)}e^{(A+\epsilon I)(n-1)T}x\|_Z^p \, dt$$

$$= \int_0^T \|Fe^{(A+\epsilon I)\tau}e^{(A+\epsilon I)(n-1)T}x\|_Z^p \, d\tau$$

$$\text{(by (7.4.2.1))} \qquad \le c_T e^{\epsilon pT} \|e^{(A+\epsilon I)(n-1)T}x\|_X^p. \qquad (7.4.2.3)$$

Thus, using (7.4.2.3), we obtain

$$
\int_0^\infty \left\| F e^{(A+\epsilon I)t} x \right\|_X^p dt = \sum_{n=1}^\infty \int_{(n-1)T}^{nT} \left\| F e^{(A+\epsilon I)t} x \right\|_Z^p dt
$$

$$
\leq c_T e^{\epsilon pT} \sum_{n=1}^\infty \left\| e^{(A+\epsilon I)(n-1)T} x \right\|_X^p
$$

$$
\text{(by (7.4.1))} \quad \leq c_T e^{\epsilon pT} M^p \|x\|_X^p \sum_{n=1}^\infty e^{-(\omega-\epsilon)(n-1)pT}
$$

$$
= c_T e^{\epsilon pT} M^p \frac{1}{1 - e^{-(\omega-\epsilon)pT}} \|x\|_X^p, \qquad (7.4.2.4)
$$

using in the last step a geometric series with $e^{-(\omega-\epsilon)pT} < 1$, that is, $\omega - \epsilon > 0$ by (7.4.1.1), as desired. Thus, (7.4.2.4) proves (7.4.2.2). $\square$

By duality on Theorem 7.4.2.1 we obtain the following result as in the proof of Theorem 7.2.1(i).

Theorem 7.4.2.2 *Assume, as stated at the outset, that X is a reflexive Banach space. Further assume the same hypotheses of Theorem 7.4.2.1: (H.1), (H.4) = (7.4.1), and (7.4.2.1). Then, with $1/p + 1/q = 1$, $1 \leq p < \infty$, $1 < q \leq \infty$, we have*

$$
\int_0^t e^{(A^*+\epsilon I)(t-\tau)} F^* v(\tau) \, d\tau : \text{continuous } L_q(0, \infty; Z^*) \to C([0, \infty]; X^*). \quad (7.4.2.5)
$$

Proof. As in the proof of Theorem 7.2.1(i), let $v \in L_q(0, \infty; Z^*)$ and $f \in L_1(0, \infty; X)$. We compute, where $\langle \;, \; \rangle$ denotes duality, by Hölder's inequality,

$$
\int_0^\infty \left\langle \int_0^t e^{(A^*+\epsilon I)(t-\tau)} F^* v(\tau) \, d\tau, \, f(t) \right\rangle dt
$$

$$
= \int_0^\infty \int_0^t \left\langle v(\tau), \, F e^{(A+\epsilon I)(t-\tau)} f(t) \right\rangle d\tau \, dt
$$

$$
\leq \int_0^\infty \left\{ \int_0^t \|v(\tau)\|_{Z^*}^q \, d\tau \right\}^{\frac{1}{q}} \left\{ \int_0^t \left\| F e^{(A+\epsilon I)(t-\tau)} f(t) \right\|_Z^p \, d\tau \right\}^{\frac{1}{p}} dt
$$

(with $t - \tau = \sigma$)

$$
= \|v\|_{L_q(0,\infty;Z^*)} \int_0^\infty \left\{ \int_0^t \left\| F e^{(A+\epsilon I)\sigma} f(t) \right\|_Z^p \, d\sigma \right\}^{\frac{1}{p}} dt
$$

(invoking (7.4.2.2) of Theorem 7.4.2.1)

$$
\leq \text{const} \|v\|_{L_q(0,\infty;Z^*)} \int_0^\infty \|f(t)\|_X \, dt
$$

$$
\leq \text{const} \|v\|_{L_q(0,\infty;Z^*)} \|f\|_{L_1(0,\infty;X)}. \qquad (7.4.2.6)
$$

[In the above estimates, either one considers $f(t) \in X$ a.e. or one takes $f \in L_1(0, \infty; X) \cap C([0, \infty]; X)$ and then extends (7.4.2.6) to all $f \in L_1(0, \infty; X)$.] Then (7.4.2.6) says that (7.4.2.5) holds true with $C([0, \infty]; X)$ replaced by $L_\infty(0, \infty; X)$. But then the approximating argument in Step 2 in the proof of Theorem 7.2.1 lifts $L_\infty(0, \infty; X)$ to $C([0, \infty]; X)$, and (7.4.2.5) is proved. $\square$

Theorem 7.4.2.3   *Assume the hypotheses of Theorem 7.4.2.1: (H.1), (H.4)$= (7.4.1)$ and (7.4.2.1). Then, the following regularity results hold true, for $1 \leq p < \infty$,*

$$
\left.
\begin{aligned}
(L_{1,F} f)(t) &\equiv \int_0^t Fe^{A(t-\tau)} f(\tau) \, d\tau \\[2mm]
(L_{2,F} f)(t) &\equiv \int_t^\infty Fe^{A(\tau-t)} f(\tau) \, d\tau
\end{aligned}
\right\}
\begin{aligned}
&: \text{ continuous } L_p(0, \infty; X) \\
&\quad \to L_p(0, \infty; Z).
\end{aligned}
\qquad
\begin{aligned}
&(7.4.2.7) \\[4mm]
&(7.4.2.8)
\end{aligned}
$$

Proof. (7.4.2.8):   As to $L_{2,F}$ we compute via Hölder's inequality

$$
\begin{aligned}
\int_0^\infty \| (L_{2,F} f)(t) \|_Z^p \, dt
&= \int_0^\infty \left\| \int_t^\infty Fe^{A(\tau-t)} f(\tau) \, d\tau \right\|_Z^p dt \\[2mm]
&= \int_0^\infty \left\| \int_t^\infty e^{-\epsilon(\tau-t)} Fe^{(A+\epsilon I)(\tau-t)} f(\tau) \, d\tau \right\|_Z^p dt \\[2mm]
&\leq \int_0^\infty \left\{ \left[\int_t^\infty e^{-q\epsilon(\tau-t)} \, d\tau \right]^{\frac{1}{q}} \right. \\[2mm]
&\qquad \left. \times \left[\int_t^\infty \left\| Fe^{(A+\epsilon I)(\tau-t)} f(\tau) \right\|_Z^p \, d\tau \right]^{\frac{1}{p}} \right\}^p dt \\[2mm]
&\leq \left(\frac{1}{q\epsilon} \right)^{\frac{p}{q}} \int_0^\infty \int_t^\infty \left\| Fe^{(A+\epsilon I)(\tau-t)} f(\tau) \right\|_Z^p \, d\tau \, dt.
\end{aligned}
$$
$$(7.4.2.9)$$

We next change the order of integration in (7.4.2.9) and obtain

$$
\begin{aligned}
\int_0^\infty \| (L_{2,F} f)(t) \|_Z^p \, dt
&\leq \left(\frac{1}{q\epsilon} \right)^{\frac{p}{q}} \int_0^\infty \int_0^\tau \left\| Fe^{(A+\epsilon I)\sigma} f(\tau) \right\|_Z^p \, d\sigma \, d\tau \\[2mm]
&\leq \left(\frac{1}{q\epsilon} \right)^{\frac{p}{q}} \int_0^\infty \int_0^\infty \left\| Fe^{(A+\epsilon I)\sigma} f(\tau) \right\|_Z^p \, d\sigma \, d\tau \\[2mm]
\text{(by (7.4.2.2))} \quad &\leq C_{\omega,\epsilon,T,p} \int_0^\infty \| f(\tau) \|_X^p \, d\tau. \qquad (7.4.2.10)
\end{aligned}
$$

Thus (7.4.2.10) proves (7.4.2.8), as desired.

(7.4.2.7)   The proof of (7.4.2.7) for $L_{1,F}$ is similar. $\square$

By duality on Theorem 7.4.2.3, we obtain

Corollary 7.4.2.4   *Assume the hypotheses of Theorem 7.4.2.1: (H.1), (H.4) $=$ (7.4.1), and (7.4.2.1), on the reflexive Banach space X. Then, the following regularity* *results hold true:*

$$\left.\begin{aligned}
(L_{1,F}^* v)(t) &= \int_t^\infty e^{A^*(\tau-t)} F^* v(t)\, dt \\
(L_{2,F}^* v)(t) &= \int_0^t e^{A^*(t-\tau)} F^* v(\tau)\, d\tau
\end{aligned}\right\} \begin{aligned} &: continuous\ L_q(0,\infty; Z^*) \\ &\to L_q(0,\infty; X^*). \end{aligned}$$

$$(7.4.2.11)$$
$$(7.4.2.12)$$

7.5 Generation and Abstract Trace Regularity under Unbounded Perturbation

In this section we collect some results on a Hilbert space Y, which will be often invoked in Volume 3, under different circumstances. More precisely, we provide a result of generation of a s.c. semigroup under a broad class of perturbations, along with a regularity result of the (abstract) "trace" of the corresponding perturbed system. Both cases $T < \infty$ and $T = \infty$ will be considered, the latter under stability assumption of the perturbed semigroup.

Theorem 7.5.1 *Consider the dynamics (7.1.1)–(7.1.3) on a Hilbert space Y, under the standing assumptions (H.1), (H.2), and (H.3) = (7.1.5) on a Hilbert space Y, with $q = 2$. Let $\Pi : Y \supset \mathcal{D}(\Pi) \to Y$ be an operator satisfying*

$$\|\Pi^* x\|_Y^2 \leq C\big[\|B^* x\|_U^2 + \|x\|_Y^2\big], \quad \forall\, x \in \mathcal{D}(B^*) \subset Y, \tag{7.5.1}$$

$\mathcal{D}(A^*) \subset \mathcal{D}(B^*)$, *so that, in particular, (7.5.1) implies $\Pi^* A^{*-1}$, $A^{-1}\Pi \in \mathcal{L}(Y)$. Then:*

(a) The perturbed closed operator

$$A_\Pi^* = A^* + \Pi^* \tag{7.5.2}$$

generates a s.c. semigroup $e^{A_\Pi^* t}$ on Y, $t > 0$.

(b) Moreover, the closable operator $B^* e^{A_\Pi^* t}$ admits a continuous extension (denoted by the same symbol) such that

$$B^* e^{A_\Pi^* t} : continuous\ Y \to L_2(0, T; U), \quad T < \infty; \tag{7.5.3a}$$

$$\int_0^T \left\| B^* e^{A_\Pi^* t} x \right\|_U^2 dt \leq C_T \|x\|_Y^2, \quad x \in Y,\ C_T > 0. \tag{7.5.3b}$$

(c) If, in addition, $e^{A_\Pi^* t}$ is uniformly stable, that is, there exist constants $M \geq 1$, $\omega > 0$ such that

$$\left\| e^{A_\Pi^* t} \right\|_{\mathcal{L}(Y)} \leq M e^{-\omega t}, \quad t \geq 0, \tag{7.5.4}$$

then (7.5.3) is strengthened to

$$\int_0^\infty \left\| B^* e^{(A_\Pi^* + \epsilon I)t} x \right\|_U^2 dt \leq C_{\epsilon,\omega} \|x\|_Y^2, \quad x \in Y, \tag{7.5.5}$$

for all $\epsilon > 0$ such that $-\omega + \epsilon < 0$ (as in (7.4.1.1)).

Proof. Parts (a) and (b):   Consider the integral equation

$$w(t) = e^{A^*t}x + \int_0^t e^{A^*(t-\tau)}\Pi^* w(\tau)\, d\tau, \quad x \in Y, \tag{7.5.6a}$$

in the Y-valued unknown $w(t) = w(t, 0; x)$, corresponding to the problem

$$\dot{w} = (A^* + \Pi^*)w \in [\mathcal{D}(A)]', \qquad w(0) = x \in Y. \tag{7.5.6b}$$

Via (7.5.1) with $\Pi^* A^{*-1} \in \mathcal{L}(Y)$, we may interpret (7.5.6a) as

$$y(t) = e^{A^*t}A^*x + A^* \int_0^t e^{A^*(t-\tau)}\Pi^* A^{*-1} y(\tau)\, d\tau \in [\mathcal{D}(A)]' \tag{7.5.6c}$$

with $y(t) = A^* w(t) \in [\mathcal{D}(A)]'$, where the s.c. semigroup e^{A^*t}, originally defined on Y, is extended while preserving the same notation, on the extrapolation space $[\mathcal{D}(A)]'$. Define the operator $\mathcal{F}$ by setting

$$(\mathcal{F}f)(t) \equiv e^{A^*t}x + \int_0^t e^{A^*(t-\tau)}\Pi^* f(\tau)\, d\tau, \quad x \in Y, \tag{7.5.7a}$$

where we note preliminarily that

$$\begin{cases} f \in L_p(0, T; \mathcal{D}(A^*)) \Rightarrow (\mathcal{F}f)(t) = \int_0^t e^{A^*(t-\tau)}\Pi^* A^{*-1} A^* f(\tau)\, d\tau \\ x = 0 \end{cases}$$

$$\in C([0, T]; Y), \tag{7.5.7b}$$

by (7.5.1), since $\Pi^* A^{*-1} \in \mathcal{L}(Y)$. We next obtain a more technical result.

Claim   We claim that $\mathcal{F}$ is well defined as an operator

$$\mathcal{F} : L_2(0, T; \mathcal{D}(B^*)) \to L_2(0, T; \mathcal{D}(B^*)), \tag{7.5.7c}$$

where

$$\|z\|_{\mathcal{D}(B^*)}^2 \equiv \|B^*z\|_U^2 + \|z\|_Y^2, \quad z \in \mathcal{D}(B^*) \subset Y. \tag{7.5.8}$$

Indeed, the term $e^{A^*t}x$ is in $L_2(0, T; \mathcal{D}(B^*))$ by the standing assumption (H.3) = (7.1.5) with $q = 2$. As to the integral term in (7.5.7), setting $v = f_1 - f_2 \in L_2(0, t_0; \mathcal{D}(B^*))$, we compute for $\mathcal{F}f_1 - \mathcal{F}f_2 = \mathcal{F}v$, via (7.5.7), Schwarz's inequality, and a change in the order of integration:

$$\int_0^{t_0} \|B^*(\mathcal{F}v)(t)\|_U^2\, dt = \int_0^{t_0} \left\| \int_0^t B^* e^{A^*(t-\tau)}\Pi^* v(\tau)\, d\tau \right\|_U^2 dt \tag{7.5.9}$$

$$\le t_0 \int_0^{t_0} \int_0^t \left\| B^* e^{A^*(t-\tau)}\Pi^* v(\tau) \right\|_U^2 d\tau\, dt$$

$$= t_0 \int_0^{t_0} \int_\tau^{t_0} \left\| B^* e^{A^*(t-\tau)}\Pi^* v(\tau) \right\|_U^2 dt\, d\tau$$

$$= t_0 \int_0^{t_0} \int_0^{t_0-\tau} \left\| B^* e^{A^*\sigma}\Pi^* v(\tau) \right\|_U^2 d\sigma\, d\tau. \tag{7.5.10}$$

Majorizing $t_0 - \tau$ by t_0 in the upper end point and using first assumption (H.3) = (7.1.5), and then assumption (7.5.1), as well as (7.5.8), we obtain from (7.5.10)

$$\int_0^{t_0} \|B^*(\mathcal{F}v)(t)\|_U^2 \, dt \leq t_0 C_{t_0} \int_0^{t_0} \|\Pi^*v(\tau)\|_Y^2 \, d\tau$$

$$\leq t_0 C_{t_0} C \int_0^{t_0} \left[\|B^*v(\tau)\|_U^2 + \|v(\tau)\|_Y^2 \right] d\tau$$

$$= t_0 C_{t_0} C \|v\|_{L_2(0,t_0; \mathcal{D}(B^*))}^2, \tag{7.5.11}$$

where C_{t_0} and C are the constants in (7.1.5c) and (7.5.1), respectively. Moreover, the same argument from (7.5.9) to (7.5.11) without B^* in front yields by (7.5.7), (7.5.1), (7.5.8), and $\|e^{A^*t}\|_{\mathcal{L}(Y)} \leq M_{t_0}$, $0 \leq t \leq t_0$:

$$\int_0^{t_0} \|(\mathcal{F}v)(t)\|_Y^2 \, dt = \int_0^{t_0} \left\| \int_0^t e^{A^*(t-\tau)} \Pi^*v(\tau) \, d\tau \right\|_U^2 dt$$

$$\leq t_0^2 M_{t_0}^2 \int_0^{t_0} \|\Pi^*v(\tau)\|_Y^2 \, d\tau$$

$$\leq t_0^2 M_{t_0}^2 C \int_0^{t_0} \left[\|B^*v(\tau)\|_U^2 + \|v(\tau)\|_Y^2 \right] d\tau$$

$$= t_0^2 M_{t_0}^2 C \|v\|_{L_2(0,t_0; \mathcal{D}(B^*))}^2. \tag{7.5.12}$$

Thus (7.5.11), (7.5.12), and (7.5.8) yield

$$\|\mathcal{F}v\|_{L_2(0,t_0; \mathcal{D}(B^*))}^2 = \int_0^{t_0} \left\{ \|B^*(\mathcal{F}v)(t)\|_U^2 + \|(\mathcal{F}v)(t)\|_Y^2 \right\} dt$$

$$\leq C_{1t_0} \|v\|_{L_2(0,t_0; \mathcal{D}(B^*))}^2, \tag{7.5.13}$$

$C_{1t_0} = t_0 C(C_{t_0} + t_0 M_{t_0}^2)$, and the claim (7.5.7c) is proved because t_0 is so far arbitrary.

Taking now t_0 sufficiently small so that $C_{1t_0} < 1$, we get that $\mathcal{F}$ is a contraction on $L_2(0, t_0; \mathcal{D}(B^*))$. Hence, the linear integral equation (7.5.6a) has a unique solution $w(t) = w(t, 0; x)$ such that $B^*w(t, 0; x) \in L_2(0, t_0; U)$, $x \in Y$. Indeed, using this result and the convolution theorem on the integral (7.5.6a) via (7.5.1), we can improve the regularity of this solution to read $w(t, 0; x) \in C([0, t_0]; Y)$, $x \in Y$. For any preassigned $T < \infty$, we can then repeat the preceding procedure a finite number of times on $[0, t_0]$, $[t_0, 2t_0]$, etc. and conclude that the linear problem (7.5.6) admits a unique global solution

$$w(t, 0; x) \in C([0, T]; Y) \quad \text{such that } B^*w(t, 0; x) \in L_2(0, T; U), \tag{7.5.14}$$

for any $x \in Y$. Moreover, one verifies from (7.5.6a) that $w(t, 0; x)$ satisfies the semigroup property. Hence, we can write $w(t, 0; x) \equiv S(t)x$, with $S(t)$ a s.c. semigroup on Y. Then, finally, $w(t, 0; x) = e^{A_\Pi^* t}x$ by (7.5.6). Both parts (a) and (b) are proved.

Part (c): We invoke Theorem 7.4.2.1 with A and F there replaced by A_Π^* and B^* now. This is permissible: A_Π^* generates a s.c. semigroup $e^{A_\Pi^* t}$ possessing the required trace regularity (7.5.3) [= assumption (7.4.2.1) in Theorem 7.4.2.1], as well as the required (uniform) stability by (7.5.4). □

7.6 Regularity of a Class of Abstract Damped Systems

The present section, though treated at the abstract level, is motivated by, and is ultimately applicable to, a large class of hyperbolic/Petrowski-type damped PDEs. A few canonical examples are given at the end to illustrate the applicability of the abstract results.

7.6.1 Mathematical Setting and Assumptions

Let X and U be two Hilbert spaces. The basic standing assumptions of the present section are:

(h.1). $\mathcal{A} : X \supset \mathcal{D}(\mathcal{A}) \to X$ is a positive, self-adjoint operator;
(h.2). $\mathcal{B} : U \to [\mathcal{D}(\mathcal{A}^{\frac{1}{2}})]'$; equivalently, $\mathcal{A}^{-\frac{1}{2}}\mathcal{B} \in \mathcal{L}(U; X)$ and $\mathcal{B}^* \mathcal{A}^{-\frac{1}{2}} \in \mathcal{L}(X; U)$.

Conservative Dynamics First we consider the conservative dynamics

$$\ddot{x} + \mathcal{A}x = \mathcal{B}u; \quad \text{or} \quad \dot{y} = Ay + Bu, \quad y = [\dot{x}, x], \quad y(0) = y_0 \in Y; \quad (7.6.1.1)$$

$$A = \begin{bmatrix} 0 & I \\ -\mathcal{A} & 0 \end{bmatrix} = -A^* : Y \supset \mathcal{D}(A) \to Y;$$

$$B = \begin{bmatrix} 0 \\ \mathcal{B} \end{bmatrix}; \quad A^{-1}B = \begin{bmatrix} -\mathcal{A}^{-1}\mathcal{B} \\ 0 \end{bmatrix} \in \mathcal{L}(U; Y); \quad (7.6.1.2)$$

$$Y \equiv \mathcal{D}(\mathcal{A}^{\frac{1}{2}}) \times X; \quad \mathcal{D}(A) = \mathcal{D}(A^*) = \mathcal{D}(\mathcal{A}) \times \mathcal{D}(\mathcal{A}^{\frac{1}{2}}); \quad (7.6.1.3)$$

$$B^*y = B^* \begin{bmatrix} y_1 \\ y_2 \end{bmatrix} = \mathcal{B}^* y_2; \quad \mathcal{B}^* \in \mathcal{L}(\mathcal{D}(\mathcal{A}^{\frac{1}{2}}); U); \quad (Bu, y)_Y = (u, B^*y)_U.$$

$$(7.6.1.4)$$

Damped Dynamics Next we consider the following damped dynamics, that is, (7.6.1.1) with feedback control $u = -\epsilon B^* \dot{x}$, where $\epsilon > 0$:

$$\ddot{x} + \mathcal{A}x + \epsilon \mathcal{B}\mathcal{B}^* \dot{x} = 0; \quad \text{or} \quad \dot{y} = A_\epsilon y, \quad y = [\dot{x}, x], \quad y(0) = y_0 \in Y; \quad (7.6.1.5)$$

$$A_\epsilon \equiv A - \epsilon B B^* = \begin{bmatrix} 0 & I \\ -\mathcal{A} & -\epsilon \mathcal{B}\mathcal{B}^* \end{bmatrix} : Y \supset \mathcal{D}(A_\epsilon) \to Y; \quad (7.6.1.6)$$

$$\mathcal{D}(A_\epsilon) = \left\{ y_1, y_2 \in \mathcal{D}(\mathcal{A}^{\frac{1}{2}}) \subset \mathcal{D}(\mathcal{B}^*) : [\mathcal{A}^{\frac{1}{2}} y_1 + \epsilon \mathcal{A}^{-\frac{1}{2}} \mathcal{B}\mathcal{B}^* y_2] \in \mathcal{D}(\mathcal{A}^{\frac{1}{2}}) \right\}. \quad (7.6.1.7)$$

where, in writing (7.6.1.7), we have made use of (h.2):

$$A_\epsilon^* = \begin{bmatrix} 0 & -I \\ \mathcal{A} & -\epsilon \mathcal{B}\mathcal{B}^* \end{bmatrix} : Y \supset \mathcal{D}(A_\epsilon^*) \to Y; \qquad (7.6.1.8)$$

$$\mathcal{D}(A_\epsilon^*) = \{y_1, y_2 \in \mathcal{D}(\mathcal{A}^{\frac{1}{2}}) \subset \mathcal{D}(\mathcal{B}^*) : [\mathcal{A}^{\frac{1}{2}} y_1 - \epsilon \mathcal{A}^{-\frac{1}{2}} \mathcal{B}\mathcal{B}^* y_2] \in \mathcal{D}(\mathcal{A}^{\frac{1}{2}})\}.$$
$$(7.6.1.9)$$

7.6.2 Main Regularity Results

The usual dissipativity arguments leading to the Lumer–Phillips theorem [Pazy, 1983, p. 14] or its corollary [Pazy, 1983, p. 15] yield a generation result.

Proposition 7.6.2.1 *Assume (h.1) and (h.2). The densely defined operators A_ϵ and A_ϵ^* in (7.6.1.6), (7.6.1.7), and (7.6.1.8), (7.6.1.9), respectively, are dissipative:*

$$\mathrm{Re}(A_\epsilon y, y)_Y = \mathrm{Re}(A_\epsilon^* y, y)_Y = -\epsilon \|\mathcal{B}^* y\|_Y^2 = -\epsilon \|\mathcal{B}^* y_2\|_X^2,$$

$$y = [y_1, y_2] \in \mathcal{D}(A_\epsilon), \qquad (7.6.2.1)$$

in fact maximally dissipative, and thus they generate s.c. contraction semigroups $e^{A_\epsilon t}$ and $e^{A_\epsilon^* t}$ on Y, $t \geq 0$.

Proof. We limit ourselves to dissipativity. First, we note that, if $y = [y_1, y_2] \in \mathcal{D}(A_\epsilon)$, then (7.6.1.7) implies that $y_1, y_2 \in \mathcal{D}(\mathcal{A}^{\frac{1}{2}}) \subset \mathcal{D}(\mathcal{B}^*)$, recalling also (h.2); and thus, by $\mathcal{A}$ in (7.6.1.2) and Y in (7.6.1.3), we have that

$$\begin{cases} \mathrm{Re}(\mathcal{A}y, y)_Y = (y_2, y_1)_{\mathcal{D}(\mathcal{A}^{\frac{1}{2}})} - (\mathcal{A}y_1, y_2)_X = \text{well defined} = 0, \\ \mathrm{Re}(\mathcal{A}^* y, y)_Y = 0, \quad y = [y_1, y_2] \in \mathcal{D}(\mathcal{A}^{\frac{1}{2}}) \times \mathcal{D}(\mathcal{A}^{\frac{1}{2}}). \end{cases} \qquad (7.6.2.2)$$

Then in view of (7.6.1.6) and (7.6.1.8), we see that (7.6.2.2) readily yields (7.6.2.1), via also (7.6.1.4). □

The main result of this Section 7.6 refers to the dynamics (7.6.1.5), under the action of a control u:

$$\ddot{x} + \mathcal{A}x + \epsilon \mathcal{B}\mathcal{B}^* \dot{x} = \mathcal{B}u; \quad \text{or} \quad \dot{y} = A_\epsilon y + Bu, \quad y(0) = y_0 \in Y, \qquad (7.6.2.3)$$

whose solution is

$$y(t) = e^{A_\epsilon t} y_0 + (L_\epsilon u)(t); \quad (L_\epsilon u)(t) = \int_0^t e^{A_\epsilon (t-\tau)} Bu(\tau)\, d\tau, \qquad (7.6.2.4)$$

where, moreover, we now assume the following hypothesis.

(h.3). With $\epsilon > 0$ fixed, $e^{A_\epsilon t}$ is exponentially stable on Y: There exist constants $M \geq 1$ and $\delta > 0$ (depending on ϵ), such that

$$\|e^{A_\epsilon t}\|_{\mathcal{L}(Y)} = \|e^{(A - \epsilon BB^*)t}\|_{\mathcal{L}(Y)} \leq Me^{-\delta t}, \quad t \geq 0. \qquad (7.6.2.5)$$

Theorem 7.6.2.2 *Assume (h.1), (h.2), and (h.3) = (7.6.2.5). Then, with reference to (7.6.2.3) or (7.6.2.4), we have:*

(i)

$$B^* e^{A_\epsilon^* t} : continuous \ Y \to L_2(0, \infty; U); \qquad (7.6.2.6a)$$

more precisely,

$$\int_0^\infty \left\| B^* e^{A_\epsilon^* t} y_0 \right\|_U^2 dt = \frac{1}{2\epsilon} \| y_0 \|_Y^2, \quad \forall \ y_0 \in Y. \qquad (7.6.2.6b)$$

(ii) With $\epsilon > 0$ fixed, and $\delta > 0$ as in (h.3) = (7.6.2.4), let $\alpha > 0$ be any number such that

$$-\delta + \alpha < 0. \qquad (7.6.2.7)$$

Then, with reference to the operator L_ϵ in (7.6.2.4), we have

$$e^{\alpha t} L_\epsilon : continuous \ L_2(0, \infty; U) \to L_2(0, \infty; Y) \cap C([0, \infty]; Y). \qquad (7.6.2.8)$$

7.6.3 Proof of Theorem 7.6.2.2: Dual Statement (7.6.2.6)

Step 1. Lemma 7.3.1 Assume (h.1) and (h.2). Then, with reference to (7.6.2.3), we have:

$$\int_0^\infty \left\| B^* e^{A_\epsilon^* t} y_0 \right\|_U^2 \leq \frac{1}{2\epsilon} \| y_0 \|_Y^2, \quad \forall \ y_0 \in \mathcal{D}(A_\epsilon^*), \qquad (7.6.3.1)$$

where the equality sign holds true in (7.6.3.1), if, moreover, $e^{A_\epsilon^* t}$ is strongly stable on Y.

Proof. Let $y_0 \in \mathcal{D}(A_\epsilon^*)$ so that, by Proposition 7.6.2.1 and (7.6.1.9),

$$\begin{cases} y(t) = e^{A_\epsilon^* t} y_0 \in C([0, T]; \mathcal{D}(A_\epsilon^*)); \ \ \mathcal{D}(A_\epsilon^*) \subset \mathcal{D}(\mathcal{A}^{\frac{1}{2}}) \times \mathcal{D}(\mathcal{A}^{\frac{1}{2}}); \\ \dot{y}(t) = A_\epsilon^* y(t) = e^{A_\epsilon^* t} A_\epsilon^* y_0 \in C([0, T]; Y), \quad \forall \ T < \infty. \end{cases} \qquad (7.6.3.2)$$

We then compute for $y(t)$ in (7.6.3.2), via (7.6.1.6) and (7.6.2.2),

$$\frac{1}{2} \frac{d}{dt} \| y(t) \|_Y^2 = \mathrm{Re}(\dot{y}(t), y(t))_Y = \mathrm{Re}(A_\epsilon^* y(t), y(t))_Y = \mathrm{Re}(y(t), A_\epsilon y(t))_Y$$

$$(7.6.3.3)$$

$$= \mathrm{Re}(y(t), (A - \epsilon B B^*) y(t))_Y$$

$$= \mathrm{Re}(y(t), A y(t))_Y - \epsilon \| B^* y(t) \|_U^2, \quad 0 \leq t \leq T, \qquad (7.6.3.4)$$

where the splitting in (7.6.3.4) is justified as in (7.6.2.2), since $y(t) \in \mathcal{D}(\mathcal{A}^{\frac{1}{2}}) \times \mathcal{D}(\mathcal{A}^{\frac{1}{2}})$, by (7.6.3.2), so that the term $(y(t), A y(t))_Y$ is well-defined, and its real part vanishes.

Thus (7.6.3.4) yields, after integration over $[0, T]$:

$$\|y(T)\|_Y^2 + 2\epsilon \int_0^T \|B^* y(t)\|_U^2 \, dt = \|y_0\|_Y^2, \quad y_0 \in \mathcal{D}(A_\epsilon^*). \qquad (7.6.3.5)$$

Dropping the first positive term in (7.6.3.5), and letting $T \to \infty$, yields (7.6.3.1) for $y_0 \in \mathcal{D}(A_\epsilon^*)$, as desired (with the equality sign, if $e^{A_\epsilon^* t}$ is strongly stable on Y, so that $y(T) \to 0$ as $T \to \infty$ via (7.6.3.2)).

Step 2   We next need to extend (7.6.3.1) to all $y_0 \in Y$. To this end, we need the following:

Lemma 7.6.3.2   *Assume (h.1) and (h.2). Let $A_\epsilon^{-1} \in \mathcal{L}(Y)$. Then*

$$A_\epsilon^{-1} B = (A - \epsilon B B^*)^{-1} B : continuous \ U \to Y = \mathcal{D}\big(A^{\frac{1}{2}}\big) \times X. \qquad (7.6.3.6)$$

Proof.   Let $u \in U$ and compute $(A - \epsilon B B^*)^{-1} Bu = z$, $z = [z_1, z_2]$ via (7.6.1.6) and (7.6.1.2), that is,

$$\begin{cases} Bu = (A - \epsilon B B^*)z = \begin{bmatrix} z_2 \\ -\mathcal{A}z_1 - \epsilon \mathcal{B}\mathcal{B}^* z_2 \end{bmatrix} = \begin{bmatrix} 0 \\ \mathcal{B}u \end{bmatrix}, \quad \text{or} & (7.6.3.7a) \\[2mm] z_2 = 0, \ -\mathcal{A}z_1 = \mathcal{B}u; \quad \text{hence } z_1 = -\mathcal{A}^{-1}\mathcal{B}u \in \mathcal{D}\big(\mathcal{A}^{\frac{1}{2}}\big), & (7.6.3.7b) \end{cases}$$

recalling (h.2). Thus, $z = [z_1, z_2] \in Y$. Hence, $A_\epsilon^{-1} B$ maps U into Y. Since A_ϵ is closed, then the closed graph theorem [Kato, 1966, p. 164] yields the desired conclusion (7.6.3.6). $\blacksquare$

Step 3   We introduce the densely defined operator J (recall Lemma 7.3.1),

$$J \equiv B^* e^{A_\epsilon^*} : Y \supset \mathcal{D}(A_\epsilon^*) \to L_2(0, \infty; U), \qquad (7.6.3.8)$$

and its adjoint J^*,

$$J^* g = \int_0^\infty e^{A_\epsilon t} B g(t) \, dt : L_2(0, \infty; U) \supset \mathcal{D}(J^*) \to Y, \qquad (7.6.3.9)$$

and show

Lemma 7.6.3.3   *Assume (h.1), (h.2), and (h.3) = (7.6.2.5). Then*

(i) The operator J^* in (7.6.3.9) is densely defined; more precisely

$$H^1(0, \infty; U) \subset \mathcal{D}(J^*). \qquad (7.6.3.10)$$

(ii) The operator J in (7.6.3.8) is closable.

Proof.   (i) Let $g \in H^1(0, \infty; U)$, so that $g(\infty) = 0$ and $\dot{g} \in L_2(0, \infty; U)$. Moreover, $A_\epsilon^{-1} \in \mathcal{L}(Y)$, a fortiori from (h.3). Then, integrating (7.6.3.9) by parts yields

$$\int_0^\infty e^{A_\epsilon t} B g(t) \, dt = \int_0^\infty \frac{d}{dt} \left(A_\epsilon^{-1} e^{A_\epsilon t} \right) B g(t) \, dt \qquad (7.6.3.11)$$

$$= -A_\epsilon^{-1} B g(0) - \int_0^\infty e^{A_\epsilon t} A_\epsilon^{-1} B \dot{g}(t) \, dt \in Y. \qquad (7.6.3.12)$$

We now invoke Lemma 7.6.3.2, Eqn. (7.6.3.6), as well as (h.3). Then, each term in (7.6.3.12) is well defined in Y. Thus J^* is well defined on $H^1(0, \infty; U)$.

(ii) It is standard [Kato, 1966, p. 168] that since J^* is densely defined, then J is closable, as claimed. $\square$

Step 4   We return to (7.6.3.1) (with equality sign, under (h.3)) and use Lemma 7.6.3.3(ii). Thus, we can extend (7.6.3.1) to (7.6.2.6b), $\forall \, y_0 \in Y$. The proof of Theorem 7.6.2.2(i) is complete. $\square$

Step 5   Then, Theorem 7.6.2.6(ii) follows from its part (i), Eqn. (7.6.2.5), by an application of Theorem 7.4.1.1 with A, ω, and ϵ there, replaced by A_ϵ, δ, and α now. (Recall that $A_\epsilon^{-1} B \in \mathcal{L}(U; Y)$ is the present assumption (H.2) in Theorem 7.4.1.1.)
                                                                                        $\square$

7.7 Illustrations of Theorem 7.6.2.2 to Boundary Damped Wave Equations

The abstract Theorem 7.6.2.2 encompasses a large class of conservative hyperbolic/Petrowski-type PDEs, once they are subject to suitable damping. Here, below we consider a few canonical cases of wave equations with boundary damping.

7.7.1 Wave Equation with Boundary Damping in the Neumann BC

Let Ω be an open, bounded domain in $\mathbb{R}^n$, $n \geq 2$, with smooth boundary $\Gamma = \Gamma_0 \cup \Gamma_1$, $\Gamma_0 \neq \phi$, $\bar{\Gamma}_0 \cap \bar{\Gamma}_1 = \phi$. On Ω we consider a wave equation with boundary dissipation on Γ_1 in the Neumann BC, subject to boundary control u here:

$$\begin{cases} w_{tt} = \Delta w & \text{in } (0, \infty) \times \Omega \equiv Q; & (7.7.1.1a) \\[2mm] w(0, \cdot) = w_0, \; w_t(0, \cdot) = w_1 & \text{in } \Omega; & (7.7.1.1b) \\[2mm] w|_{\Sigma_0} \equiv 0 & \text{in } (0, \infty) \times \Gamma_0 \equiv \Sigma_0; & (7.7.1.1c) \\[2mm] \left. \dfrac{\partial w}{\partial \nu} \right|_{\Sigma_1} = -\epsilon w_t + u & \text{in } (0, \infty) \times \Gamma_1 \equiv \Sigma_1, \; \epsilon > 0. & (7.7.1.1d) \end{cases}$$

Abstract Setting   Problem (7.7.1.1) can be brought into the abstract setting of Section 7.6.1, after introducing the following spaces and operators (see Chapter 3, Section 3.3

for details):

$$X = L_2(\Omega); \quad U = L_2(\Gamma_1);$$

$$\mathcal{A}f = -\Delta f, \quad \mathcal{D}(\mathcal{A}) = \left\{ f \in H^2(\Omega) : f|_{\Gamma_0} = 0; \left.\frac{\partial f}{\partial \nu}\right|_{\Gamma_1} = 0 \right\}; \quad (7.7.1.2)$$

$$\begin{cases} h = Ng \iff \left\{ \Delta h = 0 \text{ in } \Omega; \ h|_{\Gamma_0} = 0; \left.\frac{\partial f}{\partial \nu}\right|_{\Gamma_1} = g \right\}; & (7.7.1.3a) \\ N : L_2(\Gamma) \to H^{\frac{3}{2}}(\Omega) \subset H^{\frac{3}{2}-2\sigma}(\Omega) \equiv \mathcal{D}(\mathcal{A}^{\frac{3}{4}-\sigma}), \quad \forall \sigma > 0; & (7.7.1.3b) \end{cases}$$

$$\mathcal{B} = -\mathcal{A}N; \quad \mathcal{B}^*f = -N^*\mathcal{A}f = f|_{\Gamma}, \quad f \in \mathcal{D}(\mathcal{A}); \quad (7.7.1.4)$$

$$\mathcal{D}(\mathcal{A}^{\frac{1}{2}}) = H^1_{\Gamma_0}(\Omega); \quad Y \equiv \mathcal{D}(\mathcal{A}^{\frac{1}{2}}) \times X \equiv H^1_{\Gamma_0}(\Omega) \times L_2(\Omega), \quad (7.7.1.5)$$

where $H^1_{\Gamma_0}(\Omega) = \{ f \in H^1(\Omega) : f|_{\Gamma_0} = 0 \}$. See, in particular, Chapter 3, Eqn. (3.3.1.12) for (7.7.1.4).

Assumptions (h.1) and (h.2)   These are readily verified to hold true in the above setting; in particular,

$$\mathcal{B} = -\mathcal{A}N \in \mathcal{L}\big(U; [\mathcal{D}(\mathcal{A}^{\frac{1}{2}})]'\big) \iff \mathcal{A}^{-\frac{1}{2}}\mathcal{B} = -\mathcal{A}^{\frac{1}{2}}N \in \mathcal{L}(L_2(\Gamma); L_2(\Omega)),$$

$$(7.7.1.6)$$

where the latter relationship in (7.7.1.6) is true a fortiori by (7.7.1.3b).

Assumption (h.3)   It is well-known by now that under certain geometrical conditions (which are beyond the point to specify here) on the uncontrolled part Γ_0 of the boundary (in particular, with no geometrical conditions if Γ_0 is empty) [Bardos et al., 1992; Lasiecka, Triggiani, 1992, Lemma 7.2, p. 218], the s.c. continuous semigroup $\{w_0, w_1\} \to \{w(t), w_t(t)\}$ obtained from (7.7.1.1) after setting $u \equiv 0$ in (7.7.1.1d) is *exponentially stable in the uniform norm of Y* in (7.7.1.5), so that assumption (h.3) holds true as well. There is a large literature on these problems. The first exponential stabilization result (hence, exact controllability [Russell, 1974]) for problem (7.7.1.1) with $u \equiv 0$ was obtained in Chen [1979], using energy methods known for the corresponding exterior obstacle problem, under rather strong geometrical conditions on the triple $\{\Omega, \Gamma_0, \Gamma_1\}$. These were progressively relaxed; see Lagnese [1989]. A marked improvement was obtained in [Lasiecka, Triggiani, 1992, Lemma 7.2, p. 218], using pseudo-differential analysis, by eliminating altogether prior geometrical conditions on the *controlled* part of the boundary. The sharpest conditions are given in Bardos et al. [1992] by means of geometric optics techniques.

The point here, however, is that Theorem 7.6.2.2 is applicable to problem (7.7.1.1) and yields, for each $\epsilon > 0$ fixed, the following regularity properties:

$$\left.\begin{array}{l} u \in L_2(0, \infty; L_2(\Gamma_1)) \\ \{w_0, w_1\} \in H^1_{\Gamma_0}(\Omega) \times L_2(\Omega) \end{array}\right\} \to \begin{array}{l} \{w, w_t\} \in L_2(0, \infty; H^1_{\Gamma_0}(\Omega) \times L_2(\Omega)) \\ \cap C([0, \infty]; H^1_{\Gamma_0}(\Omega) \times L_2(\Omega)). \end{array} \quad (7.7.1.7)$$

Instead, by contrast, if $\epsilon = 0$ (and dim $\Omega \geq 2$), we know from the subsequent Chapter 8, Appendix 8A that: $\{w, w_t\}$ cannot lie in $C([0, T]; H^1(\Omega) \times L_2(\Omega))$, in fact not even in $C([0, T]; H^{\frac{3}{4}+\sigma}(\Omega) \times H^{-\frac{1}{4}+\sigma}(\Omega))$, $\forall \sigma > 0$. The *soft-analysis technique* of the relatively simple Theorem 7.6.2.2 in the damped case for $\epsilon > 0$ should be contrasted with the *pseudo-differential/micro-local analysis techniques* needed to obtain sharp regularity results for problem (7.7.1.1) in the undamped case, when $\epsilon = 0$!

7.7.2 Wave Equation with Boundary Damping in the Dirichlet BC

Let Ω be an open bounded domain in $\mathbb{R}^n$, $n \geq 1$, with smooth boundary Γ. On Ω we consider a wave equation with boundary dissipation on Γ_1 in the Dirichlet BC, subject to boundary control u here:

$$\begin{cases} w_{tt} = \Delta w & \text{in } (0, \infty) \times \Omega \equiv Q; & (7.7.2.1a) \\ w(0, \cdot) = w_0; \ w_t(0, \cdot) = w_1 & \text{in } \Omega; & (7.7.2.1b) \\ w|_\Sigma = \epsilon \dfrac{\partial(\mathcal{A}_0^{-1} w_t)}{\partial v} + u & \text{in } (0, \infty) \times \Gamma \equiv \Sigma, & (7.7.2.1c) \end{cases}$$

where the operator $\mathcal{A}_0$ is defined below in (7.7.2.2).

Abstract Setting   Problem (7.7.2.1) can be brought into the abstract setting of Section 7.6.1, after introducing the following spaces and operators [see Chapter 3, Section 3.1 for details]:

$$\mathcal{A}_0 f = -\Delta f, \quad \mathcal{D}(\mathcal{A}_0) = H^2(\Omega) \cap H^1_0(\Omega); \quad \mathcal{D}(\mathcal{A}_0^{\frac{1}{2}}) = H^1_0(\Omega); \quad (7.7.2.2)$$

$$X = [\mathcal{D}(\mathcal{A}_0^{\frac{1}{2}})]' = H^{-1}(\Omega); \quad U = L_2(\Gamma); \quad (7.7.2.3)$$

$$\begin{cases} h = Dg \Longleftrightarrow \{\Delta h = 0 \text{ in } \Omega; \ h|_\Gamma = g\}, & (7.7.2.4a) \\ D : L_2(\Gamma) \to H^{\frac{1}{2}}(\Omega) \subset H^{\frac{1}{2}-2\sigma}(\Omega) \equiv \mathcal{D}(\mathcal{A}_0^{\frac{1}{4}-\sigma}), \quad \forall \sigma > 0; & (7.7.2.4b) \end{cases}$$

$$\mathcal{B} = -\mathcal{A}_0 D, \quad \mathcal{B}^* f = -D^* \mathcal{A}_0 f = \left.\frac{\partial f}{\partial v}\right|_\Gamma, \quad f \in \mathcal{D}(\mathcal{A}) \quad (7.7.2.5)$$

(see Chapter 3, Eqn. (3.1.9) for (7.7.2.5)). We now view the $(-\Delta)$ with range in $H^{-1}(\Omega)$. Thus we define

$$\mathcal{A} = -\Delta : \mathcal{D}(\mathcal{A}_0^{\frac{1}{2}}) = H^1_0(\Omega) \to [\mathcal{D}(\mathcal{A}_0^{\frac{1}{2}})]'_{L_2(\Omega)} = H^{-1}(\Omega) = X; \quad (7.7.2.6)$$

$$\mathcal{D}(\mathcal{A}^{\frac{1}{2}}) = L_2(\Omega); \quad Y = \mathcal{D}(\mathcal{A}^{\frac{1}{2}}) \times X = L_2(\Omega) \times H^{-1}(\Omega); \quad (7.7.2.7)$$

$$[\mathcal{D}(\mathcal{A}^{\frac{1}{2}})]'_X = [\mathcal{D}(\mathcal{A}_0)]'_{L_2(\Omega)}. \quad (7.7.2.8)$$

Here $[\quad]'_X$ is the duality with respect to the pivot space X, while $[\quad]'_{L_2(\Omega)}$ is the duality with respect to the pivot space $L_2(\Omega)$.

Assumptions (h.1) and (h.3)   These are readily verified to hold true in the above setting: By (7.7.2.8),

$$\mathcal{B} = -\mathcal{A}_0 D \in \mathcal{L}\left(U; \left[\mathcal{D}(\mathcal{A}^{\frac{1}{2}})\right]'_X\right) \Longleftrightarrow \mathcal{A}_0 D \in \mathcal{L}\left(U; [\mathcal{D}(\mathcal{A}_0)]'_{L_2(\Omega)}\right) \qquad (7.7.2.9)$$

$$\Longleftrightarrow \mathcal{A}_0^{-1}\mathcal{A}_0 D = D \in \mathcal{L}(L_2(\Gamma); L_2(\Omega)),$$

$$(7.7.2.10)$$

where the latter relation in (7.7.2.10) holds true, a fortiori by (7.7.2.4b).

Assumption (h.3)   It is by now well-known that the s.c. semigroup $\{w_0, w_1\} \rightarrow \{w(t), w_t(t)\}$ obtained from problem (7.7.2.1) after setting $u \equiv 0$ in (7.7.2.1c) is *exponentially stable in the uniform norm of Y* in (7.7.2.7), so that assumption (h.3) holds true as well. One should point out that such exponential stability for system (7.7.2.1) with damped feedback in the Dirichlet BC is much more difficult than the corresponding problem of exponential decay for system (7.7.1.1) with damped feedback in the Neumann BC, as in Section 7.7.1. This is due to the different topology level of the state spaces in the two cases: the low topology $L_2(\Omega) \times H^{-1}(\Omega)$ in the Dirichlet case, as opposed to the more amenable energy level topology $H^1(\Omega) \times L_2(\Omega)$ in the Neumann case. The first exponential stability result for problem (7.7.2.1) (hence exact controllability) [Russell, 1974] with $u \equiv 0$ was obtained in Lasiecka and Triggiani [1987] under the geometrical condition that Ω is strictly convex (or the set difference $\Omega = \Omega_1 \backslash \Omega_2$ of two strictly convex sets Ω_1 and Ω_2). This geometrical condition on the *controlled* part of the boundary (in particular, with damping acting on the whole boundary as in (7.7.2.1c)) was subsequently eliminated altogether in [Lasiecka, Triggiani, 1992, Theorem 1.1, p. 191] by means of pseudo-differential calculus. The sharpest result with *damping only on a portion of the boundary* is again obtained by the geometric optics techniques of Bardos et al. [1992].

The point here, however, is that Theorem 7.6.2.2 is applicable to problem (7.7.1.1) and yields, for each $\epsilon > 0$ fixed, the following regularity properties:

$$u \in L_2(0, \infty; L_2(\Gamma)) \rightarrow \{w, w_t\} \in L_2(0, \infty; L_2(\Omega) \times H^{-1}(\Omega))$$

$$\cap \ C([0, \infty]; L_2(\Omega) \times H^{-1}(\Omega)). \quad (7.7.2.11)$$

In contrast with the Neumann illustration of Section 7.7.1, here if we let $\epsilon = 0$ in (7.7.2.1c), it remains true that

$$u \in L_2(0, T; L_2(\Gamma)) \rightarrow C([0, T]; L_2(\Omega) \times H^{-1}(\Omega)), \qquad (7.7.2.12)$$

as will be shown in Chapter 10, Section 10.5.

We finally remark that the damped coupled problem of Chapter 9, Section 9.10.4 can also be covered by Theorem 7.6.2.2.

Notes on Chapter 7

Sections 7.1–7.5.

These sections are based entirely on results established in Lasiecka and Triggiani [1983; 1988], and Flandoli et al. [1998]. More precisely, Theorem 7.2.1 and Theorem 7.3.1 were established in Lasiecka and Triggiani [1983] (a refinement of Lasiecka and Triggiani [1981] in a study of the wave equation with Dirichlet/Neumann control by abstract methods using cosine operator theory). For a general statement of Theorem 7.2.1, see also Appendix 1 of Flandoli et al. [1988]. Sections 7.3 and 7.4.1 follow closely Lasiecka and Triggiani [1988], while Sections 7.4.2 and 7.5 are taken from [Flandoli et al., 1988, pp. 344–50].

Sections 7.6–7.7

Theorem 7.6.2.2 is taken from Lasiecka [1988]. It applies to many hyperbolic/Petrowski-type PDEs We limited ourselves to the two canonical cases involving the wave equation. The first, with Neumann BC, shows the interesting feature noted at the end of Section 7.7.1: that the map: boundary control $u \to$ solution $\{w, w_t\}$ is markedly more regular (when dim $\Omega \geq 2$) for the boundary damped problem ($\epsilon > 0$) than for the undamped case ($\epsilon = 0$). This phenomenon does not occur, instead, in the case of Dirichlet BC of Section 7.7.2 for all dim $\Omega \geq 1$. This is not the place to review uniform stabilization literature, beyond the references given in the text. See the subsequent Chapter 12.

References and Bibliography

C. Bardos, G. Lebeau, and J. Rauch, Sharp sufficient conditions for the observation control and stabilization of waves from the boundary, *SIAM J. Control Optim.* **30** (1992), 1024–1065.

G. Chen, Energy decay estimates and exact boundary value controllability for the wave equation in a bounded domain, *J. Math. Pures Appl.* (9) **58** (1979), 249–274.

N. Dunford and J. Schwartz, *Linear Operators*, Part I, Wiley-Interscience, 1957.

F. Flandoli, I. Lasiecka, and R. Triggiani, Algebraic Riccati equations with nonsmoothing observating arising in hyperbolic and Euler–Bernoulli equations, *Ann. Matem. Pura Appl.*, **CLii** (1988), 307–382.

T. Kato, *Perturbation Theory for Linear Operators*, Springer-Verlag, 1966.

J. Lagnese, *Boundary Stabilization of Thin Plates*, SIAM Studies in Applied Mathematics, 1989.

I. Lasiecka, Stabilization of hyperbolic and parabolic systems with non-linearly perturbed boundary conditions, *J. Diff. Eqn.* **75** (1988), 53–87.

I. Lasiecka and R. Triggiani, A cosine operator approach to modeling $L_2(0, T; L_2(\Gamma))$ boundary input hyperbolic equations, *Appl. Math. Optim.* **7** (1981), 35–93.

I. Lasiecka and R. Triggiani, Regularity of hyperbolic equations under $L_2(0, T; L_2(\Gamma))$-boundary terms, *Appl. Math. Optim.* **10** (1983), 275–286.

I. Lasiecka and R. Triggiani, A lifting theorem for the time regularity of solutions to abstract equations with unbounded operators and applications to hyperbolic equations, *Proceedings Am. Math. Soc.* **103** (1988), 745–755.

I. Lasiecka and R. Triggiani, *Differential and Algebraic Riccati Equations with Application to Boundary/Point Control Problems: Continuous Theory and Approximation Theory*, Springer-Verlag Lecture Notes in Control and Information Sciences, **164** (1991), 160 pp.

I. Lasiecka and R. Triggiani, Uniform exponential energy decay of wave equations in a bounded region with $L_2(0, \infty; L_2(\Gamma))$-feedback control in the Dirichlet boundary conditions, *J. Diff. Eqn.* **66** (1987), 340–390.

I. Lasiecka and R. Triggiani, Uniform stabilization of the wave equation with Dirichlet or Neumann feedback control without geometrical conditions, *Appl. Math. Optim.* **25** (1992), 189–224.

J. L. Lions, *Optimal Control of Systems Governed by Partial Differential Equations*, Springer-Verlag, 1971.

R. H. Martin, *Non-Linear Operators and Differential Equations in Banach Spaces*, Wiley-Interscience, 1976.

A. Pazy, *Semigroups of Linear Operators and Applications to Partial Differential Equations*, Springer-Verlag, New York Berlin Heidelberg Tokyo, 1983.

D. L. Russell, Exact boundary value controllability theorems for wave and heat processes in star complemented regions, in *Differential Games and Control Theory*, E. Roxin, Lin, and Sternberg, eds., Dekker, 1974.

8

Optimal Quadratic Cost Problem Over a Preassigned Finite Time Interval: The Case Where the Input → Solution Map Is Unbounded, but the Input → Observation Map Is Bounded

In the present volume, beginning with (the preliminary Chapter 7 and with) this Chapter 8, the assumption of analyticity of the original s.c. semigroup e^{At} of the free dynamics is dropped altogether. Thus, Chapters 8 through 10 study, under different assumptions, the optimal quadratic cost problem over a preassigned finite time interval $[0, T]$, $T < \infty$, for the abstract dynamics $\dot{y} = Ay + Bu$, where, as before, the degree of unboundedness of B is up to that of A. (Notice that the extreme case, $\gamma = 1$ in the notation of Chapters 1 and 2, is here admitted, whereas it was excluded in Volume 1 where the s.c. semigroup e^{At} was, in addition, analytic.) The ultimate goal is, as usual, to establish existence and uniqueness of the operator differential Riccati equation that arises in the pointwise feedback synthesis of the optimal pair (optimal control and optimal trajectory). To this end, various abstract mathematical settings are possible, each distinguished by a specific set of assumptions and each motivated, in turn, by representative applications to boundary/point control problems for PDEs. In this chapter we select a general framework, marked by the key feature that the input-solution operator (L_0 below in (8.1.5)) is *unbounded* as an operator acting between the basic input space $L_2(0, T; U)$ and the basic solution space $L_2(0, T; Y)$. Although the study of this fully pathological case is deferred to the technical treatment of Chapter 16 (in Volume 3), in the present Chapter 8 we consider two different subsettings on two parallel tracks. They are characterized by two sets of assumptions: (H.1), (H.2), (H.3) and (h.1), (h.2), respectively. Our primary emphasis is on the first case where an observation operator R is present, bounded from a state space Y to an output space Z, such that the composition RL_0 is better behaved than each of its components viewed separately. Indeed, under the framework (H.1), (H.2), (H.3) of Section 8.2.1, the operator RL_0 is, in fact, continuous between the input space $L_2(0, T; U)$ and the output space $C([0, T]; Z)$. [In the more regular situation of Section 8.2.2, RL_0 induces some compactness properties on a space smoother than $L_2(0, T; U)$.]

Whereas the setting cumulatively described by (H.1), (H.2), and (H.3) couples a fully unbounded control operator B with a bounded observation operator R, an alternative parallel procedure can be identified under assumptions (h.1) and (h.2) that allows the unboundedness between the input and the output of the dynamical system to be distributed between B and R in a setting that is mathematically simpler. In particular, R may be fully unbounded and B may be, instead, bounded, in which case (h.2) is contained in (h.1). Partial differential equation applications of this case are deferred to Chapter 13 of Volume 3. To be sure, the full strength of the hypotheses of the various settings is invoked to achieve the ultimate goal of Theorem 8.2.1.2: the establishment of existence and uniqueness of the operator differential Riccati equation, which arises in the pointwise feedback synthesis of the optimal pair (optimal control and optimal trajectory).

Preliminary results (such as those of Theorem 8.2.1.1 involving the explicit formula of the optimal quantities directly in terms of the problem's data) require weaker assumptions still: (h.1) and (h.3).

We provide two approaches in establishing our main result, Theorem 8.2.1.2, which encompasses existence and uniqueness of the differential and integral Riccati equations. Our primary emphasis is on the so-called variational approach (the same approach of Chapters 1 and 2), which originates with the optimal control problem: Here, as in Volume 1, we first construct an explicit candidate Riccati operator, which is expressed solely in terms of the data of the problem, and we derive for it structural regularity properties (Subsections 8.3.3 and 8.3.4). Next, we verify that such candidate does indeed verify the DRE/IRE (Subsections 8.3.4–8.3.6). Finally, uniqueness is then established within the class of solutions possessing the aforementioned properties (Subsection 8.3.7).

We find it also instructive and enlightening to provide a second proof of our main result, by the so-called direct approach in Section 8.4, through a reverse procedure: This first settles well-posedness (existence and uniqueness) of the IRE, via a local contraction mapping near the end time T, followed by global a priori bounds (Subsections 8.4.1–8.4.4), and finally reconstructs the optimal control problem leading to the original IRE (Subsection 8.4.5).

By contrast, in the abstract setting of the subsequent Chapter 9, the input-solution operator L_0 will be continuous $L_2(0, T; U) \to C([0, T]; Y)$ and the observation operator R will be, in particular, bounded from Y to Z; yet, though specialized as far as L_0 is concerned, the setting of Chapter 9 is not generally contained in the setting of Chapter 8, because of an additional technicality that R need not be boundedly invertible. Each of the two frameworks, the one of the present Chapter 8 and the one of the subsequent Chapter 9, is inspired by special classes of mixed problems for partial differential equations. Motivation for the primary mathematical setting of this present Chapter 8 under (H.1), (H.2), and (H.3) – as well as under (H.4) and (H.5) – comes from an important and natural *purely boundary* problem (boundary control and boundary observation) for the second-order hyperbolic partial differential equations: Here, say, the wave equation on a bounded domain Ω, dim $\Omega \geq 2$, with boundary Γ is acted upon by controls exercised in the Neumann boundary conditions and the quadratic functional cost penalizes the $L_2(0, T; L_2(\Gamma))$-norm of the Neumann controls as well as the $L_2(0, T; L_2(\Gamma))$-norm of the Dirichlet *trace* of the corresponding solutions. It is this example that inspires the mathematical setting and guides the assumptions of Chapter 8 under (H.1), (H.2), and (H.3), as well as (H.4) and (H.5). It is treated in Section 8.6. (The framework (h.1), (h.2), when specialized to B bounded, whereby (h.2) is a consequence of (h.1), includes many classes of hyperbolic or "plate like" PDEs with *distributed* control, which penalize suitable *traces* of the solution; see Chapter 13 of Volume 3.) By contrast, the motivating examples of the subsequent Chapter 9 include a second-order hyperbolic partial differential equation with $L_2(0, T; L_2(\Gamma))$-control action exercised in the Dirichlet boundary conditions and penalization of the solution (position and velocity) in the *interior*; as well as many other similar problems for hyperbolic and plate like PDEs with boundary control and interior penalization of the solution. Chapter 10 will give a more refined treatment of this latter class, which under stronger assumptions (always verified in relevant PDE problems) markedly relaxes the assumptions on R. We stress that application of the abstract settings of this chapter under (H.1), (H.2), and (H.3), as well as (H.4) and (H.5), to the Neumann boundary control/Dirichlet boundary observation for general second-order hyperbolic equations, where R is the (Dirichlet) trace operator, rests critically on the sharp regularity properties (interior as well as trace regularity) for these mixed problems reported in Lasiecka and Triggiani [1983; 1989(a); 1990; 1991(b); 1994], part of which is summarized in Lemmas 8.5.1.1 and 8.5.1.2 below; see Appendix 8A.

We conclude this chapter by providing, in Section 8.7, an explicitly computable example of a one-dimensional hyperbolic equation with Dirichlet control (B unbounded) and point observation (R unbounded) that satisfies assumptions (h.1) and (h.3), but not (h.2), (H.1), (H.2), and (H.3), where the DRE is trivially satisfied as a linear (rather than quadratic) equation on $\mathcal{D}(A)$.

8.1 Mathematical Setting and Formulation of the Problem

Dynamical Model In this chapter, we consider the following abstract differential equation:

$$\dot{y} = Ay + Bu \quad \text{on, say, } [\mathcal{D}(A^*)]'; \quad y(0) = y_0 \in Y, \tag{8.1.1}$$

subject to the following preliminary assumptions *to be maintained throughout the chapter.*

(i) A is the infinitesimal generator of a strongly continuous semigroup e^{At} on the Hilbert space Y, $t \geq 0$. Without loss of generality for the problem here considered, where the dynamics (8.1.1) is studied over a finite time interval $[0, T]$, $T < \infty$, we may assume that A^{-1} exists as a bounded operator on all Y, that is, that $A^{-1} \in \mathcal{L}(Y)$.

(ii) B is a (linear) continuous operator $U \to [\mathcal{D}(A^*)]'$; equivalently,

$$A^{-1}B \in \mathcal{L}(U; Y). \tag{8.1.2}$$

In (8.1.1), A^* is the Y-adjoint of A and $[\mathcal{D}(A^*)]'$ is the Hilbert space dual to the space $\mathcal{D}(A^*) \subset Y$ with respect to the Y-topology, with norms

$$\|y\|_{\mathcal{D}(A^*)} = \|A^* y\|_Y; \quad \|y\|_{[\mathcal{D}(A^*)]'} = \|A^{-1} y\|_Y. \tag{8.1.3}$$

Via (8.1.2), we let $(B^* x, u)_U = (x, Bu)_Y$ for $u \in U$, $x \in \mathcal{D}(A^*)$, $B^* \in \mathcal{L}(\mathcal{D}(A^*); U)$, and $(A^{-1} B)^* = B^* A^{*-1} \in \mathcal{L}(Y; U)$.

The mild solution of problem (8.1.1) can be written as

$$y(t) = e^{At} y_0 + (L_0 u)(t), \tag{8.1.4}$$

$$(L_0 u)(t) = \int_0^t e^{A(t-\tau)} Bu(\tau)\, d\tau = A \int_0^t e^{A(t-\tau)} A^{-1} Bu(\tau)\, d\tau \tag{8.1.5a}$$

$$: \text{ continuous } L_1(0, T; U) \to C([0, T]; [\mathcal{D}(A^*)]'), \tag{8.1.5b}$$

where (8.1.5b) follows from (8.1.2). In contrast with (8.1.5b), L_0 is instead a closed, densely defined, *unbounded* operator, as viewed as an operator $L_2(0, T; U) \supset \mathcal{D}(L_0)$ $\to L_2(0, T; Y)$. (The subscript "0" reminds us that the action of L_0 starts at $t = 0$ and will make the notation in (8.1.5a) consistent with that in (8.1.12) below. This notation should not be confused with the notation L_T in Chapter 7.) We shall also need the adjoint operator $L_0^* : L_2(0, T; Y) \supset \mathcal{D}(L_0^*) \to L_2(0, T; U)$ in the sense that

$$(L_0 u, f)_{L_2(0,T;Y)} = (u, L_0^* f)_{L_2(0,T;U)},$$

which is explicitly given by

$$(L_0^* f)(t) = \int_t^T B^* e^{A^*(\tau - t)} f(\tau)\, d\tau \tag{8.1.6a}$$

$$: \text{ continuous } L_1(0, T; \mathcal{D}(A^*)) \to C([0, T]; U). \tag{8.1.6b}$$

Optimal Control Problem. Interval $[0, T]$ With the dynamics (8.1.1) we associate the following quadratic cost functional over a preassigned fixed time interval $[0, T]$, $0 < T < \infty$:

$$J(u, y) = \int_0^T \left[\|Ry(t)\|_Z^2 + \|u(t)\|_U^2 \right] dt, \tag{8.1.7}$$

where Z is another Hilbert (output) space and the corresponding optimal control problem is

$$\text{minimize } J(u, y) \text{ over all } u \in L_2(0, T; U), \tag{8.1.8}$$

where $y(t) = y(t; y_0)$ is the solution of (8.1.1) due to u. In (8.1.7) and (8.1.8), R is an observation operator from Y to Z, which is assumed to satisfy the hypothesis

$$\text{either} \quad R \in \mathcal{L}(Y; Z) \text{ under (H.2) below;}$$

$$\text{or else} \quad R \in \mathcal{L}(\mathcal{D}(A); Z) \text{ under (h.1) below.} \tag{8.1.9}$$

Interval $[s, T]$ In the following we shall have to consider the case where we take $t = s \geq 0$ as the new initial time for the dynamics (8.1.1) with new initial condition $y(s) = y_0$ (instead of $t = 0$ as before). The corresponding optimal control problem over the time interval $[s, T]$, $s < T < \infty$ is then:

Minimize, over all $u \in L_2(s, T; U)$, the cost functional

$$J_s(u, y) = \int_s^T \left\{ \|Ry(t)\|_Z^2 + \|u(t)\|_U^2 \right\} dt, \tag{8.1.10}$$

where now $y(t) = y(t, s; y_0)$ is the solution of Equation (8.1.1) with initial condition $y(s) = y_0$ due to u, which is therefore given by

$$y(t, s; y_0) = e^{A(t-s)}y_0 + (L_s u)(t), \quad s \leq t \leq T, \tag{8.1.11}$$

where consistently with (8.1.5a) we have set

$$(L_s u)(t) = \int_s^t e^{A(t-\tau)} Bu(\tau) \, d\tau, \quad s \leq t \leq T, \tag{8.1.12a}$$

$$: \text{continuous } L_1(s, T; U) \to C([s, T]; [\mathcal{D}(A^*)]'). \tag{8.1.12b}$$

Its L_2-adjoint L_s^*, in the same sense as in (8.1.6a) above for $s = 0$, is

$$(L_s^* f)(t) = \int_t^T B^* e^{A^*(\tau-t)} f(\tau) \, d\tau, \quad s \leq t \leq T, \tag{8.1.13a}$$

$$: \text{continuous } L_1(s, T; \mathcal{D}(A^*)) \to C([s, T]; U), \tag{8.1.13b}$$

Remark 8.1.1   In the following we shall use the notation $\mathcal{B}([0, T]; \mathcal{L}(Y))$ to indicate the Banach space of linear operators in $\mathcal{L}(Y)$ at each t, which are uniformly bounded on $[0, T]$. If, say, $P \in \mathcal{B}([0, T]; \mathcal{L}(Y))$, then the corresponding norm is

$$\|P\|_{\mathcal{B}([0,T];\mathcal{L}(Y))} \equiv \sup_{0 \leq t \leq T} \|P(t)\|_{\mathcal{L}(Y)}. \tag{8.1.14}$$

Similar considerations apply to the space $\mathcal{B}([0, T]; \mathcal{L}(Y; U))$, also to be encountered in the following.

Orientation under (H.1), (H.2), and (H.3)   In the setting of the present chapter the unbounded coefficient operators A and B give rise to an *unbounded* input-solution operator L_0 from the basic input space $L_2(0, T; U)$ to the basic solution space $L_2(0, T; Y)$, whereas the observation operator R is *bounded* from Y to Z. There are important and natural boundary control problems – such as the Neumann boundary control/Dirichlet boundary observation problem for second-order hyperbolic partial differential equations described in the subsequent Section 8.6 – that fit into this situation. It is, in fact, the desire to cover this hyperbolic problem that motivates the present chapter and provides a guiding example. Two distinct yet complementary situations will be considered: one, more general, in Section 8.2.1; and the other, more regular, in Section 8.2.2, where stronger assumptions will yield stronger conclusions. To motivate our assumptions (H.1), (H.2), and (H.3), we return to the aforementioned hyperbolic example with the Neumann boundary control and Dirichlet boundary observation. In it [see Lasiecka and Triggiani [1983; 1989(a); 1990; 1991(b); 1994], L_0

is unbounded $L_2(0, T; U) \to L_2(0, T; Y)$, $U = L_2(\Gamma)$, $Y = H^1(\Omega) \times L_2(\Omega)$, dim $\Omega \geq 2$, while the observation operator R is the Dirichlet trace, which is bounded $Y \to Z = L_2(\Gamma)$. Furthermore, the composition RL_0, which gives the maps from the control space $L_2(0, T; L_2(\Gamma))$ to the space $C([0, T]; L_2(\Gamma))$ of the Dirichlet observations of the hyperbolic solutions, is, in fact, bounded; that is, it is better behaved than its component R and L_0 viewed separately. This property is, in fact, a distinctive feature of waves and plates problems. This case is included in the framework of Section 8.2.1 (refer, in particular, to assumption (H.3) = (8.2.1.3), in its dual version (H.3*) = (8.2.1.6)).

Moreover, in the guiding hyperbolic example, if one takes the Neumann boundary controls in the smoother space

$$H^{\frac{1}{2}}(\Sigma) \equiv L_2\left(0, T; H^{\frac{1}{2}}(\Gamma)\right) \cap H^{\frac{1}{2}}(0, T; L_2(\Gamma)), \tag{8.1.15}$$

then the operator L_0 acts continuously into $C([0, T]; H^1(\Omega) \times L_2(\Omega))$, and the operators $L_t^* R^* R L_t$ form a collectively compact family of operators on $H^{\frac{1}{2}}(\Sigma)$. This situation, which yields a more regular optimal control and a more regular optimal solution, is included in the framework of Section 8.2.2. It should be stressed that it is the recent sharp regularity theory for mixed second-order hyperbolic partial differential equations (Lasiecka and Triggiani [1983; 1989(a); 1990; 1991(b); 1994]) that makes it possible to verify all the abstract assumptions (H.1), (H.2), and (H.3) of this chapter and to apply to them the resulting theory. This will be seen in Section 8.6, devoted to the applications, which will confirm that all abstract assumptions *subsume natural properties* for the hyperbolic class of problems of Section 8.6. Based on these abstract assumptions, our main theorem provides an existence and uniqueness result for the solution (given explicitly and constructively) of the differential Riccati equation (8.2.1.32) below.

As in the preceding Chapter 1, the primary strategy in Section 8.3 under the variational approach will be as follows: (i) First, we shall characterize the optimal pair $\{u^0, y^0\}$ solely in terms of the data of the problem; (ii) next, we shall construct (define) an operator $P(t)$ in terms of the original and optimal evolution, hence ultimately in terms of the original data of the problem; and finally, (iii) we shall verify by direct differentiation that such operator $P(t)$ is a global solution of the corresponding differential Riccati equation and possesses suitable regularity properties or a priori bounds. The final Subsection 8.3.6 will yield uniqueness, within the class of solutions with the aforementioned regularity properties. Section 8.4 provides a different (yet technically related) proof of the main result by the so-called direct method, a reverse procedure of the variational method.

We explicitly point out that in the general setting of Theorem 8.2.1.2, the candidate of the "evolution operator" is not well-defined pointwise on its basic space Y (only on the larger space $[\mathcal{D}(A^*)]'$), a property generally needed to carry out the existence arguments of the differential Riccati equation. To overcome this new basic difficulty, new regularity (extension) properties of the optimal solutions need to be established and exploited in Section 8.3.3.

8.2 Statement of Main Results

There are two subsections, each of which is based on three abstract assumptions for problem (8.1.1)–(8.1.9): primarily (H.1) through (H.3) or alternatively (h.1) and (h.2), for the more general setting of Subsection 8.2.1, and (H.4) through (H.6) for the more regular setting of Subsection 8.2.2. (The relationships between these two sets of assumptions will be examined in Remark 8.2.2.2.) Our goal is to develop a full theory up to, and including, a Riccati theory.

8.2.1 The General Case: Theorem 8.2.1.1, Theorem 8.2.1.2, and Theorem 8.2.1.3

Primary Set of Assumptions   Our primary standing assumptions for this subsection are as follows:

(H.1): The map $Re^{At}B$ can be extended as a map

$$Re^{At}B : \text{continuous } U \to L_1(0, T; Z); \qquad (8.2.1.1a)$$

that is,

$$\int_0^T \|Re^{At}Bu\|_Z \, dt \le c_T \|u\|_U, \quad u \in U. \qquad (8.2.1.1b)$$

(H.2): $R \in \mathcal{L}(Y; Z)$; equivalently (by taking $t = 0$),

$$Re^{At} : \text{continuous } Y \to C([0, T]; Z); \qquad (8.2.1.2a)$$

that is,

$$\max_{0 \le t \le T} \|Re^{At}x\|_Z \le c_T \|x\|_Y, \quad x \in Y. \qquad (8.2.1.2b)$$

(H.3): The map $B^*e^{A^*t}R^*$ can be extended as a map

$$B^*e^{A^*t}R^* : \text{continuous } Z \to L_2(0, T; U); \qquad (8.2.1.3a)$$

that is,

$$\int_0^T \|B^*e^{A^*t}R^*z\|_U^2 \, dt \le c_T \|z\|_Z^2, \quad z \in Z. \qquad (8.2.1.3b)$$

Remark 8.2.1.1   By duality on (H.1), (H.2), and (H.3), we obtain, respectively

(H.1*):

$$\text{The map } v \to \int_0^T B^*e^{A^*t}R^*v(t) \, dt \text{ can be extended as a map}$$
$$: \text{continuous } L_\infty(0, T; Z) \to U; \qquad (8.2.1.4a)$$

that is,

$$\left\| \int_0^T B^* e^{A^* t} R^* v(t) \, dt \right\|_U \le c_T \|v\|_{L_\infty(0,T;Z)}. \qquad (8.2.1.4b)$$

or equivalently [Chapter 7, Theorem 7.2.1 and Remark 7.2.1],

$$(L_0^* R^* v)(t) = \int_t^T B^* e^{A^*(\tau - t)} R^* v(\tau) \, d\tau \qquad (8.2.1.4c)$$

$$: \text{ continuous } L_\infty(0, T; Z) \to C([0, T]; U). \qquad (8.2.1.4d)$$

(H.2*):

$$\text{The map } v \to \int_0^T e^{A^* t} R^* v(t) \, dt : \text{continuous } L_1(0, T; Z) \to Y;$$
$$(8.2.1.5a)$$

that is,

$$\left\| \int_0^T e^{A^* t} R^* v(t) \, dt \right\|_Y \le c_T \|v\|_{L_1(0,T;Z)}. \qquad (8.2.1.5b)$$

(H.3*): The map RL_0 (L_0 as in (8.1.5)) can be extended as a map

$$RL_0 : \text{continuous } L_2(0, T; U) \to C([0, T]; Z); \qquad (8.2.1.6)$$

and by duality, the map $L_0^* R^*$ (L_0^* as in (8.1.6)) can be extended as a map

$$L_0^* R^* : \text{continuous } L_1(0, T; Z) \to L_2(0, T; U). \qquad (8.2.1.7)$$

(Equivalence between (H.3) and (H.3*) is obtained by invoking Chapter 7, Remark 7.2.2, Eqn. (7.2.7) for $q = 2$. Property (8.2.1.6) should be contrasted with the regularity (8.1.5b) of L_0 and assumption (8.2.1.2) on R.)

Remark 8.2.1.2   One readily sees, via a change of variable in the integration, that assumption (H.3*) = (8.2.1.6), or (H.3*) = (8.2.1.7), allows one to obtain the following estimates independent of s, $0 \le s \le T$: There is a constant $c_T > 0$ independent of s such that

$$\|RL_s u\|_{C([s,T];Z)} \le c_T \|u\|_{L_2(s,T;U)}; \qquad (8.2.1.8)$$

$$\|L_s^* R^* f\|_{L_2(s,T;U)} \le c_T \|f\|_{L_1(s,T;Z)}; \qquad (8.2.1.9a)$$

whereas (H.1*) = (8.2.1.4d) implies

$$\|L_s^* R^* f\|_{C([s,T];U)} \le c_T \|f\|_{L_\infty(s,T;Z)}. \qquad (8.2.1.9b)$$

Alternative Set of Assumptions   Through a parallel treatment, we shall also deal with a few combinations of the following set of alternative hypotheses:

(h.1): $R : \mathcal{D}(A) \subset \mathcal{D}(R) \to Z$ is closed, $R \in \mathcal{L}(\mathcal{D}(A); Z)$ and Re^{At} can be extended as a map

$$Re^{At} : \text{continuous } Y \to L_2(0, T; Z). \qquad (8.2.1.10)$$

(h.2): The map $Re^{At}B$ can be extended as a map

$$Re^{At}B : \text{continuous } U \to L_2(0, T; Z); \qquad (8.2.1.11a)$$

that is,

$$\int_0^T \|Re^{At}Bu\|_Z^2 \, dt \leq c_T \|u\|_U^2, \quad u \in U. \qquad (8.2.1.11b)$$

(h.3): The map RL_0 can be extended as a map

$$RL_0 : \text{continuous } L_2(0, T; U) \to L_2(0, T; Z). \qquad (8.2.1.12)$$

The dual versions are (see Chapter 7, Theorem 7.2.1 for (h.2)):

(h.1*):

$$\text{The map } v \to \int_0^T e^{A^*t} R^* v(t) \, dt \text{ can be extended as a map}$$

$$: \text{continuous } L_2(0, T; Z) \to Y; \qquad (8.2.1.13a)$$

that is,

$$\left\| \int_0^T e^{A^*t} R^* v(t) \, dt \right\|_Y \leq c_T \|v\|_{L_2(0,T;Z)}. \qquad (8.2.1.13b)$$

(h.2*):

$$\{L_0^* R^* v\}(t) = \int_t^T B^* e^{A^*(\tau - t)} R^* v(\tau) \, d\tau \text{ can be extended as} \qquad (8.2.1.14a)$$

$$: \text{continuous } L_2(0, T; Z) \to C([0, T]; U). \qquad (8.2.1.14b)$$

(h.3*):

$$L_0^* R^* \text{ can be extended as : continuous } L_2(0, T; Z) \to L_2(0, T; U).$$
$$(8.2.1.15)$$

Thus, (h.2) = (8.2.1.11) [respectively (h.2*) = (8.2.1.14)] is stronger than (h.3) = (8.2.1.12) [respectively (h.3*)= (8.2.1.15)]. Moreover, (h.1) = (8.2.1.10) is weaker than (H.2) = (8.2.1.2), whereas (h.3) = (8.2.1.12) [respectively (h.3*) = (8.2.1.15)] is weaker than (H.3) = (8.2.1.3) [respectively (H.3*) = (8.2.1.6)].

As in Remark 7.2.1.2, we then obtain that (h.3) implies the following estimates uniformly in s:

$$\|RL_s u\|_{L_2(s,T;Z)} \leq c_T \|u\|_{L_2(s,T;U)}; \quad \|L_s^* R^* f\|_{L_2(s,T;U)} \leq c_T \|f\|_{L_2(s,T;Z)}.$$
$$(8.2.1.16)$$

Assumptions (h.1)–(h.3) allow for both R and B to be unbounded operators.

We begin with a preliminary result that requires only the weaker hypotheses (h.1) and (h.3). However, a full Riccati theory is given under (h.1) and (h.2).

Theorem 8.2.1.2   *Assume hypotheses (h.1) = (8.2.1.10) and (h.3) = (8.2.1.12) (in addition to the standing assumptions (i) and (ii) of Section 8.1). Then, with reference to the optimal control problem (8.1.10) over the time interval [s, T] for the dynamics (8.1.1) with initial datum $y(s) = y_0 \in Y$, we have:*

(a) There exists a unique optimal pair $\{u^0(t, s; y_0), y^0(t, s; y_0)\}$ given explicitly in terms of the data of the problem by the representation formulas:

$$-u^0(t, s; y_0) = \left\{[I_s + (RL_s)^* RL_s]^{-1}\left[(RL_s)^*\left(Re^{A(\cdot - s)}y_0\right)\right]\right\}(t) \quad (8.2.1.17a)$$

$$\in L_2(s, T; U) \quad (8.2.1.17b)$$

(see Lemma 8.3.1.1, Eqn. (8.3.1.11));

$$y^0(t, s; y_0) = e^{A(t-s)}y_0 + \{L_s u^0(\cdot, s; y_0)\}(t) \in C([s, T]; [\mathcal{D}(A^*)]'), \quad (8.2.1.18)$$

$$Ry^0(t, s; y_0) = \left\{[I_s + RL_s(RL_s)^*]^{-1}\left(Re^{A(\cdot - s)}y_0\right)\right\}(t) \quad (8.2.1.19a)$$

$$\in L_2(s, T; Z) \quad (8.2.1.19b)$$

(see Lemma 8.3.1.1, Eqn. (8.3.1.14)), or else via (8.2.1.17a) and (8.2.1.18)

$$Ry^0(t, s; y_0) = \left\{[I_s - RL_s(I_s + (RL_s)^* RL_s)^{-1}(RL_s)^*]\left(Re^{A(\cdot - s)}y_0\right)\right\}(t)$$

$$(8.2.1.20)$$

(see proof of Lemma 8.3.1.1). The optimal pair satisfies the relation

$$u^0(t, s; y_0) = -\{(RL_s)^* Ry^0(\cdot, s; y_0)\}(t) \in L_2(s, T; U) \quad (8.2.1.21)$$

(see Lemma 8.3.1.1, Eqn. (8.3.1.6)), as well as the estimates

$$\sup_{0 \le s \le T} \left\{\|u^0(\cdot, s; y_0)\|_{L_2(s,T;U)}\right\} \le C_T \|y_0\|_Y, \quad y_0 \in Y \quad (8.2.1.22)$$

(see Lemma 8.3.2.1, Eqn. (8.3.2.1)). The optimal cost J_s^0 is given by

$$J_s^0(y_0) \equiv J(u^0(\cdot, s; y_0), y^0(\cdot, s; y_0))$$

$$= \int_s^T \left[\|Ry^0(t, s; y_0)\|_Z^2 + \|u^0(t, s; y_0)\|_U^2\right] dt$$

$$= \left\|[I_s + RL_s(RL_s)^*]^{-\frac{1}{2}}\left\{Re^{A(\cdot - s)}y_0\right\}\right\|_{L_2(s,T;Z)}^2 \quad (8.2.1.23)$$

(see Lemma 8.3.1.1, Eqn. (8.3.1.18)).

(b) There exists an operator $P(t) \in \mathcal{L}(Y)$, $0 \le t \le T$, defined explicitly in terms of the data of the problem via (8.2.1.19) by

$$P(t)x = \int_t^T e^{A^*(\tau - t)} R^* Ry^0(\tau, t; x) d\tau, \quad x \in Y, \ 0 \le t \le T, \quad (8.2.1.24)$$

such that

(i)

$$\mathcal{L}(Y) \ni P(t) = P^*(t) \geq 0, \quad 0 \leq t \leq T \ (\,^* \ in \ Y) \ (see \ (8.3.4.22)); \quad (8.2.1.25)$$

(ii)

$$P(t) : continuous \ Y \rightarrow L_\infty(0, T; Y) \ (see \ (8.3.3.5)) \qquad (8.2.1.26)$$

[i.e., in the notation of Remark 8.1.1, we have $P \in \mathcal{B}([0, T]; \mathcal{L}(Y)]$;
(iii) the following pointwise (in t) feedback representation holds true for the (unique) optimal pair u^0 and y^0 of problem (8.1.8):

$$u^0(t; s; y_0) = -B^* P(t) y^0(t; s; y_0) \in L_2(s, T; U) \ (see \ (8.3.4.18)); \quad (8.2.1.27)$$

(iv) the optimal cost of problem (8.1.10) for the dynamics (8.1.1) with initial condition $y(s) = y_0$ is given by (see (8.3.4.23))

$$J_s^0(y_0) \equiv J(u^0(\,\cdot\,, s; y_0), y^0(\,\cdot\,, s; y_0)) = (P(s)y_0, y_0)_Y; \quad y_0 \in Y \quad (8.2.1.28a)$$

$$= \left([I_s + RL_s(RL_s)^*]^{-1} Re^{A(\cdot - s)} y_0, \ Re^{A(\cdot - s)} y_0\right)_{L_2(s,T;Z)}, \quad (8.2.1.28b)$$

and thus $P(s)$ is given for $y_0 \in Y$ by

$$P(s)y_0 = \left(Re^{A(\cdot - s)}\right)^* [I_s + RL_s(RL_s)^*]^{-1} Re^{A(\cdot - s)} y_0. \qquad (8.2.1.28c)$$

Stronger assumptions are needed to claim that the operator $P(t)$ in (8.2.1.23) is the unique solution (within a specified class) of the differential Riccati equation.

Theorem 8.2.1.2 *Assume hypotheses (H.1) = (8.2.1.1), (H.2) = (8.2.1.2), and (H.3) = (8.2.1.3) (in addition to the standing assumptions (i) and (ii) of Section 8.1); or, alternatively, assume hypotheses (h.1) = (8.2.1.10) and (h.2) = (8.2.1.11). Thus, in either case, Theorem 8.2.1.1 holds true a fortiori, and in fact (8.2.1.17b) and (8.2.1.19b) can then be strengthened to*

$$u^0(\,\cdot\,, s; y_0) \in C([s, T]; U); \quad Ry^0(\,\cdot\,, s; y_0) \in C([s, T]; Z), \quad (8.2.1.29a)$$

and

$$\sup_{0 \leq s \leq T} \left\{\|Ry^0(\,\cdot\,, s; y_0)\|_{C([s,T];Z)}\right\} \leq c_T \|y_0\|_Y, \quad y_0 \in Y; \quad (8.2.1.29b)$$

$$\|u^0(\,\cdot\,, s; y_0)\|_{C([s,T];U)} \leq c_T \|y_0\|_Y \qquad (8.2.1.29c)$$

(see Lemma 8.3.2.2, Eqn. (8.3.2.4)).

Moreover, with reference to $P(t)$ defined in (8.2.1.24), the following additional results hold true:

(v)

$$V(t) \equiv B^* P(t) \in \mathcal{L}(Y; U), \quad 0 \leq t \leq T$$

and

$$V(t) = B^* P(t) : continuous \ Y \rightarrow L_\infty(0, T; U) \ (see \ (8.3.3.3)) \quad (8.2.1.30)$$

[i.e., in the notation of Remark 8.1.1, we have $V \in \mathcal{B}([0, T]; \mathcal{L}(Y; U))$].

(vi)

$$B^* P(\cdot) e^{A(\cdot - s)} B \ : \ continuous \ U \rightarrow L_2(s, T; U) \ for \ any \ s, \ 0 \leq s < T,$$
$$with \ norm \ that \ may \ be \ taken \ independent \ of \ s$$
$$(or \ U \rightarrow U_{0,T} \ in \ the \ notation \ of \ (8.3.3.1) \ below;$$
$$see \ (8.3.3.4)). \quad (8.2.1.31)$$

(vii) (Existence) The operator $P(t)$ given by (8.2.1.24) is a solution of the following differential Riccati equation (DRE):

$$\begin{cases} \left(\dfrac{d}{dt} P(t)x, y\right)_Y = -(Rx, Ry)_Z - (P(t)x, Ay)_Y - (P(t)Ax, y)_Y \\ \qquad\qquad\qquad + (B^* P(t)x, B^* P(t)y)_U, \\ \qquad\qquad\qquad\qquad\qquad\qquad\qquad \forall \, x, y \in \mathcal{D}(A), \ 0 \leq t < T, \\ \quad P(T) = 0 \end{cases}$$

$$(8.2.1.32)$$

(see Theorem 8.3.4.5, Eqn. (8.3.4.26)), as well as of the corresponding integral Riccati equation (IRE)

$$(P(t)x, y)_Y = \int_t^T \left(R e^{A(\tau - t)} x, \, R e^{A(\tau - t)} y\right)_Z d\tau$$

$$- \int_t^T \left(B^* P(\tau) e^{A(\tau - t)} x, \, B^* P(\tau) e^{A(\tau - t)} y\right)_U d\tau, \quad x, y \in Y$$

$$(8.2.1.33)$$

(see Proposition 8.3.5.1 as well as Theorem 8.3.6.1).

(viii) (Uniqueness) Finally, the operator $P(t)$ given by (8.2.1.24) is the unique solution of the IRE (8.2.1.33), hence of the DRE (8.2.1.32), to enjoy properties (i) = (8.2.1.25); (ii) = (8.2.1.26); (v) = (8.2.1.30); and (vi) = (8.2.1.31); see Section 8.3.7.

The proof of Theorems 8.2.1.1 and 8.2.1.2 will be given in Section 8.3, by the variational method (from the optimal control problem to the DRE and the IRE), and in Section 8.4, by the direct method (from the IRE to the optimal control problem).

We close this section by stating a result that provides several interesting transition properties, which, however, are not needed for the purpose of establishing the DRE or the IRE. To this end, we append a subscript T to make reference to the optimal control problem (8.1.7) on $[0, T]$, as in $u_T^0(\cdot, s; y_0)$, $\Phi_T(\cdot, \cdot)$, $P_T(t)$, etc.

Theorem 8.2.1.3   *Assume (h.1) = (8.2.1.10) and (h.3) = (8.2.1.12) as in Theorem 8.2.1.1. Then, the following transition properties hold true in the above notation, where $\Phi_T(\tau, t)x = y_T^0(\tau, t; x)$*

(i)

$$R\Phi_{T-t}(\sigma, 0)(x) = R\Phi_T(t + \sigma, t)x \underset{(\text{in } \sigma)}{\in} L_2(0, T; Z), \quad x \in Y; \quad (8.2.1.34)$$

(ii)

$$u_T^0(\tau, t; x) = u_{T-t}^0(\tau - t, 0; x) \underset{(\text{in } \tau)}{\in} L_2(t, T; U), \quad x \in Y; \quad (8.2.1.35)$$

(iii)

$$P_{T-t}(0) = P_T(t), \quad 0 \le t \le T. \quad (8.2.1.36)$$

The proof will be given in Section 8.3.8.

Remark 8.2.1.3   With reference to Theorem 8.2.1.1, if R is unbounded, then one generally has $(RL_s)^* \supset L_s^*R^*$ and $(Re^{At})^* \supset e^{A^*t}R^*$. However, the main thrust of the present chapter is to establish the DRE or the IRE and this is achieved with $R \in \mathcal{L}(Y; Z)$ by (H.2). Thus, to streamline the notation, we shall henceforth write $L_s^*R^*$ and $e^{A^*t}R^*$ for $(RL_s)^*$ and $(Re^{At})^*$, respectively, in all cases, which is correct when $R \in \mathcal{L}(Y; Z)$. We shall return to these issues in Chapter 11 in Volume 3.

8.2.2 The Regular Case: Theorem 8.2.2.1

In this subsection we study a more regular case where the optimal solution $y^0(t, 0; x)$ is, in fact, pointwise well-defined in the basic space Y (in contrast with (8.3.2.3) below). This additional regularity property is achieved under the following assumptions, which (in the next subsections) will be verified to hold true in the case of the motivating hyperbolic problem mentioned in the Orientation.

We shall postulate the existence of a Hilbert space $\mathcal{U}_{[0,T]} \subset L_2(0, T; U)$ (algebraically and topologically), with restriction $\mathcal{U}_{[t,T]} \subset L_2(t, T; U)$, such that the following assumptions hold true. (In the application to the hyperbolic problem in Section 8.6, we shall take $\mathcal{U}_{[0,T]} = H^{\frac{1}{2}}(\Sigma)$, the space defined in (8.6.4.1)).

(H.4):   The operator L_0 in (8.1.5) can be extended as a map

$$L_0 : \text{continuous } \mathcal{U}_{[0,T]} \to C([0, T]; Y); \quad (8.2.2.1a)$$

that is,

$$\|L_0 u\|_{C([0,T];Y)} \le c_T \|u\|_{\mathcal{U}_{[0,T]}}. \quad (8.2.2.1b)$$

(H.5):   For each $t \in [0, T]$, $T < \infty$, the map

$$K_t \equiv L_t^*R^*RL_t : \text{compact } \mathcal{U}_{[t,T]} \to \mathcal{U}_{[t,T]}, \quad (8.2.2.2)$$

and, moreover, $\{K_t\}$, $0 \leq t \leq T$, is a family of collectively compact operators on $\mathcal{U}_{[0,T]}$ (in the sense of [Anselone, 1971, p. 4]), that is, the set

$$\bigcup_{0 \leq t \leq T} K_t \left[\text{unit ball of } \mathcal{U}_{[0,T]}\right]$$

is pre-compact, or relatively compact, on $\mathcal{U}_{[0,T]}$. A checkable, sufficient condition – which is the one we will verify in the application of Section 8.6 – is that there exists a fixed Banach space $\mathcal{V}$ contained in $\mathcal{U}_{[0,T]}$ with compact injection, such that the following uniform estimate holds true:

$$\|K_t u\|_{\mathcal{V}} \leq c_T \|u\|_{\mathcal{U}_{[0,T]}}, \quad \forall u \in \mathcal{U}_{[0,T]}, \ \forall t \in [0, T]. \tag{8.2.2.3}$$

Remark 8.2.2.1   Assumptions (H.4) = (8.2.2.1) and (H.2) = (8.2.1.2) imply via a change of variable in the integration (as in Remark 8.2.1.2) the following estimates uniformly in t, $0 \leq t \leq T$: There is a constant $c_T > 0$ independent of t such that

$$\|L_t u\|_{C([t,T];Y)} \leq c_T \|u\|_{\mathcal{U}_{[t,T]}}; \tag{8.2.2.4}$$

$$\left\|L_t^* R^* Re^{A(\cdot -t)} x\right\|_{\mathcal{U}_{[t,T]}} \leq c_T \|x\|_Y, \tag{8.2.2.5}$$

the latter following by duality since $R^* Re^{A(\cdot -t)} x \in C([t, T]; Y)$.

Assumption (H.3*) = (8.2.1.6) gives $L_0^* R^* R L_0$: continuous $L_2(0, T; U) \rightarrow L_2(0, T; U)$, whereas (H.6) = (8.2.2.3) requires a different regularity result.

The main result of this subsection is

Theorem 8.2.2.1   (i) Assume hypothesis (H.5) = (8.2.2.2). Then, with reference to $u^0(\cdot, t; x)$ defined by (8.2.1.10) we have

$$\sup_{0 \leq t \leq T} \left\{\|u^0(\cdot, t; x)\|_{\mathcal{U}_{[t,T]}}\right\} \leq c_T \|x\|_Y. \tag{8.2.2.6}$$

(ii) Assume, in addition, hypothesis (H.4) = (8.2.2.1). Then, with reference to the optimal solution $y^0(\cdot, t; x)$ defined by (8.2.1.18) and (8.2.1.17), we have

$$\sup_{0 \leq t \leq T} \left\{\|y^0(\cdot, t; x)\|_{C([t,T];Y)}\right\} \leq c_T \|x\|_Y, \tag{8.2.2.7}$$

and, we have the standard evolution property on $\mathcal{L}(Y)$,

$$\Phi(\tau, t) = \Phi(\tau, s)\Phi(s, t), \quad t \leq s \leq \tau, \tag{8.2.2.8}$$

for the evolution operator $\Phi(\tau, t)$ defined by

$$\Phi(\tau, t)x = y^0(\tau, t; x), \quad x \in Y. \tag{8.2.2.9}$$

The proof of Theorem 8.2.2.1 will be given in Section 8.5.

8.3 The General Case. A First Proof of Theorems 8.2.1.1 and 8.2.1.2 by a Variational Approach: From the Optimal Control Problem to the DRE and the IRE Theorem 8.2.1.3

In this section we provide a proof of the main results by the variational approach.

8.3.1 Explicit Representation Formulas for the Optimal Pair $\{u^0, y^0\}$ under (h.1), (h.3)

Lemma 8.3.1.1   *Assume (h.1) = (8.2.1.10) and (h.3) = (8.2.1.12). With reference to the optimal control problem (8.1.10) over the time interval $[s, T]$ for the dynamics (8.1.1) with initial condition $y(s) = y_0 \in Y$, we have:*

There exists a unique optimal pair $\{u^0(t, s; y_0), y^0(t, s; y_0)\}$ given explicitly by the representation formulas (8.2.1.17)–(8.2.1.20) and satisfying relation (8.2.1.21).

Proof.   The existence of a unique optimal pair for the optimal problem (8.1.10) follows from standard optimization theory [Luenberger, 1969]. We present two proofs.

(a) A First Proof by Lagrange Multipliers. Proof of (8.2.1.21).   This pair satisfies the following necessary *optimality conditions.* As in Chapter 1, Section 1.4.1, we introduce the Lagrangean with $u \in L_2(s, T; U)$, $y \in L_2(s, T; \mathcal{D}(R))$, and $p \in L_2(s, T; [\mathcal{D}(R)]')$,

$$\mathcal{L}(u, y, p) \equiv \frac{1}{2}\left\{\|u\|^2_{L_2(s,T;U)} + \|Ry\|^2_{L_2(s,T;Z)}\right\} + \left(p, y - e^{A(\cdot -s)}y_0 - L_s u\right)_{L_2(s,T;Y)},$$

$$(8.3.1.1)$$

where the domain $\mathcal{D}(R)$ of the closed operator R is a Banach space under the graph norm, and $[\mathcal{D}(R)]'$ is the dual space with respect to the pivot space Y. In the notation of [Luenberger, 1969, pp. 240–3] we have $X \equiv L_2(s, T; U) \times L_2(s, T; \mathcal{D}(R))$ for $x = \{u, y\}$; $2f(u, y) = \|u\|^2_{L_2(s,T;U)} + \|Ry\|^2_{L_2(s,T;Z)}$, so that f is continuously Frechet differentiable on X; $H(u, y) = y - e^{A(\cdot -s)}y_0 - L_s u : X \to L_2(s, T; \mathcal{D}(R))$ is continuously Frechet differentiable by (h.1) and (h.3), with Frechet differential $H'(u^0, y^0)(\delta u, \delta y) = \delta y - L_s \delta u$, which is surjective from X onto $L_2(s, T; \mathcal{D}(R))$. [For $z \in L_2(s, T; \mathcal{D}(R))$, we have $\delta y - L_s \delta u = z$ with $\delta u = 0$ and $\delta y = z \in L_2(s, T; \mathcal{D}(R))$.] Liusternik's Lagrange multiplier theorem [Luenberger, 1969, p. 243] applies and gives: There exists $\{u^0, y^0, p^0\}$,

$$u^0 \in L_2(s, T; U); \quad Ry^0 \in L_2(s, T; Z); \quad p^0 \in L_2(s, T; [\mathcal{D}(R)]'),$$

such that $\mathcal{L}_u = \mathcal{L}_y = \mathcal{L}_p = 0$ at (u^0, y^0, p^0). From the Lagrangean (8.3.1.1), we compute $\mathcal{L}_p = 0$ or $y^0 = e^{A(\cdot -s)}y_0 + L_s u^0$, and moreover, $\mathcal{L}_y = 0$ and $\mathcal{L}_u = 0$ to obtain respectively,

$$(R^* R y^0, \delta y)_{L_2(s,T;Y)} + (p^0, \delta y)_{L_2(s,T;\mathcal{D}(R))} = 0, \quad \forall \delta y \in L_2(s, T; \mathcal{D}(R)), \quad (8.3.1.2)$$

or

$$p^0 = -R^* R y^0, \tag{8.3.1.3}$$

and $p^0 \in \mathcal{D}(L_s^*) \subset L_2(s, T; Y)$, $L_s^* p^0 \in L_2(0, T; U)$ and

$$(u^0 - L_s^* p^0, \delta u)_{L_2(s,T;U)} = 0, \quad \forall \, \delta u \in L_2(s, T; U), \tag{8.3.1.4}$$

or

$$u^0 = L_s^* p^0 = -L_s^* R^* R y^0 \in L_2(s, T; U), \tag{8.3.1.5}$$

where in the last step we have invoked (8.3.1.3) and (h.3) in its version (8.2.1.16). We henceforth denote u^0 and y^0 explicitly as $u^0(\,\cdot\,, s; y_0)$ and $y^0(\,\cdot\,, s; y_0)$, since they plainly depend on y_0 and on the initial time $t = s$ via (8.3.1.1). Thus, we rewrite (8.3.1.5) as

$$u^0(\,\cdot\,, s; y_0) = -L_s^* R^* R y^0(\,, s; y_0), \quad y_0 \in Y, \tag{8.3.1.6}$$

and (8.2.1.21) is proved.

Proof of (8.2.1.17) and (8.2.1.20).   As in Chapter 1, Section 1.4.1, we next use the optimal dynamics

$$y^0(\,\cdot\,, s; y_0) = e^{A(\cdot - s)} y_0 + L_s u^0(\,\cdot\,, s; y_0) \tag{8.3.1.7}$$

to eliminate y^0 via (8.3.1.6). We obtain

$$\begin{aligned} -u^0(\,\cdot\,, s; y_0) &= L_s^* R^* R y^0(\,\cdot\,, s; y_0) \\ &= L_s^* R^* R e^{A(\cdot - s)} y_0 + L_s^* R^* R L_s u^0(\,\cdot\,, s; y_0); \end{aligned} \tag{8.3.1.8}$$

$$-[I_s + L_s^* R^* R L_s] u^0(\,\cdot\,, s; y_0) = L_s^* R^* R e^{A(\cdot - s)} y_0 \tag{8.3.1.9a}$$

$$\in L_2(s, T; U). \tag{8.3.1.9b}$$

To justify the regularity claimed in (8.3.1.9b), we first see that by assumption (h.1) = (8.2.1.10), we have $R e^{A(\cdot - s)} y_0 \in L_2(s, T; Z)$ continuously in y_0; then, by invoking assumption (h.3*) = (8.2.1.15) on $L_s^* R^*$ in its version as in (8.2.1.16), we obtain the claimed regularity (8.3.1.9b), with a norm bound independent of s. But the operator $[I_s + L_s^* R^* R L_s]$ is positive self-adjoint on the space $L_2(s, T; U)$, and, in fact,

$$([I_s + L_s^* R^* R L_s] u, u)_{L_2(s,T;U)} \geq \|u\|_{L_2(s,T;U)}, \tag{8.3.1.10a}$$

$$\|[I_s + L_s^* R^* R L_s]^{-1}\|_{\mathcal{L}(L_2(s,T;U))} \leq 1, \tag{8.3.1.10b}$$

uniformly in s, $0 \leq s \leq T$. Thus, the operator on the left-hand side of (8.3.1.9a) is boundedly invertible on $L_2(s, T; U)$, and we then obtain

$$-u^0(\,\cdot\,, s; y_0) = [I_s + L_s^* R^* R L_s]^{-1} \big[L_s^* R^* \big(R e^{A(\cdot - s)} y_0 \big) \big] \in L_2(s, T; U), \tag{8.3.1.11}$$

that is, (8.2.1.17). From (8.3.1.11), one readily derives (8.2.1.20) by inserting (8.3.1.11) into the optimal dynamics (8.3.1.7).

Proof of (8.2.1.19).   In contrast, if we return to (8.3.1.6) and use this time the optimal dynamics (8.3.1.7) to eliminate u^0, we then obtain

$$[I_s + RL_s L_s^* R^*]Ry^0(\,\cdot\,, s; y_0) = Re^{A(\,\cdot\,-s)}y_0 \in L_2(s, T; Z) \qquad (8.3.1.12)$$

by assumption (h.1) = (8.2.1.10). But, again, $[I_s + RL_s L_s^* R^*]$ is a positive, self-adjoint operator on $L_2(s, T; Z)$, and

$$([I_s + RL_s L_s^* R^*]z, z)_{L_2(s,T;Z)} \geq \|z\|_{L_2(s,T;Z)}, \qquad (8.3.1.13a)$$

$$\|[I_s + RL_s L_s^* R^*]^{-1}\|_{\mathcal{L}(L_2(s,T;Z))} \leq 1, \qquad (8.3.1.13b)$$

uniformly in s, $0 \leq s \leq T$. Thus, the operator on the left-hand side of (8.3.1.12) is boundedly invertible on $L_2(s, T; Z)$, and we obtain

$$Ry^0(\,\cdot\,, s; y_0) = [I_s + RL_s L_s^* R^*]^{-1}\left(Re^{A(\,\cdot\,-s)}y_0\right) \in L_2(s, T; Z), \qquad (8.3.1.14)$$

that is, (8.2.1.19).

Proof of (8.2.1.23).   By (8.2.1.21), the optimal cost is rewritten as

$$
\begin{aligned}
J_s^0(y_0) &= (Ry^0(\,\cdot\,, s; y_0), Ry^0(\,\cdot\,, s; y_0))_{L_2(s,T;Z)} \\
&\quad + (L_s^* R^* Ry^0(\,\cdot\,, s; y_0), L_s^* R^* Ry^0(\,\cdot\,, s; y_0))_{L_2(s,T;U)}
\end{aligned}
$$
$$(8.3.1.15)$$

$$= ([I_s + RL_s L_s^* R^*]Ry^0(\,\cdot\,, s; y_0), Ry^0(\,\cdot\,, s; y_0))_{L_2(s,T;Z)}$$

$$= \left(Re^{A(\,\cdot\,-s)}y_0, Ry^0(\,\cdot\,, s; y_0)\right)_{L_2(s,T;Z)} \qquad (8.3.1.16)$$

(by (8.2.1.19)) $$= \left(Re^{A(\,\cdot\,-s)}y_0, [I_s + RL_s L_s^* R^*]^{-1} Re^{A(\,\cdot\,-s)}y_0\right)_{L_2(s,T;Z)}$$

$$(8.3.1.17)$$

$$= \left\|[I_s + RL_s L_s^* R^*]^{-\frac{1}{2}} Re^{A(\,\cdot\,-s)}y_0\right\|_{L_2(s,T;Z)}^2, \qquad (8.3.1.18)$$

where in going from (8.3.1.15) to (8.3.1.16) and from (8.3.1.16) to (8.3.1.17) we invoke (8.2.1.19) twice, and in going from (8.3.1.17) to (8.3.1.18) we use self-adjointness of the operator $[I_s + RL_s L_s^* R^*]$. Thus, (8.3.1.18) shows (8.2.1.23). $\square$

(b) A Second, Direct, Elementary Proof with Final State Penalization.   The preceding proof shows the Lagrange multiplier formalism, which is potentially applicable to more general situations. In the present quadratic cost case, an ad hoc elementary proof may be given. At little extra effort, we shall consider a more general cost functional than (8.1.10), one which penalizes also the final state

$$J_s(u, y) = \int_s^T \left\{\|Ry(t)\|_Z^2 + \|u(t)\|_U^2\right\} dt + \|Gy(T)\|_{Z_f}^2, \qquad (8.3.1.19)$$

under hypotheses (h.1) and (h.3), as well as under the following hypotheses for G:

$$Ge^{At} \in \mathcal{L}(Y; Z_f); \quad GL_{sT}u = G \int_s^T e^{A(T-\tau)} Bu(\tau) \, d\tau \in \mathcal{L}(L_2(s, T; U); Z_f),$$
$$(8.3.1.20)$$

with Z_f another Hilbert space.

Lemma 8.3.1.2 *Assume (h.1), (h.3), and (8.3.1.20). Then, omitting the dependence on y in (8.3.1.19), we have the following identity:*

$$J_s(u) - J_s(u^0) = (\Lambda_{sT}[u - u^0], [u - u^0])_{L_2(s,T;U)}, \quad (8.3.1.21)$$

where Λ_{sT} is the positive, self-adjoint operator

$$\Lambda_{sT} = [I_s + L_s^* R^* R L_s + L_{sT}^* G^* G L_{sT}] \in \mathcal{L}(L_2(s, T; U)), \quad (8.3.1.22)$$

and where u^0 is given by

$$u^0 = u^0(\cdot, s; y_0) = -\Lambda_{sT}^{-1} \left[L_s^* R^* R e^{A(\cdot - s)} y_0 + L_{sT}^* G^* G e^{A(T-s)} y_0 \right]$$
$$\in L_2(s, T; U). \quad (8.3.1.23)$$

Proof. Present assumptions justify the following computations. Eliminating y and $y(T)$ from the dynamics (8.1.1), we obtain

$$J_s(u) = \left(R \left[e^{A(\cdot - s)} y_0 + L_s u \right], R \left[e^{A(\cdot - s)} y_0 + L_s u \right] \right)_{L_2(s,T;Z)} + (u, u)_{L_2(s,T;U)}$$
$$+ \left(G \left[e^{A(T-s)} y_0 + L_{sT} u \right], G \left[e^{A(T-s)} y_0 + L_{sT} u \right] \right)_{Z_f}$$
$$= ([I_s + L_s^* R^* R L_s + L_{sT}^* G^* G L_{sT}]u, u)_{L_2(s,T;U)}$$
$$+ 2 \left(u, L_s^* R^* R e^{A(\cdot - s)} y_0 + L_{sT}^* G^* G e^{A(T-s)} y_0 \right)_{L_2(s,T;U)}$$
$$+ \left\| R e^{A(\cdot - s)} y_0 \right\|_{L_2(s,T;Z)}^2 + \left\| G e^{A(T-s)} y_0 \right\|_{Z_f}^2. \quad (8.3.1.24)$$

Recalling first (8.3.1.22) and then (8.3.1.23), we rewrite (8.3.1.24) as

$$J_s(u) = (\Lambda_{sT} u, u)_{L_2(s,T;U)}$$
$$+ 2 \left(\Lambda_{sT} u, \Lambda_{sT}^{-1} \left[L_s^* R^* R e^{A(\cdot - s)} y_0 + L_{sT}^* G^* G e^{A(T-s)} y_0 \right] \right)_{L_2(s,T;U)}$$
$$+ \left\| R e^{A(\cdot - s)} y_0 \right\|_{L_2(s,T;Z)}^2 + \left\| G e^{A(T-s)} y_0 \right\|_{Z_f}^2$$
$$= (\Lambda_{sT} u, u)_{L_2(s,T;U)} - 2(\Lambda_{sT} u, u^0)_{L_2(s,T;U)}$$
$$+ \left\| R e^{A(\cdot - s)} y_0 \right\|_{L_2(s,T;Z)}^2 + \left\| G e^{A(T-s)} y_0 \right\|_{Z_f}^2. \quad (8.3.1.25)$$

Specializing (8.3.1.25) to $u = u^0$, we obtain

$$J_s(u^0) = -(\Lambda_{sT} u^0, u^0)_{L_2(s,T;U)}$$
$$+ \left\| R e^{A(\cdot - s)} y_0 \right\|_{L_2(s,T;Z)}^2 + \left\| G e^{A(T-s)} y_0 \right\|_{Z_f}^2. \quad (8.3.1.26)$$

Subtracting (8.3.1.26) from (8.3.1.25) yields (8.3.1.21) since Λ_{sT} is self-adjoint. $\square$

We can complete the proof of optimality, since Λ_{sT} is (strictly) positive definite on $L_2(s, T; U)$, then (8.3.1.21) shows that the unique minimum of $J_s(u)$ is achieved for $u = u^0$; since this depends on s and y_0, we denote it by $u^0 = u^0(\,\cdot\,, s; y_0)$. We have re-proved (8.2.1.10) in the more general case with $G \neq 0$ as in (8.3.1.20).

Returning to the case $G = 0$, the corresponding optimal $y^0 = y^0(\,\cdot\,, s; y_0)$ satisfies, via (1.11),

$$Ry^0(\,\cdot\,, s; y_0) = Re^{A(\,\cdot\,-s)}y_0 + RL_su^0(\,\cdot\,, s; y_0) \in L_2(s, T; Z). \quad (8.3.1.27)$$

Applying $-L_s^* R^*$ on (8.3.1.27) yields

$$-L_s^* R^* Ry^0(\,\cdot\,, s; y_0) = -L_s^* R^* Re^{A(\,\cdot\,-s)}y_0 - L_s^* R^* RL_su^0(\,\cdot\,, s; y_0)$$

$$\text{(by (8.2.1.10))} \qquad = u^0(\,\cdot\,, s; y_0), \quad (8.3.1.28)$$

where in the last step we have recalled (8.2.1.10). Thus, (8.3.1.28) proves (8.2.1.14).

$\square$

8.3.2 Estimates on $u^0(\,\cdot\,, t; x)$ and $Ry^0(\,\cdot\,, t; x)$. The Operator $\Phi(\,\cdot\,,\,\cdot\,)$

Lemma 8.3.2.1 Assume $(h.1) = (8.2.1.10)$ and $(h.3) = (8.2.1.12)$. With reference to $(8.2.1.17a)$ we have

$$\sup_{0 \le t \le T} \|u^0(\,\cdot\,, t; x)\|_{L_2(t,T;U)} \le c_T \|x\|_Y. \quad (8.3.2.1)$$

Proof. The uniform estimate in (8.3.2.1) is actually contained in the argument leading to (8.2.1.17a) and based on the uniform bound (8.3.1.10b) and the uniform estimate (8.2.1.9) in s, as recognized explicitly in the paragraph below (8.3.1.9b). $\square$

With $u^0(\,\cdot\,, t; x)$ given explicitly by (8.2.1.17a) solely in terms of the data of the problem, we define the operator $\Phi(\tau, t)$ by setting for $x \in Y$:

$$\Phi(\tau, t)x = y^0(\tau, t; x) = e^{A(\tau-t)}x + \{L_t u^0(\,\cdot\,, t; x)\}(\tau) \quad (8.3.2.2a)$$

$$= e^{A(\tau-t)}x + \int_t^\tau e^{A(\tau-\sigma)}Bu^0(\sigma, t; x)\,d\sigma, \quad (8.3.2.2b)$$

and from (8.3.2.1) and the regularity property (8.1.12b) we have

$$\Phi(\,\cdot\,, t) : \text{continuous } Y \to C([t, T]; [\mathcal{D}(A^*)]'). \quad (8.3.2.3)$$

However, if we apply the observation operator R to (8.3.2.2), we obtain a more regular result.

Lemma 8.3.2.2 Assume (H.2) and (H.3). With reference to (8.3.2.2), we have

$$\sup_{0 \le t \le T} \|Ry^0(\,\cdot\,, t; x)\|_{C([t,T];Z)} = \sup_{0 \le t \le T} \|R\Phi(\,\cdot\,, t)x\|_{C([t,T];Z)} \le c_T \|x\|_Y \quad (8.3.2.4a)$$

[whereas under (h.1) and (h.3), the space $C([t, T]; Z)$ is replaced by the space $L_2(t, T; Z)$].

Moreover, under (H.1), (H.2), and (H.3) or else (h.1) and (h.2), we have

$$\|u^0(\,\cdot\,, s; x)\|_{C([s,T];U)} \leq c_T \|x\|_Y. \tag{8.3.2.4b}$$

Proof.   From (8.3.2.2), we obtain, with $x \in Y$,

$$R\Phi(\tau, t)x = Re^{A(\tau-t)}x + \{RL_t u^0(\,\cdot\,, t; x)\}(\tau) \tag{8.3.2.5}$$

$$\in C([t, T]; Z), \tag{8.3.2.6}$$

where in going from (8.3.2.5) to (8.3.2.6) we have used assumption (H.2) = (8.2.1.2)
for the first term in (8.3.2.5) and assumption (H.3*) in the form of its consequence
(8.2.1.8) for the second term in (8.3.2.5) via (8.3.2.1). In this way we also obtain
the norm bound independent of t, and (8.3.2.4a) follows. For (8.3.2.4b) we recall
$u^0(\,\cdot\,, s; x) = -L_s^* R^* R y^0(\,\cdot\,, s; x)$ from (8.2.1.21) and apply either (H.1*) in the
form of (8.2.1.9b) for $L_s^* R$, and (8.3.2.4a) or else (h.2*) = (8.2.1.14) in its form
uniform in s and (8.3.2.4a) by means also of (h.1) = (8.2.1.10).   □

We next study properties of the operator $\Phi(\,\cdot\,, \,\cdot\,)$ to be used in the following [e.g.,
in (8.3.4.19)].

Lemma 8.3.2.3   *Assume (H.2) and (H.3). With reference to $\Phi(\,\cdot\,, \,\cdot\,)$ in (8.3.2.2a),
we have:*

(i) $\Phi(t, t)x = x$, $x \in Y$.
(ii) For each s and τ fixed, $0 \leq s \leq \tau$, we have

$$Re^{A(\,\cdot\,-\tau)}\Phi(\tau, s)x \in C([t, T]; Z), \quad x \in Y \tag{8.3.2.7a}$$

continuously in x and uniformly in s and τ, that is,

$$\sup_{\tau \leq t \leq T} \|Re^{A(t-\tau)}\Phi(\tau, s)x\|_Z \leq c_T \|x\|_Y \tag{8.3.2.7b}$$

*[the space/norm of $C([t, T]; Z)$ are replaced by the space/norm of $L_2(t, T; Z)$
under (h.1) and (h.3)].*
(iii) For each s and τ fixed, $0 \leq s \leq \tau$, we have

$$R\Phi(\,\cdot\,, \tau)\Phi(\tau, s)x \in L_2(\tau, T; Z), \quad x \in Y, \tag{8.3.2.8a}$$

continuously in x and uniformly in s and τ, that is,

$$\int_\tau^T \|R\Phi(t, \tau)\Phi(\tau, s)x\|_Z^2 \, dt \leq c_T \|x\|_Y^2 \tag{8.3.2.8b}$$

[a conclusion that holds true also under (h.1) and (h.3)].

Proof.   (ii) Recalling (8.3.2.2a) we compute, with $s \leq \tau \leq t \leq T$,

$$Re^{A(t-\tau)}\Phi(\tau, s)x = Re^{A(t-\tau)}\left[e^{A(\tau-s)}x + \{L_s u^0(\,\cdot\,, s; x)\}(\tau)\right]$$

$$= Re^{A(t-s)}x + Re^{A(t-\tau)}\int_s^\tau e^{A(\tau-\sigma)}Bu^0(\sigma, s; x)\, d\sigma, \tag{8.3.2.9}$$

after invoking L_s in (8.1.12). With s, τ, and x fixed, we let $u^0_{\text{ext}}(\sigma, s; x)$ be the extension by zero of $u^0(\sigma, s; x)$ for $\tau \le \sigma$; by this, and by using the semigroup property, we obtain from (8.3.2.9)

$$Re^{A(t-\tau)}\Phi(\tau, s)x = Re^{A(t-s)}x$$
$$+ R\int_s^t e^{A(t-\sigma)}Bu^0_{\text{ext}}(\sigma, s; x)\,d\sigma. \qquad (8.3.2.10)$$

Thus, from (8.3.2.10), using (8.1.12) and recalling assumption (H.3*) in its version (8.2.1.8), we obtain

$$\left\| R\int_s^t e^{A(t-\sigma)}Bu^0_{\text{ext}}(\sigma, s; x)\,d\sigma \right\|_{C([s,T];Z)} = \left\| RL_s u^0_{\text{ext}}(\,\cdot\,, s; x) \right\|_{C([s,T];Z)}$$

$$\text{(by (8.2.1.8))} \qquad\qquad \le c_T \left\| u^0_{\text{ext}}(\,\cdot\,, s; x) \right\|_{L_2(s,T;U)}$$

$$\le c_T \| u^0(\,\cdot\,, s; x) \|_{L_2(s,T;U)}$$

$$\text{(by (8.3.2.1))} \qquad\qquad \le c_T \|x\|_Y, \quad x \in Y, \qquad (8.3.2.11)$$

after using the definition of u^0_{ext} and (8.3.2.1). As to the first term on the right-hand side of (8.3.2.10), we invoke (H.2) = (8.2.1.2) to obtain

$$\sup_{s \le t \le T} \left\| Re^{A(t-s)}x \right\|_Z \le c_T \|x\|_Y, \quad x \in Y. \qquad (8.3.2.12)$$

Then estimates (8.3.2.11) and (8.3.2.12), used in (8.3.2.10), yield the desired conclusion (8.3.2.7b). [Under (h.3), the $C([s, T]; Z)$-norm on the left-hand side of (8.3.2.11) is replaced by the $L_2(s, T; Z)$-norm. The same applies to (8.3.2.7), under the additional assumption (h.1) applied on the first term on the right-hand side of (8.3.2.10).]

(iii) We return to (8.3.1.14) = (8.2.1.19), rewritten here for convenience as to incorporate the definition (8.3.2.2a) of Φ as

$$R\Phi(\,\cdot\,, \tau)x = [I_\tau + RL_\tau L_\tau^* R^*]^{-1}[Re^{A(\cdot - \tau)}x] \in L_2(\tau, T; Z), \quad x \in Y. \qquad (8.3.2.13)$$

Replacing x with $\Phi(\tau, s)x$ in (8.3.2.13), $x \in Y$, we have that the formula

$$R\Phi(\,\cdot\,, \tau)\Phi(\tau, s)x = [I_\tau + RL_\tau L_\tau^* R^*]^{-1}[Re^{A(\cdot - \tau)}\Phi(\tau, s)x]$$
$$\in L_2(\tau, T; Z) \qquad (8.3.2.14)$$

is well defined by recalling (8.3.2.7); more precisely, from (8.3.2.14) we deduce, by (8.3.1.13b) and (8.3.2.7b),

$$\int_\tau^T \| R\Phi(t, \tau)\Phi(\tau, s)x \|_Z^2\,dt \le \int_0^T \| Re^{A(t-\tau)}\Phi(\tau, s)x \|_Z^2\,dt$$
$$\le Tc_T^2 \|x\|_Y^2, \qquad (8.3.2.15)$$

and (8.3.2.15) proves (8.3.2.8b). [The same argument applies also under (h.1) and (h.3).] $\square$

Proposition 8.3.2.4 *Assume (H.2) and (H.3) [or else, respectively, (h.1) and (h.3)].*
Let s, τ be fixed and arbitrary with $0 \leq s \leq \tau \leq t \leq T$. Then for $x \in Y$,

$$R\Phi(t, \tau)\Phi(\tau, s)x = R\Phi(t, s)x \underset{(\text{in } t)}{\in} L_2(\tau, T; Z), \qquad (8.3.2.16)$$

where the above equality is in the L_2-sense.

Proof.   It is similar to the proof of Proposition 1.4.3.1 of Chapter 1, which now,
however, takes into account (8.3.2.8) of Lemma 8.3.2.3. First, with $x \in Y$, we return
to (8.3.2.13) (with τ replaced by s) and rewrite it explicitly by recalling L_s and L_s^*
from (8.1.12) and (8.1.13), to obtain

$$R\Phi(t, s)x = Re^{A(t-s)}x - \{RL_s L_s^* R^* R\Phi(\,\cdot\,, s)x\}(t)$$

$$= Re^{A(t-s)}x - \int_s^t Re^{A(t-\sigma)}BB^* \int_\sigma^T e^{A^*(r-\sigma)}R^* R\Phi(r, s)x \, dr \, d\sigma.$$

$$(8.3.2.17)$$

Splitting the external integration in $[s, \tau]$ and $[\tau, t]$, we rewrite (8.3.2.17) as

$$R\Phi(t, s)x = Re^{A(t-\tau)}e^{A(\tau-s)}x$$

$$- Re^{A(t-\tau)} \int_s^\tau e^{A(\tau-\sigma)}BB^* \int_\sigma^T e^{A^*(r-\sigma)}R^* R\Phi(r, s)x \, dr \, d\sigma$$

$$- \int_\tau^t Re^{A(t-\sigma)}BB^* \int_\sigma^T e^{A^*(r-\sigma)}R^* R\Phi(r, s)x \, dr \, d\sigma. \quad (8.3.2.18)$$

Next, in (8.3.2.17), we replace s by τ, and x by $\Phi(\tau, s)x$, to obtain

$$R\Phi(t, \tau)\Phi(\tau, s)x = Re^{A(t-\tau)}\Phi(\tau, s)x$$

$$- \int_\tau^t Re^{A(t-\sigma)}BB^* \int_\sigma^T e^{A^*(r-\sigma)}R^* R\Phi(r, \tau)\Phi(\tau, s)x \, dr \, d\sigma, \quad x \in Y, \quad (8.3.2.19)$$

where we notice that each term in (8.3.2.19) is well defined by Lemma 8.3.2.3: the
left-hand side by (8.3.2.8); the first term on the right-hand side by (8.3.2.7); and the
second term on the right-hand side again by (8.3.2.8), along with (H.3*) in its versions
(8.2.1.8) and (8.2.1.9) [respectively, (8.2.1.16)]. Next, inserting the optimal solution
[see (8.3.2.2) and (8.3.1.8)]

$$\Phi(\tau, s)x = e^{A(\tau-s)}x - \{L_s L_s^* R^* R\Phi(\,\cdot\,, s)x\}(\tau)$$

$$= e^{A(\tau-s)}x - \int_s^\tau e^{A(\tau-\sigma)}BB^* \int_\sigma^T e^{A^*(r-\sigma)}R^* R\Phi(r, s)x \, dr \, d\sigma \quad (8.3.2.20)$$

into the first term on the right-hand side of (8.3.2.19) yields

$$R\Phi(t, \tau)\Phi(\tau, s)x$$

$$= Re^{A(t-\tau)} \left[e^{A(\tau-s)}x - \int_s^\tau e^{A(\tau-\sigma)}BB^* \int_\sigma^T e^{A^*(r-\sigma)}R^* R\Phi(r, s)x \, dr \, d\sigma \right]$$

$$- \int_\tau^t Re^{A(t-\sigma)}BB^* \int_\sigma^T e^{A^*(r-\sigma)}R^* R\Phi(r, \tau)\Phi(\tau, s)x \, dr \, d\sigma. \quad (8.3.2.21)$$

We now compare (8.3.2.21) with (8.3.2.18) and notice that the first two terms on the right of both equations are the same, and thus they cancel out if we subtract (8.3.2.21) from (8.3.2.18). We then obtain

$$R\Phi(t, s)x - R\Phi(t, \tau)\Phi(\tau, s)x$$

$$= -\int_\tau^t Re^{A(t-\sigma)} BB^* \int_\sigma^T e^{A^*(r-\sigma)} R^*[R\Phi(r, s)x - R\Phi(r, \tau)\Phi(\tau, s)x]\, dr\, d\sigma$$

$$= -RL_\tau L_\tau^* R^*[R\Phi(\cdot, s)x - R\Phi(\cdot, \tau)\Phi(\tau, s)x], \qquad (8.3.2.22)$$

recalling in the last step (8.1.12) and (8.1.13) for L_τ and L_τ^*. Thus, (8.3.2.22) can be rewritten as

$$[I + RL_\tau L_\tau^* R^*][R\Phi(\cdot, s)x - R\Phi(\cdot, \tau)\Phi(\tau, s)x] = 0, \quad x \in Y, \qquad (8.3.2.23)$$

with the function in the square brackets well defined as an $L_2(\tau, T; Z)$-function by (8.3.2.6) and (8.3.2.8). Inverting the operator in (8.3.2.33) [see (8.3.1.13)], we obtain (8.3.2.16) from (8.3.2.23).   □

8.3.3 Definition of $P(t)$ and Preliminary Properties

The main goal of this subsection, as well as of the next one, is to establish a global existence result on all of $[0, T]$ for the DRE (8.2.1.32). To this end, following the approach of Chapters 1 and 2, we shall first define in this section an operator $P(t)$, constructively in terms of the data of the problem, and exhibit some of its preliminary global regularity properties, namely (8.2.1.26), (8.2.1.30), and (8.2.1.31). In the next Subsection 8.3.4, we shall show that $P(t)$ is self-adjoint (property (8.2.1.25)) and that it is a solution of the DRE (8.2.1.32). Finally, in Section 8.3.6, we shall show that such $P(t)$ is the unique solution to possess properties (8.2.1.25), (8.2.1.26), (8.2.1.30), and (8.2.1.31).

We begin by introducing a space needed in the following. If $0 \le a \le b$, we shall denote by $X_{a,b}$ the set of all measurable mappings ϕ from $\Delta_{a,b}$ into the Banach space X, $\Delta_{a,b} = \{(t, s) \in R^2 : a \le s \le t \le b\}$, such that

$$\|\phi\|_{X_{a,b}}^2 \equiv \sup_{a \le s \le b} \int_s^b \|\phi(t, s)\|_X^2\, dt < \infty. \qquad (8.3.3.1)$$

Endowed with the norm (8.3.3.1), $X_{a,b}$ is a Banach space. In this chapter we shall take X to be the Hilbert space U and $a = s_0$ or $a = 0$, and $b = T$, in which case we write $U_{s_0,T}$ or $U_{0,T}$, respectively, a notation already recalled in (8.2.1.31).

Theorem 8.3.3.1   *Assume (H.1), (H.2), and (H.3) [or, alternatively, (h.1) and (h.2)]. The operator $P(t)$ defined constructively by*

$$P(t)x = \int_t^T e^{A^*(\tau-t)} R^* R\Phi(\tau, t)x\, d\tau, \quad x \in Y, \ 0 \le t \le T \qquad (8.3.3.2a)$$

in terms of the data of the problem, via (8.3.2.5) and (8.2.1.10), satisfies the following global properties:

   (i)

$$P(t) \in \mathcal{L}(Y), \quad 0 \le t \le T; \quad P(\cdot) : continuous\ Y \to L_\infty(0, T; Y),$$

(8.3.3.2b)

   or $P \in \mathcal{B}([0, T]; \mathcal{L}(Y))$ in the notation of Remark 8.1.1, Eqn. (8.1.14);
(ii)

$$V(t) \equiv B^* P(t) \in \mathcal{L}(Y; U), \quad 0 \le t \le T;$$
$$V(\cdot) \equiv B^* P(\cdot) : continuous\ Y \to L_\infty(0, T; U); \qquad (8.3.3.3)$$

   or $V \in \mathcal{B}([0, T]; \mathcal{L}(Y; U))$ in the notation of Remark 8.1.1;
(iii)

$$Q(t, s) \equiv B^* P(t) e^{A(t-s)} B : continuous\ U \to U_{0,T}, \qquad (8.3.3.4)$$

where the space $U_{0,T}$ is defined in (8.3.3.1) with $X = U$, $a = 0$, and $b = T$.

Proof of Theorem 8.3.3.1.

Step 1.   Lemma 8.3.3.2   *Assume (H.2) and (H.3) [or else, alternatively, (h.1) and (h.3)]. With reference to (8.3.3.2) we have*

$$P(t) \in \mathcal{L}(Y), \quad 0 \le t \le T, \quad and \quad P(\cdot) : continuous\ Y \to L_\infty(0, T; Y). \quad (8.3.3.5)$$

Proof of Lemma 8.3.3.2.   Conclusion (8.3.3.5) follows directly from (8.3.3.2a) by use of assumption (H.2) in the form $(H.2^*) = (8.2.1.5)$ followed by property (8.3.2.4a). [Alternatively, (8.3.3.5) follows by (h.1) and the corresponding version of Lemma 8.3.2.2 in the $L_2(t, T; Z)$-sense under (h.3) as well]. $\square$

We recall that (h.2) is stronger than (h.3). The operator $P(t)$ in (8.3.3.2) is our candidate, which will be shown to be the sought-after unique solution of the DRE (8.2.1.24).

Step 2   We define the operators

$$V(t) = B^* P(t), \qquad (8.3.3.6)$$
$$Q(t, s) = B^* P(t) e^{A(t-s)} B, \qquad (8.3.3.7)$$

where $P(t)$ is defined by (8.3.3.2a). The next few lemmas show that the operators V and Q satisfy the global regularity properties that we seek in connection with the uniqueness issue in the subsequent Section 8.3.6.

Proposition 8.3.3.3   *Assume (H.1), (H.2), and (H.3) [or else, alternatively, (h.1) and (h.2)]. With reference to (8.3.3.6) we have*

$$V(t) \in \mathcal{L}(Y; U), \quad 0 \le t \le T, \quad and \quad V(\cdot) : continuous\ Y \to L_\infty(0, T; U);$$

(8.3.3.8a)

that is,

$$\sup_{0 \le t \le T} \|V(t)x\|_U \le c_T \|x\|_Y. \tag{8.3.3.8b}$$

Proof of Proposition 8.3.3.3. From (8.3.3.6) and (8.3.3.2) with $\tau - t = \sigma$ and $x \in Y$, we have, using (H.1*) = (8.2.1.4b) after an innocuous extension by zero on $[T-t, T]$:

$$\|V(t)x\|_U = \|\{L_0^* R^* R \Phi(\cdot, t)x\}(t)\|_U$$

$$= \left\| \int_0^{T-t} B^* e^{A^* \sigma} R^* R \Phi(\sigma + t, t)x \, d\sigma \right\|_U$$

(by (8.2.1.4b))
$$\le c_T \sup_{0 \le \sigma \le T-t} \|R\Phi(\sigma + t, t)x\|_Z = c_T \|R\Phi(\cdot, t)x\|_{C([t,T];Z)}$$

$$\le c_T \sup_{0 \le t \le T} \|R\Phi(\cdot, t)x\|_{L_\infty(t,T;Z)} \le c_T \|x\|_Y, \quad 0 \le t \le T,$$

$$\tag{8.3.3.9}$$

where in the last step we have used (8.3.2.4). Then, (8.3.3.9) implies (8.3.3.8) by taking the sup in t. [Alternatively, we use (h.2*) = (8.2.1.14) followed by the corresponding version of Lemma 8.3.2.2 in the $L_2(t, T; Z)$-sense, valid under (h.1) and (h.3), the latter hypothesis being weaker than (h.2).]   □

Step 3   Before establishing the required regularity of Q, we need an auxiliary result on the optimal $u^0(\cdot, t; x)$ where $x = e^{A(t-s)}Bu$, $u \in U$. To this end, we recall by (8.1.2) that if $u \in U$, then $Bu \in [\mathcal{D}(A^*)]'$ by assumption. Next, the s.c. semigroup e^{At}, originally defined: $Y \to Y$, can be extended as a s.c. semigroup: $[\mathcal{D}(A^*)]' \to [\mathcal{D}(A^*)]'$ (and we keep the same symbol) by $Ae^{At}A^{-1}$. It is not clear a priori, however, that the optimal $u^0(\cdot, t; x)$ in (8.2.1.17) with $x \in Y$ can be extended to $x = e^{A(t-s)}Bu \in [\mathcal{D}(A^*)]'$.

Lemma 8.3.3.4   *Assume (H.1), (H.2), and (H.3) [or else, alternatively (h.2)]. With reference to (8.2.1.17), we have the following extension result:*

$$\sup_{0 \le t \le T} \sup_{0 \le s \le t} \left\| u^0(\cdot, t; e^{A(t-s)}Bu) \right\|_{L_2(t,T;U)} \le c_T \|u\|_U. \tag{8.3.3.10}$$

Proof.   From (8.2.1.17) we have, for $u \in U$,

$$-u^0(\tau, t; e^{A(t-s)}Bu) = \left\{ [I_t + L_t^* R^* R L_t]^{-1} L_t^* R^* R[e^{A(\cdot - t)} e^{A(t-s)}Bu] \right\}(\tau),$$

$$\tag{8.3.3.11}$$

from which we have, recalling (8.3.1.10b), (H.3*) in the form of its consequence (8.2.1.9), as well as (H.1) = (8.2.1.1):

$$\left\| u^0(\cdot, t; e^{A(t-s)}Bu) \right\|_{L_2(t,T;U)} \le \left\| L_t^* R^* [Re^{A(\cdot - s)}Bu] \right\|_{L_2(t,T;U)}$$

(by (8.2.1.9))
$$\le c_T \left\| Re^{A(\cdot - s)}Bu \right\|_{L_1(t,T;Z)}$$

(by (H.1) = (8.2.1.1))
$$\le c_T \|u\|_U, \tag{8.3.3.12a}$$

where in the last step we have used (8.2.1.1) and a change of variable. Then (8.3.3.12a) readily implies (8.3.3.10). [Alternatively, as (h.2) is stronger than (h.3),

$$\left\| u^0\left(\,\cdot\,, t; e^{A(t-s)}Bu\right)\right\|_{L_2(t,T;U)} \leq \left\| L_t^* R^* \left[Re^{(\cdot\,-s)}Bu\right]\right\|_{L_2(t,T;U)}$$

$$\text{(by (8.2.1.16))} \qquad \leq c_T \left\| Re^{A(\cdot\,-s)}Bu\right\|_{L_2(t,T;Z)}$$

$$\text{(by (8.2.1.11))} \qquad \leq c_T \|u\|_U, \tag{8.3.3.12b}$$

after invoking (h.3) in the form of (8.2.1.16) and (h.2) = (8.2.1.11).]   □

Remark 8.3.3.1   We note that the above computations will reappear at the level of estimating the term $F_1(t, s)$ in (8.4.1.15) in the direct proof of Section 8.4.   □

Step 4   The desired global result for Q is

Proposition 8.3.3.5   *Assume (H.1), (H.2), and (H.3) [or else, alternatively, (h.2)]. With reference to (8.3.3.7) and (8.3.3.1) we have*

$$\|Qu\|_{U_{0,T}}^2 \equiv \sup_{0 \leq s \leq T} \int_s^T \|Q(t, s)u\|_U^2 \, dt \leq c_T \|u\|_U^2, \tag{8.3.3.13a}$$

that is,

$$Q(t, s) : continuous \ U \to U_{0,T}, \tag{8.3.3.13b}$$

where $U_{0,T}$ is defined by (8.3.3.1) with $X = U$, $a = 0$, and $b = T$.   □

Proof of Proposition 8.3.3.5.   By (8.3.3.7) and (8.3.3.2) we have

$$Q(t, s)u = \int_t^T B^* e^{A^*(\tau-t)} R^* R\Phi(\tau, t)e^{A(t-s)}Bu \, d\tau \tag{8.3.3.14}$$

$$= (1) + (2); \tag{8.3.3.15}$$

$$(1) \quad = \int_t^T B^* e^{A^*(\tau-t)} R^* Re^{A(\tau-s)}Bu \, d\tau, \quad s \leq t \leq T \tag{8.3.3.16a}$$

$$\text{(by (8.1.13a))} \quad = \left\{ L_s^* R^* Re^{A(\cdot\,-s)}Bu\right\}(t); \tag{8.3.3.16b}$$

$$(2) \quad = \int_t^T B^* e^{A^*(\tau-t)} R^* \left\{ RL_t u^0\left(\,\cdot\,, t; e^{A(t-s)}Bu\right)\right\}(\tau) \, d\tau \tag{8.3.3.17a}$$

$$= \left\{ L_t^* R^* RL_t u^0\left(\,\cdot\,, t; e^{A(t-s)}Bu\right)\right\}(t), \tag{8.3.3.17b}$$

where in going from (8.3.3.14) to (8.3.3.15) we have used (8.3.2.5) to replace $R\Phi(\tau, t)$. As to (1), we obtain from (8.3.3.16b), after invoking assumption (H.3*) in the form (8.2.1.9),

$$\|(1)\|_{L_2(s,T;U)} \leq c_T \left\| Re^{A(\cdot\,-s)}Bu\right\|_{L_1(s,T;Z)}$$

$$\text{(by (8.2.1.1))} \qquad \leq c_T \|u\|_U, \tag{8.3.3.18}$$

where in the last step we have used (H.1) = (8.2.1.1) (i.e., exactly as in obtaining (8.3.3.12) either under (H.1), and (H.3) or under (h.2)).

As to (2), we set $\tau - t = \sigma$, and invoke (H.1*) = (8.2.1.4b) and (H.3*) in the form of its consequence (8.2.1.8) after an innocuous extension by zero on $[T - t, T]$:

$$\|(2)\|_U = \left\| \int_0^{T-t} B^* e^{A^* \sigma} R^* \{ RL_t u^0 (\cdot, t; e^{A(t-s)} Bu) \} (t + \sigma) \, d\sigma \right\|_U$$

(by (8.2.1.4b)) $\qquad \leq c_T \sup_{0 \leq \sigma \leq T-t} \left\| (RL_t u^0 (\cdot, t; e^{A(t-s)} Bu)(t + \sigma) \right\|_Z$

(by (8.2.1.8)) $\qquad \leq c_T \left\| u^0 (\cdot, t; e^{A(t-s)} Bu) \right\|_{L_2(t,T;U)}$

(by (8.3.3.10)) $\qquad \leq c_T \|u\|_U,$ $\hfill (8.3.3.19)$

where in the last step we have invoked (8.3.3.10). Thus, (8.3.3.18) and (8.3.3.19) yield readily (8.3.3.13a) via (8.3.3.15). [Alternatively, from (8.3.3.17b), since (h.2) implies (h.3), we obtain readily

$$\|(2)\|_{L_2(s,T;U)} \leq c_T \left\| u^0 (\cdot, t; e^{A(t-s)} Bu) \right\|_{L_2(s,T;U)} \leq c_T \|u\|_U,$$

by (8.2.1.16) and (8.3.3.10); and (8.3.3.13a) follows at once again.] $\quad \square$

8.3.4 $P(t)$ Solves the Differential Riccati Equation (8.2.1.32)

In this subsection, we shall show that the operator $P(t)$ defined constructively in (8.3.3.2) satisfies the DRE (8.2.1.32), is self-adjoint, and possesses the appropriate regularity properties mentioned in Theorem 8.2.1.2.

Step 1 We begin with a regularity result that extends the action of the operator $R\Phi(\ ,\ )$ in (8.3.2.2) and (8.3.2.6) from its original domain Y to the [range of B] $\subset [\mathcal{D}(A^*)]'$.

Lemma 8.3.4.1 *Assume (H.1), (H.2), and (H.3). With reference to (8.3.2.2) and (8.3.2.6), we have that the operator $R\Phi(\cdot, t)B$ can be extended as continuous, $U \to L_1(t, T; Z)$ uniformly in t, that is,*

$$\sup_{0 \leq t \leq T} \| R\Phi(\cdot, t)Bu \|_{L_1(t,T;Z)} \leq c_T \|u\|_U \qquad (8.3.4.1)$$

[or else, alternatively, under assumption (h.2), the $L_1(t, T; Z)$-norm is replaced by the $L_2(t, T; Z)$-norm].

Proof. From (8.3.2.5) we have, with $u \in U$,

$$R\Phi(\tau, t)Bu = Re^{A(\tau-t)}Bu + \{RL_t u^0(\cdot, t; Bu)\}(\tau). \qquad (8.3.4.2)$$

But from (H.3*) in the version of its consequence (8.2.1.8) and from (8.3.3.10) specialized with $t = s$ we obtain, for the second term in (8.3.4.2),

$$\|(RL_t u^0(\,\cdot\,, t; Bu)(\,\cdot\,)\|_{C([t,T];Z)} \leq c_T \|u^0(\,\cdot\,, t; Bu)\|_{L_2(t,T;U)}$$
$$\text{(by (8.3.3.10))} \qquad \leq c_T \|u\|_U. \tag{8.3.4.3}$$

Returning to (8.3.4.2) we see that (8.3.4.3) and assumption (H.1) = (8.2.1.1) on the first term on the right-hand side of (8.3.4.2) readily yield (8.3.4.1.) [Alternatively, under (h.2), the first and second term in (8.3.4.2) are in $L_2(t, T; Z)$, uniformly in t, the second term by (8.2.1.16) and, once more, by (8.3.3.10).] $\qquad\square$

Step 2   We now return to the expression of $Ry^0(\,\cdot\,, t; x)$ in (8.2.1.19) written solely in terms of the data of the problem. This, in view of the definition of $\Phi(\tau, t)x$ in (8.3.2.2a), is now rewritten as

$$R\Phi(\tau, t)x = \left\{[I_t + RL_t L_t^* R^*]^{-1}\left[Re^{A(\,\cdot\,-t)}x\right]\right\}(\tau), \quad x \in Y. \tag{8.3.4.4}$$

By (8.3.4.4), then (8.3.4.1), already proved where $x = Bu$, can be rewritten equivalently as

$$\sup_{0 \leq t \leq T} \left\{\left\|\left([I_t + RL_t L_t^* R^*]^{-1}\left[Re^{A(\,\cdot\,-t)}Bu\right]\right)(\,\cdot\,)\right\|_{L_1(t,T;Z)}\right\} \leq c_T \|u\|_U. \tag{8.3.4.5}$$

[Alternatively, under (h.2), the $L_1(t, T; Z)$-norm is replaced by the $L_2(t, T; Z)$-norm.]

Step 3.   Lemma 8.3.4.2   *Assume (H.1), (H.2), and (H.3). With reference to (8.3.2.5) or (8.3.4.4) we have*

(i)

$$\sup_{0 \leq t \leq T} \left\{\left\|\frac{dR\Phi(\tau, t)x}{dt}\right\|_{L_1(t,T;Z)}\right\} \leq c_T \|Ax\|_Y, \quad x \in \mathcal{D}(A) \tag{8.3.4.6}$$

[or else, alternatively, under (h.1) and (h.2), the $L_1(t, T; Z)$-norm is replaced by the $L_2(t, T; Z)$-norm];
(ii) for $x \in \mathcal{D}(A)$,

$$\frac{dR\Phi(\tau, t)x}{dt} = -R\Phi(\tau, t)[A - BB^* P(t)]x \text{ a.e. in } t, \quad x \in \mathcal{D}(A). \tag{8.3.4.7}$$

Proof.   We rewrite (8.3.4.4) as

$$R\Phi(\tau, t)x + \{RL_t L_t^* R^* R\Phi(\,\cdot\,, t)x\}(\tau) = Re^{A(\tau-t)}x, \tag{8.3.4.8}$$

or, explicitly by (8.1.12a) and (8.1.13a),

$$R\Phi(\tau, t)x + R \int_t^\tau e^{A(\tau-\sigma)} B \int_\sigma^T B^* e^{A^*(r-\sigma)} R^* R\Phi(r, t)x \, dr \, d\sigma = Re^{A(\tau-t)}x. \tag{8.3.4.9}$$

With $x \in \mathcal{D}(A)$ we differentiate (8.3.4.9) in t, for example, as a distributional derivative, obtaining

$$\frac{d R\Phi(\tau, t)}{dt} x - Re^{A(\tau - t)} B \int_t^T B^* e^{A^*(r-t)} R^* R\Phi(r, t)x \, dr$$

$$+ R \int_t^\tau e^{A(\tau - \sigma)} B \int_\sigma^T B^* e^{A^*(r-\sigma)} R^* \frac{d}{dt} R\Phi(r, t)x \, dr \, d\sigma$$

$$= -Re^{A(\tau - t)} Ax, \tag{8.3.4.10}$$

which in view of (8.1.12a), (8.1.13a), and (8.3.3.2) can be rewritten as

$$\left\{ [I_t + RL_t L_t^* R^*] \frac{d R\Phi(\cdot, t)x}{dt} \right\} (\tau) = -Re^{A(\tau - t)} Ax + Re^{A(\tau - t)} BB^* P(t)x.$$

$$\tag{8.3.4.11}$$

The above identity holds true (at least) in $H^{-1}(0, T; Y)$. Note that, for $x \in \mathcal{D}(A)$, the first term on the right-hand side of (8.3.4.11) is an element of $C([t, T]; Z)$ by assumption (H.2) = (8.2.1.2); and hence it may be acted upon by $[I_t + RL_t L_t^* R^*]^{-1}$. Moreover, for the second term we invoke the extension result (8.3.4.5), as well as the global estimate (8.3.3.8) for $V(t) = B^* P(t)$. Thus, from (8.3.4.11) we obtain, continuously in $x \in \mathcal{D}(A)$,

$$\frac{d R\Phi(\cdot, t)x}{dt} = -[I_t + RL_t L_t^* R^*]^{-1} \left[Re^{A(\cdot - t)} Ax \right]$$

$$+ [I_t + RL_t L_t^* R^*]^{-1} \left[Re^{A(\cdot - t)} BB^* P(t)x \right] \in L_1(t, T; Z).$$

$$\tag{8.3.4.12}$$

By (8.3.4.4), we see that (8.3.4.12) can be written equivalently as

$$\frac{d R\Phi(\tau, t)x}{dt} = -R\Phi(\tau, t)Ax + R\Phi(\tau, t)BB^* P(t)x \quad \text{a.e. in } t, \tag{8.3.4.13}$$

where, by (8.3.2.4) with $x \in \mathcal{D}(A)$, we have

$$\sup_{0 \leq t \leq T} \left\{ \|R\Phi(\cdot, t)Ax\|_{L_\infty(t,T;Z)} \right\} \leq c_T \|Ax\|_Y, \tag{8.3.4.14}$$

whereas by (8.3.4.1) we have

$$\|R\Phi(\cdot, t)BB^* P(t)x\|_{L_1(t,T;Z)} \leq c_T \|B^* P(t)x\|_U, \tag{8.3.4.15}$$

and hence by (8.3.3.8) on $V(t) = B^* P(t)$ used in (8.3.4.15) we obtain

$$\sup_{0 \leq t \leq T} \|R\Phi(\cdot, t)BB^* P(t)x\|_{L_1(t,T;Z)} \leq c_T \sup_{0 \leq t \leq T} \|B^* P(t)x\|_U \leq C_T \|x\|_Y.$$

$$\tag{8.3.4.16}$$

Then (8.3.4.13) shows (8.3.4.7), while (8.3.4.14) and (8.3.4.16) used in (8.3.4.13) yield (8.3.4.6). [Alternatively, under (h.1) and (h.2), the first term on the right-hand side of (8.3.4.11) is in $L_2(t, T; Z)$ and it can still be acted upon by $[I_t + RL_t L_t^* R^*]^{-1}$

(recall (8.3.1.10) under (h.3), weaker than (h.2)); whereas the $L_1(t, T; Z)$-norm in (8.3.4.12) and in (8.3.4.15), as well as the $L_\infty(t, T; Z)$-norm in (8.3.4.14) are all replaced by the $L_2(t, T; Z)$-norm, by quoting the corresponding results of (8.3.4.5), (8.3.2.4), and (8.3.4.1).]   □

Step 4.   Lemma 8.3.4.3   *Assume (H.2), and (H.3) [or else, alternatively, (h.1) and (h.3)]. For $x \in Y$, we have*

$$-u^0(\tau, t; x) = \{L_t^* R^* R\Phi(\,\cdot\,, t)x\}(\tau) \tag{8.3.4.17}$$

$$= B^* P(\tau)\Phi(\tau, t)x \in L_2(t, T; U)   (in \ \tau). \tag{8.3.4.18}$$

Proof.   First, note that (8.3.4.17) is nothing but (8.2.1.21), rewritten using the definition of Φ in (8.3.2.2). Next, recalling (8.1.13a) for L_t^* and (8.3.3.2a) for $P(t)$, we obtain from (8.3.4.17)

$$\{L_t^* R^* R\Phi(\,\cdot\,, t)x\}(\tau) = \int_\tau^T B^* e^{A^*(\sigma - \tau)} R^* R\Phi(\sigma, t)x \, d\sigma,   t \leq \tau \leq T,$$

$$\text{(by (8.3.2.16))}   = \int_\tau^T B^* e^{A^*(\sigma - \tau)} R^* R\Phi(\sigma, \tau)\Phi(\tau, t)x \, d\sigma \tag{8.3.4.19}$$

$$\text{(by (8.3.3.2))}   = B^* P(\tau)\Phi(\tau, t)x, \tag{8.3.4.20}$$

where in (8.3.4.19) we have recalled (8.3.2.16) of Proposition 8.3.2.4. All terms are well-defined by (H.2) and (H.3) [or else, alternatively, by (h.1) and (h.3)].   □

Step 5   We now prove that $P(t)$ is self-adjoint on Y.

Lemma 8.3.4.4   *Assume (H.2) and (H.3) [or else, alternatively, (h.1) and (h.3)].*

(i) *With reference to (8.3.3.2), we have the following identity, symmetric in $x, y \in Y$, $0 \leq t \leq T$:*

$$(P(t)x, y)_Y = \int_t^T (R\Phi(\tau, t)x, R\Phi(\tau, t)y)_Z \, d\tau$$

$$+ \int_t^T (B^* P(\tau)\Phi(\tau, t)x, B^* P(\tau)\Phi(\tau, t)y)_U \, d\tau; \tag{8.3.4.21}$$

(ii) *as a consequence*

$$P(t) = P^*(t) \geq 0,   0 \leq t \leq T; \tag{8.3.4.22}$$

(iii) *the optimal cost of the optimal control problem on $[t, T]$ initiating at the point $x \in Y$ at the initial time t is*

$$J(u^0(\,\cdot\,, t; x), y^0(\,\cdot\,, t; x)) = \int_t^T \left\{ \|R\Phi(\tau, t)x\|_Z^2 + \|B^* P(\tau)\Phi(\tau, t)x\|_U^2 \right\} d\tau$$

$$= (P(t)x, x)_Y, \tag{8.3.4.23}$$

*where the optimal $u^0(\cdot, t; x)$ is defined by (8.2.1.17) and the optimal $R\Phi(\cdot, t)x$
is defined in (8.2.1.19) via the definition (8.3.2.2a) of Φ.*

Proof. (i) As in the proof of Chapter 1, Proposition 1.4.4.8, we compute from
(8.3.3.2a), with all terms well-defined by (H.2) and (H.3) [or else, alternatively, by
(h.1) and (h.3)],

$$(P(t)x, y)_Y = \int_t^T (R\Phi(\tau, t)x, Re^{A(\tau - t)}y)_Z \, d\tau$$

$$\text{(by (8.3.2.5))} \quad = \int_t^T (R\Phi(\tau, t)x, R\Phi(\tau, t)y)_Z \, d\tau$$

$$- (R\Phi(\cdot, t)x, RL_t u^0(\cdot, t; y))_{L_2(t,T;Z)} \quad (8.3.4.24)$$

$$= \int_t^T (R\Phi(\tau, t)x, R\Phi(\tau, t)y)_Z \, d\tau$$

$$- (L_t^* R^* R\Phi(\cdot, t)x, u^0(\cdot, t; y))_{L_2(t,T;U)}$$

$$\text{(by (8.3.4.17))} \quad = (R\Phi(\cdot, t)x, R\Phi(\cdot, t)y)_{L_2(t,T;Z)}$$

$$+ (u^0(\cdot, t; x), u^0(\cdot, t; y))_{L_2(t,T;U)}, \quad (8.3.4.25)$$

after using the optimality condition (8.3.4.17). The desired conclusion (8.3.4.21)
follows now from (8.3.4.25) via (8.3.4.18) and readily implies (8.3.4.22) of part (ii).
We could also appeal directly to Eqn. (8.3.1.16).

(iii) Part (iii) follows, at once, from (8.3.4.21) with $x = y$, since the optimal
$u^0(\cdot, t; x)$ is given by (8.3.4.18) and the optimal $y^0(\tau, t; x)$ is rewritten by (8.3.2.2)
in terms of Φ.   □

Step 6 We can finally prove our desired result.

Theorem 8.3.4.5 *Assume (H.1), (H.2), and (H.3) [or else, alternatively, (h.1) and
(h.2)]. The operator $P(t)$ defined by (8.3.3.2) satisfies the following differential
Riccati equation for $x, y \in \mathcal{D}(A)$:*

$$\left(\frac{d}{dt} P(t)x, y \right)_Y = -(R^* Rx, y)_Y - (P(t)x, Ay)_Y - (P(t)Ax, y)_Y$$

$$+ (B^* P(t)x, B^* P(t)y)_U, \quad (8.3.4.26)$$

where

$$P'(t) \in \mathcal{L}(\mathcal{D}(A); [\mathcal{D}(A)]') \text{ and continuous } \mathcal{D}(A) \to L_\infty(0, T; [\mathcal{D}(A)]'),$$

$$(8.3.4.27a)$$

that is,

$$A^{*-1} P'(t) A^{-1} \in \mathcal{L}(Y) \text{ and continuous } Y \to L_\infty(0, T; Y). \quad (8.3.4.27b)$$

Proof.   Starting from (8.3.3.2a) we differentiate in t with $x, y \in \mathcal{D}(A)$:

$$
\left(\frac{d}{dt} P(t)x, y \right)_Y = -(Rx, Ry)_Z - (A^* P(t)x, y)_Y
$$
$$
+ \left(\int_t^T e^{A^*(\tau-t)} R^* \frac{dR\Phi(\tau, t)x}{dt} d\tau, y \right)_Y, \quad (8.3.4.28)
$$

after recalling (8.3.3.2) and $\Phi(t, t)x = x$. We note that the third term in (8.3.4.28) is well-defined by Lemma 8.3.4.2 and by (H.2) = (8.2.1.2) [alternatively, by (h.1) = (8.2.1.10)], for $x \in \mathcal{D}(A)$ and $y \in Y$. For this term in (8.3.4.28), we now invoke (8.3.4.7) of Lemma 8.3.4.2. We obtain for $x \in \mathcal{D}(A)$, $y \in Y$:

$$
\left(\int_t^T e^{A^*(\tau-t)} R^* \frac{dR\Phi(\tau, t)x}{dt} d\tau, y \right)_Y
$$
$$
= -\left(\int_t^T e^{A^*(\tau-t)} R^* R\Phi(\tau, t)Ax \, d\tau, y \right)_Y
$$
$$
+ \left(\int_t^T e^{A^*(\tau-t)} R^* R\Phi(\tau, t)BB^* P(t)x \, d\tau, y \right)_Y
$$
$$
\text{(by (8.3.3.2))} \quad = -(P(t)Ax, y)_Y + (P(t)BB^* P(t)x, y)_Y, \quad (8.3.4.29)
$$
$$
\text{(by (8.3.4.22))} \quad = -(P(t)Ax, y)_Y + (B^* P(t)x, B^* P(t)y)_Y, \quad (8.3.4.30)
$$

after invoking again (8.3.3.2) in the step leading to (8.3.4.29) and the self-adjointness of $P(t)$ in the step leading to (8.3.4.30). Inserting (8.3.4.30) into (8.3.4.28) yields the desired conclusion that $P(t)$ satisfies the DRE (8.3.4.26). Note that the last term in (8.3.4.29) is well-defined by (8.3.3.8) on $V(t) = B^* P(t)$. Moreover, $P(T) = 0$ by (8.3.3.2). Statement (8.3.4.27) on $P'(t)$ is contained in the above analysis of (8.3.4.28) being well-defined. $\square$

8.3.5 Differential and Integral Riccati Equations

In this section we examine the relationship between self-adjoint solutions of the differential Riccati equation (DRE) (8.3.4.26) [or (8.2.1.32)] that satisfy the regularity properties (8.3.3.3), (8.3.3.4), and (8.3.3.5) and solutions of the corresponding integral Riccati equation (IRE)

$$
(P(t)x, y)_Y = \int_t^T \left(Re^{A(\tau-t)}x, Re^{A(\tau-t)}y \right)_Z d\tau
$$
$$
- \int_t^T \left(B^* P(\tau)e^{A(\tau-t)}x, B^* P(\tau)e^{A(\tau-t)}y \right)_U d\tau \quad x, y \in Y.
$$
$$
(8.3.5.1)
$$

This analysis, besides being of interest in itself, is used in the next section in the study of uniqueness.

Proposition 8.3.5.1   *Assume (i) and (ii) of Section 8.1 on A and B, and assume (H.2) or (h.1) on R. Let $P(t) \in \mathcal{L}(Y)$, $0 \leq t \leq T$, $P(T) = 0$, be a self-adjoint solution of the DRE (8.3.4.26) that satisfies the regularity properties (8.3.3.4) and (8.3.3.5) [as is the case for the operator $P(t)$ defined by (8.3.3.2), under assumptions (H.1), (H.2), and (H.3), or else, alternatively, under assumptions (h.1) and (h.2), by virtue of Theorem 8.3.4.5, Theorem 8.3.3.1, Lemma 8.3.3.2, as well as (8.3.4.22)]. Then, such $P(t)$ also satisfies the IRE (8.3.5.1).*

Proof.   For t fixed (whose dependence is then omitted), we define the operator

$$M(\tau) = e^{A^*(\tau - t)} P(\tau) e^{A(\tau - t)} \in \mathcal{L}(Y), \quad 0 \leq t \leq \tau \leq T \qquad (8.3.5.2a)$$

$$: \text{continuous } Y \to L_\infty(0, T; Y), \qquad (8.3.5.2b)$$

where the regularity (8.3.5.2b) follows from (8.3.3.5). First, let $x, y \in \mathcal{D}(A)$. We compute

$$\frac{d}{d\tau}(M(\tau)x, y)_Y = \frac{d}{d\tau}\left(P(\tau)e^{A(\tau-t)}x, e^{A(\tau-t)}y\right)_Y$$

$$= \left(P'(\tau)e^{A(\tau-t)}x, e^{A(\tau-t)}y\right)_Y$$

$$+ \left(P(\tau)e^{A(\tau-t)}Ax, e^{A(\tau-t)}y\right)_Y$$

$$+ \left(P(\tau)e^{A(\tau-t)}x, e^{A(\tau-t)}Ay\right)_Y, \quad x, y \in \mathcal{D}(A). \quad (8.3.5.3)$$

Since $P(\tau)$ satisfies the DRE (8.3.4.26), we may rewrite the first term on the right-hand side of (8.3.5.3) as

$$\left(P'(\tau)e^{A(\tau-t)}x, e^{A(\tau-t)}y\right)_Y$$

$$= -\left(Re^{A(\tau-t)}x, Re^{A(\tau-t)}y\right)_Z - \left(P(\tau)e^{A(\tau-t)}x, Ae^{A(\tau-t)}y\right)_Y$$

$$- \left(P(\tau)Ae^{A(\tau-t)}x, e^{A(\tau-t)}y\right)_Y$$

$$+ \left(B^*P(\tau)e^{A(\tau-t)}x, B^*P(\tau)e^{A(\tau-t)}y\right)_U, \quad x, y \in \mathcal{D}(A). \quad (8.3.5.4)$$

For the left-hand side of (8.3.5.4) we recall (8.3.4.27), while for its right-hand side we note that each term is well-defined by virtue of (8.1.9) for $x, y \in \mathcal{D}(A)$ and properties (8.3.3.5) and (8.3.3.4), as assumed. Inserting the right-hand side of (8.3.5.4) into the first term on the right-hand side of (8.3.5.3) results in a cancellation of the last two terms of (8.3.5.3), so that we obtain

$$\frac{d}{d\tau}(M(\tau)x, y)_Y = -\left(Re^{A(\tau-t)}x, Re^{A(\tau-t)}y\right)_Z$$

$$+ \left(B^*P(\tau)e^{A(\tau-t)}x, B^*P(\tau)e^{A(\tau-t)}y\right)_U$$

$$x, y \in \mathcal{D}(A), \quad 0 \leq t \leq \tau \leq T. \quad (8.3.5.5)$$

Integrating now (8.3.5.5) in τ, over $[t, T]$, and using that $M(T) = 0$ (since $P(T) = 0$

by assumption), we obtain the identity

$$\int_t^T \frac{d}{d\tau}(M(\tau)x, y)_Y \, d\tau = 0 - (P(t)x, y)_Y$$

$$= -\int_t^T \left(Re^{A(\tau-t)}x, \, Re^{A(\tau-t)}y\right)_Z d\tau$$

$$+ \int_t^T \left(B^*P(\tau)e^{A(\tau-t)}x, \, B^*P(\tau)e^{A(\tau-t)}y\right)_Z d\tau,$$

$$\text{(8.3.5.6)}$$

first still for $x, y \in \mathcal{D}(A)$, and next for all $x, y \in Y$ by continuous extension using assumption (H.2) = (8.2.1.2), or (h.1) = (8.2.1.10), on the first term of the right-hand side of (8.3.5.6), and using the assumed property (8.3.3.4) on its second term. Thus, (8.3.5.6) shows that $P(t)$ is a solution of the IRE (8.3.5.1).   □

We now give the converse result.

Proposition 8.3.5.2   Let $P(t) \in \mathcal{L}(Y)$, $0 \le t \le T$, be (as in the case of the operator $P(t)$ defined by (8.3.3.2) under assumptions (H.1), (H.2), and (H.3) (or else (h.1) and (h.2)) a solution of the IRE (8.3.5.1) that satisfies properties (8.3.3.3) and (8.3.3.4), so that $P(t)$ is nonnegative and self-adjoint, $P(T) = 0$, and $P(t)$ satisfies the regularity property (8.3.3.5) as well.

Then, $P(t)$ satisfies also the DRE (8.3.4.26) for $x, y \in \mathcal{D}(A)$.

Proof.   With $x, y \in \mathcal{D}(A)$ we differentiate both sides of (8.3.5.1), thus obtaining

$$\frac{d}{dt}(P(t)x, y)_Y = -(Rx, Ry)_Z + (B^*P(t)x, B^*P(t)y)_U$$

$$- \int_t^T \left(Re^{A(\tau-t)}Ax, \, Re^{A(\tau-t)}y\right)_Z d\tau$$

$$+ \int_t^T \left(B^*P(\tau)e^{A(\tau-t)}Ax, \, B^*P(\tau)e^{A(\tau-t)}y\right)_U d\tau$$

$$- \int_t^T \left(Re^{A(\tau-t)}x, \, Re^{A(\tau-t)}Ay\right)_Z d\tau$$

$$+ \int_t^T \left(B^*P(\tau)e^{A(\tau-t)}x, \, B^*P(\tau)e^{A(\tau-t)}Ay\right)_U d\tau, \quad \text{(8.3.5.7)}$$

where all terms are well-defined by the assumed properties of $P(t)$ and by (H.2) = (8.2.1.2) or by (h.1). Using again (8.3.5.1) once with x replaced by Ax for the third and fourth terms on the right-hand side of (8.3.5.7), and once with y replaced by Ay for the fifth and sixth terms, we obtain from (8.3.5.7)

$$\frac{d}{dt}(P(t)x, y)_Y = -(Rx, Ry)_Z + (B^*P(t)x, B^*P(t)y)_U$$

$$- (P(t)Ax, y)_Y - (P(t)x, Ay)_Y, \quad 0 \le t < T, \; x, y \in \mathcal{D}(A),$$

$$\text{(8.3.5.8)}$$

as desired, while $P(T) = 0$ by (8.3.5.1). An analysis of the right-hand side of (8.3.5.8) shows property (8.3.4.27) for $P'(t)$, as in the proof of Theorem 8.3.4.5.   □

Corollary 8.3.5.3   *If the IRE (8.3.5.1) admits a unique solution within the class of nonnegative, self-adjoint $P(t) \in \mathcal{L}(Y)$ that satisfy properties (8.3.3.2b)–(8.3.3.4), then so does the DRE (8.3.4.26).*

8.3.6 The IRE without Passing through the DRE

So far, under assumptions (H.1), (H.2), and (H.3) [or, alternatively, (h.1) and (h.2)], the variational approach of the present section has proceeded along the following key steps:

  (i) from the optimality conditions to the explicit formula of the optimal solution $Ry^0(\tau, t; x) = R\Phi(\tau, t)x$ via (8.3.1.14) of Lemma 8.3.1.1, to the (constructive) definition of the operator $P(t)$ in (8.3.3.2a);
 (ii) through establishment of the critical properties of $P(t)$: nonnegative, self-adjointness in (8.3.4.22), as well as the regularity properties (8.3.3.2b)–(8.3.3.4) of Theorem 8.3.3.1;
(iii) and through proof that the operator $P(t)$ in (8.3.3.2) is a solution of the DRE (8.2.1.32) by Theorem 8.3.4.5,
(iv) and hence of the IRE (8.2.1.33) by Proposition 8.3.5.1.

Moreover, Proposition 8.3.5.2 shows that a solution of the IRE (8.2.1.33) satisfying the properties in (ii) above is also a solution of the DRE (8.2.1.32).

What is desirable is a *direct* proof that the operator $P(t)$ in (8.3.3.2a) is a solution of the IRE (8.2.1.33), without passing through the seemingly more challenging DRE (8.2.1.32). This is accomplished in the present section.

Theorem 8.3.6.1   *Assume (H.1), (H.2), and (H.3) [or else, alternatively, (h.1) and (h.2)]. The operator $P(t)$ defined by (8.3.3.2) satisfies the IRE (8.2.1.33) [or (8.3.5.1)].*

Proof.

Step 1.   Lemma 8.3.6.2   *Under the above assumptions, the following identity holds true for $x, y \in Y$:*

$$\int_t^T \left(B^* P(\tau) e^{A(\tau - t)} x, \, B^* P(\tau) e^{A(\tau - t)} y \right)_U d\tau$$

$$= - \left(RL_t u^0(\,\cdot\,, t; x), \, Re^{A(\cdot - t)} y \right)_{L_2(t, T; Z)} \tag{8.3.6.1}$$

$$= \int_t^T \left(e^{A^*(\tau - t)} R^* R \int_t^\tau e^{A(\tau - \sigma)} B B^* P(\sigma) \Phi(\sigma, t) x \, d\sigma, \, y \right)_Y d\tau. \tag{8.3.6.2}$$
□

Proof.   Via L_t given by (8.3.2.2) [or (8.1.12a)] with $u^0(\,\cdot\,, t; x)$ given by (8.3.4.18), we verify at once that the right-hand side of (8.3.6.1) coincides with (8.3.6.2). Thus, below, we establish (8.3.6.2).

Recall that $B^*P(\cdot)\Phi(\cdot,t)x \in L_2(t,T;U)$ by (8.3.4.18). Thus, (8.2.1.8) and the regularity of $V(\cdot)$ in (8.3.3.8) show that each term of (8.3.6.2) is well-defined. In fact, to this end, we compute, by use of the optimal dynamics $\Phi(\tau,t)x$ as in (8.3.4.9) [by (8.3.2.2b) and (8.3.4.18)],

$$\int_t^T \left(B^*P(\tau)e^{A(\tau-t)}x, B^*P(\tau)e^{A(\tau-t)}y\right)_U d\tau$$

$$= \int_t^T \left(B^*P(\tau)\left[\Phi(\tau,t)x + \int_t^\tau e^{A(\tau-\sigma)}BB^*P(\sigma)\Phi(\sigma,t)x\,d\sigma\right],\right.$$

$$\left. B^*P(\tau)e^{A(\tau-t)}y\right)_U d\tau$$

$$= M(t) + \int_t^T \left(B^*P(\tau)\int_t^\tau e^{A(\tau-\sigma)}BB^*P(\sigma)\Phi(\sigma,t)x\,d\sigma,\right.$$

$$\left. B^*P(\tau)e^{A(\tau-t)}y\right)_U d\tau, \qquad (8.3.6.3)$$

where we have set

$$M(t) = \int_t^T \left(B^*P(\tau)\Phi(\tau,t)x, B^*P(\tau)e^{A(\tau-t)}y\right)_U d\tau. \qquad (8.3.6.4)$$

$M(t)$ in (8.3.6.4) is well-defined by (8.3.4.18) and (8.3.3.8) with $V(t) = B^*P(t)$. The integral term in (8.3.6.3) is well-defined by (8.3.3.4) and (8.3.4.18), and by (8.3.3.8). Next, using the $L_2(\sigma,T;U)$-regularity of $B^*P(\tau)e^{A(\tau-\sigma)}Bx = Q(\tau,\sigma)x$ in τ, already established in Proposition 8.3.3.5, we interchange the order of integration in (8.3.6.3) and obtain for its R.H.S. (right-hand side):

$$\text{R.H.S. of (8.3.6.3)} = M(t) + \int_t^T \int_\sigma^T \left(B^*P(\tau)e^{A(\tau-\sigma)}BB^*P(\sigma)\Phi(\sigma,t)x,\right.$$

$$\left. B^*P(\tau)e^{A(\tau-\sigma)}e^{A(\sigma-t)}y\right)_U d\tau\,d\sigma, \qquad (8.3.6.5)$$

which is well-defined by (8.3.3.4), (8.3.4.18), and (8.3.3.8). Next, to perform the internal integration in (8.3.6.5), we use the IRE (8.3.5.1) with x there replaced by $BB^*P(\sigma)\Phi(\sigma,t)x$ (where $B^*P(\sigma)\Phi(\sigma,t)x \in U$ a.e. in σ, $x \in Y$ by (8.3.4.18)), with y there replaced by $e^{A(\sigma-t)}y$, and with t there replaced by σ. We then obtain for $x,y \in Y$:

R.H.S. of (8.3.6.3)

$$= M(t) + \int_t^T \int_\sigma^T \left(e^{A^*(\tau-\sigma)}R^*Re^{A(\tau-\sigma)}BB^*P(\sigma)\Phi(\sigma,t)x, e^{A(\sigma-t)}y\right)_Y d\tau\,d\sigma$$

$$- \int_t^T \left(P(\sigma)BB^*P(\sigma)\Phi(\sigma,t)x, e^{A(\sigma-t)}y\right)_Y d\sigma, \qquad (8.3.6.6)$$

where cancellation takes place because of (8.3.6.4). After a further change in the order of integration and use of the semigroup property, we obtain

$$\text{R.H.S. of } (8.3.6.3) = \int_t^T \left(e^{A^*(\tau-t)} R^* R \int_t^\tau e^{A(\tau-\sigma)} B B^* P(\sigma) \Phi(\sigma, t) x \, d\sigma, y \right)_Y d\tau,$$

(8.3.6.7)

which is well-defined by (8.3.4.18) and (8.2.1.8). Then, (8.3.6.3) and (8.3.6.7) verify (8.3.6.2), as desired.   □

Step 2   We return to (8.3.3.2) and compute by virtue of (8.3.2.2) and (8.3.6.1), with $x, y \in Y$:

$$(P(t)x, y)_Y = \int_t^T \left(R\Phi(\tau, t)x, Re^{A(\tau-t)}y \right)_Z d\tau$$

$$\text{(by (8.3.2.2))} \quad = \int_t^T \left(Re^{A(\tau-t)}x, Re^{A(\tau-t)}y \right)_Z d\tau$$

$$+ \int_t^T \left(RL_t u^0(\cdot, t; x), Re^{A(\tau-t)}y \right) d\tau$$

$$\text{(by (8.3.6.1))} \quad = \int_t^T \left(Re^{A(\tau-t)}x, Re^{A(\tau-t)}y \right)_Z d\tau$$

$$- \int_t^T \left(B^* P(\tau)e^{A(\tau-t)}x, B^* P(\tau)e^{A(\tau-t)}y \right)_U d\tau. \quad (8.3.6.8)$$

Thus, (8.3.6.8) is the IRE (8.3.5.1) and Theorem 8.3.6.1 is proved.   □

8.3.7 Uniqueness

The required uniqueness result in Theorem 8.2.1.2(viii) will be a consequence of the following theorem. As in the case of Chapter 1, Section 1.5.1, Theorem 1.5.1.7, it suffices by Corollary 8.3.5.3 to show uniqueness of a Riccati operator solution, within the specified class defined by properties (8.2.1.25), (8.2.1.26), (8.2.1.30), and (8.2.1.31) [or (8.3.3.2b)–(8.3.3.4)], for the corresponding integral Riccati equation (8.3.5.1). More generally, in view also of the subsequent Chapter 9, we consider the IRE of the form

$$(P(t)x, y)_Y = \left(Ge^{A(T-t)}x, Ge^{A(T-t)}y \right)_{Z_f} + \int_t^T \left(Re^{A(\tau-t)}x, Re^{A(\tau-t)}y \right)_Z d\tau$$

$$- \int_t^T \left(B^* P(\tau)e^{A(\tau-t)}x, B^* P(\tau)e^{A(\tau-t)}y \right)_U d\tau, \quad (8.3.7.1)$$

under the assumptions:

$$Ge^{At} : \text{continuous } Y \to L_2(0, T; Z_f); \quad Re^{At} : \text{continuous } Y \to L_2(0, T; Z).$$

(8.3.7.2)

Theorem 8.3.7.1   *Assume (h.1) = (8.2.1.10) and (8.3.7.2) for R and G. Then, the IRE (8.3.7.1) admits, at most, one solution satisfying the following properties:*

(i)

$$P(t) = P^*(t) \in \mathcal{L}(Y), \quad 0 \le t \le T; \qquad (8.3.7.3)$$

(ii)

$$V(\cdot) \equiv B^* P(\cdot) \in \mathcal{B}([0, T]; \mathcal{L}(Y; U)), \qquad (8.3.7.4)$$

that is, in the notation of Remark 8.1.1, Eqn. (8.1.14),

$$\begin{cases} V(t) \equiv B^* P(t) \in \mathcal{L}(Y; U), \quad 0 \le t \le T; & (8.3.7.5a) \\[2mm] V(\cdot) \equiv B^* P(\cdot) : continuous\ Y \to L_\infty(0, T; U); & (8.3.7.5b) \end{cases}$$

(iii)

$$Q(t, s) \equiv B^* P(t) e^{A(t-s)} B : continuous\ U \to U_{0,T}, \qquad (8.3.7.6a)$$

that is, in the notation of (8.3.3.1), or (8.3.3.13),

$$\|Qu\|_{U_{0,T}}^2 \equiv \sup_{0 \le s \le T} \int_s^T \|Q(t, s)u\|_U^2\, dt \le c_T \|u\|_U^2, \quad u \in U. \quad (8.3.7.6b)$$

Proof.

Step 1   Let $P_1(t)$ and $P_2(t)$ be two solutions of the IRE (8.3.7.1) within the prescribed class defined by properties (8.3.7.3)–(8.3.7.6). We set

$$\bar{P}(t) \equiv P_1(t) - P_2(t), \quad 0 \le t \le T, \qquad (8.3.7.7)$$

so that $\bar{P}(t)$ and

$$\bar{V}(t) \equiv B^* \bar{P}(t); \quad V_i(t) \equiv B^* P_i(t), \quad i = 1, 2; \qquad (8.3.7.8)$$

$$\bar{Q}(t, s) \equiv B^* \bar{P}(t) e^{A(t-s)} B; \quad Q_i(t, s) \equiv B^* P_i(t) e^{A(t-s)} B, \qquad (8.3.7.9)$$

satisfy properties (8.3.7.3)–(8.3.7.6) as well. Moreover, the difference $\bar{P}(t)$ satisfies

$$(\bar{P}(t)x, y)_Y = -\int_t^T \left(B^* P_2(\tau) e^{A(\tau-t)} x, B^* \bar{P}(\tau) e^{A(\tau-t)} y \right)_U d\tau$$

$$-\int_t^T \left(B^* \bar{P}(\tau) e^{A(\tau-t)} x, B^* P_1(\tau) e^{A(\tau-t)} y \right)_U d\tau. \quad (8.3.7.10)$$

We shall show that: There exists $0 < s_0 < T$, sufficiently near T, that is, with $T - s_0$ sufficiently small, such that

$$\bar{Q}(t, s) \equiv B^* \bar{P}(t) e^{A(t-s)} B \equiv 0, \quad s_0 \le s \le t \le T. \qquad (8.3.7.11)$$

To this end, setting both $x = e^{A(t-s)}Bu$ and $y = Bz$, $u, z \in U$, in (8.3.7.10), we obtain by virtue of (8.3.7.8) and (8.3.7.9):

$$\bar{Q}(t,s)u = -\int_t^T \bar{Q}^*(\tau,t)Q_2(\tau,s)u\,d\tau - \int_t^T Q_1^*(\tau,t)\bar{Q}(\tau,s)u\,d\tau. \quad (8.3.7.12)$$

We then obtain from (8.3.7.12)

$$\int_s^T \|\bar{Q}(t,s)u\|_U^2\,dt \leq 2\int_s^T \left\{\int_t^T \|\bar{Q}^*(\tau,t)\|^2\,d\tau\right\}\left\{\int_t^T \|Q_2(\tau,s)u\|_U^2\,d\tau\right\}dt$$

$$+2\int_s^T \left\{\int_t^T \|Q_1^*(\tau,t)\|^2\,d\tau\right\}\left\{\int_t^T \|\bar{Q}(\tau,s)u\|_U^2\,d\tau\right\}dt,$$

$$(8.3.7.13)$$

where $\|\cdot\|$ denotes the norm in $\mathcal{L}(U)$. Next, we majorize the integral terms in (8.3.7.13) involving $Q_2(\tau,s)u$ and $\bar{Q}(\tau,s)u$, by replacing the lower end point t by s, $s \leq t$, and thus making those terms independent of t. We then obtain, from (8.3.7.13)

$$\int_s^T \|\bar{Q}(t,s)u\|_U^2\,dt$$

$$\leq 2\left\{\int_s^T \|Q_2(\tau,s)u\|_U^2\,d\tau\right\}(T-s)\left\{\sup_{s\leq t\leq T}\int_t^T \|\bar{Q}^*(\tau,t)\|^2\,d\tau\right\}$$

$$+2\left\{\int_s^T \|\bar{Q}(\tau,s)u\|_U^2\,d\tau\right\}(T-s)\left\{\sup_{s\leq t\leq T}\int_t^T \|Q_1^*(\tau,t)\|^2\,d\tau\right\}. \quad (8.3.7.14)$$

Taking the sup over $s_0 \leq t \leq T$ on both sides and recalling (8.3.7.6b) we obtain from (8.3.7.14)

$$\sup_{s_0\leq s\leq T}\int_s^T \|\bar{Q}(t,s)u\|_U^2\,dt \equiv \|\bar{Q}u\|_{U_{s_0,T}}^2$$

$$\leq 2\|Q_2u\|_{U_{s_0,T}}^2(T-s_0)\|\bar{Q}\|_{\mathcal{L}(U,U_{s_0,T})}^2$$

$$+2\|\bar{Q}u\|_{U_{s_0,T}}^2(T-s_0)\|Q_1\|_{\mathcal{L}(U,U_{s_0,T})}^2. \quad (8.3.7.15)$$

Choosing s_0 near T so that $T-s_0$ is suitably small and

$$4(T-s_0)\|Q_i\|_{\mathcal{L}(U,U_{s_0,T})}^2 \leq \rho < 1, \quad i = 1, 2,$$

we obtain from (8.3.7.15)

$$\|\bar{Q}\|_{\mathcal{L}(U,U_{s_0,T})}^2 \leq \rho\|\bar{Q}\|_{\mathcal{L}(U,U_{s_0,T})}^2, \quad \rho < 1, \quad (8.3.7.16)$$

and then $\|\bar{Q}\|_{\mathcal{L}(U,U_{s_0,T})} = 0$ and (8.3.7.11) follows, as desired.

Step 2   Using (8.3.7.11), we shall now prove that

$$\bar{V}(t) \equiv B^* \bar{P}(t) \equiv 0, \quad s_0 \leq t \leq T. \tag{8.3.7.17}$$

To this end, we return to (8.3.7.10) over $s_0 \leq t \leq T$ with $y = Bz$, use (8.3.7.11), and obtain, via (8.3.7.8) and (8.3.7.9),

$$|(\bar{V}(t)x, z)_U| = |(B^* \bar{P}(t)x, z)_U|$$

$$= \left| \int_t^T \left(B^* \bar{P}(\tau) e^{A(\tau-t)} x, B^* P_1(\tau) e^{A(\tau-t)} Bz \right)_U d\tau \right|$$

$$\leq \int_t^T \left\| B^* \bar{P}(\tau) e^{A(\tau-t)} x \right\|_U \left\| B^* P_1(\tau) e^{A(\tau-t)} Bz \right\|_U d\tau$$

(by (8.3.7.8) and (8.3.7.9)) $\displaystyle \leq \left\{ \sup_{t \leq \tau \leq T} \left\| \bar{V}(\tau) e^{A(\tau-t)} x \right\|_U \right\} \sqrt{T-t}$

$$\times \left\{ \int_t^T \| Q_1(\tau, t) z \|_U^2 d\tau \right\}^{\frac{1}{2}}. \tag{8.3.7.18}$$

Taking the sup over $s_0 \leq t \leq T$ in (8.3.7.18) we obtain

$$\sup_{s_0 \leq t \leq T} |(\bar{V}(t)x, z)_U| \leq C_T \left\{ \sup_{s_0 \leq \tau \leq T} \| \bar{V}(\tau) \|_{\mathcal{L}(Y;U)} \right\} \sqrt{T - s_0} \| Q_1 \|_{\mathcal{L}(U;U_{s_0,T})} \|x\| \|z\|. \tag{8.3.7.19}$$

Taking $T - s_0$ sufficiently small yields the counterpart of (8.3.7.16), that is,

$$\left\{ \sup_{s_0 \leq t \leq T} \| \bar{V}(t) \|_{\mathcal{L}(Y;U)} \right\} \leq r \left\{ \sup_{s_0 \leq t \leq T} \| (\bar{V}(\tau) \|_{\mathcal{L}(Y;U)} \right\}, \quad r < 1, \tag{8.3.7.20}$$

with $r < 1$, and hence (8.3.7.17) follows, as desired.

Step 3   Returning to (8.3.7.10) and using here (8.3.7.17) yields, as desired,

$$\bar{P}(t) \equiv 0, \quad s_0 \leq t \leq T. \tag{8.3.7.21}$$

Step 4   Repeating the above argument, after a finite number of steps we obtain $\bar{P}(t) \equiv 0$, $0 \leq t \leq T$, as desired.   □

Completion of the Proof of Theorem 8.2.1.2 (viii).   The properties assumed in (8.3.7.3)–(8.3.7.6) are precisely those established in Sections 8.3.3 and 8.3.4 for the solution $P(t)$ in (8.3.3.2a), under assumptions (H.1), (H.2), and (H.3) [or else, alternatively, under assumptions (h.1) and (h.2)], which guarantee (8.3.7.2), where now $G \equiv 0$. Thus, Theorem 8.3.7.1 applies and yields that $P(t)$ in (8.3.3.2a) is the desired unique solution of the IRE (8.3.5.1), and hence of the DRE (8.3.4.26) by Theorem 8.3.5.2.   □

Theorem 8.2.1.2 is thus fully proved.

8.3.8 Proof of Theorem 8.2.1.3

By (h.1) and (h.3), the results of Theorem 8.2.1.1 hold true.

(i) We shall first prove (2.1.34), that is,

$$R\Phi_{T-t}(\sigma, 0)x \equiv R\Phi_T(t + \sigma, t)x \underset{(\text{in } \sigma)}{\in} L_2(0, T; Z). \tag{8.3.8.1}$$

As in the proof of Proposition 8.3.2.4, we shall use (8.2.1.19): In the notation of (8.3.2.2), we see that (8.2.1.19) is written explicitly as in (8.3.2.17), that is, for $x \in Y$,

$$R\Phi_T(t, s)x + R \int_s^t e^{A(t-\tau)} B B^* \int_\tau^T e^{A^*(r-\tau)} R^* R\Phi_T(r, s)x \, dr \, d\tau = Re^{A(t-s)}x. \tag{8.3.8.2}$$

Specializing (8.3.8.2) with $T - t$ in place of T, 0 in place of s, and σ in place of t yields

$$R\Phi_{T-t}(\sigma, 0)x + R \int_0^\sigma e^{A(\sigma-\tau)} B B^* \int_\tau^{T-t} e^{A^*(r-\tau)} R^* R\Phi_{T-t}(r, 0)x \, dr \, d\tau = Re^{A\sigma}x. \tag{8.3.8.3}$$

Specializing (8.3.8.2) with $t + \sigma$ in place of t and with t in place of s yields

$$R\Phi_T(t + \sigma, t)x + R \int_t^{t+\sigma} e^{A(t+\sigma-\tau)} B B^* \int_\tau^T e^{A^*(\alpha-\tau)} R^* R\Phi_T(\alpha, t)x \, d\alpha \, d\tau$$

$$= Re^{A(t+\sigma-t)}x. \tag{8.3.8.4}$$

We now set $\tau - t = \beta$ in the first integral of (8.3.8.4) and then $\alpha - t = r$ in the second integral of (8.3.8.4), thus obtaining

$$R\Phi_T(t + \sigma, t)x + R \int_0^\sigma e^{A(\sigma-\beta)} B B^* \int_\beta^{T-t} e^{A^*(r-\beta)} R^* R\Phi_T(t + r, t)x \, dr \, d\beta$$

$$= Re^{A\sigma}x. \tag{8.3.8.5}$$

Comparison between (8.3.8.3) and (8.3.8.5) reveals that both $R\Phi_{T-t}(\sigma, 0)$ and $R\Phi_T(t + \sigma, t)$ satisfy the same equation. But then the difference satisfies

$$[I + RL_0 L_0^* R^*][R\Phi_{T-t}(\cdot, 0)x - R\Phi_T(t + \cdot, t)x] \equiv 0, \tag{8.3.8.6}$$

from which (8.3.8.1) follows, as desired. [The present proof is similar to that of Chapter 2, Lemma 2.3.2.1.]

(ii) We next show (8.2.1.35), that is,

$$u_T^0(\tau, t; x) = u_{T-t}^0(\tau - t, 0; x) \underset{(\text{in } \tau)}{\in} L_2(0, T; U). \tag{8.3.8.7}$$

We use the characterization (8.2.1.21), that is, explicitly

$$u_T^0(\tau, t; x) = -\int_\tau^T B^* e^{A^*(\sigma - \tau)} R^* R \Phi_T(\sigma, t) x \, d\sigma \qquad (8.3.8.8)$$

$$\text{(by (8.3.8.1))} \qquad = -\int_\tau^T B^* e^{A^*(\sigma - \tau)} R^* R \Phi_{T-t}(\sigma - t, 0) x \, d\sigma$$

$$\text{(with } \sigma - t = \beta) \qquad = -\int_{\tau - t}^{T - t} B^* e^{A^*(\beta - (\tau - t))} R^* R \Phi_{T-t}(\beta, 0) x \, d\beta$$

$$\text{(by (8.3.8.8))} \qquad = u_{T-t}^0(\tau - t, 0; x), \quad \tau > t, \qquad (8.3.8.9)$$

after using (8.3.8.1), the change of variable $\sigma - t = \beta$, and (8.3.8.8). Thus, (8.3.8.7) is proved.

(iii) Finally we show (8.2.1.36), that is,

$$P_{T-t}(0) = P_T(t), \quad 0 \le t \le T. \qquad (8.3.8.10)$$

We return to (8.2.1.24) and use here (8.3.8.1) to obtain

$$P_T(t)x = \int_t^T e^{A^*(\tau - t)} R^* R \Phi_T(\tau, t) x \, d\tau$$

$$= \int_0^{T - t} e^{A^* \sigma} R^* R \Phi_T(t + \sigma, t) x \, d\sigma \qquad (8.3.8.11)$$

$$\text{(by (8.3.8.1))} \qquad = \int_0^{T - t} e^{A^* \sigma} R^* R \Phi_{T-t}(\sigma, 0) x \, d\sigma = P_{T-t}(0)x, \qquad (8.3.8.12)$$

using, in the last step, again (8.2.1.24). Thus (8.3.8.10) is proved. □

Remark 8.3.8.1   Under the case where the IRE possesses a unique solution, as guaranteed by Theorem 8.2.1.2(vii) under the hypotheses (H.1), (H.2), and (H.3) [or else, alternatively, hypotheses (h.1) = (8.2.1.10) and (h.2) = (8.2.1.11)], property (8.3.8.11) can be shown also via the IRE (8.2.1.33). This will be done explicitly in Chapter 9, Section 9.3.8 with the additional final state penalizaton.

8.4 A Second Direct Proof of Theorem 8.2.1.2: From the Well-Posedness of the IRE to the Control Problem. Dynamic Programming

In this section we provide an alternative proof of Theorem 8.2.1.2, according to the so-called direct approach. This consists in taking the well-posedness of the (differential, integral, or algebraic) Riccati equation (*not* the control problem as in Section 8.3) as the *starting* point of the investigation. Once the solution to the Riccati equation is established, one then proceeds – via a version of the dynamics programming approach – to construct the control problem that generates the original Riccati equation in the first place. Thus, a *direct* study of the Riccati equation represents, in a definite sense, a reverse procedure of the variational approach followed in Section 8.3 in the study of the optimal control problem: The former proceeds from the Riccati equation to the control problem, the latter from the control problem to the Riccati equation. The main emphasis of the present books is on the variational study from the control problem to a *constructive* solution of the Riccati equation: In this way, we can take advantage at the start of the *optimality* conditions available for the control problem. The reversed direct study to Riccati equations in infinite dimensional spaces was initiated in Da Prato [1973] and Tartar [1974]. In line with this strategy, which is a technical extension of the finite dimensional case [Lee, Markus, 1968, p. 189], our argument below in the direct study of well-posedness of the IRE will be based on two steps: (i) (unique) local solution by a contraction mapping argument and (ii) global a priori bounds to obtain a global (unique) solution. Thus, when it succeeds, this approach also provides automatically uniqueness of the Riccati solution (within a specified class). By contrast, our variational study of the control problem under the general setting of Chapter 1 in the parabolic case and of the forthcoming Chapter 10 in the hyperbolic case provides existence, but it says nothing about uniqueness.

8.4.1 Existence and Uniqueness: Preliminaries

The present Section 8.4.1, along with Sections 8.4.2 and 8.4.3 to follow, may be viewed in two ways.

On the one hand, these sections begin the *direct* study of the IRE (8.3.5.1) under the hypotheses (H.1), (H.2), and (H.3) [or else, alternatively, (h.1) and (h.2)] of the present chapter and ultimately yield local existence and uniqueness of a Riccati operator solution near T, by a contraction mapping argument. Uniqueness is asserted within the class of solutions enjoying properties (8.2.1.25), (8.2.1.26), (8.2.1.30), and (8.2.1.31). Thus, such unique solution must coincide with the operator $P(t)$, constructively defined in (8.3.3.2), which was shown in Section 8.3.4 to be a solution of the differential Riccati equation (8.2.1.32), or (8.3.4.26), and to possess properties (8.2.1.25), (8.2.1.26), (8.2.1.30), and (8.2.1.31) by Lemma 8.3.4.4 and Theorem 8.3.3.1. By Propositions 8.3.5.2 and 8.3.5.3, one then obtains existence and uniqueness (within said class) also of the DRE (8.2.1.32) near T. This local result, combined with the global a priori bounds (regularity properties) of the variables $V(t)$ and $Q(t, s)$ shown in Theorem 8.3.3.1, will yield existence and uniqueness on all of $[0, T]$. Thus, Sections 8.4.1 through 8.4.3, when combined with Sections 8.3.3 and 8.3.5, may be viewed as an alternative treatment to the (simpler and) closely related proof of global uniqueness given in Section 8.3.6.

On the other hand, Sections 8.4.1 through 8.4.3, when combined with a version of the dynamic programming approach of the forthcoming Section 8.4.4, produce an altogether independent, direct proof of Theorem 8.2.1.2.

As already mentioned, and as in the case of Chapter 1, Section 1.5.1, Theorem 1.5.1.7, it suffices by Corollary 8.3.5.3 to show existence and uniqueness within the specified class for the corresponding integral Riccati equation (8.3.5.1), rewritten here as

$$(P(t)x, y)_Y = \int_t^T \left(R^* R e^{A(\tau - t)} x, e^{A(\tau - t)} y \right)_Y d\tau$$

$$- \int_t^T \left(B^* P(\tau) e^{A(\tau - t)} x, B^* P(\tau) e^{A(\tau - t)} y \right)_U d\tau, \quad x, y \in Y,$$
(8.4.1.1)

under assumptions (H.1), (H.2), and (H.3) [or else, alternatively, (h.1) and (h.2)]. Using the new variable (8.3.3.3)

$$V(t) \equiv B^* P(t) \in \mathcal{L}(Y; U), \quad 0 \le t \le T, \tag{8.4.1.2}$$

in (8.4.1.1) with $y = Bu$, we obtain

$$V(t)x = \int_t^T B^* e^{A^*(\tau - t)} R^* R e^{A(\tau - t)} x \, d\tau$$

$$- \int_t^T B^* e^{A^*(\tau - t)} V^*(\tau) V(\tau) e^{A(\tau - t)} x \, d\tau. \tag{8.4.1.3}$$

Recalling the second new variable (8.3.3.4)

$$Q(t, s) \equiv V(t) e^{A(t - s)} B \equiv B^* P(t) e^{A(t - s)} B, \quad 0 \le s \le t \le T, \tag{8.4.1.4}$$

and setting in (8.4.1.1) both $x = e^{A(t - s)} Bu$ and $y = Bz$, we obtain the following equation for $Q(\ , \ )$:

$$Q(t, s)u = F_1(t, s)u - \int_t^T Q^*(\tau, t) Q(\tau, s)u \, d\tau, \tag{8.4.1.5}$$

where

$$F_1(t, s)u = \int_t^T B^* e^{A^*(\tau - t)} R^* R e^{A(\tau - s)} Bu \, d\tau, \quad s \le t \le T \tag{8.4.1.6a}$$

$$= \left\{ L_s^* R^* R e^{A(\cdot - s)} Bu \right\}(t), \tag{8.4.1.6b}$$

recalling (8.1.13a). Moreover, using (8.4.1.4) in (8.4.1.3), we rewrite (8.4.1.3) as

$$V(t)x = F_2(t)x - \int_t^T Q^*(\tau, t) V(\tau) e^{A(\tau - t)} x \, d\tau, \tag{8.4.1.7}$$

$$F_2(t)x = \int_t^T B^* e^{A^*(\tau - t)} R^* R e^{A(\tau - t)} x \, d\tau = \left\{ L_t^* R^* R e^{A(\cdot - t)} x \right\}(t). \tag{8.4.1.8}$$

Notice that $F_1(t, s)$ in (8.4.1.6b) compares closely with the term leading to estimate (8.3.3.12).

Equation (8.4.1.5) involves only Q, whereas Eqn. (8.4.1.7) couples V and Q. Our next task is to show local existence and uniqueness near T for the solution Q and V of (8.4.1.5) and (8.4.1.7) in appropriate function spaces.

Theorem 8.4.1.1 *Assume (H.1), (H.2), and (H.3) [or else, alternatively, (h.1) and (h.2)]. There exist unique solutions of Eqns. (8.4.1.5) and (8.4.1.3) (equivalently (8.4.1.7)):*

$$Q(t, s) : continuous \ U \to U_{s_0, T}; \quad or \ Q \in \mathcal{L}(U; U_{s_0, T}), \qquad (8.4.1.9)$$

$$V(t) \in \mathcal{L}(Y; U) : continuous \ Y \to L_\infty(s_0, T; U); \quad or \ V \in \mathcal{B}([s_0, T]; \mathcal{L}(Y; U)), \qquad (8.4.1.10)$$

provided that s_0 is sufficiently close to T, $0 < s_0 < T$; the space $U_{s_0 T}$ is defined in (8.3.3.1), while the space in (8.4.1.10) is, as in Remark 8.1.1, the space of linear operators in $\mathcal{L}(Y; U)$ that are uniformly bounded on $[s_0, T]$, with norm

$$\|V\|_{\mathcal{B}([s_0, T]; \mathcal{L}(Y; U))} \equiv \sup_{s_0 \le t \le T} \|V(t)\|_{\mathcal{L}(Y; U)}. \qquad (8.4.1.11)$$

Remark 8.4.1.1 Properties (8.4.1.9) and (8.4.1.10) are local versions of the global properties shown in Theorem 8.3.3.1 for the solution operator $P(t)$ defined by (8.3.3.2) under assumptions (H.1), (H.2), and (H.3) [or else, alternatively, (h.1) and (h.2)].

To prove Theorem 8.4.1.1, we need to establish first the following two inequalities.

Lemma 8.4.1.2 *Assume (H.1) and (H.3) [or else, alternatively, (h.2)]. With reference to (8.4.1.6) we have*

$$\sup_{0 \le s \le T} \int_s^T \|F_1(t, s)u\|_U^2 \, dt \le c_T \|u\|_U^2, \qquad (8.4.1.12)$$

that is, $F_1 \in \mathcal{L}(U; U_{0,T})$.

Lemma 8.4.1.3 *Assume (H.1) and (H.2) [or else, alternatively, (h.1) and (h.2)]. With reference to (8.4.1.8) we have $F_2(t) \in \mathcal{L}(Y; U)$ and*

$$\sup_{0 \le t \le T} \|F_2(t)x\|_U \le c_T \|x\|_Y, \qquad (8.4.1.13)$$

that is, $F_2 \in \mathcal{B}([0, T]; \mathcal{L}(Y; U))$.

Proof of Lemma 8.4.1.2. In (8.4.1.6b) we have seen that we can rewrite F_1 as

$$F_1(t, s)u = \left(L_s^* R^* R e^{A(\cdot - s)} B u\right)(t). \qquad (8.4.1.14)$$

Thus, by invoking assumption (H.3*) in the form of its consequence (8.2.1.9) and (H.1) = (8.2.1.1), we obtain from (8.4.1.14)

$$\|F_1(\cdot, s)u\|_{L_2(s,T;U)} = \left\|L_s^* R^* R e^{A(\cdot - s)} B u\right\|_{L_2(s,T;U)}$$

$$\text{(by (8.2.1.9) and (8.2.1.1))} \qquad \le c_T \|R e^{A(\cdot - s)} B u\|_{L_1(s,T;Z)} \le c_T \|u\|_U, \qquad (8.4.1.15a)$$

where in the last step we have invoked $(H.1) = (8.2.1.1)$, modulo a change of variable, to obtain readily a bound independent of s. Then, $(8.4.1.15)$ implies a fortiori $(8.4.1.12)$. [Alternatively, estimate $(8.4.1.15)$ is obtained by invoking first $(8.2.1.16)$ under $(h.2)$, which is stronger than $(h.3)$, and then $(h.2) = (8.2.1.11)$:

$$\|F_1(\,\cdot\,,s)u\|_{L_2(s,T;U)} = \|L_s^* R^* Re^{A(\cdot\,-s)} Bu\|_{L_2(s,T;U)}$$

(by $(8.2.1.16)$ and $(8.2.1.11)$) $\qquad \leq c_T \|Re^{A(\cdot\,-s)} Bu\|_{L_2(s,T;Z)} \leq c_T \|u\|_U.$

$$(8.4.1.15b)$$

Notice that if one would like to begin the estimate by just invoking $(h.3)$ in the form of $(8.2.1.16)$, one would still need $(h.2)$ – but $(h.2)$ is stronger than $(h.3)$. Here is a point where $(h.1)$ and $(h.3)$ – good for Theorem 8.2.1.1 – are not sufficient to obtain the desired estimate $(8.4.1.15)$.] $\square$

Proof of Lemma 8.4.1.3.   By $(8.4.1.8)$ with $\tau - t = \sigma$ we compute

$$\|F_2(t)x\|_U = \|\{L_t^* R^* Re^{A(\cdot\,-t)}x\}(t)\|_U$$

$$= \left\|\int_0^{T-t} B^* e^{A^*\sigma} R^* Re^{A\sigma} x \, d\sigma\right\|_U$$

$$\leq c_T \|Re^{At}x\|_{C([0,T];Z)} \leq c_T \|x\|_Y, \qquad (8.4.1.16)$$

where in the last steps we have used first assumption $(H.1^*) = (8.2.1.4)$ (after an innocuous extension by zero on $[T - t, T]$) and then assumption $(H.2) = (8.2.1.2)$. Then, $(8.4.1.16)$ implies $(8.4.1.13)$. [Alternatively, the final estimate in $(8.4.1.16)$ is achieved by first invoking $(h.2^*) = (8.2.1.14)$ for $L_t^* R^*$ uniformly in t, and next $(h.1) = (8.2.1.10)$.] $\square$

8.4.2 Unique Local Solution to Eqn. (8.4.1.5) for $Q(t,s)$

In this section we obtain a unique local solution of Eqn. $(8.4.1.5)$ in $Q(\ ,\ )$ near T, via the contraction mapping principle. First, with reference to Lemma 8.4.1.2, set for any $0 \leq s_0 < T$:

$$S_{s_0,T}(r_T) = \text{closed ball of the space } \mathcal{L}(U; U_{s_0,T}),$$
$$\text{centered at the origin, of radius } r_T = \|F_1\|_{\mathcal{L}(U;U_{s_0,T})}. \quad (8.4.2.1)$$

Thus, $F_1 \in S_{s_0,T}(r_T)$. We shall solve Eqn. $(8.4.1.5)$ in the sphere $S_{s_0,T}(2r_T)$ for s_0 sufficiently close to T. We remark that Eqn. $(8.4.1.5)$ is equivalent to

$$Q = F_1 - \Lambda(Q), \qquad (8.4.2.2)$$

where

$$\Lambda(Q)(t,s) \equiv \int_t^T Q^*(\tau,t)Q(\tau,s) \, d\tau. \qquad (8.4.2.3)$$

Proposition 8.4.2.1   *(i) Let Q, $\bar{Q} \in \mathcal{L}(U; U_{s_0,T})$. Then, the following inequality holds:*

$$\|\Lambda(Q) - \Lambda(\bar{Q})\|_{\mathcal{L}(U;U_{s_0,T})}$$

$$\leq \sqrt{T - s_0} \big[\|Q\|_{\mathcal{L}(U;U_{s_0,T})} + \|\bar{Q}\|_{\mathcal{L}(U;U_{s_0,T})} \big] \|Q - \bar{Q}\|_{\mathcal{L}(U,U_{s_0,T})}. \qquad (8.4.2.4)$$

(ii) Let Q, $\bar{Q} \in \mathcal{S}_{s_0,T}(2r_T)$ (see Definition (8.4.2.1)). Then, for all s_0 satisfying $\theta_{s_0,T} \equiv 4r_T \sqrt{T - s_0} < 1/2$, we have $\Lambda(Q) \in \mathcal{S}_{s_0,T}(r_T/2)$ and from (8.4.2.4) Λ is a contraction mapping on $\mathcal{S}_{s_0,T}(2r_T)$ with contraction constant $\theta_{s_0,T}$.

Proof.   Set $\Sigma = \Lambda(Q)$, $\bar{\Sigma} = \Lambda(\bar{Q})$, $\delta Q = Q - \bar{Q}$, and $\delta Q^* = Q^* - \bar{Q}^*$. Then, from (8.4.2.3), adding and subtracting $Q^*(\tau, t)\bar{Q}(\tau, s)$, we obtain

$$\Sigma(t, s) - \bar{\Sigma}(T, s) = \int_t^T [Q^*(\tau, t)\delta Q(\tau, s) + \delta Q^*(\tau, t)\bar{Q}(\tau, s)]\, d\tau. \qquad (8.4.2.5)$$

Next, if $v \in L_2(s, T; U)$ and $u \in U$, we compute from (8.4.2.5) in the U-inner product and in the U-norm:

$$\int_s^T (\Sigma(t, s)u - \bar{\Sigma}(t, s)u, v(t))\, dt$$

$$= \int_s^T \left\{ \int_t^T (\delta Q(\tau, s)u, Q(\tau, t)v(t))\, d\tau + \int_t^T (\bar{Q}(\tau, s)u, \delta Q(\tau, t)v(t))\, d\tau \right\} dt$$

(by Schwarz inequality and majorizing $\int_t^T$ by $\int_s^T$)

$$\leq \left[\int_s^T \|\delta Q(\tau, s)u\|^2\, d\tau \right]^{\frac{1}{2}} \int_s^T \left[\int_t^T \|Q(\tau, t)v(t)\|^2\, d\tau \right]^{\frac{1}{2}} dt$$

$$+ \left[\int_s^T \|\bar{Q}(\tau, s)u\|^2\, d\tau \right]^{\frac{1}{2}} \int_t^T \left[\int_t^T \|\delta Q(\tau, t)v(t)\|^2\, d\tau \right]^{\frac{1}{2}} dt$$

(recalling the definition (8.3.3.1) of norm for $U_{s_0,T}$)

$$\leq \|\delta Q\|_{\mathcal{L}(U;U_{s_0,T})} \|u\|\, \|Q\|_{\mathcal{L}(U;U_{s_0,T})} \int_s^T \|v(t)\|\, dt$$

$$+ \|\bar{Q}\|_{\mathcal{L}(U;U_{s_0,T})} \|u\|\, \|\delta Q\|_{\mathcal{L}(U;U_{s_0,T})} \int_s^T \|v(t)\|\, dt$$

$$\leq \sqrt{T - s_0} \big[\|Q\|_{\mathcal{L}(U;U_{s_0,T})} + \|\bar{Q}\|_{\mathcal{L}(U;U_{s_0,T})} \big] \|\delta Q\|_{\mathcal{L}(U;U_{s_0,T})} \|u\|\, \|v\|_{L_2(s_0,T;U)},$$

$$(8.4.2.6)$$

and (8.4.2.4) is proved by (8.4.2.6). Then, (i) implies (ii). $\square$

Returning to (8.4.2.2), we see by (8.4.2.1) and Proposition 8.4.2.1 that Λ in (8.4.2.2) maps $\mathcal{S}_{s_0,T}(2r_T)$ into itself and is a contraction mapping here with, say, contraction constant $\theta_{s_0,T} < 1/2$. Thus, with F_1 given by Lemma 8.4.1.2, by the contraction principle we obtain

Theorem 8.4.2.2 *Assume (H.1) and (H.3) [or else, alternatively, (h.2)]. With reference to (8.4.2.1), for s_0 satisfying*

$$\theta_{s_0,T} = 4r_T\sqrt{T - s_0} < \frac{1}{2},$$ (8.4.2.7)

Equation (8.4.1.5) has a unique solution $Q \in \mathcal{S}_{s_0,T}(2r_T) \subset \mathcal{L}(U; U_{s_0,T})$.

Remark 8.4.2.1 We point out explicitly that, under the sole alternative hypothesis (h.2), one readily obtains a *global a priori bound* for $Q(t, s)$ in $\mathcal{L}(U; U_{0,T})$ for the solution in Theorem 8.4.2.2. This is seen simply from

$$(P(t)Bu, Bu)_Y + \int_t^T \|Q(\tau, t)u\|_U^2 \, d\tau = \int_t^T \|Re^{A(\tau - t)}Bu\|_Z^2 \, d\tau$$

$$\leq c_T\|u\|_U^2,$$ (8.4.2.8)

which is obtained from the IRE (8.4.1.1) by setting $x = y = Bu$, for a nonnegative, self-adjoint $P(t)$.

8.4.3 Unique Local Solution to Eqn. (8.4.1.7) for $V(t)$. Global Solution $P(t)$ under (h.1), (h.2)

Having solved (8.4.1.5) for Q locally near T, we insert such a solution Q in (8.4.1.7) and seek a local solution of (8.4.1.7) in V near T, again by the contraction mapping principle. First, with reference to Lemma 8.4.1.3, we set for any $0 \leq s_0 < T$:

$\mathcal{S}'_{s_0,T}(r'_T) =$ closed sphere of the space $\mathcal{B}([s_0, T]; \mathcal{L}(Y; U))$,

centered at the origin, of radius $r'_T = \|F_2\|_{\mathcal{B}([s_0,T];\mathcal{L}(Y;U))}$.   (8.4.3.1)

Thus, $F_2 \in \mathcal{S}'_{s_0,T}(r'_T)$. We shall solve Eqn. (8.4.1.7) in the sphere $\mathcal{S}'_{s_0,T}(2r'_T)$ for s_0 sufficiently close to T.

We remark that Eqn. (8.4.1.7) is equivalent to

$$V = F_2 - \gamma(V),$$ (8.4.3.2)

where

$$(\gamma(V))(t) \equiv \int_t^T Q^*(\tau, t)V(\tau)e^{A(\tau - t)} \, d\tau,$$ (8.4.3.3)

and $Q(\cdot, \cdot)$ is obtained locally from Theorem 8.4.2.2.

Theorem 8.4.3.1 *Assume (H.1), (H.2), and (H.3) [or else, alternatively, (h.1) and (h.2)]. Let $\|e^{At}\|_{\mathcal{L}(Y)} \leq a_T, 0 \leq t \leq T$. With s_0 satisfying*

$$\theta'_{s_0,T} \equiv a_T4\sqrt{T - s_0}\, r_T < \frac{1}{2},$$ (8.4.3.4)

and with $Q(,)$ the local solution provided by Theorem 8.4.2.2 in $\mathcal{S}_{s_0, T}(2r_T)$, then Eqn. (8.4.1.7) admits a unique solution $V \in \mathcal{S}'_{s_0, T}(2r'_T) \subset \mathcal{B}([s_0, T]; \mathcal{L}(Y; U))$ (see definition in (8.4.1.11)).

Proof.   Let $V \in \mathcal{S}'_{s_0, T}(2r'_T)$. From (8.4.3.3),

$$|(\gamma(V)(t)x, u)| = \left| \int_t^T \left(V(\tau)e^{A(\tau - t)}x, Q(\tau, t)u\right) d\tau \right|$$

$$\leq a_T \|V\|_{\mathcal{B}; s_0, T} \sqrt{T - s_0} \|x\| \left[\int_t^T \|Q(\tau, t)u\|^2 \, d\tau \right]^{\frac{1}{2}}$$

$$\leq a_T 2r'_T \sqrt{T - s_0} \, 2r_T \|x\| \, \|u\|, \tag{8.4.3.5}$$

and by (8.4.3.4) and (8.4.3.5) we have that $\gamma(V) \in \mathcal{S}'_{s_0, T}(r'_T/2)$, so that γ is a contraction mapping on $\mathcal{S}'_{s_0, T}(2r'_T)$. Then, with F_2 given by Lemma 8.4.1.3, by (8.4.3.1), the right-hand side of (8.4.3.2) is a contraction mapping on $\mathcal{S}'_{s_0, T}(2r'_T)$ with contraction constant $\theta'_{s_0, T}$ given by (8.4.3.4) and the contraction principle applies. $\square$

We can now conclude the proof of uniqueness for the solution of the IRE (8.4.1.1). Under (H.1), (H.2), and (H.3) [or else, alternatively, (h.1) and (h.2)], once Q and hence V are found, we then have a solution $P(t)$ of the corresponding Riccati integral equation (8.4.1.1), now rewritten by (8.4.1.2) as

$$P(t) = \int_t^T e^{A^*(\tau - t)} R^* R e^{A(\tau - t)} \, d\tau - \int_t^T e^{A^*(\tau - t)} V^*(\tau) V(\tau) e^{A(\tau - t)} \, d\tau. \tag{8.4.3.6}$$

Then (8.4.3.6) reveals that $P(t)$ is self-adjoint on Y. Moreover, $P(t)$ possesses locally near T the properties of regularity corresponding to the global properties (8.2.1.25), (8.2.1.26), (8.2.1.30), and (8.2.1.31) of Theorem 8.2.1.2 [these are precisely the properties shown globally in Theorem 8.3.3.1 and Lemma 8.3.4.4 for the solution $P(t)$ defined by (8.3.3.2)]. But uniqueness of V implies uniqueness of the solution to the Riccati integral equation via (8.4.3.6), for now at least, locally near T. To extend existence and uniqueness of the solution to the IRE (8.4.1.1) over the entire interval $[0, T]$, we need to have a priori bounds. With reference to the opening paragraphs of the present subsection, we then distinguish two cases.

Case 1   If we assume the analysis of the control problem as given in Sections 8.3.1 through 8.3.5, then we can invoke the a priori bounds (8.3.3.2b)–(8.3.3.4) of Theorem 8.3.3.1, under (H.1), (H.2), and (H.3) [or, alternatively, (h.1) and (h.2)]. Thus, in this case, the unique solution $P(t)$ given by (8.4.3.6) is extended on all of $[0, T]$ and coincides here with the operator $P(t)$ defined constructively by (8.3.3.2). In this case, the present Sections 8.4.1 through 8.4.3 serve as an alternative approach to the (simpler and closely related) proof of uniqueness given in Section 8.3.6. In this case, Theorem 8.2.1.2 is thus fully proved.

Case 2 If, instead, Sections 8.4.1 through 8.4.3 are viewed as the beginning of a *direct* study of the IRE (8.4.1.1), we need to establish (independently of Sections 8.3.1–8.3.5) *global a priori estimates*. This will be done in the subsequent Section 8.4.4, under hypotheses (H.1), (H.2), and (H.3).

Instead, under the alternative set of hypotheses (h.1) and (h.2), the issue of *global* a priori estimates is far easier and rather immediate. First, under solely (h.2), a global, a priori bound for $Q(t, s)$ in $\mathcal{L}(U; U_{0,T})$ was already established in Remark 8.4.2.1. With such a priori bound for $Q(t, s)$, we return to Eqn. (8.4.1.7), which is linear in V. Then the contraction argument in Theorem 8.4.3.1 leads to a contraction constant that depends only on the length of the interval $[T - s_0]$; see (8.4.3.4). After a finite number of steps, we extend the local solution V into a global solution $V \in \mathcal{B}([0, T]; \mathcal{L}(Y; U))$. This, in turn, yields via (8.4.3.6) a global solution $P \in \mathcal{B}([0, T]; \mathcal{L}(Y))$. We have thus proved the following global existence and uniqueness result for the IRE (8.4.1.1), hence by Corollary 8.3.5.3, for the corresponding DRE (8.3.4.26).

Theorem 8.4.3.2   *Assume (h.1) = (8.2.1.10) and (h.2) = (8.2.1.11) [which allow both R and B to be unbounded]. Then the IRE (8.4.1.1), and hence the DRE (8.3.4.26), admit a unique global solution $P(t)$ satisfying properties (8.2.1.25), (8.2.1.26), (8.2.1.30), and (8.2.1.31), or else (8.3.3.2b)–(8.3.3.4). Such solution $P(t)$ is given by the explicit formula (8.3.3.20).*

8.4.4 Global A Priori Estimates for V and Q.
Global Solution P(t) under (H.1), (H.2), and (H.3)

In the present subsection we provide – under (H.1), (H.2), and (H.3) – a priori global bounds for V (and Q), thereby extending the local solution $P(t)$ of the IRE (8.4.1.1) of Section 8.4.3 near T to a global solution on $[0, T]$, while preserving its properties (8.4.1.9) and (8.4.1.10). In the process, we shall derive the explicit formula (8.3.3.2) for the solution $P(t)$, which in the approach of Sections 8.3.1 through 8.3.6 was, instead, the defining formula of an operator that only subsequently was shown to solve the IRE (8.4.1.1).

Step 1.   Proposition 8.4.4.1   *Let $V(\cdot) \in \mathcal{B}([s_0, T]; \mathcal{L}(Y; U))$ and $Q(t, s) \in \mathcal{L}(U; U_{s_0,T})$ be the operators guaranteed by Sections 8.4.2 (Theorem 8.4.2.2) and 8.4.3 (Theorem 8.4.3.1), or Proposition 8.4.1.1, near T, say for $s_0 \leq s \leq t \leq T$, under assumptions (H.1), (H.2), and (H.3) [or, alternatively, under assumptions (h.1) and (h.2); see Theorem 8.4.3.2]. Then, the integral equation*

$$\Psi(t, s)x = V(t)e^{A(t-s)}x - \int_s^t V(t)e^{A(t-\tau)}B\Psi(\tau, s)x \, d\tau \qquad (8.4.4.1)$$

admits a unique solution

$$\Psi(\cdot, s) : continuous \ Y \rightarrow L_2(s, T; U), \quad indeed \ \Psi \in \mathcal{L}(Y; U_{s_0,T}). \qquad (8.4.4.2)$$

Proof.   By property (8.4.1.10) of Proposition 8.4.1.1 we have for $s_0 \leq s \leq t \leq T$:

$$V(t)e^{A(t-s)} \in \mathcal{B}([s, T]; \mathcal{L}(Y; U)). \tag{8.4.4.3}$$

Next, we shall show that, via (8.4.1.4), the term on the right-hand side of (8.4.4.1) satisfies for $x \in Y$:

$$\int_s^T \left\| \int_s^t V(t)e^{A(t-\tau)} B\Psi(\tau, s)x \, d\tau \right\|_U^2 dt \equiv \int_s^T \left\| \int_s^t Q(t, \tau)\Psi(\tau, s)x \, d\tau \right\|_U^2 dt$$

$$\leq (T - s)\|Q\|_{\mathcal{L}(U; U_{s,T})}^2 \|\Psi(\,\cdot\,, s)x\|_{L_2(s,T;U)}^2, \tag{8.4.4.4}$$

recalling property (8.4.1.9) for Q in Proposition 8.4.1.1. In fact, by Schwarz's inequality we compute

$$\int_s^T \left\| \int_s^t Q(t, \tau)\Psi(\tau, s)x \, d\tau \right\|_U^2 dt$$

$$\leq \int_s^T \left\{ \int_s^t \|\|Q(t, \tau)\|\|^2 d\tau \right\} \left\{ \int_s^t \|\Psi(\tau, s)x\|^2 d\tau \right\} dt$$

$$\leq \left\{ \int_s^T \int_s^t \|\|Q(t, \tau)\|\|^2 d\tau \, dt \right\} \|\Psi(\,\cdot\,, s)x\|_{L_2(s,T;U)}^2$$

(changing the order of integration)

$$\leq \left\{ \int_s^T \int_\tau^T \|\|Q(t, \tau)\|\|^2 dt \, d\tau \right\} \|\Psi(\,\cdot\,, s)x\|_{L_2(s,T;U)}^2$$

$$\leq (T - s) \left\{ \sup_{s \leq \tau \leq T} \int_\tau^T \|\|Q(t, \tau)\|\|^2 dt \right\} \|\Psi(\,\cdot\,, s)x\|_{L_2(s,T;U)}^2, \tag{8.4.4.5}$$

and (8.4.4.4) follows from (8.4.4.5), via the definition of the norm (8.3.3.13) ($\|\|\quad\|\|$ denotes the norm of $\mathcal{L}(U)$). Finally, by (8.4.4.3) and (8.4.4.4), we apply the contraction mapping theorem to (8.4.4.1) and obtain $\Psi(\,\cdot\,, s)x \in L_2(s, T; U)$ for s_0 sufficiently near T.   □

Step 2   With $\Psi(t, s)$ obtained from Proposition 8.4.4.1, and $x \in Y$, we define $\Phi(t, s)$ by

$$\Phi(t, s)x = e^{A(t-s)}x - \int_s^t e^{A(t-\tau)} B\Psi(\tau, s)x \, d\tau \tag{8.4.4.6a}$$

$$= e^{A(t-s)}x - \{L_s\Psi(\,\cdot\,, s)x\}(t) \in C([s, T]; [\mathcal{D}(A^*)]'), \tag{8.4.4.6b}$$

recalling (8.1.12). Thus, applying $V(t)$ on both sides of (8.4.4.6a) and comparing the result with (8.4.4.1), we obtain, by Proposition 8.4.4.1,

$$\Psi(t, s) = V(t)\Phi(t, s) : \text{continuous:} \ Y \to L_2(s, T; U), \quad \text{indeed in } \mathcal{L}(Y; U_{s_0, T}),$$
$$(8.4.4.7)$$

which inserted in (8.4.4.6) yields for $x \in Y$:

$$\Phi(t, s)x = e^{A(t-s)}x - \int_s^t e^{A(t-\tau)} B \, V(\tau)\Phi(\tau, s)x \, d\tau \qquad (8.4.4.8a)$$

$$= e^{A(t-s)}x - \{L_s V(\cdot)\Phi(\cdot, s)x\}(t) \in C([s, T]; [\mathcal{D}(A^*)]'). \qquad (8.4.4.8b)$$

Remark 8.4.4.1   Assume (H.1), (H.2), and (H.3) [or else, alternatively, (h.1) and (h.2)], so that (8.4.4.7) holds true, and hence

$$V(t)\Phi(t, s)x = B^* P(t)\Phi(t, s)x \in U \text{ a.e. in } t \text{ for } x \in Y.$$

If we consider the IRE (8.4.1.1) with

$$x \text{ there replaced by } BB^* P(t)\Phi(t, s)x, \quad x \in Y;$$

$$y \text{ there replaced by } e^{A(t-s)}y, \quad y \in Y,$$

and t as above, we see that the resulting IRE expression, rewritten as

$$\left(B^* P(t)\Phi(t, s)x, B^* P(t)e^{A(t-s)}y\right)_U$$

$$= \int_t^T \left(Re^{A(\tau-t)} BB^* P(t)\Phi(t, s)x, Re^{A(\tau-s)}y\right)_Z d\tau$$

$$- \int_t^T \left(B^* P(\tau)e^{A(\tau-t)} BB^* P(t)\Phi(t, s)x, B^* P(\tau)e^{A(\tau-s)}y\right)_U d\tau,$$

or, by (8.4.1.2), and (8.4.1.4), as

$$\left(V(t)\Phi(t, s)x, V(t)e^{A(t-s)}y\right)_U = \int_t^T \left(Re^{A(\tau-t)} BV(t)\Phi(t, s)x, Re^{A(\tau-s)}y\right)_Z d\tau$$

$$- \int_t^T \left(Q(\tau, t)V(t)\Phi(t, s)x, V(\tau)e^{A(\tau-s)}y\right)_U d\tau$$

is well-defined a.e. in t, for all $x, y \in Y$, either under (H.1), (H.2), and (H.3) or else, alternatively, under (h.1) and (h.2). This is so because of the established regularity properties (8.4.1.9), (8.4.1.10), and (8.4.4.7). We shall use this remark in the proof of Theorem 8.4.4.2 below.

Step 3   From (8.4.4.8), recalling (8.4.4.7) and assumption (H.3) in the form of its consequence (H.3*) = (8.2.1.6), or (8.2.1.8), on RL_s, we obtain

$$R\Phi(t, s)x = Re^{A(t-s)}x - R\int_s^t e^{A(t-\tau)} B \, V(\tau)\Phi(\tau, s)x \, d\tau \qquad (8.4.4.9a)$$

$$= Re^{A(t-s)}x - \{RL_s V(\cdot)\Phi(\cdot, s)x\}(t) \in C([s, T]; Z). \qquad (8.4.4.9b)$$

Moreover, (8.4.4.8a) implies the evolution property, with $s_0 \leq s \leq \tau \leq t \leq T$:

$$V(t)\Phi(t, \tau)\Phi(\tau, s)x = V(t)\Phi(t, s)x \underset{(\text{in } t)}{\in} L_2(\tau, T; U) \qquad (8.4.4.10)$$

in the L_2-sense; indeed, (8.4.4.8a) implies after direct computations

$$V(t)[\Phi(t, \tau)\Phi(\tau, s)x - \Phi(t, s)x]$$
$$= -\int_\tau^t V(t)e^{A(t-\sigma)} B \, V(\sigma)[\Phi(\sigma, \tau)\Phi(\tau, s)x - \Phi(\sigma, s)x] \, d\sigma, \quad (8.4.4.11)$$

and an argument similar to that in Proposition 8.4.4.1 yields (8.4.4.10) from (8.4.4.11). Then (8.4.4.10), in turn, implies via (8.4.4.8)

$$R[\Phi(t, \tau)\Phi(\tau, s)x - \Phi(t, s)x]$$
$$\equiv -\int_\tau^t Re^{A(t-\sigma)} B \, V(\sigma)[\Phi(\sigma, \tau)\Phi(\tau, s)x - \Phi(\sigma, s)x] \, d\sigma \equiv 0 \quad (8.4.4.12)$$

in the $L_2(\tau, T; Z)$-sense in t.

Step 4. Theorem 8.4.4.2 *Assume (H.1) = (8.2.1.1), (H.2) = (8.2.1.2), and (H.3) = (8.2.1.3). Thus, accordingly, the IRE (8.4.1.1) has a local solution $P(t)$ near T, say $s_0 \leq t \leq T$, as guaranteed by Section 8.4.3, so that $P(t)$ is given (non-explicitly) by (8.4.3.6) and possesses locally near T the regularity properties such as (8.4.1.9) and (8.4.1.10). Moreover, (8.4.4.8) is well-defined for $\Phi(t, s)$. Then, in fact, such local solution $P(t)$ is given explicitly by*

$$P(t)x = \int_t^T e^{A^*(\tau-t)} R^* R\Phi(\tau, t)x, \quad s_0 \leq t \leq T, \ x \in Y \qquad (8.4.4.13)$$

[the proof works also under (h.1) and (h.2); recall Theorem 8.4.3.2].

Proof. We must show (8.4.4.13) as a consequence of the IRE (8.4.1.1) and of the dynamics (8.4.4.8) for Φ.

(a) Claim Under the above assumptions (H.1), (H.2), and (H.3), the following identity holds true for all $x, y \in Y$ [which corresponds to Lemma 8.3.6.2 of the variational method]:

$$\int_t^T \left(B^* P(\tau)e^{A(\tau-t)}x, \, B^* P(\tau)e^{A(\tau-t)}y\right)_U d\tau$$
$$= \int_t^T \left(e^{A^*(\tau-t)} R^* R \int_t^\tau e^{A(\tau-\sigma)} BB^* P(\sigma)\Phi(\sigma, t)x \, d\sigma, \, y\right)_Y d\tau. \qquad (8.4.4.14)$$

The proof follows closely that of Lemma 8.3.6.2, which was given within the variational approach, except that in the present development we proceed by the direct approach. Recall that $B^* P(\cdot)\Phi(\cdot, t)x \in L_2(t, T; U)$ by (8.4.4.7). This property,

(8.2.1.8) and the regularity of $V(\cdot)$ in (8.4.1.10), show that each term of (8.4.4.14) is well-defined. In fact, to this end, we compute, by use of (8.4.4.8a),

$$\int_t^T \left(B^*P(\tau)e^{A(\tau-t)}x, B^*P(\tau)e^{A(\tau-t)}y\right)_U d\tau$$

$$= \int_t^T \left(B^*P(\tau)\left[\Phi(\tau,t)x + \int_t^\tau e^{A(\tau-\sigma)}BB^*P(\sigma)\Phi(\sigma,t)x\,d\sigma\right],\right.$$

$$\left.B^*P(\tau)e^{A(\tau-t)}y\right)_U d\tau$$

$$= M(t) + \int_t^T \left(B^*P(\tau)\int_t^\tau e^{A(\tau-\sigma)}BB^*P(\sigma)\Phi(\sigma,t)x\,d\sigma,\right.$$

$$\left.B^*P(\tau)e^{A(\tau-t)}y\right)_U d\tau, \quad (8.4.4.15)$$

where we have set

$$M(t) = \int_t^T \left(B^*P(\tau)\Phi(\tau,t)x, B^*P(\tau)e^{A(\tau-t)}y\right)_U d\tau. \quad (8.4.4.16)$$

$M(t)$ in (8.4.4.16) is well-defined by (8.4.4.7) and (8.4.1.10) with $V(t) = B^*P(t)$. The integral term in (8.4.4.15) is well-defined by (8.4.1.9), (8.4.1.4), by (8.4.4.7), and by (8.4.1.10). Next, using the $L_2(\sigma, T; U)$-regularity of $B^*P(\tau)e^{A(\tau-\sigma)}Bx = Q(\tau,\sigma)x$ in τ, already established (locally) in Theorem 8.4.2.2, or in Proposition 8.4.1.1, we interchange the order of integration in (8.4.4.15) and obtain for its R.H.S. (right-hand side):

$$\text{R.H.S. of } (8.4.4.15) = M(t) + \int_t^T \int_\sigma^T \left(B^*P(\tau)e^{A(\tau-\sigma)}BB^*P(\sigma)\Phi(\sigma,t)x,\right.$$

$$\left.B^*P(\tau)e^{A(\tau-\sigma)}e^{A(\sigma-t)}y\right)_U d\tau\,d\sigma, \quad (8.4.4.17)$$

which is well-defined by (8.4.1.4), (8.4.1.9), (8.4.4.7), and (8.4.1.10). Next, to perform the internal integration in (8.4.4.17), we recall Remark 8.4.4.1 and use the IRE (8.4.1.1) with x there replaced by $BB^*P(\sigma)\Phi(\sigma,t)x$ (where $B^*P(\sigma)\Phi(\sigma,t)x \in U$ a.e. in σ, $x \in Y$ by (8.4.4.7)), with y there replaced by $e^{A(\sigma-t)}y$, and with t there replaced by σ, obtaining for $x, y \in Y$:

R.H.S. of (8.4.4.15)

$$= M(t) + \int_t^T \int_\sigma^T \left(e^{A^*(\tau-\sigma)}R^*Re^{A(\tau-\sigma)}BB^*P(\sigma)\Phi(\sigma,t)x, e^{A(\sigma-t)}y\right)_Y d\tau\,d\sigma$$

$$- \int_t^T \left(P(\sigma)BB^*P(\sigma)\Phi(\sigma,t)x, e^{A(\sigma-t)}y\right)_Y d\sigma, \quad (8.4.4.18)$$

where cancellation takes place because of (8.4.4.16). After a further change in the order of integration and use of the semigroup property, we obtain

$$\text{R.H.S. of } (8.4.4.15) = \int_t^T \left(e^{A^*(\tau-t)} R^* R \int_t^\tau e^{A(\tau-\sigma)} BB^* P(\sigma) \Phi(\sigma, t) x \, d\sigma, y \right)_Y d\tau,$$

(8.4.4.19)

which is well-defined by (8.4.4.7) and (8.2.1.8). Then, (8.4.4.15) and (8.4.4.19) verify (8.4.4.14), as desired.

The Claim is proved.

(b)   Next, we return to the IRE (8.4.1.1) and on its right-hand side we use identity (8.4.4.14) to obtain for $x, y \in Y$:

$$(P(t)x, y)_Y = \int_t^T \left(e^{A^*(\tau-t)} R^* Re^{A(\tau-t)} x, y \right)_Y d\tau$$

$$- \int_t^T \left(e^{A^*(\tau-t)} R^* R \int_t^\tau e^{A(\tau-\sigma)} BB^* P(\sigma) \Phi(\sigma, t) x \, d\sigma, y \right)_Y d\tau$$

(by (8.4.4.8)) $= \left(\int_t^T e^{A^*(\tau-t)} R^* R\Phi(\tau, t) x \, d\tau, y \right)_Y,$ (8.4.4.20)

where in the last step we have recalled (8.4.4.8a). Thus, (8.4.4.20) proves (8.4.4.13), as desired. $\square$

Step 5.   Lemma 8.4.4.3   *Assume (H.1), (H.2), and (H.3). Then, with reference to (8.4.4.13) and (8.4.4.8a), we have for $x \in Y$:*

$$B^* P(t)\Phi(t, s)x = V(t)\Phi(t, s)x = \int_t^T B^* e^{A^*(\tau-t)} R^* R\Phi(\tau, s)x \qquad (8.4.4.21a)$$

$$= \{L_s^* R^* R\Phi(\cdot, s)x\}(t) \in C([s, T]; U); \qquad s_0 \le s. \qquad (8.4.4.21b)$$

Proof.   We use (8.4.4.13) and the evolution property $R\Phi(\tau, t)\Phi(t, s)x = R\Phi(\tau, s)x$ from (8.4.4.12), and, moreover, (8.1.13). The regularity in (8.4.4.21b) is an improvement over (8.4.4.7): It follows from $(H.1^*) = (8.2.1.4d)$, or (8.2.1.9b), on $L_s^* R^*$ and (8.4.4.9b) on $R\Phi$. $\square$

Step 6.   Proposition 8.4.4.4   *Assume (H.1), (H.2), and (H.3). Then, the operator $R\Phi(t, s)$ defined via (8.4.4.6) and identified as the unique solution of (8.4.4.9), for $s_0 \le s \le t \le T$, is given explicitly by the formula*

$$R\Phi(\cdot, s) = [I_s + RL_s L_s^* R^*]^{-1} Re^{A(\cdot-s)} x \in C([s, T]; Z) \qquad (8.4.4.22)$$

and can be extended to any interval $[s, T]$, $0 \le s < T$; moreover, it satisfies the pointwise estimate

$$\|R\Phi(t, s)\|_{\mathcal{L}(Y;Z)} \le C_T, \qquad 0 \le s \le t \le T. \qquad (8.4.4.23)$$

Proof.   Inserting (8.4.4.21b) into (8.4.4.8b) yields

$$R\Phi(\cdot, s)x = Re^{A(\cdot - s)}x - RL_sL_s^*R^*R\Phi(\cdot, s)x, \tag{8.4.4.24}$$

from which (8.4.4.22) readily follows, at least first as $R\Phi(\cdot, s)x \in L_2(s, T; Z)$ (since $[I_s + RL_sL_s^*R^*]^{-1}$ is continuous $L_2(s, T; Z) \to$ itself). Next, using this preliminary regularity on the right-hand side of (8.4.4.24) and recalling assumptions (H.2) = (8.2.1.2) as well as (H.1) = (8.2.1.1) in the form of its consequence (8.2.1.8) and (8.2.1.9), we readily obtain $R\Phi(t, s)x \in C([s, T]; Z)$ and, in fact, the uniform pointwise bound (8.4.4.23).   □

Remark 8.4.4.2   At this stage, having extended $R\Phi(t, s)$ over any interval $0 \leq s \leq t \leq T$ by (8.4.4.22), we may next extend $P(t)$ over all of $[0, T]$ by (8.4.4.13). We then show that the extended $P(t)$ remains a solution of the IRE (8.4.1.1), while preserving its original properties (8.4.1.9) and (8.4.1.10).

Step 7.   Theorem 8.4.4.5   *(Global bound for $V(t)$) Assume (H.1), (H.2), and (H.3). For $V \in \mathcal{B}([s_0, T]; \mathcal{L}(Y; U))$ provided by Theorem 8.4.3.1, the following uniform global bound holds true:*

$$\begin{cases} \sup_{0 \leq t \leq T} \|V(t)x\|_U \leq C_T\|x\|_Y, \quad x \in Y; & (8.4.4.25a) \\ i.e., \ V \in \mathcal{B}([0, T]; \mathcal{L}(Y; U)). & (8.4.4.25b) \end{cases}$$

Proof.   By (8.4.4.13) and (8.4.1.2), we estimate, by invoking (H.1) in the form of its consequence (H.1*) = (8.2.1.4), as well as the uniform bound (8.4.4.23):

$$\|V(t)x\|_U = \|B^*P(t)x\|_U = \|L_t^*R^*R\Phi(\cdot, t)x\|_U$$
$$\leq C_T\|R\Phi(\cdot, t)x\|_{L_\infty(0,T;Z)} \leq \text{const}_T\|x\|_Y, \tag{8.4.4.26}$$

and (8.4.4.25) follows from (8.4.4.26).   □

Theorem 8.4.4.6   *(Global bound for $Q(t, s)$) Assume (H.1), (H.2), and (H.3). For $Q \in \mathcal{L}(U; U_{s_0,T})$ provided by Theorem 8.4.2.2, the following global bound holds true:*

$$\begin{cases} \sup_{0 \leq s \leq T} \int_s^T \|Q(t, s)u\|_U^2 \, dt \leq C_T\|u\|_U, \quad u \in U; & (8.4.4.27a) \\ or \ Q \in \mathcal{L}(U; U_{0,T}). & (8.4.4.27b) \end{cases}$$

Proof.   Since the bound (8.4.1.12) on $F_1(t, s)$ is global on $[0, T]$, that is, $F_1 \in \mathcal{L}(U; U_{0,T})$, then in (8.4.2.1) one may take the radius $r_T = \|F_1\|_{\mathcal{L}(U;U_{0,T})}$ (i.e., with $s_0 = 0$). Then, in (4.2.7) of Theorem 8.4.2.2, with constant $\theta_{s_0,T} = 4r_T\sqrt{T - s_0}$, which depends only on the interval size $(T - s_0)$, one can repeat the contraction argument of Theorem 8.4.2.2 a finite number of times and obtain a global $Q \in \mathcal{L}(U; U_{0,T})$.   □

Step 8.   Theorem 8.4.4.7 *(P(t) as a global solution) Assume (H.1), (H.2), and (H.3). Then the local solution P(t) of the IRE (8.4.1.1) defined near T, $s_0 \le t \le T$, by (8.4.4.13) is valid on all of [0, T], while preserving the regularity properties (8.4.4.25) and (8.4.4.27) and nonnegative self-adjointness.*

Proof.   We return to the Riccati integral equation solution, rewritten as a nonnegative, self-adjoint operator in (8.4.3.6), and we use the global result on V of Theorem 8.4.4.5 and on Q of Theorem 8.4.4.6.   □

8.4.5 Recovering the Optimal Control Problem under (H.1), (H.2), and (H.3) [or (h.1) and (h.2)]

Optimal dynamics   We complete the description starting from a (unique) solution of the IRE (8.4.1.1) by setting

$$u^0(t, s; x) \equiv -B^* P(t)\Phi(t, s)x \in L_2(s, T; U), \quad x \in Y \qquad (8.4.5.1a)$$
$$= -\{L_s^* R^* R\Phi(\cdot, s)x\}(t), \qquad (8.4.5.1b)$$

via (8.4.4.7) and (8.4.4.21). Thus, returning to (8.4.4.21b) and using (8.4.4.22) we obtain, via (8.4.5.1),

$$u^0(\cdot, s; x) = -B^* P(t)\Phi(\cdot, s)x = -L_s^* R^* R\Phi(\cdot, s)x \qquad (8.4.5.2)$$
$$\text{(by (8.4.4.22))} \qquad = -L_s^* R^*[I_s + RL_s L_s^* R^*]^{-1} Re^{A(\cdot - s)}x$$
$$= -[I_s + L_s^* R^* RL_s]^{-1} L_s^* R^* Re^{A(\cdot - s)}x, \qquad (8.4.5.3)$$

since, as it is easily verified,

$$[I_s + L_s^* R^* RL_s]^{-1} L_s^* R^* = L_s^* R^*[I_s + RL_s L_s^* R^*]^{-1}$$
$$: \text{continuous } L_2(s, T; Z) \to L_2(s, T; U). \qquad (8.4.5.4)$$

Thus, inserting (8.4.5.1) in (8.4.4.8), we get

$$y^0(\cdot, s; x) = \Phi(\cdot, s)x = e^{A(\cdot - s)}x + L_s u^0(\cdot, s; x) \in C([s, T]; [\mathcal{D}(A^*)]'). \qquad (8.4.5.5)$$

Finally, inserting (8.4.5.3) into (8.4.4.8b) yields

$$R\Phi(\cdot, s)x = Re^{A(\cdot - s)}x - RL_s[I_s + L_s^* R^* RL_s]^{-1} L_s^* R^* Re^{A(\cdot - s)}x$$
$$= \{I_s - RL_s[I_s + L_s^* R^* RL_s]^{-1} L_s^* R^*\}Re^{A(\cdot - s)}x. \qquad (8.4.5.6)$$

Comparing (8.4.5.5) with (8.4.4.22) we see that

$$[I_s + RL_s L_s^* R^*]^{-1} = I - RL_s[I_s + L_s^* R^* RL_s]^{-1} L_s^* R^* \in \mathcal{L}(L_2(s, T; Z)), \qquad (8.4.5.7)$$

an identity that can be readily verified directly as well.

Optimal Cost Functional   After introducing the above quantities for describing the would-be optimal dynamics – ultimately in terms of the (unique) solution $P(t)$ of the IRE (8.4.1.1), starting with Proposition 8.4.4.1 – we now define the would-be optimal cost

$$J_s^0(x) \equiv \int_s^T \left[\|R\Phi(t,s)x\|_Z^2 + \|u^0(t,s;x)\|_U^2\right]dt, \qquad (8.4.5.8)$$

with u^0 and Φ defined by (8.4.5.1) and (8.4.4.6), respectively, and explicitly given by (8.4.5.3) and (8.4.5.5). The approach of the present section proceeds from the establishment of a (unique) solution of the IRE (8.4.1.1) to the corresponding optimal control problem and culminates with the following result.

Theorem 8.4.5.1   *Assume (H.1), (H.2), and (H.3) [or, alternatively, (h.1) and (h.2); see Theorem 8.4.3.2]. Then:*

(i) *$u^0(t,s;x)$ given by (8.4.5.1), or (8.4.5.3), is the optimal control of the optimal control problem (8.1.10) for the dynamics (8.1.11);*

(ii) *$y^0(t,s;x) = \Phi(t,s)x$ given by (8.4.4.6) is the corresponding optimal trajectory, with $R\Phi(t,s)x$ given by (8.4.5.5), the corresponding optimal observation;*

(iii) *$J_s^0(x)$ given by (8.4.5.8) is the corresponding optimal cost, and, moreover,*

$$J_s^0(x) \equiv (P(s)x, x)_Y. \qquad (8.4.5.9)$$

Proof.   One proof consists simply in invoking the optimality analysis – either one of the proofs of Theorem 8.2.2.1 by Liusternik's theorem or by a more elementary method, given in Lemma 8.3.1.3 and based on (8.3.1.21)–(8.3.1.23). This shows that the quantities $u^0(\,\cdot\,,s;x)$, $R\Phi(\,\cdot\,,s)x$, and $J_s^0(x)$ are, in fact – as claimed – the optimal control, observation, and cost of the optimal control problem (8.1.10) for (8.1.11), the latter indeed satisfying (8.4.5.9); see (8.2.1.17)–(8.2.1.21) and (8.2.1.28).

We give here a second independent direct proof consistent with the present approach. It uses, as a starting point, the explicit formula (8.4.4.13) for $P(t)$ obtained in Theorem 8.4.4.2. Let $x \in Y$. By (8.4.4.13), we obtain the following identity:

$$2(P(s)x, x)_Y + \int_s^T [\|R\Phi(\tau,s)x - Ry(\tau,s;x)\|_Z^2 \, d\tau$$

$$+ \int_s^T \|u^0(\tau,s;x) - u(\tau,s;x)\|_U^2 \, d\tau = J_s(x) + J_s^0(x), \qquad (8.4.5.10)$$

where $\{u(\,\cdot\,,s;x), y(\,\cdot\,,s;x)\}$ is a pair of control and corresponding solution related to each other by the dynamics (8.1.11); moreover, here, we rewrite explicitly (8.1.10) as

$$J_s(x) = J(u(\,\cdot\,,s;x), y(\,\cdot\,,s;x))$$

$$= \int_s^T [\|Ry(\tau,s;x)\|_Z^2 + \|u(\tau,s;x)\|_U^2]d\tau. \qquad (8.4.5.11)$$

Indeed, to prove (8.4.5.10), we recall (8.4.4.13), where we then insert the general dynamics (8.1.11):

$$(P(s)x, x)_Y = \int_s^T \left(R\Phi(\tau, s)x, Re^{A(\tau-s)}x \right)_Z d\tau$$

$$\text{(by (8.1.11))} \quad = \int_s^T (R\Phi(\tau, s)x, Ry(\tau, s; x) - \{RL_s u(\cdot, s; x)\}(\tau))_Z d\tau$$

$$= \int_s^T (R\Phi(\tau, s)x, Ry(\tau, s; x))_Z d\tau$$

$$- (L_s^* R^* R\Phi(\cdot, s)x, u(\cdot, s; x))_{L_2(s,T;U)}$$

$$\text{(by (8.4.5.1))} \quad = \int_s^T (R\Phi(\tau, s)x, Ry(\tau, s; x))_Z d\tau$$

$$+ \int_s^T (u^0(\tau, s; x), u(\tau, s; x))_U d\tau. \quad (8.4.5.12)$$

Using now the identity $2(a, \alpha) + 2(b, \beta) = \|a\|^2 + \|b\|^2 + \|\alpha\|^2 + \|\beta\|^2 - \|a - \alpha\|^2 - \|b - \beta\|^2$ with $a = R\Phi(\tau, s)x$, $\alpha = Ry(\tau, s, x)$, $b = u^0(\tau, s; x)$, and $\beta = u(\tau, s; x)$, and recalling (8.4.5.8) and (8.4.5.11), we readily obtain (8.4.5.10) from (8.4.5.12). Since $J_s^0(x)$ is independent of $u(\cdot, s; x)$, then minimization of $J_s(x)$ as given by (8.4.5.11) over all $u(\cdot, s; x) \in L_2(s, T; U)$ with $y(\cdot, s; x)$ the corresponding trajectory via (8.1.4), as required by the optimal control problem (8.1.8), (8.1.10), is equivalent to such minimization of the right-hand side $[J_s(x) + J_s^0(x)]$ of (8.4.5.10), which is, in turn, equivalent to such minimization of the left-hand side of (8.4.5.10) over $u(\cdot, s; x) \in L_2(s, T; U)$. Since $(P(s)x, x) \geq 0$, then the latter minimization is obtained precisely when

$$u(\cdot, s; x) \equiv u^0(\cdot, s; x) \quad \text{and} \quad Ry(\cdot, s; x) = R\Phi(\cdot, s)x$$

$$= Ry^0(\cdot, s; x), \quad (8.4.5.13)$$

both in the $L_2(s, T; \cdot)$-sense, where $u(\cdot, s; x)$ and $y(\cdot, s; x)$ are related by (8.1.4); see (8.4.5.5). In this case we have from (8.4.5.13), (8.4.5.10), and (8.4.5.8) that

$$\min_{u \in L_2(s,T;U)} J_s(x) = J_s^0(x) = \int_s^T \left[\|Ry^0(\tau, s; x)\|_Z^2 + \|u^0(\tau, s; x)\|_U^2 \right] d\tau$$

$$= (P(s)x, x)_Y, \quad (8.4.5.14)$$

as desired. $\quad \square$

About Dynamic Programming The classical dynamic programming approach consists of two steps: (i) establishing that the DRE (8.2.1.32) is well-posed and possesses a unique solution $P(t)$ (within a specified class of nonnegative, self-adjoint solutions) and (ii) proving that the optimal pair $\{u^0(\cdot, s; x), y^0(\cdot, s; x)\}$ is related by the pointwise feedback formula (8.2.1.27) in the $L_2(s, T; U)$-sense so that the optimal dynamics $y^0(\,\cdot\,, s; x)$ is the solution of: $y^0_t(t, s; x) = Ay^0(t, s; x) - BB^*P(t)y^0(t, s; x)$, $y^0(s, s; x) = x$. Our proof of Theorem 8.4.5.1 is different from the *classical* dynamic programming proof [Balskrishnan, 1981, p. 268]. Formally, the latter goes as follows: One differentiates in t the expression $(P(t)y(t), y(t))_Y$, where $y(t) = y(t, s; x)$ and uses the DRE (8.2.1.32) to obtain the following identity, where $u(t) = u(t, s; x)$:

$$(P(s)x, x)_Y + \int_s^T \|u(t) + B^*P(t)y(t)\|^2_U \, dt$$

$$= \int_s^T \left[\|Ry(t)\|^2_Z + \|u(t)\|^2_U\right] dt = J_s(u, y). \qquad (8.4.5.15)$$

With $P(s)$ already shown to be nonnegative, self-adjoint, the optimal J^0_s is then obtained from (8.4.5.15) as given in the form $u^0(t, s; x) = -B^*P(t)y^0(t, s; x)$, as desired. The *formal* computations leading to (8.4.5.15) are as follows:

$$\frac{d}{dt}(P(t)y(t), y(t))_Y = (\dot{P}(t)y(t), y(t))_Y + (P(t)\dot{y}(t), y(t))_Y + (P(t)y(t), \dot{y}(t))$$

(by (8.2.1.32) and (8.1.1))

$$= ([-A^*P(t) - P(t)A - R^*R + P(t)BB^*P(t)]y(t), y(t))_Y$$
$$+ (P(t)[Ay(t) + Bu(t)], y(t))_Y + (P(t)y(t), Ay(t) + Bu(t))_Y \qquad (8.4.5.16)$$
$$= \|Ry(t)\|^2_Z + \|B^*P(t)y(t)\|^2_U + (P(t)Bu(t), y(t))_Y$$
$$+ (P(t)y(t), Bu(t))_Y, \qquad (8.4.5.17)$$

after a cancellation of two terms: $(A^*P(t)y(t), y(t))_Y$ and $(P(t)Ay(t), y(t))_Y$. Using the self-adjointness of $P(t)$ already established in the well-posedness step (i), we obtain from (8.4.5.17)

$$\frac{d}{dt}(P(t)y(t), y(t))_Y = \|B^*P(t)y(t)\|^2_U + 2\mathrm{Re}(u(t), B^*P(t)y(t))_Y - \|Ry(t)\|^2_Z$$

$$= \|u(t) + B^*P(t)y(t)\|^2_U - \|Ry(t)\|^2 - \|u(t)\|^2_U. \quad (8.4.5.18)$$

Integrating (8.4.5.18) from $t = s$ to $t = T$ and recalling the end condition $P(T) = 0$ leads to (8.4.5.15).

Justification   A justification of the above computations leading to identity (8.4.5.15) would require, however, that $y(t) \in \mathcal{D}(A)$ a.e. in $t \in [s, T]$, at least for trajectories corresponding to a smooth class of controls $u(t)$, dense in $L_2(s, T; U)$: In this case, then, invoking the DRE $= (8.2.1.32)$ in the step leading to (8.4.5.16) is properly justified [as required by (8.2.1.32), where $x, y \in \mathcal{D}(A)$], and the two terms to be cancelled below (8.4.5.17) are well defined. Thus, (8.4.5.15) results, first for a class of smooth controls u and later extended to all $u \in L_2(s, T; U)$. This procedure is legitimate for, say, $B \in \mathcal{L}(U; Y)$, the distributed control case. However, in the generality of the present chapter with a degree of unboundedness of the control operator B given by (8.1.2), $y(t)$ cannot belong to $\mathcal{D}(A)$, even with [$y_0 \in \mathcal{D}(A)$ and], say $u \in C^1([s, T]; U)$, $u(s) = 0$ [see Chapter 7, Eqn. (7.1.6)]:

$$y(t, s; y_0) = e^{A(t-s)} y_0 - A^{-1} B u(t) + \int_s^t e^{A(t-\tau)} A^{-1} B \dot{u}(\tau) \, d\tau. \qquad (8.4.5.19)$$

One strategy to avoid the described technical difficulties may be to approximate the system (pair) $\{A, B\}$ (say by considering the Yosida approximations $A_n = nAR(n, A)$ of the original A). This is done in [Bensoussan et al., 1993, pp. 229–37] under much more stringent assumptions than those available in the present chapter, and only to obtain an operator $P(t)$ as a strong limit of approximating Riccati operators $P_n(t)$, where $P_n^{-1}(t) = Q_n(t)$ is (the unique) solution of a corresponding *dual* Riccati equation; see Chapter 9, Section 9.6. However, there is no claim that the limit $P(t)$ is a solution of the original Riccati equation. By contrast, our approach in the present section as expressed by our proof of Theorem 8.4.5.1 (which is based on the explicit formula (8.4.4.13) for $P(t)$, obtained in Theorem 8.4.4.2) bypasses altogether the described technical difficulties, while entirely avoiding last-resort approximation arguments.

8.5 Proof of Theorem 8.2.2.1: The More Regular Case

In this section we prove Theorem 8.2.2.1, under assumptions (H.4) = (8.2.2.1) and (H.5) = (8.2.2.2). In Section 8.6, we shall show that these assumptions are automatically fulfilled in the case of the guiding example of hyperbolic second-order equations with Neumann boundary control and Dirichlet boundary observation, as a consequence of recent sharp regularity results for mixed hyperbolic problems [Lasiecka, Triggiani, 1983; 1989(a); 1990; 1991(b); 1994]; see Appendix 8A.

8.5.1 A Preliminary Lemma

Lemma 8.5.1.1   *Assumptions (H.4) = (8.2.2.1) and (H.2) = (8.2.1.2) imply the following property on the operator $K_t = L_t^* R^* R L_t$ defined in (8.2.2.3):*

$$\text{the map } t \to K_t u \text{ is continuous on } \mathcal{U}_{[0,T]} \qquad (8.5.1.1)$$

for each $u \in \mathcal{U}_{[0,T]}$ fixed, where we extend K_t by $(K_t u)(\tau) \equiv 0$, $0 \leq \tau < t$.

Proof.   Let $t \leq \tau \leq T$, and let $0 \leq t_0 < t < T$. We compute from (8.2.2.3) with $u \in \mathcal{U}_{[0,T]}$ via (8.1.12a) and (8.1.13a):

$$\left(K_{t_0} u\right)(\tau) = \int_\tau^T B^* e^{A^*(\sigma-\tau)} R^* R \left[\int_{t_0}^t e^{A(\sigma-r)} B u(r) \, dr + \int_t^\sigma e^{A(\sigma-r)} B u(r) \, dr\right] d\sigma,$$

$$\qquad (8.5.1.2)$$

and hence, from (8.5.1.2), via (8.2.2.3) and (8.1.12a), (8.1.13a),

$$
\begin{aligned}
\left(K_{t_0}u\right)(\tau) - (K_t u)(\tau) &= \int_\tau^T B^* e^{A^*(\sigma - \tau)} R^* R \int_{t_0}^t e^{A(\sigma - r)} Bu(r)\, dr\, d\sigma \\
&= \int_\tau^T B^* e^{A^*(\sigma - \tau)} R^* R e^{A(\sigma - t)} \int_{t_0}^t e^{A(t - r)} Bu(r)\, dr\, d\sigma \\
&= \left\{ L_t^* R^* \left[R e^{A(\cdot - t)} \left(L_{t_0} u\right)(t) \right] \right\}(\tau). \tag{8.5.1.3}
\end{aligned}
$$

Thus, from (8.5.1.13), invoking (H.4) and (H.2) in the form of their consequence (8.2.2.5), we obtain

$$
\begin{aligned}
\left\| K_{t_0}u - K_t u \right\|_{\mathcal{U}_{[t,T]}} &= \left\| L_t^* R^* \left[R e^{A(\cdot - t)} \left(L_{t_0} u\right)(t) \right] \right\|_{\mathcal{U}_{[t,T]}} \\
\text{(by (8.2.2.5))} &\leq c_T \left\| \left(L_{t_0} u\right)(t) \right\|_Y. \tag{8.5.1.4}
\end{aligned}
$$

But, after a change of variable $r - t_0 = \tau$, we have

$$
\begin{aligned}
\left(L_{t_0} u\right)(t) &= \int_{t_0}^t e^{A(t - r)} Bu(r)\, dr = \int_0^{t - t_0} e^{A(t - t_0 - \tau)} Bu(t_0 + \tau)\, d\tau \\
&= \{ L_0 u(t_0 + \cdot) \}(t - t_0). \tag{8.5.1.5}
\end{aligned}
$$

Using (8.5.1.5) into (8.5.1.4) and invoking (H.4) = (8.2.2.1), we obtain

$$
\left\| K_{t_0}u - K_t u \right\|_{\mathcal{U}_{[t,T]}} \leq c_T \| \{ L_0 u(t_0 + \cdot) \}(t - t_0) \|_Y \to 0 \quad \text{as } |t_0 - t| \to 0, \tag{8.5.1.6}
$$

as desired. However, with $u \in \mathcal{U}_{[0,T]}$ and $(K_t u)(\tau) \equiv 0$ for $0 \leq \tau < t$ (as defined), we have

$$
\left\| K_{t_0}u - K_t u \right\|_{\mathcal{U}_{[t_0,t]}} = \left\| K_{t_0}u \right\|_{\mathcal{U}_{[t_0,t]}} \to 0 \quad \text{as } |t_0 - t| \to 0. \tag{8.5.1.7}
$$

Then (8.5.1.6) and (8.5.1.7) prove the desired statement (8.5.1.1). $\square$

8.5.2 Completion of the Proof of Theorem 8.2.2.1

Proof of (i) = (8.2.2.6). We return to (8.2.1.17) = (8.3.1.11) for u^0. Now with $x \in Y$, we have by the version (8.2.2.5) of assumptions (H.4) = (8.2.2.1) and (H.2) that $L_t^* R^* R e^{A(\cdot - t)} x \in \mathcal{U}_{[t,T]}$ continuously in x, with norm bound uniform in t. Thus, by (8.2.1.17), to establish (8.2.2.6), we need to show that the operator $[I_t + L_t^* R^* R L_t]$ is boundedly invertible on $\mathcal{U}_{[t,T]}$ for each t and that uniformly in t we have

$$
\| [I_t + L_t^* R^* R L_t]^{-1} u \|_{\mathcal{U}_{[t,T]}} \leq c_T \| u \|_{\mathcal{U}_{[t,T]}}. \tag{8.5.2.1}
$$

First, with t fixed, since $K_t = L_t^* R^* R L_t$ is compact on $\mathcal{U}_{[t,T]}$ by (H.5) = (8.2.2.2), we have that

$$
[I_t + K_t]^{-1} \in \mathcal{L}\left(\mathcal{U}_{[t,T]}\right). \tag{8.5.2.2}
$$

In fact, this is the case if (and only if) $\lambda = -1$ is not an eigenvalue of K_t on $\mathcal{U}_{[t,T]}$, which is indeed true since $\lambda = -1$ is not an eigenvalue of the positive self-adjoint K_t on $L_2(t, T; U) \supset \mathcal{U}_{[t,T]}$.

Next, the continuity property (8.5.1.1) of Lemma 8.5.1.1 for K_t, along with the property from assumption (H.5) = (8.2.2.2) that $\{K_t\}$ is a family of collectively compact operators on $\mathcal{U}_{[0,T]}$, permits us to invoke the (proof of the) abstract Lemma 8.5.3.1 in the next section with $Z_0 = \mathcal{U}_{[0,T]}$ and to conclude that the desired estimate (8.5.2.1) holds true. We notice that Lemma 8.5.3.1 assumes a condition [property (c) given by Eqn. (8.5.3.1), which corresponds to condition (8.2.2.3) in the statement of assumption (H.5)] that is used solely to guarantee that $\{K_t\}$ is a family of collectively compact operators on $\mathcal{U}_{[0,T]} = Z_0$.

Proof of (ii) = (8.2.2.7). Estimate (8.2.2.7) now holds true by (8.3.2.2), where we use (8.2.2.6) and assumption (H.4) in the form of the uniform estimate (8.2.2.4). □

8.5.3 An Auxiliary Lemma

The following lemma is needed to complete the proof of Theorem 8.2.2.1, part (i), and, moreover, it will also be invoked in other instances throughout this book.

Lemma 8.5.3.1 *Assume hypotheses (a) through (d) below:*

(a) *Let Z_0 and Z_1 be two Banach spaces such that Z_1 is contained in Z_0, $Z_1 \subset Z_0$, with compact injection $Z_1 \to Z_0$.*

(b) *Let $\{K_t\}$, $0 \leq t \leq T$, be a family of bounded operators on Z_0, which is strongly continuous here: The map $t \to K_t x$ is continuous on Z_0, $\forall x \in Z_0$.*

(c) *Let the following uniform estimate hold true:*

$$\|K_t x\|_{Z_1} \leq c_T \|x\|_{Z_0}, \quad \forall x \in Z_0, \ \forall t \in [0, T]. \tag{8.5.3.1}$$

(d) *For each fixed $t \in [0, T]$, let the inverse $[I + K_t]^{-1}$ exist as a well-defined bounded operator on $Z_0 : [I + K_t]^{-1} \in \mathcal{L}(Z_0)$.*

Then, in fact

(i)

$$\|[I + K_t]^{-1}\|_{\mathcal{L}(Z_0)} \leq C_T, \quad \text{uniformly in } t \in [0, T]; \tag{8.5.3.2}$$

(ii) *the map $t \to [I + K_t]^{-1} x$ is continuous on Z_0, for each $x \in Z_0$.*

Proof. Assumption (8.5.3.1), along with the compact injection $Z_1 \to Z_0$ in (a), implies not only that each K_t is compact as an operator on Z_0, but in addition that $\{K_t\}$ is a family of collectively compact operators on Z_0. This means [Anselone, 1971, p. 4] that the set in Z_0:

$$\mathcal{K} \equiv \bigcup_{0 \leq t \leq T} K_t[\text{unit ball in } Z_0]$$

is precompact, or relatively compact on Z_0, where the union is made up of images of the unit ball in Z_0 under K_t. Indeed, by (8.5.3.2), $\mathcal{K}$ is contained in the ball of Z_1 of radius c_T, centered at the origin, which by assumption (a) is a compact set in Z_0. Next, by contradiction, let conclusion (8.5.3.2) be false so that there are sequences $\{t_n\}, \{x_n\}, \ t_n \in [0, T]$, and $\|x_n\|_{Z_0} = 1$ such that

$$[I + K_{t_n}]x_n \to 0 \quad \text{in } Z_0 \text{ as } n \to \infty. \tag{8.5.3.3}$$

But plainly $\{K_{t_n}x_n\} \in \mathcal{K}$, and thus there is a convergent subsequence in Z_0

$$K_{t_{n_k}}x_{n_k} \to y \in \bar{\mathcal{K}}. \tag{8.5.3.4}$$

Then, by (8.5.3.3) and (8.5.3.4), the unit vectors x_{n_k} in Z_0 converge to $-y$ in Z_0:

$$x_{n_k} = \left(I + K_{t_{n_k}}\right)x_{n_k} - K_{t_{n_k}}x_{n_k} \to -y \quad \text{in } Z_0, \tag{8.5.3.5}$$

so that $\|y\|_{Z_0} = 1$. Also, by (b), the Principle of Uniform Boundedness gives $\|K_t\|_{\mathcal{L}(Z_0)} \leq C_T$, for all $t \in [0, T]$. Thus, by (8.5.3.5),

$$K_{t_{n_k}}\left(x_{n_k} + y\right) \to 0 \quad \text{in } Z_0. \tag{8.5.3.6}$$

Also, at the price of extracting a further subsequence (denoted by the same symbol), we have $t_{n_k} \to t_0 \in [0, T]$. This, along with property (b) and (8.5.3.6), yields then

$$K_{t_{n_k}}x_{n_k} = K_{t_{n_k}}\left(x_{n_k} + y\right) - K_{t_{n_k}}y \to -K_{t_0}y \quad \text{in } Z_0. \tag{8.5.3.7}$$

From (8.5.3.3), using (8.5.3.5) and (8.5.3.7), we then obtain

$$-\left(y + K_{t_0}y\right) = 0, \quad \|y\|_{Z_0} = 1,$$

which contradicts assumption (d) at $t = t_0$. Part (i) is proved.

(ii) By the usual identity (such as Chapter 4, Eqn. (4.4.1.28)), we obtain

$$\left\|\left\{[I + K_t]^{-1} - [I + K_{t_0}]^{-1}\right\}x\right\|_{Z_0} = \left\|[I + K_t]^{-1}[K_t - K_{t_0}][I + K_{t_0}]^{-1}x\right\|_{Z_0}$$

$$\text{(by (8.5.3.2))} \qquad \leq C_T\left\|[K_t - K_{t_0}][I + K_{t_0}]^{-1}x\right\|_{Z_0} \to 0, \tag{8.5.3.8}$$

after invoking (8.5.3.2) in the last step as well as strong continuity, property (b), to conclude that the right-hand side tends to zero as $t \to t_0$.   □

8.6 Application of Theorems 8.2.1.1, 8.2.1.2, and 8.2.2.1: Neumann Boundary Control and Dirichlet Boundary Observation for Second-Order Hyperbolic Equations

8.6.1 Problem Formulation and Abstract Setting

Problem Formulation   Let Ω be an open bounded domain in R^n, $n \geq 2$, with sufficiently smooth boundary Γ (the case $n = 1$ is simpler and can be handled within the theory of the subsequent Chapter 9; see Remark 8.6.3.1 at the end of Section 8.6.3).

Consider the following canonical hyperbolic mixed problem in the unknown $w(t, x)$:

$$\begin{cases} w_{tt} = \Delta w - w & \text{in } (0, T] \times \Omega = Q; & (8.6.1.1\text{a}) \\ w(0, x) = w_0, \ w_t(0, x) = w_1 & \text{in } \Omega; & (8.6.1.1\text{b}) \\ \left. \dfrac{\partial w}{\partial v} \right|_\Sigma = u & \text{in } (0, T] \times \Gamma = \Sigma. & (8.6.1.1\text{c}) \end{cases}$$

(In what follows, one may readily replace $-(\Delta - 1)$ with a general second-order, uniformly elliptic operator on Ω, and the normal derivative $\partial/\partial v$ with the corresponding co-normal derivative; see Lasiecka and Triggiani [1990; 1994(b)].) With problem (8.6.1.1) we associate the following quadratic cost functional on a preassigned interval $[0, T]$, $T < \infty$, with $\{w_0, w_1\} \in H^1(\Omega) \times L_2(\Omega)$:

$$J(u, w) = \int_0^T \left\{ \|w(t)|_\Gamma \|_{L_2(\Gamma)}^2 + \|u(t)\|_{L_2(\Gamma)}^2 \right\} dt, \qquad (8.6.1.2)$$

which penalizes the Neumann control u, as well as the corresponding Dirichlet trace (observation) of the solution $w(t) = w(t; w_0, w_1)$, both in $L_2(0, T; L_2(\Gamma))$. The optimal control problem is then

minimize $J(u, w)$, over all $u \in L_2(0, T; L_2(\Gamma))$ with w solution of (8.6.1.1) due to u. (8.6.1.3)

The goal of this section is to show that the optimal problem (8.6.1.3) for the dynamics (8.6.1.1) is covered as a specialization of Theorems 8.2.1.1, 8.2.1.2, and 8.2.2.1. The first two will provide existence and uniqueness of the Riccati operator in $H^1(\Omega) \times L_2(\Omega)$, whereas the last will provide regularity results of the optimal control and corresponding optimal solution. For sake of convenience, we shall summarize, here, both the abstract setting for problem (8.6.1.1), (8.6.1.2), as well as the subset of the (sharp) regularity results of the mixed problem (8.6.1.1) needed in the following sections.

Abstract Setting   Let the abstract spaces Y and U of model (8.1.1) and the observation space Z be

$$Y = H^1(\Omega) \times L_2(\Omega); \quad U = L_2(\Gamma); \quad Z = L_2(\Gamma). \qquad (8.6.1.4)$$

Let $\mathcal{A} : L_2(\Omega) \supset \mathcal{D}(\mathcal{A}) \to L_2(\Omega)$ be the operator

$$-\mathcal{A}f = \Delta f - f, \quad \mathcal{D}(\mathcal{A}) = \left\{ f \in H^2(\Omega) : \left. \frac{\partial f}{\partial v} \right|_\Gamma = 0 \right\} \qquad (8.6.1.5)$$

(which is positive self-adjoint, in the canonical case (8.6.1.1), but this is not crucial). As in Chapter 3, Eqn. (3.3.7), let N be the Neumann map

$$Ng = v \iff \begin{cases} \Delta v - v = 0 & \text{in } \Omega, \\ \dfrac{\partial v}{\partial v} = g & \text{in } \Gamma, \end{cases} \qquad (8.6.1.6)$$

$$N : L_2(\Gamma) \to H^{\frac{3}{2}}(\Omega) \subset H^{\frac{3}{2} - 2\rho}(\Omega) = \mathcal{D}\big(\mathcal{A}^{\frac{3}{4} - \rho}\big), \quad \rho > 0, \qquad (8.6.1.7)$$

where the arbitrarily small constant $\rho > 0$ will be fixed once and for all. Then the operators A and B of model (8.1.1) are

$$
A = \begin{vmatrix} 0 & I \\ -\mathcal{A} & 0 \end{vmatrix}; \quad \mathcal{D}(A) = \mathcal{D}(\mathcal{A}) \times H^1(\Omega) \tag{8.6.1.8}
$$

(where A in (8.6.1.8) is skew-adjoint on Y in the canonical case (8.6.1.1); see also the identification in (8.6.1.11) for $H^1(\Omega)$) and

$$
Bu = \begin{bmatrix} 0 \\ \mathcal{A}Nu \end{bmatrix} : L_2(\Gamma) \to [\mathcal{D}(A^*)]', \tag{8.6.1.9a}
$$

where $\mathcal{D}(A^*) = \mathcal{D}(A)$ in our present case

$$
A^{-1}Bu = \begin{bmatrix} 0 & -\mathcal{A}^{-1} \\ I & 0 \end{bmatrix} \begin{bmatrix} 0 \\ \mathcal{A}Nu \end{bmatrix} = \begin{bmatrix} -Nu \\ 0 \end{bmatrix}, \tag{8.6.1.9b}
$$

where $\mathcal{A}$ in (8.6.1.9) is actually the extension, say conservatively $L_2(\Omega) \to [\mathcal{D}(\mathcal{A}^*)]'$, of the original operator defined in (8.6.1.5), where $\mathcal{A}$ is presently self-adjoint. With the above choice, problem (8.6.1.1) admits the abstract model (8.1.1). Finally, the observation operator R is defined as a Dirichlet trace of the first component by

$$
R : Y = \mathcal{D}(R) \to Z = L_2(\Gamma); \quad R\begin{bmatrix} y_1 \\ y_2 \end{bmatrix} = y_1|_\Gamma = N^*\mathcal{A}y_1, \tag{8.6.1.10}
$$

where for the last identity, which is justified by Green's theorem, we invoke Chapter 4, Lemma 4.3.1. The operator R is a fortiori in $\mathcal{L}(Y; Z)$. We also note that (algebraically and topologically) [see Chapter 3, Appendix 3A]

$$
H^1(\Omega) = \mathcal{D}(\mathcal{A}^{\frac{1}{2}}); \quad \|f\|^2_{H^1(\Omega)} = \|f\|^2_{\mathcal{D}(\mathcal{A}^{\frac{1}{2}})} = \int_\Omega \{|\nabla f|^2 + f^2\}\, d\Omega. \tag{8.6.1.11}
$$

From (8.6.1.10) we verify that for $[y_1, y_2] \in \mathcal{D}(R) = Y$ as in (8.6.1.4) and $z \in Z = L_2(\Gamma)$:

$$
\left(R\begin{bmatrix} y_1 \\ y_2 \end{bmatrix}, z \right)_{Z=L_2(\Gamma)} = (N^*\mathcal{A}y_1, z)_{L_2(\Gamma)} = (y_1, Nz)_{\mathcal{D}(\mathcal{A}^{\frac{1}{2}})}
$$

$$
= \left(\begin{bmatrix} y_1 \\ y_2 \end{bmatrix}, R^*z \right)_{Y=\mathcal{D}(\mathcal{A}^{\frac{1}{2}}) \times L_2(\Omega)}. \tag{8.6.1.12}
$$

Thus

$$
R^*z = \begin{bmatrix} Nz \\ 0 \end{bmatrix}, \quad z \in Z = L_2(\Gamma). \tag{8.6.1.13}
$$

Similarly, from (8.6.1.9), with Y as in (8.6.1.4), we compute B^*:

$$
\left(Bu, \begin{bmatrix} y_1 \\ y_2 \end{bmatrix} \right)_Y = \left(\begin{bmatrix} 0 \\ \mathcal{A}Nu \end{bmatrix}, \begin{bmatrix} y_1 \\ y_2 \end{bmatrix} \right)_Y = (u, N^*\mathcal{A}y_2)_{U=L_2(\Gamma)}
$$

$$
= \left(u, B^*\begin{bmatrix} y_1 \\ y_2 \end{bmatrix} \right)_U \tag{8.6.1.14}
$$

and hence

$$B^* \begin{bmatrix} y_1 \\ y_2 \end{bmatrix} = N^* A y_2 = y_2|_\Gamma; \quad \mathcal{D}(B^*) = H^1(\Omega) \times \mathcal{D}\big(A^{\frac{1}{4}+\rho}\big). \tag{8.6.1.15}$$

Review of Needed Regularity Results of the Mixed Problem (8.6.1.1) (See Lasiecka and Triggiani [1983; 1989(a); 1990; 1991(b); 1994] and Appendix 8A for a summary.) We shall verify that assumptions (H.1) through (H.5) hold true for the original hyperbolic problem (8.6.1.1) recast in the abstract setting of Section 8.1, as will be indicated in the present section. To this end, it will be crucial – for (H.1), (H.3), and (H.5) – to rely on the sharp regularity result of Lasiecka and Triggiani [1983; 1989(a); 1990; 1991(b); 1994], for general second-order hyperbolic mixed problems of Neumann type, of which problem (8.6.1.1) is a canonical model. The key results are collected below and invoked in the next section. To this end, we introduce the following numbers, where $\epsilon > 0$ is arbitrary:

$$\left.\begin{array}{l} \alpha = \dfrac{3}{5} - \epsilon \\[2mm] \beta = \dfrac{3}{5} \end{array}\right\} \quad \begin{array}{l} \text{for a general smooth bounded domain} \\ : \ \Omega \text{ and a general second-order uniformly} \\ \text{elliptic operator } A(\zeta, \partial); \end{array} \tag{8.6.1.16a}$$

$$\alpha = \beta = \frac{2}{3} : \text{for a sphere and the case } -A(\zeta, \partial) = \Delta \mp 1 \tag{8.6.1.16b}$$

[or when the coefficients a_{ij} of the second-order elliptic operator do not depend on the normal variable or do not depend on the tangential variable, near Γ];

$$\alpha = \beta = \frac{3}{4} : \text{for a parallelepiped and } -A(\zeta, \partial) = \Delta \mp 1. \tag{8.6.1.16c}$$

Lemma 8.6.1.1 *([Lasiecka, Triggiani, 1983; 1989(a); 1990; 1991(b); 1994], Theorem 8A.1 in Appendix 8A.) With reference to the operators L_0 in (8.1.5) and R in (8.6.1.10) we have: Given the smooth domain $\Omega \subset R^n$ ($n \geq 2$), there exists a number α as in (8.6.1.16), such that the solution to the hyperbolic mixed problem (8.6.1.1) [even with $-A(\zeta, \partial)$ replacing $\Delta - 1$] satisfies:*

(i)

$$L_0 : u \to \{w, w_t\} : continuous \ L_2(\Sigma) \to C([0, T]; H^\alpha(\Omega) \times H^{\alpha-1}(\Omega)); \tag{8.6.1.17}$$

(ii)

$$RL_0 : u \to RL_0 u = w|_\Sigma : continuous \ L_2(\Sigma) \to H^{2\alpha-1}(\Sigma); \tag{8.6.1.18}$$

$$H^{2\alpha-1}(\Sigma) = L_2(0, T; H^{2\alpha-1}(\Gamma)) \cap H^{2\alpha-1}(0, T; L_2(\Gamma)). \tag{8.6.1.19}$$

For the next lemma, we introduce the homogeneous problem corresponding to (8.6.1.1), that is,

$$
\begin{cases}
\psi_{tt} = \Delta\psi - \psi & \text{in } Q; & (8.6.1.20a) \\
\psi(0, x) = \psi_0, \; \psi_t(0, x) = \psi_1 & \text{in } \Omega; & (8.6.1.20b) \\
\left.\dfrac{\partial\psi}{\partial\nu}\right|_\Sigma \equiv 0 & \text{in } \Sigma, & (8.6.1.20c)
\end{cases}
$$

whose solution is explicitly given by

$$
\psi(t) = \psi(t, \psi_0, \psi_1) = \mathcal{C}(t)\psi_0 + \mathcal{S}(t)\psi_1, \tag{8.6.1.21}
$$

where $\mathcal{C}(\cdot)$ is the cosine operator on $L_2(\Omega)$ generated by the operator $-\mathcal{A}$ in (8.6.1.5), and $\mathcal{S}(t) = \int_0^t \mathcal{C}(\tau)\,d\tau$.

Lemma 8.6.1.2 *([Lasiecka, Triggiani, 1991(b)], Corollary 8A.6 in Appendix 8A) With reference to problem (8.6.1.20) and to (8.6.1.16), we have*

$$\{\psi_0, \psi_1\} \to \psi|_\Sigma$$

$$
: \text{ continuous} \begin{cases}
H^1(\Omega) \times L_2(\Omega) \to H^\beta(\Sigma), & (8.6.1.22) \\
L_2(\Omega) \times [H^1(\Omega)]' \to H^{\alpha-1}(\Sigma), & (8.6.1.23) \\
H^\theta(\Omega) \times [H^{1-\theta}(\Omega)]' \to H^{\alpha-1+\theta+(\beta-\alpha)\theta}(\Sigma), & (8.6.1.24)
\end{cases}
$$

where we recall from Lasiecka and Triggiani [1991(b)] that (algebraically and topologically)

$$
H^\theta(\Omega) = \mathcal{D}\big(\mathcal{A}^{\frac{\theta}{2}}\big); \quad [H^\theta(\Omega)]' = \big[\mathcal{D}\big(\mathcal{A}^{\frac{\theta}{2}}\big)\big]', \quad 0 \le \theta < \frac{3}{4}; \tag{8.6.1.25}
$$

$$
[H^\theta(\Omega)]' = H^{-\theta}(\Omega), \quad 0 \le \theta \le \frac{1}{2}. \tag{8.6.1.26}
$$

Remark 8.6.1.1 As noted before, the classical regularity theory of problems (8.6.1.1) prior to the sharp theory is as follows (Remark 8A.3 in the appendix):

(i) The version of Lemma 8.6.1.1(i) is the weaker result [Lions, Magenes, 1972, Vol. II, p. 120]

$$
u \in L_2(\Sigma) \to \{w, w_t\} \in L_2\big(0, T; H^{\frac{1}{2}}(\Omega) \times H^{-\frac{1}{2}}(\Omega)\big)
$$

instead of (8.6.1.17). Thus, in this earlier theory, we would have to take $\alpha = 1/2$ (and lose $C[0, T]$ in favor of $L_2(0, T)$).

(ii) The version of Lemma 8.6.1.2 is the weaker result

$$
\{\psi_0, \psi_1\} \in H^1(\Omega) \times L_2(\Omega) \to \psi|_\Sigma \in C\big([0, T]; H^{\frac{1}{2}}(\Gamma)\big)
$$

instead of (8.6.1.22), whereby we would have to take now $\beta = 1/2$.

The analysis below will show that to verify assumptions (H.1), (H.3), and (H.5) for problem (8.6.1.1), it is critical to have $\alpha, \beta > 1/2$; in other words, the specific values of α, β as given in (8.6.1.16) are not that important, as long as we have that $\alpha, \beta > 1/2$. Thus, invoking the sharp regularity theory of Lasiecka and Triggiani

[1983; 1989(a); 1990; 1991(b); 1994], summarized above, is crucial for the present solution of the optimal control problem (8.6.1.1)–(8.6.1.3) and related Riccati theory, whereas the earlier classical theory with $\alpha = \beta = 1/2$ would be insufficient.

8.6.2 Specialization of Theorems 8.2.1.1 and 8.2.1.2 to the Hyperbolic Problem: Theorem 8.6.2.1

Splitting the two components of $P(t)x$ as $P(t)x = \{[P(t)x]_1, [P(t)x]_2\} \in Y$, we have from (8.6.1.15) that $B^*P(t)x = [P(t)x]_2|_\Gamma$. Thus, for $x, y \in \mathcal{D}(A) = \mathcal{D}(\mathcal{A}) \times H^1(\Omega)$, by this and by (8.6.1.10), the DRE (8.2.1.32) specializes for problem (8.6.1.1)–(8.6.1.3) to

$$\frac{d}{dt}(P(t)x, y)_{H^1(\Omega) \times L_2(\Omega)}$$
$$= -(x_1|_\Gamma, y_1|_\Gamma)_{L_2(\Gamma)} - (P(t)x, Ay)_{H^1(\Omega) \times L_2(\Omega)}$$
$$- (P(t)Ax, y)_{H^1(\Omega) \times L_2(\Omega)} + ([P(t)x]_2|_\Gamma, [P(t)y]_2|_\Gamma)_{L_2(\Gamma)}, \qquad (8.6.2.1)$$

with $P(T) = 0$. In this section we shall verify that the abstract assumptions (H.1) = (8.2.1.1), (H.2) = (8.2.1.2), and (H.3) = (8.2.1.3) hold true for problem (8.6.1.1)–(8.6.1.3). As a consequence, we then obtain the following specialization of Theorems 8.2.1.1 and 8.2.1.2 as applied to problem (8.6.1.1)–(8.6.1.3).

Theorem 8.6.2.1 *With reference to the hyperbolic problem (8.6.1.1)–(8.6.1.3), there exists a unique solution (as specified in Theorem 8.2.1.2) $P(t) \in \mathcal{L}(H^1(\Omega) \times L_2(\Omega))$, $0 \le t \le T$ of the DRE (8.6.2.1) with the following properties:*

(i) $P(t) = P^*(t) \ge 0$, $0 \le t \le T$ ($*$ in $H^1(\Omega) \times L_2(\Omega)$).
(ii) $P(t)$: *continuous* $H^1(\Omega) \times L_2(\Omega) \to L_\infty(0, T; H^1(\Omega) \times L_2(\Omega))$.
(iii) $[P(t)x]_2|_\Gamma \in L_2(\Gamma)$, $0 \le t \le T$ *and*
$$\|[P(\cdot)x]_2|_\Gamma\|_{L_\infty(0,T;L_2(\Gamma))} \le c_T \|x\|_{H^1(\Omega) \times L_2(\Omega)}.$$
(iv) *By (8.6.1.15) the pointwise feedback representation (8.2.1.27) of the optimal pair specializes to*

$$u^0\left(t, \begin{bmatrix} w_0 \\ w_1 \end{bmatrix}\right) = -B^*P(t)\begin{bmatrix} w^0(t; w_0; w_1) \\ w_t^0(t; w_0; w_1) \end{bmatrix} = -\left[P(t)\begin{bmatrix} w^0(t; w_0; w_1) \\ w_t^0(t; w_0; w_1) \end{bmatrix}\right]\Bigg|_{2|\Gamma}.$$
$$(8.6.2.2)$$

(v) *By (8.2.1.28), the optimal cost specializes to*

$$J\left(u^0\left(\cdot; \begin{bmatrix} w_0 \\ w_1 \end{bmatrix}\right), \begin{bmatrix} w^0(\cdot; w_0; w_1) \\ w_t^0(\cdot; w_0; w_1) \end{bmatrix}\right) = \left(P(0)\begin{bmatrix} w_0 \\ w_1 \end{bmatrix}, \begin{bmatrix} w_0 \\ w_1 \end{bmatrix}\right)_{H^1(\Omega) \times L_2(\Omega)}.$$

8.6.3 Proof of Theorem 8.6.2.1: Verification
of Assumptions (H.1), (H.2), and (H.3)

Verification of (H.1) = (8.2.1.1)   Recalling (8.6.1.21) and (8.6.1.8) we have

$$e^{At} = \begin{bmatrix} \mathcal{C}(t) & \mathcal{S}(t) \\ -\mathcal{A}\mathcal{S}(t) & \mathcal{C}(t) \end{bmatrix}. \tag{8.6.3.1}$$

Then, by (8.6.1.10), (8.6.3.1), and (8.6.1.10),

$$Re^{At}Bu = Re^{At} \begin{bmatrix} 0 \\ \mathcal{A}Nu \end{bmatrix} = \mathcal{S}(t)\mathcal{A}Nu|_{\Gamma}. \tag{8.6.3.2}$$

By (8.6.1.7) and (8.6.1.25) we have, with $\rho > 0$ arbitrary,

$$\mathcal{A}N : \text{continuous } L_2(\Gamma) \to \left[\mathcal{D}\left(\mathcal{A}^{\frac{1}{4}+\rho}\right)\right]' = \left[H^{\frac{1}{2}+2\rho}(\Omega)\right]'. \tag{8.6.3.3}$$

Using (8.6.1.24) and (8.6.1.21) with $\psi_0 = 0$ and $\psi_1 = \mathcal{A}Nu$, $u \in L_2(\Gamma)$, with $1 - \theta = 1/2 + 2\rho$, that is, $\theta = 1/2 - 2\rho$, [see (8.6.3.3)], we obtain, via (8.6.3.2),

$$Re^{At}Bu = \mathcal{S}(t)\mathcal{A}Nu|_{\Sigma} \in H^{\alpha - \frac{1}{2} - 2\rho + (\beta - \alpha)(\frac{1}{2} - 2\rho)}(\Sigma) \tag{8.6.3.4}$$

$$\subset L_2(\Sigma) \tag{8.6.3.5}$$

continuously in u, where the last inclusion in the step from (8.6.3.4) to (8.6.3.5) attains under the sharp regularity theory where $\beta \geq \alpha > 1/2$.

Thus, from (8.6.3.5) we obtain

$$\|Re^{At}Bu\|_{L_2(0,T;L_2(\Gamma))} \leq c_T \|u\|_{L_2(\Gamma)}, \tag{8.6.3.6}$$

and assumption (H.1) = (8.2.1.1) is a fortiori verified with $Z = U = L_2(\Gamma)$ as dictated by (8.6.1.4). Note, however, that the earlier regularity theory where $\alpha = \beta = 1/2$, as recalled in Remark 8.6.1.1, would not be sufficient to obtain (H.1) = (8.2.1.1) from (8.6.3.4).   □

Verification of (H.2) = (8.2.1.2)   By trace theory, the Dirichlet trace operator R in (8.6.1.10) is plainly in $\mathcal{L}(Y; Z)$. Let $x = [x_1, x_2] \in Y = H^1(\Omega) \times L_2(\Omega)$. Then, by (8.6.3.1), (8.6.1.10), and (8.6.1.21), we have, continuously,

$$Re^{At}x = [\mathcal{C}(t)x_1 + \mathcal{S}(t)x_2]|_{\Gamma} = \psi(t; \psi_0, \psi_1)|_{\Gamma} \in C\left([0, T]; H^{\frac{1}{2}}(\Gamma)\right), \quad (8.6.3.7)$$

where $\psi_0 = x_1$, $\psi_1 = x_2$, by direct application of trace theory. Thus,

$$\|Re^{At}x\|_{C([0,T];H^{\frac{1}{2}}(\Gamma))} \leq c_T \|x\|_{H^1(\Omega) \times L_2(\Omega)}. \tag{8.6.3.8}$$

Verification of (H.3) = (8.2.1.3)   Let $z \in Z = L_2(\Gamma)$. Then, by (8.6.1.13) on R^*, (8.6.3.1) with A replaced by $A^* = -A$ (in our canonical example (8.6.1.1); but this is not crucial), and (8.6.1.15) on B^*, we obtain

$$B^*e^{A^*t}R^*z = B^*e^{A^*t} \begin{bmatrix} Nz \\ 0 \end{bmatrix} = B^* \begin{bmatrix} \mathcal{C}(t)Nz \\ \mathcal{A}\mathcal{S}(t)Nz \end{bmatrix} = \mathcal{A}\mathcal{S}(t)Nz|_{\Gamma}. \quad (8.6.3.9)$$

By (8.6.3.4) with $u = z \in L_2(\Gamma)$, and again using crucially the sharp regularity theory as in going from (8.6.3.4) to (8.6.3.6), we obtain likewise from (8.6.3.9)

$$\|B^* e^{A^* t} R^* z\|_{L_2(0,T;L_2(\Gamma))} \leq c_T \|z\|_{L_2(\Gamma)}, \tag{8.6.3.10}$$

a result that the classical theory with $\alpha = \beta = 1/2$ cannot produce. Thus, assumption (H.3) = (8.2.1.3) is again verified with $U = Z = L_2(\Gamma)$ as required.

Remark 8.6.3.1   Consider the mixed problem (8.6.1.1) for $n = 1$, with $Y = H^1(\Omega) \times L_2(\Omega)$, $\Omega = (a, b)$, a finite interval, and $\Gamma = \{a\} \cup \{b\}$. Then, in this one-dimensional case, the input-solution operator L_0 in (8.1.5) possesses the higher regularity

$$L_0 : u \to \{w, w_t\} \text{ continuous } L_2(0, T; L_2(\Gamma)) \to C([0, T]; Y = H^1(\Omega) \times L_2(\Omega)), \tag{8.6.3.11}$$

with $w_0 = w_1 = 0$, and, moreover, $R \in \mathcal{L}(Y; Z)$, $Z = L_2(\Gamma)$, as noted in (8.6.1.10) and ff. This technically simpler, more specialized situation may be handled by the treatment of the subsequent Chapter 9 [see Section 9.8.4]. We remark explicitly that the regularity expressed by (8.6.3.11) is true only in the one-dimensional case [see Chapter 9, Theorem 9.8.4.1].

8.6.4 Specialization of Theorem 8.2.2.1 to the Hyperbolic Problem: Verification of Assumptions (H.4) and (H.5)

In this section we shall specialize Theorem 8.2.2.1 to the hyperbolic problem (8.6.1.1)–(8.6.1.3) and show that, in this case, assumptions (H.4) = (8.2.2.1) and (H.5) = (8.2.2.2) hold true with

$$\mathcal{U}_{[t,T]} = H^{\frac{1}{2}}(\Sigma_t) = L_2\big(t; T; H^{\frac{1}{2}}(\Gamma)\big) \cap H^{\frac{1}{2}}(t, T; L_2(\Gamma)), \tag{8.6.4.1}$$

where $\Sigma_t = [t, T] \times \Gamma$. As a result, Theorem 8.2.2.1 specializes to

Theorem 8.6.4.1   *With reference to the hyperbolic problem (8.6.1.1)–(8.6.1.3), we have*

$$\sup_{0 \leq t \leq T} \|u^0(\,\cdot\,, t; x)\|_{H^{\frac{1}{2}}(\Sigma_t)} \leq c_T \|x\|_{H^1(\Omega) \times L_2(\Omega)}, \tag{8.6.4.2}$$

$$\sup_{0 \leq t \leq T} \|\Phi(\,\cdot\,, t)x\|_{C([t,T];H^1(\Omega) \times L_2(\Omega))} \leq c_T \|x\|_{H^1(\Omega) \times L_2(\Omega)}. \tag{8.6.4.3}$$

Proof. Verification of (H.4) = (8.2.2.1)   For problem (8.6.1.1) the regularity result with $w_0 = w_1 = 0$:

$$u \in L_2\big(0, T; H^{\frac{1}{2}}(\Gamma)\big) \to \begin{cases} w \in H^1(Q) \cap C([0, T]; H^1(\Omega)), \\ w_t \in C([0, T]; L_2(\Omega)) \end{cases} \tag{8.6.4.4}$$

is contained in [Myatake, 1973] (Theorem 8A.5 in the appendix), while a stronger result is contained in [Lasiecka, Triggiani, 1991(b), Eqn. (8.1.22)]; see Appendix 8A below Corollary 8A.4. In any case (H.4), as specified in (8.2.2.1), is a fortiori verified, and (8.6.4.4) is sufficient for this purpose, via (8.6.4.1).

Verification of (H.5) = (8.2.2.2)   It is at the level of verifying the collective compactness assumption (8.2.2.2), in its sufficient condition (8.2.2.3), that we shall make crucial use of the sharp regularity results of Lasiecka and Triggiani [1990] and [1991(b)]
in contrast, the earlier classical theory as in [Lions, Magenes, 1972, Vol. II] is inadequate. By (8.6.1.13) and (8.6.1.18), we obtain with $u \in H^{1/2}(\Sigma_0)$:

$$R^* R L_0 u = \begin{bmatrix} N R L_0 u \\ 0 \end{bmatrix} = \begin{bmatrix} N(w|_\Gamma) \\ 0 \end{bmatrix}, \tag{8.6.4.5}$$

with $w(t) = w(t; 0, 0)$ a solution of problem (8.6.1.1) with $w_0 = w_1 = 0$.

Next, by (8.6.3.1) on e^{At}, and from $A^* = -A$, we obtain by (8.6.4.5)

$$B^* e^{A^*(\sigma-\tau)} (R^* R L_0 u)(\sigma) = B^* \begin{bmatrix} C(\tau-\sigma) & C(\tau-\sigma) \\ -AS(\tau-\sigma) & S(\tau-\sigma) \end{bmatrix} \begin{bmatrix} N(w(\sigma)|_\Gamma) \\ 0 \end{bmatrix}$$

$$\text{(by (8.6.1.15))} \qquad = [AS(\sigma-\tau)N(w(\sigma)|_\Gamma)]_\Gamma, \tag{8.6.4.6}$$

where in the last step we have recalled (8.6.1.15) on B^* and the fact that S is odd.
Thus, using (8.6.4.6) in (8.1.6), we obtain

$$\{L_0^* R^* R L_0 u\}(\tau) = B^* \int_\tau^T e^{A^*(\sigma-\tau)} (R^* R L_0 u)(\sigma) \, d\sigma$$

$$= \left[\int_\tau^T AS(\sigma-\tau) N(w(\sigma)|_\Gamma) \, d\sigma \right]_\Gamma = \eta(\tau)|_\Gamma, \tag{8.6.4.7}$$

where, in view of (8.6.1.9a) and (8.6.1.6), $\eta(\tau)$ is the solution of

$$\begin{cases} \eta_{tt} = \Delta \eta - \eta & \text{in } Q; & \text{(8.6.4.8a)} \\ \eta|_{t=T} = 0, \ \eta_t|_{t=T} = 0 & \text{in } \Omega; & \text{(8.6.4.8b)} \\ \left. \dfrac{\partial \eta}{\partial \nu} \right|_\Sigma = w|_\Sigma & \text{in } \Sigma. & \text{(8.6.4.8c)} \end{cases}$$

But, by Myatake [1973] with reference to problem (8.6.1.1) with $w_0 = w_1 = 0$, we
have

$$u \in H^{\frac{1}{2}}(\Sigma_0) \to w|_\Sigma \in H^{\frac{1}{2}}(\Sigma_0) \tag{8.6.4.9}$$

continuously; and, in fact, by [Lasiecka, Triggiani, 1991(b), Corollary 4.3 with $\theta =
1/2$] [see Eqn. (8A.31) in the appendix] the following *stronger* result holds true:

$$u \in H^{\frac{1}{2}}(\Sigma_0) \to w|_\Sigma \in H^{2\alpha-\frac{1}{2}}(\Sigma_0) \tag{8.6.4.10}$$

continuously, where α is defined by (8.6.1.16). At this stage, we may still use the
conservative regularity (8.6.4.9) for $w(\sigma)|_\Gamma$ in the integral term in (8.6.4.7). However,
it is at the level of analyzing $L_0^* R^* R L_0 u$ in (8.6.4.7) and (8.6.4.8) that we crucially
use the counterpart of result (8.6.4.10),

$$w|_\Sigma \in H^{\frac{1}{2}}(\Sigma_0) \to L_0^* R^* R L_0 u = \eta|_{\Sigma_0} \in H^{2\alpha-\frac{1}{2}}(\Sigma_0), \tag{8.6.4.11}$$

as applied to problem (8.6.4.8). But $2\alpha - 1/2 > 1/2$ [see (8.6.1.16)], so that

$$\text{the injection } H^{2\alpha-\frac{1}{2}}(\Sigma_0) \rightarrow H^{\frac{1}{2}}(\Sigma_0) \text{ is compact.} \qquad (8.6.4.12)$$

Putting together (8.6.4.9) through (8.6.4.12), we conclude that

$$\begin{aligned} L_0^* R^* R L_0 &: \text{continuous } H^{\frac{1}{2}}(\Sigma_0) \rightarrow H^{2\alpha-\frac{1}{2}}(\Sigma_0) \\ &: \text{compact } H^{\frac{1}{2}}(\Sigma_0) \rightarrow H^{\frac{1}{2}}(\Sigma_0). \end{aligned} \qquad (8.6.4.13)$$

The analysis leading to the compactness property (8.6.4.14) can be repeated with L_0, L_0^* replaced by L_t, L_t^*, so that (8.2.2.3) is likewise verified. Moreover, the proof actually yields the uniform estimate

$$\|L_t^* R^* R L_t u\|_{H^{2\alpha-\frac{1}{2}}(\Sigma_0)} \leq c_T \|u\|_{H^{\frac{1}{2}}(\Sigma_0)}, \quad 0 \leq t \leq T, \qquad (8.6.4.14)$$

and since $\alpha > 1/2$, $K_t = L_t^* R^* R L_t$ is a collectively compact family of operators on $H^{1/2}(\Sigma_0) = \mathcal{U}_{[0,T]}$ as in Lemma 8.5.3.1. In the notation of (8.2.2.3) [respectively (8.5.3.1)], we have $\mathcal{V} = H^{2\alpha-1}(\Sigma_0)$, [respectively, $Z_1 = H^{2\alpha-1}(\Sigma_0)$, $Z_0 = H^{1/2}(\Sigma_0)$].

Remark 8.6.4.1 Note that, in contrast, if one has only the theory of Myatake [1973] (where then $\alpha = 1/2$ in (8.6.4.16)), the required compactness of the map in (8.6.4.19) cannot be verified. $\square$

8.7 A One-Dimensional Hyperbolic Equation with Dirichlet Control (B Unbounded) and Point Observation (R Unbounded) That Satisfies (h.1) and (h.3) but not (h.2), (H.1), (H.2), and (H.3). Yet, the DRE Is Trivially Satisfied as a Linear Equation

8.7.1 Introduction. Summary of Results

In this section we provide a simple, yet enlightening, even striking example of a one-dimensional, first-order hyperbolic equation over the infinite domain $\Omega = \mathbb{R}_x^+ = (0, \infty)$ in the unknown $y(t, x)$, possessing the following features:

(i) The scalar control acts in the Dirichlet BC at the boundary point $x = 0$.

(ii) The observation is the pointwise value $y(t, 1)$ of the solution at the observation point $x = 1$.

(iii) In the abstract model $\dot{y} = Ay + Bu$ of the problem with observation operator $Rf = f|_{x=1}$, all operators A, B, and R are unbounded.

(iv) The model satisfies the preliminary assumption (8.1.2), that is,

$$R(\lambda, A)B \in \mathcal{L}(U; Y) \quad \text{for some } \lambda \in \rho(A).$$

(v) The problem satisfies assumption (h.1) = (8.2.1.10), that is,

$$Re^{At} : \text{continuous } L_2(0, T; U) \rightarrow L_2(0, T; Z).$$

(vi) For any T the problem satisfies assumption (h.3) = (8.2.1.12), that is,

$$RL_0 : \text{continuous } L_2(0, T; U) \to L_2(0, T; Z);$$

(vii) however, for $T > 1$,

$$RL_0 \text{ is } not \text{ continuous } L_2(0, T; U) \to C([0, T]; Z),$$

and so the problem does not satisfy assumption (H.3*) = (8.2.1.6).

(viii) Indeed, with reference to (H.3) = (8.2.1.3), which is violated as noted in (vii), we have that, for $T > 1$ and $z \in Z$,

$$B^* e^{A^* t} R^* z \text{ is a distribution},$$

and so it does not belong to any $L_p(0, T; U)$, $1 \le p \le \infty$;

(ix) for $T > 1$,

$$Re^{At} Bu \text{ is a distribution},$$

and so it does not belong to any $L_p(0, T; Z)$, $1 \le p \le \infty$; so that assumption (H.1) = (8.2.1.1) is violated.

(x) For y_0 smooth,

$$F_2(t)y_0 = \left\{ L_t^* R^* Re^{A(\cdot - t)} y_0 \right\}(t) = \begin{cases} y_0(0), & t \le T - 1; \\ 0, & t > T - 1, \end{cases}$$

so that property (8.4.1.13) of Lemma 8.4.1.3,

$$\sup \| F_2(t) y_0 \|_U \le c_T \| y_0 \|_Y, \quad y_0 \in Y,$$

is false in the present case.

(xi) Similarly, property (8.4.1.12) of Lemma 8.4.1.2,

$$\sup_{0 \le s \le T} \int_s^T \| F_1(t, s) u \|_U^2 \, dt \le c_T \| u \|_U^2,$$

is false in the present case.

8.7.2 The Optimal Control Problem and Its Solution

Dynamics   We consider the following mixed problem for a first-order equation on $\Omega = (0, +\infty)$ in the unknown $y(t, x)$, subject to Dirichlet boundary control $u(t)$ exercised at the boundary point $x = 0$:

$$\begin{cases} y_t = -y_x, & t > s, \ x > 0; & (8.7.2.1a) \\ y(s, x) = y_0(x), & x \ge 0; & (8.7.2.1b) \\ y(t, 0) = u(t), & t \ge s, \ x = 0, & (8.7.2.1c) \end{cases}$$

with initial time $0 \le s$. The solution of (8.7.2.1) will be denoted by $y(t, x; s, y_0, u) = y(t, x; s)$. The letter s will always denote the initial time.

Optimal Control Problem   The observation is taken to be the value $y(t, 1; s)$ of the solution at the observation point $x = 1$. Accordingly, we take $T > 1$ throughout and define the cost functional by

$$J_{s,T}(u, y) = \int_s^T [|y(t, 1; s)|^2 + |u(t)|^2] \, dt. \qquad (8.7.2.2)$$

The corresponding optimal control problem is:

$$\begin{cases} \text{Minimize } J_{s,T}(u, y) \text{ over all } u \in L_2(s, T) \text{ where } y(t, x; s) \\ \text{is the solution of (8.7.2.1) due to } u. \end{cases} \qquad (8.7.2.3)$$

The above problem (8.7.2.3) produces the Riccati operator $P_T(s)$, $0 \le s \le T$, which is needed in the optimal control problem with $s = 0$.

Explicit Formula for the Solution of (8.7.2.1) due to $\{y_0, u\}$ The solution of problem (8.7.2.1) is given by (see Figure 8.1):

$$y(t, x; s) = y(t, x; s; y_0, u) = \begin{cases} y_0(x - (t - s)), & s \le t \le x + s; \qquad (8.7.2.4a) \\ u(t - x), & x + s < t \le T. \qquad (8.7.2.4b) \end{cases}$$

Optimal Control Solution   We note from (8.7.2.4) that along the segment $\{x = 1, \ s \le t \le T\}$ the initial condition y_0 and the control u do not interfere:

$$\text{at } x = 1 \begin{cases} \text{for } s \le t \le s + 1, & y(t, 1; s) \text{ is determined only by } y_0; \\ \text{for } s + 1 < t \le T, & y(t, 1; s) \text{ is determined only by } u. \end{cases}$$

Thus, the *optimal control* $u^0(t)$ for problem (8.7.2.1)–(8.7.2.3) is the trivial control:

$$u^0(t) \equiv 0, \quad s \le t \le T. \qquad (8.7.2.5)$$

The corresponding *optimal solution* $y^0(t, x; s)$ is, by (8.7.2.4) and (8.7.2.5),

$$y^0(t, x; s) = \begin{cases} y_0(x - (t - s)), & s \le t \le x + s; \qquad (8.7.2.6a) \\ 0, & x + s < t \le T. \qquad (8.7.2.6b) \end{cases}$$

The *optimal observed solution* $y^0(t, 1; s)$, observed at $x = 1$, is

$$y^0(t, 1; s) = \begin{cases} y_0(1 - (t - s)), & s \le t \le 1 + s; \qquad (8.7.2.7a) \\ 0, & 1 + s < t \le T. \qquad (8.7.2.7b) \end{cases}$$

The *optimal cost* $J_{s,T}^0(y_0)$ is, by (8.7.2.2), (8.7.2.5), and (8.7.2.7),

$$J_{s,T}^0(y_0) = \int_s^T [|y^0(t, 1; s)|^2 + |u^0(t)|^2] \, dt$$

$$= \int_s^{\min\{T, 1+s\}} y_0^2(1 - (t - s)) \, dt = \int_{\max\{0, 1-T+s\}}^1 y_0^2(x) \, dx \qquad (8.7.2.8)$$

$$= (P_T(s) y_0, y_0)_Y, \quad Y = L_2(0, \infty). \qquad (8.7.2.9)$$

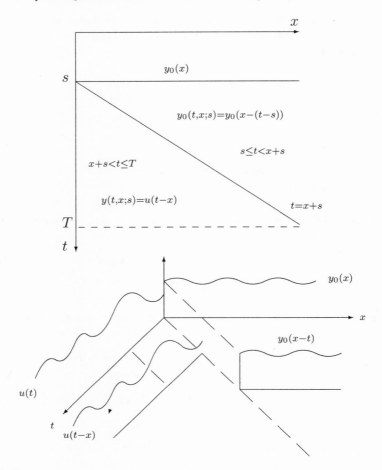

Figure 8.1   Case $s = 0$.

Thus, by (8.7.2.8) and (8.7.2.9), we obtain

$$\{P_T(s)y_0\}(x) = \begin{cases} \begin{rcases} y_0(x), & 0 \leq x \leq 1 \\ 0, & x > 1 \end{rcases}, & 1 + s \leq T; \\ \begin{rcases} 0, & 0 \leq x < 1 - T + s \\ y_0(x), & 1 - T + s \leq x \leq 1 \\ 0, & x > 1 \end{rcases}, & 1 + s > T. \end{cases}$$

$$\begin{aligned} &(8.7.2.10a) \\ &(8.7.2.10b) \\ &(8.7.2.10c) \\ &(8.7.2.10d) \\ &(8.7.2.10e) \end{aligned}$$

8.7.3 Abstract Model

Dynamics   The mixed problem (8.7.2.1) can be written abstractly as

$$y_t = Ay + Bu, \quad y(s) = y_0 \in Y \tag{8.7.3.1}$$

on the Hilbert space Y, with control space U where

(i)

$$U = \mathbb{R}^1, \quad Y = L_2(0, \infty) = L_2(\mathbb{R}_x^+); \tag{8.7.3.2}$$

(ii) A is the closed, densely defined operator $Y \supset \mathcal{D}(A) \to Y$, defined by

$$(Af)(x) = -\frac{df}{dx}, \quad \mathcal{D}(A) = \{f \in H^1(0, \infty) : f(0) = 0\}, \tag{8.7.3.3}$$

and is the generator of a s.c. semigroup e^{At} on Y, given by the right translation with speed 1:

$$\left\{e^{A(t-s)}y_0\right\}(x) = \begin{cases} y_0(x - (t - s)), & s \le t \le x + s; \tag{8.7.3.4a} \\ 0, & t > x + s; \tag{8.7.3.4b} \end{cases}$$

(iii) the operator B is given by

$$\begin{cases} Bu = u\delta(\cdot), & \delta(\cdot) = \text{Dirac measure at } x = 0, \tag{8.7.3.5a} \\ B^*\phi = \phi(0), & \phi \in H^1(0, \infty). \tag{8.7.3.5b} \end{cases}$$

In fact, multiplying Eqn. (8.7.2.1a) by a test function $\phi \in H^1(0, \infty)$ (thus, with $\phi(\infty) = 0$) and integrating by parts, one obtains

$$(y_t, \phi)_{L_2(0,\infty)} = (-y_x, \phi)_{L_2(0,\infty)} = \int_0^\infty -y_x \phi \, dx \tag{8.7.3.6}$$

$$= u\phi(0) + (y, \phi_x)_{L_2(0,\infty)} \tag{8.7.3.7}$$

$$= u\phi(0) + (Ay, \phi)_{L_2(0,\infty)}, \tag{8.7.3.8}$$

after using in (8.7.3.7) the BC (8.7.2.1c), and after using the zero BC: $y|_{x=0} = 0$ for $y \in \mathcal{D}(A)$ (see (8.7.3.3)) in going from (8.7.3.7) to (8.7.3.8). Thus, comparing (8.7.3.8) with (8.7.3.1), we must have

$$(Bu, \phi)_{L_2(0,\infty)} = (u, B^*\phi)_U = u\phi(0), \tag{8.7.3.9}$$

and (8.7.3.5) follows.

The Observation Operator R   With reference to (8.7.2.2) the cost J can be rewritten as in (8.1.10) with

$$Z = \mathbb{R}^1, \quad Rf = f(1) : Y \supset \mathcal{D}(R) \to Z, \tag{8.7.3.10a}$$

$$\mathcal{D}(R) = \{f \in L_2(0, \infty), \ f \text{ continuous around } x = 1\}, \tag{8.7.3.10b}$$

so that R is an unbounded operator on $Y = L_2(0, \infty)$. For the solution $y(t, x; s)$ of (8.7.2.1), we have, by (8.7.3.10) and (8.7.2.4),

$$
Ry(t, \cdot \, ; s) = y(t, 1; s) = \begin{cases} y_0(1 - (t - s)), & s \leq t \leq 1 + s; & (8.7.3.11a) \\ u(t - 1), & 1 + s < t \leq T. & (8.7.3.11b) \end{cases}
$$

Optimality   Because the optimal control u^0 is trivial by (8.7.2.5), the optimal dynamics and the free dynamics coincide. If $\Phi(t, s)$ denotes, as usual, the evolution operator, we have

$$
\Phi(t, s) = e^{A(t-s)};
$$
$$
R\Phi(t, s)y_0 = Re^{A(t-s)}y_0 = y^0(t, 1; s)
$$
$$
= \begin{cases} y_0(1 - (t - s)), & s \leq t \leq 1 + s; & (8.7.3.12a) \\ 0, & 1 + s < t \leq T. & (8.7.3.12b) \end{cases}
$$

Input-Solution Operator L_0 and Its Regularity   The solution of (8.7.3.1) for $s = 0$ can be written as in (8.1.4), where e^{At} is given by (8.7.3.4) and where, by (8.7.2.4),

$$
(L_0 u)(t) = \begin{cases} 0, & 0 \leq t \leq x; & (8.7.3.13a) \\ u(t - x), & t > x. & (8.7.3.13b) \end{cases}
$$

With $Y = L_2(0, \infty)$, we compute, by (8.7.3.13), the absolutely continuous map $t \to \|(L_0 u)(t)\|_Y$:

$$
\|(L_0 u)(t)\|_Y^2 = \int_0^\infty |(L_0 u)(t)|^2 \, dx = \int_0^t u^2(t - x) \, dx = \int_0^t u^2(\xi) \, d\xi \quad (8.7.3.14)
$$
$$
\leq \|u\|_{L_2(0,T)}^2. \quad (8.7.3.15)
$$

Equation (8.7.3.15) yields $(Lu)(t) \in L_\infty(0, T; Y)$, and this can be boosted to $(Lu)(t) \in C([0, T]; Y)$ by a standard approximation argument with a sequence $\{u_n\}$ smooth. Let first $u \in C[0, T]$, hence uniformly continuous on $[0, T]$. Then by (8.7.3.14) and (8.7.3.13a) with say, $0 \leq t_1 < t_2 \leq T$,

$$
\|(L_0 u)(t_1) - (L_0 u)(t_2)\|_Y^2 = \int_0^{t_1} |u(t_1 - x) - u(t_2 - x)|^2 \, dx
$$
$$
+ \int_{t_1}^{t_2} |u(t_2 - x)|^2 \, dx \to 0 \text{ as } [t_1 - t_2] \to 0,
$$

$$(8.7.3.16)$$

and thus $(Lu)(t) \in C([0, T]; Y)$ in this case. Next, given $u \in L_2(0, T)$, pick a sequence $u_n \in C[0, T]$ such that $u_n \to u$ in $L_2(0, T)$. Then, $(Lu_n)(t) \in C([0, T]; Y) \to Lu$ in $L_\infty(0, T; Y)$, by (8.7.3.15). Hence $(Lu)(t) \in C([0, T]; Y)$, as desired.

Thus, for any T:

$$L_0 : \text{continuous } L_2(0, T; U) \to C([0, T]; Y). \qquad (8.7.3.17)$$

The Adjoint Operators A^* and R^*   For $f \in \mathcal{D}(A)$ in (8.7.3.3) and $\phi \in H^1(\mathbb{R}_x^+)$, we have, with $Y = L_2(\mathbb{R}_x^+)$,

$$(Af, \phi)_Y = (f, A^*\phi)_Y, \quad A^*\phi = \frac{d\phi}{dx}, \quad \mathcal{D}(A^*) = H^1(\mathbb{R}_x^+), \quad (8.7.3.18)$$

whose corresponding s.c. semigroup e^{A^*t} on Y is

$$\{e^{A^*t}y_0\}(x) = \phi_0(x + t) = \phi(t, x), \qquad (8.7.3.19)$$

$$\phi_t(t, x) = \phi_x, \quad \phi(0, x) = \phi_0(x). \qquad (8.7.3.20)$$

From (8.7.3.10) we have

$$R^* = \delta_{x=1} = \delta(\cdot - 1), \qquad (8.7.3.21)$$

the Dirac measure concentrated at $x = 1$.

8.7.4 Analysis of the Corresponding, Integral or Differential, Riccati Equation

The Riccati Operator $P_T(x)$, $0 \le s \le T$ We have already noted that the non-negative, bounded, self-adjoint operator $P_T(s) = P_T^*(s) \ge 0$ in $\mathcal{L}(Y)$, defining the value (8.7.2.9) of the optimal control problem, is given explicitly by (8.7.2.10). Consistently with (8.3.3.2), we shall verify that $P_T(s)$ is given by

$$P_T(s)y_0 = \int_s^T e^{A^*(t-s)} R^* R\Phi(t, s) y_0 \, dt \qquad (8.7.4.1)$$

$$\text{(by (8.7.3.12))} \qquad = \int_s^T e^{A^*(t-s)} R^* R e^{A(t-s)} y_0 \, dt, \qquad (8.7.4.2)$$

so that the value of the optimal control problem is computed as

$$(P_T(s)y_0, y_0)_Y = \int_s^T \left\| R e^{A(t-s)} y_0 \right\|_Z^2 \, dt$$

$$\text{(by (8.7.3.12))} \qquad = \int_s^{\min\{T, 1+s\}} y_0^2 (1 - (t - s)) \, dt, \qquad (8.7.4.3)$$

which reproduces (8.7.2.9), hence (8.7.2.10) for $P_T(s)$.

The Operator $B^* P_T(t) e^{A(t-s)}$ Comparing (8.7.4.2) with the IRE (8.3.5.1) suggests that the quadratic term in P_T should vanish. We now verify this. That is,

$$\int_s^T \left(B^* P_T(t) e^{A(t-s)} y_0, B^* P_T(t) e^{A(t-s)} y_1 \right)_U \, dt \equiv 0, \quad 0 \le s \le T, \ \forall \, y_0, y_1 \in Y.$$

$$(8.7.4.4)$$

After this, we see that the IRE is verified and, indeed, it collapses in this case to Eqn. (8.7.4.2). By (8.7.2.10) we compute for $s \leq t \leq T$:

$$\left\{ P_T(t) e^{A(t-s)} y_0 \right\}(x)$$

$$= \begin{cases} \begin{cases} \left\{ e^{A(t-s)} y_0 \right\}(x), & 0 \leq x \leq 1 \\ 0, & x > 1 \end{cases}, & 1+t \leq T; & \begin{array}{l} (8.7.4.5a) \\ (8.7.4.5b) \end{array} \\ \begin{cases} 0, & 0 \leq x < 1-T+t \\ \left\{ e^{A(t-s)} y_0 \right\}(x), & 1-T+t \leq x \leq 1 \\ 0, & x > 1 \end{cases}, & 1+t > T. & \begin{array}{l} (8.7.4.5c) \\ (8.7.4.5d) \\ (8.7.4.5e) \end{array} \end{cases}$$

Since, by (8.7.3.5b), B^* is the evaluation at $x = 0$, we obtain from (8.7.4.5), via (8.7.3.4), for $s \leq t \leq T$,

$$B^* P_T(t) e^{A(T-s)} y_0 = \left\{ P_T(t) e^{A(t-s)} y_0 \right\}(0) \qquad (8.7.4.6)$$

$$\equiv \begin{cases} y_0(0), & s = t, \ 1+t \leq T; & (8.7.4.7a) \\ 0, & t > s, \ 1+t \leq T; & (8.7.4.7b) \\ 0, & 1+t > T. & (8.7.4.7c) \end{cases}$$

Notice that (8.7.4.7a) agrees with (8.7.2.10a) evaluated at $x = 0$. Thus (8.7.4.7) proves (8.7.4.4), as desired.

The Integral Riccati Equation   Because of (8.7.4.4), the IRE is precisely Eqn. (8.7.4.2), in the present case.

The Differential Riccati Equation   In the present case, the DRE becomes

$$\begin{cases} \dfrac{d}{ds}(P_T(s)y_0, y_1)_Y = -(Ry_0, Ry_1)_Z - (P_T(s)Ay_0, y_1)_Y \\ \qquad\qquad\qquad - (P_T(s)y_0, Ay_1)_Y, & (8.7.4.8) \\ P_T(T) = 0 \quad \forall \ y_0, y_1 \in \mathcal{D}(A), \end{cases}$$

that is, it collapses to a linear equation in $P_T(\cdot)$. Indeed, to show (8.7.4.8), we return to the IRE (8.7.4.2), where we take the Y-inner product with y_1, and we differentiate the resulting expression in s, with $y_0, y_1 \in \mathcal{D}(A)$. We obtain then

$$\frac{d}{ds}(P_T(s)y_0, y_1)_Y = -(Ry_0, Ry_1) - \int_s^T \left(R e^{A(t-s)} A y_0, R e^{A(t-s)} y_1 \right)_Y dt$$

$$- \int_s^T \left(R e^{A(t-s)} y_0, R e^{A(t-s)} A y_1 \right)_Y dt, \qquad (8.7.4.9)$$

which yields (8.7.4.8) via (8.7.4.2).

8.7.5 Further Abstract Properties

In this section we test various assumptions/conditions of Sections 8.2 and 8.3 to the present case. We shall see that some hold true, whereas others fail in the present degree of generality.

(i) Verification of Assumption (8.1.2): $R(\lambda, A)B \in \mathcal{L}(U;Y)$, $\lambda \in \rho(A)$ The resolvent operator $R(\lambda, A) = (\lambda I - A)^{-1}$ of the operator A in (8.7.3.3) is given by

$$[R(\lambda, A)f](x) = \int_0^x e^{-\lambda(x-\xi)} f(\xi)\, d\xi, \quad \lambda > 0, \ f \in Y. \qquad (8.7.5.1)$$

Recalling (8.7.3.5a) for B we obtain for $u \in \mathbb{R}^1$:

$$[R(\lambda, A)Bu](x) = \int_0^x e^{-\lambda(x-\xi)} u\delta(\xi)\, d\xi$$

$$= ue^{-\lambda x} \in L_2(0, \infty) = Y, \quad \lambda > 0, \qquad (8.7.5.2)$$

so that $\{\lambda : \lambda > 0\} \in \rho(A)$ and for such points

$$R(\lambda, A)B \in \mathcal{L}(U;Y), \quad \lambda > 0. \qquad (8.7.5.3)$$

(ii) Verification of Assumption (h.1) = (8.2.1.10): Re^{At} : continuous $Y \to L_2(0, T; Z)$ From (8.7.3.12) we obtain for Re^{At}:

$$\int_0^\infty \|Re^{At}y_0\|_Z^2\, dt = \int_0^1 y_0^2(1-t)\, dt = \int_0^1 y_0^2(x)\, dx$$

$$\leq \|y_0\|_{L_2(\mathbb{R}_x^+)}^2, \qquad (8.7.5.4)$$

and assumption (h.1) = (8.2.1.10) is verified.

(iii) Verification of Assumption (h.3) = (8.2.1.12): RL_0 : continuous $L_2(0, T; U) \to L_2(0, T; Z)$ From (8.7.3.10) and (8.7.3.13) we obtain

$$(RL_0u)(t) = \begin{cases} 0, & t \leq 1; & (8.7.5.5a) \\ u(t-1), & t > 1. & (8.7.5.5b) \end{cases}$$

Thus, by (8.7.5.5),

$$\|RL_0u\|_{L_2(0,\infty;Z)}^2 = \int_0^\infty |(RL_0u)(t)|^2\, dt = \int_1^\infty u^2(t-1)\, dt = \|u\|_{L_2(0,\infty;U)}^2, \qquad (8.7.5.6)$$

and assumption (h.3) = (8.2.1.12) is verified.

(iv) Violation of Assumption (H.3) = (8.2.1.16)*   From (8.7.5.5) it follows that, for $T > 1$,

$$RL_0 \text{ is not continuous } L_2(0, T; U) \to C([0, T]; Z),$$

and so assumption $(H.3^*) = (8.2.1.16)$ is violated. We present more on this in point (vi) below.

(v) Violation of Assumption (H.1) = (8.2.1.1): $Re^{At} B$ *: continuous $U \to L_1(0, T; Z)$*   From (8.7.3.12) and (8.7.3.5a) we have, for $u \in \mathbb{R}^1$,

$$Re^{At} Bu = \begin{cases} \delta(1-t)u, & t \le 1; & (8.7.5.7a) \\ 0, & t > 1, & (8.7.5.7b) \end{cases}$$

so that, for $T > 1$, $Re^{At} Bu$ is a distribution and cannot belong to $L_1(0, T; Z)$.

(vi) Violation of Assumption (H.3) = (8.2.1.3): $B^* e^{A^* t} R^*$ *: continuous $Z \to L_2(0, T; U)$*   From (8.7.3.19), (8.7.3.21), and (8.7.3.5b), we have, for $z \in Z = \mathbb{R}^1$,

$$B^* e^{A^* t} R^* z = \{e^{A^* t} R^* z\}_{x=0} = [\delta_1(x+t)z]_{x=0} = z\delta_1(t). \quad (8.7.5.8)$$

Thus, for $T > 1$, $B^* e^{A^* t} R^* z$ is a distribution and cannot belong to any $L_p(0, T; U)$, $1 \le p \le \infty$.

(vii) Violation of Property (8.4.1.13) of Lemma 8.4.1.3   By (8.7.3.12), (8.7.3.19), and (8.7.3.21) we obtain

$$\left\{ \int_t^T e^{A^*(\tau-t)} R^* Re^{A(\tau-t)} y_0 \, d\tau \right\}(x)$$

$$= \begin{cases} \displaystyle\int_t^{t+1} \delta(x+\tau-t-1)y_0(1-(\tau-t)) \, d\tau, & \text{if } t \le T-1; & (8.7.5.9a) \\ \displaystyle\int_t^T \delta(x+\tau-t-1)y_0(1-(\tau-t)) \, d\tau, & \text{if } t > T-1. & (8.7.5.9b) \end{cases}$$

Thus, recalling (8.7.3.5b), we obtain, by (8.7.5.9) and (8.4.1.8),

$$F_2(t)y_0 = \{L_t^* R^* Re^{A(\cdot-t)} y_0\}(t) = \int_t^T B^* e^{A^*(\tau-t)} R^* Re^{A(\tau-t)} y_0 \, d\tau$$

$$= \left\{ \int_t^T e^{A^*(\tau-t)} R^* Re^{A(\tau-t)} y_0 \, d\tau \right\}_{x=0} \quad (8.7.5.10)$$

$$= \begin{cases} y_0(0), & t \le T-1; & (8.7.5.11a) \\ 0, & t > T-1, & (8.7.5.11b) \end{cases}$$

for y_0 smooth. Then, (8.7.5.11) says that property (8.4.1.13) of Lemma 8.4.1.3,

$$\sup_{0 \le t \le T} \| F_2(t) y_0 \|_U \le c_T \| y_0 \|_Y, \quad y_0 \in Y,$$

is false.

8A Interior and Boundary Regularity of Mixed Problems for Second-Order Hyperbolic Equations with Neumann-Type BC

Throughout this appendix we take dim $\Omega \ge 2$, and we deal with the following mixed problem for second-order hyperbolic equations of Neumann type:

$$\begin{cases} w_{tt} + A(x, \partial)w = f & \text{in } Q = (0, T] \times \Omega; & (8A.1a) \\ w(0, \cdot) = w_0, \ w_t(0, \cdot) = w_1 & \text{in } \Omega; & (8A.1b) \\ \dfrac{\partial w}{\partial v_A} = u & \text{in } \Sigma = (0, T] \times \Omega, & (8A.1c) \end{cases}$$

where $A(x, \partial)$ denotes a second-order elliptic operator on Ω:

$$\begin{cases} A(x, \partial) = -\displaystyle\sum_{i,j}^{n} \frac{\partial}{\partial x_i} \left(a_{ij}(x) \frac{\partial}{\partial x_j} \right); & (8A.2a) \\ \displaystyle\sum_{i,j}^{n} a_{ij}(x)\xi_i\xi_j \ge c \sum_{i}^{n} \xi_i^2, \quad \text{constant } c > 0, & (8A.2b) \end{cases}$$

with suitably smooth coefficients $a_{ij}(x) = a_{ji}(x)$; moreover, $\partial/\partial v_A$ denotes the corresponding co-normal derivative (= normal derivative of $A(x, \partial) = -\Delta$). The following sharp regularity results for problem (A.1) are taken from the recent papers by Lasiecka and Triggiani [1990; 1991(b)], the main results of which were announced in Lasiecka [1985], Lasiecka and Triggiani [1990], and Triggiani [1987]. First, we introduce the parameters α and β, which, throughout this section, will take on only the following values for the following specified cases, where $\epsilon > 0$ arbitrary:

$$\begin{cases} \left. \begin{array}{l} \alpha = \frac{3}{5} - \epsilon \\ \beta = \frac{3}{5} \end{array} \right\} : \text{for a general smooth, bounded domain } \Omega \\ \qquad\qquad\qquad \text{and a general operator } A(x, \partial) \text{ as in (A.2)}; \quad (8A.3a) \\ \alpha = \beta = \frac{2}{3} : \text{for a sphere } \Omega \text{ and the Laplacian } -A(x, \partial) = \Delta, \\ \qquad\qquad \text{or else in the case where the coefficients } a_{ij} \text{of } A(x, \partial) \\ \qquad\qquad \text{do not depend on the normal variable or do not} \\ \qquad\qquad \text{depend on the tangential variable, near } \Gamma; \quad\quad (8A.3b) \\ \alpha = \beta = \frac{3}{4} : \text{for a parallelepiped } \Omega \text{ and the Laplacian} \\ \qquad\qquad -A(x, \partial) = \Delta. \quad\quad\quad\quad\quad\quad\quad\quad\quad\quad (8A.3c) \end{cases}$$

Throughout the results listed below we always take dim $\Omega \ge 2$, unless otherwise noted; for dim $\Omega = 1$, we refer to Remark 8A.4 below.

Some Basic Regularity Results

Theorem 8A.1   *([Lasiecka, Triggiani, 1990; 1991(b), Theorem A]) With reference to (.8A.1), assume*

$$w_0 = w_1 = 0; \quad f \equiv 0; \quad u \in L_2(\Sigma). \tag{8A.4}$$

Then, the unique solution of (8A.1) satisfies the interior regularity

$$\begin{cases} w \in C([0, T]; H^\alpha(\Omega)), & (8A.5) \\ w_t \in C([0, T]; H^{\alpha-1}(\Omega)) & (8A.6) \end{cases}$$

and the boundary regularity

$$w|_\Sigma \in H^{2\alpha-1}(\Sigma). \tag{8A.7}$$

A fortiori, the map $u \to w|_\Sigma : L_2(\Sigma) \to L_2(\Sigma)$ is compact, since $\alpha > 1/2$.

Theorem 8A.2   *([Lasiecka, Triggiani, 1991(b), Theorem A]) With reference to (8A.1), assume:*

(a) (interior regularity) $w_0 = w_1 = 0$; $f \equiv 0$; moreover,

$$u \in H^1(0, T; L_2(\Gamma)) \cap C\big([0, T]; H^{\alpha-\frac{1}{2}}(\Gamma)\big), \quad \text{and} \quad u(0) = 0 \tag{8A.8}$$

[a result that is a fortiori true, if we assume

$$u \in H^{2\alpha-1,1}(\Sigma) = L_2(0, T; H^{2\alpha-1}(\Gamma)) \cap H^1(0, T; L_2(\Gamma)), \quad u(0) = 0 \tag{8A.9}$$

by virtue of [Lions, Magenes, 1972, Theorem 3.1, $m = 1$, $j = 0$, p. 19]].
Then the unique solution of (8A.1) satisfies

$$\begin{cases} w \in C([0, T]; H^{\alpha+1}(\Omega)), & (8A.10) \\ w_t \in C([0, T]; H^\alpha(\Omega)). & (8A.11) \end{cases}$$

(b) (boundary regularity) Assume $w_0 = w_1 = 0$, $f \equiv 0$, and

$$u \in H^1(\Sigma) \quad \text{and} \quad u(0) = 0. \tag{8A.12}$$

Then the unique solution of (8A.1) satisfies

$$w|_\Sigma \in H^{2\alpha}(\Sigma). \tag{8A.13}$$

Theorem 8A.3

(a) With reference to (8A.1), assume

$$w_0 \in H^1(\Omega); \quad w_1 \in L_2(\Omega); \quad f \in L_2(Q); \quad u \equiv 0. \tag{8A.14}$$

Then, the unique solution of (8A.1) satisfies the interior regularity

$$\begin{cases} w \in C([0, T]; H^1(\Omega)), & (8A.15) \\ w_t \in C([0, T]; L_2(\Omega)) & (8A.16) \end{cases}$$

and the boundary regularity ([Lasiecka, Triggiani, 1990; 1983, Theorems B and C])

$$w|_\Sigma \in H^\beta(\Sigma). \tag{8A.17}$$

(b) With A the $L_2(\Omega)$-realization of $-A(x, \partial)$ with homogeneous Neumann boundary conditions

$$Ah = A(x, \partial)h; \quad \mathcal{D}(A) = \left\{ h \in H^2(\Omega) : \frac{\partial h}{\partial \nu_A} = 0 \right\}, \tag{8A.18}$$

assume

$$w_0 \in \mathcal{D}(A); \quad w_1 \in H^1(\Omega); \quad f \in H^1(Q); \quad u \equiv 0. \tag{8A.19}$$

Then, the unique solution of (8A.1) satisfies the interior regularity

$$\begin{cases} w \in C([0, T]; \mathcal{D}(A)), & (8A.20) \\ w_t \in C([0, T]; H^1(\Omega)) & (8A.21) \end{cases}$$

and the boundary regularity ([Lasiecka, Triggiani, 1991(b), Theorems B and C])

$$w|_\Sigma \in H^{\beta+1}(\Sigma). \tag{8A.22}$$

Remark 8A.1   In Theorem 8A.1 the boundary regularity (8A.7) does not follow from the interior regularity (8A.5), (8A.6) – it is an independent regularity result. The same holds true in Theorem 8A.2, under the assumption (8A.12), which is stronger than assumption (8A.8) since $\alpha - 1/2 < 1/2$ by (8A.3) [indeed, $u \in H^1(\Sigma)$ a fortiori implies $u \in C([0, T]; H^{1/2}(\Gamma))$ by [Lions, Magenes, 1972, Theorems 3.1, p. 19]. The boundary regularity (8A.13) is an independent result that does not follow from the interior regularity (8A.10), (8A.11). Finally, similar comments apply to Theorem 8A.3.

Remark 8A.2   In Theorem 8A.3, the interior regularity (8A.15), (8A.16) [respectively (8A.20), (8A.21)] is, in fact, a standard result under the weaker assumption that $f \in L_1(0, T; L_2(\Omega))$ [respectively $f \in L_1(0, T; H^1(\Omega))$]. It is at the level of the boundary regularity (8A.17) [respectively (8A.22)] that the full strength of the assumption (8A.14) [respectively (8A.19)] on f is used; see Lasiecka and Triggiani [1990].

Remark 8A.3   (Classical results) Classical regularity results [Lions, Magenes, 1972, Vol. II, p. 120] state, in the case of assumptions (8A.4) in Theorem 8A.1, that, in fact,

$$\{w, w_t\} \in L_2\left(0, T; H^{\frac{1}{2}}(\Omega) \times H^{-\frac{1}{2}}(\Omega)\right).$$

This result does not allow one to conclude that $w|_\Sigma \in L_2(\Sigma)$, in contrast with (8A.7). Thus, Theorem 8A.1 represents an improvement of $\alpha - 1/2$ in the space regularity of the solution, where $\alpha - 1/2$ is at least equal to $1/10 - \epsilon$ in the general case (8A.3a). Similarly, the standard interior regularity (8A.14) $\Rightarrow$ (8A.15) in Theorem 8A.3(a) would only imply by trace theory the boundary regularity $w|_\Sigma \in C([0, T]; H^{1/2}(\Gamma))$, whereby the boundary regularity (8A.17) represents then an improvement of $\beta - 1/2$ (equal at least to $1/10$ in the general case (8A.3a) in the space regularity.

Remark 8A.4   (One-dimensional case) Let now dim $\Omega = 1$ for problem (8A.1). It is a standard result, stated as Theorem 9.8.4.1 in Chapter 9 for $A(x, \partial) = -\Delta$, that then

$$\{w_0, w_1\} \in H^1(\Omega) \times L_2(\Omega); \quad u \in L_2(0, T), \ f \equiv 0 \qquad (8A.23a)$$

implies

$$\begin{cases} \{w, w_t\} \in C([0, T] : H^1(\Omega) \times L_2(\Omega)), & (8A.23b) \\ w|_\Sigma \in H^1[0, T]. & (8A.23c) \end{cases}$$

An elementary proof in the case of the wave equation $A(x, \partial) = -\Delta$ is given in Chapter 9, Section 9.9.4. A second proof, which works in the general case $A(x, \partial)$, is given in Chapter 10, Remark 10.5.10.1, as a specialization of an argument that provides regularity of second-order hyperbolic equations with Dirichlet BC [and which fails in the case of Neumann BC, except for the one-dimensional case]. The above regularity result (8A.23) is false for dim $\Omega \geq 2$; see Lasiecka [1985], Lasiecka and Triggiani [1990], or Lasiecka and Triggiani [1989(a)] for a counterexample.

Remark 8A.5   (Sharpness of results) One can show by considering the canonical Laplacian $-A(x, \partial) = \Delta$ on the half-space, and hence on a parallelepiped, that in this case, if $w_0 = w_1 = 0$, $f \equiv 0$, then [Lasiecka, Triggiani, 1989(a)]

$$u \in L_2(\Sigma) \Rightarrow w \in H^{\frac{3}{4}}(Q), \quad \text{yet } w \notin H^{\frac{3}{4}+\epsilon}(Q), \ \forall \epsilon > 0. \qquad (8A.24)$$

Remark 8A.6   It was shown in Symes [1983] by means of geometric optics techniques, and reproved in Lasiecka and Triggiani [1989(a)] by a direct Laplace–Fourier analysis, that in the canonical case of the Laplacian $-A(x, \partial) = \Delta$ on the half-space, with $u \equiv 0$, then

$$w_0 \in H^1(Q), \quad w_1 \in L_2(\Omega), \ f \in L_2(Q) \Rightarrow w|_\Sigma \in H^1(\Sigma), \qquad (8A.25)$$

provided, in addition, the data $\{w_0, w_1, f\}$ are compactly supported away from the boundary Γ.

Remark 8A.7   In the case of $A(x, \partial) = -\Delta$ on special geometries, where Ω is either a sphere or a parallelepiped, the above regularity results of Theorem 8A.1 with $\alpha = 2/3$, or $\alpha = 3/4 - \epsilon$, are proved in Lasiecka and Triggiani [1983] by direct eigenfunction expansion followed by Fourier transform. Later, Lasiecka and Triggiani [1989(a)] removed the ϵ and thus obtained $\alpha = 3/4$ by the technique on a half-space analysis, hence on a parallelepiped, recalled in Remark 8A.6. A proof by eigenfunction expansion, which reobtains $\alpha = 3/4$ for a parallelepiped, was later given in Avdonin and Ivanov [1995].

Remark 8A.8   As noted in (8A.3b) in the special case where the coefficients a_{ij} either do not depend on the normal variable or else do not depend on the tangential variable near Γ, we have the following results [Lasiecka, 1985; Lasiecka, Triggiani, 1990]:

$$w_0 = w_1 = 0; \quad f = 0; \quad u \in L_2(\Sigma) \Rightarrow \begin{cases} \{w, w_t\} \in C([0, T]; H^{\frac{2}{3}}(\Omega) \times H^{-\frac{1}{3}}(\Omega)), \\ w|_\Sigma \in H^{\frac{1}{3}}(\Sigma); \end{cases}$$

(8A.26)

$$w_0 = w_1 = 0; \quad f \in L_2(Q); \quad u = 0 \Rightarrow w|_\Sigma \in H^{\frac{2}{3}}(\Sigma). \tag{8A.27}$$

Hence we obtain the results of Theorem 8A.1 and, respectively, of Theorem 8A.3 with $\alpha = \beta = 2/3$, that is, with the same regularity obtained in the case of $A(x, \partial) = -\Delta$ on a sphere in Lasiecka and Triggiani [1983].

In these special cases, the analysis of commutators permits one to obtain a higher regularity.

Remark 8A.9 In the pseudo-differential proofs in Lasiecka and Triggiani [1990] of the two preliminary results, Theorem 8A.1 with $u \in L_2(\Sigma)$ and Theorem 8A.3 with $f \in L_2(Q)$, it is crucially used that the coefficients a_{ij} are time independent. The subsequent functional analytic approach of Lasiecka and Triggiani [1991(b)] also refers to the time-independent case.

Some Interpolation Results Used in Section 8.6 Of the several interpolation results given in Lasiecka and Triggiani [1991(b)], we shall quote only one to be invoked in Section 8.6. Interpolating between the implication (8A.9) $\Rightarrow$ (8A.10), (8A.11) and the implication (8A.4) $\Rightarrow$ (8A.5), (8A.6) of Theorem 8A.1 with $0 < \theta \leq 1/2$ so that the CR $u(0) = 0$ is irrelevant, we obtain part (a) of the following corollary. Its part (b) is instead obtained by interpolating between (8A.4) $\Rightarrow$ (8A.7) of Theorem 8A.1 and (8A.12) $\Rightarrow$ (8A.13) of Theorem 8A.2.

Corollary 8A.4 *[Lasiecka, Triggiani, 1991(b), Remark 3.6 and Corollary 4.3]*

(a) With reference to (8A.1), assume

$$w_0 = w_1 = 0; \quad f = 0; \quad u \in H^{(2\alpha-1)\theta, \theta}(\Sigma), \quad 0 \leq \theta \leq \frac{1}{2}. \tag{8A.28}$$

Then, the unique solution of (8A.1) satisfies the interior regularity

$$\{w, w_t\} \in C([0, T]; H^{\alpha+\theta}(\Omega) \times H^{\alpha+\theta-1}(\Omega)). \tag{8A.29}$$

In particular, for $\theta = 1 - \alpha < 1/2$ (by (8A.3)), we have a fortiori with $w_0 = w_1 = 0$, $f = 0$:

$$u \in H^{1-\alpha}(\Sigma) \Rightarrow \{w, w_t\} \in C([0, T]; H^1(\Omega) \times L_2(\Omega)). \tag{8A.30}$$

(b) If $w_0 = w_1 = 0$ and $f = 0$, then

$$u \in H^\theta(\Sigma) \Rightarrow w|_\Sigma \in H^{2\alpha-1+\theta}(\Sigma), \quad 0 \leq \theta \leq \frac{1}{2}. \tag{8A.31}$$

Actually, a stronger result than (8A.30) is available [see Lasiecka, Triggiani, 1991(b), Eqn. (8.1.22) or Corollary 3.4(i)]:

$$u \in H^{\theta_1}(0, T; L_2(\Gamma)) \Rightarrow \begin{cases} w \in C([0, T]; H^1(\Omega)); & (8A.32) \\ w_t \in C([0, T]; H^{\alpha-1+\theta_1}(\Omega)), & (8A.33) \end{cases}$$

for (8A.1) with $w_0 = w_1 = 0$ and $f \equiv 0$, where

$$\theta_1 = \frac{1-\alpha}{\frac{3}{2}-\alpha} \quad \left(= \frac{4}{9} + \epsilon', \ \forall \epsilon' > 0, \ \text{in the general case}\right);$$

$$\alpha - 1 + \theta_1 \approx \frac{2}{45}, \ \text{in the general case.}$$

This result with θ_1-regularity in *time* ($\theta_1 < 1/2$) for u should be contrasted with the following regularity result from Myatake [1973], with $1/2$-regularity in *space* for u, quoted in Section 8.6.

Theorem 8A.5   *[Myatake, 1973] With reference to (8A.1), assume*

$$w_0 = w_1 = 0; \quad f = 0; \ u \in L_2\left(0, T; H^{\frac{1}{2}}(\Gamma)\right). \tag{8A.34}$$

Then the unique solution of (8A.1) satisfies

$$\{w, w_t\} \in C([0, T]; H^1(\Omega) \times L_2(\Omega)). \tag{8A.35}$$

Thus the regularity in space of $1/2$ for u in Theorem 8A.5 is traded with the weaker regularity $\theta_1 < 1/2$ in time in (8A.32), (8A.33) for u. However, regularity in time and space are interchangeable, as the pseudo-differential proofs in Lasiecka and Triggiani [1990] show, in the "crucial cone" of the dual variables. Thus, the regularity in (8A.32), (8A.33) for $\{w, w_t\}$ holds true also with $u \in L_2(0, T; H^{\theta_1}(\Gamma))$.

Regularity Results with Data $\{w_0, w_1, f\}$ Less Regular than L_2

Theorem 8A.5 ([Lasiecka, Triggiani, 1991(b), Theorems E and F])   *With reference to (8A.1) with $u \equiv 0$, assume*

$$w_0 \in L_2(\Omega); \quad w_1 \in [H^1(\Omega)]'; \quad f \in [H^1(Q)]'. \tag{8A.36}$$

Then, the unique solution of (8A.1) satisfies (where $1 - \alpha < 1/2$)

$$w|_\Sigma \in H^{\alpha-1}(\Sigma) = [H^{1-\alpha}(\Sigma)]'. \tag{8A.37}$$

Corollary 8A.6   *With reference to (8A.1) with $u \equiv 0$, we have, for all $0 \le \theta \le 1$,*

$$\{w_0, w_1, f\} \in H^\theta(\Omega) \times [H^{1-\theta}(\Omega)]' \times [H^{1-\theta}(Q)]' \to H^{\alpha-1+\theta+(\beta-\alpha)\theta}(\Sigma). \tag{8A.38}$$

The proof of Corollary 8A.6 follows by interpolating Theorem 8A.3, Eqn. (8A.17), and Theorem 8A.5.

Notes on Chapter 8

Riccati Equations

This chapter follows closely the work of Lasiecka and Triggiani [1991(a)]: the variational approach of Sections 8.3.1 through 8.3.5 under (H.1), (H.2), and (H.3), as well as the more regular case of Section 8.5 under (H.4) and (H.5); the direct approach analysis of Sections 8.4.1 through 8.4.3 (which in Lasiecka and Triggiani [1991(a)] provided uniqueness); and finally, the application in Section 8.6, with the relevant regularity results taken from Lasiecka and Triggiani [1990; 1991(b)]. New contributions of the present chapter are Section 8.3.6 on uniqueness and Sections 8.4.4 and 8.4.5 on the direct approach (Section 8.4.4 being a generalization of the corresponding direct analysis of Da Prato et al. [1986]). The parallel treatment under (h.1) and (h.2) was, in part, noted in Da Prato et al. [1986].

The abstract operator formulation for second-order equations of this section was introduced by the authors in Triggiani [1977] and Lasiecka and Triggiani [1981] and used successfully since in a series of papers dealing with a variety of different topics, including regularity (see, e.g., Lasiecka and Triggiani [1981; 1983] for the Dirichlet case and Lasiecka and Triggiani [1981; 1991(b); 1994] for the Neumann case). Applicability of these abstract models includes also, mutatis mutandis, the case of plate like equations (Petrovski type), where the differential operator is of order four and two boundary conditions are available. The example of Section 7 is after Triggiani [1997].

Regularity Theory for Second-Order Mixed Hyperbolic Equations with Neumann Boundary Datum

The regularity properties of the map from the boundary datum to the solution in the interior is maximal in the case of second-order hyperbolic equations which satisfy the so-called Lopatinski conditions, e.g., Dirichlet boundary problems. This will be seen in [Chapter 10, Section 10.5.3]. This situation is quite different in the case where the Lopatinski condition is not satisfied, e.g., Neumann boundary problems. Here, the above regularity from the boundary to the interior is maximal only in the *one-dimensional case*, as will be seen in [Chapter 9, Section 9.8.4]: in this special case, an $L_2(\Sigma)$-boundary datum produces $H^1(Q)$-solutions in the interior with, actually, continuity in time; and $H^1(\Sigma)$-Dirichlet trace on the boundary. However, when dim $\Omega \geq 2$, counterexamples given in [Lasiecka, Triggiani, 1990, p. 294] and Lasiecka, Triggiani [1989(a)] show that even $H^{3/4+\epsilon}(Q)$-regularity is not achieved $\forall \epsilon > 0$. On the other hand, it is a classical result that $H^{1/2}(Q)$ regularity is always achieved. The following is a chronological summary leading to the regularity results of Appendix A, and beyond.

(i) The classical interior regularity result is that of [Lions, Magenes, 1972, Vol. II, p. 120] recalled in Remark 6.1.1: $w_0 = w_1 = 0$, $f = 0$, $u \in L_2(\Sigma)$ implies $\{w, w_t\} \in L_2$

$(0, T; H^{1/2}(\Omega) \times H^{-1/2}(\Omega))$. This result does *not* allow one to conclude that $w|_\Gamma \in L_2(\Sigma)$ for the Dirichlet trace, however.

(ii) Subsequently, Myatake [1973] proved the following result when the coefficients of the operator $A(\cdot, \partial)$ in (8A.2) are allowed to depend also on time: $w_0 = w_1 = 0$, $f = 0$, $u \in L_2(0, T; H^{1/2}(\Omega))$ implies $\{w, w_t\} \in C([0, T]; H^1(\Omega) \times L_2(\Omega))$. This result shares with that in (i) above the same philosophy of an improvement of "1/2" in Sobolev (space) regularity from the boundary datum u to the interior regularity for w. However, now, the assumption on u is not symmetric in time and space. The work of Myatake employs a pseudo-differential, micro-local analysis approch, in the style of the Japanese school of Mizohata, as in Sakamoto's work [Sakamoto, 1970; 1982]. It begins by considering the general second-order hyperbolic equation in the half-space:

$$P(x, y; D_t, D_x, D_y) \equiv -aD_t^2 + \sum_{i,j=1}^{n-1} a_{ij} D_{y_i} D_{y_j} + 2 \sum_{j=1}^{n-1} a_{nj} D_{y_j} D_x + D_x^2, \quad (8N.1)$$

$D_t = -i\partial/\partial t$, $D_x = -i\partial/\partial x$ etc., $a_{ij} = a_{ji}$, $\min a(x, y) > 0$, etc. Myatake uses only *homogeneous* pseudo-differential symbols. A later paper by Melrose [1978] solves optimally the *exterior diffractive* (concave boundary) problem, by using the technique of Airy operators.

(iii) Later, Lasiecka and Triggiani obtained a sharp regularity theory (both interior and boundary) for problem (8A.1), which includes the results given in Lemma 8.6.1.1, 8.6.1.2, as well as in Appendix 8A. They comprise the L_2-boundary case [Lasiecka, Triggiani, 1990], and, building upon [Lasiecka, Triggiani, 1990], the cases of both a more regular, and a less regular, Neumann boundary datum [Lasiecka, Triggiani, 1991(b)]. Prior announcements were given in Lasiecka [1985], Lasiecka and Triggiani [1989(b)], and Triggiani [1987]. The approach in the key foundation paper [Lasiecka, Triggiani, 1990] with $u \in L_2(\Sigma)$ is a very technical pseudo-differential/micro-local analysis proof, which is performed in the half-space on the general second-order hyperbolic equation (8N.1) above. It involves Hormander's pseudo-differential *non-homogeneous* symbols $S_{\rho,\delta}^m$, $0 < \rho, \delta < 1$, which naturally arise in the discription of the commutators between the operator P in (8N.1) above and the localization operators and other relevant operators in "the critical hyperboloid-like regions" around the bisector where the "dual time variable" and the "dual tangential derivative variable" are comparable. This analysis led to an improvement of $\alpha - 1/2$ over the classical regularity of case (i) above (see Theorem 8A.1), where $\alpha = 2/3$ in many cases, and $\alpha = 3/5 - \epsilon$ in the worst case, see (8A.3). Thus, the minimal improvement $\alpha - 1/2$ of this analysis was $[1/10 - \epsilon]$. Moreover, this analysis also provided the trace regularity $w|_\Sigma \in H^\beta(\Sigma)$ of Theorem 8A.3, again with an improvement of $\beta - 1/2$ over the classical case of Remark 8.6.1.1 (ii), where $\beta = 2/3$ again in many cases, and $\beta = 3/5$ in the worse case, see (8.A.3). As emphasized at the end of Remark 8.6.1.1, what is critical for the Riccati theory of the present chapter is to have $\alpha, \beta > 1/2$ regardless of the particular values of α and β. Thus, the classical theory is inadequate for this purpose.

(iv) Very recently, and at any rate after the writing of this Chapter 8 had been completed, Tataru [1974] managed to refine the sharp regularity results of Lasiecka and Triggiani in two directions: (a) by obtaining $\alpha = \beta = 2/3$ in the general (non-flat) case; and (b) by obtaining trace results also on general noncharacteristic surfaces, not necessarily the surface of the boundary data.

Tataru's very technical proof in Tataru [1998] relies on the full arsenal of pseudo-differential/micro-local analysis techniques, as in the monumental work of Hörmander [1983–85], Beals-Fefferman classes of pseudo-differential non-homogeneous symbols of the type used in Lasiecka and Triggiani [1990], etc., and a local change of coordinate as in Melrose and Sjöstrand [1978], which, is essence, makes the cofficients $a_{nj} \equiv 0$ in (8N.1), and thus eliminates cross terms $D_x D_{y_j}$ from (8N.1). This way, the resulting analysis of the commutators in the critical aforementioned region where "time is comparable to tangential derivative" is refined over [Lasiecka, Triggiani, 1990], and Tataru obtains $\alpha = \beta = 2/3$ in the general case.

References and Bibliography

P. M. Anselone, *Collectively Compact Operators and Approximation Theory,* Prentice-Hall, 1971.

S. Avdonin and S. A. Ivanov, *Families of Exponentials: Methods of Moments in Controllability Problems for Distributed Parameter Systems*, Cambridge University Press, 1995.

A. V. Balskrishnan, *Applied Functional Analysis*, 2nd ed., Springer-Verlag, 1981.

A. Bensoussan, G. Da Prato, M. Delfour, and S. Mitter, *Representation and Control of Infinite Dimensional Systems*, Vol. II, Birkhäuser, 1993.

G. Da Prato, Quelques résultat d'existence, unicité et regularité pour un probléme de la théorie du contrôle, *J. Math. Pures Appl.* **52** (1973), 353–375.

G. Da Prato, I. Lasiecka, and R. Triggiani, A direct study of Riccati equations arising in boundary control problems for hyperbolic equations, *J. Diff. Eqn.* **64** (1986), 26–47.

L. Hörmander, *The analysis of Linear Partial Differential Operators*, I-IV, Springer-Verlag, 1983–85.

I. Lasiecka, Sharp regularity results for mixed hyperbolic problems of second-order, Lecture Notes in Mathematics 1223, *Differential Equations in Banach Spaces*, Springer-Verlag (1986), 160–175, Proceedings of a conference held at the University of Bologna, Italy, July 1985.

I. Lasiecka and R. Triggiani, A cosine operator approach to modeling $L_2(0, T; L_2(\Gamma))$-boundary input hyperbolic equations, *Appl. Math. Optim.* **7** (1981), 35–83.

I. Lasiecka and R. Triggiani, Regularity of hyperbolic equations under $L_2(0, T; L_2(\Gamma))$-Dirichlet boundary terms, *Appl. Math. Optim.* **10** (1983), 275–286.

I. Lasiecka and R. Triggiani, Trace regularity of the solutions of the wave equation with homogeneous Neumann boundary conditions and compactly supported data, *J. Math. Anal. Appl.* **141** (1989(a)), 49–71.

I. Lasiecka and R. Triggiani, Sharp regularity theory for second order hyperbolic equations of Neumann type, *Rendiconti Classe di Scienze Fisiche, Matematiche e Naturali, Atti della Accademia Nazionale dei Lincei*, Roma, **LXXXIII** (1989(b)).

I. Lasiecka and R. Triggiani, Sharp regularity theory for mixed second-order hyperbolic equations of Neumann type, Part I: The L_2-boundary case, *Ann. Matem. Pura Appl.* (iv), **CLVII** (1990), 285–367.

I. Lasiecka and R. Triggiani, *Differential and Algebraic Riccati Equations with Applications to Boundary/Point Control Problems: Continuous Theory and Approximation Theory*, Vol. 164, Springer-Verlag Lecture Note Series, 1991, 160 pp.

I. Lasiecka and R. Triggiani, Differential Riccati equations with unbounded coefficients: Applications to boundary control/boundary observation hyperbolic problem, *J. Nonlinear Anal.* **17** (7) (1991(a)), 655–682. (Also, invited paper for special volume dedicated to E. F. Gauss; G. M. Rassias, ed., World Scientific, 9.)

I. Lasiecka and R. Triggiani, Regularity theory of hyperbolic equations with nonhomogeneous Neumann boundary conditions, Part II: General boundary data, *J. Diff. Eqn.* **94** (1991(b)), 112–164.

I. Lasiecka and R. Triggiani, Recent advances in regularity theory of second-order hyperbolic mixed problems and applications, *Dynamics Reported, Expositions in Dynamical Systems*, new series, Volume 3, C. K. R. T. Jones; U. Kirchgraber, H. O. Walther, managing edis., Springer-Verlag, 1994.

E. B. Lee and L. Markus, *Foundation of Optimal Control Theory*, Wiley, 1968.

J. L. Lions and E. Magenes, *Nonhomogeneous Boundary Value Problems, I, II*, Springer-Verlag, 1972.

D. G. Luenberger, *Optimization by Vector Space Methods*, Wiley, 1969.

R. B. Melrose, Airy operators, *Comm. Part. Differ. Eqts.* 3:1 (1978) pp. 1–76.

R. B. Melrose and J. Sjöstrand, Singularities of boundary value problems, I, *Comm. Pure Appl. Math.* **XXXI** (1978), 593–617.

S. Myatake, Mixed problems for hyperbolic equations of second order, *J. Math. Kyoto Univ.* **130-3** (1973), 435–487.

R. Sakamoto, Mixed problems for hyperbolic equations, I, II, *J. Math. Kyoto Univ.* **10-2** (1970), 343–347; and **10-3** (1970), 403–417.

R. Sakamoto, *Hyperbolic Boundary Value Problems,* Cambridge University Press, 1982.

W. W. Symes, A trace theorem for solutions of the wave equation, *Math. Methods Appl. Sci.* **5** (1983), 35–93.

L. Tartar, Sur l'étude directe d'equations non linéaires intérrenant en théorie du contrôle optimal, *J. Funct. Anal.* (1974), 1–46.

D. Tataru, On the regularity of boundary traces for the more equation, Annali Scuola Normale di Pisa, Classe Scienze (4) **26** (1998) n. 1, 185–206.

R. Triggiani, A cosine operator approach to modeling $L_2(0, T; L_2(\Gamma))$-boundary input problems for hyperbolic systems, *Lecture Notes* CIS *Springer-Verlag* (1978), 380–390. Proceedings 8th IFIP Conference, University of Würzburg, W. Germany, July 1977.

R. Triggiani, Announcement of sharp regularity theory for second-order hyperbolic equations of Neumann type, *Springer-Verlag Lecture Notes*, Proceedings IFIP Conference on Optimal Control of Systems Governed by Partial Differential Equations, University of Santiago de Compostela, Spain, July 1987.

R. Triggiani, Interior and boundary regularity with point control, Part I: Wave and Euler–Bernoulli equations, *Diff. Int. Eqn.* **6** (1993), 111–129.

R. Triggiani, An optimal control problem with unbounded control operator and unbounded observation operator where the algebraic Riccati equation is satisfied as a Lyapunov equation, *Appl. Math. Lett.* **10** (1997), 95–102.

9

Optimal Quadratic Cost Problem over a Preassigned Finite Time Interval: The Case Where the Input → Solution Map Is Bounded. Differential and Integral Riccati Equations

The present chapter studies the same problem of Chapter 8 under a still different set of assumptions. Thus, the present chapter is a companion of, and a successor to, Chapter 8. The present setting specializes the framework of Chapter 8 in one important respect: The input → solution operator L_0 is now continuous (in the sense of (9.1.10)). Yet the present setting is generally not properly contained in that of Chapter 8 because of a technicality, in that R need not be boundedly invertible. Moreover, it contains a more general cost functional, which penalizes also the final state through an observation operator G.

Nevertheless, the present treatment in Sections 9.3.1 through 9.3.4 follows the presentation of the variational approach of Chapter 8 very closely, with appropriate, and at times serious, technical modifications and additions, such as the evolution property of Lemma 9.3.2.2 and Lemma 9.3.4.2, due to the presence of G. Eventually, Chapter 8 and Chapter 9 merge at the level of showing uniqueness of the Riccati operator in Section 9.3.5. Finally, Section 9.3.6 provides a second proof of Theorem 9.2.2 by the direct method, which now takes advantage of the continuity in time on the basic state space of the optimal solution. This property – due to assumption (A.1) – was unavailable in Chapter 8. The abstract setting of each of the two Chapters 8 and 9 is motivated by different types of applications to partial differential equations. This is clearly illustrated by the examples of application presented at the end of Chapter 9 [as well as in Chapter 10]. The second part of the present chapter deals with dual Riccati equations, when A is a group generator, a topic to be further continued in Chapter 11 for $T = \infty$.

9.1 Mathematical Setting and Formulation of the Problem

Dynamical Model   In this chapter, we return to the abstract differential equation

$$\dot{y} = Ay + Bu \text{ on, say, } [\mathcal{D}(A^*)]'; \quad y(0) = y_0 \in Y, \tag{9.1.1}$$

subject to assumptions partly more specialized than, yet not fully contained in, those of Chapter 8 (see Orientation, at the end of this section). Preliminary assumptions on (9.1.1) (to be maintained throughout this chapter) are the same as those in Chapter 8:

(i) A is the infinitesimal generator of a strongly continuous semigroup e^{At} on the Hilbert space Y, $t \geq 0$. Without loss of generality for the problem considered here, where the dynamics (9.1.1) is studied over a finite time interval $[0, T]$, $T < \infty$, we may assume that A^{-1} exists as a bounded operator on all Y, that is, that $A^{-1} \in \mathcal{L}(Y)$.

(ii) B is a (linear) continuous operator $U \to [\mathcal{D}(A^*)]'$; equivalently

$$A^{-1}B \in \mathcal{L}(U; Y). \tag{9.1.2}$$

In (9.1.1), A^* is the Y-adjoint of A and $[\mathcal{D}(A^*)]'$ is the Hilbert space dual to the space $\mathcal{D}(A^*) \subset Y$ with respect to the Y-topology, with norms

$$\|y\|_{\mathcal{D}(A^*)} = \|A^*y\|_Y, \quad \|y\|_{[\mathcal{D}(A^*)]'} = \|A^{-1}y\|_Y. \tag{9.1.3}$$

Via (9.1.2), we let $(B^*x, u)_U = (x, Bu)_Y$ for $u \in U$, $x \in \mathcal{D}(A^*)$, and then $B^* \in \mathcal{L}(\mathcal{D}(A^*); U)$. Further key assumptions for the well-posedness of (9.1.1) are given below.

Optimal Control Problem. Interval $[0, T]$ With the dynamics (9.1.1), we associate the following quadratic cost functional over a preassigned fixed time interval $[0, T]$, $0 < T < \infty$:

$$J(u, y) = \int_0^T \left[\|Ry(t)\|_Z^2 + \|u(t)\|_U^2 \right] dt + \|Gy(T)\|_{Z_f}^2, \tag{9.1.4}$$

where Z and Z_f are Hilbert (output) spaces and the corresponding optimal control problem is:

$$\text{Minimize } J(u, y) \text{ over all } u \in L_2(0, T; U), \tag{9.1.5}$$

where $y(t) = y(t; y_0)$ is the solution of (9.1.1) due to u. In (9.1.4), the observation operators R and G are taken throughout this chapter to satisfy the preliminary assumption of boundedness:

(A.0)

$$R \in \mathcal{L}(Y; Z), \quad G \in \mathcal{L}(Y; Z_f). \tag{9.1.6}$$

Basic Assumptions   The present chapter rests throughout on the following basic assumptions (in addition to (i), (ii), and (A.0) above):

(A.1): (*abstract trace regularity*) The (closable) operator $B^*e^{A^*t}$ can be extended as a map

$$B^*e^{A^*t} : \text{ continuous } Y \to L_2(0, T; U), \tag{9.1.7a}$$

that is,

$$\int_0^T \| B^* e^{A^* t} x \|_U^2 \, dt \leq c_T \| x \|_Y^2, \quad x \in Y. \tag{9.1.7b}$$

[Estimate (9.1.7b) is first checked for all $x \in \mathcal{D}(A^*)$ and then extended.]

(A.2): The map $R^* R e^{At} B$ can be extended as a map

$$R^* R e^{At} B : \text{ continuous } U \to L_1(0, T; Y), \tag{9.1.8a}$$

that is,

$$\int_0^T \| R^* R e^{At} B u \|_Y \, dt \leq c_T \| u \|_U, \quad u \in U. \tag{9.1.8b}$$

(A.3): In the notation of Chapter 8, Remark 8.1.1,

$$B^* e^{A^* t} G^* G \in \mathcal{B}([0, T]; \mathcal{L}(Y; U)), \tag{9.1.9a}$$

that is,

$$\begin{cases} B^* e^{A^* t} G^* G \in \mathcal{L}(Y; U); & (9.1.9b) \\ \sup_{0 \leq t \leq T} \| B^* e^{A^* t} G^* G \|_{\mathcal{L}(Y;U)} < \infty, & (9.1.9c) \end{cases}$$

so that

$$\sup_{0 \leq t \leq T} \| B^* e^{A^* t} G^* G x \|_U \leq c_T \| x \|_Y, \quad x \in Y. \tag{9.1.9d}$$

Remark 9.1.1 By duality on (A.1), (A.2), and (A.3), we obtain

(A.1*): The map L_0 below satisfies (see Chapter 7, Theorem 7.2.1)

$$(L_0 u)(t) = \int_0^t e^{A(t-\tau)} B u(\tau) \, d\tau = A \int_0^t e^{A(t-\tau)} A^{-1} B u(\tau) \, d\tau, \tag{9.1.10a}$$

$$: \text{ continuous } L_2(0, T; U) \to C([0, T]; Y), \tag{9.1.10b}$$

so that the $L_2(0, T; \cdot)$-dual operator L_0^* satisfies

$$(L_0^* f)(t) = \int_t^T B^* e^{A^*(\tau - t)} f(\tau) \, d\tau \tag{9.1.11a}$$

$$: \text{ continuous } L_1(0, T; Y) \to L_2(0, T; U). \tag{9.1.11b}$$

(A.2*): The map

$$v \to \int_0^T B^* e^{A^* t} R^* R v(t) \, dt$$

can be extended as a map

$$: \text{continuous } L_\infty(0, T; Y) \to U, \qquad (9.1.12\text{a})$$

that is,

$$\left\| \int_0^T B^* e^{A^* t} R^* R v(t) \, dt \right\|_U \le c_T \|v\|_{L_\infty(0,T;Y)}; \qquad (9.1.12\text{b})$$

equivalently [Chapter 7, Theorem 7.2.1 and Remark 7.2.1],

$$(L_0^* R^* R v)(t) = \int_t^T B^* e^{A^*(\tau - t)} R^* R v(\tau) \, d\tau \qquad (9.1.12\text{c})$$

$$: \text{ continuous } L_\infty(0, T; Y) \to C([0, T]; U). \quad (9.1.12\text{d})$$

(A.3*): In the notation of Chapter 8, Remark 8.1.1

$$G^* G e^{At} B \in \mathcal{B}([0, T]; \mathcal{L}(U; Y)), \qquad (9.1.13\text{a})$$

that is,

$$\begin{cases} G^* G e^{At} B \in \mathcal{L}(U; Y); & (9.1.13\text{b}) \\ \sup_{0 \le t \le T} \|G^* G e^{At} B\|_{\mathcal{L}(U;Y)} < \infty, & (9.1.13\text{c}) \end{cases}$$

so that

$$\sup_{0 \le t \le T} \|G^* G e^{At} B u\|_Y \le c_T \|u\|_U; \qquad (9.1.13\text{d})$$

moreover [Chapter 7, Theorem 7.2.1 with $p = \infty$],

$$(G^* G L_0 g)(t) = G^* G \int_0^t e^{A(t - \tau)} B g(\tau) \, d\tau \qquad (9.1.13\text{e})$$

$$: \text{ continuous } L_1(0, T; U) \to C([0, T]; Y). \quad (9.1.13\text{f})$$

For $y_0 \in Y$, the mild solution of problem (9.1.1) can be written via (9.1.10) as

$$y(t) = e^{At} y_0 + (L_0 u)(t) \in C([0, T]; Y). \qquad (9.1.14)$$

Optimal Control Problem. Interval $[s, T]$ As in preceding chapters, we need to consider also the case where the initial time $t = s \ge 0$ for the dynamics (9.1.1), with corresponding initial condition $y(s) = y_0$. The resulting optimal control problem over the time interval $[s, T]$, $s < T < \infty$, is then:

Minimize over all $u \in L_2(s, T; U)$ the cost functional

$$J_s(u, y) = \int_s^T \left[\|R y(t)\|_Z^2 + \|u(t)\|_U^2 \right] dt + \|G y(T)\|_{Z_f}^2, \qquad (9.1.15)$$

where now $y(t) = y(t, s; y_0)$ is the solution of Eqn. (9.1.1), with initial condition $y(s) = y_0$. This is therefore given by

$$y(t, s; y_0) = e^{A(t-s)}y_0 + (L_s u)(t) \in C([s, T]; Y), \qquad (9.1.16)$$

where consistently with (9.1.10a) we have set

$$(L_s u)(t) = \int_s^t e^{A(t-\tau)} Bu(\tau)\, d\tau \qquad (9.1.17a)$$

$$: \text{ continuous } L_2(s, T; U) \to C([s, T]; Y). \qquad (9.1.17b)$$

Equation (9.1.17b) follows by recalling assumption (A.1*) = (9.1.10), with L_2-adjoint L_s^* given by

$$(L_s^* f)(t) = \int_t^T B^* e^{A^*(\tau - t)} f(\tau)\, d\tau \qquad (9.1.18a)$$

$$: \text{ continuous } L_1(s, T; Y) \to L_2(s, T; U). \qquad (9.1.18b)$$

In addition, as in Chapter 1, we shall need one more operator,

$$L_{sT} u = (L_s u)(T) = \int_s^T e^{A(T-t)} Bu(t)\, dt \qquad (9.1.19a)$$

$$: \text{ continuous } L_2(s, T; U) \to Y, \qquad (9.1.19b)$$

and its adjoint $L_{sT}^* : (L_{sT} u, y)_Y = (u, L_{sT}^* y)_U$,

$$(L_{sT}^* u)(t) \equiv B^* e^{A^*(T-t)} x, \quad 0 \le s \le t \le T \qquad (9.1.20a)$$

$$: \text{ continuous } Y \to L_2(s, T; U). \qquad (9.1.20b)$$

Remark 9.1.2 As in Chapter 8, Remark 8.2.1.2, one readily sees via a change of variable in the integration of (9.1.17a) and (9.1.18a) that assumption (A.1*) = (9.1.10) allows one to obtain the following estimates independent of s, $0 \le s \le T$: There is a constant $c_T > 0$ independent of s such that

$$\|L_s u\|_{C([s,T];Y)} \le c_T \|u\|_{L_2(s,T;U)}, \qquad (9.1.21)$$

$$\|L_s^* f\|_{L_2(s,T;U)} \le c_T \|f\|_{L_1(s,T;Y)}. \qquad (9.1.22)$$

Moreover, there is a constant c_T independent of s, such that

$$\|L_{sT} u\|_Y \le c_T \|u\|_{L_2(s,T;U)}, \quad \|L_{sT}^* x\|_{L_2(s,T;U)} \le c_T \|x\|_Y. \qquad (9.1.23)$$

Furthermore (A.2*) in the form (9.1.12d) and (A.3*) = (9.1.13f) allow one to obtain the uniform estimates in s:

$$\|L_s^* R^* R v\|_{C([s,T];U)} \le c_T \|v\|_{L_\infty(s,T;Y)}, \quad \|G^* G L_s g\|_{C([s,T];Y)} \le c_T \|g\|_{L_1(s,T;U)}. \qquad (9.1.24)$$

Finally, from (9.1.20a) and (9.1.9), one gets

$$\|(L_{sT}^* G^* G x)(t)\|_U = \left\| B^* e^{A^*(T-t)} G^* G x \right\|_U \leq c_T \|x\|_Y, \quad 0 \leq s \leq t \leq T.$$
(9.1.25)

Summary of Sets of Assumptions Leading to a Riccati Theory

Chapter 8: H-assumptions

$$\begin{cases} (H.1): \ Re^{At}B : U \to L_1(0, T; Z), \\ (H.2): \ R \in \mathcal{L}(Y; Z), \\ (H.3): \ B^* e^{A^* t} R^* : Z \to L_2(0, T; U); \end{cases}$$

by duality via Chapter 7, Theorem 7.2.1 and Remark 7.2.2,

$$\begin{cases} (H.1^*): L_0^* R^* : L_\infty(0, T; Y) \to C([0, T]; U) \text{ (implied by (H.1))}, \\ (H.2^*): R^* \in \mathcal{L}(Z; Y) \text{ (equivalent to (H.2))}, \\ (H.3^*): RL_0 : L_2(0, T; U) \to C([0, T]; Z) \text{ (equivalent to (H.3))}, \\ \qquad L_0^* R^* : L_1(0, T; Z) \to L_2(0, T; U) \text{ (implied)}. \end{cases}$$

Chapter 8: h-assumptions

$$\begin{cases} (h.1) \ R : Y \supset \mathcal{D}(R) \to Z, \ R \in \mathcal{L}(\mathcal{D}(A); Z), \\ \qquad Re^{At} : Y \to L_2(0, T; Z), \\ (h.2) \ Re^{At}B : U \to L_2(0, T; Z); \end{cases}$$

equivalently, by duality via Chapter 7, Theorem 7.2.1 and Remark 7.2.2,

$$\begin{cases} (h.1^*): \ v \to \displaystyle\int_0^t e^{A^*(t-\tau)} R^* v(\tau) \, d\tau \\ \qquad : L_2(0, T; Z) \to C([0, T]; U) \text{ (equivalent to (h.1))}, \\ (h.2^*): \ L_0^* R^* : L_2(0, T; Z) \to C([0, T]; U) \text{ (equivalent to (h.2))}. \end{cases}$$

Chapter 9 for $G = 0$ *(hence (A.3) is omitted)*

$$\begin{cases} (A.0) \ R \in \mathcal{L}(Y; Z), \\ (A.1) \ B^* e^{A^* t} : Y \to L_2(0, T; U), \\ (A.2) \ R^* Re^{At}B : U \to L_1(0, T; Y), \end{cases}$$

and by duration

$$\begin{cases} (\text{A.0}^*): & R^* \in \mathcal{L}(Z; Y) \text{ (equivalent to (A.0))}, \\ (\text{A.1}^*): & L_0 : L_2(0, T; U) \to C([0, T]; Y) \text{ (equivalent to (A.1))}, \\ & L_0^* : L_1(0, T; Y) \to L_2(0, T; U) \text{ (implied)}, \\ (\text{A.2}^*): & L_0^* R^* R : L_\infty(0, T; Y) \to C([0, T]; U) \text{ (implied by (A.2))}, \end{cases}$$

Remark 9.2.1 of Chapter 9

$$\left.\begin{array}{l} \left.\begin{array}{l} (\text{h.1}) \\ (\text{h.2}) \end{array}\right\} \text{ of Chapter 8} \\ (\text{A.1}) \quad \text{of Chapter 9} \end{array}\right\} \Rightarrow P(t) \in C([0, T]; Y).$$

Orientation: Comparison between the Assumptions of the Present Chapter and the Assumptions of Chapter 8, When $G = 0$ In comparing assumptions (A.1) and (A.2) of the present chapter, where $R \in \mathcal{L}(Y; Z)$ by $(A.0) = (9.1.6)$, with assumptions (H.1) and (H.3) of Chapter 8, where $R \in \mathcal{L}(Y; Z)$ by $(\text{H.2}) = (8.2.1.2)$, we see that

$$\begin{array}{ccc} & \textbf{Chapter 9} & \textbf{Chapter 8} \\ \left\{\begin{array}{l} \\ \text{while} \\ \\ \end{array}\right. & (\text{A.1}) = (9.1.7) \Rightarrow & (\text{H.3}) = (8.2.1.3) \\ & (\text{A.1}) & \Leftarrow (\text{H.3}) \text{ and } R^{-1} \in \mathcal{L}(Z; Y). \end{array}$$

Moreover,

$$\left\{\begin{array}{l} \\ \text{while} \\ \\ \end{array}\right. \quad \begin{array}{l} (\text{A.2}) = (9.1.8) \text{ and } R^{-1} \in \mathcal{L}(Z; Y) \Rightarrow (\text{H.1}) = (8.2.1.1) \\ \\ (\text{A.2}) \quad\quad\quad\quad\quad\quad\quad\quad\quad\quad \Leftarrow (\text{H.1}). \end{array}$$

In any case, we do *not* assume $R^{-1} \in \mathcal{L}(Z; Y)$. While (A.2) is weaker than (H.1), we notice that the present chapter specializes Chapter 8 in the following aspect of critical importance in applications to partial differential equations: that the input $\to$ solution operator L_0 is bounded in the sense of (9.1.10b) (unlike the setting of Chapter 8). Even though the present assumptions are not subsumed by those of Chapter 8, the solution of the optimal control problem (9.1.5) of the present chapter can be derived from the treatment of Chapter 8, with some appropriate technical modifications. First, the presence of the observation operator G is responsible for additional technical difficulties, which force a detour, for example, at the level of Lemma 9.3.3.2 and Lemma 9.3.4.2, over Chapter 8. The reason is pinpointed in Remark 9.3.2.1. Second, because of the different assumptions in L_0, the most serious departure will be at the level of obtaining the counterpart versions of Lemmas 8.3.4.1 and 8.3.4.2 of Chapter 8 [see Lemmas 9.3.4.1 and 9.3.4.2 of the present chapter], which are needed in the derivation of the differential Riccati equation. Moreover, in the present setting, it is possible to refine certain results of Chapter 8: for example, (i) formula (9.2.5) below for y^0 (not available in Chapter 8); (ii) strong continuity of the evolution operator with respect to the second variable as in (9.3.2.5) below (not available in Chapter 8); (iii) strong continuity of the Riccati operator on $[0, T]$ as in (9.3.3.3) below (while only $P \in \mathcal{B}([0, T]; \mathcal{L}(Y))$ was available in Chapter 8).

Thus this chapter relies heavily on the analysis and presentation of Chapter 8, with some additional technicalities of its own. Moreover, we shall take care to point out the technical modifications needed in the present chapter over the treatment of Chapter 8, and, further, which results, even with $G = 0$, require only [the standing assumptions (i), (ii), and (A.0) of Section 9.1, as well as] assumption (A.1), and which results, instead, hold true under the additional smoothing assumption (A.2). In particular, we shall point out that the following results are valid only under (A.1) [indeed, as we have seen in Chapter 8, only under the general assumptions (h.1) and (h.3) of that chapter, when $G = 0$]: existence and uniqueness of an optimal pair; representation formulas for optimal control and optimal solution; and a.e. pointwise feedback synthesis of the optimal pair via an explicitly defined nonnegative, self-adjoint operator $P(t)$, which satisfies a continuity regularity property, and which provides the value of the optimal control problem. In short, all these results are listed in Theorem 9.2.1 below. Instead, further results in connection with establishing that $P(t)$ is a solution, indeed a unique solution, of the differential Riccati equation will require the additional smoothing assumption on R given by (A.2), as well as the smoothing assumption on G given by (A.3); these results are comprised in Theorem 9.2.2 below. Besides being of interest in itself, such distinction between the role of assumption (A.1) alone, and the role of assumptions (A.1) and (A.2) combined, when $G = 0$, will be needed in the subsequent Chapter 11, dealing with the optimal quadratic cost problem over an infinite horizon ($T = \infty$) only under assumption (A.1) [as well as the standing assumptions (i), (ii), and (A.0) of Section 9.1].

The conceptual strategy of the present chapter will be the same as in Chapter 8. First, a variational approach explicitly constructs in Sections 9.3.1 through 9.3.4 the solution $P(t)$ of the differential Riccati equation directly in terms of the data and provides global regularity properties of this solution (a priori bounds). Uniqueness is dealt with in Section 9.3.5. In the direct method of Section 9.3.6, a local contraction mapping argument is combined with a priori bounds to establish existence and uniqueness of the solution to the corresponding integral Riccati equation, hence of the differential Riccati equation.

9.2 Statement of Main Result: Theorems 9.2.1, 9.2.2, and 9.2.3

The main results of this chapter are the following theorems, which are the counterpart of Chapter 8, Theorems 8.2.1.1 and 8.2.1.2.

Theorem 9.2.1   *Assume (in addition to the standing assumptions (i), (ii), and (A.0) of Section 9.1, as well as) (A.1) = (1.7).*

Then, with reference to the optimal control problem (9.1.15) over the time interval $[s, T]$ for the dynamics (9.1.1) with initial datum $y(s) = y_0 \in Y$, we have:

(a_1) *There exists a unique optimal pair $\{u^0(t, s; y_0), y^0(t, s; y_0)\}$ satisfying the optimality condition [Lemma 9.3.1.1 below]*

$$u^0(t, s; y_0) = -\{L_s^* R^* R y^0(\,\cdot\,, s; y_0)\}(t) - \{L_{sT}^* G^* G y^0(T, s; y_0)\}(t)$$

$$\in L_2(s, T; U) \tag{9.2.1}$$

and given explicitly in terms of the data of the problem by the representation formulas [Lemma 9.3.1.1 below]

$$u^0(t, s; y_0) = -\left\{\Lambda_{sT}^{-1}\left[L_s^* R^* R e^{A(\cdot - s)} y_0 + L_{sT}^* G^* G e^{A(T-s)} y_0\right]\right\}(t) \tag{9.2.2a}$$

$$\in L_2(s, T; U), \tag{9.2.2b}$$

$$\Lambda_{sT} = I_s + L_s^* R^* R L_s + L_{sT}^* G^* G L_{sT} \in \mathcal{L}(L_2(s, T; U)); \tag{9.2.2c}$$

$$\left\|\Lambda_{sT}^{-1}\right\|_{\mathcal{L}(L_2(s,T;U))} \leq 1; \tag{9.2.2d}$$

where I_s is the identity operator on $L_2(s, T; U)$,

$$y^0(t, s; y_0) = e^{A(t-s)} y_0 + \{L_s u^0(\,\cdot\,, s; y_0)\}(t) \in C([s, T]; Y) \tag{9.2.3}$$

$$= e^{A(t-s)} y_0 - \left\{L_s \Lambda_{sT}^{-1}\left[L_s^* R^* R e^{A(\cdot - s)} y_0 \right.\right.$$
$$\left.\left. + L_{sT}^* G^* G e^{A(T-s)} y_0\right]\right\}(t), \tag{9.2.4a}$$

$$y^0(T, s; y_0) = e^{A(T-s)} y_0 - L_{sT} \Lambda_{sT}^{-1}\left[L_s^* R^* R e^{A(\cdot - s)} y_0 \right.$$
$$\left. + L_{sT}^* G^* G e^{A(T-s)} y_0\right], \tag{9.2.4b}$$

$$y^0(t, s; y_0) = \left\{[I_s + L_s L_s^* R^* R]^{-1}\left[e^{A(\cdot - s)} y_0 - L_s L_{sT}^* G^* G y^0(T, s; y_0)\right]\right\}(t); \tag{9.2.5}$$

$$[I_s + L_s L_s^* R^* R]^{-1} = I_s - L_s[I_s + L_s^* R^* R L_s]^{-1} L_s^* R^* R \tag{9.2.6a}$$

$$= I_s - (L_s L_s^*)^{\frac{1}{2}}\left[I_s + (L_s L_s^*)^{\frac{1}{2}} R^* R (L_s L_s^*)^{\frac{1}{2}}\right]^{-1}$$
$$\times (L_s^* L_s)^{\frac{1}{2}} R^* R \tag{9.2.6b}$$

$$: \text{ continuous } L_2(s, T; Y) \to L_2(s, T; Y), \tag{9.2.6c}$$

and, in fact,

$$\|[I_s + L_s L_s^* R^* R]^{-1}\|_{\mathcal{L}(L_2(s,T;Y))} \leq 1 + \|L_s L_s^*\|_{\mathcal{L}(L_2(s,T;Y))}\|R^* R\|_{\mathcal{L}(Y)} \leq c_T \tag{9.2.7}$$

[see Lemma 9.3.1.2].

(a_2) *The optimal pair satisfies the estimates [see Lemma 9.3.2.1 below]*

$$\begin{cases} u^0(\,\cdot\,,s;y_0) \in L_\infty(s,T;U) \quad and \quad u^0(t,s;y_0) \in U \\ for\ all\ t;\ moreover,\ u^0(\,\cdot\,,s;y_0) \in C([s,T];U) \quad if\ G = 0; \end{cases} \quad (9.2.8a)$$

$$\sup_{0 \le s \le T} \|u^0(\,\cdot\,,s;y_0)\|_{L_\infty(s,T;U)} \le c_T \|y_0\|_Y; \quad y_0 \in Y,$$

$$u^0(t,s;y_0) \in U \quad for\ all\ t \ge s; \quad (9.2.8b)$$

$$\sup_{0 \le s \le T} \|y^0(\,\cdot\,,s;y_0)\|_{C([s,T];Y)} = \sup_{0 \le s \le T} \|\Phi(\,\cdot\,,s)y_0\|_{C([s,T];Y)}$$

$$\le c_T \|y_0\|_Y, \quad y_0 \in Y, \quad (9.2.9)$$

where we have set for $y_0 \in Y$ and s fixed:

$$\Phi(t,s)y_0 = y^0(t,s;y_0) : continuous\ Y \to C([s,T];Y) \quad (9.2.10)$$

(strong continuity in the first variable).

(a_3) *The operator $\Phi(\cdot,\cdot)$ is an evolution operator satisfying the following properties:*

   (i) *(transition property of optimal solution; Lemma 9.3.2.2 below)*

$$\Phi(t,t) = I \ on\ Y; \quad \Phi(t,s) = \Phi(t,\tau)\Phi(\tau,s), \quad 0 \le s \le \tau \le t \le T;$$
$$(9.2.11)$$

   (ii) *(transition property of the optimal control; Lemma 9.3.2.2 below)*

$$u^0(t,\tau;\Phi(\tau,s)x) = u^0(t,s;x)\ a.e.\ in\ t, \quad 0 \le s \le \tau \le t \le T,\ x \in Y;$$
$$(9.2.12)$$

   (iii) *for $t \le T$ fixed,*

$$the\ map\ s \to \Phi(t,s)x\ is\ continuous\ on\ Y,\ \forall x \in Y \quad (9.2.13)$$

(strong continuity in the second variable) [see Lemma 9.3.2.3 below].

(a_4) *There exists an operator $P(t) \in \mathcal{L}(Y)$, $0 \le t \le T$, defined explicitly in terms of the data of the problem via (9.2.2), (9.2.3), or (9.2.4) by*

$$P(t)x = \int_t^T e^{A^*(\tau-t)} R^* R y^0(\tau,t;x)\,d\tau + e^{A^*(T-t)} G^* G y^0(T,t;x),$$

$$x \in Y,\ 0 \le t \le T, \quad (9.2.14)$$

such that [see Proposition 9.3.3.4 and Theorem 9.3.3.1 below]

   (i)

$$P(t) = P^*(t) \ge 0, \quad 0 \le t \le T \quad (*\ in\ Y) \quad (see\ (9.3.3.33)). \quad (9.2.15)$$

   (ii)

$$P(t) : continuous\ Y \to C([0,T];Y) \quad (see\ (9.3.3.3)). \quad (9.2.16)$$

(iii) *The following pointwise (in t) feedback representation holds true for the (unique) optimal pair u^0 and y^0 of problem (9.1.15):*

$$u^0(t; s; y_0) = -B^* P(t) y^0(t; s; y_0) \in L_2(s, T; U) \quad \text{(see (9.3.3.31))}.$$
(9.2.17)

(iv) *The optimal cost of problem (9.1.15) for the dynamics (9.1.1) with initial condition $y(s) = y_0$ is given by (see (9.3.3.32))*

$$J_s(u^0(\,\cdot\,, s; y_0), y^0(\,\cdot\,, s; y_0)) = (P(s)y_0, y_0)_Y; \quad y_0 \in Y. \quad (9.2.18)$$

Theorem 9.2.2 (regular theory)   *Assume the standing hypotheses (i), (ii), and (A.0) of Section 9.1, as well as (A.1) = (9.1.7), (A.2) = (9.1.8), and (A.3) = (9.1.9). Then:*

(b₁) *The operator $P(t)$ defined in (9.2.14) satisfies the following additional regularity properties (see Theorem 9.3.3.1 below):*

   (i)

$$V(t) \equiv B^* P(t) \in \mathcal{L}(Y; U), \quad 0 \le t \le T, \quad (9.2.19a)$$

$$V(t) \equiv B^* P(t) : \text{ continuous } Y \to L_\infty(0, T; U) \quad \text{(see (9.3.3.4))}$$
(9.2.19b)

   [i.e., in the notation of Chapter 8, Remark 8.1.1, we have $V \in \mathcal{B}([0, T]; \mathcal{L}(Y; U))$].

   (ii)

$$B^* P(\cdot) e^{A(\cdot -s)} B : \text{ continuous } U \to L_2(s, T; U) \text{ for any } s, \ 0 \le s < T,$$
$$\text{with norm that may be taken independent of } s$$
$$(\text{or } U \to U_{0,T} \text{ in the notation of (9.3.3.2) below;}$$
$$\text{see (9.3.3.5))}. \quad (9.2.20)$$

(b₂) (Existence) *The operator $P(t)$ given by (9.2.14) is a solution of the following differential Riccati equation (DRE):*

$$\begin{cases} \left(\dfrac{d}{dt} P(t)x, y\right)_Y = -(Rx, Ry)_Z - (P(t)x, Ay)_Y - (P(t)Ax, y)_Y \\ \qquad\qquad\qquad + (B^* P(t)x, B^* P(t)y)_U, \\ P(T) = G^* G \qquad \forall\, x, y \in \mathcal{D}(A), \quad 0 \le t < T \end{cases}$$
(9.2.21)

(see Theorem 9.3.4.4, Eqn. (9.3.4.30)), and hence of the corresponding integral Riccati equation (IRE) (see Theorem 9.3.5.1)

$$(P(t)x, y)_Y = \int_t^T \left(Re^{A(\tau -t)}x, Re^{A(\tau -t)}y\right)_Z d\tau + \left(G^* G e^{A(T-t)}x, e^{A(T-t)}y\right)_Y$$

$$- \int_t^T \left(B^* P(\tau) e^{A(\tau -t)}x, B^* P(\tau) e^{A(\tau -t)}y\right)_U d\tau, \quad x, y \in Y.$$
(9.2.22)

(b_3) (Uniqueness) *Finally, the operator $P(t)$ given by (9.2.14) is the unique solution of the IRE (9.2.22), and hence of the DRE (9.2.21), to satisfy properties (9.2.15), (9.2.16), (9.2.19), and (9.2.20) [see Section 9.3.5 below Theorem 9.3.5.2].*

Remark 9.2.1   Adding assumption (A.1*) = (9.1.10) on L_0^* to the set of assumptions (h.1) = (9.2.1.10) and (h.2) = (9.2.1.11) of Chapter 8 permits one to take advantage of the new regularity properties (a_3) = (9.2.11) and (9.2.13), and thus to lift $P(t) \in \mathcal{B}([0, T]; \mathcal{L}(Y))$ in Chapter 8, Eqn. (8.2.1.26) under only (h.1) and (h.2) to $P(t) \in \mathcal{L}(Y; C([0, T]; Y))$ now as in (9.2.16).

We close this section by stating several interesting transition properties, which, however, are not needed for the purpose of establishing the DRE or the IRE. As in Theorem 8.2.1.3, we shall append a subscript T to quantities related to the optimal control problem (9.1.4) on $[0, T]$, as in $u_T^0(t, s; y_0)$, $\Phi_T(t, s)$, $P_T(t)$, etc.

Theorem 9.2.3   *Assume ((i), (ii), (A.0), and) (A.1) = (9.1.7), as in Theorem 9.2.1. Then, the following transition properties hold true, in the above notation:*

(i)

$$u_T^0(\tau, t; x) = u_{T-t}^0(\tau - t, 0; x) \underset{(\text{in } \tau)}{\in} L_2(t, T; U);   (9.2.23)$$

(ii)

$$\Phi_{T-t}(\sigma, 0)x = \Phi_T(t + \sigma, t)x \underset{(\text{in } \sigma)}{\in} C([0, T]; Y);   (9.2.24)$$

(iii)

$$P_{T-t}(0) = P_T(t), \quad 0 \le t \le T.   (9.2.25)$$

The proof will be given in Section 9.3.7.

9.3 Proofs of Theorem 9.2.1 and Theorem 9.2.2 (by the Variational Approach and by the Direct Approach). Proof of Theorem 9.2.3

The next few lemmas collect a series of results that involve the unique optimal pair $\{u^0(\cdot, s; y_0), y^0(\cdot, s; y_0)\}$ of problem (9.1.15). These results are the counterpart version of the statements of the following results in Chapter 8, except for the presence now of observation operator $G \in \mathcal{L}(Y; Z_f)$ in the cost functional (9.1.4): Lemmas 9.3.1.1, 9.3.2.1, 9.3.2.2, and 9.3.2.3 and Proposition 9.3.2.4, all of which were obtained solely under the more general assumptions (h.1) = (8.2.1.10) and (h.3) = (8.2.1.12) of that chapter (with no use of assumption (H.1) = (8.2.1.1) of that chapter, presently unavailable). As noted in the Orientation above, these assumptions are a fortiori in force in the setting of the present chapter, under assumption (A.1) = (9.1.7) as well as the standing assumption $R \in \mathcal{L}(Y; Z)$; see (A.0) = (9.1.6).

9.3.1 A First Proof by the Variational Method. Theorem 9.2.1: Explicit Representation Formulas for the Optimal Pair $\{u^0, y^0\}$ under (A.1)

The standing assumptions (i), (ii), and (A.0) of Section 9.1 will not be repeated.

Lemma 9.3.1.1 *Assume (A.1) = (9.1.7). Then, with reference to the optimal control problem (9.1.15), we have: There exists a unique optimal pair $\{u^0(t, s; y_0),$ $y^0(t, s; y_0)\}$ satisfying Eqns. (9.2.1) through (9.2.7) in Theorem 9.2.1.*

Proof (9.2.1)–(9.2.4).   We have already noted (at the beginning of this section) that the present assumption (A.1) = (9.1.7), along with the standing hypothesis (A.0) that $R \in \mathcal{L}(Y; Z)$, a fortiori guarantees the validity of Lemmas 8.3.1.1, 8.3.2.1, 8.3.2.2 and Proposition 8.3.2.4 of Chapter 8 when $G = 0$, under the much more general assumptions (h.1) = (8.2.1.10), (h.3) = (8.2.1.12) of that chapter, where two proofs were given, one, in fact, for $G \neq 0$ more general than in the present case; see Chapter 8, Lemma 8.3.1.1 or Chapter 2, Section 2.2.1. □

Proof of (9.2.5)–(9.2.7).   Formulas (9.2.5) and (9.2.6) are new over Chapter 8 and are proved here as a special case of Lemma 9.3.1.2 below, along with estimate (9.2.7).

The following lemma was stated and proved in Chapter 2, Appendix 2A. It is repeated here for convenience.

Lemma 9.3.1.2 *Let X be a Hilbert space and let S and V be two nonnegative, self-adjoint, bounded operators in $\mathcal{L}(X)$. Then, $[I + SV]$ is boundedly invertible on $\mathcal{L}(X)$:*

$$[I + SV]^{-1} \in \mathcal{L}(X), \tag{9.3.1.1}$$

$$[I + SV]^{-1} = I - S^{\frac{1}{2}}\left[I + S^{\frac{1}{2}}VS^{\frac{1}{2}}\right]^{-1}S^{\frac{1}{2}}V \tag{9.3.1.2}$$

[the inverse of the positive, self-adjoint operator on the right-hand side of (9.3.1.2) plainly exists as a bounded operator on X],

$$\|[I + SV]^{-1}\| \leq 1 + \|S\| \, \|V\| \tag{9.3.1.3}$$

in the norm of $\mathcal{L}(X)$.

Proof of Lemma 9.3.1.2.   (i) First, $[I + SV]$ is injective on X:

$$[I + SV]x = 0 \Rightarrow (x, Vx) + (SVx, Vx) = 0. \tag{9.3.1.4}$$

Since V and S are nonnegative, the second identity in (9.3.1.4) implies $(x, Vx) = \|V^{\frac{1}{2}}x\| = 0$; hence $V^{\frac{1}{2}}x = 0$ and $Vx = 0$. Using this in the first identity of (9.3.1.4) yields $x = 0$, as desired.

   (ii) Second, the range $\mathcal{R} = [I + SV]X$ of $[I + SV]$ is dense in X, since the adjoint $[I + SV]^* = [I + VS]$ is injective by the same argument of part (i).

(iii) Third, the identity

$$
\begin{aligned}
I &= I + SV - S^{\frac{1}{2}}\left[I + S^{\frac{1}{2}}VS^{\frac{1}{2}}\right]^{-1}\left[I + S^{\frac{1}{2}}VS^{\frac{1}{2}}\right]S^{\frac{1}{2}}V \\
&= [I + SV] - S^{\frac{1}{2}}\left[I + S^{\frac{1}{2}}VS^{\frac{1}{2}}\right]^{-1}S^{\frac{1}{2}}V[I + SV] \\
&= \left\{I - S^{\frac{1}{2}}\left[I + S^{\frac{1}{2}}VS^{\frac{1}{2}}\right]^{-1}S^{\frac{1}{2}}V\right\}[I + SV]
\end{aligned}
\tag{9.3.1.5}
$$

implies by the injectivity of part (i) that identity (9.3.1.2) holds true at least on the range $\mathcal{R}$ of $[I + SV]$; that is, for all $r \in [I + SV]X = \mathcal{R}$, we have

$$
[I + SV]^{-1}r = \left\{I - S^{\frac{1}{2}}\left[I + S^{\frac{1}{2}}VS^{\frac{1}{2}}\right]^{-1}S^{\frac{1}{2}}V\right\}r.
\tag{9.3.1.6}
$$

But the range $\mathcal{R}$ is dense in X by (ii), and the operator on the right-hand side of (9.3.1.2), or of (9.3.1.6), is bounded on X. Thus, identity (9.3.1.6) can be extended to all of X and (9.3.1.2) holds true. It then follows from (9.3.1.2) that, in the norm of $\mathcal{L}(X)$, we have

$$
\left\|[I + SV]^{-1}\right\| \leq 1 + \left\|S^{\frac{1}{2}}\right\| \cdot 1 \cdot \left\|S^{\frac{1}{2}}\right\| \|V\| = 1 + \|S\| \|V\|,
\tag{9.3.1.7}
$$

since for the nonnegative, self-adjoint S we have

$$
\left\|\left[I + S^{\frac{1}{2}}VS^{\frac{1}{2}}\right]^{-1}\right\| \leq 1; \quad \left\|S^{\frac{1}{2}}\right\| \left\|S^{\frac{1}{2}}\right\| = \|S\|.
\tag{9.3.1.8}
$$

Then (9.3.1.7) proves (9.3.1.3). □

Continuing with the proof of (9.2.5)–(9.2.7) of Theorem 9.2.1, we first insert (9.2.1) into (9.2.3), thereby obtaining

$$
[I_s + L_s L_s^* R^* R]y^0(\cdot, s; y_0) = e^{A(\cdot - s)}y_0 - L_s L_{sT}^* G^* Gy^0(T, s; y_0).
\tag{9.3.1.9}
$$

We then apply the above Lemma 9.3.1.2 to obtain the inversion of the operator in [ ] in (9.3.1.9) in $\mathcal{L}(L_2(s, T; Y))$. (Plainly, the fact that $R^* R \in \mathcal{L}(Y)$ while $L_s L_s^* \in L_2(s, T; Y)$ does not change the proof of the lemma.) We then obtain (9.2.5), (9.2.6b), and (9.2.7) directly from, respectively, (9.3.1.9), (9.3.1.2), and (9.3.1.3). Then, the final uniform estimate (9.2.8) follows from (9.2.7) by use of (A.2*) in the form of estimates (9.1.17) and (9.1.18). Finally, in the specialized case where $S = L_s L_s^*$ and $V = R^* R$, identity (9.2.6a) can be obtained with a decomposition similar to the one in (9.3.1.5), namely

$$
\begin{aligned}
I_s &= I_s + L_s L_s^* R^* R - L_s[I + L_s^* R^* R L_s]^{-1}[I_s + L_s^* R^* R L_s]L_s^* R^* R \\
&= [I_s + L_s L_s^* R^* R] - L_s[I + L_s^* R^* R L_s]^{-1}L_s^* R^* R[I_s + L_s L_s^* R^* R] \\
&= \{I - L_s[I + L_s^* R^* R L_s]^{-1}L_s^* R^* R\}[I_s + L_s L_s^* R^* R],
\end{aligned}
\tag{9.3.1.10}
$$

and the argument below (9.3.1.5) yields (9.2.6a). □

9.3.2 Theorem 9.2.1: Estimates on $u^0(\,\cdot\,, t; x)$ and $y^0(\,\cdot\,, t; x)$ under (A.1). The Operator $\Phi(\,\cdot\,, \,\cdot\,)$

We introduce the operator $\Phi(t, s)$ by

$$\Phi(t, s)x = y^0(t, s; x) = e^{A(t-s)}x + \{L_s u^0(\,\cdot\,, s; x)\}(t) \qquad (9.3.2.1a)$$

$$= e^{A(t-s)}x + \int_s^t e^{A(t-\sigma)}Bu^0(\sigma, s; x)\, d\sigma \qquad (9.3.2.1b)$$

$$: \text{continuous } Y \to C([s, T]; Y), \ x \in Y, \ s \text{ fixed.} \qquad (9.3.2.1c)$$

Lemma 9.3.2.1 *Assume (A.1) = (9.1.7). With reference to (9.2.2) and (9.3.2.1) we have*

$$\sup_{0 \le t \le T} \|u^0(\,\cdot\,, t; x)\|_{L_2(t,T;U)} \le c_T \|x\|_Y, \quad x \in Y. \qquad (9.3.2.2a)$$

In fact, $u^0(t, s; x) \in U$ for all $0 \le s \le t \le T$, and

$$\sup_{0 \le t \le T} \|u^0(\,\cdot\,, t; x)\|_{L_\infty(t,T;U)} \le c_T \|x\|_Y, \quad x \in Y; \qquad (9.3.2.2b)$$

$$\sup_{0 \le t \le T} \|y^0(\,\cdot\,, t; x)\|_{C([t,T];Y)} = \sup_{0 \le t \le T} \|\Phi(\,\cdot\,, t)x\|_{C([t,T];Y)}$$

$$\le c_T \|x\|_Y, \quad x \in Y. \qquad (9.3.2.3)$$

Proof.   Estimate (9.3.2.2a) is proved as in Chapter 8, Lemma 8.3.2.1 [where $G = 0$, however]: We start from formula (9.2.2a), and we use the uniform bound (9.2.2d) for Λ_{sT}^{-1} as well as consequences (9.1.22) for L_s^* and (9.1.23) for L_{sT}^* of assumption (A.1*). Estimate (9.3.2.3) is proved as in Chapter 8, Lemma 8.3.2.2: We return to the optimal dynamics (9.3.2.1a) and use consequence (9.1.21) for L_s of assumption (A.1*) as well as (9.3.2.2a).

Next, we may improve (9.3.2.2a) to (9.3.2.2b) as in Chapter 8, Lemma 8.3.2.2: One returns to the characterization (9.2.1a) for u^0 and invokes (A.2*) in the form of its consequence (9.1.24) for $L_s^* R^* R$, (9.3.2.3) for y^0 and, finally, (A.3*) in the form of (9.1.25) [or (9.1.9)] for $L_{sT}^* G^* G$, thus obtaining (9.3.2.2b). □

Lemma 9.3.2.2 *Assume (A.1) = (9.1.7). With reference to the operator $\Phi(\cdot, \cdot)$ in (9.3.2.1), the following transition properties hold true:*

(i)

$$u^0(t, \tau; \Phi(\tau, s)x) = u^0(t, s; x) \quad a.e. \text{ in } t, \quad 0 \le s \le \tau \le t \le T; \ x \in Y. \qquad (9.3.2.4)$$

(ii)

$$\Phi(t, t) = \text{identity on } Y; \quad \Phi(t, s) = \Phi(t, \tau)\Phi(\tau, s), \quad 0 \le s \le \tau \le t \le T. \qquad (9.3.2.5)$$

Remark 9.3.2.1 In Chapter 8, Proposition 8.3.2.4, where $G = 0$, we proved the evo-lution property (9.3.2.5) *directly*, by starting from the characterizing formula (9.2.5) for y^0, which, for $G = 0$, is directly explicit in terms of the data [see Eqn. (8.3.2.13) for $R\Phi$ in Chapter 8]. However, when $G \neq 0$, a different route as in Chapter 1 is successful: One first proves the transition property (9.3.2.4) for u^0, and next – as a consequence – one proves (9.3.2.5) for y^0. To prove u^0 we shall start with the charac-terizing formula (9.2.2a), which gives u^0 explicitly in terms of the data (while neither (9.2.4) nor (9.2.5) for $G \neq 0$ are now as convenient). A simplified proof of (9.3.2.5) for $G = 0$ is given in Remark 9.3.2.1 below. $\square$

Proof. (i) $= (9.3.2.4)$ *(refer to Chapter 1, Section 1.4.3).*   We return to Eqn. (9.2.2a), rewritten, for $x \in Y$, as

$$u^0(t, s; x) + \{L_s^* R^* R L_s u^0(\,\cdot\,, s; x)\}(t) + \{L_{sT}^* G^* G L_{sT} u^0(\,\cdot\,, s; x)\}(t)$$

$$= -\{L_s^* R^* R L_s e^{A(\cdot\,-s)} x\}(t) - \{L_{sT}^* G^* G e^{A(T-s)} x\}(t), \tag{9.3.2.6}$$

or, explicitly via (9.1.17)–(9.1.20),

$$u^0(t, s; x) + \int_t^T B^* e^{A^*(\sigma-t)} R^* R \int_s^\sigma e^{A(\sigma-r)} B u^0(r, s; x) \, dr \, d\sigma$$

$$+ B^* e^{A^*(T-t)} G^* G \int_s^T e^{A(T-r)} B u^0(r, s; x) \, dr$$

$$= -\int_t^T B^* e^{A^*(\sigma-t)} R^* R e^{A(\sigma-s)} x \, d\sigma - B^* e^{A^*(T-t)} G^* G e^{A(T-s)} x. \tag{9.3.2.7}$$

We now rewrite (9.3.2.7) with s replaced by τ, and with $x \in Y$ replaced by $\Phi(\tau, s)x \in Y$ by (9.3.2.1c):

$$u^0(t, \tau; \Phi(\tau, s)x) + \int_t^T B^* e^{A^*(\sigma-t)} R^* R \int_\tau^\sigma e^{A(\sigma-r)} B u^0(r, \tau; \Phi(\tau, s)x) \, dr \, d\sigma$$

$$+ B^* e^{A^*(T-t)} G^* G \int_\tau^T e^{A(T-r)} B u^0(r, \tau; \Phi(\tau, s)x) \, dr$$

$$= -\int_t^T B^* e^{A^*(\sigma-t)} R^* R e^{A(\sigma-\tau)} \Phi(\tau, s)x \, d\sigma - B^* e^{A^*(T-t)} G^* G e^{A(T-\tau)} \Phi(\tau, s)x.$$

$$\tag{9.3.2.8}$$

Note that for $x \in Y$, $s \leq \tau$ fixed, Eqns. (9.3.2.7) and (9.3.2.8) have all terms well-defined as $L_2(s, T; U)$-functions or $L_2(\tau, T; U)$-functions in t, by the regularity re-sults (9.1.17)–(9.1.19). Next, we return to (9.3.2.7), split the integrals $\int_s^\sigma = \int_s^\tau + \int_\tau^\sigma$ and likewise $\int_s^T = \int_s^\tau + \int_\tau^T$ in the two integral terms on its left-hand side, and finally subtract (9.3.2.8) from (9.3.2.7). We then obtain

$$u^0(t, s; x) - u^0(t, \tau; \Phi(\tau, s)x)$$

$$+ \int_t^T B^* e^{A^*(\sigma - t)} R^* R \int_\tau^\sigma e^{A(\sigma - r)} B[u^0(r, s; x) - u^0(r, \tau; \Phi(\tau, s)x)] \, dr \, d\sigma$$

$$+ \int_t^T B^* e^{A^*(\sigma - t)} R^* R \int_s^\tau e^{A(\sigma - r)} B u^0(r, s; x) \, dr \, d\sigma$$

$$+ B^* e^{A^*(T-t)} G^* G \int_\tau^T e^{A(T-r)} B[u^0(r, s; x) - u^0(r, \tau; \Phi(\tau, s)x)] \, dr$$

$$+ B^* e^{A^*(T-t)} G^* G \int_s^\tau e^{A(T-r)} B u^0(r, s; x) \, dr$$

$$= \int_t^T B^* e^{A^*(\sigma - t)} R^* R \left[e^{A(\sigma - \tau)} \Phi(\tau, s)x - e^{A(\sigma - s)}x \right] d\sigma$$

$$+ B^* e^{A^*(T-t)} G^* G \left[e^{A(T-\tau)} \Phi(\tau, s)x - e^{A(T-s)}x \right]. \tag{9.3.2.9}$$

Finally, we use the optimal dynamics (9.3.2.1b) with t replaced by τ, which yields

$$e^{A(\sigma - \tau)} \Phi(\tau, s)x - e^{A(\sigma - s)}x = \int_s^\tau e^{A(\sigma - r)} B u^0(r, s; x) \, dr, \tag{9.3.2.10}$$

$$e^{A(T-\tau)} \Phi(\tau, s)x - e^{A(T-s)}x = \int_s^\tau e^{A(T-r)} B u^0(r, s; x) \, dr. \tag{9.3.2.11}$$

Inserting (9.3.2.10) and (9.3.2.11) in the last two terms on the right-hand side of (9.3.2.9) produces a cancellation of these two terms against like terms, the third and the fifth, on the left-hand side of (9.3.2.9). Thus, (9.3.2.9) simplifies to

$$u^0(t, s; x) - u^0(t, \tau; \Phi(\tau, s)x)$$

$$+ \int_t^T B^* e^{A^*(\sigma - t)} R^* R \int_\tau^\sigma e^{A(\sigma - r)} B[u^0(r, s; x) - u^0(r, \tau; \Phi(\tau, s)x)] \, dr \, d\sigma$$

$$+ B^* e^{A^*(T-t)} G^* G \int_\tau^T e^{A(T-r)} B[u^0(r, s; x) - u^0(r, \tau; \Phi(\tau, s)x)] \, dr, \tag{9.3.2.12}$$

or, recalling (9.1.17)–(9.1.20) and (9.2.2b), to

$$[I_\tau + L_\tau^* R^* R L_\tau + L_{\tau T}^* G^* G L_{\tau T}][u^0(\cdot, s; x) - u^0(\cdot, \tau; \Phi(\tau, s)x)]$$

$$= \Lambda_{\tau T}[u^0(\cdot, s; x) - u^0(\cdot, \tau; \Phi(\tau, s)x)] = 0. \tag{9.3.2.13}$$

Since $\Lambda_{\tau T}$ is boundedly invertible on $L_2(\tau, T; U)$, then (9.3.2.13) yields (9.3.2.4), as desired.

(ii) = (9.3.2.5). We now use the transition property (9.3.2.4) for u^0 to show the transition property (9.3.2.5) for y^0. The optimal dynamics (9.3.2.1b) expressed for

$\Phi(\tau, s)x$ yields

$$\Phi(t, \tau)\Phi(\tau, s)x = e^{A(t-\tau)}\Phi(\tau, s)x + \int_{\tau}^{t} e^{A(t-\sigma)} Bu^0(\sigma, \tau; \Phi(\tau, s)x) \, d\sigma$$

$$(\text{by } (9.3.2.4)) \quad = e^{A(t-\tau)}\left[e^{A(\tau-s)}x + \int_{s}^{\tau} e^{A(\tau-\sigma)} Bu^0(\sigma, s; x) \, d\sigma\right]$$

$$+ \int_{\tau}^{t} e^{A(t-\sigma)} Bu^0(\sigma, s; x) \, d\sigma, \qquad (9.3.2.14)$$

where in the last step we have used (9.3.2.4). Thus (9.3.2.14) yields

$$\Phi(t, \tau)\Phi(\tau, s)x = e^{A(t-s)}x + \int_{s}^{t} e^{A(t-\sigma)} Bu^0(\sigma, s; x) \, d\sigma$$

$$= \Phi(t, s)x, \quad 0 \le s \le \tau \le t \le T, \ x \in Y, \qquad (9.3.2.15)$$

and property (9.3.2.5) is proved.   □

Lemma 9.3.2.3   *Assume (A.1) = (9.1.7). With reference to (9.3.2.1) we have, for $t \le T$ fixed,*

$$\text{the map } s \rightarrow \Phi(t, s)x \text{ is continuous on } Y, \quad s \le t. \qquad (9.3.2.16)$$

Proof.   The available continuity of $\Phi(\cdot, \cdot)$ in the first variable [see (9.3.2.1c)], is transferred to continuity of $\Phi(\cdot, \cdot)$ in the second variable, by virtue of the evolution properties (9.3.2.5) and of the uniform bound (9.3.2.3) of Φ in both variables, as was done in the proof of Chapter 1, Lemma 1.4.6.2(ii). Indeed, for right continuity, we choose $h > 0$ such that $s < s + h \le t \le T$. Then:

$$\|\Phi(t, s + h)x - \Phi(t, s)x\| = \|\Phi(t, s + h)[x - \Phi(s + h, s)x]\| \qquad (9.3.2.17)$$

$$\text{by } (9.3.2.3) \quad \le c_T \|x - \Phi(s + h, s)x\| \rightarrow 0 \quad \text{as } h \downarrow 0 \quad (9.3.2.18)$$

in the norm of Y, after using (9.3.2.5) in (9.3.2.17) and using (9.3.2.3) and (9.3.2.1c) in (9.3.2.18). As to the left continuity, we compute for $h > 0$, again by (9.3.2.5) and (9.3.2.3),

$$\|\Phi(t, s - h)x - \Phi(t, s)x\| = \|\Phi(t, s)[\Phi(s, s - h)x - x]\|$$

$$\le c_T \|\Phi(s, s - h)x - x\|. \qquad (9.3.2.19)$$

Next, recalling (9.3.2.1a), we have

$$\Phi(s, s - h)x - x = e^{Ah}x - x + \{L_s u^0(\cdot, s - h; x)\}(s), \qquad (9.3.2.20)$$

so that, in the norm of Y, (9.3.2.20) implies by (9.1.21)

$$\|\Phi(s, s - h)x - x\|_Y$$

$$\le \|e^{Ah}x - x\|_Y + c_T \|u^0(\cdot, s - h; x)\|_{L_2(s, s-h; U)} \rightarrow 0 \quad \text{as } h \downarrow 0, \qquad (9.3.2.21)$$

since

$$\|u^0(\,\cdot\,, s - h; x)\|^2_{L_2(s,s-h;U)} = \int_{s-h}^{s} \|u^0(\sigma, s - h; x)\|^2_U \, d\sigma$$

$$\text{(by (9.2.8))} \qquad \leq c_T \int_{s-h}^{s} 1 \, d\sigma \|x\|_Y \to 0 \text{ as } h \downarrow 0 \qquad (9.3.2.22)$$

after recalling (9.2.8). Thus, using (9.3.2.21) in (9.3.2.19) yields

$$\|\Phi(t, s - h)x - \Phi(t, s)x\| \to 0 \qquad \text{as } h \downarrow 0, \qquad (9.3.2.23)$$

as desired. Lemma 9.3.2.3 is proved. $\square$

Remark 9.3.2.1   Here we give a direct, simplified proof of (9.3.2.5) when $G = 0$, in which case formula (9.2.5) becomes directly explicit in terms of the data.

Proof of (9.3.2.4) for $G = 0$.   The present proof of the evolution properties of $\Phi(t, \tau)$ is essentially the same as the proof of Proposition 8.3.2.4 of Chapter 8, except that now we use formula (9.2.5) for Φ [rather than formula (8.3.2.13) for $R\Phi$ in Chapter 8]. (Now, however, the continuity (9.3.2.1c) makes Lemma 8.3.2.3 of Chapter 8 unnecessary.) We sketch the main steps. By (9.2.5), using the explicit formulas (9.1.17) and (9.1.18) for L_s and L_s^*, we obtain with $x \in Y$:

$$\Phi(t, s)x = e^{A(t-s)}x - \{L_s L_s^* R^* R\Phi(\,\cdot\,, s)x\}(t)$$

$$= e^{A(t-s)}x - \int_{s}^{t} e^{A(t-\sigma)} BB^* \int_{\sigma}^{T} e^{A^*(\tau-\sigma)} R^* R\Phi(r, s)x \, dr \, d\sigma$$

$$(9.3.2.24)$$

(compare with Chapter 8, Eqn. (8.3.2.18)). Similarly, recalling now (9.3.2.1),

$$\Phi(t, \tau)\Phi(\tau, s)x$$

$$= e^{A(t-\tau)} \left[e^{A(\tau-s)}x - \int_{s}^{\tau} e^{A(\tau-\sigma)} BB^* \int_{\sigma}^{T} e^{A^*(r-\sigma)} R^* R\Phi(r, s)x \, dr \, d\sigma \right]$$

$$- \int_{\tau}^{t} e^{A(t-\sigma)} BB^* \int_{\sigma}^{T} e^{A^*(r-\sigma)} R^* R\Phi(r, \tau)\Phi(\tau, s)x \, dr \, d\sigma \qquad (9.3.2.25)$$

(compare with Chapter 8, Eqn. (8.3.2.21)). Finally, subtract (9.3.2.25) from (9.3.2.24), after splitting the external integration in (9.3.2.24) in $[s, \tau]$ and $[\tau, t]$. We obtain after a cancellation of two terms:

$$\Phi(t, s)x - \Phi(t, \tau)\Phi(\tau, s)x$$

$$= -\int_{\tau}^{t} e^{A(t-\sigma)} BB^* \int_{\sigma}^{T} e^{A^*(r-\sigma)} R^* R[\Phi(r, s)x - \Phi(r, \tau)\Phi(\tau, s)x] \, dr \, d\sigma$$

$$= -L_\tau L_\tau^* R^* R[\Phi(\,\cdot\,, s)x - \Phi(\,\cdot\,, \tau)\Phi(\tau, s)x] \qquad (9.3.2.26)$$

(compare with Chapter 8, Eqn. (8.3.2.22)), or

$$[I_\tau + L_\tau L_\tau^* R^* R][\Phi(\cdot, s)x - \Phi(\cdot, \tau)\Phi(\tau, s)x] = 0 \qquad (9.3.2.27)$$

(compare with Chapter 8, Eqn. (8.3.2.23)). Finally, we invoke the bounded invertibility of $[I_\tau + L_\tau L_\tau^* R^* R]$ in $L_2(\tau, T; Y)$ as in (9.2.5) (see Lemma 9.3.1.2), and we obtain equality (9.3.2.4), first in the $L_2(\tau, T; Y)$-sense and finally in the $C([\tau, T]; Y)$-sense and by (9.3.2.1c). $\square$

9.3.3 Definition of $P(t)$ and Preliminary Properties

For $x \in Y$ we define the operator $P(t) \in \mathcal{L}(Y)$ by setting

$$P(t)x = \int_t^T e^{A^*(\tau-t)} R^* R \Phi(\tau, t)x \, d\tau + e^{A^*(T-t)} G^* G \Phi(T, t)x,$$

$$x \in Y, \ 0 \le t \le T. \quad (9.3.3.1)$$

In the development of Chapter 8 where $G = 0$, we have shown, only on the basis of the assumptions $R \in \mathcal{L}(Y; Z)$ and (H.3) = (8.2.1.3) of Chapter 8, that $P(t) \in \mathcal{L}(Y)$ and satisfies the regularity property of (9.3.3.2b) [or (9.3.3.5)] of that chapter. The same conclusion then holds true concerning the integral term of (9.3.3.1), solely under assumption (A.1) = (9.1.7) of the present chapter [as well as the standing assumption (A.0) that $R \in \mathcal{L}(Y; Z)$ in (9.1.6)], as these hypotheses imply (H.2) and (H.3) of Chapter 8, as seen in the Orientation. Moreover, the additional term in (9.3.3.1) containing G satisfies $e^{A^*(T-t)} G^* G \Phi(T, t)x \in C([0, T]; Y)$ by virtue of the assumed $G \in \mathcal{L}(Y; Z_f)$ and property (9.3.2.16) of Lemma 9.3.2.3. This is formalized in (9.3.3.3) below.

By contrast, it was at the level of establishing the regularity properties of

$$V(t) = B^* P(t) \quad \text{and} \quad Q(t, s) = B^* P(t) e^{A(t-s)} B$$

in (8.3.3.3) and (8.3.3.4) [or (8.3.3.8a) and (8.3.3.13)] of Chapter 8, where $G = 0$, that we have used, for the first time, also assumption (H.1) of that chapter [unavailable in the present chapter]. The same conclusions of $V(t)$ and $Q(t, s)$, but for different technical reasons, will continue to hold true under both assumptions (A.1) = (9.1.7) and (A.2) = (9.1.8) [and R bounded], for the integral term in (9.3.3.1), as well as under assumption (A.3) = (9.1.9) for the additional term in (9.3.3.1) containing G; see (9.3.3.4) and (9.3.3.5) below. First, we recall from Chapter 8, Eqn. (8.3.3.1)] the Banach space $U_{0,T}$ of all measurable mappings v from $\Delta_{0,T}$ into the Hilbert space U, $\Delta_{0,T} = \{(t, s) \in R^2 : 0 \le s \le t \le T\}$, such that

$$\|v\|_{U_{0,T}}^2 = \sup_{0 \le s \le T} \int_s^T \|v(t, s)\|_U^2 \, dt < \infty. \qquad (9.3.3.2)$$

Theorem 9.3.3.1 *(a) Assume (A.1) = (9.1.7). Then, the operator $P(t)$ defined by (9.3.3.1) belongs to $\mathcal{L}(Y)$ for all $0 \le t \le T$, and, in fact, it satisfies the following* *regularity property:*

(i)

$$P(t) : \ continuous \ Y \to C([0, T]; Y), \tag{9.3.3.3}$$

so that a fortiori $P(\cdot) \in \mathcal{B}([0, T]; \mathcal{L}(Y))$ *in the notation of Chapter 8, Remark 8.1.1.*

(b) Assume, in addition, $(A.2) = (9.1.8)$ *and* $(A.3) = (9.1.9)$. *Then*

(ii)

$$V(t) \equiv B^* P(t) : \ continuous \ Y \to L_\infty(0, T; U), \tag{9.3.3.4a}$$
$$V(t) \in \mathcal{L}(Y; U), \quad 0 \le t \le T; \tag{9.3.3.4b}$$

(iii)

$$Q(t, s) \equiv B^* P(t) e^{A(t-s)} B : \ continuous \ U \to U_{0,T}, \tag{9.3.3.5}$$

with the space $U_{0,T}$ *defined by (9.3.3.2).*

Proof. (i) Clearly for t fixed, the integral on the right-hand side of (9.3.3.1) is well defined, $\forall \, x \in Y$ by the continuity (9.3.2.1c) of the map $\tau \to \Phi(\tau, t)x$ and by $R \in \mathcal{L}(Y; Z)$ (these are different reasons than the ones used in the proof of the same result in Chapter 8, Proposition 8.3.3.3). Likewise, the term in (9.3.3.1) containing G is well defined, as pointed out below (9.3.3.1). Moreover, the uniform bound (9.3.2.3) for $\Phi(\cdot, \cdot)$, once applied to (9.3.3.1), plainly implies for each t:

$$\|P(t)x\|_Y \le c_T \|x\|_Y, \quad x \in Y, \tag{9.3.3.6}$$

and hence $P(t)$: continuous $Y \to L_\infty(0, T; Y)$. But, actually, the stronger statement (9.3.3.3) is true now. For this, however, we must use the continuity of the map $s \to \Phi(t, s)x$, for t fixed, expressed by (9.3.2.16), which requires only (A.1). Thus, by virtue of this property (9.3.2.16), we have

$$e^{A^*(T-t)} G^* G \Phi(T, t) : \ continuous \ Y \to C([0, T]; Y).$$

Hence, we only have to prove the same regularity for the integral term of (9.3.3.1), which corresponds to the case $G = 0$. Indeed, in this case, after adding and subtracting, we may write for t, $t_0 \in [0, T]$, say $t < t_0$, and $x \in Y$:

$$P(t)x - P(t_0)x = \int_t^{t_0} e^{A^*(\tau-t)} R^* R \Phi(\tau, t)x \, d\tau$$

$$+ \int_{t_0}^T \left[e^{A^*(\tau-t)} - e^{A^*(\tau-t_0)} \right] R^* R \Phi(\tau, t)x \, d\tau$$

$$+ \int_{t_0}^T e^{A^*(\tau-t_0)} R^* R [\Phi(\tau, t)x - \Phi(\tau, t_0)x] \, d\tau. \tag{9.3.3.7}$$

As to the first term on the right-hand side of (9.3.3.7), we invoke the uniform bound (9.3.2.3) on Φ (which requires only (A.1)) to obtain

$$\left\| \int_t^{t_0} e^{A^*(\tau-t)} R^* R\Phi(\tau, t)x \, d\tau \right\|_Y \leq |t_0 - t| c_T \|x\|_Y \to 0 \quad \text{as } t \to t_0. \quad (9.3.3.8)$$

As to the second term on the right-hand side of (9.3.3.7), we notice that, by (9.3.2.16) (which requires only (A.1)), the integrand is a Y-continuous function of t for τ fixed, $t \leq t_0 \leq \tau$, which tends to zero as $t \to t_0$; moreover, the integrand is bounded by $c_T \|x\|_Y$ by (9.3.2.3). Hence, the Lebesgue dominated convergence theorem applies and yields

$$\int_{t_0}^T \left[e^{A^*(\tau-t)} - e^{A^*(\tau-t_0)} \right] R^* R\Phi(\tau, t)x \, d\tau \to 0 \quad \text{as } t \to t_0. \quad (9.3.3.9)$$

The same (in fact, simpler) analysis, based on (9.3.2.16) and (9.3.2.3), applies to the third term and we likewise obtain

$$\int_{t_0}^T e^{A^*(\tau-t_0)} R^* R[\Phi(\tau, t)x - \Phi(\tau, t_0)]x \, d\tau \to 0 \quad \text{as } t \to t_0. \quad (9.3.3.10)$$

Invoking (9.3.3.9) and (9.3.3.10) in (9.3.3.7) gives

$$\|P(t)x - P(t_0)x\|_Y \to 0 \quad \text{as } t \to t_0, \quad (9.3.3.11)$$

and (9.3.3.3) is proved.

(ii) = (9.3.3.4) We shall also use (A.2*) = (9.1.12) and (A.3) = (9.1.9). By definition of $V(t)$ in (9.3.3.4) and by (9.3.3.1) we have

$$V(t)x = B^* P(t)x$$

$$= \int_t^T B^* e^{A^*(\tau-t)} R^* R\Phi(\tau, t)x \, d\tau + B^* e^{A^*(T-t)} G^* G\Phi(T, t)x \quad (9.3.3.12a)$$

$$= \{L_0^* R^* R\Phi(\cdot, t)x\}(t) + L_{0T}^* G^* G\Phi(T, t)x, \quad (9.3.3.12b)$$

by (9.1.18) and (9.1.20). Recalling assumption (A.2*) = (9.1.12) and (A.3) = (9.1.9) we obtain, from (9.3.3.12) with $\tau - t = \sigma$,

$$\|V(t)x\|_U \leq \left\| \int_0^{T-t} B^* e^{A^*\sigma} R^* R\Phi(\sigma + t, t)x \, d\sigma \right\|_U$$

$$+ \left\| B^* e^{A^*(T-t)} G^* G\Phi(T, t)x \right\|_U$$

$$\text{(by (9.1.12) and (9.1.9))} \quad \leq c_T \left\{ \sup_{0 \leq \sigma \leq T-t} \|\Phi(\sigma + t, t)x\|_Y + \|\Phi(T, t)x\|_Y \right\}$$

$$\text{(by (9.3.2.3))} \quad \leq c_T \|x\|_Y, \quad (9.3.3.13)$$

where in the last step we have used (9.3.2.3). Thus (9.3.3.13) proves part (ii), in particular (9.3.3.4a). (In comparing the present proof with that in Chapter 8, Eqn.

(8.3.3.9) of the same result with $G = 0$, we notice that different reasons are used in both cases.)

(iii) = (9.3.3.5) As in Chapter 8, Theorem 8.3.3.1(iii), this proof requires a preliminary extension of $u^0(\cdot, t; x)$ from $x \in Y$ to $x = e^{A(t-s)}Bu \in [\mathcal{D}(A^*)]'$ where e^{At} can be extended to a s.c. semigroup on $[\mathcal{D}(A^*)]'$. $\square$

Step 1.   Lemma 9.3.3.2 *Assume (A.1), (A.2), and (A.3). With reference to (9.2.2), we have the following extension result:*

$$\sup_{0 \le t \le T} \sup_{0 \le s \le t} \left\| u^0(\cdot, t; e^{A(t-s)}Bu) \right\|_{L_2(t,T;U)} \le c_T \|u\|_U. \tag{9.3.3.14}$$

Proof.   From (9.2.2) we have for $u \in U$

$$
\begin{aligned}
-u^0(\cdot, t; e^{A(t-s)}Bu) &= \Lambda_{tT}^{-1} L_t^* R^* R \left[e^{A(\cdot - t)} e^{A(t-s)} Bu \right] \\
&\quad + \Lambda_{tT}^{-1} L_{tT}^* G^* G e^{A(T-t)} e^{A(t-s)} Bu \\
&= (1) + (2).
\end{aligned} \tag{9.3.3.15}
$$

As to (1), recalling (9.2.2d), (A.1*) in the form of its consequence (9.1.22), as well as (A.2) = (9.1.8), we estimate

$$
\begin{aligned}
\|(1)\|_{L_2(t,T;U)} &\le \left\| L_t^* R^* R e^{A(\cdot - s)} Bu \right\|_{L_2(t,T;U)} \\
\text{(by (9.1.22))} \quad &\le c_T \left\| R^* R e^{A(\cdot - s)} Bu \right\|_{L_1(t,T;Y)} \\
\text{(by (A.2) = (9.1.8))} \quad &\le c_T \|u\|_U,
\end{aligned} \tag{9.3.3.16}
$$

where in the last step we have used (9.1.8) and a change of variable. (We notice again that the present argument uses different reasons than the ones in the proof of the corresponding result in Lemma 8.3.3.4 of Chapter 8.)

As to (2), we recall (A.1) in the form of its consequence (9.1.23), as well as (A.3) = (9.1.9), and estimate

$$
\begin{aligned}
\|(2)\|_{L_2(t,T;U)} &\le \left\| L_{tT}^* G^* G e^{A(T-s)} Bu \right\|_{L_2(t,T;U)} \\
\text{(by (9.1.23))} \quad &\le c_T \left\| G^* G e^{A(T-s)} Bu \right\|_Y \\
\text{(by (9.1.9))} \quad &\le c_T \|u\|_U.
\end{aligned} \tag{9.3.3.17}
$$

Then, (9.3.3.16) and (9.3.3.17), used in (9.3.3.15), readily imply (9.3.3.14), as desired.

Step 2   We can now show the desired global result for Q as claimed by Theorem 9.3.3.1, Eqn. (9.3.3.5).

Proposition 9.3.3.3 *Assume (A.1), (A.2), and (A.3). With reference to (9.3.3.5) we have*

$$\|Qu\|_{U_{0,T}}^2 \equiv \sup_{0 \le s \le T} \int_s^T \|Q(t,s)u\|_U^2 \, dt \le c_T \|u\|_U^2, \tag{9.3.3.18a}$$

that is,

$$Q(t, s) : \ continuous \ U \to U_{0,T}, \tag{9.3.3.18b}$$

where $U_{0,T}$ is defined by (9.3.3.2).

Proof of Proposition 9.3.3.3.   As in Chapter 8, Proposition 8.3.3.5, we compute, by (9.3.3.5) and (9.3.3.1),

$$Q(t, s)u = \int_{t}^{T} B^* e^{A^*(\tau - t)} R^* R \Phi(\tau, t) e^{A(t-s)} Bu \, d\tau$$

$$+ B^* e^{A^*(T-t)} G^* G \Phi(T, t) e^{A(t-s)} Bu \tag{9.3.3.19}$$

(by (9.3.2.1)) $= (1) + (2) + (3) + (4); \tag{9.3.3.20}$

where

$$(1) = \int_{t}^{T} B^* e^{A^*(\tau - t)} R^* R e^{A(\tau - s)} Bu \, d\tau, \quad s \leq t \leq T \tag{9.3.3.21a}$$

(by (9.1.18a)) $= \left\{ L_s^* R^* R e^{A(\cdot - s)} Bu \right\}(t); \tag{9.3.3.21b}$

$$(2) = \int_{t}^{T} B^* e^{A^*(\tau - t)} R^* \left\{ RL_t u^0 \left(\cdot, t; e^{A(t-s)} Bu \right) \right\}(\tau) \, d\tau \tag{9.3.3.22a}$$

$$= \left\{ L_t^* R^* R L_t u^0 \left(\cdot, t; e^{A(t-s)} Bu \right) \right\}(t); \tag{9.3.3.22b}$$

$$(3) = B^* e^{A^*(T-t)} G^* G e^{A(T-s)} Bu, \quad s \leq t \leq T \tag{9.3.3.23a}$$

(by (9.1.20)) $= L_{sT}^* G^* G e^{A(T-s)} Bu; \tag{9.3.3.23b}$

$$(4) = B^* e^{A^*(T-t)} G^* G \int_{t}^{T} e^{A(T-\tau)} Bu^0 \left(\tau, t; e^{A(t-s)} Bu \right) d\tau \tag{9.3.3.24a}$$

$$= L_{sT}^* G^* G L_{tT} u^0 \left(\cdot, t; e^{A^*(t-s)} Bu \right), \tag{9.3.3.24b}$$

where in going from (9.3.3.19) to (9.3.3.20) we have used (9.3.2.1) to replace $\Phi(\tau, t)$ and $\Phi(T, t)$. As to (1) in (9.3.3.21), we obtain from (9.3.3.21b), after invoking assumption (A.1*) in the form (9.1.21),

$$\|(1)\|_{L_2(s,T;U)} \leq c_T \left\| R^* R e^{A(\cdot - s)} Bu \right\|_{L_1(s,T;Y)}$$

(by (9.1.8)) $\leq c_T \|u\|_U, \tag{9.3.3.25}$

where in the last step we have used (A.2) = (9.1.8).

As to (2) in (9.3.2.22), we set $\tau - t = \sigma$ and invoke (A.2*) = (9.1.12b) and (A.1*) in the form of its consequence (9.1.21) after an innocuous extension by zero on $[T - t, T]$ to get

$$\|(2)\|_U = \left\| \int_0^{T-t} B^* e^{A^*\sigma} R^* R\{L_t u^0(\,\cdot\,, t; e^{A(t-s)} Bu)\}(t + \sigma)\, d\sigma \right\|_U$$

(by (9.1.12b)) $$\leq c_T \sup_{0 \leq \sigma \leq T-t} \left\| L_t u^0(\,\cdot\,, t; e^{A(t-s)} Bu)(t + \sigma) \right\|_Y$$

(by (9.1.21)) $$\leq c_T \left\| u^0(\,\cdot\,, t; e^{A(t-s)} Bu) \right\|_{L_2(t, T; U)}$$

(by (9.3.3.14)) $$\leq c_T \|u\|_U, \tag{9.3.3.26}$$

where in the last step we have invoked (9.3.3.14). As to (3) in (9.3.3.23), we utilize the estimate leading to (9.3.3.17) to get

$$\|(3)\|_{L_2(t, T; U)} = \left\| B^* e^{A^*(T-t)} G^* G e^{A(T-s)} Bu \right\|_{L_2(t, T; U)}$$

$$\leq c_T \|u\|_U. \tag{9.3.3.27}$$

Finally, as to (4) in (9.3.3.24b), we estimate, after invoking (A.1) in the form of its consequence (9.1.23) and $G \in \mathcal{L}(Y; Z_f)$ as in (9.1.6),

$$\|(4)\|_{L_2(s, T; U)} = \left\| L_{sT}^* G^* G L_{tT} u^0(\,\cdot\,, t; e^{A(t-s)} Bu) \right\|_{L_2(s, T; U)}$$

(by (9.1.23)) $$\leq c_T \left\| L_{tT} u^0(\,\cdot\,, t; e^{A(t-s)} Bu) \right\|_U$$

(by (9.1.23)) $$\leq c_T \left\| u^0(\,\cdot\,, t; e^{A(t-s)} Bu) \right\|_{L_2(t, T; U)}. \tag{9.3.3.28}$$

Thus, invoking estimate (9.3.3.14) in (9.3.3.28), we finally obtain

$$\sup_{0 \leq t \leq T} \sup_{0 \leq s \leq t} \|(4)\|_{L_2(s, T; U)} \leq c_T \|u\|_U. \tag{9.3.3.29}$$

Thus, estimates (9.3.3.25) for (1), (9.3.3.26) for (2), (9.3.3.27) for (3), and (9.3.3.29) for (4), used in (9.3.3.20), readily yield (9.3.3.18a), as desired. (We remark once more that the present argument, when specialized to $G = 0$, uses different reasons than the ones in the proof of the corresponding result in Chapter 8, Proposition 8.3.3.5.) Theorem 9.3.3.1 is proved.   □

We close this section by providing two additional properties of $P(t)$: (i) the pointwise feedback expression of the optimal control $u^0(\tau, t; s)$ in terms of the optimal solution $y^0(\tau, t; s)$ via the operator $P(t)$ and (ii) the value of the optimal cost in terms of $P(t)$.

Proposition 9.3.3.4 *Assume (A.1) = (9.1.7).*

(i) For $x \in Y$, we have

$$-u^0(\tau, t; x) = \{L_t^* R^* R \Phi(\,\cdot\,, t)x\}(\tau) + \{L_{tT}^* G^* G \Phi(T, t)x\}(\tau) \tag{9.3.3.30}$$

$$= B^* P(\tau) \Phi(\tau, t)x \in L_2(t, T; U) \quad (in \ \tau), \tag{9.3.3.31}$$

and the map $t \to u^0(\tau, t; s)$ is continuous in U, for τ fixed, $x \in Y$.

(ii) With reference to (9.3.3.1), we have the following identity, symmetric in x, $y \in Y$, $0 \leq t \leq T$:

$$(P(t)x, y)_Y = \int_t^T (R\Phi(\tau, t)x, R\Phi(\tau, t)y)_Z \, d\tau$$

$$+ \int_t^T (B^* P(\tau)\Phi(\tau, t)x, B^* P(\tau)\Phi(\tau, t)y)_U \, d\tau. \quad (9.3.3.32)$$

(iii) As a consequence,

$$P(t) = P^*(t) \geq 0, \quad 0 \leq t \leq T. \quad (9.3.3.33)$$

(iv) The optimal cost of the optimal control problem on $[t, T]$ initiating at the point $x \in Y$ at the initial time t is:

$$J(u^0(\cdot, t; x), y^0(\cdot, t; x))$$

$$= \int_t^T \left\{ \| R\Phi(\tau, t)x \|_Z^2 + \| B^* P(\tau)\Phi(\tau, t)x \|_U^2 \right\} d\tau + \| G\Phi(T, t)x \|_{Z_f}^2$$

$$= (P(t)x, x)_Y, \quad (9.3.3.34)$$

where the optimal $u^0(\quad, t; x)$ is defined by (9.2.2) and the optimal $\Phi(\cdot, t)x$ is defined in (9.2.3) via the definition (9.3.2.1) of Φ.

Proof.   (i) We return to (9.2.1) and write explicitly via (9.1.18), (9.1.20), and (9.3.2.1), with $x \in Y$,

$$-u^0(\tau, t; x) = \int_\tau^T B^* e^{A^*(\sigma - \tau)} R^* R\Phi(\sigma, t)x \, d\sigma + B^* e^{A^*(T - \tau)} G^* G\Phi(T, t)x$$

(by (9.3.2.5)) $= \displaystyle\int_\tau^T B^* e^{A^*(\sigma - \tau)} R^* R\Phi(\sigma, \tau)\Phi(\tau, t)x \, d\sigma$

$$+ B^* e^{A^*(T - \tau)} G^* G\Phi(T, \tau)\Phi(\tau, t)x$$

(by (9.3.3.1)) $= B^* P(\tau)\Phi(\tau, t)x \in L_2(t, T; U), \quad (9.3.3.35)$

after recalling the evolution property (9.3.2.5) for Φ and the definition (9.3.3.1) for P. [Compare with the proof of Lemma 8.3.4.3.] For τ and $x \in Y$ fixed, the map $t \to u^0(\tau, t; x)$ is continuous in U by (9.3.3.4a) and (9.3.2.16) used in (9.3.3.31).

(ii), (iii), (iv) Since (A.1) and $R \in \mathcal{L}(Y; Z)$ imply (H.2) and (H.3) of Chapter 8 (see Orientation), then parts (ii), (iii), and (iv) are proved exactly as in Chapter 8, Lemma 8.3.4.4.   □

Proposition 9.3.3.5 *Assume (A.1), (A.2), and (A.3). Then, with reference to (9.3.2.1) and (9.3.3.1) we have*

$$\frac{d\Phi(t,s)x}{dt} = [A - BB^*P(t)]\Phi(t,s)x \in L_\infty(0,T;[\mathcal{D}(A^*)]'), \quad x \in \mathcal{D}(A).$$
$$(9.3.3.36)$$

Proof. We proceed as in Chapter 1, Lemma 1.4.6.3. We return to the optimal dynamics (9.3.2.1), rewritten explicitly as

$$\Phi(t,s)x = e^{A(t-s)}x + \int_s^t e^{A(t-\tau)}Bu^0(\tau,s;x)\,d\tau. \qquad (9.3.3.37)$$

The derivative

$$\frac{d\Phi(t,s)x}{dt} = A\Phi(t,s)x + Bu^0(t,s;x)$$

exists a.e. in t in $[\mathcal{D}(A^*)]'$ by (9.1.2). We next take the Y-inner product of (9.3.3.37) with y and differentiate in t with $x \in \mathcal{D}(A)$, $y \in \mathcal{D}(A^*)$, thus obtaining a.e. in t:

$$\left(\frac{d\Phi(t,s)x}{dt}, y\right)_Y = (A\Phi(t,s)x, y)_Y + (Bu^0(t,s;x), y)_Y$$

$$\text{(by (9.3.3.1))} \qquad = ([A - BB^*P(t)]\Phi(t,s)x, y)_Y \qquad (9.3.3.38)$$

$$= (A[I - A^{-1}BB^*P(t)]\Phi(t,s)x, y)_Y, \qquad (9.3.3.39)$$

where in writing (9.3.3.38) we have recalled the pointwise feedback expression (9.3.3.31) for u^0 [compare with Chapter 1, Eqn. (1.4.6.21)]. Notice that each term of the above expressions is well defined a.e. in t by (9.1.2) and (9.3.3.4). Thus, (9.3.3.39) yields (9.3.3.36). □

As a corollary of the extension result for u^0 in Eqn. (9.3.3.14) of Lemma 9.3.3.2, we readily obtain a corresponding extension result for $\Phi(\cdot,\cdot)$.

Proposition 9.3.3.6 *Assume (A.1), (A.2), and (A.3). With reference to (9.3.2.1), we have that $\Phi(\cdot,\cdot)$ can be extended from the original setting from $Y \to Y$ to all points x of the type $x = e^{A(t-s)}Bu$, for all $u \in U$, all $s \le t$, in particular (when $s = t$), on $BU = $ range of B, in the following sense:*

$$\Phi(\tau,r)e^{A(t-s)}Bu = e^{A(\tau-r)}e^{A(t-s)}Bu$$

$$+ \int_r^\tau e^{A(\tau-\sigma)}Bu^0\left(\sigma,r;e^{A(t-s)}Bu\right)d\sigma$$

$$\in [\mathcal{D}(A^*)]', \quad \forall u \in U, \ 0 \le s \le t; \ r \le \tau. \qquad (9.3.3.40)$$

Proof. The validity of Eqn. (9.3.3.40) follows at once from (9.3.3.14) and (9.1.2). □

9.3.4 Theorem 9.2.2: $P(t)$ Solves the Differential Riccati Equation (9.2.21) under (A.1), (A.2), and (A.3)

In this subsection, we shall show that the operator $P(t)$ defined constructively in (9.3.3.1) satisfies the DRE (9.2.21), is self-adjoint, and possesses the appropriate regularity properties mentioned in Theorem 9.2.2.

Step 1.   We begin with a regularity result that complements Proposition 9.3.3.6 and extends the action of the operator $R^* R\Phi(\cdot, \cdot)$ in (9.3.2.1) from its original domain Y to the [range of B] $\subset [\mathcal{D}(A^*)]'$.

Lemma 9.3.4.1   *Assume (A.1), (A.2), and (A.3). With reference to (9.3.2.1), we have that:*

(a) *the operator $R^* R\Phi(\cdot, t)B$ can be extended as continuous $U \to L_1(t, T; Y)$ uniformly in t,*

$$\sup_{0 \leq t \leq T} \|R^* R\Phi(\cdot, t)Bu\|_{L_1(t,T;Y)} \leq c_T \|u\|_U, \quad u \in U; \qquad (9.3.4.1)$$

(b) *similarly,*

$$\sup_{0 \leq t \leq T} \|G^* G\Phi(T, t)Bu\|_Y \leq c_T \|u\|_U, \quad u \in U. \qquad (9.3.4.2)$$

Proof.   (a) From (9.3.2.1) we have, with $u \in U$,

$$R^* R\Phi(\tau, t)Bu = R^* R e^{A(\tau-t)} Bu + \{R^* R L_t u^0(\cdot, t; Bu)\}(\tau). \qquad (9.3.4.3)$$

But from (A.1*) in the version of its consequence (9.1.21), and from (9.3.3.14) specialized with $t = s$, we obtain for the second term in (9.3.4.3) since R is bounded

$$\|[R^* R L_t u^0(\cdot, t; Bu)](\cdot)\|_{C([t,T];Z)} \leq c_T \|u^0(\cdot, t; Bu)\|_{L_2(t,T;U)}$$

$$\text{(by (3.3.14))} \qquad \leq c_T \|u\|_U. \qquad (9.3.4.4)$$

Returning to (9.3.4.3) we see that (9.3.4.4) and assumption (A.2) = (9.1.8) on the first term on the right-hand side of (9.3.4.3) readily yield (9.3.4.1).

(b) Similarly, from (9.3.2.1) we have

$$G^* G\Phi(T, t)Bu = G^* G e^{A(T-t)} Bu + \int_t^T G^* G e^{A(T-\tau)} Bu^0(\tau, t; Bu)\, d\tau.$$

$$(9.3.4.5)$$

Using (A.3*) in (9.3.4.5) twice, once in the form of property (9.1.13c) and once in the form of property (9.1.13f) followed by Schwarz's inequality, we obtain from (9.3.4.5) via (9.3.3.14):

$$\|G^*G\Phi(T, t)Bu\|_Y \le c_T[\|u\|_U + \|u^0(\cdot, t; Bu)\|_{L_2(t, T; U)}] \qquad (9.3.4.6)$$

$$(\text{by } (9.3.3.14)) \qquad \le c_T\|u\|_U, \qquad (9.3.4.7)$$

and (9.3.4.2) is proved by (9.3.4.7). $\square$

Step 2. **Proposition 9.3.4.2** *Assume (A.1), (A.2), and (A.3). With reference to (9.3.2.1) we have for $x \in \mathcal{D}(A)$:*

(i)

$$\frac{d\Phi(\tau, t)x}{dt} = -\Phi(\tau, t)[A - BB^*P(t)]x \in [\mathcal{D}(A^*)]', \quad x \in \mathcal{D}(A); \quad (9.3.4.8)$$

(ii)

$$\sup_{0 \le t \le T} \left\{ \left\| \frac{dR^*R\Phi(\tau, t)x}{dt} \right\|_{L_1(t, T; Y)} \right\} \le c_T\|Ax\|_Y, \quad x \in \mathcal{D}(A); \quad (9.3.4.9)$$

(iii)

$$\sup_{0 \le t \le T} \left\| \frac{dG^*G\Phi(T, t)x}{dt} \right\|_Y \le c_T\|Ax\|_Y, \quad x \in \mathcal{D}(A). \qquad (9.3.4.10)$$

Proof. (i) = (9.3.4.8)   The present approach does not use the evolution property (9.3.2.5) for Φ [which was proved in Lemma 9.3.3.2 by means of the characterizing formula (9.2.2a) for u^0], but instead it makes use of (9.2.2a) to derive (with an argument reminiscent of that in Lemma 9.3.3.2) the following interesting, preliminary property of u^0.

Lemma 9.3.4.3   *Assume (A.1), (A.2), and (A.3). Then, with reference to (9.2.2), we obtain for $x \in \mathcal{D}(A)$:*

$$\frac{du^0(\cdot, t; x)}{dt} = -u^0(\cdot, t; Ax) - u^0(\cdot, t; Bu^0(t, t; x))$$

$$\in L_2(t, T; U); \qquad (9.3.4.11)$$

$$u^0(t, t; x) = -B^*P(t)x \in \begin{cases} U \quad \forall t, \\ L_\infty(0, T; U). \end{cases} \qquad (9.3.4.12)$$

Proof of Lemma 9.3.4.3.   We return to the characterizing formula (9.2.2a) for $u^0(\cdot, t; x)$,

$$[I_t + L_t^*R^*RL_t + L_{tT}^*G^*GL_{tT}]u^0(\cdot, t; x)$$

$$= -L_t^*R^*Re^{A(\cdot - t)}x - L_{tT}^*G^*Ge^{A(T - t)}x, \qquad (9.3.4.13)$$

rewritten explicitly for $0 \leq t \leq \tau < T$, by (9.1.17)–(9.1.20), as

$$
u^0(\tau, t; x) + \int_\tau^T B^* e^{A^*(\sigma-\tau)} R^* R \int_t^\sigma e^{A(\sigma-r)} B u^0(r, t; x) \, dr \, d\sigma
$$

$$
+ B^* e^{A^*(T-\tau)} G^* G \int_t^T e^{A(T-\sigma)} B u^0(\sigma, t; x) \, d\sigma
$$

$$
= - \int_\tau^T B^* e^{A^*(\sigma-\tau)} R^* R e^{A(\sigma-t)} x \, d\sigma - B^* e^{A^*(T-\tau)} G^* G e^{A(T-t)} x. \quad (9.3.4.14)
$$

For τ fixed, and $x \in Y$, the map $t \rightarrow u^0(\tau, t; x)$ is continuous in U by Proposition 9.3.3.4(i). Thus, $\partial u^0(\,\cdot\,, t; x)/\partial t$ exists at least in $H^{-1}(0, T; U)$. Differentiating (9.3.4.14) yields, for $x \in \mathcal{D}(A)$,

$$
\frac{du^0(\tau, t; x)}{dt} + \int_\tau^T B^* e^{A^*(\sigma-\tau)} R^* R \int_t^\sigma e^{A(\sigma-r)} B \frac{du^0(r, t; x)}{dt} \, dr \, d\sigma
$$

$$
+ B^* e^{A^*(T-\tau)} G^* G \int_t^T e^{A(T-\sigma)} B \frac{du^0(\sigma, t; x)}{dt} \, d\sigma
$$

$$
= \int_\tau^T B^* e^{A^*(\sigma-\tau)} R^* R e^{A(\sigma-t)} B u^0(t, t; x) \, d\sigma
$$

$$
+ B^* e^{A^*(T-\tau)} G^* G e^{A(T-t)} B u^0(t, t; x)
$$

$$
+ \int_\tau^T B^* e^{A^*(\sigma-\tau)} R^* R e^{A(\sigma-t)} A x \, d\sigma
$$

$$
+ B^* e^{A^*(T-\tau)} G^* G e^{A(T-t)} A x, \quad x \in \mathcal{D}(A). \quad (9.3.4.15)
$$

By (9.1.17)–(9.1.20), then Eqn. (9.3.4.15) can be rewritten as

$$
[I_t + L_t^* R^* R L_t + L_{tT}^* G^* G L_{tT}] \frac{du^0(\,\cdot\,, t; x)}{dt}
$$

$$
= \left[L_t^* R^* R e^{A(\cdot-t)} B u^0(t, t; x) + L_{tT}^* G^* G e^{A(T-t)} B u^0(t, t; x) \right]
$$

$$
+ \left[L_t^* R^* R e^{A(\cdot-t)} A x + L_{tT}^* G^* G e^{A(T-t)} A x \right], \quad x \in \mathcal{D}(A) \quad (9.3.4.16a)
$$

$$
\in L_2(t, T; U). \quad (9.3.4.16b)
$$

Membership in $L_2(t, T; U)$ of the right-hand side of (9.3.4.16a), as stated in (9.3.4.16b), will be proved below. With reference to (9.3.4.16a) we have by (9.3.3.31) and (9.3.3.4a,b):

$$
u^0(t, t; x) = -B^* P(t) \Phi(t, t) x = -B^* P(t) x; \quad (9.3.4.17)
$$

$$
u^0(t, t; x) \in U \quad \text{for each } t; \qquad \sup_{0 \leq t \leq T} \|u^0(t, t; x)\|_U \leq c_T \|x\|_Y. \quad (9.3.4.18)
$$

We now prove (9.3.4.16b): that the right-hand side of identity (9.3.4.16a), that is, of (9.3.4.15), is in $L_2(t, T; U)$. We use (A.1), (A.2), and (A.3). Indeed, for fixed t:

$$R^* Re^{A(\cdot - t)} Bu^0(t, t; x) \in L_2(t, T; Y); \qquad G^* Ge^{A(T-t)} Bu^0(t, t; x) \in Y, \tag{9.3.4.19}$$

by, respectively, (A.2) = (9.1.8) and (A.3*) = (9.1.13) via (9.3.4.8). Then, via (9.3.4.19), the regularity (9.1.18b) of L_t^* and the regularity (9.1.20b) of L_{tT}^* [both by (A.1)] show, respectively, that the first and second term on the right-hand side of (9.3.4.16a), or (9.3.4.15), are in $L_2(t, T; U)$, as desired. The remaining two terms on the right-hand side of (9.3.4.16a) are obviously in $L_2(t, T; U)$ as well, by (9.1.18b) and (9.1.20b). Finally, the operator in the square brackets on the left-hand side of identity (9.3.4.16a) is precisely Λ_{tT} in (9.2.2c), which is boundedly invertible (uniformly in t; see (9.2.2d)) in $L_2(t, T; U)$. Thus, inverting Λ_{tT} in (9.3.4.16a) yields

$$\begin{aligned}
\frac{du^0(\tau, t; x)}{dt} &= \{\Lambda_{tT}^{-1}[L_t^* R^* Re^{A(\cdot - t)} Bu^0(t, t; x) \\
&\quad + L_{tT}^* G^* Ge^{A(T-t)} Bu^0(t, t; x)]\}(\tau) \\
&\quad + \{\Lambda_{tT}^{-1}[L_t^* R^* Re^{A(\cdot - t)} Ax + L_{tT}^* G^* Ge^{A(T-t)} Ax]\}(\tau) \\
&= -u^0(\tau, t; Bu^0(t, t; x)) - u^0(\tau, t; Ax), \quad x \in \mathcal{D}(A), \tag{9.3.4.20}
\end{aligned}$$

where, in the last step, we have invoked (9.2.2a) once more. Lemma 9.3.4.3 is fully proved. □

Completion of the Proof of (i) = (9.3.4.8). We return to the optimal dynamics (9.3.2.1b),

$$\Phi(\tau, t)x = e^{A(\tau - t)}x + \int_t^\tau e^{A(\tau - \sigma)} Bu^0(\sigma, t; x) \, d\sigma,$$

and differentiate in t with $x \in \mathcal{D}(A)$, thereby obtaining by use of (9.3.4.11), (9.3.2.1), and (9.3.4.17):

$$\begin{aligned}
\frac{d\Phi(\tau, t)x}{dt} &= -e^{A(\tau - t)} Ax - e^{A(\tau - t)} Bu^0(t, t; x) \\
&\quad + \int_t^\tau e^{A(\tau - \sigma)} B \frac{\partial u^0(\sigma, t; x)}{\partial t} \, d\sigma \\
\text{(by (9.3.4.11))} \quad &= -e^{A(\tau - t)} Ax - \int_t^\tau e^{A(\tau - \sigma)} Bu^0(\sigma, t; Ax) \, d\sigma \\
&\quad - e^{A(\tau - t)} Bu^0(t, t; x) - \int_t^\tau e^{A(\tau - \sigma)} Bu^0(\sigma, t; Bu^0(t, t; x)) \, d\sigma
\end{aligned}$$

(by (9.3.2.1)) $= -\Phi(\tau, t)Ax - \Phi(\tau, t)Bu^0(t, t; x)$ (9.3.4.21)

(by (9.3.4.17)) $= -\Phi(\tau, t)[A - BB^*P(t)]x \in [\mathcal{D}(A^*)]', \quad x \in \mathcal{D}(A).$

(9.3.4.22)

In (9.3.4.21) we have used the fact that $\Phi(\tau, t)$ is well defined from [range of B] to $[\mathcal{D}(A^*)]'$ by (9.3.3.40). Thus, (9.3.4.22) shows (9.3.4.8), and the proof of (i) = (9.3.4.8) is complete. $\square$

(ii) Applying R^*R to (9.3.4.22) we obtain for $x \in \mathcal{D}(A)$:

$$\frac{dR^*R\Phi(\tau, t)x}{dt} = -R^*R\Phi(\tau, t)Ax + R^*R\Phi(\tau, t)BB^*P(t)x \quad \text{a.e. in } t,$$

(9.3.4.23)

where, by (9.3.2.3) with $x \in \mathcal{D}(A)$ and R bounded, we have

$$\sup_{0 \le t \le T} \{\|R^*R\Phi(\cdot, t)Ax\|_{C([t,T];Y)}\} \le c_T \|Ax\|_Y;$$

(9.3.4.24)

whereas by (9.3.4.1) we have using also (A.2) and (A.3):

$$\|R^*R\Phi(\cdot, t)BB^*P(t)x\|_{L_1(t,T;Y)} \le c_T \|B^*P(t)x\|_U,$$

(9.3.4.25)

and hence, by (9.3.3.4) on $V(t) = B^*P(t)$ used in (9.3.4.25), we obtain

$$\sup_{0 \le t \le T} \|R^*R\Phi(\cdot, t)BB^*P(t)x\|_{L_1(t,T;Y)} \le c_T \sup_{0 \le t \le T} \|B^*P(t)x\|_U$$

$$\le c_T \|x\|_Y.$$

(9.3.4.26)

Then (9.3.4.24) and (9.3.4.26) used in (9.3.4.23) yield (9.3.4.9). Part (i) of Proposition 9.3.4.2 is proved. $\square$

(iii) To prove (9.3.4.10) we apply G^*G to (9.3.4.22):

$$\frac{dG^*G\Phi(T, t)x}{dt} = -G^*G\Phi(T, t)Ax + G^*G\Phi(T, t)BB^*P(t)x.$$

(9.3.4.27)

By (9.3.2.3) with $x \in \mathcal{D}(A)$ and G bounded, we have

$$\sup_{0 \le t \le T} \|G^*G\Phi(T, t)Ax\|_Y \le c_T \|Ax\|_Y, \quad x \in \mathcal{D}(A),$$

(9.3.4.28)

while by (9.3.4.2) and (9.3.3.4),

$$\sup_{0 \le t \le T} \|G^*G\Phi(T, t)BB^*P(t)x\|_Y \le c_T \sup_{0 \le t \le T} \|B^*P(t)x\| \le c_T\|x\|_Y, \quad x \in Y.$$

(9.3.4.29)

Then (9.3.4.28) and (9.3.4.29), inserted in (9.3.4.27), prove (9.3.4.10), as desired. $\square$

Step 3 We can finally prove our desired result.

Theorem 9.3.4.4 *Assume (A.1), (A.2), and (A.3). The operator $P(t)$ defined by (9.3.3.1) satisfies the differential Riccati equation for $x, y \in \mathcal{D}(A)$:*

$$\begin{cases} \left(\dfrac{d}{dt}P(t)x, y\right)_Y = -(R^*Rx, y)_Y - (P(t)x, Ay)_Y - (P(t)Ax, y)_Y \\ \qquad\qquad\qquad\quad + (B^*P(t)x, B^*P(t)y)_U, \\ P(T) = G^*G, \end{cases} \tag{9.3.4.30}$$

where

$$P'(t) : \text{continuous } \mathcal{D}(A) \to L_\infty(0, T; [\mathcal{D}(A)]'); \tag{9.3.4.31}$$

that is,

$$A^{*-1}P'(t)A^{-1} : \text{continuous } Y \to L_\infty(0, T; Y). \tag{9.3.4.32}$$

Proof. (As in Chapter 8, Theorem 8.3.4.5.) Starting from (9.3.3.1) we differentiate in t with $x, y \in \mathcal{D}(A)$

$$\begin{aligned} \left(\frac{d}{dt}P(t)x, y\right)_Y = {}& -(Rx, Ry)_Z - (A^*P(t)x, y)_Y \\ & + \left(\int_t^T e^{A^*(\tau-t)}\frac{dR^*R\Phi(\tau, t)x}{dt}\, d\tau, y\right)_Y \\ & + \left(e^{A^*(T-t)}\frac{dG^*G\Phi(T, t)x}{dt}, y\right)_Y, \end{aligned} \tag{9.3.4.33}$$

after recalling (9.3.3.1) and $\Phi(t, t)x = x$. We note that the third and fourth terms in (9.3.4.33) are well defined by Proposition 9.3.4.2 for $x, y \in \mathcal{D}(A)$. For these terms we invoke (9.3.4.8) of Proposition 9.3.4.2. We obtain for $x, y \in \mathcal{D}(A)$:

$$\begin{aligned} \left(\int_t^T e^{A^*(\tau-t)}\frac{dR^*R\Phi(\tau, t)x}{dt}\, d\tau, y\right)_Y & + \left(e^{A^*(T-t)}\frac{dG^*G\Phi(T, t)x}{dt}, y\right)_Y \\ = {}& -\left(\int_t^T e^{A^*(\tau-t)}R^*R\Phi(\tau, t)Ax\, d\tau, y\right)_Y \\ & + \left(\int_t^T e^{A^*(\tau-t)}R^*R\Phi(\tau, t)BB^*P(t)x\, d\tau, y\right)_Y \\ & - \left(e^{A^*(T-t)}G^*G\Phi(T, t)Ax, y\right)_Y \\ & + \left(e^{A^*(T-t)}G^*G\Phi(T, t)BB^*P(t)x, y\right)_Y \end{aligned} \tag{9.3.4.34}$$

$$\text{(by (9.3.3.1))} \qquad = -(P(t)Ax, y)_Y + (P(t)BB^*P(t)x, y)_Y \tag{9.3.4.35}$$

$$\text{(by (9.3.3.33))} \qquad = -(P(t)Ax, y)_Y + (B^*P(t)x, B^*P(t)y)_Y, \tag{9.3.4.36}$$

after invoking again (9.3.3.1) in the step leading to (9.3.4.35) and the self-adjointness of $P(t)$ in the step leading to (9.3.4.36). Inserting (9.3.4.36) into (9.3.4.33) yields the desired conclusion that $P(t)$ satisfies the DRE (9.3.4.30). Note that the last term in (9.3.4.36) is well defined by (9.3.3.4) on $V(t) = B^*P(t) \in \mathcal{L}(Y; U)$. Moreover, $P(T) = G^*G$ by (9.3.3.1). Statement (9.3.4.31) on $P'(t)$ is contained in the above analysis.   □

Remark 9.3.4.1 If $G = 0$, the following proof, which is simpler than the one in Proposition 9.3.4.2, can be given to show (9.3.4.23), that is,

$$\frac{dR^*R\Phi(\tau, t)x}{dt} = -R^*R\Phi(\tau, t)[A - BB^*P(t)]x, \quad x \in \mathcal{D}(A). \quad (9.3.4.37)$$

To this end, we shall use the explicit formula

$$R^*R\Phi(t, s)x = \left\{[I + R^*RL_sL_s^*]^{-1}R^*Re^{A(\cdot - s)}x\right\}(t), \quad (9.3.4.38)$$

where the bounded inverse $[I + R^*RL_tL_t^*]^{-1}$ in $\mathcal{L}(L_2(t, T; Y))$ is guaranteed by Lemma 9.3.2.1. To obtain (9.3.4.38) we return to (9.2.5) (already proved) and rewrite it as

$$y^0(\cdot, s; t_0) + L_sL_s^*R^*Ry^0(\cdot, s; y_0) = e^{A(\cdot - s)}y_0. \quad (9.3.4.39)$$

Applying R^*R on both sides of (9.3.4.39) would yield (9.3.4.38) after inversion. To justify the inversion and prove (9.3.4.38), we rewrite (9.3.4.28) as

$$R^*R\Phi(\tau, t)x + \{R^*RL_tL_t^*R^*R\Phi(\cdot, t)x\}(\tau) = R^*Re^{A(\tau - t)}x, \quad (9.3.4.40)$$

or explicitly, by (9.1.17a) and (9.1.18a),

$$R^*R\Phi(\tau, t)x + R^*R\int_t^\tau e^{A(\tau - \sigma)}B\int_\sigma^T B^*e^{A^*(r - \sigma)}R^*R\Phi(r, t)x\, dr\, d\sigma$$

$$= R^*Re^{A(\tau - t)}x. \quad (9.3.4.41)$$

With $x \in \mathcal{D}(A)$ we differentiate (9.3.4.41) in t, for example, as a distributional derivative in $H^{-1}(0, T; Y)$ by (9.3.2.16), obtaining

$$\frac{dR^*R\Phi(\tau, t)x}{dt} = R^*Re^{A(\tau - t)}B\int_t^T B^*e^{A^*(r - t)}R^*R\Phi(r, t)x\, dr$$

$$+ R^*R\int_t^\tau e^{A(\tau - \sigma)}B\int_\sigma^T B^*e^{A^*(r - \sigma)}\frac{d}{dt}R^*R\Phi(r, t)x\, dr\, d\sigma$$

$$= -R^*Re^{A(\tau - t)}Ax, \quad (9.3.4.42)$$

which in view of (9.1.17a), (9.1.18a), and (9.3.3.1) can be rewritten as

$$\left\{[I_t + R^*RL_tL_t^*]\frac{dR^*R\Phi(\cdot, t)x}{dt}\right\}(\tau)$$

$$= -R^*Re^{A(\tau - t)}Ax + R^*Re^{A(\tau - t)}BB^*P(t)x. \quad (9.3.4.43)$$

The above identity holds true (at least) in $H^{-1}(0, T; Y)$. Note that, for $x \in \mathcal{D}(A)$, the first term on the right-hand side of (9.3.4.43) is an element of $C([t, T]; Y)$, and hence it may be acted upon by $[I_t + R^*RL_t L_t^*]^{-1}$. As to the second term on the right-hand side of (9.3.4.43), we proceed as follows. By (9.3.4.1), we may extend (9.3.4.38) for $x = Bu$, to obtain

$$R^*R\Phi(\tau, t)Bu = \{[I_t + R^*RL_t L_t^*]^{-1}R^*Re^{A(\cdot -t)}Bu\}(\tau),$$

where

$$\sup_{0 \le t \le T} \left\{ \left\| \left([I_t + R^*RL_t L_t^*]^{-1}[R^*Re^{A(\cdot -t)}Bu] \right)(\cdot) \right\|_{L_1(t,T;Z)} \right\} \le c_T \|u\|_U.$$

$$(9.3.4.44)$$

Thus, recalling the global regularity (9.3.3.4) for $V(t) = B^*P(t)$, we obtain from (9.3.4.33) continuously in $x \in \mathcal{D}(A)$:

$$\begin{aligned} \frac{dR^*R\Phi(\cdot, t)x}{dt} &= -[I_t + R^*RL_t L_t^*]^{-1}[R^*Re^{A(\cdot -t)}Ax] \\ &\quad + [I_t + R^*RL_t L_t^*]^{-1}[R^*Re^{A(\cdot -t)}BB^*P(t)x] \\ &\in L_1(t, T; Y). \end{aligned} \qquad (9.3.4.45)$$

By (9.3.4.38), we see that (9.3.4.45) can be written equivalently as

$$\frac{dR^*R\Phi(\tau, t)x}{dt} = -R^*R\Phi(\tau, t)Ax + R^*R\Phi(\tau, t)BB^*P(t)x$$

$$\text{a.e. in } t, \quad x \in \mathcal{D}(A), \quad (9.3.4.46)$$

and (9.3.4.37) is established. □

9.3.5 Merger with the Proof of Chapter 8: Uniqueness and the IRE

Having shown that the Riccati operator $P(t)$ defined by (9.3.3.1) possesses the global regularity properties (9.3.3.3)–(9.3.3.5) and satisfies the Riccati equation (9.3.4.30), we are left with the statement of uniqueness to complete the proof of Theorem 9.2.2. But to this end, we can invoke Chapter 8, Theorem 8.3.7.1, which, under more general assumptions than presently available (see Chapter 8, assumption (8.3.7.2)), provides the desired uniqueness of the IRE. The properties in Chapter 8, (8.3.7.3)–(8.3.7.6) defining the class within which uniqueness is asserted, were proved in Theorem 9.3.3.1 above. The results in Chapter 8, Section 8.3.6 concerning the relationship between the DRE and the IRE extend to the present case of $G \ne 0$ [see also Chapter 1, Proposition 1.5.3.4]. Similarly, the proof of Chapter 8, Theorem 8.3.6.1 showing *directly* that $P(t)$ defined by (9.3.3.1) satisfies the IRE without passing through the DRE carries over to the present context.

To summarize the relevant results. We begin with the counterpart of Chapter 8, Lemma 8.3.6.2.

Lemma 9.3.5.1   *Assume (A.1), (A.2), and (A.3). Then, the following identity holds true for all $x, y \in Y$:*

$$
\int_t^T \left(B^* P(\tau) e^{A(\tau-t)} x, \, B^* P(\tau) e^{A(\tau-t)} y \right)_U dt
$$

$$
= - \left(R L_t u^0(\,\cdot\,, t; x), \, R e^{A(\,\cdot\,-t)} y \right)_{L_2(t,T;Z)}
$$

$$
- \left(G L_{tT} u^0(\,\cdot\,, t; x), \, G e^{A(T-t)} y \right)_{Z_f} \tag{9.3.5.1}
$$

$$
= \int_t^T \left(e^{A^*(\tau-t)} R^* R \int_t^\tau e^{A(\tau-\sigma)} B B^* P(\sigma) \Phi(\sigma, t) x \, d\sigma, \, y \right)_Y d\tau
$$

$$
+ \int_t^T \left(G^* G e^{A(T-\sigma)} B B^* P(\sigma) \Phi(\sigma, t) x \, d\sigma, \, e^{A(T-t)} y \right)_Y. \tag{9.3.5.2}
$$

[A proof of Lemma 9.3.5.1 in the *direct method* will be given in the proof of Theorem 9.3.6.5 below.]

As a corollary of Lemma 9.3.5.1, we obtain the counterpart of Chapter 8, Theorem 8.3.6.1.

Theorem 9.3.5.2   *Assume (A.1), (A.2), and (A.3). The operator $P(t)$ defined by (9.3.3.1) satisfies the IRE (9.2.2.2), that is,*

$$
(P(t)x, y)_Y = \left(G e^{A(T-t)} x, \, G e^{A(T-t)} y \right)_{Z_f} + \int_t^T \left(R e^{A(\tau-t)} x, \, R e^{A(\tau-t)} y \right)_Z d\tau
$$

$$
- \int_t^T \left(B^* P(\tau) e^{A(\tau-t)} x, \, B^* P(\tau) e^{A(\tau-t)} y \right)_U d\tau, \quad \forall \, x, y \in Y. \tag{9.3.5.3}
$$

We now turn to the counterpart of Propositions 8.3.5.1 and 8.3.5.2 of Chapter 8.

Theorem 9.3.5.3 [Relationship between DRE and IRE]   *Assume hypotheses (i), and (ii) of Section 9.1 on A and B, as well as (A.0) = (1.6). (However, the weaker assumption*

$$
G e^{At}, \, R e^{At} : \text{continuous } Y \to L_2(0, T; \,\cdot\,) \tag{9.3.5.4}
$$

on G and R would suffice.)

Let $P(t) \in \mathcal{L}(Y)$, $0 \le t \le T$, $P(T) = G^* G$, *be a self-adjoint operator, which satisfies the regularity properties (9.3.3.3)–(9.3.3.5) [as is the case for the operator $P(t)$ defined by (9.3.3.1), under assumptions (A.1), (A.2), and (A.3), by virtue of Theorem 9.3.3.1, as well as (9.3.3.33)].*

(a) *Let $P(t)$ be a solution of the DRE (9.2.21), that is, (9.3.4.30) [as is the case for the operator $P(t)$ defined by (9.3.3.1), by Theorem 9.3.4.4]. Then, such $P(t)$ also satisfies the IRE (9.2.22), that is, (9.3.5.3).*

(b) *Conversely, let $P(t)$ be a solution of the IRE (9.3.5.3) [as is the case for the operator $P(t)$ defined by (9.3.3.1), by Theorem 9.3.5.2.]*

Then, such $P(t)$ is also a solution of the DRE (9.3.4.30).

Theorem 9.3.5.4 (Uniqueness) *Assume hypotheses (i) and (ii) of Section 9.1 on A and B, as well as hypothesis (9.3.5.4) on R and G. Then, the IRE (9.3.5.3) admits at most one nonnegative, self-adjoint solution $P(t) = P^*(t) \in \mathcal{L}(Y)$ satisfying the regularity properties (9.3.3.3)–(9.3.3.5).*

9.3.6 A Second Direct Proof of Theorem 9.2.2: From the Well-Posedness of the IRE to the Control Problem. Dynamic Programming

Preliminaries The present subsection is the counterpart of Chapter 8, Section 8.4, under the present set of assumptions (A.0), (A.1), (A.2), and (A.3). It seeks to provide an alternate proof of Theorem 9.2.2, this time by the direct method: from the well-posedness of the IRE (9.2.22) to the optimal control problem. As in Chapter 8, Sections 8.4.1–8.4.5, well-posedness of the IRE (9.2.22) is established by combining (i) local well-posedness near T, by an application of the contraction mapping theorem, with (ii) global a priori bounds. The line of the arguments is exactly the same as the one carried out in Chapter 8, except that now the new set of hypotheses (A.0)–(A.3) will make it work (rather than hypotheses (H.1), (H.2), and (H.3) or (h.1) and (h.2) of Chapter 8). Thus, only some technical differences will be noted below.

With $V(t)$ and $Q(t, s)$ defined by (9.3.3.4) and (9.3.3.5), we return to the IRE (9.2.22), first with $x = e^{A(t-s)}Bu$ and $y = Bz$ and next with $y = Bu$, and obtain, respectively, the following Q- and V-equations:

$$
\begin{cases}
Q(t, s)u = F_{1,G}(t, s)u - \int_t^T Q^*(\tau, t)Q(\tau, s)u \, d\tau, & (9.3.6.1) \\[2mm]
F_{1,G}(t, s)u = F_1(t, s)u + B^*e^{A^*(T-t)}G^*Ge^{A(T-s)}Bu, & (9.3.6.2) \\[2mm]
F_1(t, s)u = \int_t^T B^*e^{A^*(\tau-t)}R^*Re^{A(\tau-s)}Bu \, d\tau \\[2mm]
\quad\quad\quad\quad = \{L_s^*R^*Re^{A(\cdot-s)}Bu\}(t), \quad s \leq t \leq T; & (9.3.6.3)
\end{cases}
$$

$$
\begin{cases}
V(t)x = F_{2,G}(t)x - \int_t^T Q^*(\tau, t)V(\tau)e^{A(\tau-t)}x \, d\tau, & (9.3.6.4) \\[2mm]
F_{2,G}(t)x = F_2(t)x + B^*e^{A^*(T-t)}G^*Ge^{A(T-t)}x, & (9.3.6.5) \\[2mm]
F_2(t)x = \int_t^T B^*e^{A^*(\tau-t)}R^*Re^{A(\tau-t)}x \, d\tau = \{L_t^*R^*Re^{A(\cdot-t)}x\}(t), & (9.3.6.6)
\end{cases}
$$

which are the present counterpart versions of [Eqns. (8.4.1.5)–(8.4.1.8)], with the same terms F_1 and F_2. We note that $F_{1,G}(t, s)$ compares closely with the terms leading to estimates (9.3.3.16) and (9.3.3.17), since $B^*e^{A^*(T-t)}x = \{L_{sT}^*x\}(t)$ by (9.1.20).

The counterpart versions of Chapter 8, Theorem 8.4.1.1 and Lemmas 8.4.1.1–8.4.1.3 are given next.

9.3.6.1 Local Well-Posedness of the IRE (9.2.22)

Theorem 9.3.6.1   *Assume (the standing hypotheses (i), (ii), and (A.0) of Section 9.1 as well as) (A.2) and (A.3). Then there exists a unique solution of Eqn. (9.3.6.1) for Q:*

$$Q(t, s) : \text{ continuous } U \to U_{s_0, T}; \quad \text{or } Q \in \mathcal{L}(U, U_{s_0, T}), \qquad (9.3.6.7)$$

and, under the additional hypothesis (A.1), a unique solution of Eqn. (9.3.6.4) for V:

$$V(t) \in \mathcal{L}(Y; U) : \text{ continuous } Y \to L_\infty(s_0, T; U);$$

$$\text{or } V \in \mathcal{B}([s_0, T]; \mathcal{L}(Y; U)), \qquad (9.3.6.8)$$

provided that s_0 is sufficiently close to T : $0 < s_0 < T$; the space $U_{s_0, T}$ is defined by (9.3.3.2), and the space $\mathcal{B}([s_0, T]; \cdot)$ by Chapter 8, Remark 8.1.1.

Remark 9.3.6.1   Properties (9.3.6.7) and (9.3.6.8) are local versions of the global properties shown in Theorem 9.3.3.1 above, for the solution operator $P(t)$ defined by (9.3.3.1), under (A.1), (A.2), and (A.3).

As in Chapter 8, the proof of Theorem 9.3.6.1 is based on the following lemmas.

Lemma 9.3.6.2   *Assume (A.1), (A.2), and (A.3). With reference to (9.3.6.2), we have*

$$\sup_{0 \leq s \leq T} \int_s^T \|F_{1,G}(t, s)u\|_U^2 \, dt \leq c_T \|u\|_U^2, \qquad (9.3.6.9)$$

that is, $F_{1,G} \in \mathcal{L}(U; U_{0,T})$.

Lemma 9.3.6.3   *Assume (A.2) and (A.3). With reference to (9.3.6.5), we have $F_{2,G}(t) \in \mathcal{L}(Y; U)$ and*

$$\sup_{0 \leq t \leq T} \|F_{2,G}(t)x\|_U \leq c_T \|x\|_Y, \qquad (9.3.6.10)$$

that is, $F_{2,G} \in \mathcal{B}([0, T]; \mathcal{L}(Y; U))$.

Proof of Lemma 9.3.6.2   With reference to (9.3.6.3), we estimate by invoking assumption (A.1*) in the form of (9.1.22) and (A.2) = (9.1.8):

$$\|F_1(\cdot, s)u\|_{L_2(s,T;U)} = \left\| L_s^* R^* Re^{A(\cdot - s)} Bu \right\|_{L_2(s,T;U)}$$

$$\text{(by (9.1.22))} \qquad \leq c_T \left\| R^* Re^{A(\cdot - s)} Bu \right\|_{L_1(s,T;Y)}$$

$$\text{(by (9.1.8b))} \qquad \leq c_T \|u\|_U. \qquad (9.3.6.11)$$

[The above steps are different from those used in the proof of Chapter 8, Lemma 8.4.1.2 under either (H.1) and (H.3) or else (h.2).] Moreover, with reference to (9.3.6.2), we estimate by invoking assumption (A.1) in the form of its consequence (9.1.20b) and (A.3*) = (9.1.13c):

$$\left\| B^* e^{A^*(T-t)} G^* G e^{A(T-s)} B u \right\|_{L_2(s,T;U)}$$

(by (9.1.20b)) $\leq C_T \left\| G^* G e^{A(T-s)} u \right\|_Y$

(by (9.1.13)) $\leq c_T \| u \|_U.$ (9.3.6.12)

Using (9.3.6.11) and (9.3.6.12) in (9.3.6.2) readily yields (9.3.6.9), as desired. □

Proof of Lemma 9.3.6.3   With reference to (9.3.6.6), we estimate by invoking assumption (A.2*) = (9.1.12) (after an innocuous extension by zero on $[T - t, T]$):

$$\| F_2(t) x \|_U = \left\| \left\{ L_t^* R^* R e^{A(\cdot - t)} x \right\} (t) \right\|_U$$

$$= \left\| \int_0^{T-t} B^* e^{A^* \sigma} R^* R e^{A\sigma} x \, d\sigma \right\|_U$$

(by (9.1.12b)) $\leq c_T \| e^{At} x \|_{L_\infty(0,T;Y)} \leq c_T \| x \|_Y.$ (9.3.6.13)

[The above steps are different from those used in the proof of Chapter 8, Lemma 8.4.1.3 under either (H.1) and (H.2) or else (h.1) and (h.2).] Moreover, with reference to (9.3.6.5), we estimate, by invoking assumption (A.3) in (9.1.9),

$$\left\| B^* e^{A^*(T-t)} G^* G e^{A(T-t)} x \right\|_U \leq C_T \left\| e^{A(T-t)} x \right\|_Y \leq c_T \| x \|_Y.$$ (9.3.6.14)

Using (9.3.6.13) and (9.3.6.14) in (9.3.6.5) yields (9.3.6.10), as desired. □

Starting from Lemmas 9.3.6.2 and 9.3.6.3, the proofs by contraction mapping to obtain both conclusions of Theorem 9.3.6.1 are exactly the same as in Chapter 8, Sections 8.4.2 for Q and 8.4.3 for V and will not be repeated. Theorems 8.4.2.2 and 8.4.3.1 in Chapter 8 hold true now, with F_1 and F_2 replaced by $F_{1,G}$ and $F_{2,G}$ now; thus the contraction constant is now $\theta_{s_0,T} = 4 r_T \sqrt{T - s_0}$, with $r_T = \| F_{1,G} \|_{\mathcal{L}(U;U_{0,T})}$ for Q (compare with [Chapter 8, Eqns. (8.4.2.1) and (8.4.2.7)]).

We can now conclude the proof of uniqueness for the solution of the IRE (9.2.22). Once Q and V are found, we then have a solution $P(t)$ of the corresponding Riccati integral equation (9.2.22), now rewritten by (9.3.3.4) as

$$P(t) x = \int_t^T e^{A^*(\tau-t)} R^* R e^{A(\tau-t)} x \, d\tau + e^{A^*(T-t)} G^* G e^{A(T-t)} x$$

$$- \int_t^T e^{A^*(\tau-t)} V^*(\tau) V(\tau) e^{A(\tau-t)} x \, d\tau, \quad x \in Y, \quad s_0 \leq t \leq T.$$ (9.3.6.15)

Then (9.3.6.15) reveals that $P(t)$ is self-adjoint on Y. Moreover, $P(t)$ possesses locally near T the properties of regularity corresponding to the global properties (9.2.16), (9.2.19a,b), and (9.2.20) of Theorems 9.2.1 and 9.2.2. These are precisely the properties shown globally in Theorem 9.3.3.1 for the solution $P(t)$ defined by

(9.3.3.1). But uniqueness of (Q and) V implies uniqueness of the solution of the IRE via (9.3.6.15), for now at least near T, $s_0 \leq t \leq T$.

Global A Priori Estimates   The treatment of Chapter 8, Section 8.4.4 may be simplified under the present assumption (A.1) = (9.1.7), to lead to stronger regularity results. Although Proposition 8.4.4.1 continues to hold true, with the same proof, a leaner and more direct approach may be given: Instead of seeking a fixed point solution of the candidate *optimal control* equation [Chapter 8, Eqn. (8.4.4.1)] and next obtaining the corresponding candidate *optimal trajectory* [Chapter 8, Step 2, Eqn. (8.4.4.6)], one may now obtain the candidate optimal trajectory directly by fixed point.

Step 1.   ***Proposition 9.3.6.4***   Let $V(t) \in \mathcal{B}([s_0, T]; \mathcal{L}(Y; U))$ be the operator guaranteed by Theorem 9.3.6.1 near T, say for $s_0 \leq s \leq t \leq T$, under assumptions (A.0), (A.2), and (A.3). Assume further (A.1). Then the integral equation

$$\Phi(t, s)x = e^{A(t-s)}x - \int_s^t e^{A(t-\tau)}BV(\tau)\Phi(\tau, s)x \, d\tau, \quad x \in Y, \quad (9.3.6.16)$$

admits (for s_0 sufficiently near T) a unique solution

$$\Phi(\cdot, s) : \text{ continuous } Y \rightarrow C([s, T]; Y), \quad (9.3.6.17)$$

which possesses the evolution properties

$$\Phi(\tau, \tau) = I; \quad \Phi(t, \sigma)\Phi(\sigma, \tau) = \Phi(t, \tau), \quad s_0 \leq s \leq \tau \leq \sigma \leq t \leq T. \quad (9.3.6.18)$$

Proof.   Let $f \in L_2(s, T; Y)$ and $\Phi(\cdot, \cdot) \in \mathcal{L}(Y; Y_{s_0,T})$, with the space $Y_{s_0,T}$ defined as in (9.3.3.2). By use of (A.1) = (9.1.7), and (9.3.3.2), we see that the term on the right-hand side of (9.3.6.16) satisfies the following estimate:

$$\left| \int_s^T \left(\int_s^t e^{A(t-\tau)}BV(\tau)\Phi(\tau, s)x \, d\tau, f(t) \right)_Y dt \right|$$

$$= \left| \int_s^T \int_s^t \left(V(\tau)\Phi(\tau, s)x, B^* e^{A^*(t-\tau)}f(t) \right)_U d\tau \, dt \right|$$

$$\leq \|\|V\|\|_{\mathcal{B},s_0,T} \int_s^T \left\{ \left[\int_s^t \|\Phi(\tau, s)x\|_Y^2 \, d\tau \right]^{\frac{1}{2}} \left[\int_0^{t-s} \|B^* e^{A^*\sigma}f(t)\|_U^2 \, d\sigma \right]^{\frac{1}{2}} \right\} dt$$

$$\text{(by (A.1))} \qquad \leq c_T \|\|V\|\|_{\mathcal{B},s_0,T} \|\Phi\|_{\mathcal{L}(Y;Y_{s_0,T})} \|x\|_Y \int_s^T \|f(t)\|_Y \, dt$$

$$\leq c_T \sqrt{T - s_0} \|\|V\|\|_{\mathcal{B},s_0,T} \|\Phi\|_{\mathcal{L}(Y;Y_{s_0,T})} \|x\|_Y \|f\|_{L_2(s_0,T;Y)}, \quad (9.3.6.19)$$

where the norm $\|\| \ \|\|$ for V is the norm of $\mathcal{B}([s_0, T]; \mathcal{L}(Y; U))$. By the contraction principle applied to the linear equation (9.3.6.16) via (9.3.6.19), we preliminarily get a unique solution $\Phi(\cdot, s) \in L_2(s, T; Y)$ for s_0 sufficiently close to T. Rewriting (9.3.6.16) as

$$\Phi(t, s)x = e^{A(t-s)}x - \{L_s[V(\cdot)\Phi(\cdot, s)x]\}(t), \qquad (9.3.6.20)$$

and invoking (A.1*) in the form of (9.1.17b) on L_s, with $V(\cdot)\Phi(\cdot, s)x \in L_2(s, T; U)$, by the assumption on V and the preliminary regularity on Φ, we then improve the regularity of Φ to $\Phi(\cdot, s)x \in C([s, T]; Y)$, as desired. Thus, (9.3.6.17) is proved. One can readily show then the evolution property (9.3.6.18) for the solution of the linear integral equation (9.3.6.16). $\quad\square$

Remark 9.3.6.2 (Counterpart of Chapter 8, Remark 8.4.4.1)   Under assumptions (A.1), (A.2), and (A.3), we have

$$V(t)\Phi(t, s)x = B^*P(t)\Phi(t, s)x \in U \text{ for all } t \in [s, T], \quad x \in Y,$$

by (9.3.6.8) and (9.3.6.17). If we consider the IRE = (9.2.22) with

$$x \text{ there replaced by } BB^*P(t)\Phi(t, s)x, \quad x \in Y;$$
$$y \text{ there replaced by } e^{A(t-s)}y, \quad y \in Y,$$

we see that the resulting IRE expression, rewritten as

$$\left(B^*P(t)\Phi(t, s)x, B^*P(t)e^{A(t-s)}y\right)_U$$

$$= \int_t^T \left(Re^{A(\tau-t)}BB^*P(t)\Phi(t, s)x, Re^{A(\tau-s)}y\right)_Z d\tau$$

$$- \int_t^T \left(B^*P(\tau)e^{A(\tau-t)}BB^*P(t)\Phi(t, s)x, B^*P(\tau)e^{A(\tau-s)}y\right)_U d\tau$$

$$+ \left(Ge^{A(T-t)}BB^*P(t)\Phi(t, s)x, Ge^{A(T-s)}y\right)_{Z_f},$$

or, by (9.3.3.4) and (9.3.3.5),

$$\left(V(t)\Phi(t, s)x, V(t)e^{A(t-s)}y\right)_U = \int_t^T \left(Re^{A(\tau-t)}BV(t)\Phi(t, s)x, Re^{A(\tau-s)}y\right)_Z d\tau$$

$$- \int_t^T \left(Q(\tau, t)V(t)\Phi(t, s)x, V(\tau)e^{A(\tau-s)}y\right)_U d\tau$$

$$+ \left(Ge^{A(T-t)}BV(t)\Phi(t, s)x, Ge^{A(T-s)}y\right)_{Z_f}$$

is well-defined for all t, x, $y \in Y$. This is so because of the established regularity properties (9.3.6.7) for Q, (9.3.6.8) for V, and (9.3.6.17) for Φ, along with hypothesis (A.3) = (9.1.9). We shall use this remark in the proof of Theorem 9.3.6.5 below.

Step 2.   (Corresponding to Chapter 8, Section 8.4.4, Theorem 8.4.4.2)

Theorem 9.3.6.5   *Assume (in addition to (i), (ii), and (A.0) of Section 9.1) (A.1), (A.2), and (A.3). Thus, according to the above analysis, the IRE (9.2.22) has a local solution $P(t)$ near T, say $s_0 \leq t \leq T$, given nonexplicitly by (9.3.6.15). Moreover, the operator $\Phi(t,s)$ is well-defined, as given by Proposition 9.3.6.4. Then, in fact, such local solution $P(t)$ is given explicitly by*

$$P(t)x = \int_t^T e^{A^*(\tau-t)} R^* R \Phi(\tau,t)x \, d\tau + e^{A^*(T-t)} G^* G \Phi(T,t)x,$$

$$s_0 \leq t \leq T, \ x \in Y \quad (9.3.6.21)$$

$$\in C([s_0, T]; Y).$$

Proof.   We must show (9.3.6.21) as a consequence of the IRE (9.2.22) and of the dynamics (9.3.6.16). This is done exactly as in the proof of Chapter 8, Theorem 8.4.4.2; the presence of $G \neq 0$ does not introduce any new essential difficulty; it merely complicates the resulting expressions.

Claim   Under the above assumptions (A.1), (A.2), and (A.3), the following identity holds true for all $x, y \in Y$ [which corresponds to Lemma 9.3.5.1 in the variational method]:

$$\int_t^T \left(B^* P(\tau) e^{A(\tau-t)} x, B^* P(\tau) e^{A(\tau-t)} y \right)_U d\tau$$

$$= \int_t^T \left(e^{A^*(\tau-t)} R^* R \int_t^\tau e^{A(\tau-\sigma)} B B^* P(\sigma) \Phi(\sigma,t)x \, d\sigma, y \right)_Y d\tau$$

$$+ \left(\int_t^T G^* G e^{A(T-\sigma)} B B^* P(\sigma) \Phi(\sigma,t)x \, d\sigma, e^{A(T-t)} y \right)_Y, \quad (9.3.6.22)$$

which is the counterpart of Eqn. (8.4.4.14) of Chapter 8; we notice that each term of (9.3.6.22) is well-defined by (9.3.6.8), (9.3.6.17), (A.1*) = (9.1.10b), and (A.3*) = (9.1.13).

Verification of (9.3.6.22) proceeds in two steps, precisely as in Chapter 8, Eqns. (8.4.4.14) through (8.4.4.20). In the first step, as in Chapter 8, Eqns. (8.4.4.15) and (8.4.4.17), we use (9.3.6.16) to replace $e^{A(\tau-t)}x$ on the left-hand side term of (9.3.6.22), we change the order of integration, and we obtain

$$\int_t^T \left(B^* P(\tau) e^{A(\tau-t)} x, B^* P(\tau) e^{A(\tau-t)} y \right)_U d\tau$$

$$= M(t) + \int_t^T \int_\sigma^T \left(B^* P(\tau) e^{A(\tau-\sigma)} B B^* P(\sigma) \Phi(\sigma,t)x, \right.$$

$$\left. B^* P(\tau) e^{A(\tau-\sigma)} e^{A(\sigma-t)} y \right)_U d\tau \, d\sigma; \quad (9.3.6.23)$$

$$M(t) = \int_t^T \left(B^* P(\tau) \Phi(\tau, t) x, \, B^* P(\tau) e^{A(\tau - t)} y \right)_U d\tau. \qquad (9.3.6.24)$$

The change of the order of integration is justified by the regularity property (9.3.6.7), while all terms in (9.3.6.23) and (9.3.6.24) are well-defined by (9.3.6.8), (9.3.6.17), and $(A.1^*) = (9.1.10b)$. In the second step, we perform the internal integration in (9.3.6.23) by recalling Remark 9.3.6.2, and thus using the IRE (9.2.22) with x there replaced by $BB^* P(\sigma) \Phi(\sigma, t) x$ (where $B^* P(\sigma) \Phi(\sigma, t) x \in U$ for all $\sigma, x \in Y$ by (9.3.6.8) and (9.3.6.17)), with y there replaced by $e^{A(\sigma - t)} y$, and with t there replaced by σ. We then obtain for all $x, y \in Y$:

$$\text{R.H.S. of } (9.3.6.23) = M(t)$$

$$+ \int_t^T \int_\sigma^T \left(R e^{A(\tau - \sigma)} BB^* P(\sigma) \Phi(\sigma, t) x, \, R e^{A(\tau - t)} y \right)_Z d\tau \, d\sigma$$

$$+ \int_t^T \left(G e^{A(T - \sigma)} BB^* P(\sigma) \Phi(\sigma, t) x, \, G e^{A(T - t)} y \right)_{Z_f} d\sigma$$

$$- \int_t^T \left(P(\sigma) BB^* P(\sigma) \Phi(\sigma, t) x, \, e^{A(\sigma - t)} y \right)_Y d\sigma, \qquad (9.3.6.25)$$

where cancellation takes place because of (9.3.6.24). All terms are well-defined by (9.3.6.8), (9.3.6.17), $(A.3^*) = (9.1.13)$, and $(A.1^*) = (9.1.10)$. After a further change in the order of integration, we obtain from (9.3.6.25)

$$\text{R.H.S. of } (9.3.6.23) = \int_t^T \left(e^{A^*(\tau - t)} R^* R \int_t^\tau e^{A(\tau - \sigma)} BB^* P(\sigma) \Phi(\sigma, t) x \, d\sigma, \, y \right)_Y$$

$$+ \left(\int_t^T G^* G e^{A(T - \sigma)} B^* B P(\sigma) \Phi(\sigma, t) x \, d\sigma, \, e^{A(T - t)} y \right)_Y. \qquad (9.3.6.26)$$

All terms are well-defined by (9.3.6.8), (9.3.6.17), $(A.1^*) = (9.1.10b)$, and $(A.3^*) = (9.1.13)$. Then (9.3.6.23) and (9.3.6.26) verify (9.3.6.22), as desired.

Next, we return to the IRE (9.2.22) and on its right-hand side we use identity (9.3.6.22), to obtain for $x, y \in Y$:

$$(P(t)x, y)_Y = \int_t^T \left(e^{A^*(\tau - t)} R^* R e^{A(\tau - t)} x, \, y \right)_Y d\tau + \left(e^{A^*(T - t)} G^* G e^{A(T - t)} x, \, y \right)_Y$$

$$- \int_t^T \left(e^{A^*(\tau - t)} R^* R \int_t^\tau e^{A(\tau - \sigma)} BB^* P(\sigma) \Phi(\sigma, t) x \, d\sigma, \, y \right)_Y$$

$$- \left(e^{A^*(T - t)} G^* G \int_t^T e^{A(T - \sigma)} BB^* P(\sigma) \Phi(\sigma, t) x \, d\sigma, \, y \right)_Y$$

$$\text{(by (9.3.6.16))} = \int_t^T \left(e^{A^*(\tau-t)} R^* R \Phi(\tau, t)x \, d\tau, y\right)_Y$$

$$+ \left(e^{A^*(T-t)} G^* G \Phi(T, t)x, y\right)_Y, \tag{9.3.6.27}$$

where in the last step we have recalled the dynamics (9.3.6.16), once with (t, s) replaced by (τ, t), and once replaced by (T, t). Thus, (9.3.6.27) proves (9.3.6.21), as desired. The regularity property $P(t)$: continuous $Y \to C([s_0, T]; Y)$ follows as in Theorem 9.3.3.1(i), Eqn. (9.3.3.3). $\square$

Step 3   (corresponding to Chapter 8, Section 8.4.4, Step 5)

Lemma 9.3.6.6   *Assume (A.1), (A.2), and (A.3). Then, with reference to (9.3.6.21) and (9.3.6.16), we have for $x \in Y$, $s_0 \le s \le t \le T$:*

$$-u^0(t, s; x) \stackrel{\text{def}}{=} B^* P(t) \Phi(t, s)x = V(t) \Phi(t, s)x \tag{9.3.6.28}$$

$$= \int_t^T B^* e^{A^*(\tau-t)} R^* R \Phi(\tau, s)x \, d\tau$$
$$+ B^* e^{A^*(T-t)} G^* G \Phi(T, s)x \tag{9.3.6.29a}$$

$$= \{L_s^* R^* R \Phi(\cdot, s)x\}(t) + \{L_{sT}^* G^* G \Phi(T, s)x\}(t), \tag{9.3.6.29b}$$

$$B^* P(t) \Phi(t, s) \in \mathcal{B}([s, T]; \mathcal{L}(Y; U)). \tag{9.3.6.29c}$$

Proof.   We use (9.3.6.21) and the evolution property $\Phi(\tau, t)\Phi(t, s)x = \Phi(\tau, s)x$ and the same with $\tau = T$ from (9.3.6.18), and, moreover, (9.1.18) and (9.1.20). The regularity in (9.3.6.29c) follows from $(A.2^*) = (9.1.12d)$ on $L_s^* R^* R$, (9.3.6.17) on Φ and $(A.3) = (9.1.9)$. $\square$

Step 4.   Proposition 9.3.6.7   *Assume (A.1), (A.2), and (A.3). Then, the operator $\Phi(t, s)$ defined as the unique solution of Eqn. (9.3.6.16) by Proposition 9.3.6.4 satisfies the following relations for $s_0 \le s \le t \le T$; $x \in Y$:*

$$\Phi(t, s)x = e^{A(t-s)}x + \int_s^t e^{A(t-\tau)} B u^0(\tau, s; x) \, d\tau \tag{9.3.6.30a}$$

$$= e^{A(t-s)}x + \{L_s u^0(\cdot, s; x)\}(t) \in C([s, T]; Y), \tag{9.3.6.30b}$$

$$\Phi(T, s)x = e^{A(T-s)}x + L_{sT} u^0(\cdot, s; x) \in Y, \tag{9.3.6.31}$$

$$\Phi(t, s)x = \left\{[I_s + L_s L_s^* R^* R]^{-1} \left[e^{A(\cdot-s)}x - L_s L_{sT}^* G^* G \Phi(T, s)x\right]\right\}(t). \tag{9.3.6.32}$$

Moreover, the quantity $u^0(\,\cdot\,, s; x)$ defined by (9.3.6.28) is given solely in terms of the problem data by the formula

$$u^0(t, s; x) = -\left\{\Lambda_{sT}^{-1}\left[L_s^* R^* Re^{A(\cdot -s)}x + L_{sT}^* G^* Ge^{A(T-s)}x\right]\right\}(t); \quad (9.3.6.33)$$

$$u^0(t, s; x) \in U \quad \text{for all } t; \quad u^0(t, s; x) \in L_\infty(s, T; U); \quad (9.3.6.34)$$

$$\Lambda_{sT} = [I_s + L_s^* R^* RL_s + L_{sT}^* G^* GL_{sT}] \in \mathcal{L}(L_2(s, T; U)); \quad (9.3.6.35)$$

$$\left\|\Lambda_{sT}^{-1}\right\|_{\mathcal{L}(L_2(s,T;U))} \le 1, \quad (9.3.6.36)$$

and can be extended therefore to any interval $[s, T]$, $0 \le s \le T$. Then, $\Phi(t, s)$ can, via (9.3.6.30), likewise be extended to any such interval $[s, T]$, $0 \le s \le T$, and we have here:

$$\|\Phi(\,\cdot\,, s)x\|_{C([s,T];Y)} \le c_T\|x\|_Y; \quad (9.3.6.37)$$

$$\|u^0(t, s; x)\|_U + \|u^0(\,\cdot\,, s; x)\|_{L_\infty(s,T;U)} \le c_T\|x\|_Y, \quad \text{for all } s \le t \le T. \quad (9.3.6.38)$$

Proof. Equation (9.3.6.30) follows from using (9.3.6.28) in (9.3.6.16), and then it specializes to (9.3.6.31). Inserting (9.3.6.29b) into (9.3.6.30b) yields (9.3.6.32) via Lemma 9.3.1.2. Conversely, inserting (9.3.6.30b) and (9.3.6.31) into (9.3.6.29b) yields (9.3.6.33) and (9.3.6.35). The regularity properties in (9.3.6.30b) for Φ and in (9.3.6.34) for u^0 were established in (9.3.6.17) and (9.3.6.29c), respectively. Next, (9.3.6.36) used in (9.3.6.33), along with (9.1.23) for L_{sT}^* and (9.1.22) for L_s^*, yield first

$$\|u^0(\,\cdot\,, s; x)\|_{L_2(s,T;U)} \le c_T\|x\|_Y, \quad x \in Y,$$

and using this preliminary result in (9.3.6.30b) along with (9.1.21) for L_s, we then obtain (9.3.6.37) for Φ. This, in turn, used in (9.3.6.29a) along with (9.1.24) for $L_s^* R^* R$ and (9.1.9), yields (9.3.6.38) for u^0, which is the same regularity statement as the one in (9.3.6.29c), or (9.3.6.34), with the added uniformity in s. $\square$

Remark 9.3.6.3   At this stage, having extended the operator $\Phi(t, s)$ over any interval $0 \le s \le t \le T$ by the explicit formula (9.3.6.33) for $u^0(t, s; x)$ used in (9.3.6.30b) [thereby removing the original constraint $s_0 \le s$], we may next extend $P(t)$ over all of $[0, T]$ by (9.3.6.21). We then show that the extended $P(t)$ continues to remain a solution of the IRE (9.2.22), which preserves its original properties (9.3.6.7) and (9.3.6.8).

Step 5.   **Theorem 9.3.6.8** (Global bound for $V(t)$)   *Assume (A.1), (A.2), and (A.3). For $V \in \mathcal{B}([s_0, T]; \mathcal{L}(Y; U))$ provided by Theorem 9.3.6.1, the following uniform global bound holds true:*

$$\begin{cases} \sup_{0 \le t \le T} \|V(t)x\|_U \le c_T\|x\|_Y, \quad x \in Y; & (9.3.6.39a) \\ \text{i.e., } V \in \mathcal{B}([0, T]; \mathcal{L}(Y; U)). & (9.3.6.39b) \end{cases}$$

Proof.   By (9.3.6.21) and (9.3.3.4) we estimate, by invoking (A.2*) in the form of its version (9.1.24), as well as (A.3) = (9.1.9),

$$
\|V(t)x\|_U = \|B^* P(t)x\|_U
$$

$$
= \left\| L_t^* R^* R \Phi(\,\cdot\,, t)x + B^* e^{A^*(T-t)} G^* G \Phi(T, t)x \right\|_U
$$

$$
\leq \left\| L_t^* R^* R \Phi(\,\cdot\,, t)x \right\|_U + \left\| B^* e^{A^*(T-t)} G^* G \Phi(T, t)x \right\|_U
$$

(by (9.1.24) and (9.1.9)) $\leq c_T \left\{ \|\Phi(\,\cdot\,, t)x\|_{L_\infty(t,T;Y)} + \|\Phi(T, t)x\|_Y \right\}$

$$
\leq c_T \|x\|_Y, \quad x \in Y, \quad 0 \leq t \leq T, \qquad (9.3.6.40)
$$

where in the last step we have invoked (9.3.6.37) via (A.1) as well. $\square$

Theorem 9.3.6.9 (Global bound for $Q(t, s)$)   *Assume (A.1), (A.2), and (A.3). For $Q \in \mathcal{L}(U; U_{s_0,T})$ provided by Theorem 9.3.6.1, the following global bound holds true:*

$$
\begin{cases} \sup_{0 \leq s \leq T} \int_s^T \|Q(t, s)u\|_U^2 \, dt \leq c_T \|u\|_U^2, \quad u \in U; & (9.3.6.41a) \\[2mm] or \; Q \in \mathcal{L}(U; U_{0,T}). & (9.3.6.41b) \end{cases}
$$

Proof.   Since the bound (9.3.6.9) on $F_{1,G}$ is global on $[0, T]$, that is, $F_{1,G} \in \mathcal{L}(U; U_{0,T})$, then the contraction constant $\theta_{s_0,T} = 4r_T \sqrt{T - s_0}, r_T = \|F_{1,G}\|_{\mathcal{L}(U;U_{0,T})}$, noted in the concluding discussion below (9.3.6.14) (following the proof of Lemma 9.3.6.3) for the fixed point argument for Q depends only on the interval size $(T - s_0)$. Thus, one can repeat the contraction argument of Chapter 8, Theorem 8.4.2.2 a finite number of times and obtain a global $Q \in \mathcal{L}(U; U_{0,T})$. $\square$

Step 6.   Theorem 9.3.6.10 ($P(t)$ as a global solution)   *Assume (A.1), (A.2), and (A.3). Then, the local solution $P(t)$ of the IRE (9.2.22) defined near T, $s_0 \leq t \leq T$, by (9.3.6.21) is valid on all of $[0, T]$, while preserving the regularity properties (9.3.6.39), (9.3.6.41), and (9.3.6.21) with $s_0 = 0$, as well as nonnegative self-adjointness.*

Proof.   We return to the Riccati integral equation solution, rewritten as a nonnegative, self-adjoint operator in (9.3.6.15), and we use the global results on V of Theorem 9.3.6.8 and on Q of Theorem 9.3.6.9. $\square$

Recovering the Optimal Control Problem   It remains to show that the quantities

$$
u^0(\,\cdot\,, s; y_0) \equiv -B^* P(t)\Phi(\,\cdot\,, s)y_0 \quad \text{and} \quad y^0(\,\cdot\,, s; y_0) \overset{\text{def}}{\equiv} \Phi(\,\cdot\,, s)y_0 \quad (9.3.6.42)
$$

obtained above in (9.3.6.28), (9.3.6.34), and (9.3.6.30) on any interval $0 \leq s \leq t^* \leq T$ are, in fact, the optimal control in feedback form and the optimal trajectory, respectively, of the optimal control problem (9.1.15) for (9.1.1), whereby the optimal cost is then

$$J_s(u^0) = \int_s^T \left[\|Ry^0(t, s; y_0)\|_Z^2 + \|u^0(t, s; y_0)\|_U^2 \right] dt + \|Gy^0(T, s; y_0)\|_{Z_f}^2$$

(9.3.6.43a)

$$= (P(s)y_0, y_0)_Y.$$

(9.3.6.43b)

This can be done as in Chapter 8, Section 8.4.5, in fact in two ways. One proof is to rely on the formula

$$J_s(u) - J_s(u^0) = (\Lambda_{sT}[u - u^0], [u - u^0])_{L_2(s,T;U)}$$

(9.3.6.44)

of Chapter 8, Lemma 8.3.1.3, shown there for an even more general context than the one presently available, with u^0 and the positive self-adjoint operator Λ_{sT} given by (9.3.6.33)–(9.3.6.35) in Proposition 9.3.6.7. Thus, such control $u^0 = u^0(\cdot, s; y_0)$ is the unique optimal control, as claimed, with feedback form and corresponding optimal trajectory given by (9.3.6.42).

A second independent proof, consistent with the approach of the present section, relies instead on the explicit formula (9.3.6.21) for $P(t)$ in Theorem 9.3.6.5, extended to $0 \leq s \leq t \leq T$ in Theorem 9.3.6.10. As in Chapter 8, Section 8.4.5, the key is the following:

Lemma 9.3.6.11 *Assume (A.1), (A.2), and (A.3). Then the following identity holds true:*

$$2(P(s)x, x)_Y + \int_s^T \left[\|R\Phi(\tau, s)x - Ry(\tau, s; x)\|^2 \right]_Z d\tau$$

$$+ \int_s^T \|u^0(\tau, s; x) - u(\tau, s; x)\|_U^2 d\tau + \|G\Phi(T, s)x - Gy(T, s; x)\|_{Z_f}^2$$

$$= J_s(x) + J_s^0(x),$$

(9.3.6.45)

where $P(\cdot)$ is given by (9.3.6.21). Moreover, $J_s(x)$ and $J_s^0(x)$ are the cost functionals (9.1.15) and the optimal cost (9.3.6.43), in a new notation that emphasizes only the dependence on the initial point $x \in Y$:

$$J_s(x) = \int_s^T \left[\|Ry(\tau, s; x)\|_Z^2 + \|u(\tau, s; x)\|_U^2 \right] d\tau + \|Gy(T, s; x)\|_{Z_f}^2,$$

(9.3.6.46)

$$J_s^0(x) = \int_s^T \left[\|R\Phi(\tau, s)x\|_Z^2 + \|u^0(\tau, s; x)\|_U^2 \right] d\tau + \|G\Phi(T, s)x\|_{Z_f}^2.$$

(9.3.6.47)

Finally, $\{u(\cdot, s; x), y(\cdot, s; x)\}$ is a pair consisting of a control and its corresponding solution which are related to each other by the dynamics (9.1.16).

Proof. We return to the basic formula (9.3.6.21) for $P(t)$, and we insert here the general dynamics (9.1.16) to replace $e^{A(\tau-s)}x$ and $e^{A(T-s)}x$:

$$(P(s)x, x)_Y \quad = \int_s^T \left(R\Phi(\tau, s)x, Re^{A(\tau-s)}x\right)_Z d\tau$$

$$+ \left(G\Phi(T, s)x, Ge^{A(T-s)}x\right)_{Z_f}$$

$$\text{(by (9.1.16))} \quad = \int_s^T (R\Phi(\tau, s)x, Ry(\tau, s; x))_Z d\tau$$

$$- (L_s^* R^* R\Phi(\cdot, s)x, u(\cdot, s; x))_{L_2(s,T;U)}$$

$$- (L_{sT}^* G^* G\Phi(T, s)x, u(\cdot, s; x))_{L_2(s,T;U)}$$

$$+ (G\Phi(T, s)x, Gy(T, s; x))_{Z_f}$$

$$\text{(by (9.2.1))} \quad = \int_s^T (R\Phi(\tau, s)x, Ry(\tau, s; x))_Z d\tau$$

$$+ \int_s^T (u^0(\tau, s; x), u(\tau, s; x))_U d\tau$$

$$+ (G\Phi(T, s)x, Gy(T, s; x))_{Z_f}. \tag{9.3.6.48}$$

9.3.7 Proof of Theorem 9.2.3

By (A.0) and (A.1) the results of Theorem 9.2.1 hold true. Because of the presence of $G \neq 0$, we shall first prove the transition property for u^0 and then use this to prove the transition property for Φ. This strategy (used already in the proof of Lemma 9.3.2.2) is meant to take advantage of the explicit formula (9.2.2a) for u^0, solely in terms of the problem data. A reverse strategy was used in Chapter 8, Section 8.3.7 in the case $G = 0$.

(i) We shall prove (9.2.23), that is,

$$u_T^0(\tau, t; x) = u_{T-t}^0(\tau - t, 0; x) \underset{(\text{in } \tau)}{\in} L_2(t, T; U). \tag{9.3.7.1}$$

We shall use (9.2.2a),

$$\left\{[I_t + L_t^* R^* RL_t + L_{tT}^* G^* GL_{tT}]u_T^0(\cdot, t; x)\right\}(\tau)$$

$$= -\left\{L_t^* R^* Re^{A(\cdot-t)}x\right\}(\tau) - \left\{L_{tT}^* G^* Ge^{A(T-t)}x\right\}(\tau), \tag{9.3.7.2}$$

which explicitly becomes, via (9.1.17)–(9.1.20),

$$u_T^0(\tau, t; x) + \int_\tau^T B^* e^{A^*(\sigma - \tau)} R^* R \int_t^\sigma e^{A(\sigma - \beta)} B u_T^0(\beta, t; x) \, d\beta \, d\sigma$$

$$+ B^* e^{A^*(T - \tau)} G^* G \int_t^T e^{A(T - r)} B u_T^0(r, t; x) \, dr$$

$$= - \int_\tau^T B^* e^{A^*(\sigma - \tau)} R^* R e^{A(\sigma - t)} x \, d\sigma - B^* e^{A^*(T - \tau)} G^* G e^{A(T - t)} x. \qquad (9.3.7.3)$$

In (9.3.7.3), writing $T - t$ in place of T, $\tau - t$ in place of τ, and 0 in place of t, we obtain

$$u_{T-t}^0(\tau - t, 0; x) + \int_{\tau - t}^{T - t} B^* e^{A^*(\alpha - (\tau - t))} R^* R \int_0^\alpha e^{A(\alpha - r)} B u_{T-t}^0(r, 0; x) \, dr \, d\alpha$$

$$+ B^* e^{A^*(T - \tau)} G^* G \int_0^{T - t} e^{A(T - t - \beta)} B u_{T-t}^0(\beta, 0; x) \, d\beta$$

$$= - \int_{\tau - t}^{T - t} B^* e^{A^*(\alpha - (\tau - t))} R^* R e^{A(\alpha - 0)} x \, d\alpha$$

$$- B^* e^{A^*(T - t - (\tau - t))} G^* G e^{A(T - t - 0)} x. \qquad (9.3.7.4)$$

Making the changes of variable: $\sigma = \alpha + t$ in the external integral and then $\beta = r + t$ in the internal integral on the left-hand side of (9.3.7.4), then $\beta + t = r$ and $\alpha + t = \sigma$, respectively, on the first and second integral on the right-hand side:

$$u_{T-t}^0(\tau - t, 0; x) + \int_\tau^T B^* e^{A^*(\sigma - \tau)} R^* R \int_t^\sigma e^{A(\sigma - \beta)} B u_T^0(\beta - t, 0; x) \, d\beta \, d\sigma$$

$$+ B^* e^{A^*(T - \tau)} G^* G \int_t^T e^{A(T - r)} B u_{T-t}^0(r - t, 0; x) \, dr$$

$$= - \int_\tau^T B^* e^{A^*(\sigma - \tau)} R^* R e^{A(\sigma - t)} x \, d\sigma - B^* e^{A^*(T - \tau)} G^* G e^{A(T - t)} x. \qquad (9.3.7.5)$$

Comparison between (9.3.7.5) and (9.3.7.3) reveals the following fact: that both $u_T^0(\cdot, t; x)$ and $u_{T-t}^0(\cdot - t, 0; x)$ satisfy the same equation. But then the difference satisfies

$$[I_t + L_t^* R^* R L_t + L_{tT}^* G^* G L_{tT}] \left[u_T^0(\cdot, t; x) - u_{T-t}^0(\cdot - t, 0; x) \right] = 0, \qquad (9.3.7.6)$$

from which (9.3.7.1) follows, as desired.

(ii) We next prove (9.2.24), that is,

$$\Phi_{T-t}(\sigma, 0)x = \Phi_T(t + \sigma, t) \underset{(\text{in } \sigma)}{\in} C([0, T]; Y). \tag{9.3.7.7}$$

We use (9.3.7.1) in the optimal dynamics (9.2.3):

$$\Phi_T(t + \sigma, t)x = e^{A(t+\sigma-t)}x + \int_t^{t+\sigma} e^{A(t+\sigma-\tau)}u_T^0(\tau, t; x)\, d\tau$$

$$\text{(by (9.3.7.1))} \qquad = e^{A\sigma}x + \int_t^{t+\sigma} e^{A(t+\sigma-\tau)}u_{T-t}^0(\tau - t, 0; x)\, d\tau$$

$$\text{(with } \tau - t = \beta) \qquad = e^{A\sigma}x + \int_0^{\sigma} e^{A(\sigma-\beta)}u_{T-t}^0(\beta, 0; x)\, d\beta$$

$$\text{(by (9.2.3))} \qquad = \Phi_{T-t}(\sigma, 0)x, \tag{9.3.7.8}$$

and (9.3.7.7) is proved.

(iii) Finally we prove (9.2.25), that is,

$$P_{T-t}(0) = P_T(t), \quad 0 \le t \le T. \tag{9.3.7.9}$$

We use (9.3.7.7) in (9.2.14):

$$P_T(t)x = \int_t^T e^{A^*(\tau-t)}R^*R\Phi_T(\tau, t)x\, d\tau + e^{A^*(T-t)}G^*G\Phi_T(T, t)x$$

$$\text{(by (9.3.7.7))} \quad = \int_t^T e^{A^*(\tau-t)}R^*R\Phi_{T-t}(\tau - t, 0)x$$

$$+ e^{A^*(T-t)}G^*G\Phi_{T-t}(T - t, 0)x$$

$$\text{(with } \tau - t = \beta) \quad = \int_0^{T-t} e^{A^*\beta}R^*R\Phi_{T-t}(\beta, 0)x + e^{A^*(T-t)}G^*G\Phi_{T-t}(T - t, 0)x$$

$$\text{(by (9.2.14))} \quad = P_{T-t}(0)x, \tag{9.3.7.10}$$

recalling (9.2.14) in the last step. Thus (9.3.7.9) is proved.

Remark 9.3.7.1   In the situation where existence and uniqueness of the IRE (9.2.22) has already been established as in Theorem 9.2.2(b$_3$), under the stronger set of assumptions (A.0), (A.1), (A.2), and (A.3) – then an alternative proof of (9.3.7.9) may be given, as follows. From (9.2.22) with $x, y \in Y$,

$$(P_T(t)x, y)_Y = \int_t^T \left(Re^{A(r-t)}x, Re^{A(r-t)}y\right)_Z dr + \left(Ge^{A(T-t)}x, Ge^{A(T-t)}y\right)_{Z_f}$$

$$- \int_t^T \left(B^*P_T(r)e^{A(r-t)}x, B^*P_T(r)e^{A(r-t)}y\right)_U dr. \tag{9.3.7.11}$$

Next, we rewrite (9.3.7.11) with $T + \tau$ in place of T, and with $t + \tau$ in place of t; we then use the change of variable $\sigma = r - \tau$. We obtain for $x, y \in Y$:

$$(P_{T+\tau}(t + \tau)x, y)_Y$$

$$= \int_{t+\tau}^{T+\tau} \left(Re^{A(r-(t+\tau))}x, Re^{A(r-(t+\tau))}y \right)_Z dr$$

$$+ \left(Ge^{A(T+\tau-(t+\tau))}x, Ge^{A(T+\tau-(t+\tau))}y \right)_{Z_f}$$

$$- \int_{t+\tau}^{T+\tau} \left(B^*P_{T+\tau}(r)e^{A(r-(t+\tau))}x, B^*P_{T+\tau}(r)e^{A(r-(t+\tau))}y \right)_U dr$$

$$= \int_t^T \left(Re^{A(\sigma-t)}x, Re^{A(\sigma-t)}y \right)_Z d\sigma + \left(Ge^{A(T-t)}x, Ge^{A(T-t)}y \right)_{Z_f}$$

$$- \int_t^T \left(B^*P_{T+\tau}(\tau + \sigma)e^{A(\sigma-t)}x, B^*P_{T+\tau}(\tau + \sigma)e^{A(\sigma-t)}y \right)_U d\sigma. \quad (9.3.7.12)$$

By Theorem 9.2.2(b_3), Eqn. (9.3.7.12) has the unique solution $P_T(t)$. Thus

$$P_T(t) = P_{T+\tau}(t + \tau), \quad \tau + t \geq 0, \; 0 \leq t \leq T, \quad (9.3.7.13)$$

and (9.3.7.9) is proved for $\tau = -t$. $\quad\square$

9.4 Isomorphism of $P(t)$, $0 \leq t < T$, and Exact Controllability of $\{A^*, R^*\}$ on $[0, T - t]$ When $G = 0$

In this section we complete the description of the optimal control problem (9.1.13) on $[s, T]$ by providing an explicit formula for the optimal cost (9.2.18) in $[s, T]$. Besides being of interest in itself, such formula reveals, as an immediate corollary, an interesting and important equivalence between the property that $P(t)$ be an isomorphism on Y and the exact controllability property on $[0, T - t]$ (from the origin) of the dynamical system on Y,

$$\dot\zeta(t) = A^*\zeta(t) + Sg(t), \quad \zeta(0) = 0, \quad (9.4.1)$$

where we may indifferently choose

$$\text{either } S = R^* \quad \text{or} \quad S = (R^*R)^{\frac{1}{2}}, \quad (9.4.2)$$

and what follows holds true for either choice.

Definition 9.4.1   The dynamical system (9.4.1), in short the pair $\{A^*, S\}$, is exactly controllable from the origin on the space Y over the time interval $[0, T]$, $T < \infty$, with controls $g \in L_2(0, T; Y)$ in the case where the totality of all solutions points $\zeta(T)$ of (9.4.1) [with initial condition $\zeta(0) = 0$] coincides with all of Y when g runs over

$L_2(0, T; Y)$. Equivalently, in case the (obviously bounded) input-solution operator of (9.4.1),

$$O_T g = \int_0^T e^{A^*(T-t)} Sg(t)\, dt, \qquad (9.4.3)$$

is surjective:

$$O_T : L_2(0, T; Y) \quad \text{onto } Y. \qquad (9.4.4)$$

Recalling a standard result in operator theory, in particular for bounded operators on Hilbert spaces [Taylor, Lay, 1980, pp. 235–6], we find that condition (9.4.4) admits, in turn, the following equivalent formulation: There exists a constant $C_T > 0$ such that

$$\|O_T^* y\|_{L_2(0,T;Y)} \geq C_T \|y\|_Y, \quad \forall\, y \in Y. \qquad (9.4.5)$$

Here O_T^* is the adjoint of $O_T : (O_T g, y)_Y = (g, O_T^* y)_{L_2(0,T;Y)}$ and is given explicitly by

$$(O_T^* y)(t) = S^* e^{A(T-t)} y, \quad 0 \leq t \leq T. \qquad (9.4.6)$$

Thus, by (9.4.5) and (9.4.6), exact controllability from the origin of (9.4.1) within the class of $L_2(0, T; Y)$-controls is equivalent to the condition: There exists $C_T > 0$ such that

$$\int_0^T \|S^* e^{At} y\|_Y^2\, dt \geq C_T^2 \|y\|_Y^2, \quad \forall\, y \in Y, \qquad (9.4.7)$$

or, explicitly for either choice in (9.4.2),

$$\int_0^T \left\|(R^* R)^{\frac{1}{2}} e^{At} y\right\|_Y^2 dt = \int_0^T \|R e^{At} y\|_Y^2\, dt \geq C_T^2 \|y\|_Y^2. \qquad (9.4.8)$$

Theorem 9.4.1   *Assume (in addition to the standing assumptions (i), (ii) = (9.1.2), and (A.0) = (9.1.6) of Section 9.1) hypothesis (A.1) = (9.1.7) and $G = 0$. Then:*

(i) *The optimal cost of the optimal control problem (9.1.15) with $G = 0$ over the interval $[t, T]$ is given by*

$$J_{tT}^0(y_0) \equiv J(u^0(\cdot\,, t; y_0), y^0(\cdot\,, t; y_0)) = (P_T(t) y_0, y_0)_Y$$

$$= \left\| \left[I_t + (R^* R)^{\frac{1}{2}} L_t L_t^* (R^* R)^{\frac{1}{2}} \right]^{-\frac{1}{2}} (R^* R)^{\frac{1}{2}} \left[e^{A(\cdot - t)} y_0 \right] \right\|_{L_2(t,T;Z)}^2 \qquad (9.4.9a)$$

$$= \left\| \left[I_t + R L_t L_t^* R^* \right]^{-\frac{1}{2}} R e^{A(\cdot - t)} y_0 \right\|_{L_2(t,T;Z)}^2 \qquad (9.4.9b)$$

$$= \left([I_t + R L_t L_t^* R^*]^{-1} R e^{A(\cdot - t)} y_0, \, R e^{A(\cdot - t)} y_0 \right)_{L_2(t,T;Z)}. \qquad (9.4.9c)$$

(ii) The operator $P(t)$, $0 \leq t < T$, defined by (9.2.14) with $G = 0$, is an isomorphism on Y if and only if the following estimate holds true:

$$\int_t^T \left\| Re^{A(\tau - t)} y_0 \right\|_Y^2 d\tau = \int_t^T \left\| (R^*R)^{\frac{1}{2}} e^{A(\tau - t)} y_0 \right\|_Y^2 d\tau$$

$$\equiv \int_0^{T-t} \| S^* e^{A\sigma} y_0 \|_Y^2 \, d\sigma$$

$$\geq c_{T-t} \| y_0 \|_Y^2, \quad y_0 \in Y, \tag{9.4.10}$$

for some constant $c_{T-t} > 0$, that is, if and only if the dynamical system $\{A^*, S\}$ in (9.4.1) is exactly controllable on Y (from the origin) over the interval $[0, T - t]$, within the class of $L_2(0, T - t; Y)$-controls, for either choice of S in (9.4.2).

Proof.

(i) Step 1   Equations (9.4.9b,c) were already proved in Chapter 8, Eqn. (8.2.1.23) of Theorem 8.2.1.1, with proof given in Chapter 8, Lemma 8.3.1.1, Eqns. (8.3.1.17) and (8.3.1.18), indeed only under the weaker assumptions [Chapter 8, (h.1) = (8.2.1.10) and (h.2) = (8.2.1.12)]. These identities were also noted in Chapter 1, Remark 1.2.1.3. Here, we reproduce the short proof for sake of convenience. We compute from the definition of optimal cost and the optimal dynamics (9.2.3), via the characterization (9.2.1),

$$J_{tT}^0(y_0) \equiv J(u^0(\cdot, t; y_0), y^0(\cdot, t; y_0))$$

$$= (Ry^0(\cdot, t; y_0), Ry^0(\cdot, t; y_0))_{L_2(t,T;Y)} + \| u^0(\cdot, t; y_0) \|_{L_2(t,T;U)}^2$$

(by (9.2.1)) $\quad = (Ry^0(\cdot, t; y_0), Ry^0(\cdot, t; y_0))_{L_2(t,T;Z)}$

$$+ (L_t^* R^* Ry^0(\cdot, t; y_0), L_t^* R^* Ry^0(\cdot, t; y_0))_{L_2(t,T;U)}$$

$$= ([I_t + RL_t L_t^* R^*] Ry^0(\cdot, t; y_0), Ry^0(\cdot, t; y_0))_{L_2(t,T;Z)} \tag{9.4.11}$$

$$= \left(Re^{A(\cdot - t)} y_0, Ry^0(\cdot, t; y_0) \right)_{L_2(t,T;Z)} \tag{9.4.12}$$

$$= \left(Re^{A(\cdot - t)} y_0, [I_t + RL_t L_t^* R^*]^{-1} Re^{A(\cdot - t)} y_0 \right)_{L_2(t,T;Z)}, \tag{9.4.13}$$

where in the last step we have used

$$[I_t + RL_t L_t^* R^*] Ry^0(\cdot, t; y_0) = Re^{A(\cdot - t)} y_0 \tag{9.4.14}$$

obtained from

$$y^0(\cdot, t; y_0) = e^{A(\cdot - t)} y_0 - L_t L_t^* R^* Ry^0(\cdot, t; y_0), \tag{9.4.15}$$

upon applying R, where (9.4.13) is in turn the result of inserting (9.2.1a) with $G = 0$ in (9.2.3); see Chapter 8, Eqn. (8.2.1.20).

Then (9.4.13) is (9.4.9c). The equality with (9.4.9a) can be readily verified.

(ii) With $R \in \mathcal{L}(Y; Z)$ by the standing assumption (A.0) = (9.1.6) and L_t continuous as in (9.1.17b) by the assumption (A.1*) = (9.1.10), we have that the operators

$$\text{either } \mathcal{T}_t \equiv [I_t + RL_t L_t^* R^*] \quad \text{or} \quad \mathcal{T}_t \equiv \left[I_t + (R^*R)^{\frac{1}{2}} L_t L_t^* (R^*R)^{\frac{1}{2}}\right]$$

$$\text{are isomorphisms on } L_2(t, T; Y), \quad (9.4.16)$$

with norms that may be made independent of t by (9.1.21) and (9.1.22). Thus, using (9.4.16) into (9.4.9), we have

$$J_{tT}^0(y_0) = (P(t)y_0, y_0)_Y \sim \left\|(R^*R)^{\frac{1}{2}} e^{A(\cdot - t)} y_0\right\|_{L_2(t,T;Y)}^2 = \left\|Re^{A(\cdot - t)} y_0\right\|_{L_2(t,T;Z)}^2,$$

or

$$c_{tT} \int_0^{T-t} \|S^* e^{A\sigma} y_0\|^2 \, d\sigma \leq (P(t)y_0, y_0)_Y \leq C_{tT} \int_0^{T-t} \|S^* e^{A\sigma} y_0\|_Y^2 \, d\sigma, \quad (9.4.17)$$

for either choice $S^* = R$ or $S^* = (R^*R)^{\frac{1}{2}}$ as in (9.4.2). In (9.4.17), $\sim$ means that the left-hand side is bounded from above and from below by the right-hand side modulo positive constants c_{tT}, C_{tT}; in fact, respectively, these constants are the square of the $\mathcal{L}(L_2(t, T; Y))$-norm of $\mathcal{T}_t^{-\frac{1}{2}}$ in one direction and the reciprocal of the square of the $\mathcal{L}(L_2(t, T; Y))$-norm of $\mathcal{T}_t^{\frac{1}{2}}$ in the other direction, recalling (9.4.16). Then, the double inequality (9.4.17) readily proves the equivalence between the isomorphism of $P(t)$, that is, essentially $(P(t)y_0, y_0) \geq k_{tT} \|y_0\|^2$ for $k_{tT} > 0$, since $P(t) \in \mathcal{L}(Y)$ is self-adjoint, and the exact controllability condition for $\{A^*, S\}$,

$$\int_0^{T-t} \|S^* e^{A\sigma} y_0\|_Y^2 \, d\sigma \geq c_{tT} \|y_0\|^2, \quad y_0 \in Y, \quad (9.4.18)$$

over $[0, T - t]$ within the class of $L_2(0, T - t; U)$-controls, according to the characterization (9.4.8) with T replaced by $T - t$, for either choice $S = R$ or $S = (R^*R)^{\frac{1}{2}}$ in (9.4.2). $\square$

Remark 9.4.1   As a matter of fact, Theorem 9.4.1 requires only the assumptions (h.1) = (8.2.1.10) and (h.3) = (8.2.1.12) of Chapter 8, that is,

$$Re^{At} : \text{ continuous } Y \to L_2(0, T; Z); \quad RL_t : \text{ continuous } L_2(t, T; U)$$

$$\to L_2(t, T; Z), \quad (9.4.19)$$

in which case (9.4.16) is satisfied; see also Chapter 8, Lemma 8.3.1.1, Eqns. (8.3.1.17) and (8.3.1.18).

Remark 9.4.2   Part (i) of Theorem 9.4.1 (i.e., formula (9.4.9)), holds true, of course, also in the analytic case of Chapter 1: However, in the framework of Chapter 1, $P(t)$ cannot be an isomorphism, and indeed, via the smoothing property (1.2.1.5) in Chapter 1, $P(t)$ is compact if A has compact resolvent. In other words: in the framework of the analytic case of Chapter 1, the pair $\{A^*, R^*\}$ cannot be exactly controllable in finite time (in the sense of Definition 9.4.1), and so part (ii) of Theorem 9.4.1 is not applicable.

9.5 Nonsmoothing Observation R: "Limit Solution" of the Differential Riccati Equation under the Sole Assumption (A.1) When $G = 0$

9.5.1 Introduction

Let $G = 0$ in (9.1.15). Under the standing assumptions (i), (ii) $= (9.1.2)$, and (A.0) $=$ (9.1.6) of Section 9.1, hence with $R \in \mathcal{L}(Y; Z)$ generally nonsmoothing, Theorem 9.2.1 provides – under the sole main abstract trace regularity assumption (A.1) $=$ (9.1.7) – a series of results, (9.2.1) through (9.2.18), of the optimal control problem (9.1.15) for (9.1.1). These, in fact, hold true only under the assumptions in (9.4.19), as pointed out in Chapter 8, Theorem 8.2.1.1. Among them, we cite the existence of the nonnegative, self-adjoint operator $P(t)$ in (9.2.14), which satisfies the pointwise synthesis (9.2.17) of the optimal pair, and which provides the optimal cost via (9.2.18). What is conspicuously absent from the statement of Theorem 9.2.1 is the claim that, in the present generality, with $R \in \mathcal{L}(Y; Z)$ only, such $P(t)$ is a solution of a differential (or integral) Riccati equation, in some sense. When R is only in $\mathcal{L}(Y; Z)$, lack of (proof of) regularity properties of the gain operator $B^*P(t)$ prevents one from justifying the formal steps leading to the desired conclusion [as in Theorem 9.2.2 under the smoothing hypothesis (A.2) $= (9.1.8)$ on R] that such $P(t)$ satisfies the differential Riccati equation (9.2.21) or the integral Riccati equation (9.2.22) with $G = 0$. We recall from Theorem 9.4.1 that, at least when the pair $\{A^*, R^*\}$ is exactly controllable from the origin on $[0, T]$ (in the sense of Definition 9.4.1), the operator $P(t)$, $0 \leq t < T$, is an isomorphism on Y, and hence $B^*P(t)$ is bounded from Y to U if and only if B is bounded from U to Y, the trivial case. Thus, in general, $B^*P(t)$ is unbounded [see Chapter 8, Section 8.7], and it is an issue whether, for example, $B^*P(t)$ is even densely defined. In the setting of Chapter 8, as well as in the present setting of Chapter 9, it turned out that $B^*P(t) \in \mathcal{L}(Y; U)$ for all $0 \leq t \leq T$.

Under these circumstances – with $R \in \mathcal{L}(Y; Z)$ only, and subject to (A.1) $=$ (9.1.7) – the issue remains to be settled whether the operator $P(t)$ of Theorem 9.2.1(a) may be regarded as a "solution" of a corresponding differential Riccati equation in a suitably weaker sense. Two approaches are presented in the remainder of this chapter to answer the above question: (i) one, in the present Section 9.5, which is based on the idea of regularizing the observation operator R, and then taking a limit process, as the parameter of regularization tends to zero, and (ii) another, in the subsequent Section 9.6, which is based on the idea of a dual Riccati equation. Finally, in the subsequent Chapter 10, we shall consider the case where the operator $P(t)$ does satisfy the differential Riccati equation, under a weak smoothing hypothesis on R, as well as the application of this result to second-order hyperbolic equations, first-order hyperbolic systems, Euler–Bernoulli equations, Kirchhoff equations, Schrödinger equations, etc.

9.5.2 Regularization of R. Statement of Main Result: Theorem 9.5.2.2

Throughout this section we assume only the standing assumptions (i), (ii) = (9.1.2), (A.0) = (9.1.6) [i.e., $R \in \mathcal{L}(Y; Z)$], and (A.1) = (9.1.7).

The basic idea of the present section is to show that the operator $P(t)$ of Theorem 9.2.1 may be seen as the limit, as the parameter of regularization $\epsilon \downarrow 0$, of appropriate, bonafide Riccati operators $P_\epsilon(t)$ of regularizing problems with smoothing observation R_ϵ which fit into the framework of Theorem 9.2.2, whereby, in particular, all such $P_\epsilon(t)$ satisfy the differential Riccati equation (9.2.21).

Regularizing Approach. Hypotheses   We introduce a parameter of regularization $\epsilon \downarrow 0$, $\epsilon_0 \geq \epsilon > 0$, and consider the family $\{R_\epsilon\}$ of observation operators, $R_\epsilon \in \mathcal{L}(Y; Z)$ satisfying hypothesis (A.2) = (9.1.8) of Section 9.1:

(A.2$_\epsilon$): The map $R_\epsilon^* R_\epsilon e^{At} B$ can be extended as a map

$$R_\epsilon^* R_\epsilon e^{At} B : \text{ continuous } U \to L_1(0, T; Y), \qquad (9.5.2.1a)$$

that is,

$$\int_0^T \|R_\epsilon^* R_\epsilon e^{At} Bu\|_Y \, dt \leq C_{T,\epsilon} \|u\|_U, \quad u \in U, \qquad (9.5.2.1b)$$

and, moreover,

$$R_\epsilon^* R_\epsilon \to R^* R \text{ strongly: } R_\epsilon^* R_\epsilon x \to R^* R x, \quad \text{as } \epsilon \downarrow 0, x \in Y. \quad (9.5.2.2)$$

Remark 9.5.2.1   Given $R \in \mathcal{L}(Y; Z)$, such family $\{R_\epsilon\}$ always exists: We may, in fact, take

$$R_\epsilon = R \left[\frac{1}{\epsilon} R \left(\frac{1}{\epsilon}, A \right) \right], \qquad (9.5.2.3)$$

$R(\cdot, A)$ being the resolvent operator of A, satisfying the well-known strong limit $\lambda R(\lambda, A)x \to x$, $x \in Y$, as $\lambda \to +\infty$ for a generator A, in order to comply with (A.2$_\epsilon$) = (9.5.2.1) and (9.5.2.2), as well. In fact, from (9.5.2.3),

$$\|R_\epsilon^* R_\epsilon e^{At} Bu\|_Y = \left\| \frac{1}{\epsilon^2} R \left(\frac{1}{\epsilon}, A^* \right) (R^* R) R \left(\frac{1}{\epsilon}, A \right) Bu \right\|_Y$$

$$\leq \frac{c}{\epsilon} c_\epsilon \|u\|_U, \quad u \in U, \qquad (9.5.2.4)$$

since by (ii) = (9.1.2), $\|R \left(\frac{1}{\epsilon}, A \right) B\|_{\mathcal{L}(U;Y)} \leq c_\epsilon$; and (9.5.2.1) then follows at once from (9.5.2.4).

Regularizing Optimal Control Problems   With each R_ϵ we associate the corresponding optimal control problem OCP$_{T,\epsilon}$ on $[0, T]$, $T < \infty$:

Minimize the cost functional

$$J_\epsilon(u, y) \equiv \int_s^T \left[\|R_\epsilon y(t)\|_Z^2 + \|u(t)\|_U^2 \right] dt \qquad (9.5.2.5)$$

over all $u \in L_2(s, T; U)$ where $y(t) = y(t; s; y_0)$ is the solution of (9.1.1) subject to u, with initial condition $y(s) = y_0$.

Theorem 9.5.2.1 *Under [the standing assumptions (i), (ii) = (9.1.2), and (A.0) = (9.1.6) of Section 9.1, as well as] the main assumption (A.1) = (9.1.7), and in view of (A.2$_\epsilon$) = (9.5.2.1), Theorems 9.2.1 and 9.2.2 apply and provide for $G = 0$:*

(i) *a unique optimal pair $\{u_\epsilon^0(\cdot, s; y_0), y_\epsilon^0(\cdot, s; y_0)\}$ of* OCP$_{T,\epsilon}$ = (9.5.2.5) *given by*

$$-u_\epsilon^0(\cdot, s; y_0) = [I_s + L_s^* R_\epsilon^* R_\epsilon L_s]^{-1} L_s^* R_\epsilon^* R_\epsilon e^{A(\cdot - s)} y_0, \quad (9.5.2.6)$$

$$y_\epsilon^0(\cdot, s; y_0) = [I_s + L_s L_s^* R_\epsilon^* R_\epsilon]^{-1} e^{A(\cdot - s)} y_0 \qquad (9.5.2.7)$$

$$= e^{A(\cdot - s)} y_0 + L_s u_\epsilon^0(\cdot, s; y_0); \qquad (9.5.2.8)$$

(ii) *a nonnegative, self-adjoint operator $P_\epsilon(t) \in \mathcal{L}(Y)$ given by*

$$P_\epsilon(t)x = \int_t^T e^{A^*(\tau - t)} R_\epsilon^* R_\epsilon \Phi_\epsilon(\tau, t)x \, d\tau \qquad (9.5.2.9a)$$

$$: \ continuous \ Y \to C([s, T]; Y) \ uniformly \ in \ s, \quad (9.5.2.9b)$$

where

$$\Phi_\epsilon(t, \tau)x = y_\epsilon^0(t, \tau; x), \quad 0 \leq s \leq \tau \leq t \leq T, x \in Y, \quad (9.5.2.10)$$

defines an evolution operator enjoying the properties (9.2.10) through (9.2.13). Moreover:

(iii) *The following pointwise (in t) feedback representation holds true for the optimal pair:*

$$u_\epsilon^0(t, s; y_0) = -B^* P_\epsilon(t) y_\epsilon^0(t, s; y_0) \in C([s, T]; U). \quad (9.5.2.11)$$

(iv) *The optimal cost is*

$$J\left(u_\epsilon^0(\cdot, s; y_0), y_\epsilon^0(\cdot, s; y_0)\right) = (P_\epsilon(s)y_0, y_0)_Y, \quad y_0 \in Y. \quad (9.5.2.12)$$

(v)

$$B^* P_\epsilon(t) : continuous \ Y \to L_\infty(s, T; U) \ uniformly \ in \ s, \quad (9.5.2.13)$$

$$B^* P_\epsilon(\cdot) e^{A(\cdot - s)} B : continuous \ U \to L_2(s, T; U) \ uniformly \ in \ s. \quad (9.5.2.14)$$

(vi) (Existence) For $x, y \in \mathcal{D}(A)$, $P_\epsilon(t)$ satisfies the differential Riccati equation DRE$_\epsilon$

$$
\begin{cases}
\left(\dfrac{d}{dt} P_\epsilon(t)x, y \right)_Y = -(R_\epsilon x, R_\epsilon y)_Z - (P_\epsilon(t)x, Ay)_Y - (P_\epsilon(t)Ax, y)_Y \\
\qquad\qquad\qquad\qquad + (B^* P_\epsilon(t)x, B^* P_\epsilon(t)y)_U, \\
P_\epsilon(T) = 0,
\end{cases}
$$

$$(9.5.2.15)$$

and hence, by Theorem 9.3.5.3, it also satisfies the corresponding integral Riccati equation IRE$_\epsilon$

$$
(P_\epsilon(t)x, y)_Y = \int_t^T \left(R_\epsilon e^{A(\tau-t)}x, R_\epsilon e^{A(\tau-t)}y \right)_Z d\tau
$$

$$
\qquad\qquad - \int_t^T \left(B^* P_\epsilon(\tau) e^{A(\tau-t)}x, B^* P_\epsilon(\tau) e^{A(\tau-t)}y \right)_U d\tau, \quad x, y \in Y.
$$

$$(9.5.2.16)$$

(viii) (Uniqueness) The operator $P_\epsilon(t)$ given by (9.5.2.9) is the unique, non-negative, self-adjoint solution of the IRE$_\epsilon$ = (9.5.2.16), and hence of the DRE$_\epsilon$ = (9.5.2.15) to satisfy properties (9.5.2.9b), (9.5.2.13), and (9.5.2.14).

Limit Solution of Original DRE Definition 9.5.2.1   With $G = 0$, under [the standing assumptions (i), (ii) = (9.1.2), and (A.0) = (9.1.6) of Section 9.1, as well as] the main assumption (A.1) = (9.1.7), we say that the nonnegative, self-adjoint operator $P(t)$ defined by (9.2.14) with $G = 0$ of the applicable Theorem 9.2.1 is a "limit solution" of the DRE (9.2.21), in case there exists a regularizing sequence $\{R_\epsilon\}$ satisfying conditions (A.2$_\epsilon$) = (9.5.2.1) and (9.5.2.2), such that the corresponding nonnegative, self-adjoint operators $P_\epsilon(t)$ defined by (9.5.2.9) of the applicable Theorem 9.5.2.1, and hence satisfying the DRE (9.2.21), converge strongly to $P(t)$ as $\epsilon \downarrow 0$, uniformly on $[0, T]$.

The desired connection between the original optimal control problem of Theorem 9.2.1 and the constructed regularizing family of optimal control problems OCP$_\epsilon$ of Theorem 9.5.2.1 is given by the following main result of the present section, which in particular yields the nonnegative, self-adjoint operator $P(t)$ of Theorem 9.2.1 as a limit solution of the DRE (9.2.21), that is, as a strong limit of bonafide Riccati operators $P_\epsilon(t)$ satisfying the DRE$_\epsilon$ = (9.5.2.15). This result is a specialized version of a more sophisticated approximating result to be given in Chapter 14, Theorem 14.1.3.1, which was originally envisioned for numerical purposes.

Theorem 9.5.2.2   *Assume [(i), (ii), and (A.0) of Section 9.1 and] (A.1) = (9.1.7). Then, with reference to the optimal control problem (9.1.15) of Theorem 9.2.1 and to the (arbitrary) regularizing optimal control problems OCP$_{T,\epsilon}$ = (9.5.2.5) of Theorem 9.5.2.1, subject to hypotheses (A.2$_\epsilon$) = (9.5.2.1) and (9.5.2.2), we have the following* *convergence results, as $\epsilon \downarrow 0$ and $y_0 \in Y$:*

(i)

$$\left\| u_\epsilon^0(\,\cdot\,, s; y_0) - u^0(\,\cdot\,, s; y_0) \right\|_{L_2(s,T;U)} \to 0 \ \textit{uniformly in } s; \qquad (9.5.2.17)$$

(ii)

$$\left\| y_\epsilon^0(\,\cdot\,, s; y_0) - y^0(\,\cdot\,, s; y_0) \right\|_{C([s,T];Y)} \to 0 \ \textit{uniformly in } s; \qquad (9.5.2.18)$$

(iii)

$$\left\| J\left(u_\epsilon^0(\,\cdot\,, s; y_0), y_\epsilon^0(\,\cdot\,, s; y_0) \right) - J(u^0(\,\cdot\,, s; y_0), y^0(\,\cdot\,, s; y_0)) \right\| \to 0$$
$$\textit{uniformly in } s; \qquad (9.5.2.19)$$

(iv)

$$\left\| P_\epsilon(t)x - P(t)x \right\|_{C([0,T];Y)} \to 0, \quad x \in Y. \qquad (9.5.2.20)$$

Thus, a fortiori, $P(t)$ is a limit solution of the DRE (9.2.21), and it is unique within the class of nonnegative, self-adjoint limit solutions, satisfying properties (9.2.15) and (9.2.16).

Remark 9.5.2.2 One of the advantages of Theorem 9.5.2.2 is that it leads to a numerical algorithm for the computation of the limit solution $P(t)$ [under the sole assumption (A.1) = (9.1.7), with R merely in $\mathcal{L}(Y; Z)$], by means of solving finite-dimensional differential Riccati equations, which approximate the regularizing problem. More specifically, to compute $P(t)$, one first considers a regularizing family of problems subject to the hypotheses (9.5.2.1) and (9.5.2.2). Next, one operates on these an approximation (with parameter of approximation $h \downarrow 0$) by means of the finite-dimensional $\mathrm{DRE}_{h\epsilon}$

$$\dot{P}_{\epsilon,h} = -A_h^* P_{\epsilon,h} - P_{\epsilon,h} A_h - R_{\epsilon,h}^* R_{\epsilon,h} + P_{\epsilon,h} B_h B_h^* P_{\epsilon,h},$$
$$P_{\epsilon,h}(T) = 0, \qquad (9.5.2.21)$$

where A_h, B_h, and $R_{\epsilon,h}$ are suitable approximations of A, B, and R_ϵ, as described in detail in the subsequent Chapter 14, Section 14.1 of Volume III: In Chapter 14 it is then shown, under suitable, natural hypotheses of approximation, that we obtain

$$\lim_{\epsilon \downarrow 0} \lim_{h \downarrow 0} \left\| P_{\epsilon,h}(t)x - P(t)x \right\|_{C([0,T];Y)} = 0, \quad x \in Y, \qquad (9.5.2.22)$$

as desired. More details are given in Chapter 14.

9.5.3 Proof of Theorem 9.5.2.2

(i) Step 1 We first note, with reference to (9.2.2a) and (9.5.2.6) that, as $\epsilon \downarrow 0$, $y_0 \in Y$:

$$\left\| L_s^* [R_\epsilon^* R_\epsilon - R^* R] e^{A(\,\cdot\,-s)} y_0 \right\|_{L_2(s,T;Y)} \to 0 \quad \textit{uniformly in } s, \ 0 \le s \le T,$$
$$(9.5.3.1)$$

as it follows from the strong convergence (9.5.2.2) of $R_\epsilon^* R_\epsilon$ to $R^* R$, as well as from (9.1.18).

Step 2 Similarly, for $v \in L_2(s, T; U)$, as $\epsilon \downarrow 0$,

$$\|L_s^*[R_\epsilon^* R_\epsilon - R^* R] L_s v\|_{L_2(s,T;U)} \to 0 \text{ uniformly in } s, \qquad (9.5.3.2)$$

via (9.1.17) and (9.1.18) as well.

Step 3 Setting, then, consistently with (9.2.2c)

$$\Lambda_{sT,\epsilon} = I_s + L_s^* R_\epsilon^* R_\epsilon L_s \in \mathcal{L}(L_2(s, T; Y)), \quad \text{uniformly in } s; \qquad (9.5.3.3)$$

$$\Lambda_{sT} = I_s + L_s^* R^* R L_s \in \mathcal{L}(L_2(s, T; Y)), \quad \text{uniformly in } s, \qquad (9.5.3.4)$$

and recalling the second resolvent equation

$$\Lambda_{sT,\epsilon}^{-1} - \Lambda_{sT}^{-1} = \Lambda_{sT,\epsilon}^{-1}(\Lambda_{sT} - \Lambda_{sT,\epsilon})\Lambda_{sT}^{-1}, \qquad (9.5.3.5)$$

we obtain, via (9.5.3.1)–(9.5.3.5), a standard composition theorem, and the Principle of Uniform Boundedness on $\Lambda_{sT,\epsilon}$, as $\epsilon \downarrow 0$, $y_0 \in Y$:

$$\left\| \{[I_s + L_s^* R_\epsilon^* R_\epsilon L_s]^{-1} L_s^* R_\epsilon^* R_\epsilon - [I_s + L_s^* R^* R L_s]^{-1} L_s^* R^* R\} e^{A(\cdot - s)} y_0 \right\|_{L_2(s,T;U)}$$

$$\to 0 \quad \text{uniformly in } s. \qquad (9.5.3.6)$$

Thus, recalling (9.2.2a) and (9.5.2.6), we have that (9.5.3.6) is the desired convergence (9.5.1.17) of the optimal controls.

(ii) Having proved (9.5.2.17) in part (i), we return to (9.2.3) and (9.5.2.8) to obtain, via the regularity (9.1.17) of L_s, that (9.5.2.18) of part (ii) holds true as well.

(iii) Conclusion (9.5.2.19) is then a ready consequence of the two convergence results in (9.5.2.17) and (9.5.2.18), via the cost formula.

(iv) By formulas (9.2.14) and (9.5.2.9), we compute, after adding and subtracting

$$\|P_\epsilon(t)x - P(t)x\|_Y \leq \left\| \int_t^T e^{A^*(\tau-t)}[R_\epsilon^* R_\epsilon - R^* R] y^0(\tau, t; x) \, d\tau \right\|_Y$$

$$+ \left\| \int_t^T e^{A^*(\tau-t)} R_\epsilon^* R_\epsilon \left[y_\epsilon^0(\tau, t; x) - y^0(\tau, t; x) \right] d\tau \right\|_Y. \qquad (9.5.3.7)$$

Then, using the strong convergence (9.5.2.2), as well as the convergence (9.5.2.18) uniform in the second parameter on the first and second term of (9.5.3.7), respectively, one can readily conclude with the desired convergence in (9.5.2.20). $\quad\square$

9.6 Dual Differential and Integral Riccati Equations When A is a Group Generator under (A.1) and $R \in \mathcal{L}(Y; Z)$ and $G = 0$. (Bounded Control Operator, Unbounded Observation)

9.6.1 Motivation. Orientation.

Motivation   In Theorem 9.4.1(ii) of Section 9.4, we have seen that under [the standing assumptions (i), (ii) = (9.1.2), and (A.0) = (9.1.6) of Section 9.1, so that $R \in \mathcal{L}(Y; Z)$ only, as well as] the main trace regularity assumption (A.1) = (9.1.7) and $G = 0$, the nonnegative, self-adjoint operator $P(t)$, $0 \le t < T$, defined by (9.2.14) and satisfying the feedback synthesis (9.2.17) of the optimal pair and the value (9.2.18) of the optimal cost, possesses the following interesting property: that $P(t)$ is an isomorphism on Y if and only if the dynamical system (9.4.1), say the pair $\{A^*, R^*\}$, is exactly controllable in Y over $[0, T - t]$ from the origin, using the class of $L_2(0, T - t, U)$-controls (Definition 9.4.1). Under these circumstances, setting

$$Q(t) = P^{-1}(t), \quad 0 \le t < T, \tag{9.6.1.1}$$

we see that $Q(t)$ is a nonnegative, self-adjoint operator. Formally, for the purpose of motivation, differentiating (formally) $P(t)P^{-1}(t) \equiv I$ yields

$$\frac{dP^{-1}(t)}{dt} = -P^{-1}(t) \frac{dP(t)}{dt} P^{-1}(t).$$

Thus, if $P(t)$ is taken to satisfy the DRE (9.2.21), this formal procedure leads to the conclusion that $Q(t) = P^{-1}(t)$ in (9.6.1.1) satisfies the Riccati equation

$$\frac{d}{dt}(Q(t)x, y) = (Q(t)x, A^*y) + (A^*x, Q(t)y) + (RQ(t)x, RQ(t)y) - (B^*x, B^*y), \tag{9.6.1.2a}$$

where we would then require $x, y \in \mathcal{D}(A^*)$. Equation (9.6.1.2a) will have to be supplemented with a terminal condition

$$Q(T) = Q_T \ge 0. \tag{9.6.1.2b}$$

One then verifies that (9.6.1.2) is the differential Riccati equation corresponding to the optimal control problem: Minimize the cost functional

$$\int_0^T \left[\|B^*z(t)\|_U^2 + \|v(t)\|_Z^2 \right] dt + (Q_T z(T), z(T))_Y \tag{9.6.1.3}$$

over all $v \in L_2(0, T; Z)$ for the dynamics

$$\dot{z} = -A^*z + R^*v, \quad z(0) = z_0 \in Y. \tag{9.6.1.4}$$

This then requires the additional assumption that A (equivalently, A^*) be the generator of a s.c. *group* of operators e^{At}, $t \in \mathbb{R}$, on Y. Notice that, in this optimal control problem, the *control operator R^* is bounded*, whereas *the observation operator B^* is unbounded*.

9.6.2 Dual Optimal Control Problem; Dual Differential Riccati Equation, When A is a Group Generator

Mathematical Setting   Throughout Section 9.6, we shall assume the [standing hypotheses (i), (ii) = (9.1.2), and (A.0) = (9.1.6) of Section 9.1, so that $R \in \mathcal{L}(Y; Z)$, as well as] abstract trace regularity hypothesis (A.1) = (9.1.7); in addition, we shall henceforth assume that A (equivalently, A^*) is a generator of a s.c. group on Y.

With motivation coming from Section 9.6.1, in this section we consider the following dual dynamics:

$$\dot{z}(t) = -A^* z(t) + R^* v, \quad z(0) = z_0 \in Y, \tag{9.6.2.1}$$

and pose for it the following

Dual Optimal Control Problem on $[0, T]$ Introduce the cost functional

$$J_{T;\mathcal{G}}(v, z) = \int_0^T \left[\|B^* z(t)\|_U^2 + \|v(t)\|_Z^2 \right] dt + \|\mathcal{G}z(T)\|_{Z_f}^2, \tag{9.6.2.2}$$

where throughout this Section 9.6,

$$\mathcal{G} \in \mathcal{L}(Y; Z_f), \tag{9.6.2.3}$$

where Z_f is another Hilbert space. The Dual Optimal Control Problem (DOCP$_T$) is:

$$\left.\begin{array}{c} \text{Minimize } J_{T;\mathcal{G}}(z, v) \text{ over all } v \in L_2(0, T; Z) \\ \text{where } z \text{ is the solution of (9.6.2.1) due to } v \end{array}\right\}. \tag{9.6.2.4}$$

Preliminaries   The solution of the dual dynamics is

$$z(t) = e^{-A^* t} z_0 + (Wv)(t), \tag{9.6.2.5a}$$

$$(Wv)(t) = \int_0^t e^{-A^*(t-\tau)} R^* v(\tau)\, d\tau. \tag{9.6.2.5b}$$

Lemma 9.6.2.1   *Assume (A.1) = (9.1.7) and also that A is a s.c. group generator on Y. Then:*

(i) The corresponding solution z of (9.6.2.1) given by (9.6.2.5) satisfies

$$z_0 \in Y, \quad v \in L_1(0, T; Z) \Rightarrow B^* z \in L_2(0, T; U). \tag{9.6.2.6a}$$

In particular,

$$B^* W : \text{ continuous } L_1(0, T; Z) \to L_2(0, T; U). \tag{9.6.2.6b}$$

(ii) The pair $\{-A, B\}$ is exactly controllable on Y from the origin over the interval $[0, T]$, within the class of $L_2(0, T; U)$-controls, if and only if the pair $\{A, B\}$ is also similarly controllable.

Proof. (i) By $(A.1) = (9.1.7)$ and the group property of e^{A^*t}, we readily compute with reference to the two terms in (9.6.2.5):

(i_1) For $x \in Y$:

$$\int_0^T \left\| B^* e^{-A^*t} x \right\|_U^2 dt = \int_0^T \left\| B^* e^{A^*(T-t)} e^{-A^*T} x \right\|_U^2 dt \qquad (9.6.2.7)$$

$$\text{(by (9.1.7))} \qquad \leq c_T \left\| e^{-A^*T} x \right\|_Y^2 \leq c_T' \|x\|_Y^2. \qquad (9.6.2.8)$$

(i_2) For $g \in L_1(0, T; Y)$ and $w \in L_2(0, T; U)$, interchanging the order of integration gives

$$\left| \int_0^T \left(\int_0^t B^* e^{-A^*(t-\tau)} g(\tau) \, d\tau, \, w(t) \right)_U dt \right|$$

$$= \left| \int_0^T \int_\tau^T \left(B^* e^{-A^*(t-\tau)} g(\tau), \, w(t) \right)_U dt \, d\tau \right|$$

$$\leq \int_0^T \left\{ \int_\tau^T \left\| B^* e^{-A^*(t-\tau)} g(\tau) \right\|_U^2 dt \right\}^{\frac{1}{2}} d\tau \left\{ \int_\tau^T \|w(t)\|_U^2 \, dt \right\}^{\frac{1}{2}}$$

(by (9.6.2.7) and $t - \tau = \sigma$)

$$\leq (c_T')^{\frac{1}{2}} \|w\|_{L_2(0,T;U)} \int_0^T \|g(\tau)\|_Y \, d\tau$$

$$\leq (c_T')^{\frac{1}{2}} \|w\|_{L_2(0,T;U)} \|g\|_{L_1(0,T;Y)}. \qquad (9.6.2.9)$$

Then (9.6.2.9) says that

$$g \to \int_0^t B^* e^{-A^*(t-\tau)} g(\tau) \, d\tau : \text{ continuous } L_1(0, T; Y) \to L_2(0, T; U).$$

$$(9.6.2.10)$$

[The above argument is similar to the proof of Theorem 7.2.1 in Chapter 7.]

Then, (9.6.2.8) and (9.6.2.10), used in (9.6.2.5), prove (9.6.2.6), since $R \in \mathcal{L}(Y; Z)$.

(ii) With reference to identity (9.6.2.7), if $\{A, B\}$ is exactly controllable, then by the characterization as in (9.4.7) for $\{A^*, S\}$ given by (9.4.1), the right-hand side of (9.6.2.7) is greater than or equal to $c_T \|e^{-A^*T} x\|^2 \geq c_T'' \|x\|^2$; and then, by (9.4.7), this time applied to the left-hand side of (9.6.2.7), we obtain that $\{-A, B\}$ is exactly controllable. The argument is reversible from $\{-A, B\}$ to $\{A, B\}$. $\square$

Dual Differential and Integral Riccati Equations With reference to [the primal problem (9.1.1), (9.1.15), and] the dual problem (9.6.2.1), (9.6.2.2), the dual differential Riccati equation, DDRE, is

$$
\begin{cases}
\dfrac{d}{dt}(Q(t)x, y)_Y = (Q(t)x, A^*y)_Y + (A^*x, Q(t)y)_Y - (B^*x, B^*y)_U \\
\qquad\qquad + (RQ(t)x, RQ(t)y)_Z, \quad x, y \in \mathcal{D}(A^*), \\
Q(T) = \mathcal{G}^*\mathcal{G},
\end{cases} \tag{9.6.2.11}
$$

whose corresponding dual integral Riccati equation, DIRE, is

$$
(Q(t)x, y)_Y = \int_t^T \left(B^* e^{-A^*(s-t)}x, \, B^* e^{-A^*(s-t)}y\right)_U ds
$$

$$
- \int_t^T \left(RQ(s)e^{-A^*(s-t)}x, \, RQ(s)e^{-A^*(s-t)}y\right)_Z ds
$$

$$
+ \left(\mathcal{G}e^{-A^*(T-t)}x, \, \mathcal{G}e^{-A^*(T-t)}y\right)_{Z_f}, \quad x, y \in Y. \tag{9.6.2.12}
$$

We begin by stating the relationship between the DDRE (9.6.2.11) and the DIRE (9.6.2.12).

Theorem 9.6.2.2   *Assume [(i), (ii) = (9.1.2), (A.0) = (9.1.6) of Section 9.1 and] (A.1) = (9.1.7), and, moreover, that A is a s.c. group generator.*

With reference to the DDRE (9.6.2.11) and the DIRE (9.6.2.12), the following statements are equivalent:

(a) *$Q(\cdot)$: continuous $Y \to C([0, T]; Y)$ satisfies the DIRE (9.6.2.12) [so that $Q^*(t) = Q(t)$].*

(b) *$Q(\cdot)$: continuous $Y \to C([0, T]; Y)$ is such that $Q(t) = Q^*(t)$; $(Q(t)x, z)_Y$ is continuously differentiable in t for each $x, z \in \mathcal{D}(A^*)$; and $Q(\cdot)$ satisfies the DDRE (9.6.2.11).*

Next, we state existence and uniqueness for the DIRE (9.6.2.12).

Theorem 9.6.2.3   (Existence and uniqueness of the DIRE)   *Assume [(i), (ii) = (9.1.2), (A.0) = (9.1.6) of Section 9.1 and] (A.1) = (9.1.7) and that A is a s.c. group generator.*

Then, there exists a unique solution $Q(\cdot) \in \mathcal{L}(Y; C([0, T]; Y))$ of the DIRE (9.6.2.12), which then satisfies $Q^*(t) = Q(t) \geq 0$.

Finally, from Lemma 9.6.2.1 and Theorem 9.6.2.2, we then obtain at once existence and uniqueness of the DDRE (9.6.2.11).

Theorem 9.6.2.4   (Existence and uniqueness of the DDRE)   *Assume [(i), (ii) = (9.1.2), (A.0) = (9.1.6) of Section 9.1 and] (A.1) = (9.1.7) and that A is a s.c. group generator.*

Then, with reference to the DDRE $= (9.6.2.11)$, we have:

(i) *(Existence) There exists*

$$Q(\cdot) : \quad continuous \ Y \rightarrow C([0, T]; Y), \qquad (9.6.2.13)$$

such that
(i_1)

$$Q(t) = Q^*(t) \geq 0, \quad 0 \leq t \leq T; \qquad (9.6.2.14)$$

(i_2) $(Q(t)x, z)_Y$ *is continuously differentiable in t, for each $x, z \in \mathcal{D}(A^*)$;*
(i_3) $Q(t)$ *satisfies the DDRE (9.6.2.11).*
(ii) *(Uniqueness) The DDRE (9.6.2.11) admits a unique solution, the one asserted in (i), within the class of solutions satisfying properties (i_1) and (i_2).*

Solution of the Dual Optimal Control Problem   Finally, via dynamic programming, we shall recover the solution of the dual optimal control problem (9.6.2.1), (9.6.2.2).

Theorem 9.6.2.5   *Assume $[(i), (ii), (A.0)$ and] $(A.1) = (9.1.7)$ and that A is a s.c. group generator. Then, with reference to the dual optimal control problem (9.6.2.1), (9.6.2.2), we have:*

(i) *There exists a unique optimal pair $\{v^0 = v^0(t, 0; z_0), z^0 = z^0(t, 0; z_0)\}$ such that*

$$v^0 \in C([0, T]; Z), \quad z^0 \in C([0, T]; Y). \qquad (9.6.2.15)$$

(ii) *The optimal pair satisfies the following pointwise feedback synthesis:*

$$v^0(t, 0; z_0) = - R Q(t) z^0(t, 0; z_0) \in C([0, T]; Z). \qquad (9.6.2.16)$$

(iii) *The optimal cost in (9.6.2.2) is*

$$J_{T,\mathcal{G}}(v^0(\cdot, 0; z_0), z^0(\cdot, 0; z_0)) = (Q(0)z_0, z_0)_Y, \quad z_0 \in Y. \qquad (9.6.2.17)$$

9.6.3 Proof of Theorem 9.6.2.2

This is the counterpart of the proof of Proposition 8.3.5.2 and Proposition 8.3.5.3 of Chapter 8 regarding the primal differential and integral Riccati equations.

   (a) $\rightarrow$ (b). This follows by direct differentiation of the DIRE (9.6.2.12), which is justified for $x, y \in \mathcal{D}(A^*)$. We obtain:

$$\frac{d}{dt}(Q(t)x, y)_Y = -(B^*x, B^*y)_U + \int_t^T \left(B^* e^{-A^*(\tau - t)} A^* x, B^* e^{-A^*(\tau - t)} y \right)_U d\tau$$

$$+ \int_t^T \left(B^* e^{-A^*(\tau - t)} x, B^* e^{-A^*(\tau - t)} A^* y \right)_U d\tau + (R Q(t)x, R Q(t)y)_Z$$

$$- \int_t^T \left(RQ(\tau) e^{-A^*(\tau - t)} A^* x, \, RQ(\tau) e^{-A^*(\tau - t)} y \right)_Z d\tau$$

$$- \int_t^T \left(RQ(\tau) e^{-A^*(\tau - t)} x, \, RQ(\tau) e^{-A^*(\tau - t)} A^* y \right)_Z d\tau$$

$$+ \left(Ge^{-A^*(T - t)} A^* x, \, Ge^{-A^*(T - t)} y \right)_{Z_f}$$

$$+ \left(Ge^{-A^*(T - t)} x, \, Ge^{-A^*(T - t)} A^* y \right)_{Z_f}. \tag{9.6.3.1}$$

We note that all terms in (9.6.3.1) are well-defined for $x, y \in \mathcal{D}(A^*)$, via (ii) $= (9.1.2)$, (A.1) $= (9.1.7)$, and the given hypothesis on $Q(\cdot)$. We next note that, by (9.6.2.12), the second, fifth, and seventh terms on the right-hand side of (9.6.3.1) sum up to $(Q(t)A^* x, y)_Y$, while the third, sixth, and eighth terms sum up to $(Q(t)x, A^* y)_Y$. Thus, (9.6.3.1) becomes for $x, y \in \mathcal{D}(A^*)$, $0 \le t < T$,

$$\frac{d}{dt}(Q(t)x, y)_Y = -(B^* x, B^* y)_U + (Q(t)A^* x, y)_Y$$

$$+ (Q(t)x, A^* y)_Y + (RQ(t)x, RQ(t)y)_Z, \tag{9.6.3.2}$$

which along with $Q(T) = \mathcal{G}^* \mathcal{G}$, from (9.6.2.12), yields the DDRE (9.6.2.11), as desired. The self-adjoint property $Q^*(t) = Q(t)$ is built in (9.6.2.12).

(b) $\to$ (a). By assumption (b), we compute for $x, y \in \mathcal{D}(A^*)$:

$$\frac{d}{ds} \left(Q(s) e^{-A^*(s - t)} x, \, e^{-A^*(s - t)} y \right)_Y$$

$$= \left[\frac{d}{dr} \left(Q(r) e^{-A^*(s - t)} x, \, e^{-A^*(s - t)} y \right)_Y \right]_{r = s} - \left(Q(s) e^{-A^*(s - t)} A^* x, \, e^{-A^*(s - t)} y \right)_Y$$

$$- \left(Q(s) e^{-A^*(s - t)} x, \, e^{-A^*(s - t)} A^* y \right)_Y. \tag{9.6.3.3}$$

Using (9.6.2.11) in the first term on the right-hand side of (9.6.3.3), and recalling $Q^*(t) = Q(t)$ by assumption (b), results in a cancellation of four terms, and we then obtain from (9.6.3.3):

$$\frac{d}{ds} \left(Q(s) e^{-A^*(s - t)} x, \, e^{-A^*(s - t)} y \right)_Y$$

$$= -\left(B^* e^{-A^*(s - t)} x, \, B^* e^{-A^*(s - t)} y \right)_U + \left(RQ(s) e^{-A^*(s - t)} x, \, RQ(s) e^{-A^*(s - t)} y \right)_Z,$$

$$x, y \in \mathcal{D}(A^*). \tag{9.6.3.4}$$

All terms in (9.6.3.4) are well defined also via (A.1) $= (9.1.7)$. After integration of (9.6.3.4) on $[0, T]$, we finally obtain the DIRE (9.6.2.12), originally for $x, y \in \mathcal{D}(A^*)$ and next for all $x, y \in Y$ by continuous extension, using (A.1) $= (9.1.7)$ and the assumption on $Q(\cdot)$ in (b). Theorem 9.6.2.1 is proved. $\square$

9.6.4 Proof of Theorem 9.6.2.3

The proof of existence and uniqueness of a nonnegative, self-adjoint solution to the DIRE (9.6.2.12) is based on the argument, already employed in Chapter 8, Section 8.4, consisting of a two-step procedure: (1) a local contraction principle, combined with (2) a priori bounds or estimates. Indeed, the technicalities of the present proof are contained in those encountered in the more demanding proof of Chapter 8, Section 8.4, which we shall parallel in a more sketchy manner.

Step 1   We return to the DIRE (9.6.2.12) and introduce the linear operators on Y:

$$(M_1(t)x, y)_Y = \int_t^T \left(B^* e^{-A^*(\tau-t)}x, B^* e^{-A^*(\tau-t)}y\right)_U d\tau \qquad (9.6.4.1a)$$

$$= \int_0^{T-t} \left(B^* e^{-A^*r}x, B^* e^{-A^*r}y\right)_U dr, \qquad (9.6.4.1b)$$

$$(M_2(t)x, y)_Y = \left(\mathcal{G} e^{-A^*(T-t)}x, \mathcal{G} e^{-A^*(T-t)}y\right)_{Z_f}, \qquad (9.6.4.2)$$

$$M(t) = M_1(t) + M_2(t). \qquad (9.6.4.3)$$

Furthermore, we set

$$\Psi(Q)(t) = \int_t^T e^{-A(\tau-t)} Q^*(\tau) R^* R Q(\tau) e^{-A^*(\tau-t)} d\tau \qquad (9.6.4.4a)$$

$$: \ \mathcal{L}(Y; C([0, T]; Y)) \to \ \text{itself} \qquad (9.6.4.4b)$$

so that, by (9.6.4.1)–(9.6.4.4), the DIRE (9.6.2.12) can be rewritten as

$$Q = M - \Psi(Q), \qquad (9.6.4.5)$$

for which we seek a fixed point solution on the Banach space $C([0, T]; \mathcal{L}(Y))$. Here, for any $0 \le T_0 < T$ we denote by $C([T_0, T]; \mathcal{L}(Y))$ the space $\mathcal{L}(Y; C([T_0, T]; Y))$ endowed with the norm

$$\|\|Q(\cdot)\|\|_{T_0} = \max_{T_0 \le t \le T} \|Q(t)\|_{\mathcal{L}(Y)} \qquad (9.6.4.6)$$

[Eqn. (9.6.4.5) is the counterpart of Eqn. (8.4.2.2) in Chapter 8.]

Step 2.   **Lemma 9.6.4.1**   *Assume $(A.1) = (9.1.7)$ and that A is a s.c. group generator. Then, with reference to $(9.6.4.1)$ and $(9.6.4.3)$, we have*

$$M_1, M_2, M : \ \text{continuous } Y \to C([0, T]; Y). \qquad (9.6.4.7)$$

Proof.   The conclusion (9.6.4.7) is plainly true for M_2 in (9.6.4.2), and so we only need to verify it for M_1 in (9.6.4.1). By Schwarz's inequality on (9.6.4.1b) followed by an application of (9.6.2.8), we obtain that $M_1(t) \in \mathcal{L}(Y)$ and indeed

$$|(M_1(t)x, y)_Y| \le c_T' \|x\|_Y \|y\|_Y, \quad x, y \in Y, \ 0 \le t \le T, \qquad (9.6.4.8)$$

or $M_1 \in \mathcal{L}(Y; L_\infty(0, T; Y))$. Indeed, $M_1(t)$ is strongly continuous on Y: From (9.6.4.1b), Schwarz's inequality, and (9.6.2.8), we obtain more precisely

$$
\begin{aligned}
&|(M_1(t)x - M_1(s)x, y)_Y| \\
&= \left| \int_{T-s}^{T-t} \left(B^* e^{-A^* r} x, B^* e^{-A^* r} y \right)_U dr \right| \\
&\leq \left[\int_{T-s}^{T-t} \left\| B^* e^{-A^* r} x \right\|_U^2 dr \right]^{\frac{1}{2}} (c_T')^{\frac{1}{2}} \|y\|_Y \to 0 \quad \text{as } s \to t, \quad (9.6.4.9)
\end{aligned}
$$

and this shows (9.6.4.7) for M_1. $\square$

Step 3.   Lemma 9.6.4.2 (Local existence and uniqueness)   *Assume* $(A.1) = (9.1.7)$ *and that A is a s.c. group generator. Then, with reference to (9.6.4.6), we have that: For all $T_0 < T$ sufficiently close to T [deduced from (9.6.4.10) and (9.6.4.11) below], the DIRE (9.6.2.12), equivalently (9.6.4.5), has a unique solution $Q(\cdot)$ in $C([T_0, T]; \mathcal{L}(Y))$.*

Proof.   Let Q, Q_1, and Q_2 be in $C([T_0, T]; \mathcal{L}(Y))$. Then, the definition (9.6.4.4) for $\Psi(Q)$ provides the following bounds in the norm of (9.6.4.6):

(a)

$$
\|\|\Psi(Q(\cdot))\|\|_{T_0} \leq (T - T_0) c_{1T}^2 \|\|Q(\cdot)\|\|_{T_0}^2 \|R^* R\|_{\mathcal{L}(Y)}; \quad (9.6.4.10)
$$

(b)

$$
\begin{aligned}
&\|\|\Psi(Q_1(\cdot)) - \Psi(Q_2(\cdot))\|\|_{T_0} \\
&\leq 2(T - T_0) c_{1T}^2 \|R^* R\|_{\mathcal{L}(Y)} k_{T_0} \|\|Q_1(\cdot) - Q_2(\cdot)\|\|_{T_0}, \quad (9.6.4.11)
\end{aligned}
$$

$$
k_{T_0} = \max\{\|\|Q_1(\cdot)\|\|_{T_0}, \|\|Q_2(\cdot)\|\|_{T_0}\}, \quad C_{1T} = \|\|e^{-A \cdot}\|\|_0.
$$

The bound (9.6.4.11) is readily obtained by adding and subtracting. Let now

$$
\begin{aligned}
\mathcal{S}_{T_0}(\eta) = &\text{ closed ball in the space } C([T_0, T]; \mathcal{L}(Y)) \text{ centered} \\
&\text{ at the origin and of radius } \eta. \quad (9.6.4.12)
\end{aligned}
$$

Fix

$$
\eta > \|\|M(\cdot)\|\|_0, \quad (9.6.4.13)
$$

so that $M(\cdot) \in \mathcal{S}_{T_0}(\eta)$. Moreover, from (9.6.3.10) and (9.6.4.11), we readily see that: For all $T_0 < T$ sufficiently close to T, we have:

$$
\begin{cases}
\Psi \text{ maps } \mathcal{S}_{T_0}(\eta) \text{ into itself;} \\
\Psi \text{ is a contraction on } \mathcal{S}_{T_0}(\eta) \text{ with} \\
\quad \text{contraction constant less than, say, } \frac{1}{2}.
\end{cases} \quad (9.6.4.14)
$$

The suitable value of T_0 is readily read off from (9.6.4.10) and (9.6.4.11) via (9.6.4.13) and (9.6.4.14), respectively. Thus, the contraction principle applies and yields a unique solution of the DIRE (9.6.2.12), that is, of (9.6.4.5), on $C([T_0, T]; \mathcal{L}(Y))$, for $T_0 < T$ suitably close to T. Lemma 9.6.4.2 is proved.   □

Step 4.   Lemma 9.6.4.3 *Assume (A.1)$=$(9.1.7) and that A is a s.c. group generator. Then, the unique solution $Q(\cdot) \in \mathcal{L}(Y; C([T_0, T]; Y))$ of the DIRE (9.6.2.12), guaranteed by Lemma 9.6.4.2 for a $T_0 < T$ suitably close to T, is nonnegative, self-adjoint: $Q^*(t) = Q(t) \geq 0$, $t_0 \leq t \leq T$.*

Proof. The symmetricity of (9.6.2.12) guarantees at once that if

$$Q(\cdot) \in \mathcal{L}(Y; C([T_0, T]; Y))$$

is a solution of (9.6.2.12), so is $Q^*(t)$ and then, by uniqueness of Lemma 9.6.4.2, we have $Q^*(t) = Q(t)$, $T_0 \leq t \leq T$. We next prove that $Q(t)$ is nonnegative definite. To this end, let $t_0 \in [T_0, T]$, and let $z_0 \in Y$ be fixed. Given $v \in L_2(t_0, T; Y)$, let $z(t)$ be defined by

$$z(t) = e^{-A^*(t-t_0)} z_0 + \int_{t_0}^{t} e^{-A^*(t-\tau)} R^* v(\tau) \, d\tau, \tag{9.6.4.15}$$

which is a mild solution of Eqn. (9.6.2.1), with $z(t_0) = z_0$. Indeed,

$$\begin{cases} z_0 \in \mathcal{D}(A^*) \\ v \in H^1(t_0, T; Z) \end{cases} \Rightarrow \begin{cases} z(t) \in C([t_0, T]; \mathcal{D}(A^*)), & (9.6.4.16a) \\ \dot{z}(t) \in C([t_0, T]; Y), & (9.6.4.16b) \\ \dot{z}(t) = -A^* z(t) + R^* v(t). & (9.6.4.16c) \end{cases}$$

Next, we have seen in the proof of the implication (a) $\to$ (b) in Section 9.6.3 that $(Q(t)x, y)_Y$ is continuously differentiable in t for each $x, y \in \mathcal{D}(A^*)$ and that, then, (9.6.2.11) is satisfied. It then follows by (9.6.4.16) that the map $t \to (Q(t)z(t), z(t))_Y$ is differentiable and

$$\frac{d}{dt}(Q(t)z(t), z(t))_Y = (\dot{Q}(t)z(t), z(t))_Y + (Q(t)\dot{z}(t), z(t))_Y + (Q(t)z(t), \dot{z}(t))_Y. \tag{9.6.4.17}$$

Thus, using the DDRE (9.6.2.11) on the first term on the right-hand side of (9.6.4.17) – which is legal by (9.6.4.16a) – and using (9.6.4.16c) on the other two terms, we obtain for $t_0 \leq t \leq T$:

$$\begin{aligned} \frac{d}{dt}(Q(t)z(t), z(t))_Y = \ & (Q(t)z(t), A^* z(t))_Y + (A^* z(t), Q(t)z(t))_Y \\ & - (B^* z(t), B^* z(t))_U + (RQ(t)z(t), RQ(t)z(t))_Z \\ & + (Q(t)[-A^* z(t) + R^* v(t)], z(t))_Y \\ & + (Q(t)z(t), -A^* z(t) + R^* v(t))_Y. \end{aligned} \tag{9.6.4.18}$$

Recalling the self-adjointness $Q^*(t) = Q(t)$, $t_0 \leq t \leq T$, in (9.6.4.18), we obtain after a double cancellation

$$
\frac{d}{dt}(Q(t)z(t), z(t))_Y = -(B^*z(t), B^*z(t))_U + (RQ(t)z(t), RQ(t)z(t))_Z
$$
$$
+ 2(Q(t)R^*v(t), z(t))_Y
$$
$$
= -\|B^*z(t)\|_U^2 + \|v(t) + RQ(t)z(t)\|_Z^2 - \|v(t)\|_Z^2,
$$

(9.6.4.19)

with data $z_0 \in \mathcal{D}(A^*)$ and $v \in H^1(t_0, T; Z)$ as in (9.6.4.16). Hence, integrating (9.6.4.19) over $[t_0, T]$ and using the terminal condition $Q(T) = \mathcal{G}^*\mathcal{G}$ in (9.6.2.11) and $z(t_0) = z_0$ yields

$$
(Q(t_0)z_0, z_0)_Y = \int_0^T \left[\|B^*z(t)\|_U^2 + \|v(t)\|_Z^2 \right] dt + (\mathcal{G}z(T), \mathcal{G}z(T))_{Z_f}
$$
$$
- \int_{t_0}^T \|v(t) + RQ(t)z(t)\|_Z^2 \, dt, \quad \forall \, z_0 \in \mathcal{D}(A^*), \; v \in H^1(t_0, T; Z).
$$

(9.6.4.20)

Next, the above identity (9.6.4.20) can be extended by a density argument to all $z_0 \in Y$ and all $v \in L_2(t_0, T; Z)$, by virtue also of Lemma 9.6.2.1(i), Eqn. (9.6.2.6), on $B^*z(t)$.

Next, consider the (closed-loop) integral equation

$$
z(t) = e^{-A^*(t-t_0)}z_0 - \int_{t_0}^t e^{-A^*(t-\tau)} R^* RQ(\tau)z(\tau) \, d\tau.
$$

(9.6.4.21)

It has a unique solution, denoted henceforth by $z^0(t, t_0; z_0)$, in $C([t_0, T]; Y)$: This is so since $RQ(\cdot)$, is a strongly continuous perturbation of the infinitesimal generator $-A^*$ (see, e.g. [Balakrishnan, 1981, Section 4.13]). [Here, $Q(\cdot)$ is the unique solution in $\mathcal{L}(Y; C([T_0, T]; Y))$ of the DIRE (9.6.2.12) guaranteed by Lemma 9.6.4.2.] Then, let $v^0(t, t_0; z_0)$ be defined by the feedback formula

$$
v^0(t, t_0; z_0) = -RQ(t)z^0(t, t_0; z_0) \in C([t_0, T]; Z),
$$

(9.6.4.22)

so that then $z^0(t, t_0; z_0)$ is the solution of (9.6.4.15) due to $v^0(t, t_0; z_0)$. Inserting the pair $\{v^0, z^0\}$ in (9.6.4.20) results in

$$
(Q(t_0)z_0, z_0)_Y = \int_{t_0}^T \left[\|B^*z^0(t, t_0; z_0)\|_U^2 + \|v^0(t, t_0; z_0)\|_Z^2 \right] dt
$$
$$
+ (\mathcal{G}z^0(T, t_0; z_0), \mathcal{G}z^0(T; t_0; z_0))_{Z_f} \geq 0,
$$

(9.6.4.23)

which proves $Q(t_0) \geq 0$, as desired. Lemma 9.6.4.3 is established.   □

Step 5. Theorem 9.6.4.4 (Global existence and uniqueness) *Assume (A.1) = (9.1.7) and that A is a s.c. group generator. Then, there exists a unique solution $Q(\cdot) \in \mathcal{L}(Y; C([0, T]; Y))$ of the DIRE (9.6.2.12), that is, of (9.6.4.5), such that $Q(t) = Q^*(t) \geq 0$.*

Proof. Having a unique local solution of (9.6.2.12) with the required properties, we now need to establish an a priori bound. Since $Q(t)$ has been established in Lemma 9.6.4.3 to be nonnegative definite, we then drop the second negative term on the right-hand side of (9.6.2.12) with $x = y$ and estimate from (9.6.2.12), with $x \in Y$, for all t on the maximal interval of existence I_M:

$$0 \leq (Q(t)x, x)_Y = |(Q(t)x, x)_Y|$$

$$\leq \int_0^{T-t} \left\| B^* e^{-A^* r} x \right\|_U^2 \, dr + \|\mathcal{G}^* \mathcal{G}\|_{\mathcal{L}(Y)} C_{1T}^2 \|x\|_Y^2$$

(by (9.6.2.8)) $\leq [C_T' + \|\mathcal{G}^* \mathcal{G}\| C_{1T}^2] \|x\|_Y^2, \quad t \in I_M,$ (9.6.4.24)

recalling (9.6.2.8) in the last step.

With the a priori estimate $\|Q(t)\|_{\mathcal{L}(Y)} \leq C_T$ at hand from (9.6.4.24), for all t on the maximal interval of existence I_M, we can now extend the local solution $Q(\cdot)$ of Lemma 9.6.4.2, satisfying the nonnegative, self-adjointness of Lemma 9.6.4.3, to a global solution $Q(\cdot) \in C([0, T]; \mathcal{L}(Y))$, in finitely many steps. Theorem 9.6.4.4 is proved. $\square$

Theorem 9.6.4.4 is a restatement of Theorem 9.6.2.2.

9.6.5 Proof of Theorem 9.6.2.4

Existence   We shall use an argument commonly referred to as dynamic Programming [see Chapter 8, end of Section 8.4.5]. It consists in differentiating $(Q(t)z(t), z(t))_Y$ in t, for $z(t)$ the solution of the differential equation (9.6.2.1) due to $z_0 \in \mathcal{D}(A^*)$ at first, followed by extension to all $z_0 \in Y$ and $v \in L_2(0, T; Z)$, as done already in (9.6.4.18) through (9.6.4.19), followed by an integration leading to (9.6.4.20). We rewrite (9.6.4.20) at $t_0 = 0$ by virtue of (9.6.2.2) as

$$(Q(0)z_0, z_0)_Y = J_{T,\mathcal{G}}(v, z) - \int_0^T \|v(t) + RQ(t)z(t)\|_Y^2 \, dt. \quad (9.6.5.1)$$

Furthermore, we recall the argument and the results obtained following (9.6.4.21), once specialized to $t_0 = 0$: The (closed-loop) integral equation

$$z(t) = e^{-A^* t} z_0 - \int_0^t e^{-A^*(t-\tau)} R^* RQ(\tau)z(\tau) \, d\tau, \quad (9.6.5.2)$$

with $Q(\cdot)$ provided by Theorem 9.6.2.2, or Theorem 9.6.2.3, has a unique solution denoted by $z^0(t, 0; z_0)$ in $C([0, T]; Y)$. Define next $v^0(t, 0; z_0)$ by the feedback formula

$$v^0(t, 0; z_0) = -RQ(t)z^0(t, 0; z_0) \in C([0, T]; Z),$$

so that $z^0(t, 0; z_0)$ is the solution of (9.6.4.15), or (9.6.2.1), to $v^0(t, 0; z_0)$. Then (9.6.5.1) yields

$$(Q(0)z_0, z_0)_Y = J_{T,\mathcal{G}}(v^0, z^0)$$

$$= \int_0^T \left[\|B^*z^0(t, 0; z_0)\|_U^2 + \|v^0(t)\|_Z^2 \right] dt + \|\mathcal{G}z^0(T, 0; z_0)\|_{Z_f}^2,$$

(9.6.5.3)

recalling (9.6.2.2). Moreover, (9.6.5.1) implies also

$$(Q(0)z_0, z_0)_Y \le J_{T,\mathcal{G}}(v, z), \quad \forall\, v \in L_2(0, T; Z). \tag{9.6.5.4}$$

Putting (9.6.5.3) and (9.6.5.4) together, we then conclude that $\{v^0(\cdot, 0; z_0), z^0(\cdot, 0; z_0)\}$ is the optimal pair of problem (9.6.2.4) for (9.6.2.1).

Uniqueness   Conversely, if $\hat{v}$ is an optimal control of problem (9.6.2.4) for (9.6.2.1), with corresponding optimal solution $\hat{z}$, then by (9.6.5.4) and (9.6.5.3), we have

$$J_{T,\mathcal{G}}(\hat{v}, \hat{z}) = (Q(0)z_0, z_0) = J_{T,\mathcal{G}}(v^0, z^0), \tag{9.6.5.5}$$

and from (9.6.5.1),

$$\hat{v}(t) = -RQ(t)\hat{z}(t) \quad \text{a.e. in } [0, T]. \tag{9.6.5.6}$$

This implies that $\hat{z}$ is a solution of the integral equation (9.6.5.2) in $C([0, T]; Y)$. From the uniqueness result for (9.6.5.2), we have $\hat{z} = z^0$, and hence $\hat{v} = v^0$ via (9.6.5.6). The proof of Theorem 9.6.2.4 is complete. $\quad\square$

Remark 9.6.5.1   Through a constructive argument such as the one of Theorem 9.3.6.5 for the original problem (9.1.1), (9.1.15), it is also possible to give an explicit formula for the Riccati solution operator for the dual problem (9.6.2.1), (9.6.2.2):

$$Q_T(t)x = \int_t^T e^{-A(\tau-t)}BB^*z^0(\tau, t; x)\, d\tau + e^{-A(T-t)}\mathcal{G}^*\mathcal{G}z^0(T, t; x), \quad x \in Y. \tag{9.6.5.7}$$

9.6.6 A Transition Property of $Q(\cdot)$

In this section, we find it convenient to write $Q_T(\cdot)$, instead of simply $Q(\cdot)$, by adding a subscript T to emphasize the time interval of the optimal problem. As a corollary of the uniqueness result of Theorem 9.6.2.3 for the DIRE (9.6.2.12), we shall obtain a transition property for $Q_T(\cdot)$, analogous to property (9.2.25) for $P_T(\cdot)$.

We shall employ the argument of Remark 9.3.7.1, that is, we shall use this time the DIRE.

Proposition 9.6.6.1 *Assume [(i), (ii) = (9.1.2), and (A.0) = (9.1.6) of Section 9.1 and] (A.2) = (9.1.8) and, moreover, assume that A is an s.c. group generator. Then, with reference to $Q_T(\cdot)$ satisfying the DIRE (9.6.2.12) we have*

$$Q_T(t) = Q_{T+\tau}(t + \tau), \quad T, \ t + \tau \geq 0, \ 0 \leq t \leq T. \tag{9.6.6.1}$$

Proof. We rewrite (9.6.2.12) with $T + \tau$ in place of T and $t + \tau$ in place of t, and then using the change of variable $\sigma = s - \tau$, we obtain

$$
(Q_{T+\tau}(t + \tau)x, y)_Y
$$
$$
= \int_t^T \left(B^* e^{-A^*(\sigma - t)} x, \ B^* e^{-A^*(\sigma - t)} y \right)_U d\sigma
$$
$$
- \int_t^T \left(R Q_{T+\tau}(\tau + \sigma) e^{-A^*(\sigma - t)} x, \ R Q_{T+\tau}(\tau + \sigma) e^{-A^*(\sigma - t)} y \right)_Z d\sigma
$$
$$
+ \left(\mathcal{G} e^{-A^*(T - t)} x, \ \mathcal{G} e^{-A^*(T - t)} y \right)_{Z_f}. \tag{9.6.6.2}
$$

This equation (9.6.6.2) has the unique solution $Q_T(t)$ by Theorem 9.6.2.3. Thus, we obtain $Q_{T+\tau}(t + \tau) \equiv Q_T(t)$, and (9.6.6.1) is proved. $\quad\square$

9.6.7 Isomorphism of $Q_T(t)$, $0 \leq t \leq T$ and Exact Controllability of $\{A, B\}$ on $[0, T - t]$

In this section we report a characterization that the dual Riccati operator $Q_T(t)$, $0 \leq t \leq T$ be an isomorphism on Y, which is the counterpart of the characterization, given by Theorem 9.4.1, that the primal Riccati operator $P_T(t)$, $0 \leq t \leq T$, be an isomorphism on Y. Implications of the present result will be given in Chapter 11, Section 11.7. We recall, from Definition 9.4.1, the definition of exact controllability on Y from the origin of the dynamical system (9.4.1), over the finite interval $[0, T]$, within the class of $L_2(0, T; Z)$-controls, which applies, of course, also to the z-dynamics (9.6.2.1), defined by the pair $\{-A^*, R^*\}$. By Lemma 9.6.2.1(ii), when A is the generator of a s.c. group, then the pair $\{A, B\}$ is exactly controllable on $[0, T]$ in the above sense if and only if $\{-A, B\}$ is as well; and this occurs precisely when inequality (9.6.7.3) below holds true.

The present dual counterpart of Theorem 9.4.1 is the following:

Theorem 9.6.7.1 *Assume [(i), (ii), and (A.0) of Section 1 and] (A.1) = (9.1.7), and, moreover, assume that A is a s.c. group generator. Then:*

(i) The optimal cost of the dual optimal control problem (9.6.2.2)–(9.6.2.4) over $[t, T]$ (rather than $[0, T]$ as is (9.6.2.2)) with $\mathcal{G} = 0$, for the dynamics (9.6.2.1),

is given by

$$J^0_{t,T}(z_0) = J_{t,T}(v^0(\cdot, t; z_0), z^0(\cdot, t; z_0))$$

$$= \int_t^T \left[\|B^* z^0(\tau, t; z_0)\|^2_U + \|v^0(\tau, t; z_0)\|^2_Z \right] d\tau = (Q_T(t) z_0, z_0)_Y$$

$$= \left\| [I_t + B^* W_t W_t^* B]^{-\frac{1}{2}} B^* e^{-A^*(\cdot - t)} z_0 \right\|^2_{L_2(t,T;U)} \tag{9.6.7.1}$$

$$= \left\| \left[I_t + (BB^*)^{\frac{1}{2}} W_t W_t^* (BB^*)^{\frac{1}{2}} \right]^{-\frac{1}{2}} (BB^*)^{\frac{1}{2}} \left[e^{-A^*(\cdot - t)} z_0 t \right] \right\|^2_{L_2(t,T;Y)}, \tag{9.6.7.2}$$

where, refining (9.6.2.5), we have set

$$(W_t v)(\tau) = \int_t^\tau e^{-A^*(\tau - s)} R^* v(s) \, ds. \tag{9.6.7.3}$$

(ii) The operator $Q_T(t)$, $0 \le t \le T$, guaranteed by Theorem 9.6.2.4, is an isomorphism on Y if and only if the following estimate holds true:

$$\int_t^T \left\| (BB^*)^{\frac{1}{2}} e^{-A^*(\tau - t)} z_0 \right\|^2_Y d\tau = \int_t^T \left\| B^* e^{-A^*(\tau - t)} z_0 \right\|^2_U \, d\tau$$

$$= \int_0^{T-t} \left\| B^* e^{-A^* \sigma} z_0 \right\|^2_U \, d\tau$$

$$\ge c_{T-t} \|z_0\|^2_Y \tag{9.6.7.4}$$

for some constant $c_{T-t} > 0$, that is, if and only if the dynamical system $\{-A, B\}$ [equivalently, by Lemma 9.6.2.1(ii), the dynamical system $\{A, B\}$ in (9.1.1)] is exactly controllable on Y (from the origin) over the interval $[0, T - t]$, within the class of $L_2(0, T - t; Z)$-controls.

Proof.   The validity of (9.6.7.1) was already asserted in (9.6.2.17). Formula (9.6.7.2) for the dynamics (9.6.2.1), in particular, the input-solution operator W_t in (9.6.7.3), and with observation operator B^* in the cost (9.6.2.2) [with $\mathcal{G} = 0$], is exactly formula (9.4.9) for the dynamics (9.1.1), in particular, the input solution operator L_t in (9.1.17), and with observation R in the cost (9.1.15) and $G = 0$. Next, we recall from (9.6.2.6) of Lemma 9.6.2.1, that, a fortiori,

$$B^* W_t : \text{ continuous } L_2(t, T; Z) \to L_2(t, T; U) \tag{9.6.7.5}$$

(indeed, with a norm that may be made dependent only on T, not on t, as in (9.1.17) for L_t). Hence, from (9.6.7.5), the operators

$$[I_t + B^* W_t W_t^* B]^{-\frac{1}{2}}, \quad \left[I_t + (BB^*)^{\frac{1}{2}} W_t W_t^* (BB^*)^{\frac{1}{2}} \right]^{-\frac{1}{2}}$$

are isomorphisms on $L_2(t, T; Y)$, respectively,   (9.6.7.6) with norms that may be made dependent only on T, not on t [counterpart of (9.4.16)]. Thus, using (9.6.7.6) in (9.6.7.2), we obtain

$$c_{tT} \int_t^T \left\| (BB^*)^{\frac{1}{2}} e^{-A^*(\tau - t)} z_0 \right\|_Y^2 d\tau$$

$$\leq (Q_T(t) z_0, z_0)_Y \leq C_{tT} \int_t^T \left\| (BB^*)^{\frac{1}{2}} e^{-A^*(\tau - t)} z_0 \right\|_Y^2 d\tau \qquad (9.6.7.7)$$

[the counterpart of (9.4.17)]. Then, arguing as below (9.4.17), we conclude from (9.6.7.7) that the nonnegative, self-adjoint operator $Q_T(t)$ is an isomorphism on Y if and only if (9.4.7.4) holds true, which is the stated exact controllability characterization (recall (9.4.7)). $\square$

9.7 Optimal Control Problem with Bounded Control Operator and Unbounded Observation Operator

Section 9.6 deals with the dual optimal control problem with quadratic cost (9.6.2.2) for the dynamics (9.6.2.1). As noted there, in the formulation of this dual problem, the *control operator R^** is *bounded*, whereas the *observation operator B^** is *unbounded* as in (9.1.2) and subject to the trace regularity property $(A.1) = (9.1.7)$. Regarding the fact that the dual z-dynamics (9.6.2.1) has the free dynamics operator $-A^*$, which then requires the assumption on the original y-dynamics (9.1.1) that A be the generator of a s.c. group, we note explicitly that this is relevant only in connection with the objective of Section 9.6: which is to derive a (dual) differential or integral Riccati equation for the inverse $P^{-1}(t) = Q(t)$ of the Riccati operator $P(t)$ of the original optimal control problem (9.1.1), (9.1.4) [under exact controllability assumption of the pair $\{A^*, R^*\}$]. Had the goal been to derive a full solution of the optimal control problem with bounded control operator R^* and unbounded observation B^*, including the derivation of the corresponding differential or integral Riccati equation, the assumption that A be a group generator would not be needed. Thus, if starting from the z-problem (9.6.2.1), (9.6.2.2), we view it not as a dual problem of the original problem (9.1.1), (9.1.4), but as a bonafide problem in the usual notation of these books, according to the correspondences,

$$z \longrightarrow \text{state variable } y,$$
$$-A^* \longrightarrow \text{free dynamics generator of } A \text{ of a s.c. semigroup},$$
$$R^* \longrightarrow \text{control operator } B,$$
$$B^* \longrightarrow \text{observation operator } R,$$
$$\mathcal{G} \longrightarrow \text{final state observation } G,$$
$$\mathcal{J} \longrightarrow \text{cost functional } J,$$

we can conclude that Section 9.6 contains, in fact, the treatment of the optimal control problem with *bounded control* and *unbounded observation* operators, according to the precise statement given below. Such case, at least for $G = 0$, is also contained as a special case in the treatment of Chapter 8, under assumptions (h.1) $= (8.2.1.10)$ and (h.2) $= (8.2.1.11)$ of that chapter, possibly complemented by assumption (A.1) $=$ (9.1.7) of the present Chapter 9 (see Remark 9.2.1). In fact, for $B \in \mathcal{L}(U; Y)$, hypothesis (h.2) is superfluous as it is contained in hypothesis (h.1). Indeed, (h.1) of Chapter 8 is nothing but (a₃) below.

Consider the dynamics

$$\dot{y} = Ay + Bu, \quad y(0) = y_0 \in Y \tag{9.7.1}$$

on the Hilbert space Y, for which we pose the corresponding optimal control problem:

$$\text{Minimize } J(u, y) \text{ over all } u \in L_2(0, T; U), \tag{9.7.2}$$

where

$$J(u, y) \equiv \int_0^T \left[\|Ry(t)\|_Z^2 + \|u(t)\|_U^2 \right] dt + \|Gy(T)\|_{Z_f}^2, \tag{9.7.3}$$

where U, Z, and Z_f are also Hilbert spaces, subject to the following assumptions:

(a.1): $A : Y \supset \mathcal{D}(A) \to Y$ is the infinitesimal generator of a s.c. semigroup e^{At} on Y, $t \geq 0$;

(a.2): B is bounded from U to Y,

$$B \in \mathcal{L}(U; Y); \tag{9.7.4}$$

(a.3): $R : Y \supset \mathcal{D}(R) \supset \mathcal{D}(A) \to Z$,

$$R \in \mathcal{L}(\mathcal{D}(A); Z) \quad \text{or} \quad RA^{-1} \in \mathcal{L}(Y; Z); \tag{9.7.5}$$

and the (closable) operator Re^{At} can be extended as a map

$$Re^{At} : \text{continuous } Y \to L_2(0, T; Z); \tag{9.7.6a}$$

that is,

$$\int_0^T \|Re^{At}x\|_Z^2 \, dt \leq c_T \|x\|_Y^2, \quad x \in Y. \tag{9.7.6b}$$

Then, Theorems 9.6.2.3, 9.6.2.4, and 9.6.2.5, expressed for problem (9.7.1)–(9.7.6) [where A is however only the generator of a s.c. semigroup] yields the following result.

Theorem 9.7.1 *Assume the above hypotheses (a.1) through (a.3) for problem (9.7.1)–(9.7.3). Then:*

(i) There exists a unique optimal pair $\{u^0(t, 0; y_0), y^0(t, 0; y_0\}$, $y_0 \in Y$, such that

$$u^0(t, 0; y_0) \in C([0, T]; U); \quad y^0(t, 0; y_0) \in C([0, T]; Y). \tag{9.7.7}$$

(ii) *There exists*

$$P(\cdot): \ continuous \ Y \to C([0, T]; Y). \tag{9.7.8}$$

such that

(ii₁)

$$P(t) = P^*(t) \geq 0, \quad 0 \leq t \leq T. \tag{9.7.9}$$

(ii₂) $(P(t)x, y)_Y$ *is continuously differentiable in* t, *for* $x, y \in \mathcal{D}(A)$.

(ii₃) *(Existence)* $P(t)$ *satisfies the differential Riccati equation*

$$\begin{cases} \dfrac{d}{dt}(P(t)x, y)_Y = -(Rx, Ry)_Z - (P(t)x, Ay)_Y - (P(t)Ax, y)_Y \\ \qquad\qquad\qquad + (B^*P(t)x, B^*P(t)y)_U, \\ \quad P(T) = G^*G, \end{cases} \tag{9.7.10}$$

and hence the integral Riccati equation

$$(P(t)x, z)_Y = \int_t^T \left(Re^{A(\tau-t)}x, \ Re^{A(\tau-t)}z \right)_Z d\tau$$

$$- \int_t^T \left(B^*P(\tau)e^{A(\tau-t)}x, \ B^*P(\tau)e^{A(\tau-t)}z \right)_U d\tau$$

$$+ \left(Ge^{A(T-t)}x, \ Ge^{A(T-t)}z \right)_{Z_f}, \quad x, z \in Y. \tag{9.7.11}$$

(ii₄) *Indeed,* $P(t)$ *is given explicitly by*

$$P(t)x = \int_t^T e^{A^*(\tau-t)}R^*Ry^0(\tau, t; x)\,d\tau + e^{A^*(T-t)}G^*Gy^0(T, t; x).$$

$$\tag{9.7.12}$$

(iii) *The optimal pair satisifies the following pointwise feedback synthesis:*

$$u^0(t, 0; y_0) = -B^*P(t)y^0(t, 0; y_0) \in C([0, T]; U). \tag{9.7.13}$$

(iv) *The optimal cost in (9.7.3) is*

$$J(u^0(\cdot, 0; y_0), y^0(\cdot, 0; y_0)) = (P(0)y_0, y_0)_Y, \quad y_0 \in Y. \tag{9.7.14}$$

(v) *(Uniqueness) The operator* $P(t)$ *in (9.7.12) is the unique solution of the IRE (9.7.11) and hence of the DRE (9.7.10) within the class of solutions satisfying properties (9.7.8) and (9.7.9).*

Relevant examples of partial differential equations with *distributed* control and with *trace* observation, which fit into the framework of the present section, are deferred to Chapter 13 of Volume 3.

9.8 Application to Hyperbolic Partial Differential Equations with Point Control. Regularity Theory

This section provides an illustration of the abstract theory of Theorems 9.2.1 and 9.2.2, as it applies to some hyperbolic partial differential equations (PDEs). By choice, which is explained below, emphasis will be put on *interior point-control* problems [rather than on *boundary control* problems], where a scalar control function acts on the hyperbolic PDE through a Dirac mass $+1$ concentrated at an interior pont of the bounded spatial domain Ω, for dim $\Omega = 1, 2, 3$. The first illustration involves the wave equation (Subsection 9.8.1); this is then followed by the Kirchhoff equation under one choice of BC (others are possible as well) in Subsection 9.8.2. In Subsection 9.8.3, we present a *boundary control* exception, which centers on the wave equation with Neumann boundary control and Dirichlet boundary observation in *one space dimension*, dim $\Omega = 1$, thus complementing Chapter 8, Section 8.6, which treats the same problem for dim $\Omega \geq 2$. When dim $\Omega = 1$, a definitely higher interior and trace regularity of the wave equation with Neumann boundary control is available; this permits inclusion of this case in the present more regular theory of Chapter 9, where the input-solution operator L is continuous $L_2(0, T; U) \rightarrow C([0, T]; Y)$, as in (A.1*) = (1.10). The reason for deemphasizing boundary control problems for PDEs in this chapter is that they will be more properly handled by the richer theory of the subsequent Chapter 10. By requiring – in addition to (A.1) = (9.1.7) – regularity results for the input-solution operator L (and its adjoint L^*) under controls smoother than L_2 in time/space, the general abstract theory of the subsequent Chapter 10 manages to reduce drastically (to a "minimum" $R \sim A^{-\epsilon}$) the smoothing requirements of the observation operator R, over the abstract theory of the present Chapter 9. Needless to say, property (A.1) = (9.1.7), as well as the additional regularity requirements for L and L^* are, in fact, *intrinsic properties* enjoyed by numerous classes of PDE dynamics, of hyperoblic or Petrowski-type: second-order hyperbolic equations with Dirichlet control; first-order hyperbolic systems; Kirchhoff equations and Euler–Bernoulli equations under a variety of boundary controls; Schrödinger equation with Dirichlet control, etc. Accordingly, illustrations involving these mixed problems are deferred to Chapter 10, where they can be handled under drastically reduced requirements of smoothing by the observation operator R.

9.8.1 Applications: Wave Equation with Interior Point Control and Dirichlet Boundary Conditions

Dynamics   We consider the following interior point control problem for the wave equation:

$$
\begin{cases}
w_{tt} = \Delta w + \delta(x)u(t) & \text{in } (0, T] \times \Omega = Q; & (9.8.1.1a) \\[2mm]
w(0, \,\cdot\,) = w_0, \, w_t(0, \,\cdot\,) = w_1 & \text{in } \Omega; & (9.8.1.1b) \\[2mm]
w|_\Sigma \equiv 0 & \text{in } (0, T] \times \Gamma = \Sigma, & (9.8.1.1c)
\end{cases}
$$

where $\delta(x)$ is the Dirac mass $+1$ at the point 0 (origin), assumed to be an interior point of the open bounded domain $\Omega \subset R^n$, $n = 1, 2, 3$. The control u is assumed in $L_2(0, T)$.

Consistently with established (optimal) regularity theory, described below in (9.8.1.11), we take

$$\{w_0, w_1\} \in Y = Y_1 \times Y_2 \equiv \begin{cases} L_2(\Omega) \times H^{-1}(\Omega) & n = 3; & (9.8.1.2a) \\ H_{00}^{\frac{1}{2}}(\Omega) \times \left[H_{00}^{\frac{1}{2}}(\Omega) \right]' & n = 2; & (9.8.1.2b) \\ H_0^1(\Omega) \times L_2(\Omega) & n = 1. & (9.8.1.2c) \end{cases}$$

The space $H_{00}^{\frac{1}{2}}(\Omega)$ may also be defined as the interpolation space $H_{00}^{\frac{1}{2}}(\Omega) = [H_0^1(\Omega), L_2(\Omega)]_{\frac{1}{2}}$ or intrinsically, as in [Lions, Magenes, 1972, p. 66]. Accordingly, the cost functional we seek to minimize over all $u \in L_2(0, T)$ is

$$J(u, w) = \int_0^T \left[\left\| R \begin{bmatrix} w(t) \\ w_t(t) \end{bmatrix} \right\|_Y^2 + |u(t)|^2 \right] dt + \left\| G \begin{bmatrix} w(T) \\ w_t(T) \end{bmatrix} \right\|_{Z_f}^2 \quad (9.8.1.3)$$

for a preassigned finite time $0 < T < \infty$, where the space Y is defined by (9.8.1.2), while the space Z_f and the operators R and G are selected below. We set

$$\mathcal{A}h = -\Delta h; \quad \mathcal{A} : \mathcal{D}(\mathcal{A}) = H^2(\Omega) \cap H_0^1(\Omega) \rightarrow L_2(\Omega), \quad (9.8.1.4)$$

so that $\mathcal{A}$ is a positive, self-adjoint operator on $L_2(\Omega)$, and we recall that, with equivalent norms:

$$\mathcal{D}(\mathcal{A}^{\frac{1}{2}}) = H_0^1(\Omega); \quad \mathcal{D}(\mathcal{A}^{\frac{1}{4}}) = H_{00}^{\frac{1}{2}}(\Omega). \quad (9.8.1.5)$$

Abstract Setting To put problem (9.8.1.1)–(9.8.1.3) into the abstract model (9.1.1), (9.1.15), we take

$$A = \begin{vmatrix} 0 & I \\ -\mathcal{A} & 0 \end{vmatrix}; \quad B = \begin{vmatrix} 0 \\ \delta \end{vmatrix}; \quad (9.8.1.6)$$

$$y(t) = \begin{bmatrix} w(t) \\ w_t(t) \end{bmatrix}; \quad Z = Z_f = Y \text{ (as defined in (9.8.1.2))}; \quad U = \mathbb{R}^1. \quad (9.8.1.7)$$

Thus, A is the generator of a s.c. group on Y.

Assumption (1.2) $A^{-1}B \in \mathcal{L}(U, Y)$. From (9.8.1.6) we compute with $u \in \mathbb{R}^1$:

$$A^{-1}Bu = \begin{bmatrix} 0 & -\mathcal{A}^{-1} \\ I & 0 \end{bmatrix} \begin{bmatrix} 0 \\ \delta u \end{bmatrix} = \begin{bmatrix} -\mathcal{A}^{-1}\delta u \\ 0 \end{bmatrix} \in Y, \quad (9.8.1.8)$$

where it remains to show membership to Y, as stated in (9.8.1.8). To this end, we recall that for the second-order differential operator $\mathcal{A}$ we have $\mathcal{D}(\mathcal{A}^{\frac{\theta}{2}}) \subset H^\theta(\Omega)$;

hence

$$\delta \in [H^\theta(\Omega)]' \subset [\mathcal{D}(A^{\frac{\theta}{2}})]', \quad \text{or} \quad A^{-\frac{\theta}{2}}\delta \in L_2(\Omega), \tag{9.8.1.9}$$

after invoking Sobolev embedding, where with $\epsilon > 0$ arbitrary: $\theta = 3/2 + 2\epsilon$ for $n = 3$; $\theta = 1 + 2\epsilon$ for $n = 2$; $\theta = 1/2 + 2\epsilon$ for $n = 1$. Thus, setting $z = A^{-\theta/2}\delta \in L_2(\Omega)$, we have, by (9.8.1.9),

$$A^{-1}\delta = A^{\frac{\theta}{2}-1}z$$

$$= \begin{cases} A^{-\frac{1}{4}+\epsilon}z \in \mathcal{D}(A^{\frac{1}{4}-\epsilon}) \subset L_2(\Omega) & n = 3; \quad (9.8.1.10a) \\ A^{-\frac{1}{2}+\epsilon}z \in \mathcal{D}(A^{\frac{1}{2}-\epsilon}) \subset \mathcal{D}(A^{\frac{1}{4}}) = H^{\frac{1}{2}}_{00}(\Omega) & n = 2 \quad (9.8.1.10b) \\ A^{-\frac{3}{4}+\epsilon}z \in \mathcal{D}(A^{\frac{3}{4}-\epsilon}) \subset \mathcal{D}(A^{\frac{1}{2}}) = H^1_0(\Omega) & n = 1. \quad (9.8.1.10c) \end{cases}$$

Thus, comparing (9.8.1.10) with (9.8.1.2), we see that (9.8.1.8) holds true. Thus, we conclude that assumption (9.1.2) is verified.

Assumption (A.1) = (9.1.7)   This is equivalent to its dual version (A.1*) = (9.1.10), which is guaranteed by the first part (a) of the following result. Its second part (b) is included for completeness.

Theorem 9.8.1.1   *With reference to problem (9.8.1.1), let $w_0 = w_1 = 0$. Then:*

(a) Interior regularity. *The following interior regularity holds true, with Y defined by (9.8.1.2):*

$$L_0 : u \rightarrow \{w(t), w_t(t)\} = y(t) : \quad continuous \ L_2(0, T) \rightarrow C([0, T]; Y). \tag{9.8.1.11}$$

(b) Boundary regularity. *The following boundary regularity holds true:*

$$u \rightarrow \left.\frac{\partial w}{\partial v}\right|_\Sigma : \quad continuous \ L_2(0, T) \rightarrow \begin{cases} H^{-1}(\Sigma) & n = 3; \quad (9.8.1.12a) \\ H^{-\frac{1}{2}}(\Sigma) & n = 2; \quad (9.8.1.12b) \\ L_2(\Sigma) & n = 1. \quad (9.8.1.12c) \end{cases}$$

Proof.   A proof will be given in Section 9.9.1 below.   □

Remark 9.8.1.1   The *interior* regularity (9.8.1.11) is "$1/2 + \epsilon$" sharper (in the space variable), measured in Sobolev space order, than the regularity that one would obtain by using only property (9.8.1.9) of membership of δ, without exploiting the individual properties of the Dirac measure δ. Moreover, as in the case of the mixed problem in Chapter 10, Section 10.5, Remark 10.5.3.1, the *boundary* regularity in (9.8.1.12) is "$1/2$" sharper (in the space variable) than the one that follows from the interior regularity (9.8.1.11) by a formal application of the trace theory.

Remark 9.8.1.2 For $n = 3$, several proofs have been given of the interior regularity (9.8.1.11), with (9.8.1.2a). Three different proofs are reported in J. L. Lions [1988], one proof, due to Y. Meyer [1985], uses harmonic analysis; the proofs of J. L. Lions and L. Niremberg are based on the explicit solution formula (generally refered to as Kirchhoff formula) for the corresponding free space problem in R^3. Some proofs (e.g., Niremberg's) refer to the dual problem. Throughout this section, for both the wave equation and the Kirchhoff equation, we follow closely the general method given in R. Triggiani [1991; 1993(a)], which was explicitly used for $n = 1, 2, 3$, and which, in principle, may be applied to any dimension of n. This approach, which studies the original problem (9.8.1.1), hinges on the Laplace–Fourier transforms for the corresponding free space problem, followed by suitable changes of variables leading to standard problems. This method has been subsequently used also in the case of control concentrated on interior curves by S. Jafford and M. Tucsnak [1995].

Remark 9.8.1.3 In Eqn. (9.8.1.1a), if one replaces the Dirac distribution δ (supported at the interior point $x_0 = 0$) with its derivative δ', then the corresponding regularity results are one unit *less*, in Sobolev space regularity, over those of Theorem 9.8.1.1(a), with Y as in (9.8.1.2).

Case #1 (nonsmoothing observations)   For any $R \in \mathcal{L}(Y; Z)$ and $G \in \mathcal{L}(Y; Z_f)$, Theorem 9.2.1 holds true for problem (9.8.1.1), (9.8.1.3), yielding, in particular, a *pointwise synthesis of the optimal pair.*

Case #2 (smoothing observations)   In preparation for the verification of assumption $(A.2) = (9.1.8)$ on R, we let $C(t)$ be the cosine operator on $L_2(\Omega)$ generated by $-\mathcal{A}$, and let $S(t) = \int_0^t C(\tau)\,d\tau$ be its corresponding sine operator. Then, with $u \in \mathbb{R}^1$, we compute by (9.8.1.6)

$$
e^{\mathcal{A}t} Bu = \begin{bmatrix} C(t) & S(t) \\ -\mathcal{A}S(t) & C(t) \end{bmatrix} \begin{bmatrix} 0 \\ \delta u \end{bmatrix} = \begin{bmatrix} S(t)\delta u \\ C(t)\delta u \end{bmatrix}
$$
$$
= \begin{bmatrix} \mathcal{A}^{\frac{\theta-1}{2}} \mathcal{A}^{\frac{1}{2}} S(t)zu \\ \mathcal{A}^{\frac{\theta}{2}} C(t)zu \end{bmatrix}, \tag{9.8.1.13}
$$

where $z = \mathcal{A}^{-\theta/2}\delta \in L_2(\Omega)$ as in (9.8.1.9). Recalling the values of θ for $n = 1, 2, 3$ given above (9.8.1.10), as well as the standard regularity properties: $C(t)z$, $\mathcal{A}^{\frac{1}{2}}S(t)z \in C([0, T]; L_2(\Omega))$, we then obtain from (9.8.1.13)

$$
e^{\mathcal{A}t} Bu \in \begin{cases} C\big([0, T]; \big[\mathcal{D}(\mathcal{A}^{\frac{1}{4}+\epsilon})\big]' \times \big[\mathcal{D}(\mathcal{A}^{\frac{3}{4}+\epsilon})\big]'\big) & n = 3; \quad (9.8.1.14\text{a}) \\ C\big([0, T]; \big[\mathcal{D}(\mathcal{A}^{\epsilon})\big]' \times \big[\mathcal{D}(\mathcal{A}^{\frac{1}{2}+\epsilon})\big]'\big) & n = 2; \quad (9.8.1.14\text{b}) \\ C\big([0, T]; \mathcal{D}(\mathcal{A}^{\frac{1}{4}-\epsilon}) \times \big[\mathcal{D}(\mathcal{A}^{\frac{1}{4}+\epsilon})\big]'\big) & n = 1. \quad (9.8.1.14\text{c}) \end{cases}
$$

Assumption (A.2) = (9.1.8). From (9.8.1.14) we see that assumption (A.2) = (9.1.8) on R : $R^* Re^{At} B$: $U \rightarrow L_1(0, T; Y)$, continuously, is satisfied, provided that

$R^* R$: continuous :

$$
\begin{cases}
[\mathcal{D}(\mathcal{A}^{\frac{1}{4}+\epsilon})]' \times [\mathcal{D}(\mathcal{A}^{\frac{3}{4}+\epsilon})]' \\
[\mathcal{D}(\mathcal{A}^{\epsilon})]' \times [\mathcal{D}(\mathcal{A}^{\frac{1}{2}+\epsilon})]' \\
\mathcal{D}(\mathcal{A}^{\frac{1}{4}-\epsilon}) \times [\mathcal{D}(\mathcal{A}^{\frac{1}{4}+\epsilon})]'
\end{cases}
\rightarrow
\begin{cases}
L_2(\Omega) \times [\mathcal{D}(\mathcal{A}^{\frac{1}{2}})]', & n = 3; & (9.8.1.15a) \\
\mathcal{D}(\mathcal{A}^{\frac{1}{4}}) \times [\mathcal{D}(\mathcal{A}^{\frac{1}{4}})]', & n = 2; & (9.8.1.15b) \\
\mathcal{D}(\mathcal{A}^{\frac{1}{2}}) \times L_2(\Omega), & n = 1. & (9.8.1.15c)
\end{cases}
$$

This requires that for $n = 1, 2, 3$:

$$
R^* R \sim
\begin{bmatrix}
\mathcal{A}^{-\frac{1}{4}-\epsilon} & 0 \\
0 & \mathcal{A}^{-\frac{1}{4}-\epsilon}
\end{bmatrix},
\tag{9.8.1.16}
$$

meaning that $R^* R$ has to produce a smoothing comparable to $\mathcal{A}^{-\frac{1}{4}-\epsilon}$ on each coordinate space.

Assumption (A.3) = (9.1.9) The requirements for $G^* G$ to satisfy equivalently (A.3*) = (9.1.13):

$$
\sup_{0 \leq t \leq T} \| G^* G e^{At} B \|_{\mathcal{L}(U;Y)} < \infty,
$$

are seen from (9.8.1.14) to be the same as those in (9.8.1.15) for $R^* R$.

We conclude that Theorem 9.2.2 applies to (9.8.1.1)–(9.8.1.3) with $R^* R$ and $G^* G$ satisfying (9.8.1.16).

Duality results In order to state corresponding duality results, obtained by a duality argument on those of Theorem 9.8.1.1, we introduce the following system in the unknown $z(t, x)$:

$$
\begin{cases}
z_{tt} = \Delta z & \text{in } (0, T] \times \Omega \equiv Q; & (9.8.1.17a) \\
z(T, \cdot) = z_0; \, z_t(T, \cdot) = z_1 & \text{in } \Omega; & (9.8.1.17b) \\
z|_{\Sigma} \equiv g & \text{in } (0, T] \times \Gamma \equiv \Sigma. & (9.8.1.17c)
\end{cases}
$$

Then, with reference to problems (9.8.1.1) and (9.8.1.17) we readily have that:

$$
\text{the map } v(t) \rightarrow \left\{ w(T, \cdot), w_t(T, \cdot), \frac{\partial w}{\partial v}\bigg|_{\Sigma} \right\}
\tag{9.8.1.18}
$$
$$
\text{is dual to the map } \{z_1, z_0, g\} \rightarrow z(t, 0).
$$

Applying the duality relationship (9.8.1.18) on the results of Theorem 9.8.1.1 we obtain

Theorem 9.8.1.2 *With reference to problem (9.8.1.17) we have that*

$$
z(t, 0) \in L_2(0, T)
\tag{9.8.1.19}
$$

continuously on $\{z_0, z_1, g\}$, where

(a) for $n = \dim \Omega = 3$,

$$\{z_0, z_1\} \in H_0^1(\Omega) \times L_2(\Omega); \quad g \in H^1(\Sigma); \qquad (9.8.1.20)$$

(b) for $n = \dim \Omega = 2$,

$$\{z_0, z_1\} \in H_{00}^{1/2}(\Omega) \times \left[H_{00}^{1/2}(\Omega)\right]' = \mathcal{D}(A^{1/4}) \times [\mathcal{D}(A^{1/4})]'; \quad g \in H^{1/2}(\Sigma);$$
$$(9.8.1.21)$$

(c) for $n = \dim \Omega = 1$,

$$\{z_0, z_1\} \in L_2(\Omega) \times H^{-1}(\Omega) = L_2(\Omega) \times [\mathcal{D}(A^{1/2})]'; \quad g \in L_2(\Sigma).$$
$$(9.8.1.22)$$

9.8.2 Applications: Wave Equation with Interior Point Control and Boundary Observation

Dynamics We consider the following interior point control problem for the wave equation:

$$\begin{cases} w_{tt} = \Delta w + \delta(x)u(t) & \text{in } (0, T] \times \Omega = Q; & (9.8.2.1a) \\ w(0, \cdot) = w_0, \ w_t(0, \cdot) = w_1 & \text{in } \Omega; & (9.8.2.1b) \\ w|_{\Sigma_0} \equiv 0, \ \dfrac{\partial w}{\partial \nu}\bigg|_{\Sigma_1} \equiv 0 & \text{in } (0, T] \times \Gamma_i \equiv \Sigma_i, & (9.8.2.1c) \end{cases}$$

where $\delta(x)$ is the Dirac mass $+1$ at the point 0 (origin), assumed to be an interior point of the open bounded domain $\Omega \subset R^n$, $n = 1, 2, 3$, with boundary $\Gamma = \Gamma_0 \cup \Gamma_1$, Γ_i open in Γ. We assume for notational convenience (but this is not critical) that Γ_0 is nonempty. The scalar control u is assumed in $L_2(0, T)$.

Regularity of (9.8.2.1) With reference to problem (9.8.2.1), we have the following regularity result, taken from Triggiani [1991; 1993(a)]. First, as in Chapter 8, Appendix 8A, or Eqn. (8.6.1.16), we set

$$\left.\begin{array}{l} \alpha = \dfrac{3}{5} - \epsilon \\[2mm] \beta = \dfrac{3}{5} \end{array}\right\} : \text{for a general smooth domain}, \qquad (9.8.2.2a)$$

$$\alpha = \beta = \dfrac{2}{3} \quad : \text{for a sphere}, \qquad (9.8.2.2b)$$

$$\alpha = \beta = \dfrac{3}{4} \quad : \text{for a parallelepiped}, \qquad (9.8.2.2c)$$

for a domain Ω with $\dim \Omega = n \geq 2$. [See Notes in Chapter 8 for refinements due to Tataru.]

Theorem 9.8.2.1   *With reference to problem (9.8.2.1) and to the numbers α and β in (9.8.2.2), let*

$$w_0 = w_1 = 0 \quad \text{and} \quad u \in L_2(0, T).$$

Then, continuously,

(a) for $n = \dim \Omega = 3$,

$$\begin{cases} w \in C([0, T]; L_2(\Omega)), & (9.8.2.3a) \\ w_t \in C([0, T]; H^{-1}(\Omega)), & (9.8.2.3b) \\ w_{tt} \in L_2(0, T; H^{-2}(\Omega)), & (9.8.2.3c) \end{cases}$$

and

$$w|_{\Sigma_1} \in H^{\alpha-1}(\Sigma); \qquad (9.8.2.4)$$

(b) for $n = \dim \Omega = 2$,

$$\begin{cases} w \in C[0, T]; H^{\frac{1}{2}}(\Omega)), & (9.8.2.5a) \\ w_t \in C([0, T]; H^{-\frac{1}{2}}(\Omega)), & (9.8.2.5b) \\ w_{tt} \in L_2\big(0, T; \big[H_{00}^{\frac{3}{2}}(\Omega)\big]'\big), & (9.8.2.5c) \end{cases}$$

and

$$w|_{\Sigma_1} \in H^{\frac{\alpha+\beta-1}{2}}(\Sigma); \qquad (9.8.2.6)$$

(c) for $n = \dim \Omega = 1$,

$$\begin{cases} w \in C([0, T]; H^1(\Omega)), & (9.8.2.7a) \\ w_t \in C([0, T]; L_2(\Omega)), & (9.8.2.7b) \\ w_{tt} \in L_2(0, T; H^{-1}(\Omega)), & (9.8.2.7c) \end{cases}$$

and

$$w|_{\Sigma_1} \in H^1(\Sigma). \qquad (9.8.2.8)$$

Proof.   A proof will be given in Section 9.9.2. $\square$

Here we shall use only a subset of Theorem 9.8.2.1.

Remark 9.8.2.1   In Eqn. (9.8.2.1a), if one replaces the Dirac distribution δ (supported at the interior point $x_0 = 0$) with its derivative δ', then the corresponding regularity results are one unit less, in Sobolev space regularity in Ω, over those of Theorem 9.8.2.1. See Remark 9.8.1.3.

Remark 9.8.2.2   If the homogeneous BC (9.8.2.1c) is replaced with the damped BC $\frac{\partial w}{\partial \nu}|_{\Sigma_1} = -w_t$, the interior regularity results of Theorem 9.8.2.1 continue to hold true.

Control Problem   The cost functional we seek to minimize is defined by

$$J(u, w) = \int_0^T \left[\|w|_{\Gamma_1}\|_{L_2(\Gamma_1)}^2 + u^2(t) \right] dt \tag{9.8.2.9}$$

for a fixed preassigned $0 < T < \infty$. According to (9.8.2.2), (9.8.2.4), (9.8.2.6), and (9.8.2.8), J in (9.8.2.9) is well-defined for all $u \in L_2(0, T)$ and say, zero initial data $\{w_0, w_1\}$ only for $n = 1, 2$. Henceforth, we shall investigate the applicability of Theorems 9.2.1 and 9.2.2 of Chapter 9 and Theorem 8.2.1.2 of Chapter 8 to problem (9.8.2.1), (9.8.2.9) for $n = 1, 2$.

Abstract Setting   To put problem (9.8.2.1), (9.8.2.9) into the abstract form (9.1.1), (9.1.15), we introduce the following operators and spaces:

(i) $\mathcal{A} : L_2(\Omega) \supset \mathcal{D}(\mathcal{A}) \to L_2(\Omega)$ positive (with Γ_0 nonempty) and self-adjoint, defined by

$$\mathcal{A}h = -\Delta h; \quad \mathcal{D}(\mathcal{A}) = \left\{ h \in H^2(\Omega) : h|_{\Gamma_0} = \left.\frac{\partial h}{\partial \nu}\right|_{\Gamma_1} = 0 \right\}; \tag{9.8.2.10}$$

$$\mathcal{D}\left(\mathcal{A}^{\frac{1}{2}}\right) = H^1(\Omega); \quad \mathcal{D}(\mathcal{A}^\theta) = H^{2\theta}(\Omega), \quad 0 \le \theta < \frac{3}{4} \text{ (equivalent norms).} \tag{9.8.2.11}$$

(ii)

$$A = \begin{bmatrix} 0 & I \\ -\mathcal{A} & 0 \end{bmatrix}; \quad B = \begin{bmatrix} 0 \\ \delta \end{bmatrix}; \quad y = \begin{bmatrix} w \\ w_t \end{bmatrix}; \tag{9.8.2.12}$$

$$Z = L_2(\Gamma_1); \quad U = \mathbb{R}^1. \tag{9.8.2.13}$$

Then the operator A, with domain $\mathcal{D}(A) = \mathcal{D}(\mathcal{A}^{s+1}) \times \mathcal{D}(\mathcal{A}^{s+\frac{1}{2}})$ is the generator of a s.c. semigroup, in fact a unitary group, on the scale of spaces $\mathcal{D}(\mathcal{A}^{s+\frac{1}{2}}) \times \mathcal{D}(\mathcal{A}^s)$, with the convention that $\mathcal{D}(\mathcal{A}^s) = [\mathcal{D}(\mathcal{A}^{-s})]'$, duality with respect to $L_2(\Omega)$, if $s < 0$. Out of this scale, we shall pick the space Y according to the dimension of Ω (see below).

(iii) We define the observation operator R by

$$R \begin{bmatrix} w \\ w_t \end{bmatrix} = w|_{\Gamma_1}. \tag{9.8.2.14}$$

By proceeding exactly as in (9.8.1.8)–(9.8.1.10) of Section 9.8.1, we compute from (9.8.1.12) with $u \in \mathbb{R}^1$:

$$A^{-1}Bu = \begin{bmatrix} 0 & -\mathcal{A}^{-1} \\ I & 0 \end{bmatrix} \begin{bmatrix} 0 \\ \delta u \end{bmatrix} = \begin{bmatrix} -\mathcal{A}^{-1}\delta u \\ 0 \end{bmatrix}; \qquad (9.8.2.15)$$

$$\mathcal{A}^{-1}\delta \in \begin{cases} \mathcal{D}(\mathcal{A}^{\frac{1}{2}-\epsilon}) & \text{for } n = 2; & (9.8.2.16a) \\ \mathcal{D}(\mathcal{A}^{\frac{3}{4}-\epsilon}) & \text{for } n = 1. & (9.8.2.16b) \end{cases}$$

In preparation for the verification of assumption (A.2) = (9.1.8), we compute, via (9.8.2.12) and (9.8.2.14), with $u \in \mathbb{R}^1$ [as in (9.8.1.13) of Section 9.8.1]:

$$Re^{At}Bu = R \begin{bmatrix} C(t) & S(t) \\ -\mathcal{A}S(t) & C(t) \end{bmatrix} \begin{bmatrix} 0 \\ \delta u \end{bmatrix} = R \begin{bmatrix} S(t)\delta u \\ C(t)\delta u \end{bmatrix}$$

$$= S(t)\delta u|_{\Gamma_1} = \psi(t; \psi_0, \psi_1)|_{\Gamma_1}, \qquad (9.8.2.17)$$

where $C(t)$ and $S(t)$ are the cosine and sine operators generated by $-\mathcal{A}$, and where ψ is, therefore, the solution of the problem

$$\begin{cases} \psi_{tt} = \Delta\psi & \text{in } (0, T] \times \Omega; & (9.8.2.18a) \\ \psi(0, \cdot) = \psi_0 = 0, \ \psi_t(0, \cdot) = \psi_1 = \delta u & \text{in } \Omega; & (9.8.2.18b) \\ \psi|_{\Sigma_0} \equiv 0 & \text{in } (0, T] \times \Gamma_0 \equiv \Sigma_0, & (9.8.2.18c) \\ \left. \dfrac{\partial\psi}{\partial\nu} \right|_{\Sigma_1} \equiv 0 & \text{in } (0, T] \times \Gamma_1 \equiv \Sigma_1. & (9.8.2.18d) \end{cases}$$

We now seek to verify the required assumptions of Theorems 9.2.1 and 9.2.2 or, possibly, of Theorem 8.2.1.1 of Chapter 8.

Case $n = \dim \Omega = 2$ In this case, in order to have $R \in \mathcal{L}(Y; Z)$, $Z = L_2(\Gamma_1)$ as in (9.8.2.13), as required by (9.1.6) [or (8.2.1.2) of Chapter 8], we must take the space $Y \equiv Y_1 \times Y_2$ with first component $Y_1 = \mathcal{D}(\mathcal{A}^{\frac{1}{4}+\epsilon}) = H^{\frac{1}{2}+\epsilon}(\Omega)$, $\epsilon > 0$, in view of the definition (9.8.2.14) of R and trace theory. With such choice $Y = \mathcal{D}(\mathcal{A}^{\frac{1}{4}+\epsilon}) \times [\mathcal{D}(\mathcal{A}^{\frac{1}{4}-\epsilon})]'$, however, the operator $L_0 u = [w(t), w_t(t)]$ for problem (9.8.2.1) with $w_0 = w_1 = 0$ fails to be continuous $L_2(0, T) \to C([0, T]; Y)$, in view of the sharp interior regularity (9.8.2.5) of Theorem 9.8.2.1. Thus, (A.1) = (9.1.7) of Chapter 9 fails to be true, and the results of the present chapter are not applicable for $n = 2$.

Worse, even the weaker requirement (H.3*) = (8.2.1.6) of Chapter 8: $RL_0 u = w(t)|_{\Gamma_1}$: continuous $L_2(0, T) \to C([0, T]; L_2(\Gamma_1))$ cannot be verified, in view of the sharp boundary regularity (9.8.2.6), (9.8.2.2). Similarly, assumption (H.1) = (8.2.2.1) of Chapter 8 cannot be verified to hold true, as it would mean via (9.8.2.17) that the solution ψ of problem (9.8.2.18) would satisfy $\psi(t; \psi_0 = 0, \ \psi_1 = \delta u)|_{\Gamma_1} \in L_1(0, T; L_2(\Gamma_1))$. By contrast, with $\psi_1 = \delta u \in [H^{1+2\epsilon}(\Omega)]'$, the sharp regularity result from [Lasiecka, Triggiani, 1991, Corollary 2.5, Eqn. (9.2.37)], a companion result of

Chapter 8, Lemma 8.5.1.2, gives only for dim $\Omega \geq 2$,

$$\{\psi_0, \psi_1\} \in \mathcal{D}(A^{\frac{1-\alpha}{2}}) \times [\mathcal{D}(A^{\frac{\alpha}{2}})]' \to \psi(t; \psi_0, \psi_1)|_{\Sigma_1} \in L_2(\Sigma_1),$$

a result inadequate for our purposes. [Compare with Theorem 9.8.2.2(b) below for dim $\Omega = 1$.]

Thus, in the case $n = \dim \Omega = 2$, we cannot apply either the theory of Chapter 8 or the theory of Chapter 9 to problem (9.8.2.1), (9.8.2.9).

Case $n = \dim \Omega = 1$ Here we have some positive choices.

Choice #1   We select

$$Y = \mathcal{D}(A^{\frac{1}{2}}) \times L_2(\Omega) = H^1(\Omega) \times L_2(\Omega). \tag{9.8.2.19}$$

Then, we proceed to verify the required assumptions.

Assumption (ii) $= (9.1.2)$ $A^{-1}Bu \in Y$. This is satisfied via (9.8.2.16b), (9.8.2.19), and (9.8.2.15).

Assumption $(A.1) = (9.1.7)$ This is plainly verified by Theorem 9.8.2.1, Eqn. (9.8.2.7): with $w_0 = w_1 = 0$,

$$L : u \to \{w, w_t\} \text{ continuous } L_2(0, T) \to C([0, T]; H^1(\Omega) \times L_2(\Omega)). \tag{9.8.2.20}$$

Assumption $(A.2) = (9.1.8)$ With reference to the boundary-homogeneous problem (9.8.2.18a–d) in its one-dimensional version, the following result holds true:

$$\{\psi_0, \psi_1\} \in L_2(\Omega) \times [H^1(\Omega)]' \to \psi(t)|_{\xi=1} = \psi(t)|_{\Gamma_1} \in L_2(0, T) \tag{9.8.2.21}$$

continuously. Thus, considering the initial conditions in (9.8.2.18b): ψ_0, $\psi_1 = \delta u \in [H^{\frac{1}{2}+\epsilon}(\Omega)]' \subset [H^1(\Omega)]'$, we have a fortiori, via (9.8.2.17),

$$Re^{At}Bu = \psi(t; \psi_0 = 0; \ \psi_1 = \delta u)|_{\Gamma_1} \in L_2(0, T), \tag{9.8.2.22}$$

and thus assumption $(H.1) = (8.2.1.2)$ of Chapter 8 is verified; a fortiori, since presently $R \in \mathcal{L}(Y; Z)$, we have that assumption $(A.2) = (9.1.8)$ of Chapter 9 is fulfilled for dim $\Omega = 1$, via (9.8.2.22).

We conclude: under the present choice $Y = H^1(\Omega) \times L_2(\Omega)$ as in (9.8.2.19), Theorem 9.2.1 and Theorem 9.2.2 [or Theorem 8.2.1.2 of Chapter 8] are applicable to problem (9.8.2.1), (9.8.2.9) for dim $\Omega = 1$.

Choice #2   Here we select a smoother space

$$Y = \mathcal{D}(A^{\frac{3}{4}-\epsilon}) \times \mathcal{D}(A^{\frac{1}{4}-\epsilon}), \quad \epsilon > 0, \tag{9.8.2.23}$$

for a fixed $\epsilon > 0$. Then, assumption (ii) $= (9.1.2)$, $A^{-1}Bu \in Y$, is again satisfied by (9.8.2.15), (9.8.2.16b), and (9.8.2.23). Moreover, assumptions (H.1) $= (8.2.2.1)$ and (H.2) $= (8.2.2.2)$ of Chapter 8 are satisfied as well by the above analysis carried out under choice #1. Now, however, the operator L_0 is not continuous $L_2(0, T) \rightarrow C([0, T]; Y)$, with Y selected as in (9.8.2.23), by the sharp result (9.8.2.7) of Theorem 9.8.2.1. Hence, assumption (A.1) $= (9.1.7)$ [or (A.1*) $= (9.1.10)$] of the present Chapter 9 does not hold true. On the other hand, assumption (H.3) $= (8.2.1.3)$ of Chapter 8 is verified. With reference to problem (9.8.2.1), we have

$$(RL_0u)(t) = R \begin{bmatrix} w(t) \\ w_t(t) \end{bmatrix} = w(t)|_{\Gamma_1} \in C[0, T] = C([0, T]; L_2(\Gamma_1)), \quad (9.8.2.24)$$

by (9.8.2.8) of Theorem 9.8.2.1. We conclude: Under the choice $Y = \mathcal{D}(\mathcal{A}^{\frac{3}{4}-\epsilon}) \times \mathcal{D}(\mathcal{A}^{\frac{1}{4}-\epsilon})$ as in (9.8.2.23), Theorem 8.2.1.2 of Chapter 8 applies to problem (9.8.2.1), (9.8.2.9), while Theorems 9.2.1 and 9.2.2 do not apply. Thus, choice #2 fits better into the spirit of Chapter 8 and yields a Riccati equation on a smoother space over choice #1.

Duality results   In order to state corresponding duality results, obtained by a transposition (duality) argument on those of Theorem 9.8.2.1, we introduce the following system in the unknown $z(t, x)$:

$$\begin{cases} z_{tt} = \Delta z & \text{in } Q, & (9.8.2.25a) \\ z(T, \cdot) = z_0, z_t(T, \cdot) = z_1 & \text{in } \Omega, & (9.8.2.25b) \\ z|_{\Sigma_0} \equiv 0 & \text{in } \Sigma_0, & (9.8.2.25c) \\ \left. \dfrac{\partial z}{\partial \nu} \right|_{\Sigma_1} \equiv g & \text{in } \Sigma_1. & (9.8.2.25d) \end{cases}$$

The duality between the w-problem (9.8.2.1) and the z-problem (9.8.2.25) is described by the correspondence of the maps

$$v \rightarrow \{w(T, \cdot), w_t(T, \cdot), w|_\Sigma\} \text{ is dual of } \{z_1, z_0, g\} \rightarrow z(t, 0) \quad (9.8.2.26)$$

as in the Dirichlet case, see (9.8.1.19). Duality on Theorem 9.8.2.1 using (9.8.2.26) yields then

Theorem 9.8.2.2   *With reference to problem (9.8.2.25) we have that*

$$z(t, 0) \in L_2(0, T) \quad (9.8.2.27)$$

continuously in $\{z_0, z_1, g\}$ *where*

(a) for $n = \dim \Omega = 3$,

$$\{z_0, z_1\} \in H^1(\Omega) \times L_2(\Omega); \quad g \in H^{1-\alpha}(\Sigma); \quad (9.8.2.28)$$

(b) for $n = \dim \Omega = 2$,

$$\{z_0, z_1\} \in H^{1/2}(\Omega) \times H^{-1/2}(\Omega); \quad g \in \left[H^{\frac{\alpha+\beta-1}{2}}(\Sigma)\right]'; \tag{9.8.2.29}$$

(c) for $n = \dim \Omega = 1$,

$$\{z_0, z_1\} \in L_2(\Omega) \times [H^1(\Omega)]'; \quad g \in [H^1(\Sigma)]'. \tag{9.8.2.30}$$

9.8.3 Applications: Kirchhoff Equation with Interior Point Control

Dynamics   We consider the following interior point control problem for the Kirchhoff equation:

$$\begin{cases} w_{tt} - \rho \Delta w_{tt} + \Delta^2 w = \delta(x) u(t) & \text{in } (0, T] \times \Omega = Q; & (9.8.3.1a) \\ w(0, \cdot) = w_0, \ w_t(0, \cdot) = w_1 & \text{in } \Omega; & (9.8.3.1b) \\ w|_\Sigma \equiv 0 & \text{in } (0, T] \times \Gamma = \Sigma; & (9.8.3.1c) \\ \Delta w|_\Sigma \equiv 0 & \text{in } \Sigma, & (9.8.3.1d) \end{cases}$$

where $\rho > 0$ is a constant, and where δ is the Dirac mass $+1$ exercised at the origin, assumed to be an interior point of the open bounded domain $\Omega \subset R^n$, $n = 1, 2, 3$. Again, the control u is assumed to be in $L_2(0, T)$.

Consistently with (optimal) regularity theory, described below in (9.8.3.17), we take

$$\{w_0, w_1\} \in Y = Y_1 \times Y_2 \equiv \begin{cases} \mathcal{D}(A^{\frac{1}{2}}) \times \mathcal{D}(A^{\frac{1}{4}}), & n = 3; & (9.8.3.2a) \\ \mathcal{D}(A^{\frac{5}{8}}) \times \mathcal{D}(A^{\frac{3}{8}}), & n = 2; & (9.8.3.2b) \\ \mathcal{D}(A^{\frac{3}{4}}) \times \mathcal{D}(A^{\frac{1}{2}}), & n = 1. & (9.8.3.2c) \end{cases}$$

In (9.8.3.2), we have that $A : L_2(\Omega) \supset \mathcal{D}(A) \to L_2(\Omega)$ is the positive self-adjoint operator defined by

$$\mathcal{A}h = \Delta^2 h; \quad \mathcal{D}(A) = \{h \in H^4(\Omega) : h|_\Gamma = \Delta h|_\Gamma = 0\}. \tag{9.8.3.3}$$

Then [Grisvard, 1967, see also Chapter 3, Appendix 3A]

$$\mathcal{D}(A^{\frac{3}{4}}) = \{h \in H^3(\Omega) : h|_\Gamma = \Delta h|_\Gamma = 0\}; \quad \mathcal{D}(A^{\frac{1}{4}}) = H_0^1(\Omega); \tag{9.8.3.4}$$

$$A^{\frac{1}{2}}h = -\Delta h; \quad \mathcal{D}(A^{\frac{1}{2}}) = H^2(\Omega) \cap H_0^1(\Omega). \tag{9.8.3.5}$$

According to (9.8.3.2), the cost functional we seek to minimize over all $u \in L_2(0, T; L_2(\Gamma))$ is

$$J(u, w) = \int_0^T \left[\left\| R \begin{bmatrix} w(t) \\ w_t(t) \end{bmatrix} \right\|_Y^2 + |u(t)|_{L_2(\Gamma)}^2 \right] dt + \left\| G \begin{bmatrix} w(T) \\ w_t(T) \end{bmatrix} \right\|_{Z_f}^2, \tag{9.8.3.6}$$

for a preassigned finite time $0 < T < \infty$, where the space Y is defined by (9.8.3.2), while the operators R and G and the space Z_f are selected below.

Abstract Setting   To put problems (9.8.3.1)–(9.8.3.6) into the abstract model (9.1.1), (9.1.15), we first observe, via (9.8.3.4) and (9.8.3.5), that the dynamics (9.8.3.1) can be rewritten abstractly as

$$(I + \rho \mathcal{A}^{\frac{1}{2}})w_{tt} + \mathcal{A}w = \delta u, \tag{9.8.3.7}$$

and then we take

$$A = \begin{bmatrix} 0 & I \\ -\mathbb{A} & 0 \end{bmatrix}; \quad B = \begin{bmatrix} 0 \\ (I + \rho \mathcal{A}^{\frac{1}{2}})^{-1}\delta \end{bmatrix}; \quad \mathbb{A} = (I + \rho \mathcal{A}^{\frac{1}{2}})^{-1}\mathcal{A}; \quad R = I. \tag{9.8.3.8}$$

$$y(t) = \begin{bmatrix} w(t) \\ w_t(t) \end{bmatrix}; \quad Z = Z_f = Y \text{ (as defined in (9.8.3.1))}; \quad U = \mathbb{R}^1. \tag{9.8.3.9}$$

We notice that $\mathbb{A}$ is a positive self-adjoint operator on each of the following spaces:

(i) the space $\mathcal{D}(\mathcal{A}_\rho^{\frac{1}{4}})$, norm equivalent to $\mathcal{D}(\mathcal{A}^{\frac{1}{4}})$, with inner product

$$(x, y)_{\mathcal{D}(\mathcal{A}_\rho^{\frac{1}{4}})} = ((I + \rho \mathcal{A}^{\frac{1}{2}})x, y)_{L_2(\Omega)}, \quad x, y \in \mathcal{D}(\mathcal{A}^{\frac{1}{4}}); \tag{9.8.3.10}$$

(ii) the space $\mathcal{D}(\mathcal{A}_\rho^{\frac{3}{8}})$, norm equivalent to $\mathcal{D}(\mathcal{A}^{\frac{3}{8}})$, with inner product

$$(x, y)_{\mathcal{D}(\mathcal{A}_\rho^{\frac{3}{8}})} = ((I + \rho \mathcal{A}^{\frac{1}{2}})\mathcal{A}^{\frac{1}{4}}x, y)_{L_2(\Omega)}, \quad x, y \in \mathcal{D}(\mathcal{A}^{\frac{3}{8}}); \tag{9.8.3.11}$$

(iii) the space $\mathcal{D}(\mathcal{A}_\rho^{\frac{1}{2}})$, norm equivalent to $\mathcal{D}(\mathcal{A}^{\frac{1}{2}})$, with inner product

$$(x, y)_{\mathcal{D}(\mathcal{A}_\rho^{\frac{1}{2}})} = ((I + \rho \mathcal{A}^{\frac{1}{2}})\mathcal{A}^{\frac{1}{2}}x, y)_{L_2(\Omega)}, \quad x, y \in \mathcal{D}(\mathcal{A}^{\frac{1}{2}}). \tag{9.8.3.12}$$

Thus, as a consequence of (9.8.3.10)–(9.8.3.12), the operator A in (9.8.3.8) (with the obvious maximal domain) is *skew-adjoint* on each of the spaces

$$Y_\rho = \begin{cases} \mathcal{D}(\mathcal{A}^{\frac{1}{2}}) \times \mathcal{D}(\mathcal{A}_\rho^{\frac{1}{4}}), & n = 3; & (9.8.3.13a) \\ \mathcal{D}(\mathcal{A}^{\frac{5}{8}}) \times \mathcal{D}(\mathcal{A}_\rho^{\frac{3}{8}}), & n = 2; & (9.8.3.13b) \\ \mathcal{D}(\mathcal{A}^{\frac{3}{4}}) \times \mathcal{D}(\mathcal{A}_\rho^{\frac{1}{2}}), & n = 1, & (9.8.3.13c) \end{cases}$$

and thus A is the *generator of a s.c. unitary group on* Y_ρ, norm equivalent to Y.

Assumption (1.2) $A^{-1}B \in \mathcal{L}(U; Y)$ With $u \in \mathbb{R}^1$ we compute from (9.8.3.8)

$$A^{-1}Bu = \begin{bmatrix} 0 & -\mathbb{A}^{-1} \\ I & 0 \end{bmatrix}\begin{bmatrix} 0 \\ (I + \rho \mathcal{A}^{\frac{1}{2}})^{-1}\delta \end{bmatrix} = \begin{bmatrix} -\mathcal{A}^{-1}\delta u \\ 0 \end{bmatrix} \in Y, \tag{9.8.3.14}$$

and we have to show membership to Y, as stated in (9.8.3.14). To this end, we recall that for the fourth-order operator $\mathcal{A}$ in (9.8.3.3) we have $\mathcal{D}(\mathcal{A}^{\theta/4}) \subset H^\theta(\Omega)$; hence invoking Sobolev embedding we have

$$\delta \in [H^\theta(\Omega)]' \subset [\mathcal{D}(\mathcal{A}^{\frac{\theta}{4}})]' \quad \text{or} \quad \mathcal{A}^{-\theta/4}\delta \in L_2(\Omega), \tag{9.8.3.15}$$

where $\theta = 3/2 + 4\epsilon$ for $n = 3$; $\theta = 1 + 4\epsilon$ for $n = 2$; and $\theta = 1/2 + 4\epsilon$ for $n = 1$. Thus, setting $z = \mathcal{A}^{-\frac{\theta}{4}}\delta \in L_2(\Omega)$, we have by (9.8.3.15)

$$\mathcal{A}^{-1}\delta = \mathcal{A}^{\frac{\theta}{4}-1}z = \begin{cases} \mathcal{A}^{-\frac{5}{8}+\epsilon}z \in \mathcal{D}(\mathcal{A}^{\frac{5}{8}-\epsilon}) \subset \mathcal{D}(\mathcal{A}^{\frac{1}{2}}), & n = 3; & (9.8.3.16a) \\ \mathcal{A}^{-\frac{3}{4}+\epsilon}z \in \mathcal{D}(\mathcal{A}^{\frac{3}{4}-\epsilon}) \subset \mathcal{D}(\mathcal{A}^{\frac{5}{8}}), & n = 2; & (9.8.3.16b) \\ \mathcal{A}^{-\frac{7}{8}+\epsilon}z \in \mathcal{D}(\mathcal{A}^{\frac{7}{8}-\epsilon}) \subset \mathcal{D}(\mathcal{A}^{\frac{3}{4}}), & n = 1. & (9.8.3.16c) \end{cases}$$

Thus, by (9.8.3.14), (9.8.3.16), and (9.8.3.2), we see that (9.8.3.14) holds true. We conclude that assumption (9.1.2) is verified.

Assumption (A.1) = (9.1.7)   This is equivalent to its dual version (A.1*) = (9.1.10), which is guaranteed by the first part of the following result, taken from Triggiani [1993(b)].

Theorem 9.8.3.1   *With reference to problem (9.8.3.1), let $w_0 = w_1 = 0$. Then, with Y defined by (9.8.3.2), we have*

$$L_0 : u \to \{w(t), w_t(t)\} = y(t) : \text{continuous}$$

$$L_2(0, T; L_2(\Gamma)) \to C([0, T]; Y). \tag{9.8.3.17}$$

Proof.   A proof will be given in Section 9.9.3. $\square$

Remark 9.8.3.1   The interior regularity (9.8.3.17) is sharper by "$1/8 + \epsilon$" in space regularity, measured in fractional powers of $\mathcal{A}$ (essentially, by "$1/2+4\epsilon$" measured in Sobolev space order), than the regularity that one would obtain by using only property (9.8.3.15) of membership of δ, rather than its distinctive properties.

Case #1   (nonsmoothing observations)   For any $R \in \mathcal{L}(Y; Z)$ and $G \in \mathcal{L}(Y; Z_f)$, Theorem 9.2.1 applies to problem (9.8.3.1), (9.8.3.6), yielding, in particular, *a pointwise synthesis of the optimal pair.*

Case #2   (smoothing observations)   In preparation for the verification of assumption (A.2) = (9.1.8) on R, we notice that the operator

$$-\mathbb{A} = -\left(I + \rho\mathcal{A}^{\frac{1}{2}}\right)^{-1}\mathcal{A} = \frac{-\mathcal{A}^{\frac{1}{2}}}{\rho} + \frac{I}{\rho^2} - \frac{1}{\rho^2}\left(I + \rho\mathcal{A}^{\frac{1}{2}}\right)^{-1}, \quad \mathcal{D}(\mathbb{A}) = \mathcal{D}(\mathcal{A}^{\frac{1}{2}}), \tag{9.8.3.18}$$

being a bounded perturbation of $-\mathcal{A}^{\frac{1}{2}}/\rho$, generates a s.c. cosine operator $\mathbb{C}(t)$ on $L_2(\Omega)$ with $\mathbb{S}(t) = \int_0^t \mathbb{C}(t)\,d\tau$: continuous $L_2(\Omega) \to C([0, T]; \mathcal{D}(\mathcal{A}^{1/4}))$. Identity (9.8.3.18) is obtained by applying twice the identity $R(\lambda, \mathcal{A}^{1/2})\mathcal{A}^{1/2} = -I +$

$\lambda R(\lambda, \mathcal{A}^{\frac{1}{2}})$. From (9.8.3.7) and (9.8.3.8), we then obtain, with $u \in \mathbb{R}$,

$$
\begin{aligned}
e^{At} Bu &= \begin{bmatrix} \mathbb{C}(t) & \mathbb{S}(t) \\ -A\mathbb{S}(t) & \mathbb{C}(t) \end{bmatrix} \begin{bmatrix} 0 \\ (I + \rho \mathcal{A}^{\frac{1}{2}})^{-1} \delta \end{bmatrix} \\
&= \begin{bmatrix} \mathcal{A}^{\frac{\theta-1}{4}} (I + \rho \mathcal{A}^{\frac{1}{2}})^{-1} \mathcal{A}^{\frac{1}{4}} \mathbb{S}(t) zu \\ \mathcal{A}^{\frac{\theta}{4}} (I + \rho \mathcal{A}^{\frac{1}{2}})^{-1} \mathbb{C}(t) zu \end{bmatrix},
\end{aligned}
\tag{9.8.3.19}
$$

where $z = \mathcal{A}^{-\theta/4} \delta \in L_2(\Omega)$ by (9.8.3.15). Recalling the values of θ given above and the above property of $\mathbb{S}(t)$, we obtain from (9.8.3.19)

$$
e^{At} Bu \in \begin{cases} C([0, T]; \mathcal{D}(\mathcal{A}^{\frac{3}{8}-\epsilon}) \times \mathcal{D}(\mathcal{A}^{\frac{1}{8}-\epsilon})), & n = 3; & (9.8.3.20a) \\ C([0, T]; \mathcal{D}(\mathcal{A}^{\frac{1}{2}-\epsilon}) \times \mathcal{D}(\mathcal{A}^{\frac{1}{4}-\epsilon})), & n = 2; & (9.8.3.20b) \\ C([0, T]; \mathcal{D}(\mathcal{A}^{\frac{5}{8}-\epsilon}) \times \mathcal{D}(\mathcal{A}^{\frac{3}{8}-\epsilon})), & n = 1. & (9.8.3.20c) \end{cases}
$$

Assumption (A.2) = (9.1.8)   Thus, by (9.8.3.20), we see that a fortiori assumption (A.2) = (9.1.8) is satisfied provided that the observation operator $R \in \mathcal{L}(Y; Z)$, Y as in (9.8.3.2), satisfies

$R^* R$: continuous

$$
\begin{cases} \mathcal{D}(\mathcal{A}^{\frac{3}{8}-\epsilon}) \times \mathcal{D}(\mathcal{A}^{\frac{1}{8}-\epsilon}) \\ \mathcal{D}(\mathcal{A}^{\frac{1}{2}-\epsilon}) \times \mathcal{D}(\mathcal{A}^{\frac{1}{4}-\epsilon}) \\ \mathcal{D}(\mathcal{A}^{\frac{5}{8}-\epsilon}) \times \mathcal{D}(\mathcal{A}^{\frac{3}{8}-\epsilon}) \end{cases} \longrightarrow \begin{cases} \mathcal{D}(\mathcal{A}^{\frac{1}{2}}) \times \mathcal{D}(\mathcal{A}^{\frac{1}{4}}), & n = 3; & (9.8.3.21a) \\ \mathcal{D}(\mathcal{A}^{\frac{5}{8}}) \times \mathcal{D}(\mathcal{A}^{\frac{3}{8}}), & n = 2; & (9.8.3.21b) \\ \mathcal{D}(\mathcal{A}^{\frac{3}{4}}) \times \mathcal{D}(\mathcal{A}^{\frac{1}{2}}), & n = 1. & (9.8.3.21c) \end{cases}
$$

This requires that for $n = 1, 2, 3$:

$$
R^* R \sim \begin{bmatrix} \mathcal{A}^{-\frac{1}{8}+\epsilon} & 0 \\ 0 & \mathcal{A}^{-\frac{1}{8}+\epsilon} \end{bmatrix},
\tag{9.8.3.22}
$$

meaning that $R^* R$ has to produce a smoothing comparable to $\mathcal{A}^{-\frac{1}{8}+\epsilon}$ on each coordinate space.

Assumption (A.3) = (9.1.9)   The requirements for $G^* G$ to satisfy equivalently (A.3*) = (9.1.13):

$$
\sup_{0 \leq t \leq T} \|G^* G e^{At} B\|_{\mathcal{L}(U; Y)} < \infty,
$$

are seen from (9.8.3.20) to be the same as those in (9.8.3.21) for $R^* R$.

We conclude that Theorem 9.2.2 applies to problem (9.8.3.1), (9.8.3.6) with $R^* R$ and $G^* G$ satisfying (9.8.3.22).

Remark 9.8.3.2   If the BC $\Delta w|_\Sigma \equiv 0$ in (9.8.3.1d) is replaced by $\frac{\partial w}{\partial \nu}|_\Sigma \equiv 0$, then the corresponding Kirchhoff problem has the same regularity property (9.8.3.17), (9.8.3.2), in terms of the spaces of fractional powers of the new operator $\mathcal{A}$, which incorporates the new BC. This is proved in Triggiani [1993(b)] to which we refer for the new technical issues that appear in this case.

Remark 9.8.3.3   In Eqn. (9.8.3.1a), if one replaces the Dirac distribution δ (supported at the interior point $x_0 = 0$) with its derivative δ', then the corresponding regularity results are one unit less in Sobolev space regularity over those of Theorem 9.8.3.1, with Y as in Eqn. (9.8.3.2). This (readily established) fact will be used in Section 9.10.3, Theorem 9.10.3.1. See corresponding Remark 9.8.1.3 for the wave equation.

Remark 9.8.3.4   If the homogeneous BC (9.8.3.1d) is replaced with the damped BC $\Delta w = -\frac{\partial w_t}{\partial \nu}$ on Σ, the regularity results of Theorem 9.8.3.1 continue to hold true (via, essentially, the same proof). See corresponding Remark 9.8.2.2 for the wave equation.

9.8.4 One-Dimensional Wave Equation with Neumann Boundary Control

The present subsection illustrates the more regular theory [over that of Chapter 8, Section 8.6, where dim $\Omega \geq 2$] that arises when the wave equation with Neumann boundary control is defined on a one-dimensional bounded domain, say $\Omega = (0, 1)$. The resulting higher regularity theory available when dim $\Omega = 1$ allows one to treat, within the framework of the more regular Chapter 9, even the case with Neumann boundary control and Dirichlet boundary observation.

Dynamics   Let $\Omega = (0, 1)$ with boundary $\Gamma = \{0, 1\}$ consisting of the two endpoints, and consider the following mixed problem for the wave equation in the unknown $w(t, x)$:

$$
\begin{cases}
w_{tt} = w_{xx} & \text{in } (0, T] \times \Omega; & (9.8.4.1a) \\
w(0, \cdot) = w_0, \ w_t(0, \cdot) = w_1 & \text{in } \Omega; & (9.8.4.1b) \\
w_x|_{x=0} = u, \ w|_{x=1} \equiv 0 & \text{in } (0, T] \times \Gamma_0, \ \Gamma_0 = \{0\}, & (9.8.4.1c)
\end{cases}
$$

with $L_2(0, T)$-control u acting in the Neumann BC at $x = 0$. Consistently with the regularity theory stated below in Theorem 9.8.4.1, we associate to (9.8.4.1) the following cost functional:

$$
J(u, w) = \int_0^T \left[\left\| \begin{bmatrix} R_1 & 0 \\ 0 & R_2 \end{bmatrix} \begin{bmatrix} w(t) \\ w_t(t) \end{bmatrix} \right\|_{H^1(\Omega) \times L_2(\Omega)}^2 + (w(t)|_{x=0})^2 + |u(t)|^2 \right] dt,
$$
$$(9.8.4.2)$$

which penalizes the Neumann boundary control at $x = 0$, the observed solution in the interior in the energy norm, as well as the Dirichlet trace (boundary) observation $w(t)|_{x=0}$ at $x = 0$.

Abstract Setting   To put problem (9.8.4.1), (9.8.4.2) into the abstract model (9.1.1), (9.1.15), we take the following spaces and operators:

$$Y = H^1(\Omega) \times L_2(\Omega) = \mathcal{D}\big(A^{\frac{1}{2}}\big) \times L_2(\Omega); \tag{9.8.4.3}$$

$$y(t) = \begin{bmatrix} w(t) \\ w_t(t) \end{bmatrix}; \quad U = L_2(\Gamma_0) = \mathbb{R}^1; \quad Z = Y \times L_2(\Gamma_0); \tag{9.8.4.4}$$

$$A = \begin{bmatrix} 0 & I \\ -\mathcal{A} & 0 \end{bmatrix} : \mathcal{D}(\mathcal{A}) \times \mathcal{D}\big(\mathcal{A}^{\frac{1}{2}}\big) \to Y, \tag{9.8.4.5}$$

where $\mathcal{A} : L_2(\Omega) \supset \mathcal{D}(\mathcal{A}) \to L_2(\Omega)$ is the positive self-adjoint operator defined by

$$-\mathcal{A}h = h_{xx}, \quad \mathcal{D}(\mathcal{A}) = \{h \in H^2(\Omega) : h|_{x=1} = h_x|_{x=0} = 0\}, \tag{9.8.4.6}$$

so that A is the generator of a s.c. unitary group on $\mathcal{D}(\mathcal{A}^{\frac{1}{2}}) \times L_2(\Omega)$;

$$Bu = \begin{bmatrix} 0 \\ \mathcal{A}Nu \end{bmatrix}, \quad u \in \mathbb{R}^1; \quad Ng = v \iff \begin{cases} \Delta v = 0 \text{ in } \Omega; \\ v_x|_{x=0} = g; \ v|_{x=1} \equiv 0; \end{cases} \tag{9.8.4.7}$$

$$N : \text{ continuous } L_2(\Gamma_1) \to H^{\frac{3}{2}}(\Omega) \subset H^{\frac{3}{2}-2\rho}(\Omega) = \mathcal{D}\big(\mathcal{A}^{\frac{3}{4}-\rho}\big), \quad \rho > 0 \tag{9.8.4.8}$$

(compare with Chapter 8, Section 8.6.1);

$$R \begin{bmatrix} w \\ w_t \end{bmatrix} = \{R_1 w; \ R_2 w_t; \ w|_{x=0}\} \in Z = Y \times L_2(\Gamma_0), \tag{9.8.4.9}$$

where we recall from Chapter 3, Section 3.3, Eqn. (3.3.9) that the Dirichlet trace on $x = 0$ may be expressed as:

$$N^* \mathcal{A}w = w|_{x=0} = w|_{\Gamma_0}, \tag{9.8.4.10}$$

and we have

$$R : \text{ continuous } Y = H^1(\Omega) \times L_2(\Omega) \to Y \times L_2(\Gamma_0) = Z \text{ for } \begin{bmatrix} R_1 & 0 \\ 0 & R_2 \end{bmatrix} \in \mathcal{L}(Y), \tag{9.8.4.11}$$

so that assumption (9.1.6) is satisfied, via trace theory.

Assumption (9.1.2)   With $u \in \mathbb{R}^1$, we verify from (9.8.4.5) and (9.8.4.7) that

$$A^{-1}Bu = \begin{bmatrix} 0 & -\mathcal{A}^{-1} \\ I & 0 \end{bmatrix} \begin{bmatrix} 0 \\ \mathcal{A}Nu \end{bmatrix} = \begin{bmatrix} -Nu \\ 0 \end{bmatrix} \in \mathcal{D}\big(\mathcal{A}^{\frac{1}{2}}\big) \times L_2(\Omega), \tag{9.8.12}$$

invoking in the last step (9.8.4.8). Thus assumption (9.1.2) is verified.

Assumption (A.1) = (9.1.7) This assumption is verified, as a consequence of the following regularity result. To this end, we consider the corresponding homogeneous problem:

$$\begin{cases} \psi_{tt} = \psi_{xx} & \text{in } (0, T] \times \Omega; & (9.8.4.13a) \\ \psi(0, \cdot) = \psi_0, \ \psi_t(0, \cdot) = \psi_1 & \text{in } \Omega; & (9.8.4.13b) \\ \psi_x|_{x=0} \equiv 0, \ \psi|_{x=1} \equiv 0 & \text{in } (0, T] \times \Gamma, & (9.8.4.13c) \end{cases}$$

which would be the dual problem of (9.8.4.1), if the initial data were given at $t = T$, rather than at $t = 0$, an inessential modification since the problem is time reversible. The solution of (9.8.4.13) may be written as

$$\psi(t; \psi_0, \psi_1) = C(t)\psi_0 + S(t)\psi_1, \qquad (9.8.4.14)$$

where $C(t)$ is the s.c. cosine operator on $L_2(\Omega)$ generated by $-\mathcal{A}$, while $S(t) = \int_0^t C(\tau) \, d\tau$ is the corresponding sine operator.

Theorem 9.8.4.1 *With reference to the one-dimensional problem (9.8.4.1), we have:*

(a) Let

$$\{w_0, w_1\} \in H^1(\Omega) \times L_2(\Omega); \quad u \in L_2(0, T). \qquad (9.8.4.15)$$

Then, continuously,

$$\begin{cases} \{w, w_t\} \in C([0, T]; H^1(\Omega) \times L_2(\Omega)), & (9.8.4.16a) \\ \text{as well as} \\ w|_{\Gamma_0} = w|_{x=0} \in H^1(0, T). & (9.8..4.16b) \end{cases}$$

(b) Equivalently to (9.8.4.16a), by duality, with reference to problem (9.8.4.13), we have, continuously,

$$\{\psi_0, \psi_1\} \in L_2(\Omega) \times [H^1(\Omega)]' \to \psi(t; \psi_0, \psi_1)|_{\Gamma_1} \in L_2(0, T), \quad (9.8.4.17a)$$

or else

$$\{\psi_0, \psi_1\} \in H^1(\Omega) \times L_2(\Omega) \to \psi_t(t; \psi_0, \psi_1)|_{\Gamma_0} \in L_2(0, T), \quad (9.8.4.17b)$$

or in the operator norm

$$\left.\begin{array}{l} N^* \mathcal{A}^{1+\frac{1}{2}} S(t) \\ N^* \mathcal{A} C(t) \end{array}\right\} : \textit{continuous } L_2(\Omega) \to L_2(0, T). \qquad (9.8.4.18)$$

Proof. One elementary proof of part (a) is given in Section 9.9.4. Another proof of the equivalent part (b) is given in Chapter 10, Remark 10.5.10.1. □

Thus, the interior regularity (9.8.4.15) in part (a) of Theorem 9.8.4.1, once specialized to $w_0 = w_1 = 0$, yields

$$(L_0 u)(t) = \begin{bmatrix} w(t) \\ w_t(t) \end{bmatrix} : \text{ continuous } L_2(0, T) \to C([0, T]; Y), \qquad (9.8.4.19)$$

with $Y = H^1(\Omega) \times L_2(\Omega)$ as in (9.8.4.11). Thus (9.8.4.19) verifies assumption $(A.1^*) = (9.1.10)$. Moreover, from (9.8.4.7) we compute, via (9.8.4.3),

$$\left(Bu, \begin{bmatrix} y_1 \\ y_2 \end{bmatrix} \right)_Y = (ANu, y_2)_{L_2(\Omega)} = (u, N^* A y_2)_{L_2(\Gamma_1)}, \qquad (9.8.4.20)$$

and thus we see that

$$B^* \begin{bmatrix} y_1 \\ y_2 \end{bmatrix} = N^* A y_2 = y_2|_{\Gamma_1}. \qquad (9.8.4.21)$$

Furthermore, since $A^* = -A$, and so $e^{A^* t} = e^{A(-t)}$, we compute from (9.8.4.14), since $C(\cdot)$ is even and $S(\cdot)$ odd, with $x = [x_1, x_2] \in Y$,

$$B^* e^{A^* t} \begin{bmatrix} x_1 \\ x_2 \end{bmatrix} = B^* \begin{bmatrix} C(t)x_1 - S(t)x_2 \\ AS(t)x_1 + C(t)x_2 \end{bmatrix} \qquad (9.8.4.22)$$

$$= N^* A[AS(t)x_1 + C(t)x_2]$$

$$= N^* A C(t)x_2 + N^* A^{1+\frac{1}{2}} S(t) A^{\frac{1}{2}} x_1 \qquad (9.8.4.23)$$

$$\text{(by (9.8.4.18))} \qquad = \psi(t; \psi_0, \psi_1)|_{\Gamma_1} \in L_2(0, T); \qquad (9.8.4.24)$$

$$\psi_0 = x_2 \in L_2(\Omega); \quad \psi_1 = Ax_1 \in \left[\mathcal{D}\left(A^{\frac{1}{2}} \right) \right]' = [H^1(\Omega)]'. \qquad (9.8.4.25)$$

Thus, part (b) of Theorem 9.8.4.1 yields the dual version,

$$B^* e^{A^* t} : Y \to L_2(0, T) = L_2(0, T; L_2(\Gamma_1)), \qquad (9.8.4.26)$$

which verifies assumption $(A.1) = (9.1.7)$.

Assumption (A.2) = (9.1.8) We similarly compute from (9.8.4.5) and (9.8.4.7), with $u \in \mathbb{R}^1$,

$$e^{At} Bu = \begin{bmatrix} C(t) & S(t) \\ -AS(t) & C(t) \end{bmatrix} \begin{bmatrix} 0 \\ ANu \end{bmatrix} = \begin{bmatrix} S(t)ANu \\ C(t)ANu \end{bmatrix}, \qquad (9.8.4.27)$$

and thus by (9.8.4.9) applied to (9.8.4.27), we obtain

$$Re^{At} Bu = [R_1 S(t)ANu; R_2 C(t)ANu; S(t)ANu|_{\Gamma_1}]. \qquad (9.8.4.28)$$

Since $t \to C(t)x$ and $A^{\frac{1}{2}} S(t)x$ are $L_2(\Omega)$-continuous, for $x \in L_2(\Omega)$, we have, recalling (9.8.4.8),

$$R_1 S(t) ANu = R_1 \mathcal{A}^{-\frac{1}{4}+\epsilon} \mathcal{A}^{\frac{1}{2}} S(t) \mathcal{A}^{\frac{3}{4}-\epsilon} Nu \in C\big([0, T]; \mathcal{D}\big(\mathcal{A}^{\frac{1}{2}}\big)\big), \quad (9.8.4.29)$$

$$R_2 C(t) ANu = R_2 \mathcal{A}^{\frac{1}{4}+\epsilon} C(t) \mathcal{A}^{\frac{3}{4}-\epsilon} Nu \in C\left([0, T]; L_2(\Omega)\right), \quad (9.8.4.30)$$

as desired, provided that

$$\mathcal{A}^{\frac{1}{2}} R_1 \mathcal{A}^{-\frac{1}{4}+\epsilon} \in \mathcal{L}(L_2(\Omega)); \quad R_2 \mathcal{A}^{\frac{1}{4}+\epsilon} \in \mathcal{L}(L_2(\Omega)), \quad (9.8.4.31)$$

which means that both R_1 and R_2 are required to provide a smoothing comparable to $\mathcal{A}^{-\frac{1}{4}-\epsilon}$. Moreover, as to the third entry in (9.8.4.28), recalling (9.8.4.10):

$$\begin{aligned} S(t) ANu|_{\Gamma_1} &= N^* A S(t) ANu \\ &= N^* \mathcal{A}^{1+\frac{1}{2}} S(t) \mathcal{A}^{\frac{1}{2}} Nu \in L_2(0, T), \end{aligned} \quad (9.8.4.32)$$

by (9.8.4.18) and (9.8.4.8). Using (9.8.4.29), (9.8.4.30), and (9.8.4.32) in (9.8.4.28), we see that

$$Re^{At} Bu : U = \mathbb{R}^1 \to C([0, T]; Y) \times L_2(0, T), \quad (9.8.4.33)$$

which, a fortiori, verifies assumption (A.2) = (9.1.8), provided that the operators R_1 and R_2 satisfy the smoothing requirements in (9.3.4.31).

We conclude that Theorem 9.2.1 applies to problem (9.8.4.1), (9.8.4.2) for any

$$\begin{bmatrix} R_1 & 0 \\ 0 & R_2 \end{bmatrix} \in \mathcal{L}(Y),$$

whereas Theorem 9.2.2 applies if, in addition, the smoothing requirements (9.8.4.31) are fulfilled.

9.9 Proof of Regularity Results Needed in Section 9.8

In this section we give a proof of the regularity Theorems 9.8.1.1, 9.8.2.1, 9.8.3.1, and 9.8.4.1 needed in Section 9.8. Regarding the first three regularity results of the wave equation problems (9.8.1.1) and (9.8.2.1) [with homogeneous Dirichlet or Neumann BC respectively], and of the Kirchhoff equation (9.8.3.1), all with interior point control, we follow a general approach [Triggiani, 1991; 1993a, b]. This strategy applies in principle to any dimension and has the virtue of not requiring the explicit solution of the corresponding free space problem. Thus, it has been more recently applied successfully also to situations where the latter is not available, as in Jafford and Tucsnak [1995], where the control is concentrated on interior curves. The key steps of the approach are:

(i) the analysis of (sharp) regularity of the corresponding free space problem by means of Laplace–Fourier transform techniques, which hinges on a sharp a priori estimate (Lemma 9.9.1.1 in the case of the wave equation, and Lemma 9.9.3.1 in the case of the Kirchhoff equation);

(ii) suitable changes of variables, which lead to a corresponding standard problem on Ω, homogeneous on the boundary but with nonhomogeneous forcing term on Ω, which does not, however, involve the Dirac distribution δ.

As noted already in Remarks 9.8.1.1 and 9.8.3.1, this sharp approach produces interior regularity results that are "$1/2 + \epsilon$" stronger in space regularity, measured in Sobolev space order, over those that can be directly obtained by simply using the Sobolev regularity that $\delta \in [H^\alpha(\Omega)]'$; $\alpha = 3/2 + \epsilon$, $1 + \epsilon$; $1/2 + \epsilon$; for $n = 3, 2, 1$, respectively.

9.9.1 Proof of Theorem 9.8.1.1

With reference to the wave equation problem (9.8.1.1), with $w_0 = w_1 = 0$, and control $u \in L_2(0, T)$, we must show – by the closed graph theorem – the following interior and boundary regularity results:

(a) for $n = \dim \Omega = 3$,

$$
\begin{cases}
w \in C([0, T]; L_2(\Omega)), & (9.9.1.1) \\
w_t \in C([0, T]; H^{-1}(\Omega)), & (9.9.1.2) \\
w_{tt} \in L_2(0, T; H^{-2}(\Omega)), & (9.9.1.3)
\end{cases}
$$

and

$$
\left. \frac{\partial w}{\partial \nu} \right|_\Sigma \in H^{-1}(\Sigma); \qquad (9.9.1.4)
$$

(b) for $n = \dim \Omega = 2$,

$$
\begin{cases}
w \in C\big([0, T]; H_{00}^{\frac{1}{2}}(\Omega) = \mathcal{D}(\mathcal{A}^{\frac{1}{4}})\big), & (9.9.1.5) \\
w_t \in C\big([0, T]; \big[H_{00}^{\frac{1}{2}}(\Omega)\big]' = \big[\mathcal{D}(\mathcal{A}^{\frac{1}{4}})\big]'\big), & (9.9.1.6) \\
w_{tt} \in L_2\big(0, T; \big[\mathcal{D}(\mathcal{A}^{\frac{3}{4}})\big]'\big) \subset L_2\big(0, T; \big[H_{00}^{\frac{3}{2}}(\Omega)\big]'\big), & (9.9.1.7)
\end{cases}
$$

and

$$
\left. \frac{\partial w}{\partial \nu} \right|_\Sigma \in H^{-\frac{1}{2}}(\Sigma); \qquad (9.9.1.8)
$$

(c) for $n = \dim \Omega = 1$,

$$
\begin{cases}
w \in C\big([0, T]; H_0^1(\Omega) = \mathcal{D}(\mathcal{A}^{\frac{1}{2}})\big), & (9.9.1.9) \\
w_t \in C\big([0, T]; L_2(\Omega)\big), & (9.9.1.10) \\
w_{tt} \in L_2\big(0, T; H^{-1}(\Omega) = \big[\mathcal{D}(\mathcal{A}^{\frac{1}{2}})\big]'\big), & (9.9.1.11)
\end{cases}
$$

and

$$
\left. \frac{\partial w}{\partial \nu} \right|_\Sigma \in L_2(\Sigma). \qquad (9.9.1.12)
$$

In the above results, $\mathcal{A}$ is the positive, self-adjoint operator in (9.8.1.4) and (9.8.1.5):

$$\mathcal{A}h = -\Delta h; \quad \mathcal{D}(\mathcal{A}) = H^2(\Omega) \cap H_0^1(\Omega); \quad \mathcal{D}\big(\mathcal{A}^{\frac{1}{2}}\big) = H_0^1(\Omega); \quad (9.9.1.13)$$

$$\mathcal{D}\big(\mathcal{A}^{\frac{1}{4}}\big) = \big[\mathcal{D}\big(\mathcal{A}^{\frac{1}{2}}\big), L_2(\Omega)\big]_{\frac{1}{2}} = \big[H_0^1(\Omega), L_2(\Omega)\big]_{\frac{1}{2}} = H_{00}^{\frac{1}{2}}(\Omega); \quad (9.9.1.14)$$

$$\mathcal{D}\big(\mathcal{A}^{\frac{3}{4}}\big) = \big[\mathcal{D}(\mathcal{A}), \mathcal{D}\big(\mathcal{A}^{\frac{1}{2}}\big)\big]_{\frac{1}{2}} \supset \big[H_0^2(\Omega), H_0^1(\Omega)\big]_{\frac{1}{2}} \equiv H_{00}^{\frac{3}{2}}(\Omega) \quad (9.9.1.15)$$

(with equivalent norms); see [Lions, Magenes, p. 66] for $H_{00}^{\frac{1}{2}}(\Omega)$.

Step 1 In studying the corresponding free space problem (9.9.1.17), we shall need the following uniform estimate:

Lemma 9.9.1.1 *Let $\gamma > 0$ be a fixed constant, and let $\omega \in R^1$ be a parameter. Then, there exists a constant C_γ depending on γ such that*

$$I(\omega) \equiv \int_0^\infty \frac{y^2\,dy}{(y^2 + \gamma^2 - \omega^2)^2 + 4\gamma^2\omega^2} \le C_\gamma < \infty, \quad (9.9.1.16)$$

for all $\omega \in R^1$.

Proof. The proof is relegated to Appendix 9A, where it is also pointed out that Lemma 9.9.1.1 is sharp. □

Step 2 (Free space problem) We consider the first free space problem corresponding to the original problem (9.8.1.1), in the unknown $\phi(t, x)$:

$$\begin{cases} \phi_{tt} = \Delta\phi + \delta(x)u(t) & \text{in } R_{t+}^1 \times R_x^n; & (9.9.1.17a) \\ \phi(0, x) \equiv \phi_t(0, x) \equiv 0 & \text{in } R_x^n, & (9.9.1.17b) \end{cases}$$

after extending $u(t)$ by zero for $t > T$, where $R_{t+}^1 = (0, \infty)$ in t and $R_x^n = n$-dimensional Euclidean space in x.

Proposition 9.9.1.2 *With reference to problem (9.9.1.17), let $u \in L_2(0, T)$. Then, continuously,*

(a) for $n = \dim \Omega = 3$,

$$\begin{cases} \phi \in C([0, T]; L_2(R^3)), & (9.9.1.18) \\ \phi_t \in C([0, T]; H^{-1}(R^3)), & (9.9.1.19) \\ \phi_{tt} \in L_2(0, T; H^{-2}(R^3)); & (9.9.1.20) \end{cases}$$

(b) for $n = \dim \Omega = 2$, and for any $\psi \in C_0^\infty(\Omega)$, $\psi(0) = 1$,

$$\begin{cases} \phi \in C([0, T]; H^{\frac{1}{2}}(R^2)); \quad \psi\phi \in C([0, T]; H_{00}^{\frac{1}{2}}(\Omega) = \mathcal{D}\big(\mathcal{A}^{\frac{1}{4}}\big)), & (9.9.1.21) \\ \phi_t \in C([0, T]; H^{-\frac{1}{2}}(R^2)); & (9.9.1.22) \\ \phi_{tt} \in L_2\big(0, T; H^{-\frac{3}{2}}(R^2)\big); & (9.9.1.23) \end{cases}$$

(c) for n = dim Ω = 1,

$$
\begin{cases}
\phi \in C([0, T]; H^1(R^1)), & (9.9.1.24) \\
\phi_t \in C([0, T]; L_2(R^1)), & (9.9.1.25) \\
\phi_{tt} \in L_2(0, T; H^{-1}(R^1)). & (9.9.1.26)
\end{cases}
$$

Proof. Let $\hat{\phi}(\lambda, \xi)$, with $\lambda = \gamma + i\omega$, $\gamma > 0$, and $\omega \in R^1$, $\xi \in R^n$, be the Laplace (in t)–Fourier (in x) transform of $\phi(t, x)$:

$$
\hat{\phi}(\lambda, \xi) = \frac{1}{(2\pi)^{N/2}} \int_0^\infty e^{-\lambda t} \int_{R^N} e^{-ix\cdot\xi} \phi(t, x)\, dx\, dt, \qquad (9.9.1.27)
$$

so that the transformed version of problem (9.9.1.17) is

$$
\hat{\phi}(\lambda, \xi) = \frac{\hat{u}(\lambda)}{\lambda^2 + |\xi|^2}. \qquad (9.9.1.28)
$$

It suffices to show the results for $\{\phi, \phi_t\}$ in (9.9.1.18)–(9.9.1.25) with $C[0, T]$ replaced by $L_2(0, T)$ and then appeal to the general result [Chapter 7, Theorem 7.3.1] for time-reversible dynamics (groups of operators) to lift $L_2(0, T)$ to $C[0, T]$, while preserving the space regularity.

Proof of (9.9.1.18). $n = 3$. Accordingly, it suffices to show that

$$
e^{-\gamma t}\phi(t, x) \in L_2\left(R^1_{t+} \times R^n_x\right), \qquad n = 3, \qquad (9.9.1.29)
$$

or, equivalently, by the Parseval identity [Doetsch, 1974, p. 212] that

$$
\hat{\phi}(\gamma + i\omega, \xi) \in L_2\left(R^1_\omega \times R^n_\xi\right), \qquad n = 3. \qquad (9.9.1.30)
$$

To this end we compute by (9.9.1.28)

$$
\|\hat{\phi}\|^2_{L_2(R^1_\omega \times R^n_\xi)} = \int_{R^1_\omega} \int_{R^n_\xi} |\hat{\phi}(\gamma + i\omega, \xi)|^2\, d\xi\, d\omega
$$

$$
\text{(by (9.9.1.28))} \qquad = \int_{R^1_\omega} |\hat{u}(\gamma + i\omega)|^2 \left(\int_{R^n_\xi} \frac{d\xi}{|\lambda^2 + |\xi|^2|^2} \right) d\omega. \qquad (9.9.1.31)
$$

Since $|\lambda^2 + |\xi|^2|^2 = (|\xi|^2 + \gamma^2 - \omega^2)^2 + 4\gamma^2\omega^2$, we see that (9.9.1.31) yields

$$
\|\hat{\phi}\|^2_{L_2(R^1_\omega \times R^n_\xi)} \leq C_\gamma \int_{R^1_\omega} |\hat{u}(\gamma + i\omega)|^2\, d\omega
$$

$$
= \frac{C_\gamma}{2\pi} \int_0^\infty e^{-2\gamma t} |u(t)|^2\, dt \leq \text{const} \|u\|^2_{L_2(0,T)}, \qquad (9.9.1.32)
$$

as desired, as soon as we prove that for $n = 3$, there exists C_γ such that

$$
\int_{R^n_\xi} \frac{d\xi}{|\lambda^2 + |\xi|^2|^2} = \int_{R^n_\xi} \frac{d\xi}{(|\xi|^2 + \gamma^2 - \omega^2)^2 + 4\gamma^2\omega^2} \leq C_\gamma < \infty, \qquad (9.9.1.33)
$$

uniformly in $\omega \in R^1$. To this end, we use spherical coordinates (ρ, Φ, θ) so that $d\xi = d\xi_1 \, d\xi_2 \, d\xi_3 = \rho^2 \sin \Phi \, d\Phi \, d\rho \, d\theta$. We then see that estimate (9.9.1.33) holds true provided

$$\int_0^\infty \frac{\rho^2 \, d\rho}{(\rho^2 + \gamma^2 - \omega^2)^2 + 4\gamma^2 \omega^2} \le C_\gamma < \infty, \quad \text{uniformly in } \omega \in R^1, \qquad (9.9.1.34)$$

which is precisely estimate (9.9.1.16) asserted by Lemma 9.9.1.1. Then (9.9.1.33), hence (9.9.1.32) and (9.9.1.30), (9.9.1.29) are all proved.

Proof of (9.9.1.21). $n = 2$.   Here, it suffices to show that

$$e^{-\gamma t} \phi(t, x) \in L_2\big(R_{t+}^1; H^{\frac{1}{2}}(R_x^n)\big), \quad n = 2,$$

or, equivalently,

$$|\xi|^{\frac{1}{2}} \hat{\phi}(\gamma + i\omega, \xi) \in L_2\big(R_\omega^1 \times R_\xi^n\big), \quad n = 2. \qquad (9.9.1.35)$$

To this end we compute, via (9.9.1.28),

$$\big\| |\xi|^{\frac{1}{2}} \hat{\phi}(\gamma + i\omega, \xi) \big\|_{L_2(R_\omega^1 \times R_\xi^n)}^2 = \int_{R_\omega^1} |\hat{u}(\gamma + i\omega)|^2 \left(\int_{R_\xi^n} \frac{|\xi| \, d\xi}{|\lambda^2 + |\xi|^2|^2} \right) d\omega$$

$$\le C_\gamma \int_{R_\omega^1} |\hat{u}(\gamma + i\omega)|^2 \, d\omega$$

$$= \frac{C_\gamma}{2\pi} \int_0^\infty e^{-2\gamma t} |u(t)|^2 \, dt = C_\gamma \|u\|_{L_2(0,T)}^2,$$

$$(9.9.1.36)$$

as desired, as soon as we prove that for $n = 2$, there exists C_γ such that

$$\int_{R_\xi^n} \frac{|\xi| \, d\xi}{|\lambda^2 + |\xi|^2|^2} = \int_{R_\xi^n} \frac{|\xi| \, d\xi}{(|\xi|^2 + \gamma^2 - \omega^2)^2 + 4\gamma^2 \omega^2} \le C_\gamma < \infty, \qquad (9.9.1.37)$$

uniformly in $\omega \in R^1$. Using now, for $n = 2$, polar coordinates (r, ω) so that $d\xi = d\xi_1 \, d\xi_2 = r \, dr \, d\theta$, we see that (9.9.1.37) leads to the same integral estimate (9.9.1.34) with ρ replaced by r, that is, to estimate (9.9.1.16). The regularity of $\psi\phi$ in (9.9.1.21) uses that $\psi \equiv 0$ near Γ.

Proof of (9.9.1.24). $n = 1$.   Now, it suffices to show that

$$e^{-\gamma t} \phi(t, x) \in L_2\big(R_{t+}^1; H^1(R_x^n)\big), \quad n = 1,$$

or, equivalently,

$$|\xi| \hat{\phi}(\gamma + i\omega, \xi) \in L_2\big(R_\omega^1 \times R_\xi^n\big), \quad n = 1, \qquad (9.9.1.38)$$

which is seen to hold true precisely as in the two preceding cases, by virtue of the same estimate (9.9.1.16) of Lemma 9.9.1.1.

Proof of (9.9.1.20). It follows from (9.9.1.18) already proved in the $L_2(0, T)$-sense, via [Lions, Magenes, 1972, p. 85], that

$$\Delta\phi \in L_2\left(0, T; H^{-2}\left(R_x^n\right)\right).$$ (9.9.1.39)

Moreover, by Sobolev embedding with $n = 3$, $H_0^2(\Omega) \subset H^{\frac{3}{2}+\epsilon}(\Omega) \subset C^0(\Omega)$; hence

$$\delta \in H^{-2}(\Omega) \quad \text{and} \quad \delta u \in L_2(0, T; H^{-2}(\Omega)), \quad n = 3.$$ (9.9.1.40)

Hence, (9.9.1.39) and (9.9.1.40), used on the right-hand side of (9.9.1.17a), yield

$$\phi_{tt} \in L_2\left(0, T; H^{-2}\left(R_x^n\right)\right),$$

as desired.

Proof of (9.9.1.19). Applying the intermediate derivative theorem [Lions, Magenes, 1972, p. 15] to (9.9.1.18) and (9.9.1.20) yields $\phi_t \in L_2(0, T; H^{-1}(R_x^n))$. By virtue of the lifting property in time recalled below (9.9.1.28), the regularity of $\{\phi, \phi_t\}$ in the $L_2(0, T)$-sense is sufficient to establish the regularity (9.9.1.18), (9.9.1.19).

Proof of (9.9.1.26). It follows from (9.9.1.24), which was already proved in the $L_2(0, T)$-sense, via [Lions, Magenes, 1972, p. 85] that

$$\Delta\phi \in L_2\left(0, T; H^{-1}\left(R_x^n\right)\right).$$ (9.9.1.41)

Moreover, by Sobolev embedding with $n = 1$, $H_0^1(\Omega) \subset H^{\frac{1}{2}+\epsilon}(\Omega) \subset C^0(\Omega)$; hence

$$\delta \in H^{-1}(\Omega) \quad \text{and} \quad \delta u \in L_2(0, T; H^{-1}(\Omega)), \quad n = 1.$$ (9.9.1.42)

Hence, (9.9.1.41) and (9.9.1.42), used on the right-hand side of (9.9.1.17a), yield

$$\phi_{tt} \in L_2\left(0, T; H^{-1}\left(R_x^n\right)\right),$$

as desired.

Proof of (9.9.1.25). The intermediate derivative theorem [Lions, Magenes, 1972, p. 15] applied to (9.9.1.24) and (9.9.1.26) yields $\phi_t \in L_2(0, T; L_2(R_x^n))$. The regularity of $\{\phi, \phi_t\}$ in the $L_2(0, T)$-sense is then lifted in time to the regularity (9.9.1.24), (9.9.1.25), as recalled below (9.9.1.28).

Proof of (9.9.1.23). It follows from (9.9.1.21), which was already proved in the $L_2(0, T)$-sense, via [Lions, Magenes, 1972, p. 85] that

$$\Delta\phi \in L_2\left(0, T; H^{-\frac{3}{2}}\left(R_x^n\right)\right); \quad \Delta\phi \in L_2\left(0, T; \left[H_{00}^{\frac{3}{2}}\left(R_x^n\right)\right]'\right).$$ (9.9.1.43)

Moreover, by Sobolev embedding with $n = 2$, $H_{00}^{\frac{3}{2}}(\Omega) \subset H_0^{\frac{3}{2}}(\Omega) \subset H^{1+\epsilon}(\Omega) \subset C^0(\Omega)$; hence

$$\delta \in H^{-\frac{3}{2}}(\Omega) \subset \left[H_{00}^{\frac{3}{2}}(\Omega)\right]'; \quad \delta u \in L_2\left(0, T; \left[H_{00}^{\frac{3}{2}}(\Omega)\right]'\right); \quad n = 2,$$ (9.9.1.44)

and (9.9.1.43) and (9.9.1.44), used on the right-hand side of (9.9.1.17a), yield

$$\phi_{tt} \in L_2\big(0, T; H^{-\frac{3}{2}}(R_x^n)\big), \quad \phi_{tt} \in L_2\big(0, T; \big[H_{00}^{\frac{3}{2}}(\Omega)\big]'\big),$$

as desired.

Moreover, to see that $\psi\phi_{tt} \in L_2(0, T; [\mathcal{D}(\mathcal{A}^{\frac{3}{4}})]')$, we note that multiplying (9.9.1.17a) by ψ, $\psi \in C_0^\infty(\Omega)$, $\psi(0) = 1$, we obtain $\psi\phi_{tt} = \psi\Delta\phi + \delta u$. Since $\mathcal{D}(\mathcal{A}^{\frac{3}{4}}) \subset H^{\frac{3}{2}}(\Omega) \subset H^{1+\epsilon}(\Omega) \subset C^0(\Omega)$ for $n = 2$, we have $\delta \in ([H^{1+\epsilon}(\Omega)]' \subset [\mathcal{D}(\mathcal{A}^{\frac{3}{4}})]')$. It remains therefore to show that $\psi\Delta\phi \in L_2(0, T; [\mathcal{D}(\mathcal{A}^{\frac{3}{4}})]')$. To this end, we pick $g \in L_2(0, T; \mathcal{D}(\mathcal{A}^{\frac{3}{4}})) \subset L_2(0, T; H^{\frac{3}{2}}(\Omega))$, so that $\psi g \in H_{00}^{\frac{3}{2}}(R_x^2) \subset H_{00}^{\frac{3}{2}}(\Omega)$, almost everywhere in t, since $\psi \equiv 0$ near Γ. Moreover, $\phi \in L_2(0, T; H^{\frac{1}{2}}(R_x^2))$ by (9.9.1.21), which was already proved in the $L_2(0, T)$-sense, implies $\Delta\phi \in L_2(0, T; [H_{00}^{1/2}(\Omega)]')$ by [Lions, Magenes, 1972, p. 85], and so the integral

$$\int_0^T \int_\Omega \Delta\phi \, g\psi \, d\Omega \, dt$$

is well defined, as desired.

Proof of (9.9.1.22). The intermediate derivative theorem [Lions, Magenes, 1972, p. 15], applied on (9.9.1.21) and (9.9.1.23), yields (9.9.1.22) with $L_2(0, T)$-time regularity. This, as mentioned below (9.9.1.28), is all we need to show. The proof of Proposition 9.9.1.2 is complete.

Step 3 (Auxiliary problems) We return from the free space ϕ-problem (9.9.1.17) to the original w-problem (9.8.1.1), via two intermediary problems. We introduce two new variables,

$$\phi_c(t, x) = \psi(x)\phi(t, x) \quad \text{and} \quad h(t, x) = \phi_c(t, x) - w(t, x), \quad (9.9.1.45)$$

where $\psi(x)$ is a C^∞-function with compact support in Ω, $\psi \in C_0^\infty(\Omega)$, so that ϕ_c is a "cutoff" of ϕ. Moreover, we impose $\psi(0) = 1$, since $0 \in \Omega$. Then, multiplying (9.9.1.17) by ψ yields ($' = \frac{d}{dt}$):

$$\begin{cases} \psi\phi'' = \psi\Delta\phi + \delta(x)u(t) & \text{in } (0, T] \times \Omega \equiv Q, & (9.9.1.46a) \\ \psi(x)\phi(0, x) \equiv \psi(x)\phi_t(0, x) \equiv 0 & \text{in } \Omega, & (9.9.1.46b) \\ \psi\phi|_\Sigma \equiv 0 & \text{in } (0, T] \times \Gamma \equiv \Sigma. & (9.9.1.46c) \end{cases}$$

Then, via (9.9.1.46), ϕ_c satisfies the problem

$$\begin{cases} \phi_c'' = \Delta\phi_c + \delta(x)u(t) + f(t, x) & \text{in } Q, & (9.9.1.47a) \\ \phi_c(0, x) \equiv \phi_c'(0, x) \equiv 0 & \text{in } \Omega, & (9.9.1.47b) \\ \phi_c|_\Sigma \equiv 0 & \text{in } \Sigma, & (9.9.1.47c) \end{cases}$$

$$f \equiv -(\Delta\psi)\phi - 2\nabla\psi \cdot \nabla\phi. \quad (9.9.1.48)$$

Finally, via (9.9.1.47), $h(t, x)$ satisifes the homogeneous (on Σ) problem

$$\begin{cases} h'' = \Delta h + f & \text{in } Q, & (9.9.1.49a) \\ h(0, x) = h'(0, x) \equiv 0 & \text{in } \Omega, & (9.9.1.49b) \\ h|_\Sigma \equiv 0 & \text{in } \Sigma. & (9.9.1.49c) \end{cases}$$

Since the regularity of ϕ (Proposition 9.9.1.2) determines that of ϕ_c by (9.9.1.43) (left), in order to obtain the regularity of w via (9.9.1.43) (right), it remains to establish the regularity of $h(t, x)$. From the regularity of ϕ in (9.9.1.18), and (9.9.1.21), and (9.9.1.24), it follows via [Lions, Magenes, 1972, Eqn. (12.64), p. 85] and $\psi \in C_0^\infty$ that

$$\nabla \psi \cdot \nabla \phi, \text{ hence by (9.9.1.48)}, f \in \begin{cases} C([0, T]; H^{-1}(\Omega)), & n = 3; & (9.9.1.50) \\ C\big([0, T]; \big[H_{00}^{\frac{1}{2}}(\Omega)\big]'\big), & n = 2; & (9.9.1.51) \\ C([0, T]; L_2(\Omega)), & n = 1. & (9.9.1.52) \end{cases}$$

Remark 9.9.1.1   The results in (9.9.1.50) and (9.9.1.51) are all that is needed in the present Dirichlet case. They will be strengthened in Section 9.9.2 to a form needed for the Neumann case; see (9.9.2.15) and (9.9.2.16) below.

Then (9.9.1.50) and (9.9.1.51) mean, equivalently, via (9.9.1.13) and (9.9.1.14),

$$\mathcal{A}^{-\frac{1}{2}} f \in C([0, T]; L_2(\Omega)), \quad n = 3; \tag{9.9.1.53}$$

$$\mathcal{A}^{-\frac{1}{4}} f \in C([0, T]; L_2(\Omega)), \quad n = 2. \tag{9.9.1.54}$$

Let now $C(t)$ and $S(t) = \int_0^t C(\tau) \, d\tau$ denote the cosine and sine operators generated by the negative, self-adjoining operator $-\mathcal{A}$ on $L_2(\Omega)$ in (9.9.1.13), $t \in \mathbb{R}$. We recall that the maps: $t \to C(t)y$ and $t \to \mathcal{A}^{\frac{1}{2}} S(t)y$, $y \in L_2(\Omega)$, are (strongly) continuous, a fact that will be used freely below. The solution of the h-problem (9.9.1.49) is (abstractly)

$$h(t) = \int_0^t S(t - \tau) f(\tau) \, d\tau; \tag{9.9.1.55}$$

$$h'(t) = \int_0^t C(t - \tau) f(\tau) \, d\tau; \quad h'' = -\mathcal{A}h + f. \tag{9.9.1.56}$$

Proposition 9.9.1.3   *With reference to problem (9.9.1.49), equivalently (9.9.1.55), (9.9.1.56), we have*

(a) for $n = \dim \Omega = 3$,

$$\begin{cases} h \in C([0, T]; L_2(\Omega)), & (9.9.1.57) \\ h_t \in C\big([0, T]; \big[\mathcal{D}\big(\mathcal{A}^{\frac{1}{2}}\big)\big]'\big) = H^{-1}(\Omega)), & (9.9.1.58) \\ h_{tt} \in C([0, T]; [\mathcal{D}(\mathcal{A})]'), & (9.9.1.59) \end{cases}$$

and

$$\frac{\partial h}{\partial v}\bigg|_{\Sigma} \in H^{-1}(\Sigma); \tag{9.9.1.60}$$

(b) for $n = \dim \Omega = 2$,

$$\begin{cases} h \in C\big([0, T]; \mathcal{D}(A^{\frac{1}{4}}) = H_{00}^{1/2}(\Omega)\big), & (9.9.1.61) \\[2mm] h_t \in C\big([0, T]; [\mathcal{D}(A^{\frac{1}{4}})]' = [H_{00}^{1/2}(\Omega)]'\big), & (9.9.1.62) \\[2mm] h_{tt} \in C\big([0, T]; [\mathcal{D}(A^{\frac{3}{4}})]'\big) \subset C\big([0, T]; [H_{00}^{1/2}(\Omega)]'\big), & (9.9.1.63) \end{cases}$$

and

$$\frac{\partial h}{\partial v}\bigg|_{\Sigma} \in H^{-\frac{1}{2}}(\Sigma); \tag{9.9.1.64}$$

(c) for $n = \dim \Omega = 1$,

$$\begin{cases} h \in C\big([0, T]; \mathcal{D}(A^{\frac{1}{2}}) = H_0^1(\Omega)\big), & (9.9.1.58) \\[2mm] h_t \in C\big([0, T]; L_2(\Omega)\big); & (9.9.1.66) \\[2mm] h_{tt} \in C\big([0, T]; [\mathcal{D}(A^{\frac{1}{2}})]' = H^{-1}(\Omega)\big), & (9.9.1.67) \end{cases}$$

and

$$\frac{\partial h}{\partial v}\bigg|_{\Sigma} \in L_2(\Sigma). \tag{9.9.1.68}$$

Proof. Interior Regularity.   We use convolution properties between C-spaces and L_2-spaces. For $n = 3$, via (9.9.1.53a), we obtain

$$h(t) = \int_0^t A^{\frac{1}{2}} S(t - \tau) A^{-\frac{1}{2}} f(\tau)\, d\tau \in C([0, T]; L_2(\Omega)), \quad (9.9.1.69)$$

$$A^{-\frac{1}{2}} h_t(t) = \int_0^t C(t - \tau) A^{-\frac{1}{2}} f(\tau)\, d\tau \in C([0, T]; L_2(\Omega)), \quad (9.9.1.70)$$

and, equivalently, (9.9.1.57) and (9.9.1.58) follow. For $n = 2$, via (9.9.1.53), we obtain

$$A^{\frac{1}{4}} h(t) = \int_0^t A^{\frac{1}{2}} S(t - \tau) A^{-\frac{1}{4}} f(\tau)\, d\tau \in C([0, T]; L_2(\Omega)), \quad (9.9.1.71)$$

$$A^{-\frac{1}{4}} h_t(t) = \int_0^t C(t - \tau) A^{-\frac{1}{4}} f(\tau)\, d\tau \in C([0, T]; L_2(\Omega)), \quad (9.9.1.72)$$

or, equivalently, (9.9.1.61) and (9.9.1.62) by (9.9.1.14). For $n = 1$, via (9.9.1.52),

$$A^{\frac{1}{2}} h(t) = \int_0^t A^{\frac{1}{2}} S(t - \tau) f(\tau)\, d\tau \in C([0, T]; L_2(\Omega)), \quad (9.9.1.73)$$

$$h_t(t) = \int_0^t C(t - \tau) f(\tau)\, d\tau \in C([0, T]; L_2(\Omega)), \quad (9.9.1.74)$$

or, equivalently, (9.9.1.65) and (9.9.1.66). Finally, the regularity (9.9.1.59), (9.9.1.63), and (9.9.1.67) for h_{tt} follows from that of h in (9.9.1.57) and (9.9.1.61), via (9.9.1.56) (right) and that of f in (9.9.1.53), (9.9.1.54), and (9.9.1.52), respectively.

Boundary Regularity   The regularity results (9.9.1.60), (9.9.1.64), and (9.9.1.68) – which are not needed in the development of Section 9.8 – are contained a fortiori [i.e., with the regularity in time of f in L_1 rather than in C] in Lasiecka et al. [1986].

Step 4   (Return to w-Problem). ***Interior Regularity***   We now return to the w-problem (9.8.1.1), via $w = \psi\phi - h$ (see (9.9.1.45)). Employing the regularity of ϕ in Proposition 9.9.1.2 and the regularity of h in Proposition 9.9.1.3, we obtain the regularity (9.9.1.1)–(9.9.1.12) of w in Theorem 9.8.1.1. In analyzing w, note that for $n = 2$, $\phi \in C([0, T]; H^{\frac{1}{2}}(R^2))$ by (9.9.1.21) implies $\psi\phi \in C([0, T]; H^{\frac{1}{2}}_{00}(\Omega))$ since $\psi \equiv 0$ near Γ [see Lions, Magenes, 1972, p. 66]; similarly, for $n = 1$, $\phi \in C([0, T]; H^1(R^1))$ by (9.9.1.24) implies $\psi\phi \in C([0, T]; H^1_0(\Omega))$. Finally, for w_{tt} in case $n = 3$, note that $H^2_0(\Omega) \subset \mathcal{D}(A)$ yields $[\mathcal{D}(A)]' \subset H^{-2}(\Omega)$.

Boundary Regularity   Since $\psi(x)$ has compact support in Ω, Eqn. (9.9.1.45) implies that $\frac{\partial w}{\partial \nu}|_\Sigma = -\frac{\partial h}{\partial \nu}|_\Sigma$. Thus, (9.9.1.4), (9.9.1.8), and (9.9.1.12) follow from (9.9.1.60), (9.9.1.64), and (9.9.1.68).

The proof of Theorem 9.8.1.1 is complete.

9.9.2 Proof of Theorem 9.8.2.1

With reference to the wave equation problem (9.8.2.1), with $w_0 = w_1 = 0$ and control $u \in L_2(0, T)$, we must show – by the closed graph theorem – the following interior and boundary regularity results:

(a) for $n = \dim \Omega = 3$,

$$
\begin{cases}
w \in C([0, T]; L_2(\Omega)), & (9.9.2.1) \\[2mm]
w_t \in C([0, T]; H^{-1}(\Omega)), & (9.9.2.2) \\[2mm]
w_{tt} \in L_2(0, T; H^{-2}(\Omega)), & (9.9.2.3)
\end{cases}
$$

and

$$
w|_{\Sigma_1} \in H^{\alpha-1}(\Sigma); \qquad (9.9.2.4)
$$

(b) for $n = \dim \Omega = 2$,

$$
\begin{cases}
w \in C([0, T]; H^{\frac{1}{2}}(\Omega)), & (9.9.2.5) \\[2mm]
w_t \in C([0, T]; H^{-\frac{1}{2}}(\Omega)), & (9.9.2.6) \\[2mm]
w_{tt} \in L_2(0, T; [H^{\frac{3}{2}}_{00}(\Omega)]'), & (9.9.2.7)
\end{cases}
$$

and

$$w|_{\Sigma_1} \in H^{\frac{\alpha+\beta-1}{2}}(\Sigma);$$

(9.9.2.8)

(c) for $n = \dim \Omega = 1$,

$$\begin{cases} w \in C([0, T]; H^1(\Omega), & (9.9.2.9) \\ w_t \in C([0, T]; L_2(\Omega)), & (9.9.2.10) \\ w_{tt} \in L_2(0, T; H^{-1}(\Omega)), & (9.9.2.11) \end{cases}$$

and

$$w|_{\Sigma_1} \in H^1(\Sigma).$$

(9.9.2.12)

The constants α and β in (9.9.2.4) and (9.9.2.8) are defined in (9.8.2.2). We follow the same approach of Section 9.9.1.

Step 1 This step on the free space problem (9.9.1.17) is the same as in Section 9.9.1, Proposition 9.9.1.2.

Step 2 This then consists in considering the regularity of the following homogeneous problem:

$$\begin{cases} h'' = \Delta h + f & \text{in } Q, & (9.9.2.13a) \\ h(0, x) = h'(0, x) \equiv 0 & \text{in } \Omega, & (9.9.2.13b) \\ h|_{\Sigma_0} \equiv 0 & \text{in } \Sigma_0, & (9.9.2.13c) \\ \left. \dfrac{\partial h}{\partial \nu} \right|_{\Sigma_1} \equiv 0 & \text{in } \Sigma_1, & (9.9.2.13d) \end{cases}$$

where again the changes of variables (9.9.1.45) are used, with, now, w being the solution of (9.8.2.1). The function f is the same as the one in (9.9.1.48), that is,

$$f = -(\Delta \psi)\phi - 2\nabla \psi \cdot \nabla \phi.$$

(9.9.2.14)

We now strengthen the regularity results for f, as given by (9.9.1.50) for $n = 3$ and by (9.9.1.51) for $n = 2$, to suit our present needs.

Proposition 9.9.2.1 *With reference to (9.9.2.14), where ϕ solves (9.9.1.17) and $\psi \in C_0^\infty(\Omega)$, we have*

(a) *for $n = \dim \Omega = 3$,*

$$f \in C([0, T]; [H^1(\Omega)]');$$

(9.9.2.15)

(b) *for $n = \dim \Omega = 2$,*

$$f \in C\big([0, T]; H^{-\frac{1}{2}}(\Omega)\big).$$

(9.9.2.16)

Proof.   (a) Let $n = 3$. We have $\phi \in L_2(\Omega)$ by (9.9.1.18) so that $|\nabla \phi| \in H^{-1}(\Omega)$ by [Lions, Magenes, 1972, p. 85]. Pick $g \in H^1(\Omega)$. Then $g|\nabla \psi| \in H_0^1(\Omega)$, since $\psi \equiv 0$ near Γ. Therefore the integral

$$\int_\Omega g \nabla \psi \cdot \nabla \phi \, d\Omega \tag{9.9.2.17}$$

is well defined, and then $\nabla \psi \cdot \nabla \phi \in [H^1(\Omega)]'$, as desired.

(b) Let $n = 2$. We have $\phi \in H^{\frac{1}{2}}(\Omega)$ by (9.9.1.21), so that $|\nabla \phi| \in [H_{00}^{\frac{1}{2}}(\Omega)]'$ by [Lions, Magenes, 1972, p. 85]. Pick $g \in H^{\frac{1}{2}}(\Omega)$. Then $g|\nabla \psi| \in H_{00}^{\frac{1}{2}}(\Omega)$ since $\psi \equiv 0$ near Γ. Therefore the integral in (9.9.2.17) is well-defined also in this case, and then $\nabla \psi \cdot \nabla \phi \in [H^{\frac{1}{2}}(\Omega)]' = H^{-\frac{1}{2}}(\Omega)$, as desired. $\square$

Continuing with the proof of Theorem 9.8.2.1, we now recall the operator $\mathcal{A}$ in (9.8.2.10), so that (with Γ_0 nonempty):

$$\mathcal{A}\psi = -\Delta\psi; \quad \mathcal{D}(\mathcal{A}) = \left\{ \psi \in H^2(\Omega) : \left. \psi \right|_{\Gamma_0} = \left. \frac{\partial \psi}{\partial \nu} \right|_{\Gamma_1} = 0 \right\}; \tag{9.9.2.18}$$

$$\mathcal{D}(\mathcal{A}^{\frac{1}{2}}) = H^1(\Omega); \quad \mathcal{D}(\mathcal{A}^{\frac{1}{4}}) = H^{\frac{1}{2}}(\Omega) = H_0^{\frac{1}{2}}(\Omega); \tag{9.9.2.19a}$$

$$\left[\mathcal{D}(\mathcal{A}^{\frac{1}{2}})\right]' = \left[H^1(\Omega)\right]'; \quad \left[\mathcal{D}(\mathcal{A}^{\frac{1}{4}})\right]' = H^{-\frac{1}{2}}(\Omega). \tag{9.9.2.19b}$$

By (9.9.2.15) and (9.9.2.19b), we then obtain

$$\mathcal{A}^{-\frac{1}{2}} f \in C([0, T]; L_2(\Omega))$$
$$\left[\text{or } (\mathcal{A} + I)^{-\frac{1}{2}} f \in C([0, T]; L_2(\Omega)) \text{ if } \Gamma_0 = \phi\right], \quad n = 3. \tag{9.9.2.20}$$

Similarly, by (9.9.2.16) and (9.9.2.19b), we obtain

$$\mathcal{A}^{-\frac{1}{4}} f, \quad \text{or} \quad (\mathcal{A} + I)^{-\frac{1}{4}} f \in C([0, T]; L_2(\Omega)), \quad n = 2, \tag{9.9.2.21}$$

according to whether $\Gamma_0 \neq \phi$ or $\Gamma_0 = \phi$. Instead, for $n = 1$, we recall that

$$f \in C([0, T]; L_2(\Omega)), \quad n = 1, \tag{9.9.2.22}$$

from (9.9.1.52). Finally, (9.9.2.20)–(9.9.2.22) permit us to establish the regularity of the h-problem (9.9.2.13) by proceeding exactly as in the proof of Proposition 9.9.1.3, using (9.9.2.19). We obtain the interior regularity results of the following proposition.

Proposition 9.9.2.2   *With reference to problem (9.9.2.13), equivalently to (9.9.1.55), (9.9.1.56) (where now $C(t)$ is the cosine operator generated on $L_2(\Omega)$ by the operator $-\mathcal{A}$ defined in (9.9.2.18)), we have*

(a) *for* $n = \dim \Omega = 3$,

$$\begin{cases} h \in C([0, T]; L_2(\Omega)), & (9.9.2.23) \\ h_t \in C([0, T]; [H^1(\Omega)]'), & (9.9.2.24) \\ h_{tt} \in C([0, T]; [\mathcal{D}(\mathcal{A})]'), & (9.9.2.25) \end{cases}$$

and

$$h|_{\Sigma_1} \in H^{\alpha-1}(\Sigma); \qquad (9.9.2.26)$$

(b) *for* $n = \dim \Omega = 2$,

$$\begin{cases} h \in C\left([0, T]; H^{\frac{1}{2}}(\Omega)\right), & (9.9.2.27) \\ h_t \in C\left([0, T]; H^{-\frac{1}{2}}(\Omega)\right), & (9.9.2.28) \\ h_{tt} \in C\left([0, T]; \left[\mathcal{D}\left(\mathcal{A}^{\frac{3}{4}}\right)\right]'\right), & (9.9.2.29) \end{cases}$$

and

$$h|_{\Sigma_1} \in H^{\frac{\alpha+\beta-1}{2}}(\Sigma); \qquad (9.9.2.30)$$

(c) *for* $n = \dim \Omega = 1$,

$$\begin{cases} h \in C([0, T]; H^1(\Omega)), & (9.9.2.31) \\ h_t \in C([0, T]; L_2(\Omega)), & (9.9.2.32) \\ h_{tt} \in C([0, T]; [H^1(\Omega)]'), & (9.9.2.33) \end{cases}$$

and

$$h|_{\Sigma_1} \in H^1(\Sigma). \qquad (9.9.2.34)$$

Proof. The interior regularity results are proved just above the statement of Proposition 9.9.2.2. As to the boundary regularity results (9.9.2.26) and (9.9.2.30) (which are not needed in the treatment of Section 9.8) they are contained in Lasiecka and Triggiani [1990] and Lasiecka and Triggiani [1991(b)]; see Triggiani [1993(a)] for details. The case $n = 1$, that is (9.9.2.34), is shown in Chapter 10, Remark 10.5.10.1.

Step 3 (Return to the w-Problem). *Interior Regularity* Recalling from (9.9.1.45), $w = \psi\phi - h$, and $\psi \in C_0^\infty(\Omega)$, we see that the following results follow.

For $n = 3$, we use (9.9.1.18) for $\{\phi, \phi_t\}$ and (9.9.2.23) and (9.9.2.24) for $\{h, h_t\}$ to obtain the desired regularity (9.9.2.1), (9.9.2.2), for $\{w, w_t\}$ since $[H^1(\Omega)]' \subset H^{-1}(\Omega)$. Then (9.9.2.3) follows for $w_{tt} = \Delta w + \delta(x)u(t)$ with $\Delta w \in C([0, T]; H^{-2}(\Omega))$ by [Lions, Magenes, 1972, p. 85] and (9.9.1.40) for δv; or else (9.9.2.3) follows from (9.9.1.20) and (9.9.2.25) with $[\mathcal{D}(\mathcal{A})]' \subset H^{-2}(\Omega)$.

For $n = 2$, we use (9.9.1.21) and (9.9.1.22) for $\{\phi, \phi_t\}$ and (9.9.2.27) and (9.9.2.28) for $\{h, h_t\}$ to obtain the desired regularity (9.9.2.5), (9.9.2.6) for $\{w, w_t\}$. Then (9.9.2.7) follows from $w_{tt} = \Delta w + \delta u$, with $\Delta w \in C([0, T]; [H_{00}^{\frac{3}{2}}(\Omega)]')$ by [Lions, Magenes, 1972, Eqn. (12.64), p. 85] and (9.8.1.44) for δu; or else (9.9.2.7) follows from (9.9.1.23) (left) and (9.9.2.29) with

$$\mathcal{D}(A^{\frac{3}{4}}) = [\mathcal{D}(A), L_2(\Omega)]_{\frac{1}{4}} \supset [H_0^2(\Omega), L_2(\Omega)]_{\frac{1}{4}} = H_{00}^{\frac{3}{2}}(\Omega),$$

and hence $[\mathcal{D}(A^{\frac{3}{4}})]' \subset [H_{00}^{\frac{3}{2}}(\Omega)]'$.

Finally, for $n = 1$, we use (9.9.1.24) and (9.9.1.25) for $\{\phi, \phi_t\}$ and (9.9.2.31) and (9.9.2.32) for $\{h, h_t\}$ to obtain the desired regularity (9.9.2.9), (9.9.2.10), for $\{w, w_t\}$. Then (9.9.2.11) follows for $w_{tt} = \Delta w + \delta u$ with $\Delta w \in C([0, T]; H^{-1}(\Omega))$ and (9.9.1.42) for δu; or else (9.9.2.11) follows from (9.9.1.26) and (9.9.2.33).

Boundary Regularity   Since $\psi(x)$ has compact support in Ω, Eqn. (9.9.1.45) implies that $w|_{\Sigma_1} = -h|_{\Sigma_1}$. Thus, (9.9.2.4), (9.9.2.8), and (9.9.2.12) follow from (9.9.2.26), (9.9.2.30), and (9.9.2.34).

The proof of Theorem 9.8.2.1 is complete.   □

9.9.3   Proof of Theorem 9.8.3.1

With reference to the Kirchhoff equation problem (9.8.3.1), with $w_0 = w_1 = 0$ and control $u \in L_2(0, T)$, we must show – by the closed graph theorem – the following interior and boundary regularity results:

(a) for $n = \dim \Omega = 3$,

$$\begin{cases} w \in C([0, T]; \mathcal{D}(A^{\frac{1}{2}}) = H^2(\Omega) \cap H_0^1(\Omega)), & (9.9.3.1) \\[2mm] w_t \in C([0, T]; \mathcal{D}(A^{\frac{1}{4}}) = H_0^1(\Omega)), & (9.9.3.2) \\[2mm] w_{tt} \in L_2(0, T; L_2(\Omega)); & (9.9.3.3) \end{cases}$$

(b) for $n = \dim \Omega = 2$,

$$\begin{cases} w \in C([0, T]; \mathcal{D}(A^{\frac{5}{8}})) \subset C([0, T]; H^{\frac{5}{2}}(\Omega)), & (9.9.3.4) \\[2mm] w_t \in C([0, T]; = \mathcal{D}(A^{\frac{3}{8}})) \subset C([0, T]; H^{\frac{3}{2}}(\Omega)), & (9.9.3.5) \\[2mm] w_{tt} \in L_2(0, T; \mathcal{D}(A^{\frac{1}{8}}) = H_{00}^{\frac{1}{2}}(\Omega)); & (9.9.3.6) \end{cases}$$

(c) for $n = \dim \Omega = 1$,

$$\begin{cases} w \in C([0, T]; \mathcal{D}(A^{\frac{3}{4}})), & (9.9.3.7) \\[2mm] w_t \in C([0, T]; \mathcal{D}(A^{\frac{1}{2}}) = H^2(\Omega) \cap H_0^1(\Omega)), & (9.9.3.8) \\[2mm] w_{tt} \in L_2(0, T; \mathcal{D}(A^{\frac{1}{4}}) = H_0^1(\Omega)). & (9.9.3.9) \end{cases}$$

In the above results $\mathcal{A}$ is the positive, self-adjoint operator in (9.8.3.3), so that

$$\mathcal{A}h = \Delta^2 h; \quad \mathcal{D}(\mathcal{A}) = \{h \in H^4(\Omega) : h|_\Gamma = \Delta h|_\Gamma = 0\}, \quad (9.9.3.10)$$

and we recall that (with equivalent norms)

$$\mathcal{D}(\mathcal{A}^{\frac{3}{4}}) = \{h \in H^3(\Omega) : h|_\Gamma = \Delta h|_\Gamma = 0\}; \quad \mathcal{D}(\mathcal{A}^{\frac{1}{4}}) = H_0^1(\Omega); \quad (9.9.3.11)$$

$$\mathcal{A}^{\frac{1}{2}}h = -\Delta h; \quad \mathcal{D}(\mathcal{A}^{\frac{1}{2}}) = H^2(\Omega) \cap H_0^1(\Omega); \quad (9.9.3.12)$$

$$\mathcal{D}(\mathcal{A}^{\frac{1}{8}}) = [\mathcal{D}(\mathcal{A}^{\frac{1}{4}}), L_2(\Omega)]_{\frac{1}{2}} = [H_0^1(\Omega), L_2(\Omega)]_{\frac{1}{2}} = H_{00}^{\frac{1}{2}}(\Omega). \quad (9.9.3.13)$$

Step 1   In studying below the corresponding free space problem (9.9.3.15), we shall need the following uniform estimate:

Lemma 9.9.3.1   *Let $\gamma > 0$ and $\rho > 0$ be two fixed constants, and let $\omega \in R^1$ be the parameter. Then there exists a constant C_γ depending on γ such that*

$$I(\omega) = \int_0^\infty \frac{y^{\frac{5}{2}}\, dy}{[y^2 + (1 + \rho y)(\gamma^2 - \omega^2)]^2 + 4\gamma^2\omega^2(1 + \rho y)^2} \leq C_\gamma < \infty, \quad (9.9.3.14)$$

for all $\omega \in R^1$.

Proof.   The proof is relegated to Appendix 9B, where it is pointed out that Lemma 9.9.3.1 is sharp.   □

Step 2   (Free space problem)   As in preceding sections, we consider first the free space problem corresponding to problem (9.8.3.1) in the unknown $\phi(t, x)$,

$$\begin{cases} \phi_{tt} - \rho\Delta\phi_{tt} + \Delta^2\phi = \delta(x)u(t) & \text{in } R_{t+}^1 \times R_x^n; & (9.9.3.15a) \\ \phi(0, x) \equiv \phi_t(0, x) \equiv 0 & \text{in } R_x^n, & (9.9.3.15b) \end{cases}$$

after extending $u(t)$ by zero for $t > T$.

Proposition 9.9.3.2   *With reference to problem (9.9.3.15), let $u \in L_2(0, T)$. Then, continuously,*

(a) for $n = \dim \Omega = 3$,

$$\begin{cases} \phi \in C([0, T]; H^2(R_x^n)), & (9.9.3.16) \\ \phi_t \in C([0, T]; H^1(R_x^n)), & (9.9.3.17) \\ \phi_{tt} \in L_2(0, T; L_2(R_x^n)); & (9.9.3.18) \end{cases}$$

(b) for n = $\dim \Omega = 2$,

$$
\begin{cases}
\phi \in C\big([0, T]; H^{\frac{5}{2}}\big(R_x^n\big)\big), & (9.9.3.19) \\[2mm]
\phi_t \in C\big([0, T]; H^{\frac{3}{2}}\big(R_x^n\big)\big), & (9.9.3.20) \\[2mm]
\phi_{tt} \in L_2\big(0, T; H^{\frac{1}{2}}\big(R_x^n\big)\big); & (9.9.3.21)
\end{cases}
$$

(c) for n = $\dim \Omega = 1$,

$$
\begin{cases}
\phi \in C\big([0, T]; H^3\big(R_x^n\big)\big), & (9.9.3.22) \\[2mm]
\phi_t \in C\big([0, T]; H^2\big(R_x^n\big)\big), & (9.9.3.23) \\[2mm]
\phi_{tt} \in L_2\big(0, T; H^1\big(R_x^n\big)\big). & (9.9.3.24)
\end{cases}
$$

Proof. Let $\hat{\phi}(\lambda, \xi)$, with $\lambda = \gamma + i\omega$, $\gamma > 0$, and $\omega \in R^1$, $\xi \in R^n$, be the Laplace (in t)-Fourier (in x) transform of $\phi(t, x)$, so that the transformed version of problem (9.9.3.15) is

$$
\hat{\phi}(\lambda, \xi) = \frac{\hat{u}(\lambda, \xi)}{\lambda^2(1 + \rho|\xi|^2) + |\xi|^4}. \tag{9.9.3.25}
$$

As in the proof of Proposition 9.9.1.2, it suffices to show the results (9.9.3.16), (9.9.3.23) for $\{\phi, \phi_t\}$ with $C[0, T]$ replaced by $L_2(0, T)$ and then invoke the general result [Chapter 7, Theorem 7.3.1] for time-reversible dynamics (groups of operators) to lift $L_2(0, T)$ to $C[0, T]$, while preserving the space regularity.

Proof of (9.9.3.16), (9.9.3.19), and (9.9.3.22). Accordingly, it suffices to show that

$$
e^{-\gamma t}\phi(t, x) \in L_2\big(R_{t+}^1; H^\alpha\big(R_x^n\big)\big),
$$

or, equivalently, that

$$
|\xi|^{2\alpha}\hat{\phi}(\gamma + i\omega, \xi) \in L_2\big(R_\omega^1 \times R_\xi^n\big), \tag{9.9.3.26}
$$

where $\alpha = 2$ for $n = 3$; $\alpha = 5/2$ for $n = 2$; and $\alpha = 3$ for $n = 1$. Since

$$
|\lambda^2(1 + \rho|\xi|^2) + |\xi|^4|^2
$$
$$
= [|\xi|^4 + (\gamma^2 - \omega^2)(1 + \rho|\xi|^2)]^2 + 4\gamma^2\omega^2(1 + \rho|\xi|^2)^2, \tag{9.9.3.27}
$$

we see that, via (9.9.3.25) and Parseval's identity [Doetsch, 1974, p. 212],

$$
\||\xi|^{2\alpha}\hat{\phi}\|^2_{L_2(R_\omega^1 \times R_\xi^n)} = \int_{R_\omega^1} |\hat{u}(\gamma + i\omega)|^2 \left(\int_{R_\xi^n} \frac{|\xi|^{2\alpha}\, d\xi}{|\lambda^2(1 + \rho|\xi|^2) + |\xi|^4|^2}\right) d\omega
$$

$$
\leq C \int_{R_\omega^1} |\hat{u}(\gamma + i\omega)|^2\, d\omega = \frac{C}{2\pi}\int_0^\infty e^{-2\gamma t}|v(t)|^2\, dt
$$

$$
= \text{const}|u|^2_{L_2(0,T)}, \tag{9.9.3.28}
$$

as desired, as soon as we prove via (9.9.3.27) that

$$
\int_{R_\xi^n} \frac{|\xi|^{2\alpha}\, d\xi}{|\lambda^2(1 + \rho|\xi|^2) + |\xi|^4|^2}
$$

$$
\leq C_\gamma \int_{R_\xi^n} \frac{|\xi|^{2\alpha}\, d\xi}{[|\xi|^4 + (1 + \rho|\xi|^2)(\gamma^2 - \omega^2)]^2 + 4\gamma^2\omega^2(1 + \rho|\xi|^2)^2}
$$

$$
\leq C_\gamma < \infty, \quad \text{uniformly in } \omega \in R^1, \tag{9.9.3.29}
$$

in each of the three cases. For $n = 3$ and thus $\alpha = 2$, we use spherical coordinates $d\xi = x^2 \sin \Phi\, dx\, d\Phi\, d\theta$, where $|\xi| = x$, and see that the uniform bound (9.9.3.29) holds true, provided that there exists C_γ such that

$$
\int_0^\infty \frac{x^6\, dx}{[x^4 + (1 + \rho x^2)(\gamma^2 - \omega^2)]^2 + 4\gamma^2\omega^2(1 + \rho x^2)^2} \leq C_\gamma, \tag{9.9.3.30}
$$

uniformly in $\omega \in R^1$. But upon setting $x^2 = y$, estimate (9.9.3.30) becomes precisely estimate (9.9.3.14) asserted by Lemma 9.9.3.1.

For $n = 2$ and thus $\alpha = 5/2$, we use polar coordinates $d\xi = r\, dr\, d\theta$, $|\xi| = r$, and thus obtain that estimate (9.9.3.29) is again reduced to estimate (9.9.3.30) with, now, x replaced by r. We can use a similar approach, for $n = 1$ and $\alpha = 3$.

Proof of (9.9.3.18). $n = 3$. It follows from (9.9.3.16), which was already proved in the $L_2(0, T)$-sense, that

$$
\Delta^2\phi \in L_2\left(0, T; H^{-2}\left(R_x^n\right)\right); \quad \text{while } \delta \in \left[H^{\frac{3}{2}+\epsilon}(\Omega)\right]' \subset H^{-\frac{3}{2}-\epsilon}(\Omega). \tag{9.9.3.31}
$$

Then, using Eqn. (9.9.3.15a), we obtain that

$$
(1 - \rho\Delta)\phi_{tt} = -\Delta^2\phi + \delta u \in L_2\left(0, T; H^{-2}\left(R_x^n\right)\right). \tag{9.9.3.32}
$$

Consider the domain of the Laplacian operator to be $\mathcal{D}(\Delta) = \{f \in L_2(R_x^n) : \Delta f \in L_2(R_x^n)\}$, where Δu is understood in the sense of distributions. Then, $(I - \rho\Delta)$ is a positive, self-adjoint operator on $L_2(R^n)$ and $(I - \rho\Delta)^{-1} : H^{-2}(R_x^n) \to L_2(R_x^n)$, so that (9.9.3.31) then implies (9.9.3.18).

Proof of (9.9.3.21). $n = 2$.   The proof is the same with, now, $\Delta^2\phi \in L_2(0, T; H^{-\frac{3}{2}}(R_x^n))$ by (9.9.3.19), while $\delta u \in L_2(0, T; H^{-1-\epsilon}(\Omega))$. Thus, the right-hand side of identity (9.9.3.32) is now in $L_2(0, T; H^{-\frac{3}{2}}(R_x^n))$, while $(I - \rho\Delta)^{-1} : H^{-\frac{3}{2}}(R_x^n) \to H^{\frac{1}{2}}(R_x^n)$. Hence (9.9.3.21) follows.

Proof of (9.9.3.24). $n = 1$.   Now $\Delta^2\phi \in L_2(0, T; H^{-1}(R_x^n))$ by (9.9.3.22), which was already proved in the $L_2(0, T)$-sense. Then the right-hand side of the identity in (9.9.3.32) is in $L_2(0, T; H^{-1}(R_x^n))$, and thus (9.9.3.24) follows.

Proof of (9.9.3.17), (9.9.3.20), and (9.9.3.23).   This is by the intermediate derivative theorem on ϕ and ϕ_{tt} [Lions, Magenes, 1972, p. 16], at least in $L_2(0, T)$. But then $\{\phi, \phi_t\}$ are lifted to $C[0, T]$, as remarked below (9.9.3.25).

Step 3 (Auxiliary problems)   As in preceding sections, we introduce new variables

$$\phi_c(t, x) = \psi(x)\phi(t, x) \quad \text{and} \quad h(t, x) = \phi_c(t, x) - w(t, x), \quad (9.9.3.33)$$

where $\psi \in C_0^\infty(\Omega)$, $\psi(0) = 1$, and ϕ, w solve problems (9.9.3.15) and (9.8.3.1), respectively. Multiplying (9.9.3.15) by ψ yields

$$\begin{cases} \psi\phi'' - \psi\rho\,\Delta\phi'' + \psi\Delta^2\phi = \delta(x)u(t) & \text{in } Q, & (9.9.3.34a) \\ \psi(x)\phi(0, x) = \psi(x)\phi'(0, x) \equiv 0 & \text{in } \Omega, & (9.9.3.34b) \\ \psi\phi|_\Sigma \equiv 0 & \text{in } \Sigma, & (9.9.3.34c) \\ \psi(\Delta\phi)|_\Sigma \equiv 0 & \text{in } \Sigma. & (9.9.3.34d) \end{cases}$$

Then, using (9.9.3.34), we see that $\phi_c = \psi\phi$ satisfies the problem

$$\begin{cases} \psi_c'' - \rho\,\Delta\phi_c'' + \Delta^2\phi_c = \delta(x)u(t) + F(t, x), & (9.9.3.35a) \\ \phi_c(0, x) \equiv \phi_c'(0, x) \equiv 0, & (9.9.3.35b) \\ \phi_c|_\Sigma \equiv 0, & (9.9.3.35c) \\ \Delta\phi_c|_\Sigma \equiv 0; & (9.9.3.35d) \end{cases}$$

$$F(x, t) \equiv f(x, t) - \rho(\Delta\psi)\phi'' - \rho\nabla\psi \cdot \nabla\phi''; \quad (9.9.3.36)$$

$$f = 4\nabla\psi \cdot \nabla(\Delta\phi) + 2(\Delta\psi)\Delta\phi + 4\nabla(\Delta\psi) \cdot \nabla\phi + 4\sum_i^n \nabla(\psi_{x_i}) \cdot \nabla(\phi_{x_i}) + (\Delta^2\psi)\phi. \quad (9.9.3.37)$$

Finally, using (9.9.3.35), we see that $h(t, x)$ in (9.9.3.33) satisfies the homogeneous (on Σ) problem

$$\begin{cases} h_{tt} - \rho\,\Delta h_{tt} + \Delta^2 h = F & \text{in } Q, & (9.9.3.38a) \\ h(0, x) = h_t(0, x) \equiv 0 & \text{in } \Omega, & (9.9.3.38b) \\ h|_\Sigma \equiv 0 & \text{in } \Sigma, & (9.9.3.38c) \\ \Delta h|_\Sigma \equiv 0 & \text{in } \Sigma, & (9.9.3.38d) \end{cases}$$

or, abstractly, using (9.9.3.10) and (9.9.3.12),

$$\left(I + \rho\mathcal{A}^{\frac{1}{2}}\right)h'' + \mathcal{A}h = F; \quad \text{or} \quad h'' = -\left(I + \rho\mathcal{A}^{\frac{1}{2}}\right)^{-1}\mathcal{A}h + \left(I + \rho\mathcal{A}^{\frac{1}{2}}\right)^{-1}F. \quad (9.9.3.39)$$

The regularity of ϕ (Proposition 9.9.3.2) determines that of ϕ_c by (9.9.3.33) (left). Thus, to determine the regularity of w via (9.9.3.33) (right), it remains to establish the regularity of h. To this end, we need the following lemma.

Lemma 9.9.3.3   *With reference to F defined by (9.9.3.36) and (9.9.3.37), we have*

$$
\left.\begin{array}{ll}
\mathcal{A}^{-\frac{1}{4}}f, & \mathcal{A}^{-\frac{1}{4}}F \\[4pt]
\mathcal{A}^{-\frac{1}{8}}f, & \mathcal{A}^{-\frac{1}{8}}F \\[4pt]
f, & F
\end{array}\right\} \in C([0,T];L_2(\Omega))
\quad
\begin{array}{ll}
n=3, & (9.9.3.40) \\[4pt]
n=2, & (9.9.3.41) \\[4pt]
n=1. & (9.9.3.42)
\end{array}
$$

Proof. From the regularity (9.2.17a), (9.2.18a), and (9.2.19a) for ϕ and (9.2.17c), (9.2.18c), and (9.2.19c) for ϕ_{tt}, we obtain, with $\psi \equiv 0$ near Γ (recall (9.2.3) and (9.2.5)),

$$
\nabla\psi \cdot \nabla\phi_{tt}, \; \nabla\psi \cdot \nabla(\Delta\phi)
$$

$$
\in
\begin{cases}
C([0,T]; H^{-1}(\Omega) = [\mathcal{D}(\mathcal{A}^{\frac{1}{4}})]'), & n=3; & (9.9.3.43) \\[6pt]
C([0,T]; [H_{00}^{\frac{1}{2}}(\Omega)]' = [\mathcal{D}(\mathcal{A}^{\frac{1}{8}})]'), & n=2; & (9.9.3.44) \\[6pt]
C([0,T]; L_2(\Omega)), & n=1. & (9.9.3.45)
\end{cases}
$$

For $n=2$, note that $|\nabla\phi_{tt}|$, $|\nabla(\Delta\phi)| \in [H_{00}^{\frac{1}{2}}(\Omega)]'$ by [Lions, Magenes, 1972, p. 85]. The terms in (9.9.3.43) and (9.9.3.45) are the critical terms in the definition (9.9.3.37) of f, and they are the critical additional terms in the definition (9.9.3.36) of F. Thus (9.9.3.40)–(9.9.3.45) follow.   □

The following result forms the basis upon which the regularity of the h-problem (9.9.3.28), equivalently (9.9.3.39), will be established by virtue of Lemma 9.9.3.3. Consider (with $\rho > 0$ a constant) the problem in the unknown $y(t,x)$:

$$
\begin{cases}
y_{tt} - \rho\,\Delta y_{tt} + \Delta^2 y = g & \text{in } Q, & (9.9.3.46a) \\[4pt]
y(0,x) \equiv y_t(0,x) \equiv 0 & \text{in } \Omega, & (9.9.3.46b) \\[4pt]
y|_\Sigma \equiv \Delta y|_\Sigma \equiv 0 & \text{in } \Sigma, & (9.9.3.46c)
\end{cases}
$$

which we write abstractly, via (9.9.3.10) and (9.9.3.12), as

$$
\ddot{y} + \rho\mathcal{A}^{\frac{1}{2}}\ddot{y} + \mathcal{A}y = g; \quad \ddot{y} = -\left(I + \rho\mathcal{A}^{\frac{1}{2}}\right)^{-1}\mathcal{A}y + \left(I + \rho\mathcal{A}^{\frac{1}{2}}\right)^{-1}g. \quad (9.9.3.47)
$$

Lemma 9.9.3.4   *With reference to problem (9.9.3.46), let*

$$
g \in L_1(0,T;L_2(\Omega)). \quad (9.9.3.48)
$$

Then, continuously,

$$
\begin{cases}
y \in C([0,T]; \mathcal{D}(\mathcal{A}^{\frac{3}{4}})), & (9.9.3.49) \\[4pt]
y_t \in C([0,T]; \mathcal{D}(\mathcal{A}^{\frac{1}{2}})), & (9.9.3.50) \\[4pt]
y_{tt} \in C([0,T]; \mathcal{D}(\mathcal{A}^{\frac{1}{4}})), & (9.9.3.51)
\end{cases}
$$

where the domains of fractional powers are identified in (9.9.3.11) and (9.9.3.12).

Proof. (First proof by energy methods)   We multiply Eqn. (9.9.3.46a) by Δy_t, and integrate by parts using (9.9.3.46b, c). We obtain after standard computations

$$\||\nabla y_t\||^2_{L_\infty(0,T;L_2(\Omega))} + \rho \, \|\Delta y_t\|^2_{L_\infty(0,T;L_2(\Omega))} + \|\nabla(\Delta y)\|^2_{L_\infty(0,T;L_2(\Omega))}$$
$$\leq C_T \|g\|_{L_1(0,T;L_2(\Omega))}, \tag{9.9.3.52}$$

where by Green's first and second theorems we obtain

$$\left\|\mathcal{A}^{\frac{3}{4}} y\right\|^2_{L_2(\Omega)} = \int_\Omega |\nabla(\Delta y)|^2 \, d\Omega; \tag{9.9.3.53}$$

$$\left\|\mathcal{A}^{\frac{1}{2}} y_t\right\|^2_{L_2(\Omega)} = \int_\Omega |\Delta y_t|^2 \, d\Omega; \quad \left\|\mathcal{A}^{\frac{1}{4}} y_t\right\|^2_{L_2(\Omega)} = \int_\Omega |\nabla y_t|^2 \, d\Omega. \tag{9.9.3.54}$$

Thus, by (9.9.3.53) and (9.9.3.54), we see that estimate (9.9.3.52) implies (9.9.3.49) and (9.9.3.50) for $\{y, y_t\}$, at least with $C[0, T]$ replaced by $L_\infty(0, T)$. The improvement to $C[0, T]$ can be done by approximating arguments, as usual, starting with smooth datum g. Then (9.9.3.51) for y_{tt} is obtained from (9.9.3.47) via (9.9.3.49).

(Second proof via operator techniques) On $\mathcal{D}(\mathcal{A}^{\frac{1}{2}})$ we have that

$$-(I + \rho \mathcal{A}^{\frac{1}{2}})^{-1} \mathcal{A} = -\frac{\mathcal{A}^{\frac{1}{2}}}{\rho} + \frac{1}{\rho}(I + \rho \mathcal{A}^{\frac{1}{2}})^{-1} \mathcal{A}^{\frac{1}{2}}, \tag{9.9.3.55}$$

so that the operator is a bounded perturbation of $-\mathcal{A}^{\frac{1}{2}}/\rho$, and then the regularity of the y-problem (9.9.3.46), or (9.9.3.47), is the same as the regularity of the following abstract η-problem:

$$\ddot{\eta} = -\tilde{A}\eta + \mathcal{A}^{-\frac{1}{2}}g, \quad \eta_0 = 0, \tag{9.9.3.56}$$

where $-\tilde{A} = -\mathcal{A}^{\frac{1}{2}}$ generates a s.c. cosine operator $\tilde{C}(t)$ with $\tilde{S}(t) = \int_0^t \tilde{C}(\tau) \, d\tau$ on $L_2(\Omega)$. Then the map $t \to \tilde{A}^{\frac{1}{2}} \tilde{S}(t) = \mathcal{A}^{\frac{1}{4}} \tilde{S}(t)$ is strongly continuous on $L_2(\Omega)$, and $\tilde{S}(t)$ and $\tilde{C}(t)$ commute with powers of $\tilde{A}$ and of $\mathcal{A}$. The solution of (9.9.3.56) is

$$\eta(t) = \int_0^t \tilde{S}(t - \tau) \mathcal{A}^{-\frac{1}{2}} g(\tau) \, d\tau = \int_0^t \tilde{A}^{-1} \tilde{S}(t - \tau) g(\tau) \, d\tau, \tag{9.9.3.57}$$

$$\eta_t(t) = \int_0^t \tilde{C}(t - \tau) \mathcal{A}^{-\frac{1}{2}} g(\tau) \, d\tau = \int_0^t \tilde{A}^{-1} \tilde{C}(t - \tau) g(\tau) \, d\tau. \tag{9.9.3.58}$$

Thus, by convolution properties using (9.9.3.48),

$$\tilde{A}^{\frac{3}{2}} \eta(t) = \mathcal{A}^{\frac{3}{4}} \eta(t) = \int_0^t \tilde{A}^{\frac{1}{2}} \tilde{S}(t - \tau) g(\tau) \, d\tau \in C([0, T]; L_2(\Omega)), \tag{9.9.3.59}$$

$$\tilde{A} \eta_t(t) = \mathcal{A}^{\frac{1}{2}} \eta_t(t) = \int_0^t \tilde{C}(t - \tau) g(\tau) \, d\tau \in C([0, T]; L_2(\Omega)), \tag{9.9.3.60}$$

or, equivalently,

$$y, \eta \in C\big([0, T]; \mathcal{D}(A^{\frac{3}{4}})\big), \tag{9.9.3.61}$$

$$y_t, \eta_t \in C\big([0, T]; \mathcal{D}(A^{\frac{1}{2}})\big), \tag{9.9.3.62}$$

and (9.9.3.49) and (9.9.3.50) are proved. $\quad\square$

Corollary 9.9.3.5 *With reference to the h-problem (9.9.3.38), equivalently (9.9.3.39), where F satisfies (9.9.3.40)–(9.9.3.42), we have:*

(a) for $n = \dim \Omega = 3$,

$$\begin{cases} h \in C\big([0, T]; \mathcal{D}(A^{\frac{1}{2}}) = H^2(\Omega) \cap H_0^1(\Omega)\big), & (9.9.3.63) \\ h_t \in C\big([0, T]; \mathcal{D}(A^{\frac{1}{4}}) = H_0^1(\Omega)\big), & (9.9.3.64) \\ h_{tt} \in C\big([0, T]; L_2(\Omega))\big); & (9.9.3.65) \end{cases}$$

(b) for $n = \dim \Omega = 2$,

$$\begin{cases} h \in C\big([0, T]; \mathcal{D}(A^{\frac{5}{8}})\big), & (9.9.3.66) \\ h_t \in C\big([0, T]; \mathcal{D}(A^{\frac{3}{8}})\big), & (9.9.3.67) \\ h_{tt} \in C\big([0, T]; \mathcal{D}(A^{\frac{1}{8}}) = H_{00}^{\frac{1}{2}}(\Omega)\big); & (9.9.3.68) \end{cases}$$

(c) for $n = \dim \Omega = 1$,

$$\begin{cases} h \in C\big([0, T]; \mathcal{D}(A^{\frac{3}{4}})\big), & (9.9.3.69) \\ h_t \in C\big([0, T]; \mathcal{D}(A^{\frac{1}{2}})\big), & (9.9.3.70) \\ h_{tt} \in C\big([0, T]; \mathcal{D}(A^{\frac{1}{4}})\big). & (9.9.3.71) \end{cases}$$

Proof. Equation (9.9.3.39) (left) for h yields, upon applying $A^{-\theta}$,

$$\big(I + \rho A^{\frac{1}{2}}\big)(A^{-\theta}h)_{tt} + A(A^{-\theta}h) = A^{-\theta}F \in C([0, T]; L_2(\Omega)) \tag{9.9.3.72}$$

by (9.9.3.40), where $\theta = 1/4, \ 1/8, \ 0$ for $n = 3, 2, 1$, respectively. Thus problem (9.9.3.72) becomes Eqn. (9.9.3.47) (left) with $y = A^{-\theta}h$ and $g = A^{-\theta}F$. Applying Lemma 9.8.3.4 on y yields a fortiori the desired conclusions. $\quad\square$

Step 4 (Return to the w-problem)　We now return to the w-problem (9.8.3.1) via $w = \psi\phi - h$ [see (9.9.3.33)] with the regularity of ϕ, $\psi\phi$ given by Proposition 9.9.3.2 and the regularity of h given by Corollary 9.9.3.5.

Proof of (9.9.3.1), (9.9.3.4), and (9.9.3.7).　For $n = 3$ we have

$$\psi\phi \in C\big([0, T]; H^2(\Omega) \cap H_0^1(\Omega) = \mathcal{D}(A^{\frac{1}{2}})\big)$$

by (9.9.3.16) since $\psi \equiv 0$ near Γ. Comparing this with (9.9.3.63) yields $w \in C([0, T]; \mathcal{D}(A^{\frac{1}{2}}))$, that is, (9.9.3.1). For $n = 2$ we have similarly by (9.9.3.19), $\psi\phi \in C([0, T];$

$H^{\frac{5}{2}}(\Omega))$. But $\psi \equiv 0$ near Γ, so that $\psi\phi \in C([0, T]; \mathcal{D}(A^{\frac{5}{8}}))$, that is, $A^{\frac{1}{2}}(\psi\phi) = -\Delta(\psi\phi) \in \mathcal{D}(A^{\frac{1}{8}}) = H_{00}^{\frac{1}{2}}(\Omega)$ see [Lions, Magenes, 1972, p. 85]. Comparing this with (9.9.3.66) yields $w \in C([0, T]; \mathcal{D}(A^{\frac{5}{8}}))$, that is, (9.9.3.4). For $n = 1$, we have $\psi\phi \in C([0, T]; H^3(\Omega))$ by (9.9.3.22) while $[\psi\phi]_\Gamma \equiv \Delta(\psi\phi)|_\Gamma \equiv 0$ since $\psi \equiv 0$ near Γ. Thus by (9.9.3.11),

$$\psi\phi \in \mathcal{D}(A^{\frac{3}{4}}). \tag{9.9.3.73}$$

Comparing (9.9.3.73) with (9.9.3.69) for h, we conclude that $w \in C([0, T]; \mathcal{D}(A^{\frac{3}{4}}))$, that is, (9.9.3.7).

Proof of (9.9.3.2), (9.9.3.5), and (9.9.3.8).   We use $w_t = \psi\phi_t - h_t$. For $n = 3$, we have $\psi\phi_t \in H_0^1(\Omega)$ by (9.9.3.17) and $\psi \equiv 0$ near Γ. Comparing this with (9.9.3.64) for h_t yields (9.9.3.2) for w_t. For $n = 2$, we have $\psi\phi_t \in H_{00}^{\frac{3}{2}}(\Omega)$ by (9.9.3.20) and $\psi \equiv 0$ near Γ. Comparing this with (9.9.3.68) for h_t yields (9.9.3.5) for w_t, since

$$\mathcal{D}(A^{\frac{3}{8}}) = \left[\mathcal{D}(A^{\frac{1}{2}}), \mathcal{D}(A^{\frac{1}{4}})\right]_{\frac{1}{2}} = \left[H^2(\Omega) \cap H_0^1(\Omega), H_0^1(\Omega)\right]_{\frac{1}{2}}$$

$$\supset \left[H_0^2(\Omega), H_0^1(\Omega)\right]_{\frac{1}{2}} = H_{00}^{\frac{3}{2}}(\Omega). \tag{9.9.3.74}$$

For $n = 1$, we have $\psi\phi_t \in H^2(\Omega) \cap H_0^1(\Omega) = \mathcal{D}(A^{\frac{1}{2}})$ by (9.9.3.23) and $\psi \equiv 0$ near Γ. Comparing this with (9.9.3.70) for h_t yields (9.9.3.8).

Proof of (9.9.3.3), (9.9.3.6), and (9.9.3.9).   We use $w_{tt} = \psi\phi_{tt} - h_{tt}$ with ϕ_{tt} given by (9.9.3.18), (9.9.3.21), and (9.9.3.24), for $n = 3, 2, 1$, and h_{tt} given by (9.9.3.65), (9.9.3.68), and (9.9.3.71), respectively. For $n = 1$ we must use that, in fact, $\psi\phi_{tt} \in H_0^1(\Omega)$, since $\psi \equiv 0$ near Γ.

The proof of Theorem 9.8.3.1 is complete.   □

9.9.4 Proof of Theorem 9.8.4.1

(i) An elementary proof of the (direct) part (a) of Theorem 9.8.4.1 is as follows. Define the function

$$U(r) = \begin{cases} -\displaystyle\int_0^r u(\sigma)d\sigma, & r \geq 0; \\ 0, & r < 0. \end{cases} \tag{9.9.4.1}$$

Then the solution of problem (9.8.4.1) with $w_0 = w_1 = 0$ is given explicitly by

$$w(t, x) = U(t - x), \quad 0 \leq t < 1; \tag{9.9.4.2a}$$

$$w(t, x) = U(t - x) - U(t - 2 + x), \quad 1 \leq t < 2; \tag{9.9.4.2b}$$

$$w(t, x) = U(t - x) - U(t - 2 + x) - U(t - 2 - x), \quad 2 \leq t < 3, \tag{9.9.4.2c}$$

and generally

$$\begin{cases} w(t, x) = \sum_{\substack{k=0 \\ k \text{ even}}}^{K} a_k U(t - k - x) - \sum_{\substack{k=1 \\ k \text{ odd}}}^{K} a_k U(t - (k + 1) + x); \\ a_k \equiv 1 \text{ for } k = 0, 3, 4, 7, 8, \ldots; \\ a_k \equiv -1 \text{ for } k = 1, 2, 5, 6, 9, 10, \ldots, \quad K \leq t < K + 1. \end{cases}$$

(9.9.4.2d)

Let, first, $0 \leq t < 1$. We thus obtain, by (9.9.4.1) and (9.9.4.2a),

$$w_t(t, x) = \dot{U}(t - x) = -u(t - x) = -w_x(t, x), \qquad (9.9.4.3)$$

$$w|_{x=0} = U(t) = -\int_0^t u(\sigma) \, d\sigma \in H^1(0, 1), \qquad (9.9.4.4)$$

where we extend $u(t) \equiv 0$ for $t < 0$. Using this and (9.9.4.3), we obtain

$$\|w_t(t, \cdot)\|_{L_2(\Omega)}^2 = \|w_x(t, \cdot)\|_{L_2(\Omega)}^2 = \int_0^1 u^2(t - x) \, dx = \int_{t-1}^t u^2(\sigma) \, d\sigma \quad (9.9.4.5)$$

$$= \int_0^t u^2(\sigma) d\sigma \leq \|u\|_{L_2(0,1)}^2, \qquad (9.9.4.6)$$

and hence

$$u \to \{w_t, w_x\} : \text{ continuous } L_2(0, 1) \to L_\infty(0, 1; L_2(\Omega) \times L_2(\Omega)). \quad (9.9.4.7)$$

Next, let $u \in C[0, 1]$, so that u is uniformly continuous on finite intervals. Then, by (9.9.4.3),

$$\|w_t(t_1, \cdot) - w_t(t_2, \cdot)\|_{L_2(\Omega)}^2 = \|w_x(t_1, \cdot) - w_x(t_2, \cdot)\|_{L_2(\Omega)}^2$$

$$= \int_0^1 |u(t_1 - x) - u(t_2 - x)|^2 \, dx \to 0 \text{ as } [t_1 - t_2] \to 0, \quad (9.9.4.8)$$

and hence

$$u \in C[0, 1] \to \{w_t, w_x\} \in C([0, 1]; L_2(\Omega) \times L_2(\Omega)). \qquad (9.9.4.9)$$

Finally, let $u \in L_2(0, 1)$. Pick $u_n \in C[0, 1]$ such that $u_n \to u$ in $L_2(0, 1)$. If w_n and w denote the solutions of problem (9.8.4.1) with $w_0 = w_1 = 0$, due to u_n and u, respectively, then

$$\{w_{n,t}, w_{n,x}\} \in C([0, 1]; L_2(\Omega) \times L_2(\Omega)) \qquad (9.9.4.10)$$

by (9.9.4.9). Moreover, by (9.9.4.7),

$$\{w_{n,t}, w_{n,x}\} \to \{w_t, w_x\} \text{ in } L_\infty(0, 1; L_2(\Omega) \times L_2(\Omega)). \qquad (9.9.4.11)$$

Hence, $\{w_t, w_x\} \in C([0, 1]; L_2(\Omega) \times L_2(\Omega))$, as desired. This argument a fortiori yields $w \in C([0, 1]; L_2(\Omega))$. Thus, the desired regularity (9.8.4.16) is proved for $T = 1$, and can be proven similarly, for any other time interval. The regularity (9.8.4.16) of Theorem 9.8.4.1 is proved.

   (ii) In Chapter 10, Remark 10.5.10.1 we shall point out another (more complicated) proof of the dual regularity (9.8.4.17) of Theorem 9.8.4.1. This will be a specialization of an argument that provides regularity for second-order hyperbolic equations with Dirichlet BC [and that fails in the case of Neumann BC, except for the one-dimensional case dim $\Omega = 1$]. $\square$

9.10 A Coupled System of a Wave and a Kirchhoff Equation with Point Control, Arising in Noise Reduction. Regularity Theory

In the present section, as well as in the next section, we consider two mathematical models that arise in the problem of noise reduction in a chamber, with piezo-ceramic patches attached to a moving flat wall. The model of the present section couples a wave equation, which describes the acoustic waves in the chamber, with a (hyperbolic) Kirchhoff equation on the moving flat wall, which describes its elastic displacement. The model of the next section replaces the hyperbolic Kirchhoff equation on the moving, flat wall with a parabolic Euler–Bernoulli equation with Kelvin Voigt (structural) damping. The moving, flat wall is subject to the action of, say, a piezo-ceramic patch (smart material), which is mathematically modeled by the derivative of a Dirac measure concentrated at a point of the wall, through which the scalar control influences the system's dynamics. [Piezo-ceramic patches are wired pairwise: the voltage imposed upon them causes the elastic wall to produce a bending moment, which, if properly modulated, reduces the noise pressure within the chamber.] In either case, our goal is to establish a sharp regularity result, verifying the abstract trace regularity (9.1.7) or its dual version (9.1.10).

9.10.1 Statement of Problem for dim $\Omega = 2$. Main Results

Let Ω be a two-dimensional domain ("the chamber"). We consider explicitly two cases:

(i) either Ω is a two-dimensional rectangle with three consecutive hard walls comprising the boundary Γ_1 and one vibrating wall comprising the boundary Γ_0 fixed at its extremes, where $\Gamma_0 \cup \Gamma_1 = \partial\Omega$ (Figure 9.1);

(ii) or else Ω is a general two-dimensional bounded domain, whose smooth boundary Γ is divided into two parts Γ_0 and Γ_1, $\Gamma = \Gamma_0 \cup \Gamma_1$, with Γ_1 the portion of the boundary acting as the hard wall, and Γ_0 the flat portion acting as the moving wall fixed at its extremes (Figure 9.2).

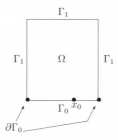

Figure 9.1   A rectangular Ω; $\alpha = 3/4$.

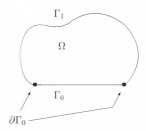

Figure 9.2   A general Ω; $\alpha = 3/5 - \epsilon$.

If $z(t, x)$ denotes the acoustic wave (unwanted noise) in the chamber and $v(t, x)$, $x \in \Gamma_0$, denotes the displacement of Γ_0, then the relevant system of partial differential equations describing the given problem is

$$
\begin{cases}
\begin{cases}
z_{tt} = \Delta z & \text{on } (0, T] \times \Omega \equiv Q; \quad\quad\quad\quad\quad\quad\quad (9.10.1.1a) \\[2mm]
\left.\dfrac{\partial z}{\partial \nu}\right|_{\Sigma_1} \equiv 0 & \text{on } (0, T] \times \Gamma_1 \equiv \Sigma_1; \quad\quad\quad\quad (9.10.1.1b) \\[2mm]
\left.\dfrac{\partial z}{\partial \nu}\right|_{\Sigma_0} \equiv -v_t & \text{on } (0, T] \times \Gamma_0 \equiv \Sigma_0; \quad\quad\quad\quad (9.10.1.1c)
\end{cases} \\[14mm]
\begin{cases}
v_{tt} - \gamma \Delta v_{tt} + \Delta^2 v - z_t = \delta'(x_0)u(t) & \text{on } \Sigma_0; \\[1mm]
\text{either clamped BC on } \partial\Gamma_0, & \quad\quad\quad\quad\quad\quad\quad (9.10.1.1d) \\[1mm]
v \equiv \dfrac{\partial v}{\partial \nu} \equiv 0 & \text{on } (0, T] \times \partial\Gamma_0, \quad (9.10.1.1e_c) \\[1mm]
\text{or else hinged BC on } \partial\Gamma_0, \\[1mm]
v = \Delta v \equiv 0 & \text{on } (0, T] \times \partial\Gamma_0; \quad (9.10.1.1e_h)
\end{cases} \\[2mm]
z(0, \cdot) = z_0;\ z_t(0, \cdot) = z_1 \text{ in } \Omega;\ v(0, \cdot) = v_0;\ v_t(0, \cdot) = v_1 \text{ in } \Gamma_0, \quad (9.10.1.1f)
\end{cases}
$$

where $\nu(x) = $ unit outward normal vector at $x \in \Gamma$ and $\gamma > 0$ is a constant. In (9.10.1.1d), x_0 is a chosen point on Γ_0, $u(t)$ is the scalar control function, and $\delta'(x_0)$

denotes the derivative of the Dirac measure at x_0, which models the action of the bending moment caused by the piezo-ceramic patch.

The main goal of the present section is twofold, as expressed by the following two main results.

Theorem 9.10.1.1   *Let $y(t) = [z(t), z_t(t), v(t), v_t(t)]$. Then the coupled PDE system (9.10.1.1) can be rewritten abstractly as the equation (9.1.1), that is, as*

$$\dot{y} = Ay + Bu \in [\mathcal{D}(A^*)]', \quad y(0) = y_0 = [z_0, z_1, v_0, v_1] \in Y, \quad (9.10.1.2)$$

where A and B are suitable operators, given by (9.10.2.16) and (9.10.2.19) below, and where Y is the Hilbert space

$$Y = H^1(\Omega) \times L_2(\Omega) \times H^2(\Gamma_0) \times H^1(\Gamma_0). \quad (9.10.1.3)$$

[Indeed, the last two components of (9.10.1.3) are refined as $H_0^2(\Gamma_0) \times H_0^1(\Gamma_0)$ for clamped BC and $[H^2(\Gamma_0) \cap H_0^1(\Gamma_0)] \times H_0^1(\Gamma_0)$ for hinged BC; see (9.10.2.10) (9.10.2.11), and (9.10.2.28) below.] When Y is topologized by an equivalent norm as in (9.10.2.10) below, then the operator A is skew-adjoint, $A^* = -A$, and thus generates a s.c. unitary group e^{At} on Y, $t \geq 0$ (conservative homogeneous problem); see (9.10.2.21) and (9.10.2.22).

As a consequence of Theorem 9.10.1.1, the solution of the abstract version (9.10.1.2) of the coupled PDE problem (9.10.1.1) is then as in (9.1.16) and (9.1.17), that is,

$$y(t) = e^{At}y_0 + (Lu)(t); \quad (Lu)(t) = \int_0^t e^{A(t-\tau)}Bu(\tau)\,d\tau. \quad (9.10.1.4)$$

Since the operator B given by (9.10.2.19) below is *not* bounded from $U = \mathbb{R}$ to Y, but instead satisfies the regularity $B : U \to [\mathcal{D}(A^*)]'$ where [ ]' denotes duality of $\mathcal{D}(A^*)$ with respect to the pivot space Y, as required by (9.1.2) [see (9.10.2.24) below], it is necessary to specify the regularity of the operator L in (9.10.1.4). This is given next. The main result of this section is the following regularity Theorem for (9.10.1.2), or (9.10.1.4), which verifies the abstract trace regularity (A.1) = (9.1.7), or its dual version (A.1*) = (9.1.10).

Theorem 9.10.1.2   *With reference to the abstract version (9.10.1.2), or (9.10.1.4), of the coupled PDE system (9.10.1.1) and to the space Y in (9.10.1.3), we have that*

(i) For each $0 < T < \infty$, the operator L in (9.10.1.4) satisfies the property

$$L : continuous \ L_2(0, T) \to C([0, T]; Y). \quad (9.10.1.5)$$

In PDE terms, the meaning of (9.10.1.5) is as follows. Set $z_0 = z_1 = v_0 = v_1 = 0$ in (9.10.1.1f), then the corresponding solution of (9.10.1.1) satifies

$$u \in L_2(0, T) \to [z(t), z_t(t), v(t), v_t(t)] \in C([0, T]; Y). \quad (9.10.1.6)$$

(ii) Equivalently, by duality [Chapter 7, Theorem 7.2.1], then the following "abstract trace regularity" property holds true: For each $0 < T < \infty$, there exists a constant $C_T > 0$ such that

$$\int_0^T |B^* e^{A^* t} y|^2 \le C_T \|y\|_Y^2, \quad \forall\, y \text{ first in } \mathcal{D}(A^*), \text{ next extended to all of } Y.$$
$$(9.10.1.7)$$

In PDE terms the meaning of (9.10.1.7) is as follows. Let $u \equiv 0$ in Eqn. (9.10.1.1d), then, the corresponding homogeneous *problem (9.10.1.1) satisfies the estimate*

$$\int_0^T |v_{tx}(t, x_0; y_0)|^2 \, dt \le C_T \|y_0\|_Y^2, \quad y_0 \in Y, \qquad (9.10.1.8)$$

$y_0 = [z_0, z_1, v_0, v_1]$, where $v_{tx}(t, x_0; y_0)$ is the second partial derivative of the solution v in t and x, evaluated at the point $x = x_0 \in \Gamma_0$ (point observation), and due to the initial condition y_0 and to $u \equiv 0$.

Remark 9.10.1.1 Theorem 9.10.1.2 is sharp. The regularity $[v, v_t] \in C([0, T]; H^2(\Gamma_0) \times H^1(\Gamma_0))$ of the Kirchhoff component (the one subject directly to the control action u) of the coupled problem (9.10.1.1) is exactly the same as the regularity described by Theorem 9.10.3.1 for the *uncoupled* Kirchhoff problem (9.10.3.1) below. As noted in Remark 9.10.3.1, such regularity is "$1/2 + \epsilon$" higher in Sobolev space units, in the space variable, over the regularity obtained by the variation of parameter formula of the corresponding semigroup, based simply on the membership property that $\delta'(x_0) \in [H^{\frac{3}{2}+\epsilon}(\Gamma_0)]' \subset [\mathcal{D}(\mathcal{A}^{\frac{3}{8}+\frac{\epsilon}{4}})]'$, or $\mathcal{A}^{-\frac{3}{8}-\frac{\epsilon}{4}} \delta'(x_0) \in L_2(\Gamma_0)$, where $\dim \Gamma_0 = 1$, and where $\mathcal{A}$ is the biharmonic operator defined by (9.10.2.1) below (in the hinged case). A similar loss of "$1/2 + \epsilon$" would occur, if one used directly and analogously the abstract model (9.10.1.4) with e^{At} the s.c. semigroup of Theorem 9.10.1.1 and with B given by (9.10.2.19) below. Similarly, (9.10.1.8) does not follow from energy regularity. In conclusion, Theorem 9.10.1.2 cannot be obtained by standard methods: It requires a combination of sharp regularity results for the uncoupled Kirchhoff part (Theorem 9.8.3.1 plus Remark 9.8.3.3) and for the uncoupled wave part (Chapter 8, Appendix 8A, Theorem 8A.2).

The present section gives a direct proof of the regularity statement (9.10.1.5) following [Camurdan, Triggiani, 1997]. A different proof of the equivalent dual trace regularity statement (9.10.1.8) may be given Triggiani [1997]. The present direct proof is preferable, as it is simpler and more streamlined than that in Triggiani [1997] for the dual version in (9.1.8). Both proofs, however, rely critically on the sharp regularity results of the two basic dynamical components of the noise reduction model: Triggiani [1993(b)] for the Kirchhoff equation with point control and Lasiecka and Triggiani [1991(b)], or related results, for the wave equation with Neumann control.

9.10.2 Abstract Model of the Original Problem (9.10.1.1) in $\{z, v\}$ (Hinged BC). Theorem 9.10.1.1

In this section we introduce the relevant functional analytic setting for problem (9.10.1.1), which culminates with the proof of Theorem 9.10.1.1. For simplicity of notation, we restrict to the case of hinged BC. The case of clamped BC can be handled similarly by using [Triggiani, 1993(b), Section 3]. Some relevant operators, in this latter case, are given explicitly at the end of this section.

Hinged BC

(i) Let $\mathcal{A} : L_2(\Gamma_0) \supset \mathcal{D}(\mathcal{A}) \to L_2(\Gamma_0)$ be the positive self-adjoint operator

$$\mathcal{A}f = \Delta^2 f, \quad \mathcal{D}(\mathcal{A}) = \{f \in H^4(\Gamma_0) : f|_{\partial\Gamma_0} = \Delta f|_{\partial\Gamma_0} = 0\}. \quad (9.10.2.1)$$

(ii) Next, introduce the operator $\mathbb{A}$ as in (9.8.3.8):

$$\mathbb{A} = (I + \gamma \mathcal{A}^{\frac{1}{2}})^{-1} \mathcal{A} : L_2(\Gamma_0) \supset \mathcal{D}(\mathbb{A}) = \mathcal{D}(\mathcal{A}^{\frac{1}{2}}) \to L_2(\Gamma_0). \quad (9.10.2.2)$$

The operator $\mathbb{A}$ is positive self-adjoint on the space $\mathcal{D}(\mathcal{A}_\gamma^{\frac{1}{4}})$ topologized by the inner product as in (9.8.3.10),

$$(x, y)_{\mathcal{D}(\mathcal{A}_\gamma^{\frac{1}{4}})} = ((I + \gamma \mathcal{A}^{\frac{1}{2}})x, y)_{L_2(\Gamma_0)}, \quad x, y \in \mathcal{D}(\mathcal{A}^{\frac{1}{4}}), \quad (9.10.2.3a)$$

$$\mathcal{A}_\gamma^{\frac{1}{4}} = (I + \gamma \mathcal{A}^{\frac{1}{2}})^{\frac{1}{2}}. \quad (9.10.2.3b)$$

We have, as in (9.8.3.4) and (9.8.3.5) [G.1; and Chapter 3, Appendix 3B],

$$\mathcal{D}(\mathcal{A}^{\frac{1}{2}}) = H^2(\Gamma_0) \cap H_0^1(\Gamma_0); \quad \mathcal{D}(\mathcal{A}^{\frac{1}{4}}) = H_0^1(\Gamma_0). \quad (9.10.2.4)$$

(iii) Define the Neumann map N as in Chapter 3, Section 3.3, Eqn. (3.3.1.7) by

$$L_2^0(\Omega) \ni h = Ng \iff \begin{cases} \Delta h = 0 \quad \text{in } \Omega; & (9.10.2.5a) \\ \dfrac{\partial h}{\partial \nu} = 0 \text{ on } \Gamma_1; \quad \dfrac{\partial h}{\partial \nu} = g \text{ on } \Gamma_0. & (9.10.2.5b) \end{cases}$$

(iv) Let $\mathcal{A}_N : L_2^0(\Omega) \equiv L_2(\Omega)/\mathcal{N}(\mathcal{A}_N) \to L_2^0(\Omega)$ be the positive self-adjoint operator

$$\mathcal{A}_N f = -\Delta f; \quad \mathcal{D}(\mathcal{A}_N) = \left\{f \in H^2(\Omega) : \frac{\partial f}{\partial \nu}\Big|_\Gamma = 0\right\}, \quad (9.10.2.6)$$

where $\mathcal{N}(\mathcal{A}_N)$ is the one-dimensional nullspace of $\mathcal{A}_N$ of constant functions. We have, as in Chapter 3, Section 3.3, Lemma 3.3.1.1,

$$N^* \mathcal{A}_N f = \begin{cases} 0 & \text{on } \Gamma_1, \\ -f|_{\Gamma_0} & \text{on } \Gamma_0, \end{cases} \quad f \in \mathcal{D}(\mathcal{A}_N). \quad (9.10.2.7)$$

Elliptic theory and Grisvard [1967] yield, as in Chapter 3, Section 3.3, Eqn. (3.3.1.8b), that for any $\epsilon > 0$:

$$N : \text{continuous } L_2(\Gamma) \rightarrow H^{\frac{3}{2}}(\Omega) \supset H^{\frac{3}{2}-2\epsilon}(\Omega) \equiv \mathcal{D}\big(A_N^{\frac{3}{4}-\epsilon}\big); \quad (9.10.2.8)$$

$$A_N^{\frac{3}{4}-\epsilon} N : \text{ continuous } L_2(\Gamma) \rightarrow L_2^0(\Omega). \quad (9.10.2.9)$$

(v) Finally, we introduce the space (norm equivalent to (9.10.1.3))

$$Y \equiv \mathcal{D}\big(A_N^{\frac{1}{2}}\big) \times L_2(\Omega) \times \mathcal{D}\big(A^{\frac{1}{2}}\big) \times \mathcal{D}\big(A_\gamma^{\frac{1}{4}}\big) \equiv Y_W \times Y_K; \quad (9.10.2.10)$$

$$Y_W = H^1(\Omega) \times L_2(\Omega); \quad Y_K = \big[H^2(\Gamma_0) \cap H_0^1(\Gamma_0)\big] \times H_0^1(\Gamma_0) \quad (9.10.2.11)$$

(norm equivalence). Thus, by (9.10.2.1), (9.10.2.6), and (9.10.2.7), the coupled problem (9.10.1.1) in $\{z, v\}$ can be rewritten by (9.10.2.7) and [Chapter 3, Section 3, Eqn. (3.3.1.6)] as

$$\begin{cases} z_{tt} = -A_N z + A_N N(v_t|_{\Gamma_0}) & \text{on } [\mathcal{D}(A_N)]'; \quad (9.10.2.12) \\ \big(I + \gamma A^{\frac{1}{2}}\big)v_{tt} + Av = -N^* A_N z_t + \delta'(x_0)u(t) & \text{on } L_2(\Gamma_0), \quad (9.10.2.13) \end{cases}$$

and the second equation (9.10.2.13) becomes via (9.10.2.2)

$$v_{tt} = -\mathbb{A}v - \big(I + \gamma A^{\frac{1}{2}}\big)^{-1} N^* A_N z_t + \big(I + \gamma A^{\frac{1}{2}}\big)^{-1} \delta'(x_0)u(t). \quad (9.10.2.14)$$

The first-order system corresponding to (9.10.2.12), (9.10.2.14) with $y(t) = [z(t), z_t(t), v(t), v_t(t)]$ is

$$\dot{y} = Ay + Bu, \quad y(0) = y_0 = [z_0, z_1, v_0, v_1] \in Y. \quad (9.10.2.15)$$

Here

$$A = \begin{bmatrix} 0 & I & 0 & 0 \\ -A_N & 0 & 0 & A_N N(\cdot|_{\Gamma_0}) \\ 0 & 0 & 0 & I \\ 0 & -\big(I + \gamma A^{\frac{1}{2}}\big)^{-1} N^* A_N & -\mathbb{A} & 0 \end{bmatrix}; \quad (9.10.2.16)$$

$$Y \supset \mathcal{D}(A) = \big\{ y \in Y : y_2 \in \mathcal{D}\big(A_N^{\frac{1}{2}}\big), y_3 \in \mathcal{D}\big(A^{\frac{1}{4}}\big), y_4 \in \mathcal{D}\big(A^{\frac{1}{2}}\big),$$

$$A_N^{\frac{1}{2}}[y_1 - N(y_4|_{\Gamma_0})] \in \mathcal{D}\big(A_N^{\frac{1}{2}}\big)\big\} \rightarrow Y; \quad (9.10.2.17)$$

and the Y-adjoint of A is

$$A^* = \begin{bmatrix} 0 & -I & 0 & 0 \\ A_N & 0 & 0 & -A_N N(\cdot|_{\Gamma_0}) \\ 0 & 0 & 0 & -I \\ 0 & \big(I + \gamma A^{\frac{1}{2}}\big)^{-1} N^* A_N & \mathbb{A} & 0 \end{bmatrix} = -A; \quad (9.10.2.18)$$

$\mathcal{D}(A^*) = \mathcal{D}(A)$. The operator $B : U \to [\mathcal{D}(A^*)]'$, [here $[\ ]'$ is the dual of $\mathcal{D}(A^*)$ with respect to the pivot space Y], $U = \mathbb{R}$, and its L_2-adjoint B^* are

$$Bu = \begin{bmatrix} 0 \\ 0 \\ 0 \\ \left(I + \gamma \mathcal{A}^{\frac{1}{2}}\right)^{-1} \delta'(x_0)u \end{bmatrix}; \quad B^* \begin{bmatrix} y_1 \\ y_2 \\ y_3 \\ y_4 \end{bmatrix} = -\frac{d}{dx} y_4|_{x=x_0}, \quad y \in \mathcal{D}(A^*).$$

(9.10.2.19)

Indeed, to find B^*, by (9.10.2.10) and (9.10.2.3), we compute via (9.10.2.1)

$$(Bu, y)_Y = \left(\left(I + \gamma \mathcal{A}^{\frac{1}{2}}\right)^{-1} \delta'(x_0)u, y_4\right)_{\mathcal{D}(\mathcal{A}^{\frac{1}{4}}_Y)} = (\delta'(x_0)u, y_4)_{L_2(\Gamma_0)}$$

$$= u\left(-\frac{d}{dx} y_4|_{x=x_0}\right) = (u, B^*y)_{U=\mathbb{R}}, \quad (9.10.2.20)$$

and (9.10.2.19) follows for B^*.

Next, we establish the claimed regularity $B : U \to [\mathcal{D}(A^*)]'$, as required by (9.1.2). First, we readily find that the inverse $A^{-1} \in \mathcal{L}(Y)$ of A in (9.10.2.16) is

$$A^{-1} = \begin{bmatrix} 0 & -\mathcal{A}_N^{-1} & N(\cdot|_{\Gamma_0}) & 0 \\ I & 0 & 0 & 0 \\ -\mathcal{A}^{-1}N^*\mathcal{A}_N & 0 & 0 & -\mathcal{A}^{-1}\left(I + \gamma \mathcal{A}^{\frac{1}{2}}\right) \\ 0 & 0 & I & 0 \end{bmatrix} \in \mathcal{L}(Y).$$

(9.10.2.21)

Next, for $\dim \Gamma_0 = 1$, Sobolev embedding yields

$$\delta'(x_0) \in \left[H^{\frac{3}{2}+4\epsilon}(\Gamma_0)\right]' \subset \left[\mathcal{D}\left(\mathcal{A}^{\frac{3}{8}+\epsilon}\right)\right]'; \quad \mathcal{A}^{-\frac{3}{8}-\epsilon}\delta'(x_0) \in L_2(\Gamma_0), \quad (9.10.2.22)$$

since $\mathcal{D}(\mathcal{A}^{\frac{3}{8}+\epsilon}) \subset H^{\frac{3}{2}+4\epsilon}(\Gamma_0)$ for the fourth-order operator $\mathcal{A}$ in (9.10.2.1), with duality with respect to $L_2(\Gamma_0)$. A fortiori from (9.10.2.22),

$$\mathcal{A}^{-1}\delta'(x_0) = \mathcal{A}^{-\frac{5}{8}+\epsilon}\mathcal{A}^{-\frac{3}{8}-\epsilon}\delta'(x_0) \in \mathcal{D}\left(\mathcal{A}^{\frac{5}{8}-\epsilon}\right) \subset \mathcal{D}\left(\mathcal{A}^{\frac{1}{2}}\right). \quad (9.10.2.23)$$

Thus, recalling B from (9.10.2.19) and A^{-1} from (9.10.2.21), we obtain for $u \in U \equiv \mathbb{R}$:

$$A^{-1}Bu = \begin{bmatrix} 0 \\ 0 \\ -\mathcal{A}^{-1}\delta'(x_0)u \\ 0 \end{bmatrix} \in Y = \mathcal{D}\left(\mathcal{A}_N^{\frac{1}{2}}\right) \times L_2(\Omega) \times \mathcal{D}\left(\mathcal{A}^{\frac{1}{2}}\right) \times \mathcal{D}\left(\mathcal{A}_Y^{\frac{1}{4}}\right),$$

(9.10.2.24)

where in the last step we have recalled (9.10.2.23) and (9.10.2.10). Thus, (9.10.2.24) shows (9.1.2), as desired. [Of course, we do not need to compute A^{-1} as in (9.10.2.21) to obtain (9.10.2.24): Because of the special form of A in (9.10.2.16) and B in

(9.10.2.19), we have that $Bu = Ag$ readily yields g given by the right-hand side of (9.10.2.24). Equation (9.10.2.21) is an additional piece of information.]

Finally, from (9.10.2.16)–(9.10.2.18) we have that A generates a s.c. unitary group e^{At} on Y:

$$e^{A^*t} = e^{-At}, \quad \|e^{A^*t}\|_{\mathcal{L}(Y)} \equiv \|e^{-At}\|_{\mathcal{L}(Y)} \equiv 1, \qquad (9.10.2.25)$$

$$\mathrm{Re}(Ax, x)_Y = \mathrm{Re}(A^*x, x)_Y \equiv 0, \quad x \in \mathcal{D}(A) = \mathcal{D}(A^*), \quad (9.10.2.26)$$

with Y as in (9.10.2.10). With A and B given by (9.10.2.16) and (9.10.2.19), the solution of problem (9.10.1.1) is given by (9.10.1.4). The equivalence between (9.10.1.7) and (9.10.1.8) follows from (9.10.2.4) for B^* and (9.10.2.25).

Theorem 9.10.1.1 is proved (in the hinged case).

Clamped BC Here we introduce the following positive, self-adjoint operators [Triggiani, 1993(b), Section 3]:

$$\mathcal{A}f = \Delta^2 f, \quad \mathcal{D}(\mathcal{A}) = \left\{ f \in H^4(\Omega) : f|_{\Gamma_0} = \frac{\partial f}{\partial \nu}\bigg|_{\partial \Gamma_0} = 0 \right\}; \qquad (9.10.2.27)$$

$$\mathcal{D}(\mathcal{A}^{\frac{1}{2}}) = H_0^2(\Gamma_0); \quad \mathcal{D}(\mathcal{A}^{\frac{1}{4}}) = H_0^1(\Gamma_0) \qquad (9.10.2.28)$$

(which are the counterparts of (9.10.2.2) and (9.10.2.4)). Moreover, let

$$\mathcal{A}_D f = -\Delta f, \quad \mathcal{D}(\mathcal{A}_D) = H^2(\Gamma_0) \cap H_0^1(\Gamma_0), \quad \mathbb{A} = (I + \gamma \mathcal{A}_D)^{-1}\mathcal{A}, \qquad (9.10.2.29)$$

which is the counterpart of (9.10.2.2).

9.10.3 Proof of Theorem 9.10.1.2: Regularity of L in (9.10.1.5)

In this section we prove the regularity of L in (9.10.1.5).

Step 1 (Uncoupled Nonhomogeneous Kirchhoff Equation on Γ_0) Henceforth, $\phi(t, x)$ will denote the solution of the following mixed problem for the Kirchhoff equation (which is problem (9.10.1.1d,e), this time uncoupled):

$$
\begin{cases}
\phi_{tt} - \gamma\Delta\phi_{tt} + \Delta^2\phi = \delta'(x_0)u(t) & \text{in } (0, T] \times \Gamma_0; & (9.10.3.1a) \\[4pt]
\phi(0, \cdot) = \phi_0, \quad \phi_t(0, \cdot) = \phi_1 & \text{in } \Gamma_0; & (9.10.3.1b) \\[4pt]
\text{either clamped BC on } \partial\Gamma_0, \\[4pt]
\phi \equiv \dfrac{\partial\phi}{\partial\nu} \equiv 0 & \text{on } (0, T] \times \partial\Gamma_0, & (9.10.3.1c_c) \\[4pt]
\text{or else hinged BC on } \partial\Gamma_0, \\[4pt]
\phi \equiv \Delta\phi \equiv 0 & \text{in } (0, T] \times \partial\Gamma_0. & (9.10.3.1c_h)
\end{cases}
$$

An optimal regularity result for problem (9.10.3.1) is given next (recall Remark 9.8.3.1).

Theorem 9.10.3.1 (ϕ-problem)   *Recall that* dim $\Gamma_0 = 1$. *Consider the ϕ-problem in (9.10.3.1) with, say, hinged BC, and thus with*

$$\{\phi_0, \phi_1\} \in \left[H^2(\Gamma_0) \cap H_0^1(\Gamma_0)\right] \times H_0^1(\Gamma_0), \quad u \in L_2(0, T). \qquad (9.10.3.2)$$

Then, continuously,

$$\{\phi, \phi_t, \phi_{tt}\} \in C\left([0, T]; \left[H^2(\Gamma_0) \cap H_0^1(\Gamma_0)\right] \times H_0^1(\Gamma_0)\right) \times L_2(0, T; L_2(\Gamma_0)).$$
$$(9.10.3.3)$$

For clamped BC the first component space is $H_0^2(\Gamma_0)$.   □

Proof.   It suffices to take zero Initial Condition by Theorem 9.10.1.1. Then, the result follows directly, at least in the hinged case, from Theorem 9.8.3.1, Eqn. (9.8.3.17) referring to δ, along with Remark 9.8.3.3 referring to δ', that is, the space of regularity in Theorem 9.10.3.1 with δ' is one Sobolev unit less than the space of regulairty in Theorem 9.8.3.1 with δ, which is given by (9.8.3.2c) for dim $\Gamma_0 = 1$. The same regularity in the clamped case is given in [Triggiani, 1993(b), Section 3]; see Remark 9.8.3.2.   □

Step 2 (Uncoupled Nonhomogeneous Wave Equation on Ω)   Henceforth, $\psi(t, x)$ will denote the solution of the following mixed problem for the wave equation (which is problem (9.10.1.1a–c), this time uncoupled):

$$\begin{cases} \psi_{tt} = \Delta\psi & (0, T] \times \Omega \equiv Q; & (9.10.3.4a) \\[2mm] \psi(0, \cdot) = 0, \ \psi_t(0, \cdot) = 0 & \text{in } \Omega; & (9.10.3.4b) \\[2mm] \dfrac{\partial\psi}{\partial\nu}\bigg|_\Sigma \equiv g & (0, T] \times \partial\Omega \equiv \Sigma. & (9.10.3.4c) \end{cases}$$

A sharp regularity theory for problem (9.10.3.4) is given in Chapter 8, Appendix 8A, Theorem 8A.2 after Lasiecka and Triggiani [1986; 1991(b)], from which we shall quote below.

Henceforth, α is a constant taking up the following values (where $\epsilon > 0$ is arbitrary):

$$\begin{cases} \alpha = \dfrac{3}{5} - \epsilon & \text{for a general smooth domain } \Omega \text{ of } R^n, \ n \geq 2; & (9.10.3.5a) \\[3mm] \alpha = \dfrac{3}{4} & \text{for a parallelepiped } \Omega \text{ of } R^n, \ n \geq 2; & (9.10.3.5b) \\[3mm] \alpha = \dfrac{2}{3} & \text{for a sphere } \Omega \text{ of } R^n, \ n \geq 2, & (9.10.3.5c) \end{cases}$$

as in Chapter 8, Appendix 8A, Eqn. (8A.3). We next quote [Chapter 8, Appendix 8A, Theorem 8A.2] from Lasiecka and Triggiani [1990; 1991(b)].

Theorem 9.10.3.2   *With reference to problem (9.10.3.4) (where, actually,* dim $\Omega = 2$), *we have:*

(i) (interior regularity) Let

$$g \in H^1\left(0, T; L_2(\Gamma)\right) \cap C\left([0, T]; H^{\alpha - \frac{1}{2}}(\Gamma)\right), \quad g|_{t=0} = 0. \qquad (9.10.3.6)$$

Then, continuously,

$$\{\psi, \psi_t, \psi_{tt}\} \in C\left([0, T]; H^{\alpha+1}(\Omega) \times H^{\alpha}(\Omega) \times H^{\alpha-1}(\Omega)\right). \qquad (9.10.3.7)$$

(ii) (boundary regularity) Let

$$g \in H^1(\Sigma) \equiv L_2(0, T; H^1(\Gamma)) \cap H^1(0, T; L_2(\Gamma)), \quad g|_{t=0} = 0. \quad (9.10.3.8)$$

Then, continuously,

$$\psi|_\Sigma \in H^{2\alpha}(\Sigma) = L_2(0, T; H^{2\alpha}(\Gamma)) \cap H^{2\alpha}(0, T; L_2(\Gamma)). \qquad (9.10.3.9)$$

The case of interest for the present acoustic model is captured next. With ϕ_t provided by Theorem 9.10.3.1 for the Kirchhoff problem (9.10.3.1), consider the specialization of the wave equation problem (9.10.3.4) given by

$$\begin{cases} w_{tt} = \Delta w & \text{in } Q; & (9.10.3.10a) \\[4pt] w(0, \cdot) = 0, \quad w_t(0, \cdot) = 0 & \text{in } \Omega; & (9.10.3.10b) \\[4pt] \left. \dfrac{\partial w}{\partial \nu} \right|_{\Sigma_1} \equiv 0 & \text{in } \Sigma_1; & (9.10.3.10c) \\[4pt] \left. \dfrac{\partial w}{\partial \nu} \right|_{\Sigma_0} = -\phi_t & \text{in } \Sigma_0, & (9.10.3.10d) \end{cases}$$

where, by Theorem 9.10.3.1, Eqn. (9.10.3.3), we have

$$\begin{cases} -\phi_t \in C\left([0, T]; H_0^1(\Gamma_0)\right) \cap H^1(0, T; L_2(\Gamma_0)) & (9.10.3.11a) \\[4pt] \qquad \subset H^1(\Sigma_0). & (9.10.3.11b) \end{cases}$$

Moreover, by (9.10.3.1b), we verify that

$$\text{on } \Gamma_0: \ -\phi_t|_{t=0} = -\phi_1 = 0, \qquad (9.10.3.12)$$

so that the compatibility relation $g|_{t=0} = 0$ required by Theorem 9.10.3.2 is satisfied with $g = 0$ on Σ_1, and $g = -\phi_t$ on Σ_0. As a corollary of Theorem 9.10.3.2, we thus obtain the result of interest.

Corollary 9.10.3.3   *With reference to the wave problem (9.10.3.10) we have:*

(i) (interior regularity)

$$\{w, w_t, w_{tt}\} \in C\left([0, T]; H^{\alpha+1}(\Omega) \times H^{\alpha}(\Omega) \times H^{\alpha-1}(\Omega)\right); \qquad (9.10.3.13)$$

(ii) (boundary regularity)

$$w_t \in H^{2\alpha-1}(0, T; L_2(\Gamma_0)). \qquad (9.10.3.14)$$

Proof. (i) Since $H^1(\Gamma_0) \subset H^{\alpha-\frac{1}{2}}(\Gamma_0)$ by (9.10.3.5), we see by (9.10.3.11) and (9.10.3.12) that we can invoke Theorem 9.10.3.2 for problem (9.10.3.10) and obtain (9.10.3.13) from (9.10.3.7), and (9.10.3.14) from (9.10.3.9), that is, from $w \in H^{2\alpha}(0, T; L_2(\Gamma))$. $\square$

Step 3   With w_t given by (9.10.3.14), we next consider the following coupled system of two PDEs, in the variables $\zeta(t, x)$ and $h(t, x)$:

$$
\begin{cases}
\begin{cases}
\zeta_{tt} = \Delta\zeta \quad \text{on } (0, T] \times \Omega = Q; & \text{(9.10.3.15a)} \\[2mm]
\left.\dfrac{\partial\zeta}{\partial\nu}\right|_{\Sigma_1} \equiv 0 \quad \text{on } (0, T] \times \Gamma_1 \equiv \Sigma_1; & \text{(9.10.3.15b)} \\[2mm]
\left.\dfrac{\partial\zeta}{\partial\nu}\right|_{\Sigma_0} \equiv -h_t \text{ on } (0, T] \times \Gamma_0 \equiv \Sigma_0; & \text{(9.10.3.15c)}
\end{cases} \\[10mm]
\begin{cases}
h_{tt} - \gamma\Delta h_{tt} + \Delta^2 h - \zeta_t = w_t \quad \text{on } \Sigma_0; & \text{(9.10.3.15d)} \\[2mm]
\text{either clamped BC on } \partial\Gamma_0, \\
h \equiv \dfrac{\partial h}{\partial\nu} \equiv 0 \quad\quad\quad\quad\quad \text{on } (0, T] \times \partial\Gamma_0; & (9.10.3.15e_c) \\
\text{or else hinged BC on } \partial\Gamma_0, \\
h \equiv \Delta h \equiv 0 \quad\quad\quad\quad\quad \text{on } (0, T] \times \partial\Gamma_0; & (9.10.3.15e_h)
\end{cases} \\[10mm]
\zeta(0, \cdot) = 0; \ \zeta_t(0, \cdot) = 0 \text{ in } \Omega; \ h(0, \cdot) = 0; \ h_t(0, \cdot) = 0 \quad \text{in } \Gamma_0. \\
\hfill \text{(9.10.3.15f)}
\end{cases}
$$

Remark 9.10.3.1   As a motivation, we have that problem (9.10.3.15) is obtained by setting

$$\zeta(t, x) \equiv z(t, x) - w(t, x), \quad h(t, x) \equiv v(t, x) - \phi(t, x), \quad \text{(9.10.3.16)}$$

differentiating formally and using problems (9.10.1.1), (9.10.3.1), and (9.10.3.10); this procedure is justified in Step 5 below.

Remark 9.10.3.2   The coupled problem (9.10.3.15) in the new variables $\{\zeta, h\}$ is the same as the original coupled problem (9.10.1.1) in the variables $\{z, v\}$, except that: On Σ_0, the point control term $\delta'(x_0)u(t)$ in (9.10.1.1d) is replaced with the nonhomogeneous term w_t in (9.3.15d), for which the a priori regularity (9.10.3.14) is available. Thus, problem $\{\zeta, h\}$ is easier to analyze than problem $\{z, v\}$. In fact, we shall obtain a regularity result for problem $\{\zeta, h\}$ by semigroup methods, using the unitary group obtained in Theorem 9.10.1.1. By contrast, as we know, semigroup methods applied to even a single hyperbolic equation – such as the wave equation or the Kirchhoff equation – acted upon by interior point control, are definitely nonoptimal, as they lose "$1/2 + \epsilon$" regularity in Sobolev space units, in the space variable; see Remark 9.8.1.1 for the wave equation and Remark 9.8.3.1 for the Kirchhoff equation.

According to (9.10.2.1), (9.10.2.6), and (9.10.2.7), problem (9.10.3.15) can be rewritten abstractly as

$$\zeta_{tt} = -A_N\zeta + A_N N\left(h_t|_{\Gamma_0}\right) \quad \text{on } [\mathcal{D}(A_N)]', \quad (9.10.3.17\text{a})$$

$$\left(I + \gamma A^{\frac{1}{2}}\right)h_{tt} + Ah = -N^* A_N \zeta_t + w_t \quad \text{on } L_2(\Gamma_0), \quad (9.10.3.17\text{b})$$

at least in the hinged case: Compare with Eqns. (9.10.2.12) and (9.10.2.13). [In the clamped case, we use the operators in (9.10.2.27)–(9.10.2.29).] The corresponding first-order system of (9.10.3.17) and (9.10.3.18) is

$$\frac{d}{dt}\begin{bmatrix} \zeta \\ \zeta_t \\ h \\ h_t \end{bmatrix} = A\begin{bmatrix} \zeta \\ \zeta_t \\ h \\ h_t \end{bmatrix} + \begin{bmatrix} 0 \\ 0 \\ 0 \\ \left(I + \gamma A^{\frac{1}{2}}\right)^{-1}w_t \end{bmatrix}, \quad (9.10.3.18)$$

where the operator A is given by (9.10.2.16) and (9.10.2.17). Thus, according to Theorem 9.10.1.1, the solution of (9.10.3.18) with zero initial condition (by (9.10.3.15f)) is given by

$$\eta(t) = \begin{bmatrix} \zeta(t) \\ \zeta_t(t) \\ h(t) \\ h_t(t) \end{bmatrix} = \int_0^t e^{A(t-\tau)}\begin{bmatrix} 0 \\ 0 \\ 0 \\ \left(I + \gamma A^{\frac{1}{2}}\right)^{-1}w_t(\tau) \end{bmatrix} d\tau, \quad (9.10.3.19)$$

where e^{At} is the unitary group on Y [see (9.10.2.10)] of Theorem 9.10.1.1. We next establish the regularity of $\eta(t)$ in (9.10.3.19).

Step 4. ***Proposition 9.10.3.4*** With reference to (9.10.3.19), and Y in (9.10.2.10), we have that

$$\eta(t) \equiv [\zeta(t), \zeta_t(t), h(t), h_t(t)] \in C([0, T]; Y). \quad (9.10.3.20)$$

Proof. We return to the a priori regularity (9.10.3.14) for w_t and obtain

$$\left(I + \gamma A^{\frac{1}{2}}\right)^{-1}w_t \in H^{2\alpha-1}\left(0, T; \mathcal{D}(A^{\frac{1}{2}})\right) \quad (9.10.3.21)$$

$$\subset L_2\left(0, T; \mathcal{D}(A_\gamma^{\frac{1}{4}})\right), \quad (9.10.3.22)$$

since $2\alpha - 1 > 0$ by (9.10.3.5). It will suffice to proceed with the loss of space regularity of one Sobolev unit of (9.10.3.22) over (9.10.3.21); see (9.10.2.4). Using

(9.10.3.22) in (9.10.3.19) and recalling that $\mathcal{D}(\mathcal{A}_\gamma^{\frac{1}{4}})$ is the last component space of Y in (9.10.2.10), we readily obtain (9.10.3.20).   □

Step 5   We now return from the $\{\zeta, h\}$ problem to the $\{z, v\}$ original problem. It is readily checked (recall Remark 9.10.3.1) that, having $\{\zeta, h\}$ from problem (9.10.3.15) [Proposition 9.10.3.4], and having ϕ and w from problem (9.10.3.1) [Theorem 9.10.3.1] and problem (9.10.3.10) [Corollary 9.10.3.3], respectively, the functions $\{\zeta(t, x) + w(t, x), h(t, x) + \phi(t, x)\}$ satisfy problem (9.10.1.1), so that we, in fact, have

$$z(t, x) \equiv \zeta(t, x) + w(t, x), \quad v(t, x) \equiv h(t, x) + \phi(t, x). \quad (9.10.3.23)$$

Thus, via (9.10.3.23), we can obtain the regularity of the original variables $\{z(t, x), v(t, x)\}$ from the regularity of the new variables $\{\zeta(t, x), h(t, x)\}$ given by Proposition 9.10.3.4, Eqn. (9.10.3.20), and from the regularity of $w(t, x)$ and $\phi(t, x)$ given by Corollary 9.10.3.3, Eqn. (9.10.3.13), and Theorem 9.10.3.1, Eqn. (9.10.3.3), respectively. We obtain, recalling (9.10.3.20) with Y as in (9.10.2.10), in the hinged case:

$$\begin{bmatrix} \zeta \\ \zeta_t \\ h \\ h_t \end{bmatrix} \in C \left([0, T]; \begin{bmatrix} \mathcal{D}(\mathcal{A}_N^{\frac{1}{2}}) = H^1(\Omega) \\ L_2(\Omega) \\ \mathcal{D}(\mathcal{A}^{\frac{1}{2}}) = H^2(\Gamma_0) \cap H_0^1(\Gamma_0) \\ \mathcal{D}(\mathcal{A}_\gamma^{\frac{1}{4}}) = H_0^1(\Gamma_0) \end{bmatrix} \right), \quad (9.10.3.24)$$

and recalling Eqns. (9.10.3.13) and (9.10.3.3),

$$\begin{bmatrix} w \\ w_t \end{bmatrix} \in C \left([0, T]; \begin{bmatrix} H^{\alpha+1}(\Omega) \\ H^\alpha(\Omega) \end{bmatrix} \right);$$

$$\begin{bmatrix} \phi \\ \phi_t \end{bmatrix} \in C \left([0, T]; \begin{bmatrix} H^2(\Gamma_0) \cap H_0^1(\Gamma_0) \\ H_0^1(\Gamma_0) \end{bmatrix} \right). \quad (9.10.3.25)$$

Thus, via (9.10.3.23), using (9.10.3.24) and (9.10.3.25), we finally obtain

$$\begin{bmatrix} z \\ z_t \\ v \\ v_t \end{bmatrix} \in C \left([0, T]; \begin{bmatrix} \mathcal{D}(\mathcal{A}_N^{\frac{1}{2}}) = H^1(\Omega) \\ L_2(\Omega) \\ H^2(\Gamma_0) \cap H_0^1(\Gamma_0) \\ H_0^1(\Gamma_0) \end{bmatrix} \right) = C([0, T]; Y), \quad (9.10.3.26)$$

and Theorem 9.10.1.2 is proved, at least in the hinged case.

For the clamped case, we use instead the operators in (9.10.2.27) and (9.10.2.29). The proof of Theorem 9.10.1.2 is complete.

9.10.4 A Corresponding Damped Model

The Damped PDE Model As we have seen, the original model (9.10.1.1) in $\{z, v\}$ is *conservative*: The dynamics operator A explicitly given by (9.10.2.16) is skew-adjoint, $A^* = -A$, on the space Y topologized by (9.10.2.10); see Theorem 9.10.1.1.

In this section, we briefly consider a correspondingly *damped* model, say in the hinged case:

$$
\begin{cases}
\quad z_{tt} = \Delta z - d_0 z_t \quad \text{on } Q, & (9.10.4.1a) \\[2mm]
\quad \dfrac{\partial z}{\partial v}\bigg|_{\Sigma_1} \equiv -d_1 z_t \quad \text{on } \Sigma_1, & (9.10.4.1b) \\[2mm]
\quad \dfrac{\partial z}{\partial v}\bigg|_{\Sigma_0} \equiv -d_1 z_t - v_t \quad \text{on } \Sigma_0; & (9.10.4.1c) \\[2mm]
\quad \begin{cases} v_{tt} - \gamma \Delta v_{tt} + \Delta^2 v - z_t = \delta'(x_0) u(t) - k_0 \Delta v_t & \text{on } \Sigma_0, \quad (9.10.4.1d) \\[2mm] v = 0, \ \Delta v = -k_1 \dfrac{\partial v_t}{\partial v} & \text{on } (0, T] \times \partial \Gamma_0; \end{cases} \\
\hspace{11cm} (9.10.4.1e_h) \\[2mm]
\quad z(0, \cdot) = z_0; \quad z_t(0, \cdot) = z_1 \text{ in } \Omega; \quad v(0, \cdot) = v_0, \quad v_t(0, \cdot) = v_1 \quad \text{in } \Gamma_0, \\
\hspace{11cm} (9.10.4.1f)
\end{cases}
$$

with constant $\gamma > 0$ throughout where the damping constants d_i and k_i satisfy the conditions

$$d_i \geq 0, \ k_i \geq 0; \quad d_0 + d_1 > 0; \quad k_0 + k_1 > 0. \tag{9.10.4.2}$$

To streamline the exposition, it will suffice to carry out the analysis in the most demanding *boundary damping* case, where

$$d_0 = 0, \ d_1 = 1; \quad k_0 = 0, \ k_1 = 1. \tag{9.10.4.3}$$

Abstract Model of (9.10.4.1) under (9.10.4.3) To put problem (9.10.4.1), subject to (9.10.4.3), into an abstract framework, we need two additional operators over those in Section 9.10.2.

(i) Define the following Green map (recall Chapter 3, Section 3.6, Eqn. (3.6.3)):

$$h = G_2 g \iff \{\Delta^2 h = 0 \text{ on } \Gamma_0; \ h|_{\partial \Gamma_0} = 0, \ \Delta h|_{\partial \Gamma_0} = g\}, \tag{9.10.4.4}$$

so that, if $\mathcal{A}$ is the elastic operator in (9.10.2.1), we then know from Chapter 3, Section 3.6, Eqn. (3.6.6) that

$$G_2 = -\mathcal{A}^{-\frac{1}{2}} D, \quad \text{where } r = Dg \iff \{\Delta r = 0 \text{ on } \Gamma_0, r|_{\partial \Gamma_0} = g\}. \tag{9.10.4.5}$$

Thus, if G_2^* is the L_2-adjoint of $G_2 : (G_2 g, f)_{L_2(\Gamma_0)} = (g, G_2^* f)_{L_2(\partial\Gamma_0)}$, then (recall Chapter 3, Section 3.1, Lemma 3.1.1, Eqn. (3.1.9))

$$G_2^* \mathcal{A} h = -D^* \mathcal{A}^{\frac{1}{2}} h = \frac{\partial h}{\partial \nu} \quad \text{on } \partial\Gamma_0. \tag{9.10.4.6}$$

(ii) Complementing the operator N in (9.10.2.5), we introduce another Neumann map N_1 by

$$h = N_1 g \iff \left\{ \Delta h = 0 \text{ in } \Omega; \ \left.\frac{\partial h}{\partial \nu}\right|_\Gamma = g \right\}, \quad h \in L^0(\Omega) \equiv L_2(\Omega)/\mathcal{N}(\mathcal{A}_N),$$

so that, by Chapter 3, Section 3.3, Lemma 3.3.1.1, we have

$$N_1^* \mathcal{A}_N f = -f|_\Gamma, \quad f \in \mathcal{D}(\mathcal{A}_N), \tag{9.10.4.7}$$

where $\mathcal{A}_N$ is defined by (9.10.2.6).

Then, recalling the corresponding trace property of $N^* \mathcal{A}_N$ in (9.10.2.7), we see, as in Section 9.10.2, that the damped system (9.10.4.1), subject to (9.10.4.3), can be rewritten abstractly, by means of (9.10.2.1), (9.10.2.6), (9.10.2.7), and (9.10.4.7), as

$$
\begin{cases}
z_{tt} = -\mathcal{A}_N z - \mathcal{A}_N N_1 N_1^* \mathcal{A}_N z_t + \mathcal{A}_N N(v_t|_{\Gamma_0}) & \text{on } [\mathcal{D}(\mathcal{A}_N)]'; \quad (9.10.4.8) \\
v_{tt} = -\mathbb{A}v - \left(I + \gamma \mathcal{A}^{\frac{1}{2}}\right)^{-1} N^* \mathcal{A}_N z_t - \mathbb{A} G_2 G_2^* \mathcal{A} v_t \\
\quad + \left(I + \gamma \mathcal{A}^{\frac{1}{2}}\right)^{-1} \delta'(x_0) u(t) & \text{on } L_2(\Gamma_0), \quad (9.10.4.9)
\end{cases}
$$

corresponding to Eqn. (9.10.2.12) and Eqn. (9.10.2.14), respectively, in the conservative (undamped) case of Section 9.10.2. The corresponding first-order system of Eqns. (9.10.4.8) and (9.10.4.9), with $y(t) = [z(t), z_t(t), v(t), v_t(t)]$, is

$$\dot{y} = A_F y + Bu, \quad y(0) = y_0 = [z_0, z_1, v_0, v_1], \tag{9.10.4.10}$$

where the operator B is given by (9.10.2.19), and the operator A_F (the subindex F reminds us that the new damping terms may be viewed as "feedbacks" to the original conservative problem (9.10.1.1)) is given by

$$
A_F = \begin{bmatrix}
0 & I & 0 & 0 \\
-\mathcal{A}_N & -\mathcal{A}_N N_1 N_1^* \mathcal{A}_N & 0 & \mathcal{A}_N N(\cdot|_{\Gamma_0}) \\
0 & 0 & 0 & I \\
0 & -\left(I + \gamma \mathcal{A}^{\frac{1}{2}}\right)^{-1} N^* \mathcal{A}_N & -\mathbb{A} & -\mathbb{A} G_2 G_2^* \mathcal{A}
\end{bmatrix} \tag{9.10.4.11}
$$

[we could write $(\cdot|_{\Gamma_0}) = -N^* \mathcal{A}_N$ and $\partial/\partial\nu = G_2^* \mathcal{A}$ by (9.10.2.7) and (9.10.4.6), respectively];

$$Y = \mathcal{D}(\mathcal{A}_N^{\frac{1}{2}}) \times L_2^0(\Omega) \times \mathcal{D}(\mathcal{A}^{\frac{1}{2}}) \times \mathcal{D}(\mathcal{A}_\gamma^{\frac{1}{2}}) \tag{9.10.4.12}$$

$$\supset \mathcal{D}(A_F) = \left\{ y \in Y : \; y_2 \in \mathcal{D}(\mathcal{A}_N^{\frac{1}{2}}), \; y_3 \in \mathcal{D}(\mathcal{A}^{\frac{3}{4}}), \; y_4 \in \mathcal{D}(\mathcal{A}^{\frac{1}{2}}), \right.$$

$$- \mathcal{A}_N^{\frac{1}{2}}[y_1 - N_1(y_2|_\Gamma) - N(y_4|_{\Gamma_0})] \in \mathcal{D}(\mathcal{A}_N^{\frac{1}{2}}),$$

$$\left. - (I + \gamma\mathcal{A}^{\frac{1}{2}})^{-\frac{1}{2}} \mathcal{A}[y_3 + G_2 G_2^* \mathcal{A} y_4] \in L_2(\Omega) \right\}. \tag{9.10.4.13}$$

The Y-adjoint A_F^* of A_F is

$$A_F^* = \begin{bmatrix} 0 & -I & 0 & 0 \\ \mathcal{A}_N & -\mathcal{A}_N N_1 N_1^* \mathcal{A}_N & 0 & -\mathcal{A}_N N(\cdot|_{\Gamma_0}) \\ 0 & 0 & 0 & -I \\ 0 & (I + \gamma\mathcal{A}^{\frac{1}{2}})^{-1} N^* \mathcal{A}_N & \mathbb{A} & -\mathbb{A} G_2 G_2^* \mathcal{A} \end{bmatrix}, \tag{9.10.4.14}$$

with $\mathcal{D}(A_F^*)$ written analogously as (9.10.4.13) for $\mathcal{D}(A_F)$. As in Section 9.10.2, by virtue of a corollary of the Lumer–Phillips theorem [Pazy, 1983, p. 15], we obtain

Theorem 9.10.4.1 *The damped system (9.10.4.1), subject to (9.10.4.2), is modeled abstractly by (9.10.4.10) with B as in (9.10.2.19) and A_F as in (9.10.4.11). The operator A_F and its Y-adjoint A_F^* in (9.10.4.14) are maximally dissipative on the space Y defined by (9.10.4.12), and thus they generate s.c. contraction semigroups $e^{A_F t}$ and $e^{A_F^* t}$ on Y, $t \geq 0$.*

The solution to problem (9.10.4.1), that is, to problem (9.10.4.10), is given by

$$y(t) = e^{A_F t} y_0 + (L_F u)(t); \quad (L_F u)(t) = \int_0^t e^{A_F(t-\tau)} B u(\tau) \, d\tau. \tag{9.10.4.15}$$

Theorem 9.10.4.2 *The operator L_F in (9.10.4.15) providing the input-solution map of the damped problem (9.10.4.1), subject to (9.10.4.2), has the same regularity as the operator L of the conservative problem (9.10.1.1), that is,*

$$L_F : \text{continuous } L_2(0, T) \rightarrow C([0, T]; Y), \tag{9.10.4.16}$$

with Y as in (9.10.4.12).

Proof. (Sketch) We limit ourselves to point out the main changes that are needed in the proof of Section 9.10.3, when the conservative problem (9.10.1.1) is replaced by the damped problem (9.10.4.1), to obtain the same regularity as in Theorem 9.10.1.2.

(i) First, problem (9.10.3.1) now replaces the BC (9.10.3.1c$_h$) with the BC

$$\phi \equiv 0, \quad \Delta\phi = -\frac{\partial\phi_t}{\partial\nu}. \tag{9.10.4.17}$$

As noted in Remark 9.8.3.4, the proof of Section 9.9.3 readily adapts to this case and thus Theorem 9.10.3.1 continues to hold true.

(ii) Second, problem (9.10.3.4) now replaces the BC (9.10.3.4c) with the BC

$$\left.\frac{\partial \psi}{\partial v}\right|_{\Sigma} = -\psi_t + g. \tag{9.10.4.18}$$

This new damped problem (9.10.3.4a,b), (9.10.4.18) provides the regularity [Chapter 7, Section 7.6]

$$g \in L_2(0, T; L_2(\Gamma)) \Rightarrow \{\psi, \psi_t\} \in C([0, T]; H^1(\Omega) \times L_2(\Omega)) \tag{9.10.4.19}$$

[rather than the regularity (Chapter 8, Appendix 8A, Theorem 8A.1)

$$g \in L_2(0, T; L_2(\Gamma)) \Rightarrow \{\psi, \psi_t\} \in C([0, T]; H^\alpha(\Omega) \times H^{\alpha-1}(\Omega)), \tag{9.10.4.20}$$

$\alpha < 1$ as in (9.10.3.5), for problem (9.10.3.4)]. Thus, not only the damped BC (9.10.4.18) is responsible now for the higher regularity result (9.10.4.19), which is false in the case of problem (9.10.3.4) with dim $\Omega \geq 2$, but, moreover, the proof of (9.10.4.19) is far easier than that of (9.10.4.20), as it only requires a direct (operator) energy method of Chapter 7, Section 7.6, based on [Lasiecka, 1997, Section 4] rather than pseudo-differential analysis of Chapter 8, Appendix 8A, based on Lasiecka and Triggiani [1981; 1983].   □

Theorem 9.10.4.3 *[Camurdan, 1997] Consider the damped problem (9.10.4.1), say in the damping case of Eqn. (9.10.4.3), whose dynamics is described by the s.c. contraction semigroup $e^{A_F t}$ of Theorem 9.10.4.1. Let the initial condition $y_0 = \{z_0, z_1, v_0, v_1\} \in Y$ satisfy the condition*

$$\int_{\Gamma_0} z_0 \, d\Gamma_0 + \int_{\Omega} z_1 \, d\Omega + \int_{\Gamma_0} v_0 \, d\Gamma_0 = 0. \tag{9.10.4.21}$$

Then, the s.c. semigroup $e^{A_F t}$ is uniformly stable in Y: there exists constants $M \geq 1$, $\delta > 0$ such that

$$\left\| e^{A_F t} \right\|_{\mathcal{L}(Y)} \leq M e^{-\delta t}, \quad t \geq 0; \quad \text{equivalently, } E(t) \leq M e^{-\delta t} E(0), \tag{9.10.4.22}$$

where $E(t)$ is the energy of the system (9.10.4.1),

$$E(t) \equiv \left\| e^{A_F t} y_0 \right\|_Y^2 = \int_{\Omega} \left[|\nabla z(t)|^2 + z_t^2(t) \right] d\Omega$$
$$+ \int_{\Gamma_0} \left[|\Delta v(t)|^2 + v_t^2(t) + \gamma |\nabla v_t(t)|^2 \right] d\Gamma_0. \tag{9.10.4.23}$$

9.11 A Coupled System of a Wave and a Structurally Damped Euler–Bernoulli Equation with Point Control, Arising in Noise Reduction. Regularity Theory

9.11.1 Statement of Problem for dim $\Omega = 2$. *Main Results*

Let Ω be a two-dimensional domain ("the chamber") as in Section 9.10; see Figures 9.1 and 9.2 there. We shall preserve the same symbols for the boundary $\partial\Omega = \Gamma_0 \cup \Gamma_1$, for the acoustic wave $z(t, x)$ in Ω and for the displacement $v(t, x)$ of Γ_0, as in Section 9.10. However, the present section studies the model

$$
\begin{cases}
\begin{cases}
z_{tt} = \Delta z & \text{on } (0, T] \times \Omega \equiv Q, & (9.11.1.1a) \\[2mm]
\dfrac{\partial z}{\partial \nu}\Big|_{\Sigma_1} \equiv 0 & \text{on } (0, T] \times \Gamma_1 \equiv \Sigma_1, & (9.11.1.1b) \\[2mm]
\dfrac{\partial z}{\partial \nu}\Big|_{\Sigma_0} \equiv -v_t & \text{on } (0, T] \times \Gamma_0 \equiv \Sigma_0; & (9.11.1.1c)
\end{cases} \\[8mm]
\begin{cases}
v_{tt} - \gamma \Delta v_{tt} + \Delta^2 v + \Delta^2 v_t - z_t = \delta'(x_0)u(t) & \text{on } \Sigma_0; & (9.11.1.1d) \\
\text{either clamped BC on } \partial\Gamma_0, & \\[2mm]
v \equiv \dfrac{\partial v}{\partial \nu} \equiv 0 & \text{on } (0, T] \times \partial\Gamma_0, & (9.11.1.1e_c) \\[2mm]
\text{or else hinged BC on } \partial\Gamma_0, & \\[2mm]
v \equiv \Delta v \equiv 0 & \text{on } (0, T] \times \partial\Gamma_0; & (9.11.1.1e_h)
\end{cases} \\[8mm]
z(0, \cdot) = z_0; \ z_t(0, \cdot) = z_1 \ \text{in } \Omega; \ v(0, \cdot) = v_0; \ v_t(0, \cdot) = v_1 \ \text{in } \Gamma_0 \quad (9.11.1.1f)
\end{cases}
$$

where $\nu(x) = $ unit outward normal vector at $x \in \Gamma$, and $\gamma \geq 0$ is a constant. We shall treat simultaneously the case $\gamma = 0$ (Euler–Bernoulli model in v) and the case $\gamma > 0$ (Kirchhoff model in v, accounting for rotational forces). As in Section 9.10, x_0 is a chosen point on Γ_0. Apart from the possibility now allowed to have $\gamma = 0$ as well, the main difference between model (9.10.1.1) of Section 9.10 and model (9.11.1.1) of the present section is the presence of the new damping term $\Delta^2 v_t$ in Eqn. (9.11.1.1d), which is of the same order as the basic elastic term $\Delta^2 v$ (Kelvin–Voigt damping). Thus, regardless of whether $\gamma = 0$ or $\gamma > 0$, we have that: If $u \equiv 0$ and the term z_t is missing from Eqn. (9.11.1.1d), then the corresponding v-problem is *parabolic*, that is, it gives rise to an analytic semigroup by virtue of Chapter 3, Appendix 3B, Theorem 3B.1. Thus, the overall system (9.11.1.1) with $\gamma \geq 0$ couples a *hyperbolic* equation in z with a *parabolic* equation in v, while the overall system (9.10.1.1) of Section 9.10 with $\gamma > 0$ couples two hyperbolic equations. The main goal of the present section is two fold, as expressed by the following two main results.

Theorem 9.11.1.1   Let $y(t) = [z(t), z_t(t), v(t), v_t(t)]$. Then the coupled PDE system (9.11.1.1) can be rewritten abstractly as equation (9.1.1), that is, as

$$\dot{y} = Ay + Bu \in [\mathcal{D}(A^*)]', \quad y(0) = y_0 = [z_0, z_1, v_0, v_1] \in Y, \qquad (9.11.1.2)$$

where A is a suitable operator given explicitly by (9.11.2.6) below, and B is the same operator (9.10.2.19) of Section 9.10 (rewritten in (9.11.2.8) below). Moreover, Y is the Hilbert space

$$Y = \mathcal{D}\big(A_N^{\frac{1}{2}}\big) \times L_2(\Omega) \times \mathcal{D}\big(A^{\frac{1}{2}}\big) \times \mathcal{D}\big(\mathcal{M}_\gamma^{\frac{1}{2}}\big) \qquad (9.11.1.3)$$

$$= \begin{cases} H^1(\Omega) \times L_2(\Omega) \times H_0^2(\Gamma_0) \times H_0^1(\Gamma_0) \\ \quad : \textit{clamped BC } \gamma > 0; & (9.11.1.4a) \\ H^1(\Omega) \times L_2(\Omega) \times H_0^2(\Gamma_0) \times L_2(\Gamma_0) \\ \quad : \textit{clamped BC } \gamma = 0; & (9.11.1.4b) \\ H^1(\Omega) \times L_2(\Omega) \times \big[H^2(\Gamma_0) \cap H_0^1(\Gamma_0)\big] \times H_0^1(\Gamma_0) \\ \quad : \textit{hinged BC } \gamma > 0; & (9.11.1.4c) \\ H^1(\Omega) \times L_2(\Omega) \times \big[H^2(\Gamma_0) \cap H_0^1(\Gamma_0)\big] \times L_2(\Gamma_0) \\ \quad : \textit{hinged BC } \gamma = 0 & (9.11.1.4d) \end{cases}$$

(norm equivalence). In (9.11.1.3), A_N is the operator defined by (9.10.2.6) $(-\Delta$ with Neumann BC on Ω), whereas A is the elastic operator Δ^2 with hinged BC on Γ_0 as in (9.10.2.1) or with clamped BC on Γ_0 as in (9.10.2.27). Finally,

$$\mathcal{M}_\gamma = \begin{cases} I + \gamma A^{\frac{1}{2}} & \textit{for hinged BC, } A \textit{ as in (9.10.2.1),} & (9.11.1.5a) \\ I + \gamma A_D & \textit{for clamped BC, } A_D \textit{ as in (9.10.2.29),} & (9.11.1.5b) \end{cases}$$

so that $\mathcal{D}(A_\gamma^{\frac{1}{2}}) = H_0^1(\Gamma_0)$ in both cases of hinged or clamped BC for $\gamma > 0$, whereas $\mathcal{D}(\mathcal{M}_\gamma^{\frac{1}{2}}) = L_2(\Gamma_0)$ if $\gamma = 0$. When Y is topologized by the norm in (9.11.1.3), then the operator A is maximally dissipative and thus generates a s.c. contraction semigroup e^{At} on Y, $t \geq 0$.

As a consequence of Theorem 9.11.1.1, the solution of the abstract version (9.11.1.2) of the coupled PDE system (9.11.1.1) is then as in (9.1.16) and (9.1.17), that is,

$$y(t) = e^{At}y_0 + (Lu)(t); \quad (Lu)(t) = \int_0^t e^{A(t-\tau)}Bu(\tau)\,d\tau. \qquad (9.11.1.6)$$

The main result of this section is the following regularity theorem for (9.11.1.2), or (9.11.1.6), which verifies the "abstract trace regularity" (A.1) = (9.1.7), or its dual version (A.1*) = (9.1.10).

Theorem 9.11.1.2   *With reference to the abstract version (9.11.1.2), or (9.11.1.6), of the coupled PDE system (9.11.1.1), and to the space Y defined in (9.11.1.3) and (9.11.1.4), we have that:*

(i) *For each $0 < T < \infty$, the operator L in (9.11.1.6) satisfies the property*

$$L : \ continuous \ L_2(0, T) \to C([0, T]; Y). \qquad (9.11.1.7)$$

In PDE terms, the meaning of (9.11.1.6) is as follows. Set $z_0 = z_1 = v_0 = v_1 = 0$ in (9.11.1.1f); then the corresponding solution of (9.11.1.1) satisfies

$$u \in L_2(0, T) \to [z(t), z_t(t), v(t), v_t(t)] \in C([0, T]; Y). \quad (9.11.1.8)$$

(ii) *Equivalently, by duality [Chapter 7, Theorem 7.2.1], then the following "abstract trace regularity" property holds true: For each $0 < T < \infty$, there exists a constant $C_T > 0$ such that*

$$\int_0^T \left| B^* e^{A^* t} y \right| dt \leq C_T \|y\|_Y^2, \quad \forall y \ first \ in \ \mathcal{D}(A^*), \ next \ extended \ to \ all \ of \ Y.$$

$$(9.11.1.9)$$

In PDE terms the meaning of (9.11.1.9) is as follows. Let $[\tilde{z}, \tilde{z}_t, \tilde{v}, \tilde{v}_t]$ be the solution of problem (9.11.2.11) below, which is the adjoint of problem (9.11.1.1) for $u \equiv 0$. Then:

$$\int_0^T |\tilde{v}_{tx}(t, x_0; y_0)|^2 \, dt \leq C_T \|y_0\|_Y^2, \quad y_0 \in Y, \qquad (9.11.1.10)$$

$y_0 = [\tilde{z}_0, \tilde{z}_1, \tilde{v}_0, \tilde{v}_1]$, *where $\tilde{v}_{tx}(t, x_0; y_0)$ is the second partial derivative of the solution $\tilde{v}$ in t and x, evaluated at the point $x = x_0 \in \Gamma_0$ (point observation), and due to the initial condition y_0 in (9.11.2.11) below.*

9.11.2 Abstract Model of Original Problem (9.11.1.1) in $\{z, v\}$. Theorem 9.11.1.1

The present section provides for problem (9.11.1.1) a counterpart treatment of Section 9.10.2 for problem (9.10.1.1). Thus, let $\mathcal{A}_N$ be the operator in (9.10.2.6) ($-\Delta$ with Neumann BC); let $\mathcal{A}$ be the elastic operator either in (9.10.2.1) (Δ^2 with hinged BC) or in (9.10.2.27) (Δ^2 with clamped BC); let $\mathcal{A}_D$ be the operator in (9.10.2.29) ($-\Delta$ with Dirichlet BC). Finally, let N be the Neumann map defined by (9.10.2.5), with regularity (9.10.2.8), (9.10.2.9). Then, the abstract model of the coupled system (9.11.1.1) is given by

$$\begin{cases} z_{tt} = -\mathcal{A}_N z + \mathcal{A}_N N(v_t|_{\Gamma_0}) & \text{on } [\mathcal{D}(\mathcal{A}_N)]', & (9.11.2.1) \\ \mathcal{M}_\gamma v_{tt} + \mathcal{A} v + \mathcal{A} v_t = -N^* \mathcal{A}_N z_t + \delta'(x_0) u(t) & \text{on } L_2(\Gamma_0). & (9.11.2.2) \end{cases}$$

Thus it is the same as the abstract model (9.10.2.12), (9.10.2.13), with the addition now of the damping term $\mathcal{A}v_t$ in (9.11.2.2), where the operator $\mathcal{M}_\gamma$ is defined by (9.11.1.5) and we have

$$(x, y)_{\mathcal{D}(\mathcal{M}_\gamma^{\frac{1}{2}})} = (\mathcal{M}_\gamma x, y)_{L_2(\Gamma_0)}, \quad x, y \in \mathcal{D}(\mathcal{M}_\gamma^{\frac{1}{2}}) = H_0^1(\Gamma_0), \quad \gamma > 0, \quad (9.11.2.3)$$

[compare with (9.10.2.3)], where for $\gamma > 0$ (equivalent norms)

$$\left. \begin{array}{c} H^2(\Gamma_0) \cap H_0^1(\Gamma_0) \\[2mm] H_0^2(\Gamma_0) \end{array} \right\} = \mathcal{D}(\mathcal{A}^{\frac{1}{2}}) \subset H_0^1(\Gamma_0) = \mathcal{D}(\mathcal{M}_\gamma^{\frac{1}{2}}) = \left\{ \begin{array}{l} \mathcal{D}(\mathcal{A}^{\frac{1}{4}}) \text{ hinged BC,} \\[2mm] \hspace{2cm} (9.11.2.4a) \\[2mm] \mathcal{D}(\mathcal{A}_D^{\frac{1}{2}}) \text{ clamped BC.} \\[2mm] \hspace{2cm} (9.11.2.4b) \end{array} \right.$$

Thus, the corresponding first-order system of (9.11.2.1), (9.11.2.2) with $y(t) = [z(t), z_t(t), v(t), v_t(t)]$ is

$$\dot{y} = Ay + Bu, \quad y(0) = y_0 = [z_0, z_1, v_0, v_1] \in Y, \quad (9.11.2.5)$$

where Y is given by (9.11.1.3) and (compare with (9.10.2.16)–(9.10.2.18)):

(i)

$$A = \begin{bmatrix} 0 & I & 0 & 0 \\ -\mathcal{A}_N & 0 & 0 & \mathcal{A}_N N(\cdot |_{\Gamma_0}) \\ 0 & 0 & 0 & I \\ 0 & -\mathcal{M}_\gamma^{-1} N^* \mathcal{A}_N & -\mathcal{M}_\gamma^{-1}\mathcal{A} & -\mathcal{M}_\gamma^{-1}\mathcal{A} \end{bmatrix}; \quad (9.11.2.6)$$

$$Y \supset \mathcal{D}(A) = \{ y \in Y : y_2 \in \mathcal{D}(\mathcal{A}_N^{\frac{1}{2}}), \ y_4 \in \mathcal{D}(\mathcal{A}^{\frac{1}{2}}); \ \mathcal{A}_N^{\frac{1}{2}}[y_1 - N(y_4|_{\Gamma_0})] \in \mathcal{D}(\mathcal{A}_N^{\frac{1}{2}}),$$

$$\mathcal{A}^{\frac{1}{2}} y_3 + \mathcal{A}^{\frac{1}{2}} y_4 \in \mathcal{D}(\mathcal{M}_\gamma^{-\frac{1}{2}} \mathcal{A}^{\frac{1}{2}})\}. \quad (9.11.2.7)$$

The Y-adjoint of A is

$$A^* = \begin{bmatrix} 0 & -I & 0 & 0 \\ \mathcal{A}_N & 0 & 0 & -\mathcal{A}_N N(\cdot |_{\Gamma_0}) \\ 0 & 0 & 0 & -I \\ 0 & \mathcal{M}_\gamma^{-1} N^* \mathcal{A}_N & \mathcal{M}_\gamma^{-1}\mathcal{A} & -\mathcal{M}_\gamma^{-1}\mathcal{A} \end{bmatrix}, \quad (9.11.2.8)$$

$$\mathcal{D}(A^*) = \mathcal{D}(A).$$

Because of (9.11.2.3), the operator obtained from A in (9.10.2.6) by removing the bottom-right corner element $-\mathcal{M}_\gamma^{-1}\mathcal{A}$ is skew-adjoint on Y given by (9.11.1.3), so that A is dissipative on Y and, moreover,

$$\mathrm{Re}(Ax; x)_Y = \mathrm{Re}(A^*x, x)_Y = -\left\|A^{\frac{1}{2}}x_4\right\|_{L_2(\Gamma_0)} \leq 0, \qquad (9.11.2.9)$$

for $x = [x_1, x_2, x_3, x_4] \in \mathcal{D}(A) = \mathcal{D}(A^*)$. Since A is densely defined on Y, then (9.11.2.9) allows one to invoke a standard corollary of the Lumer–Phillips theorem [Pazy, 1983, p. 15] and conclude that the operators A and A^* in (9.11.2.6), (9.11.2.7), and in (9.11.2.8) generate s.c. contraction semigroups e^{At} and e^{A^*t} on Y, $t \geq 0$, as desired. Theorem 9.11.1.1 is proved.

We notice also, for future use in Section 9.11.3, that

$$y(t) = e^{A^*t}y_0 \text{ means } y(t) = [\tilde{z}(t), \tilde{z}_t(t), \tilde{v}(t), \tilde{v}_t(t)], \qquad (9.11.2.10)$$

where

$$
\begin{cases}
\begin{cases}
\tilde{z}_{tt} = \Delta \tilde{z} & \text{on } (0, T] \times \Omega \equiv Q, & (9.11.2.11a) \\
\left.\dfrac{\partial \tilde{z}}{\partial v}\right|_{\Sigma_1} \equiv 0 & \text{on } (0, T] \times \Gamma_1 \equiv \Sigma_1, & (9.11.2.11b) \\
\left.\dfrac{\partial \tilde{z}}{\partial v}\right|_{\Sigma_0} \equiv \tilde{v}_t & \text{on } (0, T] \times \Gamma_0 \equiv \Sigma_0; & (9.11.2.11c)
\end{cases} \\[2em]
\begin{cases}
\tilde{v}_t - \gamma \Delta \tilde{v}_{tt} + \Delta^2 \tilde{v} + \Delta^2 \tilde{v}_t + \tilde{z}_t = 0 & \text{on } \Sigma_0; & (9.11.2.11d) \\
\text{either clamped BC on } \partial\Gamma_0, & \\
\tilde{v} \equiv \dfrac{\partial \tilde{v}}{\partial v} \equiv 0 & \text{on } (0, T] \times \partial\Gamma_0, & (9.11.2.11e_c) \\
\text{or else hinged BC on } \partial\Gamma_0, & \\
\tilde{v} \equiv \Delta \tilde{v} \equiv 0; & & (9.11.2.11f_h)
\end{cases} \\[2em]
\tilde{z}(0, \cdot) = \tilde{z}_0; \ \tilde{z}_t(0, \cdot) = \tilde{z}_1 \text{ in } \Omega; \ \tilde{v}(0, \cdot) = \tilde{v}_0; \ \tilde{v}_t(0, \cdot) = \tilde{v}_1 \text{ in } \Gamma_0. & (9.11.2.11f)
\end{cases}
$$

Thus, the adjoint $[\tilde{z}, \tilde{z}_t, \tilde{v}, \tilde{v}_t]$ problem in (9.11.2.11) has the coupling with opposite sign in (9.11.2.11c) and in (9.11.2.11d) with respect to the original $[z, z_t, v, v_t]$-problem in (9.11.1.1).

(ii) $B : U \to [\mathcal{D}(A^*)]'$, duality with respect to the pivot space Y, $U = \mathbb{R}$, and its L_2-adjoint B^* are

$$Bu = \begin{bmatrix} 0 \\ 0 \\ 0 \\ \mathcal{M}_\gamma^{-1}\delta'(x_0)u \end{bmatrix}; \quad B^* \begin{bmatrix} y_1 \\ y_2 \\ y_3 \\ y_4 \end{bmatrix} = -\frac{d}{dx}y_4|_{x=x_0}, \quad y \in \mathcal{D}(A^*).$$

$$(9.11.2.12)$$

Indeed, to obtain B^*, we compute from (9.11.2.12), via (9.11.1.3) and (9.11.2.3),

$$(Bu, y)_Y = \left(M_\gamma^{-1} \delta'(x_0)u, y_4 \right)_{\mathcal{D}(M_\gamma^{\frac{1}{2}})} = (\delta'(x_0)u, y_4)_{L_2(\Gamma_0)}$$

$$= u \left(-\frac{d}{dx} y_4 |_{x=x_0} \right) = (u, B^*y)_U = \mathbb{R},$$ (9.11.2.13)

and B^* in (9.11.2.12) is proved. Finally, the inverse $A^{-1} \in \mathcal{L}(Y)$ of A in (9.11.2.6) is given by

$$A^{-1} = \begin{bmatrix} 0 & -A_N^{-1} & N(\cdot |_{\Gamma_0}) & 0 \\ I & 0 & 0 & 0 \\ -A^{-1}N^*A_N & 0 & -I & -A^{-1}M_\gamma \\ 0 & 0 & I & 0 \end{bmatrix} \in \mathcal{L}(Y). \quad (9.11.2.14)$$

Thus, by (9.11.2.12) and (9.10.2.14), we have

$$A^{-1}Bu = [0, \ 0, \ -A^{-1}\delta'(x_0)u, \ 0] \in Y,$$ (9.11.2.15)

as in Section 9.10, Eqn. (9.10.2.4), and (9.1.2) is verified. Theorem 9.11.1.1 is proved. [Again, as in Section 9.10, there is no need to compute A^{-1} as in (9.11.2.14) to obtain (9.11.2.15): Writing $Bu = Ag$, the special form of B as in (9.11.2.12) and A as in (9.11.2.6), readily yields that g is given by the right-hand side of (9.11.2.15). The expression of A^{-1} in (9.11.2.14) is an additional piece of information.]

9.11.3 Proof of Theorem 9.11.1.2: Abstract Trace Condition (9.11.1.9)

Whereas in Section 9.10 we gave a direct proof of the regularity of L as in (9.10.1.5), in the present section we prove instead the dual statement (9.11.1.9), or (9.11.1.10), not the direct statement (9.11.1.7), or (9.11.1.8), on L. In either case, we have chosen the simplest proof available.

We begin by defining the "energy" of problem (9.11.1.1):

$$E_\gamma(t) \equiv E_\gamma(t; y_0) = \|e^{At}y_0\|_Y^2 = \|[z(t), z_t(t), v(t), v_t(t)]\|_Y^2$$

$$= \left\| A_N^{\frac{1}{2}}z(t) \right\|_{L_2(\Omega)}^2 + \|z_t(t)\|_{L_2(\Omega)}^2 + \left\| A^{\frac{1}{2}}v(t) \right\|_{L_2(\Gamma_0)}^2 + \left\| M_\gamma^{\frac{1}{2}}v_t(t) \right\|_{L_2(\Gamma_0)}^2,$$

(9.11.3.1)

recalling Y in (9.11.1.3), for $y_0 = [z_0, z_1, v_0, v_1] \in Y$, valid for both $\gamma > 0$ and $\gamma = 0$.

Step 1.   Lemma 9.11.3.1   With reference to problem (9.11.1.1) and to (9.11.3.1), we have the following energy identity, for $y_0 \in Y$ and any $0 < T$ finite:

$$E_\gamma(T) + 2 \int_0^T \left\| A^{\frac{1}{2}}v_t(t) \right\|_{L_2(\Gamma_0)}^2 dt = E_\gamma(0).$$ (9.11.3.2)

Proof. Let $y_0 = [z_0, z_1, v_0, v_1] \in \mathcal{D}(A)$, so that, by Theorem 9.11.1.1, we have

$$\begin{cases} y(t) = [z(t), z_t(t), v(t), v_t(t)] \in C([0, T]; \mathcal{D}(A)), \text{ so that by (9.11.1.7)} \\ z_t(t) \in C([0, T]; \mathcal{D}(A_N^{\frac{1}{2}}) = H^1(\Omega)) \text{ and } z_t(t)|_{\Gamma_0} \in C([0, T]; L_2(\Gamma_0)). \\ \text{Also, } v_t(t) \in C([0, T]; \mathcal{D}(A^{\frac{1}{2}})); \quad \mathcal{D}(A^{\frac{1}{2}}) \subset \mathcal{D}(\mathcal{M}_\gamma^{\frac{1}{2}}) \text{ by (9.11.2.4).} \end{cases}$$
$$(9.11.3.3)$$

Next, we take the $L_2(\Omega)$-inner product of Eqn. (9.11.2.1) with $z_t(t)$ and the $L_2(\Gamma_0)$-inner product of Eqn. (9.11.2.2) with $v_t(t)$, integrate in t over $[0, T]$, use (9.10.2.7) for N^*A_N, and obtain, respectively:

$$\|z_t(T)\|_{L_2(\Omega)}^2 + \|A_N^{\frac{1}{2}}z(T)\|_{L_2(\Omega)}^2 + 2\int_0^T (v_t(t), z_t(t))_{L_2(\Gamma_0)}\, dt$$

$$= \|z_t(0)\|_{L_2(\Omega)}^2 + \|A_N^{\frac{1}{2}}z(0)\|_{L_2(\Omega)}^2, \qquad (9.11.3.4)$$

$$\|\mathcal{M}_\gamma^{\frac{1}{2}}v_t(T)\|_{L_2(\Gamma_0)}^2 + \|A^{\frac{1}{2}}v(T)\|_{L_2(\Gamma_0)}^2 + 2\int_0^T \|A^{\frac{1}{2}}v_t(t)\|_{L_2(\Gamma_0)}^2\, dt$$

$$- 2\int_0^T (z_t(t), v_t(t))_{L_2(\Gamma_0)}\, dt$$

$$= \|\mathcal{M}_\gamma^{\frac{1}{2}}v_t(0)\|_{L_2(\Gamma_0)}^2 + \|A^{\frac{1}{2}}v(0)\|_{L_2(\Gamma_0)}^2. \qquad (9.11.3.5)$$

Summing up (9.11.3.4) and (9.11.3.5) produces a cancellation of the $L_2(\Gamma_0)$-integral term for v_t and z_t and yields (9.11.3.2), by virtue of (9.11.3.1), at least for $y_0 \in \mathcal{D}(A)$. Identity (9.11.3.2) is then extended to all $y_0 \in Y$ by density. $\square$

Step 2 We recall that dim $\Gamma_0 = 1$. Hence, we have as in (9.10.2.22):

$$\delta'(x_0) \in \left[H^{\frac{3}{2}+4\epsilon}(\Gamma_0)\right]' \subset \left[\mathcal{D}(A^{\frac{3}{8}+\epsilon})\right]', \quad \text{or} \quad A^{-\frac{3}{8}-\epsilon}\delta'(x_0) \in L_2(\Gamma_0), \quad (9.11.3.6)$$

since $\mathcal{D}(A^\theta) \subset H^{4\theta}(\Gamma_0)$, $\theta > 0$ for the fourth-order operator A, with duality with respect to $L_2(\Gamma_0)$. Then, by recalling (9.11.2.9), we have

$$|(u, B^*y)_{U=\mathbb{R}}| = |(\delta'(x_0)u, y_4)_{L_2(\Gamma_0)}| = \left|\left(A^{-\frac{3}{8}-\epsilon}\delta'(x_0)u, A^{\frac{3}{8}+\epsilon}y_4\right)_{L_2(\Gamma_0)}\right|$$
$$(9.11.3.7)$$

$$\text{(by (9.11.3.6))} \qquad \leq \text{const}\|A^r y_4\|_{L_2(\Gamma_0)}|u|, \quad r = \frac{3}{8} + \epsilon. \qquad (9.11.3.8)$$

Lemma 9.11.3.2   *(i) With reference to the original problem (9.11.1.1) and to (9.11.3.1), as well as to A in (9.11.2.6), we have*

$$\int_0^T |B^* e^{At} y_0|^2 \, dt \le C_T \|y_0\|_Y^2, \quad y_0 \in Y. \tag{9.11.3.9}$$

(ii) With reference to A^* in (9.11.2.8), or to the adjoint problem (9.11.2.10), (9.11.2.11), we have

$$\int_0^T |B^* e^{A^*t} y_0|^2 \, dt \le C_T \|y_0\|_Y^2, \quad y_0 \in Y. \tag{9.11.3.10}$$

Proof.   (i) Let $y(t) = [z(t), z_t(t), v(t), v_t(t)] = e^{At} y_0$, $y_0 \in Y$ by Theorem 9.11.1.1. Recalling (9.11.3.8), we compute

$$\int_0^T |B^* y(t)|^2 \, dt = \int_0^T |B^* e^{At} y_0|^2 \, dt$$

$$= \int_0^T |B^*[z(t), z_t(t), v(t), v_t(t)]|^2 \, dt$$

(by (9.11.3.8)) $\displaystyle \le \text{const} \int_0^T \|\mathcal{A}^r v_t(t)\|_{L_2(\Gamma_0)}^2 \, dt \tag{9.11.3.11}$

$\displaystyle \le \text{const} \int_0^T \left\|\mathcal{A}^{\frac{1}{2}} v_t(t)\right\|_{L_2(\Gamma_0)}^2 \, dt \tag{9.11.3.12}$

(by (9.11.3.2)) $\displaystyle \le \text{const} \, E_\gamma(0) = \text{const} \|y_0\|_Y^2, \tag{9.11.3.13}$

where in going from (9.11.3.11) to (9.11.3.12) we have critically used that $r = 3/8 + \epsilon \le 1/2$. Moreover, the identity in (9.11.3.13) follows from (9.11.3.1). Thus, (9.11.3.13) proves (9.11.3.9).

(ii) The proof of (9.11.3.10) is similar; this time we use A^* in (9.11.2.8) rather than A in (9.11.2.6); that is, we use the adjoint problem (9.11.2.11). Lemma 9.11.3.1 still holds true for the adjoint problem, with the same proof, which still produces a cancellation as noted below (9.11.3.5). The original problem (9.11.1.1) in $[z, z_t, v, v_t]$ and the adjoint problem (9.11.2.11) in $[\tilde{z}, \tilde{z}_t, \tilde{v}, \tilde{v}_t]$ only differ in the sign on both coupling terms. $\square$

The proof of Theorem 9.11.1.2 is complete. $\square$

9A Proof of (9.9.1.16) in Lemma 9.9.1.1

(i) We first take $\gamma^2 - \omega^2 < 0$. Set $a_\omega^2 = \omega^2 - \gamma^2 > 0$, and split the integral $I(\omega)$ in (9.9.1.16) as

$$I(\omega) = \int_0^{\frac{a_\omega}{2}} + \int_{\frac{a_\omega}{2}}^{2a_\omega} + \int_{2a_\omega}^\infty = I_1(\omega) + I_2(\omega) + I_3(\omega). \tag{9A.1}$$

For $I_1(\omega)$ we compute

$$I_1(\omega) = \int_2^{\frac{a_\omega}{2}} \frac{y^2 \, dy}{\left(y^2 - a_\omega^2\right)^2 + 4\gamma^2\omega^2} \leq \int_0^{\frac{a_\omega}{2}} \frac{y^2 \, dy}{\left(\frac{3}{4}a_\omega^2\right)^2 + 4\gamma^2\omega^2}$$

$$= \frac{1}{3} \frac{\left(\frac{a_\omega}{2}\right)^3}{\left(\frac{3}{4}a_\omega^2\right)^2 + 4\gamma^2\omega^2} \leq \text{const.} \tag{9A.2}$$

For $I_3(\omega)$ we compute with $-a_\omega^2 \geq -\frac{y^2}{4}$, $\gamma^2 < \omega^2$:

$$I_3(\omega) = \int_{2a_\omega}^\infty \frac{y^2 \, dy}{\left(y^2 - a_\omega^2\right)^2 + 4\gamma^2\omega^2} \leq \int_{2a_\omega}^\infty \frac{y^2 \, dy}{\left(\frac{3}{4}y^2\right)^2 + 4\gamma^4} \leq \text{const.} \tag{9A.3}$$

Finally, for $I_2(\omega)$, we set $t = y^2 - a_\omega^2$ and compute

$$I_2(\omega) = \int_{\frac{a_\omega}{2}}^{2a_\omega} \frac{y^2 \, dy}{\left(y^2 - a_\omega^2\right)^2 + 4\gamma^2\omega^2} = \frac{1}{2} \int_{-\frac{3}{4}a_\omega^2}^{3a_\omega^2} \frac{\sqrt{t + a_\omega^2} \, dt}{t^2 + 4\gamma^2\omega^2}$$

$$\left(\text{by } t \leq 3a_\omega^2\right) \quad \leq a_\omega \int_{-\frac{3}{4}a_\omega^2}^{3a_\omega^2} \frac{1 \, dt}{t^2 + 4\gamma^2\omega^2} = \frac{a_\omega}{2\gamma\omega} \left[\arctan \frac{t}{2\gamma\omega}\right]_{t=-3/4a_\omega^2}^{t=3a_\omega^2}$$

$$\left(\text{by } a_\omega \leq \omega\right) \quad \leq \frac{\omega}{2\gamma\omega} \left(\frac{\pi}{2} + \frac{\pi}{2}\right) \leq \text{const.} \tag{9A.4}$$

Thus, estimates (9A.2)–(9A.4) used in (9A.1) yield

$$I(\omega) \leq \text{const for all } \omega^2 > \gamma^2. \tag{9A.5}$$

(ii) Next, we take $\gamma^2 - \omega^2 \geq 0$ and split $I(\omega)$ as

$$I(\omega) = \int_0^1 + \int_1^\infty = J_1(\omega) + J_2(\omega). \tag{9A.6}$$

For $J_2(\omega)$ we drop $\gamma^2 - \omega^2 \geq 0$ and obtain

$$J_2(\omega) = \int_1^\infty \frac{y^2 \, dy}{(y^2 + \gamma^2 - \omega^2)^2 + 4\gamma^2\omega^2} \leq \int_1^\infty \frac{1}{y^2} \, dy = 1. \tag{9A.7}$$

For $J_1(\omega)$ we first restrict to $0 \leq \omega^2 \leq \gamma^2 - \epsilon$, $\epsilon > 0$ and obtain

$$J_1(\omega) = \int_0^1 \frac{y^2 \, dy}{(y^2 + \gamma^2 - \omega^2)^2 + 4\gamma^2\omega^2} \leq \int_0^1 \frac{y^2 \, dy}{(y^2 + \epsilon)^2} \leq \frac{1}{\epsilon^2} \int_0^1 y^2 \, dy = \frac{1}{3\epsilon^2};$$
$$\tag{9A.8}$$

we next consider $\gamma^2 - \epsilon < \omega^2 \le \gamma^2$ and obtain

$$J_1(\omega) = \int_0^1 \frac{y^2 \, dy}{(y^2 + \gamma^2 - \omega^2)^2 + 4\gamma^2\omega^2} \le \int_0^1 \frac{y^2 \, dy}{4\gamma^2\omega^2}$$

$$\le \int_0^1 \frac{y^2 \, dy}{4\gamma^2(\gamma^2 - \epsilon)} = \frac{1}{12\gamma^2(\gamma^2 - \epsilon)}. \tag{9A.9}$$

Using estimates (9A.7), (9A.8), and (9A.9) in (9A.6) yields

$$I(\omega) \le \text{const} \quad \text{for all } \omega^2 \le \gamma^2. \tag{9A.10}$$

Thus (9A.5) and (9A.10) prove (9.9.1.16) and Lemma 9.9.1.1 is established.   □

Remark 9A.1   Estimate (9.9.1.16) is sharp (optimal) in the sense that, say for $\omega^2 > \gamma^2$ and $\epsilon > 0$, we have instead, where $t = y^2 - a_\omega^2$:

$$\int_0^\infty \frac{y^{2+\epsilon} \, dy}{(y^2 + \gamma^2 - \omega^2)^2 + 4\gamma^2\omega^2} \ge \int_{\frac{a_\omega}{2}}^{2a_\omega} \frac{y^{2+\epsilon} \, dy}{(y^2 + \gamma^2 - \omega^2)^2 + 4\gamma^2\omega^2}$$

$$= \frac{1}{2} \int_{-3/4a_\omega^2}^{3a_\omega^2} \frac{\left(t + a_\omega^2\right)^{\frac{1+\epsilon}{2}} dt}{t^2 + 4\gamma^2\omega^2}$$

$$\ge \frac{1}{2} \left(\frac{a_\omega^2}{4}\right)^{\frac{1+\epsilon}{2}} \int_{-3/4a_\omega^2}^{3a_\omega^2} \frac{dt}{t^2 + 4\gamma^2\omega^2}$$

$$= \frac{1}{2} \left(\frac{1}{4}\right)^{\frac{1+\epsilon}{2}} \frac{a_\omega^{1+\epsilon}}{2\gamma\omega} \left[\arctan \frac{t}{2\gamma\omega}\right]_{t \,=\, -3/4a_\omega^2}^{t \,=\, 3a_\omega^2} \to \infty,$$

$$\tag{9A.11}$$

since $a_\omega^{1+\epsilon}/(2\gamma\omega) \sim \omega^{1+\epsilon}/\omega \to +\infty$ as $\omega \to +\infty$, while the arctan term tends to $\pi/2 + \pi/2$ as $\omega \to \infty$, since $a_\omega^2/\omega \sim \omega$.   □

9B   Proof of (9.9.3.14) in Lemma 9.9.3.1

The proof is a generalization of that of Lemma 9.9.1.1.
   (i) We first take $\gamma^2 - \omega^2 < 0$. Set $a_\omega^2 = \omega^2 - \gamma^2 > 0$. Then

$$y^2 + (1 + \rho y)(\gamma^2 - \omega^2) = y^2 - \rho a_\omega^2 y - a_\omega^2 = (y - y_{1\omega})(y + |y_{2\omega}|), \tag{9B.1}$$

where the roots $y_{1\omega} > 0$ and $y_{2\omega} < 0$ are

$$2y_{1,2\omega} = \rho a_\omega^2 \pm \sqrt{\rho^2 a_\omega^4 + 4a_\omega^2}. \tag{9B.2}$$

Remark 9B.1 Note that (where $a_n \sim b_n$ means $c|b_n| \le |a_n| \le C|b_n|$ as $n \to \infty$, with $0 < c < C < \infty$, as usual)

$$\text{for } \rho > 0, \quad y_{1\omega}, |y_{2\omega}| \sim a_\omega^2 \sim \omega^2; \tag{9B.3a}$$

$$\text{for } \rho = 0, \quad y_{1\omega} = |y_{2\omega}| = a_\omega \sim \omega. \tag{9B.3b}$$

The first case (which will apply to Kirchhoff equations) accounts for the improvement of $y^{\frac{5}{2}}$ in the numerator of (9B.1) over y^2 in the numerator of (9.9.1.16) in Section 9.9.1 when $\rho = 0$, which was applied to the wave equation.

We split the integral $I(\omega)$ in (9B.1) as

$$I(\omega) = \int_0^{(1/2)y_{1\omega}} + \int_{(1/2)y_{1\omega}}^{2y_{1\omega}} + \int_{2y_{1\omega}}^\infty = I_1(\omega) + I_2(\omega) + I_3(\omega). \tag{9B.4}$$

For $I_1(\omega)$ we compute, using (9B.1) and recalling (9B.3a),

$$I_1(\omega) \equiv \int_0^{(1/2)y_{1\omega}} \frac{y^{\frac{5}{2}}\,dy}{(y - y_{1\omega})^2(y + |y_{2\omega}|)^2 + 4\gamma^2\omega^2(1 + \rho y)^2}$$

$$\le \int_0^{(1/2)y_{1\omega}} \frac{y^{\frac{5}{2}}\,dy}{\left(\frac{y_{1\omega}}{2}\right)^2 (y_{2\omega})^2} = \frac{\frac{2}{7}\left(\frac{y_{1\omega}}{2}\right)^{\frac{7}{2}}}{\left(\frac{y_{1\omega}}{2}\right)^2 (y_{2\omega})^2} \sim \frac{(\omega^2)^{\frac{7}{2}}}{(\omega^2)^4} \le \text{Const.} \tag{9B.5}$$

For $I_3(\omega)$ we compute, by (9B.1) and $-y_{1\omega} \ge -y/2$,

$$I_3(\omega) = \int_{2y_{1\omega}}^\infty \frac{y^{\frac{5}{2}}\,dy}{(y - y_{1\omega})^2(y + |y_{2\omega}|)^2 + 4\gamma^2\omega^2(1 + \rho y)^2}$$

$$\le \int_0^\infty \frac{y^{\frac{5}{2}}\,dy}{\left(\frac{y}{2}\right)^2 y^2 + 4\gamma^2\omega^2} = \int_0^1 + \int_1^\infty$$

$$\le \frac{1}{4\gamma^2\omega^2} \int_0^1 y^{\frac{5}{2}}\,dy + \int_1^\infty \frac{y^{\frac{5}{2}}\,dy}{\left(\frac{y}{2}\right)^2 y^2} \le \text{Const}, \tag{9B.6}$$

after recalling that $\omega^2 > \gamma^2$.

Finally, for $I_2(\omega)$, we have, recalling (9B.1),

$$I_2(\omega) \equiv \int_{\frac{y_{1\omega}}{2}}^{2y_{1\omega}} \frac{y^{\frac{5}{2}}\,dy}{(y - y_{1\omega})^2(y + |y_{2\omega}|)^2 + 4\gamma^2\omega^2(1 + \rho y)^2}$$

$$\equiv \int_{\frac{y_{1\omega}}{2}}^{2y_{1\omega}} \frac{y^{\frac{5}{2}}\,dy}{(y - y_{1\omega})^2 y^2 + 4\gamma^2\omega^2\rho^2 y^2} \tag{9B.7}$$

(cancelling y^2 and majorizing y with $2y_{1\omega}$ in the numerator)

$$\le (2y_{1\omega})^{\frac{1}{2}} \int_{\frac{y_{1\omega}}{2}}^{2y_{1\omega}} \frac{dy}{(y - y_{1\omega})^2 + 4\gamma^2\omega^2\rho^2}$$

(setting $t = y - y_{1\omega}$)

$$= (2y_{1\omega})^{\frac{1}{2}} \int_{-\frac{y_{1\omega}}{2}}^{y_{1\omega}} \frac{dt}{(t^2 + 4\gamma^2\omega^2\rho^2)}$$

$$= (2y_{1\omega})^{\frac{1}{2}} \frac{1}{2\gamma\omega\rho} \left\{ \arctan\frac{t}{2\gamma\omega\rho} \right\}_{t=-\frac{y_{1\omega}}{2}}^{t=y_{1\omega}}$$

(recalling (9B.3a))

$$I_2(\omega) \le \text{Const}\, \frac{\omega}{\omega}\left(\frac{\pi}{2} + \frac{\pi}{2}\right) \le \text{Const}. \tag{9B.8}$$

Thus, estimates (9B.5) and (9.B6), used in (9B.4), yield

$$I(\omega) \le \text{Const} \quad \text{for all } \omega^2 > \gamma. \tag{9B.9}$$

(ii) Next, we take $\gamma^2 - \omega^2 \ge 0$ and split $I(\omega)$ as

$$I(\omega) = \int_0^1 + \int_1^\infty = J_1(\omega) + J_2(\omega). \tag{9B.10}$$

For $J_2(\omega)$ we obtain, dropping positive terms,

$$J_2(\omega) \le \int_1^\infty \frac{y^{\frac{5}{2}}}{y^4}\, dy < \infty. \tag{9B.11}$$

For $J_1(\omega)$ we first restrict to $0 \le \omega^2 \le \gamma^2 - \epsilon, \epsilon > 0$, and obtain

$$J_1(\omega) \le \int_0^1 \frac{y^{\frac{5}{2}}\, dy}{(y^2 + \epsilon)^2} \le \frac{1}{\epsilon^2} \int_0^1 y^{\frac{5}{2}}\, dy < \infty, \tag{9B.12}$$

and we next consider $\gamma^2 - \epsilon < \omega^2 \le \gamma^2$ and obtain

$$J_1(\omega) \le \int_0^1 \frac{y^{\frac{5}{2}}\, dy}{4\gamma^2\omega^2} \le \frac{1}{4\gamma^2(\gamma^2 - \epsilon)} \int_0^1 y^{\frac{5}{2}}\, dy < \infty. \tag{9B.13}$$

Using estimates (9B.11), (9B.12), and (9B.13) in (9B.10) yields

$$I(\omega) \le \text{Const} \quad \text{for all } \omega^2 \le \gamma^2. \tag{9B.14}$$

Thus (9B.10) and (9B.14) prove (9.9.3.14) and Lemma 9.9.3.1 is established. $\square$

Remark 9B.2   Estimate (9.3.14) is sharp (optimal) in the sense that, say for $\omega^2 > \gamma^2$, we have instead with $\epsilon > 0$, recalling identity (9B.1),

$$I_{2,\epsilon}(\omega) \equiv \int_{(1/2)y_{1\omega}}^{2y_{1\omega}} \frac{y^{\frac{5}{2}+\epsilon}\, dy}{(y - y_{1\omega})^2(y + |y_{2\omega}|)^2 + 4\gamma^2\omega^2(1 + \rho y)^2} \to \infty$$

$$\text{as } \omega \to \infty. \tag{9B.15}$$

In fact, replacing y with $y_{1\omega}/2$ in the numerator of (9B.15), and y with $2y_{1\omega}$ in two spots in the denominator, we obtain

$$I_{2,\epsilon}(\omega) \geq \left(\frac{y_{1\omega}}{2}\right)^{\frac{5}{2}+\epsilon} \int_{(1/2)y_{1\omega}}^{2y_{1\omega}} \frac{dy}{(y - y_{1\omega})^2 (2y_{1\omega} + |y_{2\omega}|)^2 + 4\gamma^2 \omega^2 (1 + \rho 2 y_{1\omega})^2}$$

(setting $t = y - y_{1\omega}$, $b_\omega = 2y_{1\omega} + |y_{2\omega}|$, and $a_\omega^2 = 4\gamma^2\omega^2(1 + \rho 2 y_{1\omega})^2$)

$$= \left(\frac{y_{1\omega}}{2}\right)^{\frac{5}{2}+\epsilon} \int_{-(1/2)y_{1\omega}}^{y_{1\omega}} \frac{dt}{t^2 b_\omega^2 + a_\omega^2}$$

$$= \left(\frac{y_{1\omega}}{2}\right)^{\frac{5}{2}+\epsilon} \frac{1}{a_\omega b_\omega} \left[\arctan \frac{b_\omega t}{a_\omega}\right]_{t=-(1/2)y_{1\omega}}^{t=y_{1\omega}} \to \infty, \qquad (9B.16)$$

as $\omega \to \infty$, since

$$\left(\frac{y_{1\omega}}{2}\right)^{\frac{5}{2}+\varepsilon} \frac{1}{a_\omega b_\omega} = \left(\frac{y_{1\omega}}{2}\right)^{\frac{5}{2}+\varepsilon} \frac{1}{2\gamma\omega(1 + \rho^2 y_{1\omega})(2y_{1\omega})(2y_{1\omega} + |y_{2\omega}|)}$$

$$\sim \frac{(\omega^2)^{\frac{5}{2}+\epsilon}}{\omega\omega^2\omega^2} \to \infty \quad \text{as } \omega \to \infty,$$

while the arctan term in (9B.16) tends to $(\pi/2 + \pi/2)$ as $\omega \to \infty$ since $b_\omega y_{1\omega}/a_\omega \sim \omega^2\omega^2/\omega\omega^2$. □

Notes on Chapter 9

Sections 9.1–9.3

The *variational* approach of Sections 9.3.1 through 9.3.5 follows, in the present new set of assumptions (A.0), (A.1), (A.2), and (A.3) inspired by Da Prato, et al. [1986], the treatment of Chapter 8, except for the additional presence of the finite state penalization via the operator G. The detours required by $G \neq 0$ can be exported also to the original setting of Chapter 8, and we leave this to the reader. Further comments on the few differences of approach between $G = 0$ and $G \neq 0$ are given below. The *direct* approach of Sections 9.3.6 follows closely that of Da Prato, et al. [1986] except for adding the final state penalization operator G. This requires some changes of strategy, illustrated, for example, by the proof of Lemma 9.3.2.2 and by the proof of Theorem 9.2.3 in Section 9.3.7. Qualitatively, the guiding philosophy is as follows: If one seeks to show certain properties, say the transition properties (9.3.2.4), or (9.2.24), for $y^0(\cdot, s; x) = \Phi(\cdot, s)x$ [the latter being needed to establish the transition property (9.2.25) for $P(\cdot)$], then this can be done by working directly on $y^0(\cdot, s; x) = \Phi(\cdot, s)x$ in the case $G = 0$, by use of formula (9.2.3), which for $G = 0$ is expressed solely in terms of the problem data. Instead, in the case $G \neq 0$, one begins by working on u^0 and first shows corresponding properties for $u^0(\cdot, s; x)$, say the transition properties (9.3.24), or (9.2.23), by use of formula (9.2.2a), which is explicit also for $G \neq 0$; next one transfers to y^0 said properties of u^0, via (9.2.3).

Section 9.4

The interesting relationship in Section 9.4 between the isomorphism of $P_T(t)$ and the exact controllability of $\{A^*, R^*\}$ was established by each of the authors of Flandoli, et al. [1988] (see also Flandoli [1987(a)]). The proof given in Section 9.4, based on the explicit formula (9.4.9) for the cost functional, is new.

Section 9.5

The regularization procedure of R as given in Section 9.5 much expands on an original idea in [Chang, Lasiecka, 1986, Corollary 9.2.6] and was also reported in Lasiecka and Triggiani [1991(a)].

Section 9.6

The idea of introducing the dual (integral) Riccati equation (9.6.2.12) for $Q(t)$ (easier to use than the primal equation (9.2.22) for $P(t)$, since the quadratic term of (9.6.2.12) involves a bounded operator R), as a way of gaining insight in the original, primal, more demanding Riccati equation (9.2.22) via $P(t) = Q^{-1}(t)$, originated in Flandoli [1987(b)]. More precisely, with G^*G assumed initially an isomorphism, Flandoli [1987(b)] then considers the dual IRE (9.6.2.12) with $\mathcal{G}^*\mathcal{G} = (G^*G)^{-1}$, shows by the direct method existence and uniqueness and, moreover, shows that $Q(t)$ is an isomorphism on $0 \leq t \leq T$. Then, setting $P(t) = Q^{-1}(t)$, Flandoli [1987(b)] shows that such $P(t)$ possesses the following properties: (i) It satisfies the optimal cost formula (9.2.18); (ii) it satisfies the explicit formula (9.2.14); and (iii) it satisfies the synthesis (9.2.17). However, there is no claim that $P(t)$ satisfies the original Riccati equation (9.2.21) or (9.2.22) (except formally). The case where G^*G is not an isomorphism is reduced to the latter by considering $(G^*G + \epsilon I)$, obtaining $Q_\epsilon(t)$, $P_\epsilon(t) = Q_\epsilon^{-1}(t)$, and taking the limit $\lim_\epsilon P_\epsilon(t) = P(t)$. Our discussion in Section 9.6 of the dual Riccati equation (9.6.2.12) is primarily directed toward providing the necessary background for the limit case $T = \infty$ to be treated in Chapter 11, Sections 11.7 and 11.8 of Volume 3. Here the relationships between the algebraic Riccati operator P_∞ and the dual algebraic Riccati operator Q_∞ will be analyzed.

A different dual Riccati equation, say for $Q_1(t)$, is considered in Barbu and Da Prato [1992], through a different approach that relates $Q_1(t)$ to the original $P(t)$.

However, again, there is no claim that $P(t)$ is the solution of the original Riccati equation.

Sections 9.8 and 9.9

On the application side, the regularity theory for wave and Kirchhoff equations of Sections 9.8 and 9.9 is taken directly from Triggiani [1991; 1993a, b], where additional results are given (different boundary conditions for the Kirchhoff equation, etc.).

As noted already in Remark 9.8.1.2, in the case of the wave equation for dim $\Omega = n = 3$ of Section 8.1, several proofs have been given of the interior regularity of Theorem 9.8.1.1. Three different proofs are reported in Lions [1988]. One proof, due to Meyer [1985], uses harmonic analysis; the proofs of Lions [1992] and Nirenberg are based on the explicit solution formula (generally referred to as Kirchhoff formula) for the corresponding free space problem in R^3. Nirenberg's proof refers to the dual problem. Here we have opted for the proofs of Triggiani [1991; 1993a, b], since they are more flexible, as they apply to wave, Kirchhoff equations, etc. in principle on any dimension, as no explicit solution formula is needed. The method of Triggiani [1991; 1993a, b], and some of his key preliminary estimates (such as Lemma 9.9.1.1 for wave equations and Lemma 9.9.3.1 for Kirchhoff equations) have subsequently been used by Jafford and Tucsnak [1995] to study regularity in the case of control concentrated on interior curves.

Section 9.10

This section follows closely the work of Camurdan and Triggiani [1999], where a direct proof of Theorem 9.10.1.2 is given, which is reported here. A different proof of the dual statement was previously given in Triggiani [1997]. The direct proof is preferable. Theorem 9.10.4.3 for the damped model is due to Camurdan [1997]. A generalization to the three-dimentional chamber appears in Camurdan and Ji [1998].

The basic structure of acoustic flow models has been known for a long time; see, for example [Mores, Ingard, 1968, example on p. 263]. Some related mathematical questions regarding spectral properties of such model are studied in Beale [1976], Balakrishnan [1981]. The key contribution of smart materials technology to the modeling issue, as supplied, for example, by NASA related groups [Banks, et al., 1993; 1995], appears to be the presence of (finitely many) δ' on the moving wall, caused by the action of wired piezo-ceramic patches. The models considered by these groups, however, included a (Kelvin–Voigt) high structural damping on the moving wall, such as occurs in Eqn. (9.11.1.1d) with $\gamma = 0$ (Euler–Bernoulli equation with Kelvin–Voigt damping). This strong damping then causes the dynamics of the moving wall to be *parabolic* (by an immediate use of the results of Chen and Triggiani reported in Chapter 3, Appendix 3B, as specialized to this easy case). The model of Section 9.10 considers a more realistic *hyperbolic* Kirchhoff equation for the moving wall.

Section 9.11

This section where the model of the moving wall has a Kelvin–Voigt damping is taken from Lasiecka [1998] and Lasiecka and Triggiani [1999].

References and Bibliography

A. V. Balakrishnan, *Applied Functional Analysis*, 2nd ed., Springer-Verlag, 1981.

H. T. Banks, R. J. Silcox, and R. C. Smith, The modeling and control of acoustic/structure interaction problems via piezo-ceramic actuators: 2-D numerical examples, *ASME J. Vibration Acoustics* **2** (1993), 343–390.

H. T. Banks, R. C. Smith, and Y. Wang, The modeling of piezo-ceramic patch interactions with shells, plates and beams, *Quarterly of Applied Mathematics* **53**(2) (1995), 353–381.

V. Barbu and G. Da Prato, A representation formula for the solutions to the operator Riccati equation, *Diff. Int. Eqn.* **5** (4) (1992), 821–829.

J. T. Beale, Spectral properties of an acoustic boundary condition, *Indiana Univ. Math. J.* **25** (1976), 895–917.

M. Camurdan, Uniform stabilization of a coupled structural acoustic system with boundary dissipation, presented at Conference held at University of Tennessee, Knoxville, March 1997, *Abstract and Applied Analysis*, To appear. Part of a Ph.D. thesis, University of Virginia, May 1999.

M. Camurdan and G. Ji, Uniform stabilization of a 3-d structural acoustic model via boundary dissipation, *Proceedings CDC Conference*, Dec 1998, Tampa, FL. Decision & Control Conference. To appear as a full paper.

M. Camurdan and R. Triggiani, Sharp regularity of a coupled system of a wave and a Kirchhoff equation with point control, arising in noise reduction, March 1997 (presented at SPIE Conference, San Diego, March 1997), *Diff. Int. Eqns.*, Vol. 12, No. 1 (Jan 1999), 101–118.

S. Chang and I. Lasiecka, Riccati equations for non-symmetric and non-dissipative hyperbolic systems with L_2-boundary control, *J. Math. Analy. Appl.* **116** (1986), 378–414.

G. Da Prato, I. Lasiecka, and R. Triggiani, A direct study of Riccati equations arising in boundary control problems for hyperbolic equations, *J. Diff. Eqn.* **64** (1986), 26–47.

G. Doetsch, *Introduction to the Theory and Applications of the Laplace Transformation*, Springer-Verlag, 1974.

F. Flandoli, Invertibility of Riccati operators and controllability of related systems, *Systems Control Lett.* **9** (1987)(a), 65–72.

F. Flandoli, A new approach to the LQR problem for hyperbolic dynamics with boundary control, Springer-Verlag LNCIS **102** (1987)(b), 89–111.

F. Flandoli, I. Lasiecka, and R. Triggiani, Algebraic Riccati equations with non-smoothing observation arising in hyperbolic and Euler–Bernoulli boundary control problems, *Ann. Matemat. Pura Appli.* (iv), **CLIII** (1988), 307–382.

P. Grisvard, A characterization de quelques espaces d'interpolation, *Arch. Rat. Mech. Anal.* **25** (1967), 40–63.

S. Jafford and M. Tucsnak, Regularity of plate equations with control concentrated on interior curves, *Centre de Mathematiques Appliques* R. I., No. 325, June 1995, Ecole Polytechnique.

I. Lasiecka, Mathematical control theory in structural acoustic problems, *Mathematical Methods in Applied Science*, Vol. 8, No. 7, pp. 1119–1153, 1998 (based on the invited Barret lectures given at the University of Tennessee, Knoxville, March 1997).

I. Lasiecka, Boundary stabilization of a 3-d structural acoustic model, *J. Math. Pures et Appl.* **78** (1999), 203–232.

I. Lasiecka and R. Triggiani, A cosine operator approach to modelling $L_2(0, T; L_2(\Gamma))$-boundary input hyperbolic equations, *Appl. Math. Optim.* **7** (1981), 35–83.

I. Lasiecka and R. Triggiani, Regularity of hyperbolic equations under $L_2(0, T; L_2(\Gamma))$-Dirichlet boundary terms, *Appl. Math. Optim.* **10** (1983), 275–286.

I. Lasiecka and R. Triggiani, Riccati equations for hyperbolic partial differential equations with $L_2(\Sigma)$-Dirichlet boundary terms, *SIAM J. Control Optim.* **24** (1986), 884–926.

I. Lasiecka and R. Triggiani, A lifting theorem for the time regularity of solutions to abstract equations with unbounded operators and applications to hyperbolic equations, *Proc. Am. Math. Soc.* **10** (1988), 745–755.

I. Lasiecka and R. Triggiani, Trace regularity of the solutions of the wave equation with homogeneous Neumann boundary conditions and compactly supported data, *J. Math. Analy. Appl.* **141** (1989), 49–71.

I. Lasiecka and R. Triggiani, Sharp regularity theory for second-order hyperbolic equations of Neumann type, Part I: L_2-nonhomogeneous data, *Ann. Matem. Pura Appl.* **CLVII** (1990), 267–285.

I. Lasiecka and R. Triggiani, *Differential and Algebraic Riccati Equations with Applications to Boundary/Point Control Problems: Continuous Theory and Approximation Theory*, Vol. 164, Springer-Verlag Lecture Notes Series, 1991(a); 160 pp.

I. Lasiecka and R. Triggiani, Regularity theory of hyperbolic equations with nonhomogeneous Neumann boundary conditions, Part II: General boundary data, *J. Diff. Eqn.* **94** (1991(b)), 112–164.

I. Lasiecka and R. Triggiani, Feedback noise control in an acoustic chamber: Mathematical theory, invited paper in *Nonlinear Problems in Aviation and Aerospace*, Stability and Control Theory Methods and Applications: Volume 12, Gordon and Breach Science Publishers, S. Sivasundaram, ed., to appear in 1999.

I. Lasiecka, J. L. Lions, and R. Triggiani, Nonhomogeneous boundary value problems for second order hyperbolic operators, *J. Math. Pures Appl.* **65** (1986), 149–192.

J. L. Lions, Exact controllability, stabilization and perturbations for distributed systems, *SIAM Review* **30** (1988), 1–68.

J. L. Lions, Pointwise control for distributed systems, in *Control and Estimation in Distributed Parameter Systems*, H. T. Banks, ed., SIAM, 1992, 1–41.

J. L. Lions and E. Magenes, *Nonhomogeneous Boundary Value Problems, I,* Springer-Verlag, 1972.

Y. Meyer, Etude a un modele mathematique issu du contrôle des structures spatiales déformables, in *Nonlinear Partial Differential Equations and Their Applications*, College de France Seminar, Vol. II, H. Brezis and J. L. Lions eds., Research Notes in Mathematics, Pitman, 1985, 234–242.

P. M. Mores and K. Ingard, *Theoretical Acoustics*, McGraw-Hill, 1968.

A. Pazy, *Semigroups of linear operators and applications to partial differential equations*, Springer-Verlag, New York Berlin Heidelberg Tokyo, 1983

R. Triggiani, A cosine operator approach to modeling boundary input problems for hyperbolic systems, Proceedings of 8th IFIP Conference on Optimization Techniques, University of Wurzburg, West Germany, September 1977, Springer-Verlag Lecture Notes on Control Sciences N.6 (1978), 380–390.

R. Triggiani, Regularity with point control, Part I: Wave equations and Bernoulli–Euler equations, Springer-Verlag LNCIS, No. 178, J. P. Zolesio ed., (1991), 321–355.

R. Triggiani, Interior and boundary regularity of the wave equation with interior point control, *Diff. Int. Eqn.* **6** (1993(a)), 111–129.

R. Triggiani, Regularity with point control, Part II: Kirchhoff equations, *J. Diff. Eqn.* **103** (1993(b)), 394–420.

R. Triggiani, Control problems in noise reduction: The case of two coupled hyperbolic equations, Proceedings of SPIE Conference, San Diego, March 1997, **3039**, 382–392.

A. E. Taylor and D. C. Lay, *Introduction to Functional Analysis,* 2nd ed., Wiley, 1980.

10

Differential Riccati Equations under Slightly Smoothing Observation Operator. Applications to Hyperbolic and Petrowski-Type PDEs. Regularity Theory

The present chapter continues once more the study of the quadratic optimal control problem over a finite horizon, $T < \infty$, this time with no final state penalization. This investigation was already initiated in Chapter 8 and pursued further in Chapter 9 (where $G \neq 0$) under two different abstract settings, each motivated by classes of applications to PDEs of hyperbolic and Petrowski type, with boundary/point control. The present Chapter 10 provides yet another abstract framework, which in fact specializes, and is properly contained in, that of Chapter 9 (with $G = 0$). By way of providing a justification, we recall that the treatment of Chapter 9, with no final state penalization ($G = 0$), rests on just one *dynamical* assumption imposed on the pair $\{A, B\}$ of the free dynamics operator and the control operator, that is, the "abstract trace regularity" assumption stated in Chapter 9, (A.1) = (9.1.7). In return, that study assumes a requirement of "smoothing" on the part of the (bounded) observation operator R, namely, the L_1-condition that: $R^* Re^{At} Bu \in L_1(0, T; Y)$, $u \in U$ [see Chapter 9, (A.2) = (9.1.8)]. To appreciate more explicitly the quantitative degree of smoothing of $R^* R$ included in such a condition, it suffices to consider the canonical case of the wave equation (or more generally a second-order hyperbolic equation) with Dirichlet boundary control. This is a typical case where the abstract trace condition is satisfied, as we shall see in Section 10.5 of the present chapter. Then, in this case, the aforementioned L_1-condition in Chapter 9, (A.2) = (9.1.8) is satisfied when the smoothing of $R^* R$ is comparable to $\mathcal{A}^{-(\frac{1}{4}+\epsilon)}$ for each of the two coordinates (position and velocity), $R^* R \sim \mathcal{A}^{-(\frac{1}{4}+\epsilon)} \times \mathcal{A}^{-(\frac{1}{4}+\epsilon)}$, $\epsilon > 0$ arbitrary, where $\mathcal{A} = -\Delta$ with homogeneous Dirichlet BC. *The goal of the present chapter is then to markedly reduce the requirement of smoothing on $R^* R$ imposed in Chapter 9.* In the wave equation example in question, the present chapter will require only the smoothing $R^* R \sim \mathcal{A}^{-\epsilon} \times \mathcal{A}^{-\epsilon}$, $\epsilon > 0$ arbitrary.

Chapter 9 gives additional PDE examples with point/boundary control, where the L_1-condition of assumption (A.2) = (9.1.8) requires a finite, quantitatively identified degree of smoothing on $R^* R$. These include: (i) the wave equation with interior point control and homogeneous Dirichlet boundary conditions [Chapter 9, Section 9.8.1, Eqn. (9.8.1.16)]; (ii) the Kirchoff equation with interior point control and

919

homogeneous boundary conditions in the position and the "moment" [Chapter 9, Section 9.8.3, Eqn. (9.8.3.22)]; and (iii) the one-dimensional wave equation with Neumann boundary control [Chapter 9, Section 9.8.4, Eqn. (9.8.4.31)].

To deal with the nonsmoothing observation R, Chapter 9, Section 9.5 introduced the notion of limit solution of the differential Riccati equations.

To relax the smoothing of R^*R from the L_1-condition of Chapter 9 to a condition of "ϵ-smoothing" of the present chapter (technically, (H.8) = (10.1.23) below), the treatment of the present chapter extracts and singles out further *dynamical* properties on the pair $\{A, B\}$, namely a set of seven dynamical hypotheses, (H.1) through (H.7) below, (H.1) being the abstract trace regularity of Chapter 9. Accordingly, Chapter 10 is conceptually divided into two parts. The first part provides the abstract study, by the variational approach, and culminates with the final, sought-after goal of asserting that the operator $P(t)$ of the optimal cost, given constructively by Eqn. (10.2.3) below, is indeed a solution of the differential Riccati equation (10.2.6) or of the integral Riccati equation (10.2.7) below. The second part then verifies that all the above abstract hypotheses (H.1) through (H.7) are nothing but *intrinsic dynamical properties* of the class of hyperbolic and Petrowski-type PDE problems with boundary control, which motivate this chapter in the first place. Numerous PDE illustrations are given, to which we show that the abstract theory applies. Compared with the abstract framework of Chapter 9, it will be clear that the present Chapter 10 makes use also of additional *higher level* "interior" and "boundary" regularity properties of the dynamics. The resulting abstract theory leads to a drastic reduction on the smoothing of R^*R and thus represents a quantitative jump over the abstract theory of Chapter 9. This is perhaps best illustrated by the few main new features distinctive of the present theory: (i) The gain operator $B^*P(t)$ is no longer bounded (as in Chapters 8 and 9) from the state space Y to the control space U; however, $B^*P(t)$ is densely defined in Y, with a domain that is independent of t, and which is an explicitly identified subspace Y_δ^- of Y (see conclusion (10.2.4) of Theorem 10.2.2 below), where Y_δ^- is related to the assumption (H.8) = (10.1.23) on R^*R; (ii) the usual candidate $P(t)$ of the optimal value satisfies the differential Riccati equation for all x, y in Y_δ^- (rather than for all $x, y \in \mathcal{D}(A)$ as in Chapters 8 and 9), where $\mathcal{D}(A) \subset Y_\delta^-$ by (H.6); (iii) no claim of uniqueness of the (nonnegative) solution of the DRE or IRE is made in the present degree of generality, unlike the situation of both Chapters 8 and 9, where uniqueness was asserted for the solution $P(t)$.

10.1 Mathematical Setting and Problem Statement

Dynamical Model   As in preceding chapters, we consider the abstract differential equation

$$\dot{y} = Ay + Bu \quad \text{on, say,} \ [\mathcal{D}(A^*)]'; \quad y(s) = y_0 \in Y, \tag{10.1.1}$$

or its mild version

$$y(t, s; y_0) = e^{A(t-s)}y_0 + (L_s u)(t), \tag{10.1.2a}$$

$$(L_s u)(t) = \int_s^t e^{A(t-\tau)}Bu(\tau)\,d\tau, \tag{10.1.2b}$$

where $0 \le s \le T < \infty$, subject to the abstract hypotheses listed below.

Optimal Control Problem on the Interval $[s, T]$ We introduce the cost functional

$$J(u, y) = \int_s^T \left[\|Ry(t)\|_Z^2 + \|u(t)\|_U^2\right] dt, \tag{10.1.3}$$

as in Chapter 9, Eqn. (9.1.13), and the corresponding optimal control problem (OCP) is then:

Minimize $J(u, y)$ over all $u \in L_2(s, T; U)$,   where $y(t) = y(t, s; y_0)$
is the solution of Eqn. (10.1.1) with initial condition $y(s) = y_0$.   (10.1.4)

We now list the abstract assumptions of the present chapter.

Abstract Assumptions   We first group together in (i) below some standing preliminary basic assumptions:

(i) U, Y, and Z are Hilbert spaces; A is the generator of a s.c. semigroup e^{At} on Y, $t \ge 0$; B is a (linear) continuous operator $U \to [\mathcal{D}(A^*)]'$, equivalently,

$$A^{-1}B \in \mathcal{L}(U; Y). \tag{10.1.5a}$$

[Without loss of generality, we take $A^{-1} \in \mathcal{L}(Y)$.] In (10.1.1), A^* is the Y-adjoint of A, and $[\mathcal{D}(A^*)]'$ is the Hilbert space dual to the space $\mathcal{D}(A^*) \subset Y$ with respect to the Y-topology, with norms

$$\|y\|_{\mathcal{D}(A^*)} = \|A^*y\|_Y, \quad \|y\|_{[\mathcal{D}(A^*)]'} = \|A^{-1}y\|_Y. \tag{10.1.5b}$$

Via (10.1.5a), we let $(B^*x, u)_U = (x, Bu)_Y$ for $u \in U$, $x \in \mathcal{D}(A^*)$, and then $B^* \in \mathcal{L}(\mathcal{D}(A^*); U)$.

(H.1):   (Abstract trace regularity) The (closable) operator $B^*e^{A^*t}$ can be extended as a map

$$B^*e^{A^*t} : \text{continuous } Y \to L_2(0, T; U); \tag{10.1.6a}$$

$$\int_0^T \|B^*e^{A^*t}x\|_U^2\,dt \le c_T\|x\|_Y^2, \quad x \in Y. \tag{10.1.6b}$$

Consequently, as seen in Chapter 7, Theorem 7.2.1, it follows equivalently that the operator L_s in (10.1.2b) satisfies:

(H.1*):

$$L_s : \text{continuous } L_2(s, T; U) \to C([s, T]; Y)$$
$$\text{with a norm that may be made independent of } s, \qquad (10.1.7a)$$

that is,

$$\|L_s u\|_{C([s,T];Y)} \le c_T \|u\|_{L_2(s,T;U)} \text{ uniformly in } s. \qquad (10.1.7b)$$

Then, the operator L_s^*, adjoint of L_s in the sense that $(L_s u, f)_{L_2(s,T;Y)} = (u, L_s^* f)_{L_2(s,T;U)}$, and thus given by

$$(L_s^* f)(t) = \int_t^T B^* e^{A^*(\tau - t)} f(\tau) \, d\tau, \qquad (10.1.8)$$

satisfies

$$L_s^* : \text{continuous } L_1(s, T; Y) \to L_2(s, T; U)$$
$$\text{with a norm that may be made independent of } s, \qquad (10.1.9a)$$

that is,

$$\|L_s^* f\|_{L_2(s,T;U)} \le c_T \|f\|_{L_1(s,T;Y)} \text{ uniformly in } s. \qquad (10.1.9b)$$

Distinctive new hypotheses of the present chapter are as follows: There exist families of Hilbert spaces (which in applications to PDEs are Sobolev spaces)

$$\begin{cases} U_\theta; \ Y_\theta, \ 0 \le \theta \le 1/2 + \delta, \text{ for some } \delta > 0, \ \theta \ne 1/2 \text{ for } Y_\theta; \\ U_0 = U; Y_0 = Y; \delta \text{ henceforth kept fixed,} \\ \text{with the property that:} \\ \text{injection } U_{\theta_2} \to U_{\theta_1} \text{ and } Y_{\theta_2} \to Y_{\theta_1} \text{ is compact, } 0 \le \theta_1 < \theta_2 \le 1/2 + \delta, \\ \hspace{8cm} (10.1.10) \\ \text{and the interpolating property} \\ \left[Y_{\frac{1}{2} - \delta}, \left[Y_{\frac{1}{2} + \delta} \right]' \right]_{\theta = \frac{1}{2} - \delta} = Y, \quad \delta > 0, \end{cases}$$

with duality of $[Y_{\frac{1}{2} + \delta}]'$ with respect to Y, such that, setting

$$\mathcal{U}^\theta[s, T] \equiv L_2(s, T; U_\theta) \cap H^\theta(s, T; U), \qquad (10.1.11)$$

$$\mathcal{Y}^\theta[s, T] \equiv L_2(s, T; Y_\theta) \cap H^\theta(s, T; Y), \qquad (10.1.12)$$

where $u \in H^\theta(s, T; U)$ means that the fractional time derivative $D_t^\theta u \in L_2(s, T; U)$, as usual, with norm

$$\|u\|_{\mathcal{U}^\theta[s,T]}^2 \equiv \|u\|_{L_2(s,T;U_\theta)}^2 + \|u\|_{H^\theta(s,T;U)}^2, \qquad (10.1.13)$$

and similarly for $\mathcal{Y}^\theta[s, T]$, then:

(H.2):

$$L_s : \text{continuous } \mathcal{U}^\theta[s, T] \to \mathcal{Y}^\theta[s, T] \cap C([s, T]; Y_\theta), \quad 0 \le \theta < 1/2,$$

with a norm that may be made independent of s, $\qquad$ (10.1.14a)

that is,

$$\|L_s u\|_{\mathcal{Y}^\theta[s,T] \cap C([s,T],Y_\theta)} \le c_{T,\theta} \|u\|_{\mathcal{U}^\theta[s,T]} \text{ uniformly in } s, \quad 0 \le \theta < 1/2.$$
$$(10.1.14b)$$

[For $\theta = 0$, (H.2) = (10.1.14) specializes to (H.1*) = (10.1.7), via (10.1.10).]

Remark 10.1.1 In applications to mixed problems for PDEs [see Sections 10.5 through 10.9 below], one first establishes (H.1) = (10.1.6), hence the regularity (H.1*) = (10.1.7) for L_s [case $\theta = 0$]; next, one establishes a regularity result for L_s for $\theta = 1$ involving the spaces U_1 and Y_1, which, however, requires a compatibility condition. In interpolating between the two above cases $\theta = 0$ and $\theta = 1$ for $\theta < 1/2$, the compatibility condition is irrelevant, and one thus obtains (H.2) = (10.1.14).

We also assume

(H.3):

$$L_s^* : \text{continuous } L_2(s, T; Y_\theta) \to \mathcal{U}^\theta[s, T], \quad 0 \le \theta \le 1/2 + \delta, \quad \theta \ne 1/2;$$

with a norm that may be made independent of s, $\qquad$ (10.1.15a)

that is,

$$\|L_s^* f\|_{\mathcal{U}^\theta[s,T]} \le C_{T,\theta} \|f\|_{L_2(s,T;Y_\theta)}, \quad \text{uniformly in } s \quad (10.1.15b)$$

[for $\theta = 0$, (H.3) = (10.1.15) is contained in (10.1.9)]. Henceforth, we shall fix once and for all a number $\delta > 0$ arbitrarily small and set for convenience

$$Y_\delta^- \equiv Y_{\theta=\frac{1}{2}-\delta} \supset Y_\delta^+ \equiv Y_{\theta=\frac{1}{2}+\delta}, \qquad (10.1.16)$$

$$U_\delta^- \equiv U_{\theta=\frac{1}{2}-\delta} \supset U_\delta^+ \equiv U_{\theta=\frac{1}{2}+\delta}. \qquad (10.1.17)$$

[The values $\theta = 1/2 \pm \delta$ and $\theta = 1/2 - \delta/2$ will be the only values of θ where the assumptions (H.2) = (10.1.14) and (H.3) = (10.1.15) will be used.]

Remark 10.1.2 In applications to mixed problems for PDEs, passage from $Y_\delta^- = Y_{\frac{1}{2}-\delta}$ to $Y_\delta^+ = Y_{\frac{1}{2}+\delta}$ may represent a jump across compatibility conditions, as in the applications of Sections 10.5–10.7. The meaning of the assumption on R^*R, made in (H.8) = (10.1.23) below, may be precisely this: to perform a passage to bypass compatibility conditions.

Our additional hypotheses are:

(H.4)   (complementing (10.1.6)):

$$B^* e^{A^* t} : \text{continuous } Y_\delta^+ \to C([0, T]; U); \qquad (10.1.18)$$

(H.5):

$$e^{At} \text{ is also a s.c. semigroup on } Y_\delta^- \text{ and } \mathcal{D}(A) \text{ is dense}$$
$$\text{in } Y_\delta^- \text{ in the } Y_\delta^- \text{-topology;} \qquad (10.1.19)$$

(H.6):

$$A : \text{continuous } Y_\delta^- \to [Y_\delta^+]', \qquad (10.1.20)$$

where the duality $[Y_\delta^+]'$ of Y_δ^+ is with respect to the space Y.

(H.7)   (complementing (10.1.5a)):

$$B : \text{continuous } U_\delta^- \to [Y_\delta^+]' \qquad (10.1.21)$$

[which is automatically implied by

$$A^{-1} B : \text{continuous } U_\delta^- \to Y_\delta^- \qquad (10.1.22)$$

via assumption (H.6) = (10.1.20)];

(H.8)   (assumption on smoothing observation):

$$R \in \mathcal{L}(Y; Z) \quad \text{and} \quad R^* R : \text{continuous } Y_\delta^- \to Y_\delta^+, \qquad (10.1.23)$$

which then, by duality, implies

$$R^* R : \text{continuous } [Y_\delta^+]' \to [Y_\delta^-]'. \qquad (10.1.24)$$

Remark 10.1.3   As already noted, in the applications in the subsequence sections to PDEs with boundary control, the spaces Y_δ^- and Y_δ^+ are Sobolev spaces (which may coincide with domains of appropriate fractional powers of the basic differential operator) invariant under the action of the semigroups e^{At} and $e^{A^* t}$. More insight on the impact of the smoothing assumption $R^* R \in \mathcal{L}(Y_\delta^-; Y_\delta^+)$ in (10.1.23) is provided in the orientation below. Needless to say, for the class of boundary control problems for PDEs for which this chapter is intended [hyperbolic dynamics, platelike equations; see Sections 10.5–10.9 below], all basic assumptions (H.1) through (H.7) on A and B are nothing but *intrinsic dynamical properties*.

Preliminary, Direct Consequences of the Assumptions   Some preliminary, direct consequences of the abstract assumptions, to be invoked in the following, are listed next.

(C.1): Putting together (H.2) = (10.1.14) and (H.3) = (10.1.15) for $\theta = 1/2 - \delta/2$, we obtain, via (10.1.11), (10.1.12), and (10.1.16),

$$L_s L_s^* : \text{continuous } L_2\left(s, T; Y_{\frac{\delta}{2}}^-\right) \to C\left([s, T]; Y_{\frac{\delta}{2}}^-\right) \cap H^{\frac{1}{2}-\frac{\delta}{2}}(s, T; Y)$$

$$\subset \mathcal{Y}^{\frac{1}{2}-\frac{\delta}{2}}[s, T] \text{ with a norm that may be made independent of } s,$$

$$\text{(10.1.25a)}$$

that is,

$$\|L_s L_s^* f\|_{C([s,T];Y_{\frac{\delta}{2}}^-)} + \|L_s L_s^* f\|_{\mathcal{Y}^{\frac{1}{2}-\frac{\delta}{2}}[s,T]} \le C_{T,\delta} \|f\|_{L_2(s,T;Y_{\frac{\delta}{2}}^-)}, \quad \text{uniformly in } s.$$

$$\text{(10.1.25b)}$$

(C.2): By differentiating (10.1.2b) in t, we obtain

$$\left(\frac{dL_s u}{dt}\right)(t) = A\left[\int_s^t e^{A(t-\tau)} Bu(\tau)\, d\tau + A^{-1} Bu(t)\right] \quad \text{(10.1.26a)}$$

$$= A(L_s u)(t) + Bu(t). \quad \text{(10.1.26b)}$$

Of the possible regularity results that may be given on (10.1.26), we point out the following one, to be invoked below. By recalling (H.2) = (10.1.14) on L_s for $\theta = 1/2 - \delta$ and (H.6) = (10.1.20) and (H.7) = (10.1.21) on A and B, respectively, we obtain, via (10.1.26b) and (10.1.11),

$$\frac{dL_s}{dt} = AL_s + B : \text{continuous } \mathcal{U}^{\frac{1}{2}-\delta}[s, T] \to L_2(s, T; [Y_\delta^+]')$$

$$\text{with a norm that may be made independent of } s, \quad \text{(10.1.27a)}$$

that is,

$$\left\|\frac{dL_s}{dt} u\right\|_{L_2(s,T;[Y_\delta^+]')} \le C_{T,\delta} \|u\|_{\mathcal{U}^{\frac{1}{2}-\delta}[s,T]}, \quad \text{uniformly in } s. \quad \text{(10.1.27b)}$$

(C.3): The assumption that $\mathcal{D}(A)$ is dense in Y_δ^-, made in (H.5) = (10.1.19), implies that: Given $x \in Y_\delta^-$, there exists $x_n \in \mathcal{D}(A)$ such that $\|x_n - x\|_{Y_\delta^-} \to 0$ and

$$\|e^{At} Ax_n\|_{C([0,T];[Y_\delta^+]')} = \|Ae^{At} x_n\|_{C([0,T];[Y_\delta^+]')} \le C_T \|x_n\|_{Y_\delta^-}, \quad \text{(10.1.28)}$$

by recalling (H.6) = (10.1.20) for A, and that e^{At} is a s.c. semigroup on Y_δ^- by (H.5) = (10.1.19). Then, by continuous extension, (10.1.28) yields

$$e^{At} A : \text{continuous } Y_\delta^- \to C([0, T]; [Y_\delta^+]') \quad \text{(10.1.29)}$$

under assumptions (H.5) and (H.6).
(C.4): Assumption (H.4) = (10.1.18), by duality, is equivalent to

$$e^{At} B : \text{continuous } U \to C([0, T]; [Y_\delta^+]'). \quad \text{(10.1.30)}$$

Indeed, for $u \in U$ and $y \in [Y_\delta^+]'$ we compute

$$|(e^{At}Bu, y)_Y| = |(u, B^*e^{A^*t}y)_U| \leq C_T\|u\|_U\|y\|_{Y_\delta^+}, \qquad (10.1.31)$$

and (10.1.30) follows then from (10.1.31).

We then see that assumption (10.1.30), equivalently (H.4) = (10.1.18), for $e^{At}B = e^{At}AA^{-1}B$ is implied by the property

$$A^{-1}B : \text{continuous } U \to Y_\delta^-, \qquad (10.1.32)$$

along with (H.5) and (H.6), since these, in turn, imply (10.1.29).

10.2 Statement of the Main Results

Our starting point is Theorem 9.2.1 of Chapter 9, which applies by virtue of the assumptions (i), (H.1) = (10.1.6), and $R \in \mathcal{L}(Y; Z)$. In the present setting, a far richer and complete theory becomes available.

Theorem 10.2.1 (Regularity of the optimal pair)   *Assume hypotheses (i), (H.1) = (10.1.6), (H.2) = (10.1.14), (H.3) = (10.1.15), (H.5) = (10.1.19), (H.6) = (10.1.20), (H.7) = (10.1.21), and (H.8) = (10.1.23) [actually the weaker requirement R^*R : $Y_{\frac{1}{2}-\delta} \to Y_{\frac{1}{2}}$ will suffice]. Then, the unique optimal pair $\{u^0(\cdot, s; y_0), y^0(\cdot, s; y_0)\}$ of the OCP (10.1.4) for (10.1.1) [guaranteed by Theorem 9.2.1 of Chapter 9] satisfies the following regularity properties: With $y_0 \in Y_\delta^-$ defined in (10.1.16), we have (see Chapter 9, Theorem 9.2.1, with $G = 0$):*

(i)

$$y^0(\cdot, s; y_0) = [I_s + L_sL_s^*R^*R]^{-1}\left[e^{A(\cdot - s)}y_0\right] \qquad (10.2.1a)$$
$$\in C\left([s, T]; Y_\delta^- = Y_{\frac{1}{2}-\delta}\right) \cap H^{\frac{1}{2}-\delta}(s, T; Y) \subset \mathcal{Y}^{\frac{1}{2}-\delta}[s, T]; \qquad (10.2.1b)$$

(ii)

$$u^0(\cdot, s; y_0) = -L_s^*R^*Ry^0(\cdot, s; y_0) \qquad (10.2.2a)$$
$$\in L_2\left(s, T; U_\delta^+ = U_{\frac{1}{2}+\delta}\right) \cap H^{\frac{1}{2}+\delta}(s, T; U) \equiv \mathcal{U}^{\frac{1}{2}+\delta}[s, T], \qquad (10.2.2b)$$

a fortiori,

$$u^0(\cdot, s; y_0) \in C([s, T]; U) \qquad (10.2.2c)$$

(see Theorem 10.3.1.2 below).

All the above results are with norms that may be made independent of s.

Theorem 10.2.2 (Regularity of the gain operator $B^*P(t)$) *Assume hypotheses (i) and (H.1) = (10.1.6) through (H.8) = (10.1.23). Then, the operator $P(t)$ defined by Eqn. (9.2.14) of Chapter 9 with $G = 0$,*

$$P(t)x = \int_t^T e^{A^*(\tau-t)} R^* R y^0(\tau, t; x)\, d\tau \qquad (10.2.3a)$$

$$: \ continuous \ Y \to C([0, T]; Y), \qquad (10.2.3b)$$

satisfies the following regularity property

$$B^*P(t) : continuous \ Y_\delta^- \to C([0, T]; U) \qquad (10.2.4)$$

(see Theorem 10.3.2.1 below).

Theorem 10.2.3 (DRE) *Assume (i) and (H.1) = (10.1.6) through (H.8) = (10.1.23). Then, the operator $P(t)$ defined by Eqn. (9.2.14) of Chapter 9 satisfies*

$$(P(t)x, Ay)_Y, \qquad (P(t)Ax, y)_Y \in C[0, T], \quad \forall x, y \in Y_\delta^-, \qquad (10.2.5)$$

in the sense that the above quantities, originally defined on $\mathcal{D}(A)$, can be extended to Y_δ^-; and, moreover, $P(t)$ satisfies the following differential Riccati equation for all $0 \le t < T$:

$$\begin{cases} \dfrac{d}{dt}(P(t)x, y)_Y = -(Rx, Ry)_Z - (P(t)x, Ay)_Y - (P(t)Ax, y)_Y \\ \qquad\qquad\qquad + (B^*P(t)x, B^*P(t)y)_U, \quad \forall x, y \in Y_\delta^-; \\ P(T) = 0; \end{cases} \qquad (10.2.6)$$

as well as the corresponding integral Riccati equation for all $0 \le t \le T$:

$$(P(t)x, y)_Y = \int_t^T \left(Re^{A(\tau-t)}x, Re^{A(\tau-t)}y \right)_Z d\tau$$

$$- \int_t^T \left(B^*P(\tau)e^{A(\tau-t)}x, B^*P(\tau)e^{A(\tau-t)}y \right)_U d\tau, \quad x, y \in Y_\delta^- \qquad (10.2.7)$$

(see Lemma 10.4.2.1 and Theorem 10.4.2.2 below).

Orientation Theorem 9.2.1 of Chapter 9 applies to the present situation (by virtue of assumptions (i) and (H.1), and $R \in \mathcal{L}(Y; Z)$), and we seek to go beyond these preliminary results. Now, Eqns. (10.1.14) for L_s and (10.1.15) for L_s^* show, by (10.1.11) and (10.1.12), that – in the present setting, and unlike the case of analytic semigroups of Volume 1 – the operators L_s and L_s^* do not provide any smoothing in the (Sobolev spaces) Y_θ and U_θ, that is, in what in PDE applications will be "the space variable."

Thus, to achieve a complete theory, which in particular includes the derivation of a differential Riccati equation, two main problems of similar nature arise:

First, in seeking regularity properties for the optimal trajectory $y^0(\cdot, s; y_0)$ with a "regular" initial datum $y_0 \in Y_\delta^-$, one needs to perform a critical bounded inversion of the operator $[I_s + L_s L_s^* R^* R]$, which describes $y^0(\cdot, s; y_0)$ (see Chapter 9, Eqn. (9.2.5), that is, Eqn. (10.2.1a) above) on the smoother space $L_2(s, T; Y_\delta^-)$, in fact on its subspace $\mathcal{Y}^{\frac{1}{2}-\delta}[s, T] \equiv L_2(s, T; Y_\delta^-) \cap H^{\frac{1}{2}-\delta}(s, T; Y)$; see (10.1.12). This bounded inversion would, however, pose a serious problem, unless $L_s L_s^* R^* R$ could be asserted to be compact on $\mathcal{Y}^{\frac{1}{2}-\delta}[s, T]$. It is to this end that a "minimal" smoothing assumption on $R^* R$, such as, for example, (H.8) = (10.1.23), is then invoked [but even the weaker requirement $R^* R$: continuous $Y_{\theta = \frac{1}{2} - \delta} \to Y_{\theta = \frac{1}{2}}$ would do it, of course]. Once the bounded inversion of $[I_s + L_s L_s^* R^* R]$ on $\mathcal{Y}^{\frac{1}{2}-\delta}[s, T]$ is performed, one obtains, along with assumptions (H.5) = (10.1.19), regularity properties of $y^0(\cdot, s; y_0)$, and hence, via (H.3) = (10.1.15) applied to the optimality condition [Eqn. (9.2.1) of Chapter 9], that is, Eqn. (10.2.2a) above, regularity properties of $y^0(\cdot, s; y_0)$, for s fixed. Next, however, to obtain the regularity property of the gain operator $B^* P(t)$: continuous $Y_\delta^- \to C([0, T]; U)$ (via Eqn. (10.2.3) above), we see that we need to refine the preceding result by asserting that, in fact, $[I_s + L_s L_s^* R^* R]$ is boundedly invertible on $\mathcal{Y}^{\frac{1}{2}-\delta}[s, T]$, *uniformly in* s. The aforementioned regularity of $B^* P(t)$ then justifies the well-posedness of the critical quadratic term, which occurs in the differential Riccati equation. All this summarizes the content of Section 10.3, which provides the proof of the regularity Theorems 10.2.1 and 10.2.2.

Second, in seeking to derive the differential Riccati equation on Y_δ^-, one encounters the obstacle of performing the bounded inversion of the operator $[I_s + L_s L_s^* R^* R]$, this time, however, on the weaker space $L_2(s, T; [Y_\delta^+]')$; equivalently, by duality, one faces the bounded inversion of $[I_s + R^* R L_s L_s^*]$ on the space $L_2(s, T; Y_\delta^+)$. This task would, however, be again a serious problem, unless $R^* R L_s L_s^*$ could be asserted to be compact on $L_2(s, T; Y_\delta^+)$. It is at this level that the smoothing assumption $R^* R$: continuous $Y_\delta^- \to Y_\delta^+$ in (H.8) = (10.1.23) is used in full force, as the operator L_s, by assumption (H.2) = (10.1.14), has a known regularity property only on $Y_\theta, \theta < 1/2$. Accordingly, the bounded inversion of $[I_s + L_s L_s^* R^* R]$ on $L_2(s, T; [Y_\delta^+]')$ is then performed for each s, a result sufficient in the derivation of the DRE in Section 10.4.

10.3 Proof of Theorems 10.2.1 and 10.2.2

10.3.1 *Bounded Inversion of* $[I_s + L_s L_s^* R^* R]$ *on the Space* $\mathcal{Y}^{\frac{1}{2}-\delta}[s, T]$, *Uniformly in* s. *Proof of Theorem 10.2.1*

The key preliminary result is the following.

Theorem 10.3.1.1   *Assume (H.1), (H.2) and (H.3) only for* $\theta < 1/2$, *(H.5), (H.6), (H.7), and (H.8) [though the weaker requirement* $R^* R : Y_{\frac{1}{2}-\delta} \to Y_{\frac{1}{2}}$ *will suffice]. Then:*

(i) With reference to the spaces in (10.1.12) for $\theta = 1/2 - \delta$ and $\theta = 1/2 - \delta/2$, we have the following estimate:

$$\|L_s L_s^* R^* R f\|_{\mathcal{Y}^{\frac{1}{2}-\frac{\delta}{2}}[s,T]} \leq C_{T,\delta} \|f\|_{\mathcal{Y}^{\frac{1}{2}-\delta}[s,T]}, \quad \text{uniformly in } s. \tag{10.3.1.1}$$

(ii) For fixed s, the operator $L_s L_s^* R^* R : \mathcal{Y}^{\frac{1}{2}-\delta}[s, T] \to$ itself, is compact for fixed s, and, in fact, $\{L_s L_s^* R^* R\}$ is a family (in s) of collectively compact operators on $\mathcal{Y}^{\frac{1}{2}-\delta}[0, T]$, once extended by zero on $[0, s)$ (in the sense of [Anselone, 1971, p. 3]).

(iii) For $f \in \mathcal{Y}^{\frac{1}{2}-\delta}[0, T]$, indeed, even $f \in L_2(0, T; Y_\delta^-)$, the map $s \to L_s L_s^* R^* R f$ is continuous in $\mathcal{Y}^{\frac{1}{2}-\delta}[0, T]$.

(iv) The operator $[I_s + L_s L_s^* R^* R]$ is boundedly invertible on $\mathcal{Y}^{\frac{1}{2}-\delta}[s, T]$, indeed uniformly with respect to s:

$$\|[I_s + L_s L_s^* R^* R]^{-1}\|_{\mathcal{L}(\mathcal{Y}^{\frac{1}{2}-\delta}[s,T])} \leq C_{T,\delta}, \quad \text{uniformly in } s. \tag{10.3.1.2}$$

Proof. (i) The proof of estimate (10.3.1.1) is a consequence of part of the following diagram, where all continuity maps are uniform in s:

$$\mathcal{Y}^{\frac{1}{2}-\delta}[s, T] \xrightarrow[\text{continuous} \atop \text{by (10.1.23)}]{R^* R} L_2(s, T; Y_\delta^+) \xrightarrow[\text{injection}]{\text{continuous}} L_2(s, T; Y_{\frac{\delta}{2}}^-)$$

$$L_s L_s^* \Big\downarrow \text{continuous by (10.1.25)}$$

$$\mathcal{Y}^{\frac{1}{2}-\delta}[s, T] \xleftarrow[\text{injection} \atop \text{by (10.3.1.3)}]{\text{compact}} \mathcal{Y}^{\frac{1}{2}-\frac{\delta}{2}}[s, T].$$

In the first step, we use (H.8) = (10.1.23) for $R^* R$ [but $R^* R$: continuous $Y_{\frac{1}{2}-\delta} \to Y_{\frac{1}{2}}$ would suffice], followed by [the combination of (H.2) = (10.1.14) and (H.3) = (10.1.15) for $\theta = 1/2 - \delta/2$ culminating in] the regularity (C.1) = (10.1.25), followed in the last step by the

compact injection $\mathcal{Y}^{\frac{1}{2}-\frac{\delta}{2}}[s, T] \to \mathcal{Y}^{\frac{1}{2}-\delta}[s, T]$, $\tag{10.3.1.3}$

a consequence, via (10.1.12), of the compact injection $Y_{\frac{1}{2}-\frac{\delta}{2}} \to Y_{\frac{1}{2}-\delta}$ in (10.1.10) and of $T < \infty$.

(ii) A fortiori from the diagram, $L_s L_s^* R^* R$, extended by zero on $[0, s)$ is a compact operator on $\mathcal{Y}^{\frac{1}{2}-\delta}[0, T]$, and the family $\{L_s L_s^* R^* R\}$ is collectively compact (in s) on $\mathcal{Y}^{\frac{1}{2}-\delta}[0, T]$, by estimate (10.3.1.1).

(iii)

Step 1   Let $g \in \mathcal{U}^{\frac{1}{2}-\delta}[0, T] \equiv L_2(0, T; U_\delta^-) \cap H^{\frac{1}{2}-\delta}(0, T; U)$. We shall first show that, when $L_s g$ is extended by zero on $[0, s)$, then

$$\text{the map } s \to L_s g \text{ is continuous from } [0, T] \text{ to } L_2(0, T; Y_\delta^-). \qquad (10.3.1.4)$$

In fact, with, say, $t > s_1 > s$, recalling (10.1.2b),

$$\|(L_s g)(t) - (L_{s_1} g)(t)\|_{Y_\delta^-} = \left\| \int_s^{s_1} e^{A(t-\tau)} B g(\tau) \, d\tau \right\|_{Y_\delta^-}$$

$$= \left\| e^{A(t-s_1)} \int_s^{s_1} e^{A(s_1-\tau)} B g(\tau) \, d\tau \right\|_{Y_\delta^-}$$

$$(\text{by } (10.1.19)) \qquad \leq C_T \|(L_s g)(s_1)\|_{Y_\delta^-}$$

$$\leq C'_{T,\delta} \|g\|_{\mathcal{U}^{\frac{1}{2}-\delta}[s,s_1]} \to 0 \quad \text{as } [s - s_1] \to 0, \quad (10.3.1.5)$$

where in the last steps we have recalled (10.1.2b) as well as assumption (H.5) = (10.1.19) on e^{At}, and (H.2) = (10.1.14) with $\theta = 1/2 - \delta$. Thus, by (10.3.1.5),

$$\|L_s g - L_{s_1} g\|_{C([s_1,T];Y_\delta^-)} \to 0 \quad \text{as } [s_1 - s] \to 0, \qquad (10.3.1.6a)$$

as well as

$$\|L_s g - L_{s_1} g\|_{C([s,s_1];Y_\delta^-)} = \|L_s g\|_{C([s,s_1];Y_\delta^-)} \to 0 \quad \text{as } [s - s_1] \to 0. \qquad (10.3.1.6b)$$

Then, (10.3.1.6a) and (10.3.1.6b) a fortiori imply (10.3.1.4).

Step 2   Next, let $f \in L_2(0, T; Y_\delta^-)$. We then show that

$$\text{the map } s \to L^* f \text{ is continuous from } [0, T] \text{ to } \mathcal{U}^{\frac{1}{2}-\delta}[0, T]. \qquad (10.3.1.7)$$

In fact, the definition (10.1.8) implies, still with $s_1 > s$,

$$\|(L_s^* f - L_{s_1}^* f\|_{\mathcal{U}^{\frac{1}{2}-\delta}[0,T]} = \|L_s^* f\|_{\mathcal{U}^{\frac{1}{2}-\delta}[s,s_1]}$$

$$(\text{by } (10.1.15b)) \qquad \leq C_{T,\delta} \|f\|_{L_2(s,s_1;Y_\delta^-)} \to 0 \quad \text{as } [s - s_1] \to 0, \qquad (10.3.1.8)$$

after using (H.3) = (10.1.15b), and (10.3.1.7) is proved.

Step 3   Next, with $g = L_s^* R^* R f \in \mathcal{U}^{\frac{1}{2}-\delta}[0, T]$ (conservatively) with $f \in L_2(0, T; Y_\delta^-)$, via (H.8) = (10.1.23) and (H.3) = (10.1.15), we recall (10.1.26b) and write

$$\frac{d(L_s g)}{dt} = \frac{d(L_s L_s^* R^* R f)}{dt} = A L_s L_s^* R^* R f + B L_s^* R^* R f \qquad (10.3.1.9)$$

$$= A L_s L^* R^* R f + B L_s^* R^* R f, \qquad (10.3.1.10)$$

since, by the definitions (10.1.2b) and (10.1.8), we have readily $L_s L_s^* = L_s L^*$. With reference to (10.3.1.10), and with $f \in L_2(0, T; Y_\delta^-)$, we then have that

$$\text{the map } s \to A L_s L^* R^* R f \text{ is continuous in } L_2(0, T; [Y_\delta^+]'), \quad (10.3.1.11)$$

by combining (10.3.1.4) with $g = L^* R^* R f$ and (H.6) = (10.1.20) on A. Also, again with $f \in L_2(0, T; Y_\delta^-)$, hence $R^* R f \in L_2(0, T; Y_\delta^-)$, a fortiori

$$\text{the map } s \to B L_s^* R^* R f \text{ is continuous in } L_2(0, T; [Y_\delta^+]'), \quad (10.3.1.12)$$

by combining (10.3.1.7) and (H.7) = (10.1.21) on B. Using (10.3.1.11) and (10.3.1.12) in (10.3.1.10) we conclude that:

If $f \in L_2(0, T; Y_\delta^-)$, then:

$$\text{the map } s \to \frac{d}{dt}(L_s L_s^* R^* R f) \text{ is continuous in } L_2(0, T; [Y_\delta^+]'), \quad (10.3.1.13)$$

as well as

$$\text{the map } s \to L_s L_s^* R^* R f \text{ is continuous in } L_2(0, T; Y_\delta^-), \quad (10.3.1.14)$$

by (3.1.4).

Step 4 Hence, by interpolation between (10.3.1.13) and (10.3.1.14), we obtain, recalling the interpolation property in (10.1.10),

$$\text{the map } s \to \left(D_t^{\theta=\frac{1}{2}-\delta}\right) L_s L_s^* R^* R f \text{ is continuous in}$$
$$L_2\left(0, T; [Y_\delta^-, [Y_\delta^+]']_{\theta=\frac{1}{2}-\delta}\right) = L_2(0, T; Y), \quad (10.3.1.15)$$

via [Lions, Magenes, 1972, pp. 15, 23]. Then (10.3.1.14) and (10.3.1.5) together mean:

If $f \in L_2(0, T; Y_\delta^-)$, then:

$$\text{the map } s \to L_s L_s^* R^* R f \text{ is continuous in } \mathcal{Y}^{\frac{1}{2}-\delta}[0, T], \quad (10.3.1.16)$$

which proves the desired part (iii).

(iv) We first show that $[I_s + L_s L_s^* R^* R]$ is boundedly invertible on $\mathcal{Y}^{\frac{1}{2}-\delta}[s, T]$ for each s fixed,

$$[I_s + L_s L_s^* R^* R]^{-1} \in \mathcal{L}\left(\mathcal{Y}^{\frac{1}{2}-\delta}[s, T]\right). \quad (10.3.1.17)$$

Indeed, since $L_s L_s^* R^* R$ is a compact operator on $\mathcal{Y}^{\frac{1}{2}-\delta}[s, T]$ by part (ii), then a (necessary and) sufficient condition for (10.3.1.17) to hold true is that $\lambda = 1$ not be an eigenvalue of $L_s L_s^* R^* R$ on $\mathcal{Y}^{\frac{1}{2}-\delta}[s, T]$, which is certainly the case, for otherwise $\lambda = 1$ would also be an eigenvalue of $L_s L_s^* R^* R$ on $L_2(s, T; Y)$, thus contradicting Lemma 8.3.1.2 of Chapter 8, which asserts that $[I_s + L_s L_s^* R^* R]^{-1} \in \mathcal{L}(L_2(s, T; Y))$ [see Eqn. (9.2.6) of Chapter 9]. Thus, (10.3.1.7) is proved.

Finally, to assert the uniform estimate (10.3.1.2), we simply invoke Lemma 8.5.3.1 of Chapter 8 with $Z_1 \equiv \mathcal{Y}^{\frac{1}{2}-\frac{\delta}{2}}[0, T]$ with compact injection into $Z_0 \equiv \mathcal{Y}^{\frac{1}{2}-\delta}[0, T]$ (see (10.3.1.3)). This is legal by virtue also of (10.3.1.1) of part (i), (10.3.1.16) of part (iii), and (10.3.1.17) of part (iv). Theorem 10.3.1.1 is proved. □

Remark 10.3.1.1   In the preceding diagram the weaker requirement R^*R: continuous $Y_{\frac{1}{2}-\delta} \to Y_{\frac{1}{2}}$ would suffice.

Remark 10.3.1.2   With reference to (10.2.1a), setting

$$\Gamma_s = [I_s + L_s L_s^* R^* R], \quad \text{we obtain} \quad \Gamma_s^{-1} - \Gamma_{s_1}^{-1} = \Gamma_s^{-1}[\Gamma_{s_1} - \Gamma_s]\Gamma_{s_1}^{-1} \quad (10.3.1.18)$$

by the second resolvent equation. Hence, estimate (10.3.1.2) of Theorem 10.3.1.1 applied to (10.3.1.18) readily implies that, for each $f \in \mathcal{Y}^{\frac{1}{2}-\delta}[s, T]$ fixed,

$$\text{the map } s \to \Gamma_s^{-1} f \text{ is continuous in } \mathcal{Y}^{\frac{1}{2}-\delta}[0, T], \quad (10.3.1.19)$$

a result we shall not need, however, for $y^0(\cdot, s; y_0)$. See also Remark 10.3.1.3 below.

As a corollary of Theorem 10.3.1.1, we shall prove Theorem 10.2.1 on the regularity of the optimal pair.

Theorem 10.3.1.2   *Assume the hypotheses of Theorem 10.3.1.1: (H.1) through (H.3) and (H.5) through (H.8). Then, the optimal pair $\{u^0(\cdot, s; y_0), y^0(\cdot, s; y_0)\}$ guaranteed by Theorem 9.2.1 of Chapter 9 satisfies the following regularity properties for $y_0 \in Y_\delta^-$:*

(i)

$$y^0(\cdot, s; y_0) \equiv \Phi(\cdot, s)y_0 \in C([s, T]; Y_\delta^-) \cap H^{\frac{1}{2}-\delta}(s, T; Y) \subset \mathcal{Y}^{\frac{1}{2}-\delta}[s, T], \quad (10.3.1.20a)$$

with norms that may be made independent of s,

$$\|\Phi(\cdot, s)\|_{\mathcal{L}(C([s,T];Y_\delta^-);Y_\delta^-)} + \|\Phi(\cdot, s)\|_{\mathcal{L}(\mathcal{Y}^{\frac{1}{2}-\delta}[s,T];Y_\delta^-)} \leq C_{T,\delta}, \quad \text{uniformly in } s; \quad (10.3.1.20b)$$

(ii) still for $y_0 \in Y_\delta^-$,

$$u^0(\cdot, s; y_0) \in \mathcal{U}^{\frac{1}{2}+\delta}[s, T], \quad (10.3.1.21a)$$

with a norm that may be made independent of s,

$$\|u^0(\cdot, s; y_0)\|_{\mathcal{L}(\mathcal{U}^{\frac{1}{2}+\delta}[s,T];Y_\delta^-)} \leq C_{T,\delta}, \quad \text{uniformly in } s. \quad (10.3.1.21b)$$

Proof.

Step 1   We recall (10.2.1a) (i.e., Eqn. (9.2.5) in Theorem 9.2.1 of Chapter 9) with $G = 0$:

$$y^0(\cdot, s; y_0) \equiv \Phi(\cdot, s)y_0 = [I_s + L_s L_s^* R^* R]^{-1}\big[e^{A(\cdot - s)}y_0\big]. \quad (10.3.1.22)$$

With $y_0 \in Y_\delta^-$, we apply (H.5) = (10.1.19), which gives that $e^{A(\cdot - s)}$ is a s.c. semi-group on Y_δ^-, and finally we invoke Theorem 10.3.1.1(iv), Eqn. (10.3.1.2), to obtain (10.3.1.20b) for $\mathcal{L}(\mathcal{Y}^{\frac{1}{2}-\delta}[s, T]; Y_\delta^-)$.

Step 2   We now recall the optimality condition

$$u^0(\cdot, s; y_0) = -L_s^* R^* R y^0(\cdot, s; y_0), \tag{10.3.1.23}$$

from (10.2.2a) (i.e., from Eqn. (9.2.1) in Theorem 9.2.1 in Chapter 9), to which we apply the diagram

$$\mathcal{Y}^{\frac{1}{2}-\delta}[s, T] \overset{R^* R}{\underset{\text{by }(10.1.23)}{\longrightarrow}} L_2(s, T; Y_\delta^+) \overset{L_s^*}{\underset{\text{by }(10.1.15b)}{\longrightarrow}} \mathcal{U}^{\frac{1}{2}+\delta}[s, T], \tag{10.3.1.24}$$

with $y^0(\cdot, s; y_0) \in \mathcal{Y}^{\frac{1}{2}-\delta}[s, T]$ uniformly in s by (10.3.1.20b), which was just proved in Step 1. All the maps in the diagram are uniform with respect to s, the last one, L_s^*, by (H.3) = (10.1.15b) with $\theta = 1/2 + \delta$. Then the above diagram and (10.3.1.23) prove part (ii), that is, (10.3.1.21).

Step 3   It remains to complete the proof of part (i), by showing the statement for $C([s, T]; Y_\delta^-)$. To this end, we use the optimal dynamics

$$y^0(\cdot, s; y_0) = e^{A(\cdot - s)} y_0 + L_s u^0(\cdot, s; y_0), \tag{10.3.1.25}$$

with $y_0 \in Y_\delta^-$, hence $e^{A(\cdot - s)} y_0 \in C([s, T]; Y_\delta^-)$ by (H.5) = (10.1.19), and finally $L_s u^0(\cdot, s; y_0) \in C([s, T]; Y_\delta^-)$ by (H.2) = (10.1.14a) with $\theta = 1/2 - \delta$, since a fortiori from part (ii), $u^0(\cdot, s; y_0) \in \mathcal{U}^{\frac{1}{2}-\delta}[s, T]$. Then, $y^0(\cdot, s; y_0) \in C([s, T]; Y_\delta^-)$ by (10.3.1.25). Moreover, all results are uniform in s. Theorem 10.3.1.2 is fully proved. $\square$

Remark 10.3.1.3   As we have seen [e.g., in the proof of Lemma 9.3.2.2, Eqn. (9.3.2.16) of Chapter 9 or Lemma 1.4.6.2(ii) of Chapter 1] continuity of $\Phi(t, s)x$ in the first variable, as established by (10.3.1.20a),

$$t \to \Phi(t, s)x \text{ continuous in } Y_\delta^-, \quad \text{for } x \in Y_\delta^-, \quad T \geq t \geq s, \tag{10.3.1.26}$$

for s fixed, combined with the uniform bound obtained in (10.3.2.20b),

$$\|\Phi(t, s)\|_{\mathcal{L}(Y_\delta^-)} \leq C_T \text{ uniformly in } s \leq t \leq T, \tag{10.3.1.27}$$

implies continuity of $\Phi(t, s)x$ in the second variable,

$$s \to \Phi(t, s)x \text{ continuous in } Y_\delta^-, \quad \text{for } x \in Y_\delta^-, \quad s \leq t. \tag{10.3.1.28}$$

10.3.2 Proof of Theorem 10.2.2

We restate Theorem 10.2.2 as

Theorem 10.3.2.1 *Assume hypotheses (H.1)=(10.1.6) through (H.8)=(10.1.23). Then, the operator $P(t)$ defined by Eqn. (10.2.3a), that is, by Eqn. (9.2.14) in Theorem 9.2.1 of Chapter 9, satisfies*

$$B^*P(t) : continuous\ Y_\delta^- \to C([0, T]; U); \qquad \max_{0 \le t \le T} \|B^*P(t)x\|_U \le C_T \|x\|_{Y_\delta^-}.$$

$$(10.3.2.1)$$

Remark 10.3.2.1 The weaker statement

$$B^*P(t) : continuous\ Y_\delta^- \to L_\infty(0, T; U) \qquad (10.3.2.2)$$

can be immediately proved, by applying (H.4) = (10.1.18), (H.8) = (10.1.23), and (10.3.1.20) of Theorem 10.3.1.2 (or (10.3.1.27)) to

$$B^*P(t)x = \int_t^T B^*e^{A^*(\tau-t)}R^*R\Phi(\tau, t)x\, d\tau. \qquad (10.3.2.3)$$

We obtain with $x \in Y_\delta^-$:

$$\|B^*P(t)x\|_U \le C_T \int_t^T \|\Phi(\tau, t)x\|_{Y_\delta^-}\, d\tau \le C'_{T,\delta}\|x\|_{Y_\delta^-}, \qquad (10.3.2.4)$$

and (10.3.2.2) is proved.

Proof of Theorem 10.3.2.1 Let $t_1 \in [0, T)$ and let $t > t_1$. From (10.3.2.3), we compute after a change of variable, with $x \in Y_\delta^-$:

$$B^*P(t)x - B^*P(t_1)x = B^* \int_0^{T-t} e^{A^*\sigma}R^*R\Phi(t + \sigma, t)x\, d\sigma$$

$$- B^* \int_0^{T-t_1} e^{A^*\sigma}R^*R\Phi(t_1 + \sigma, t_1)x\, d\sigma$$

$$= I_1(t)x - I_2(t)x, \qquad (10.3.2.5)$$

where, after adding and subtracting, we get

$$I_1(t)x = \int_0^{T-t} B^*e^{A^*\sigma}R^*R[\Phi(t + \sigma, t)x - \Phi(t_1 + \sigma, t_1)x]\, d\sigma, \quad (10.3.2.6)$$

$$I_2(t)x = \int_{T-t}^{T-t_1} B^*e^{A^*\sigma}R^*R\Phi(t_1 + \sigma, t_1)x\, d\sigma. \qquad (10.3.2.7)$$

As to $I_2(t)x$, we apply (H.4) = (10.1.18), (H.8) = (10.1.23), and (10.3.1.21b) of Theorem 10.3.1.2, or (10.3.1.27), to obtain

$$\|I_2(t)x\|_U \le C_T \int_{T-t}^{T-t_1} \|\Phi(t_1 + \sigma, t_1)x\|_{Y_\delta^-}\, d\sigma$$

$$\le C'_{T,\delta}(t - t_1)\|x\|_{Y_\delta^-} \to 0 \quad \text{as } t \downarrow t_1. \qquad (10.3.2.8)$$

As to $I_1(t)x$, we again apply (H.4) = (10.1.18) and (H.8) = (10.1.23) to obtain, after adding and subtracting $\Phi(t+\sigma, t_1)x = \Phi(t+\sigma, t)\Phi(t, t_1)x$ [recall (9.2.11) in Theorem 9.2.1 of Chapter 9]:

$$\|I_1(t)x\|_U \le C_T \int_0^{T-t} \|\Phi(t+\sigma, t)x - \Phi(t_1+\sigma, t_1)x\|_{Y_\delta^-} \, d\sigma$$

$$\le C_T \left\{ \int_0^{T-t} \|\Phi(t+\sigma, t)x - \Phi(t+\sigma, t)\Phi(t, t_1)x\|_{Y_\delta^-} \, d\sigma \right.$$

$$\left. + \int_0^{T-t} \|\Phi(t+\sigma, t_1)x - \Phi(t_1+\sigma, t_1)x\|_{Y_\delta^-} \, d\sigma \right\}. \quad (10.3.2.9)$$

As to the first term on the right-hand side of (10.3.2.9), we compute

$$\int_0^{T-t} \|\Phi(t+\sigma, t)[x - \Phi(t, t_1)x]\|_{Y_\delta^-} \, d\sigma$$

$$\le \int_0^{T-t} \|\Phi(t+\sigma, t)\|_{\mathcal{L}(Y_\delta^-)} \|x - \Phi(t, t_1)x\|_{Y_\delta^-} \, d\sigma$$

(by (10.3.1.27)) $\le C_{T,\delta} T \|x - \Phi(t, t_1)x\|_{Y_\delta^-} \to 0 \quad \text{as } t \downarrow t_1, \quad x \in Y_\delta^-,$

$$(10.3.2.10)$$

after recalling the uniform bound (10.3.1.27), that is, (10.3.1.21b), where convergence to zero attains because of the continuity property in (10.3.1.17) or (10.3.1.26). As to the second term in (10.3.2.9), the integrand, with $[t+\sigma] - [t_1+\sigma] = t - t_1$, is uniformly continuous and hence arbitrarily small as $t - t_1$ is sufficiently small. Thus

$$\lim_{t \downarrow t_1} \int_0^{T-t} \|\Phi(t+\sigma, t_1)x - \Phi(t_1+\sigma, t_1)x\|_{Y_\delta^-} \, d\sigma = 0. \quad (10.3.2.11)$$

Using (10.3.2.10) and (10.3.2.11) on the right-hand side of (10.3.2.9) then yields

$$\lim_{t \downarrow t_1} I_1(t)x = 0, \quad x \in Y_\delta^-, \quad (10.3.2.12)$$

as desired. Then, (10.3.2.8) for $I_2(t)x$ and (10.3.2.12) for $I_1(t)x$, used in (10.3.2.5), complete the proof that

$$\lim_{t \downarrow t_1} \|B^* P(t)x - B^* P(t_1)x\|_U = 0, \quad x \in Y_\delta^-. \quad (10.3.2.13)$$

A similar argument applies if $t < t_1$ and $t \uparrow t_1$. We then obtain that

$$B^* P(t)x \in C([0, T]; Y_\delta^-), \quad x \in Y_\delta^-. \quad (10.3.2.14)$$

This, along with (10.3.2.4), shows (10.3.2.1), as desired.   □

Remark 10.3.2.2   Recalling the pointwise relationship

$$u^0(t, 0; y_0) = -B^* P(t) y^0(t, 0; y_0), \quad y_0 \in Y_\delta^-, \quad (10.3.2.15)$$

from Eqn. (10.2.17) in Theorem 9.2.1 of Chapter 9, and applying to it the continuity $y^0(t, 0; y_0) \in C([0, T]; Y_\delta^-)$ via (10.3.1.20) of Theorem 10.3.1.2, as well as (10.3.2.1) of Theorem 10.3.2.1, we reobtain that $u^0(t, 0; y_0) \in C([0, T]; U)$, a result a fortiori contained in (10.2.2b), or (10.3.1.21); see (10.2.2c).

10.4 Proof of Theorem 10.2.3

10.4.1 Bounded Inversion of $[I_s + L_s L_s^* R^* R]$ on the Space $L_2(s, T; [Y_\delta^+]')$. Consequences on $\Phi(t, s)$

We begin with the result that will serve our purposes in what follows.

Theorem 10.4.1.1   *Assume (i), (H.1) = (10.1.6), (H.2) = (10.1.14), (H.3) = (10.1.15), and (H.8) = (10.1.23). Then, for s fixed:*

(i) The operator $R^* R L_s L_s^*$ is compact on $L_2(s, T; Y_\delta^+)$.
(ii) The operator $[I_s + R^* R L_s L_s^*]$ is boundedly invertible on $L_2(s, T; Y_\delta^+)$:

$$[I_s + R^* R L_s L_s^*]^{-1} \in \mathcal{L}(L_2(s, T; Y_\delta^+)). \tag{10.4.1.1}$$

(iii) The operator $[I_s + L_s L_s^* R^* R]$ is boundedly invertible on $L_2(s, T; [Y_\delta^+]')$:

$$[I_s + L_s L_s^* R^* R]^{-1} \in \mathcal{L}(L_2(s, T; [Y_\delta^+]'). \tag{10.4.1.2}$$

Proof.   (i) The proof of part (i) is a consequence of the following diagram:

$$
\begin{array}{ccccc}
L_2(s, T; Y_\delta^+) & \xrightarrow[\substack{\text{continuous}\\ \text{by (10.1.15)}}]{L_s^*} & \mathcal{U}^{\frac{1}{2}+\delta}[s, T] & \xrightarrow[\substack{\text{compact}\\ \text{by (10.4.1.3)}}]{\text{injection}} & \mathcal{U}^{\frac{1}{2}-\delta}[s, T] \\
 & & & & \Big\downarrow {\substack{L_s \\ \text{continuous}\\ \text{by (10.1.14)}}} \\
 & & L_2(s, T; Y_\delta^+) & \xleftarrow[\substack{\text{continuous}\\ \text{by (10.1.23)}}]{R^* R} & \mathcal{Y}^{\frac{1}{2}-\delta}[s, T].
\end{array}
$$

The above diagram uses (H.3) = (10.1.15) for $\theta = 1/2 + \delta$ on L_s^*, followed by the

injection $\mathcal{U}^{\frac{1}{2}+\delta}[s, T] \to \mathcal{U}^{\frac{1}{2}-\delta}[s, T]$ compact, $\qquad\qquad (10.4.1.3)$

as a consequence of the compactness property $U_{\frac{1}{2}+\delta} \to U_{\frac{1}{2}-\delta}$ of the injection contained in (10.1.10) and of $T < \infty$; followed by (H.2) = (10.1.14) for $\theta = 1/2 - \delta$ on

L_s; followed by (H.8) = (10.1.23) on R^*R. Thus, as a result, $R^*RL_sL_s^*$ is a compact operator on $L_2(s, T; Y_\delta^+)$, as desired.

(ii) Since $R^*RL_sL_s^*$ is compact on $L_2(s, T; Y_\delta^+)$ by part (i), then a (necessary and) sufficient condition for (10.4.1.1) to hold true is that $\lambda = 1$ not be an eigenvalue of $R^*RL_sL_s^*$ on $L_2(s, T; Y_\delta^+)$, which is certainly the case, for otherwise $\lambda = 1$ would also be an eigenvalue of $R^*RL_sL_s^*$ on $L_2(s, T; Y)$, thus contradicting Lemma 8.3.1.2 of Chapter 8, which asserts that $[I_s + R^*RL_sL_s^*]^{-1} \in \mathcal{L}(L_2(s, T; Y))$. Thus, (10.4.1.1) is proved.

(iii) Part (iii), Eqn. (10.4.1.2), follows from part (ii), Eqn. (10.4.1.1), by duality.

$\square$

We can now draw some consequences of Theorem 10.4.1.1 on properties of the evolution operator $\Phi(t, s)$ in (10.3.1.20).

Corollary 10.4.1.2 *Assume preliminarily (i), (H.1) = (10.1.6) through (H.3) = (10.1.15), and (H.8) = (10.1.23).*

(i) Assume (H.5) = (10.1.19) on e^{At} and (H.6) = (10.1.20). Then

$$\Phi(\cdot, s)A : continuous \ Y_\delta^- \to L_2(s, T; [Y_\delta^+]'), \qquad (10.4.1.4a)$$

which holds true if one knows that

$$\Phi(\cdot, s) : continuous \ [Y_\delta^+]' \to L_2(s, T; [Y_\delta^+]'). \qquad (10.4.1.4b)$$

(ii) Assume (H.4) = (10.1.18). Then

$$\Phi(t, s)B : continuous \ U \to L_2(s, T; [Y_\delta^+]'). \qquad (10.4.1.5)$$

(iii) Assume (H.4) = (H.7). Then

$$\Phi(t, s)[A - BB^*P(s)] : continuous \ Y_\delta^- \to L_2(s, T; [Y_\delta^+]'). \qquad (10.4.1.6)$$

Proof. (i) Recalling (10.3.1.22), we have

$$\Phi(\cdot, s)Ax = [I_s + L_sL_s^*R^*R]^{-1}[e^{A(\cdot - s)}Ax], \qquad (10.4.1.7)$$

where, for $x \in Y_\delta^-$ we have $e^{A(\cdot - s)}Ax \in C([s, T]; [Y_\delta^+]')$ by (C.3) = (10.1.29), that is, by (H.5) and (H.6) continuously in $x \in Y_\delta^-$. Finally, we invoke (10.4.1.2) of Theorem 10.4.1.1 and obtain $\Phi(\cdot, s)Ax \in L_2(s, T; [Y_\delta^+]')$, continuously in $x \in Y_\delta^-$, from (10.4.1.7), as desired.

(ii) Similarly, we have for $u \in U$, via (10.3.1.22),

$$\Phi(\cdot, s)Bu = [I_s + L_sL_s^*R^*R]^{-1}[e^{A(\cdot - s)}Bu], \qquad (10.4.1.8)$$

where now $e^{A(\cdot - s)}Bu \in C([s, T]; [Y_\delta^+]')$ by consequence (C.4) = (10.1.30), that is, duality on (H.4) = (10.1.18). Again, (10.4.1.2) then yields $\Phi(\cdot, s)Bu \in L_2(s, T; [Y_\delta^+]')$, continuously in $u \in U$, from (10.4.1.8), as desired.

(iii) Regularity (10.4.1.6) is an immediate consequence of (10.4.1.4) and (10.4.1.5), via (10.3.2.1) of Theorem 10.3.2.1.   □

Corollary 10.4.1.3   *Assume (H.1) through (H.8). Then:*

(i) for $x \in Y_\delta^-$, $s \leq t \leq T$,

$$\frac{d\Phi(t, s)x}{dt} = [A - BB^*P(t)]\Phi(t, s)x \in C([s, T]; [Y_\delta^+]'), \quad x \in Y_\delta^-;$$
(10.4.1.9)

(ii) for $x \in Y_\delta^-$, $s \leq t \leq T$,

$$\frac{d\Phi(t, s)x}{ds} = -\Phi(t, s)[A - BB^*P(s)]x \in L_2(s, T; [Y_\delta^+]'), \quad x \in Y_\delta^-.$$
(10.4.1.10)

Proof.   (i) Equation (10.4.1.9) is simply the optimal dynamics in differential form via (10.3.2.15) [or Eqn. (9.2.17) in Theorem 9.2.1 of Chapter 9] and may be obtained by differentiation on its integral version (10.3.1.25), that is,

$$\Phi(t, s)x = e^{A(t-s)}x - \int_s^t e^{A(t-\tau)}BB^*P(\tau)\Phi(\tau, s)x \, d\tau, \quad (10.4.1.11)$$

where the regularity in $C([s, T]; [Y_\delta^+]')$ in (10.4.1.9) is obtained by use of assumptions (H.6) = (10.1.20) and (H.7) = (10.1.21) on A and B, respectively, combined with the regularity properties of $\Phi(t, s)x \in C([s, T]; Y_\delta^-)$ in (10.3.1.20) and (10.3.2.1) on $B^*P(t)$.

(ii) One way to derive (10.4.1.10) [in line with the derivation of Eqn. (8.3.4.7) in Chapter 8 and of (9.3.4.4) in Chapter 9] is to start from (10.3.1.22) rewritten as

$$\Phi(t, s)x + \{L_s L_s^* R^* R\Phi(\cdot, s)x\}(t) = e^{A(t-s)}x, \quad x \in Y_\delta^-, \quad (10.4.1.12a)$$

or, explicitly via (10.1.2b) and (10.1.8), as

$$\Phi(t, s)x + \int_s^t e^{A(t-\tau)}BB^* \int_\tau^T e^{A^*(\sigma-\tau)}R^*R\Phi(\sigma, s)x \, d\sigma \, d\tau = e^{A(t-s)}x,$$
(10.4.1.12b)

and then take the distributional derivative in s, to obtain

$$[I_s + L_s L_s^* R^* R]\frac{d\Phi(\cdot, s)x}{ds} = -e^{A(\cdot-s)}[A - BB^*P(s)]x$$
$$\in C([s, T]; [Y_\delta^+]') \subset L_2(s, T; [Y_\delta^+]'), \quad x \in Y_\delta^-,$$
(10.4.1.13)

after invoking the definition of $P(s)$ from Eqn. (10.2.3a). The regularity displayed at the right-hand side of (10.4.1.13) is a consequence of (10.3.2.1) for $B^*P(t)$; (C.3) = (10.1.28) for $e^{At}A$; (C.4) = (10.1.30) for $e^{At}B$. Then, applying (10.4.1.2) of Theorem 10.4.1.1 on (10.4.1.13) yields

$$\frac{d\Phi(\cdot, s)x}{ds} = -[I_s + L_s L_s^* R^* R]^{-1} e^{A(\cdot - s)}[A - BB^* P(s)]x$$

$$\text{(by (10.3.1.22))} \quad = -\Phi(\cdot, s)[A - BB^* P(s)]x \in L_2(s, T; [Y_\delta^+]'), \quad x \in Y_\delta^-,$$

$$\text{(10.4.1.14)}$$

and (10.4.1.10) is proved.

Alternatively, writing

$$\Phi(t, \tau)x = \Phi(t, s)\Phi(s, \tau)x, \quad \tau \le s \le t, \quad x \in Y_\delta^-, \quad \text{(10.4.1.15)}$$

by the evolution property of Eqn. (9.2.11) in Theorem 9.2.1, we differentiate both sides of (10.4.1.15) in s, for example, as a distributional derivative, obtaining

$$0 = \frac{d\Phi(t, \tau)x}{ds} = \frac{d\Phi(t, s)}{ds}\Phi(s, \tau)x + \Phi(t, s)\frac{d\Phi(s, \tau)x}{ds}, \quad \text{(10.4.1.16)}$$

or using (10.4.1.9)

$$\frac{d\Phi(t, s)}{ds}\Phi(s, \tau)x = -\Phi(t, s)[A - BB^* P(s)]\Phi(s, \tau)x$$

$$\underset{\text{in } t}{\in} L_2(s, T; [Y_\delta^+]'), \quad x \in Y_\delta^-, \quad \text{(10.4.1.17)}$$

recalling the regularity of (10.4.1.9) combined with that of $\Phi(t, s)$ on $[Y_\delta^+]'$ given by (10.4.1.4b). Since (10.4.1.17) is valid for all $\tau \le s$, setting $\tau = s$ yields (10.4.1.10), as desired. $\square$

10.4.2 Derivation of the Differential and Integral Riccati Equations

Lemma 10.4.2.1   *Assume (i), (H.1) = (10.1.6) through (H.3) = (10.1.15), and (H.5) = (10.1.19) through (H.8) = (10.1.23). Then, with reference to the nonnegative, self-adjoint operator $P(t) \in \mathcal{L}(Y)$ defined by Eqn. (10.2.3a), we have*

(i)

$$A^* P(t) : continuous \ Y_\delta^- \to C([0, T]; [Y_\delta^-]'), \quad \text{(10.4.2.1)}$$

so that, for $x, y \in Y_\delta^-$, we have the duality pairings

(ii)

$$(P(t)x, Ay)_Y, \quad (P(t)Ax, y)_Y \in C[0, T]. \quad \text{(10.4.2.2)}$$

Proof.   (i) We examine

$$A^* P(t)x = \int_t^T A^* e^{A^*(\tau - t)} R^* R\Phi(\tau, t)x \, d\tau \quad \text{(10.4.2.3)}$$

for $x \in Y_\delta^-$. Then, by (10.3.1.20), or (10.3.1.26), and by (H.8) = (10.1.23), we have

$$R^* R \Phi(\tau, t) x \in C([t, T]; Y_\delta^+), \quad x \in Y_\delta^-, \qquad (10.4.2.4)$$

and by duality on (C.3) = (10.1.29), we have

$$A^* e^{A^* t} : \text{continuous } Y_\delta^+ \to C([0, T]; [Y_\delta^-]'). \qquad (10.4.2.5)$$

Using (10.4.2.4) and (10.4.2.5) in (10.4.2.3) yields $A^* P(t) x \in [Y_\delta^-]'$. Actually, since by (10.3.1.28) and (10.4.2.5), respectively,

$$t \to \Phi(\tau, t) x \text{ continuous in } Y_\delta^- \text{ for } x \in Y_\delta^-; \qquad (10.4.2.6)$$

$$t \to A^* e^{A^*(\tau - t)} y \text{ continuous in } [Y_\delta^-]' \text{ for } y \in Y_\delta^+, \qquad (10.4.2.7)$$

then, in fact, $A^* P(t) x \in C([0, T]; [Y_\delta^-]')$, $x \in Y_\delta^-$, as desired. The closed graph theorem then yields (10.4.2.1).

Part (ii), Eqn. (10.4.2.2), is an immediate consequence of part (i) and of $P(t)$ being self-adjoint on Y. $\quad \square$

Remark 10.4.2.1 Notice that we would have:

$$P(t) : \text{continuous } Y_\delta^- \to C([0, T]; Y_\delta^+), \qquad (10.4.2.8)$$

$$\text{if and only if } A \text{ is an isomorphism } Y_\delta^- \text{ onto } [Y_\delta^+]', \qquad (10.4.2.9)$$

a property for A which is not always true (see PDE illustrations below). Property (10.4.2.9) is true for the examples in Sections 10.7 through 10.9 [see (10.7.9.7), (10.8.9.7), and (10.9.9.1)]; and it is false for the example in Section 10.5 [see (10.5.9.4)–(10.5.9.6)].

We can finally establish that $P(t)$ satisfies the DRE.

Theorem 10.4.2.2 *Assume (i) and (H.1) through (H.8). Then the nonnegative, self-adjoint operator $P(t)$ defined by Eqn. (10.2.3a) satisfies the following differential Riccati equation for all $0 \leq t < T$:*

$$\begin{cases} \dfrac{d}{dt}(P(t)x, y)_Y = -(Rx, Ry)_Z - (P(t)x, Ay)_Y - (P(t)Ax, y)_Y \\ \qquad\qquad\qquad + (B^* P(t)x, B^* P(t)y)_U, \quad \forall\, x, y \in Y_\delta^-, \qquad (10.4.2.10) \\ P(T) = 0. \end{cases}$$

Proof. Let $x, y \in Y_\delta^-$. We differentiate in t the expression

$$(P(t)x, y)_Y = \int_t^T \left(R^* R \Phi(\tau, t)x, e^{A(\tau - t)} y \right)_Y d\tau \qquad (10.4.2.11)$$

obtained from Eqn. (10.2.3a), thus obtaining

$$\frac{d}{dt}(P(t)x, y)_Y = -(R^*Rx, y)_Y + \int_t^T \left(R^*R\frac{\partial\Phi(\tau, t)x}{\partial t}, e^{A(\tau-t)}y\right)_Y d\tau$$

$$-\int_t^T \left(R^*R\Phi(\tau, t)x, e^{A(\tau-t)}Ay\right)_Y d\tau \qquad (10.4.2.12)$$

(by (10.4.1.10)) $= -(R^*Rx, y)_Y - \int_t^T \left(R^*R\Phi(\tau, t)[A - BB^*P(t)]x,\right.$

$$\left. e^{A(\tau-t)}y\right)_Y d\tau - (P(t)x, Ay)_Y, \quad x, y \in Y_\delta^-, \quad (10.4.2.13)$$

after substituting (10.4.1.10) in the second term on the right-hand side of (10.4.2.12), as well as substituting (10.4.2.11) [with y replaced by Ay] in the third term on the right-hand side of (10.4.2.12). We notice explicitly that each term in (10.4.2.12), or (10.4.2.13), is well-defined at each t: the last term by (10.4.2.2), and the critical second term on the right-hand side of (10.4.2.12), or (10.4.2.13), by the regularity in (10.4.1.10) for $d\phi(\tau, t)x/dt$, combined with R^*R: continuous $[Y_\delta^+]' \to [Y_\delta^-]'$ by (10.1.24), as well as with

$$e^{A(\tau-t)}y \in C([t, T]; Y_\delta^-) \quad \text{for } y \in Y_\delta^-, \quad \text{by (H.5)} = (10.1.19).$$

Thus, invoking again (10.4.2.11) on the second term on the right-hand side of (10.4.2.13), we obtain

$$\frac{d}{dt}(P(t)x, y)_Y = -(R^*Rx, y)_Y - (P(t)[A - BB^*P(t)]x, y)_Y - (P(t)x, Ay)_Y$$

$$= -(R^*Rx, y)_Y - (P(t)Ax, y)_Y - (P(t)x, Ay)_Y$$

$$+ (B^*P(t)x, B^*P(t)y)_U, \quad x, y \in Y_\delta^-, \qquad (10.4.2.14)$$

where each term is well-defined, by virtue of (10.3.2.1) and (10.4.2.2). Then, (10.4.2.14) proves (10.4.2.10), as desired.   □

As a consequence of Theorem 10.4.2.2, we obtain that the operator $P(t)$ satisfies the integral Riccati equation as well.

Theorem 10.4.2.3   *Assume (i) and (H.1) = (10.1.6) through (H.8) = (10.1.23). Then, the nonnegative, self-adjoint operator $P(t)$ of Theorem 10.4.2.2 satisfies the following integral Riccati equation for all $0 \le t < T$:*

$$(P(t)x, y)_Y = \int_t^T \left(Re^{A(\tau-t)}x, Re^{A(\tau-t)}y\right)_Z d\tau$$

$$-\int_t^T \left(B^*P(\tau)e^{A(\tau-t)}x, B^*P(\tau)e^{A(\tau-t)}y\right)_U d\tau, \quad x, y \in Y_\delta^-,$$

$$(10.4.2.15)$$

where all terms are well defined by (10.1.19), (10.3.2.1), and (10.1.23).

Proof.   We proceed as in the proof of Proposition 8.3.5.1 of Chapter 8, in the present context. For $x, y \in Y_\delta^-$, we compute

$$
\frac{d}{d\tau} \left(e^{A^*(\tau-t)} P(\tau) e^{A(\tau-t)} x, y \right)_Y = \frac{d}{d\tau} \left(P(\tau) e^{A(\tau-t)} x, e^{A(\tau-t)} y \right)_Y
$$

$$
= \frac{\partial}{\partial r} \left(P(r) e^{A(\tau-t)} x, e^{A(\tau-t)} y \right)_Y \Big|_{r=\tau} + \left(P(\tau) e^{A(\tau-t)} A x, e^{A(\tau-t)} y \right)_Y
$$

$$
+ \left(P(\tau) e^{A(\tau-t)} x, e^{A(\tau-t)} A y \right)_Y , \tag{10.4.2.16}
$$

where by using the DRE (10.4.2.10), we have

$$
\frac{\partial}{\partial r} \left(P(r) e^{A(\tau-t)} x, e^{A(\tau-t)} y \right) \Big|_{r=\tau} = - \left(R^* R e^{A(\tau-t)} x, e^{A(\tau-t)} y \right)_Y
$$

$$
- \left(P(\tau) e^{A(\tau-t)} x, A e^{A(\tau-t)} y \right)_Y - \left(P(\tau) A e^{A(\tau-t)} x, e^{A(\tau-t)} y \right)_Y
$$

$$
+ \left(B^* P(\tau) e^{A(\tau-t)} x, B^* P(\tau) e^{A(\tau-t)} y \right)_U , \quad x, y \in Y_\delta^- . \tag{10.4.2.17}
$$

We note explicitly that each term of (10.4.2.16) and (10.4.2.17) is well defined; indeed, we have $e^{A(\tau-t)} x, e^{A(\tau-t)} y$ in $C([t, T]; Y_\delta^-)$, for $x, y \in Y_\delta^-$, and hence:

$$
A^* P(\tau) e^{A(\tau-t)} x \in C([t, T]; [Y_\delta^-]') , \quad \text{by (10.4.2.1)}; \tag{10.4.2.18}
$$

$$
A e^{A(\tau-t)} x \in C([0, T]; [Y_\delta^+]') , \quad \text{by (10.1.19) and (10.1.20)}; \tag{10.4.2.19}
$$

$$
B^* P(\tau) e^{A(\tau-t)} x \in C([t, T]; U) , \quad \text{by (10.3.2.1) and (10.1.19)}; \tag{10.4.2.20}
$$

and similarly for y_0. Thus, (10.4.2.18)–(10.4.2.20) and (10.1.23) show that each term in (10.4.2.16) and (10.4.2.17) is well-defined. Inserting (10.4.2.17) into (10.4.2.16) results in a cancellation of the last two terms of (10.4.2.16); hence

$$
\frac{d}{d\tau} \left(e^{A^*(\tau-t)} P(\tau) e^{A(\tau-t)} x, y \right)_Y = - \left(R e^{A(\tau-t)} x, R e^{A(\tau-t)} y \right)_Z
$$

$$
+ \left(B^* P(\tau) e^{A(\tau-t)} x, B^* P(\tau) e^{A(\tau-t)} y \right)_U , \quad x, y \in Y_\delta^- . \tag{10.4.2.21}
$$

Integrating (10.4.2.21) in τ over $[t, T]$ and using $P(T) = 0$ from the DRE (10.4.2.10) results in (10.4.2.15), as desired. $\square$

10.5 Application: Second-Order Hyperbolic Equations with Dirichlet Boundary Control. Regularity Theory

In this section we consider an optimal quadratic cost problem over a finite horizon for a second-order hyperbolic equation with control function acting in the Dirichlet boundary conditions. The equation is defined on a sufficiently smooth, bounded domain of arbitrary dimension. We show that all abstract system's assumptions (H.1) = (10.1.6) through (H.7) = (10.1.21) of Section 10.1 are automatically satisfied in the natural mathematical setting. Accordingly, Theorems 10.2.1, 10.2.2, and 10.2.3 are then applicable to the present class, for any observation operator R with "minimal" smoothing as in (H.8) = (10.1.23).

10.5.1 Problem Formulation

The Dynamics Let $\Omega \subset R^n$ be an open bounded domain with sufficiently smooth boundary, say, of class C^1, or else a parallelopiped. Let $-\mathcal{A}(x, \partial)$ be a general second-order operator with uniformly elliptic principal part and with variable (real) coefficients, depending smoothly on the space variable x:

$$-\mathcal{A}(x, \partial) = \sum_{i,j=1}^{n} \frac{\partial}{\partial x_i}\left(a_{ij}(x)\frac{\partial}{\partial x_j}\right) + \sum_{j=1}^{n} b_j(x)\frac{\partial}{\partial x_j} + c(x) \quad (10.5.1.1)$$

[canonically, $-\mathcal{A}(x, \partial) = \Delta =$ the Laplacian in x],

$$\sum_{i,j=1}^{n} a_{ij}(x)\eta_i\eta_j \geq \alpha \sum_{j=1}^{n} \eta_j^2, \quad \alpha > 0, \quad \forall \eta \in R^n, \quad (10.5.1.2)$$

where the matrix $[a_{ij}]$ is symmetric: $a_{ij} = a_{ji}$ [as in Eqn. (3.1.14) of Chapter 3]. We consider the mixed hyperbolic problem

$$\begin{cases} w_{tt} = -\mathcal{A}(x, \partial)w & \text{in } (0, T] \times \Omega \equiv Q; & (10.5.1.3a) \\ w(0, x) = w_0(x), \quad w_t(0, x) = w_1(x) & \text{in } \Omega; & (10.5.1.3b) \\ w|_{\Sigma} = u & \text{in } (0, T] \times \Gamma \equiv \Sigma, & (10.5.1.3c) \end{cases}$$

where the Dirichlet control function $u \in L_2(0, T; L_2(\Gamma)) = L_2(\Sigma)$.

The Optimal Control Problem on $[s, T]$ Consistently with the (optimal) regularity theory for problem (10.5.1.3) presented in Theorem 10.5.3.2 below, the cost functional we seek to minimize over all $u \in L_2(s, T; L_2(\Gamma)) = L_2(\Sigma_{sT})$ is taken to be

$$J(u, \{w, w_t\}) = \int_s^T \left\{ \left\| R\begin{bmatrix} w(t) \\ w_t(t) \end{bmatrix} \right\|_{L_2(\Omega) \times H^{-1}(\Omega)}^2 + \|u(t)\|_{L_2(\Gamma)}^2 \right\} dt, \quad (10.5.1.4)$$

with initial data $\{w_0, w_1\} \in L_2(\Omega) \times H^{-1}(\Omega)$ at the initial time $s \geq 0$, where the observation operator $R \in \mathcal{L}(L_2(\Omega) \times H^{-1}(\Omega))$ will be further specified below in (10.5.2.3).

10.5.2 Main Results

As a specialization to problem (10.5.1.3), (10.5.1.4) of the abstract theory presented in Theorems 10.2.1, 10.2.2, and 10.2.3 of this chapter, the present section states the following results, to be proved below.

Theorem 10.5.2.1   *(a) With the observation operator R in (10.5.1.4) only assumed to satisfy*

$$R \in \mathcal{L}(L_2(\Omega) \times H^{-1}(\Omega)), \tag{10.5.2.1}$$

Theorem 9.2.1 of Chapter 9 applies to the optimal control problem (10.5.1.3), (10.5.1.4), with

$$y(t) = \begin{bmatrix} w(t) \\ w_t(t) \end{bmatrix}; \quad U = L_2(\Gamma); \quad Y \equiv L_2(\Omega) \times H^{-1}(\Omega), \tag{10.5.2.2}$$

and yields a unique optimal pair $\{u^0(\,\cdot\,, s; y_0), y^0(\,\cdot\,, s; y_0)\}$, $y_0 = [w_0, w_1] \in Y$, satisfying, in particular, the pointwise feedback synthesis (9.2.17) of Chapter 9, and the optimal cost relation (9.2.18) of Chapter 9, as well as the other properties listed there.

   (b) Assume, in addition to (10.5.2.1), that R satisfies the smoothing assumption

$$R^*R : continuous \; H^{\frac{1}{2}-\delta}(\Omega) \times H^{-\frac{1}{2}-\delta}(\Omega) \to H_0^{\frac{1}{2}+\delta}(\Omega) \times H^{-\frac{1}{2}+\delta}(\Omega), \tag{10.5.2.3}$$

with R^* the adjoint of R in Y, for a fixed arbitrarily small $\delta > 0$. Then, Theorems 10.2.1, 10.2.2, and 10.2.3 hold true, with

$$Y_\delta^- \equiv H^{\frac{1}{2}-\delta}(\Omega) \times H^{-\frac{1}{2}-\delta}(\Omega); \quad Y_\delta^+ = H_0^{\frac{1}{2}+\delta}(\Omega) \times H^{-\frac{1}{2}+\delta}(\Omega);$$

$$U_\delta^- = H^{\frac{1}{2}-\delta}(\Gamma); \quad U_\delta^+ = H^{\frac{1}{2}+\delta}(\Gamma) \tag{10.5.2.4}$$

(see Section 10.5.6 below for details). In particular, explicitly:

   (b_1) (Regularity of optimal pair) The optimal pair satisfies, for $y_0 = [w_0, w_1] \in H^{\frac{1}{2}-\delta}(\Omega) \times H^{-\frac{1}{2}-\delta}(\Omega)$, the following regularity properties:

$$\begin{bmatrix} w^0(\,\cdot\,, s; y_0) \\ w_t^0(\,\cdot\,, s; y_0) \end{bmatrix} = y^0(\,\cdot\,, s; y_0) \in C\big([s, T]; H^{\frac{1}{2}-\delta}(\Omega) \times H^{-\frac{1}{2}-\delta}(\Omega)\big)$$

$$\cap \, H^{\frac{1}{2}-\delta}(s, T; L_2(\Omega) \times H^{-1}(\Omega)), \tag{10.5.2.5}$$

$$u^0(\,\cdot\,, s; y_0) \in L_2\big(s, T; H^{\frac{1}{2}+\delta}(\Gamma)\big) \cap H^{\frac{1}{2}+\delta}(s, T; L_2(\Gamma)) \equiv H^{\frac{1}{2}+\delta, \frac{1}{2}+\delta}(\Sigma_{sT}),$$

$$\tag{10.5.2.6}$$

$\Sigma_{sT} = [s, T] \times \Gamma$, *a fortiori,*

$$u^0(\,\cdot\,, s; y_0) \in C([s, T]; L_2(\Gamma)). \tag{10.5.2.7}$$

All the above results hold true uniformly in s, that is, with norms that may be made independent of s.

(b_2) (Gain operator) The gain operator $B^* P(t)$, with $P(t)$ defined by (9.2.14) of Chapter 9, satisfies the following regularity property (see (10.2.4)):

$$B^* P(t)x = B^* \begin{bmatrix} P_{11}(t) & P_{12}(t) \\ P_{21}(t) & P_{22}(t) \end{bmatrix} \begin{bmatrix} x_1 \\ x_2 \end{bmatrix} \tag{10.5.2.8a}$$

$$= \frac{\partial}{\partial v_A} \mathcal{A}^{*-\frac{1}{2}} \mathcal{A}^{-\frac{1}{2}} [P_{21}(t)x_1 + P_{22}(t)x_2] \tag{10.5.2.8b}$$

$$: \text{continuous } H^{\frac{1}{2}-\delta}(\Omega) \times H^{-\frac{1}{2}-\delta}(\Omega) \to C([0, T]; L_2(\Gamma)) \tag{10.5.2.8c}$$

(see (10.5.5.14), generalizing the self-adjoint case in (10.5.4.27)), where $\mathcal{A} = $ realization of $\mathcal{A}(x, \partial)$ on $L_2(\Omega)$ with Dirichlet homogeneous boundary conditions [see (10.5.4.1) below] and $\mathcal{A}^*$ is its $L_2(\Omega)$-adjoint.

(b_3) (Differential Riccati equation) With

$$A = \begin{bmatrix} 0 & I \\ -\mathcal{A} & 0 \end{bmatrix}$$

as in (10.5.4.18) below, we have that $P(t)$ satisfies the following DRE:

$$\begin{cases} \dfrac{d}{dt}(P(t)x, y)_Y = -(Rx, Ry)_Y - (P(t)x, Ay)_Y - (P(t)Ax, y)_Y \\ \qquad\qquad + \left(\dfrac{\partial}{\partial v_A} \mathcal{A}^{*-\frac{1}{2}} \mathcal{A}^{-\frac{1}{2}} [P_{21}(t)x_1 + P_{22}(t)x_2], \right. \\ \qquad\qquad \left. \dfrac{\partial}{\partial v_A} \mathcal{A}^{*-\frac{1}{2}} \mathcal{A}^{-\frac{1}{2}} [P_{21}(t)y_1 + P_{22}(t)y_2] \right)_{L_2(\Gamma)}, \\ P(T) = 0, \forall x, y \in H^{\frac{1}{2}-\delta}(\Omega) \times H^{-\frac{1}{2}-\delta}(\Omega). \end{cases} \tag{10.5.2.9}$$

Furthermore, $P(t)$ satisfies a corresponding integral Riccati equation, as in (10.2.7), for all $x, y \in H^{\frac{1}{2}-\delta}(\Omega) \times H^{-\frac{1}{2}-\delta}(\Omega)$.

10.5.3 Regularity Theory for Problem (10.5.1.3) with $u \in L_2(\Sigma)$

The following well-posedness Theorem 10.5.3.2 provides the critical regularity result, which justifies as natural the selection of the cost functional J in (10.5.1.4), and which will permit us to verify assumption (H.1) = (10.1.6) below. To this end, as we shall see, it is expedient to associate with problem (10.5.1.3) the following boundary-homogeneous version (which would be the corresponding adjoint problem (10.5.5.7) below, if the initial conditions were given at $t = T$, an inessential modification since the problem is time reversible):

$$\begin{cases} \phi_{tt} = -\mathcal{A}^*(x, \partial)\phi + f & \text{in } Q; & (10.5.3.1a) \\ \phi(0, x) = \phi_0(x), \quad \phi_t(0, x) = \phi_1(x) & \text{in } \Omega; & (10.5.3.1b) \\ \phi|_\Sigma = 0 & \text{in } \Sigma, & (10.5.3.1c) \end{cases}$$

where, with sufficiently smooth coefficients, say $a_{ij} = a_{ji} \in L_\infty(Q)$

$$-\mathcal{A}^*(x, \partial)v = \sum_{i,j=1}^{n} \frac{\partial}{\partial x_i}\left(a_{ij}(x)\frac{\partial v}{\partial x_j}\right) - \sum_{j=1}^{n} \frac{\partial}{\partial x_j}(b_j(x)v) + c(x)v. \qquad (10.5.3.2)$$

The following key results are taken from Lasiecka and Triggiani [1981; 1983], Lions [1983], and Lasiecka et al. [1986]; see also Lasiecka and Triggiani [1994] for a review and an historical account (summarized in the Notes).

Theorem 10.5.3.1   *With reference to (10.5.3.1), assume $\{\phi_0, \phi_1\} \in H_0^1(\Omega) \times L_2(\Omega)$; $f \in L_1(0, T; L_2(\Omega))$. Then, the unique solution of (10.5.3.1) satisfies*

$$\{\phi, \phi_t\} \in C\big([0, T]; H_0^1(\Omega) \times L_2(\Omega)\big); \qquad (10.5.3.3)$$

$$\phi_{tt} \in C([0, T]; H^{-1}(\Omega)) \quad if\ f = 0, \quad \phi_{tt} \in L_1(0, T; H^{-1}(\Omega)) \quad if\ f \not\equiv 0;$$

and, assuming the boundary Γ of class C^1:

$$\frac{\partial \phi}{\partial \nu_A} \in L_2(0, T; L_2(\Gamma)) \equiv L_2(\Sigma) \qquad (10.5.3.4)$$

continuously, where $\partial/\partial \nu_A$ is the co-normal derivative relative to the self-adjoint principal part of $\mathcal{A}(x, \partial)$,

$$\frac{\partial \phi}{\partial \nu_A} = \sum_{i,j=1}^{n} a_{ij}\nu_i \frac{\partial \phi}{\partial x_j}, \quad \nu_i = \begin{array}{l} ith\ direction\ cosine\ of\ \nu,\ \nu\ being \\ the\ normal\ at\ \Gamma,\ exterior\ to\ \Omega, \end{array} \qquad (10.5.3.5)$$

where, moreover [Lasiecka, Triggiani, 1994, Eqn. (4.17)],

$$c\left|\frac{\partial \phi}{\partial \nu_A}\right| \leq \left|\frac{\partial \phi}{\partial \nu}\right| \leq (\alpha - \epsilon)\left|\frac{\partial \phi}{\partial \nu_A}\right|, \quad c > 0,\ \alpha\ as\ in\ (10.5.1.2). \qquad (10.5.3.6)$$

Proof.   The proof of (10.5.3.4) of Theorem 10.5.3.1 is by PDE energy methods and will be given in Section 10.5.10 below. $\square$

Remark 10.5.3.1   The trace regularity (10.5.3.4) is, of course, the key result of Theorem 10.5.3.1. It does not follow from the optimal interior regularity (10.5.3.3); indeed, (10.5.3.4) is "1/2 higher" in Sobolev space regularity on Ω over a formal application of trace theory to ϕ in (10.5.3.3).

Theorem 10.5.3.2   *With reference to the nonhomogeneous problem (10.5.1.3), with sufficiently smooth coefficients, assume*

$$\{w_0, w_1\} \in L_2(\Omega) \times H^{-1}(\Omega); \quad u \in L_2(\Sigma). \qquad (10.5.3.7)$$

Then, the unique solution of (10.5.1.3) satisfies

$$\{w, w_t, w_{tt}\} \in C([0, T]; L_2(\Omega) \times H^{-1}(\Omega) \times H^{-2}(\Omega)) \qquad (10.5.3.8)$$

continuously, that is, more precisely, there is $C_T > 0$ such that

$$\|\{w, w_t, w_{tt}\}\|^2_{C([0,T];L_2(\Omega) \times H^{-1}(\Omega) \times H^{-2}(\Omega))}$$

$$\leq C_T \left\{ \|u\|^2_{L_2(\Sigma)} + \|\{w_0, w_1\}\|^2_{L_2(\Omega) \times H^{-1}(\Omega)} \right\}. \qquad (10.5.3.9)$$

Proof. The proof of Theorem 10.5.3.2 is a consequence of Theorem 10.5.3.1 and will be given in Section 10.5.11 below. □

10.5.4 Abstract Setting for Problem (10.5.1.3)

We follow [Triggiani, 1977], [Lasiecka, Triggiani, 1981; 1982], and [Lasiecka et al., 1986, Sect. 3]. To put problem (10.5.1.3), (10.5.1.4) into the abstract model (10.1.1), (10.1.3), we introduce the following operators and spaces:

(i)

$$\mathcal{A}h = A(x, \partial)h; \quad \mathcal{A} : \mathcal{D}(\mathcal{A}) \to L_2(\Omega); \quad \mathcal{D}(\mathcal{A}) = H^2(\Omega) \cap H^1_0(\Omega), \qquad (10.5.4.1)$$

where without (essential) loss of generality we shall henceforth take $\mathcal{A}^{-1} \in \mathcal{L}(L_2(\Omega))$, and, moreover, that the *fractional powers of $\mathcal{A}$ are well-defined*. Furthermore, as in the case of Chapter 3, say Section 3.1, we shall freely extend the $\mathcal{A}$ originally defined in (10.5.4.1) [while maintaining the same symbol, with no fear of confusion] as $\mathcal{A} : L_2(\Omega) \to [\mathcal{D}(\mathcal{A}^*)]'$, duality with respect to the pivot space $L_2(\Omega)$, with $\mathcal{A}^*$ the $L_2(\Omega)$-adjoint of $\mathcal{A}$ in (10.5.4.1), that is, the $L_2(\Omega)$-realization of $\mathcal{A}^*(x, \partial)$ in (10.5.3.2). In the absence of first-order terms in (10.5.1.1), $\mathcal{A}$ is self-adjoint in $L_2(\Omega)$, and in fact we are presently taking it [without essential loss of generality] to be positive self-adjoint. We shall use freely below that [Lions, Magenes, 1972, p. 196 with Remark 102.6, p. 121; Lions, 1962, Remark 10.5.1, p. 238]

$$\mathcal{D}(\mathcal{A}) = \mathcal{D}(\mathcal{A}^*), \text{ hence } \mathcal{D}(\mathcal{A}^\theta) = \mathcal{D}(\mathcal{A}^{*\theta}), \quad 0 \leq \theta \leq 1, \qquad (10.5.4.2)$$

as in Eqns. (3.1.16) and (3.1.17) of Chapter 3. Moreover, the following identifications hold true [Grisvard, 1967; Lasiecka, 1980, Appendix; Fujiwara, 1967; see also Chapter 3, Appendix 3A]:

$$\mathcal{D}(\mathcal{A}^\theta) = \mathcal{D}(\mathcal{A}^{*\theta}) = \begin{cases} H^{2\theta}_0(\Omega), & 2\theta < \frac{3}{2}, \quad \theta \neq \frac{1}{4}; & (10.5.4.3) \\ H^{2\theta}(\Omega), & 2\theta < \frac{1}{2}. & (10.5.4.4) \end{cases}$$

Problem (10.5.3.1) can then abstractly be rewritten as

$$\ddot{\phi} + \mathcal{A}^*\phi = f; \quad \phi(0) = \phi_0; \quad \dot{\phi}(0) = \phi_1. \qquad (10.5.4.5)$$

In any case, $-\mathcal{A}^*$ is the generator of a s.c. cosine operator $C^*(t)$ on $L_2(\Omega)$ [Fattorini, 1985], with corresponding "sine" operator $S^*(t) = \int_0^t C^*(\tau)\, d\tau$. [Technically, $\mathcal{A}^{*\frac{1}{2}} S^*(t)$ has the right credentials for being called a sine operator.] Accordingly, the solution of problem (10.5.3.1), or (10.5.4.5), can be written abstractly as

$$\phi(t) = \phi(t; \phi_0, \phi_1; f) = C^*(t)\phi_0 + S^*(t)\phi_1 + \int_0^t S^*(t-\tau)f(\tau)\,d\tau, \quad (10.5.4.6)$$

$$\dot{\phi}(t) = \dot{\phi}(t; \phi_0, \phi_1; f) = -\mathcal{A}^* S^*(t)\phi_0 + C^*(t)\phi_1 + \int_0^t C^*(t-\tau)f(\tau)\,d\tau,$$
$$(10.5.4.7)$$

in appropriate function spaces, depending on $\{\phi_0, \phi_1, f\}$. To this end, we note that:

$$\left\{ \begin{array}{l} \text{The maps } t \to C^*(t)x, \ \mathcal{A}^{*\frac{1}{2}}S^*(t)x \text{ are well defined and continuous} \\ \text{on } L_2(\Omega) \text{ for } x \in L_2(\Omega) \text{ and } \|C^*(t)\|, \ \|\mathcal{A}^{*\frac{1}{2}}S^*(t)\| \leq M_\gamma e^{\gamma t} \\ \text{in the } \mathcal{L}(L_2(\Omega))\text{-norm, for some constants } \gamma, M_\gamma. \end{array} \right. \quad (10.5.4.8)$$

(ii) As in Eqns. (3.1.6) and (3.1.7) of Chapter 3, let D be the Dirichlet map defined by

$$h = Dg \quad \text{iff} \quad -\mathcal{A}(x,\partial)h = 0 \text{ in } \Omega; \quad h|_\Gamma = g \text{ on } \Gamma; \quad (10.5.4.9)$$

$$D : \text{continuous } H^s(\Gamma) \to H^{s+\frac{1}{2}}(\Omega); \text{ in particular,} \quad (10.5.4.10)$$

$$: \text{continuous } L_2(\Gamma) \to H^{\frac{1}{2}}(\Omega) \subset H^{\frac{1}{2}-2\epsilon}(\Omega) = \mathcal{D}\big(\mathcal{A}^{\frac{1}{4}-\epsilon}\big), \forall \epsilon > 0, \quad (10.5.4.11)$$

with the property that [Eqn. (3.1.15) of Chapter 3]

$$D^*\mathcal{A}^*h = -\frac{\partial h}{\partial \nu_\mathcal{A}}, \quad h \in \mathcal{D}(\mathcal{A}^*). \quad (10.5.4.12)$$

(iii) Problem (10.5.1.3) can be written abstractly as in (10.1.1), that is, as

$$y_t = Ay + Bu \quad \text{on } [\mathcal{D}(A^*)]', \quad y(0) = y_0 \in Y; \quad (10.5.4.13)$$

$$Z \equiv Y \equiv L_2(\Omega) \times H^{-1}(\Omega); \quad U = L_2(\Gamma); \quad (10.5.4.14)$$

$$H_0^1(\Omega) = \mathcal{D}\big(\mathcal{A}^{\frac{1}{2}}\big) = \mathcal{D}\big(\mathcal{A}^{*\frac{1}{2}}\big); \quad H^{-1}(\Omega) = \big[\mathcal{D}\big(\mathcal{A}^{\frac{1}{2}}\big)\big]' = \big[\mathcal{D}\big(\mathcal{A}^{*\frac{1}{2}}\big)\big]'$$
$$\text{(equivalent norms)} \quad (10.5.4.15)$$

(duality with respect to $L_2(\Omega)$) with norms

$$\|x\|_{\mathcal{D}(\mathcal{A}^{\frac{1}{2}})} = \big\|\mathcal{A}^{\frac{1}{2}}x\big\|_{L_2(\Omega)}; \quad \|x\|_{\mathcal{D}(\mathcal{A}^{*\frac{1}{2}})} = \big\|\mathcal{A}^{*\frac{1}{2}}x\big\|_{L_2(\Omega)}; \quad (10.5.4.16)$$

$$\|x\|_{[\mathcal{D}(\mathcal{A}^{\frac{1}{2}})]'} = \big\|\mathcal{A}^{*-\frac{1}{2}}x\big\|_{L_2(\Omega)}; \quad \|x\|_{[\mathcal{D}(\mathcal{A}^{*\frac{1}{2}})]'} = \big\|\mathcal{A}^{-\frac{1}{2}}x\big\|_{L_2(\Omega)}; \quad (10.5.4.17)$$

$$y = \begin{bmatrix} w \\ w_t \end{bmatrix}; \quad A = \begin{bmatrix} 0 & I \\ -\mathcal{A} & 0 \end{bmatrix} : \mathcal{D}(A) = \mathcal{D}\big(\mathcal{A}^{\frac{1}{2}}\big) \times L_2(\Omega) \to Y; \quad (10.5.4.18)$$

$$Bu = \begin{bmatrix} 0 \\ \mathcal{A}Du \end{bmatrix} : \text{continuous } U \to [\mathcal{D}(A^*)]' \text{ (duality with respect to } Y),$$
$$(10.5.4.19a)$$

or, equivalently,

$$A^{-1}B : \text{continuous } U \to Y, \tag{10.5.4.19b}$$

and, in fact, even the stronger property

$$A^{-1}Bu = \begin{bmatrix} 0 & -\mathcal{A}^{-1} \\ I & 0 \end{bmatrix} \begin{bmatrix} 0 \\ \mathcal{A}Du \end{bmatrix} = \begin{bmatrix} -Du \\ 0 \end{bmatrix},$$

$$u \in U, \quad Du \in \mathcal{D}(\mathcal{A}^{\frac{1}{4}-\epsilon}) \subset L_2(\Omega) \tag{10.5.4.20}$$

holds true in the present case.

The solution of problem (10.5.1.3), or (10.5.4.13), with initial condition $y(s) = y_0 = [w_0, w_1]$ is written abstractly as

$$\begin{bmatrix} w(t, s; y_0) \\ w_t(t, s; y_0) \end{bmatrix} = e^{A(t-s)} \begin{bmatrix} w_0 \\ w_1 \end{bmatrix} + (L_s u)(t), \tag{10.5.4.21}$$

$$(L_s u)(t) = \int_s^t e^{A(t-\tau)} Bu(\tau)\, d\tau = \begin{bmatrix} \mathcal{A} \int_s^t S(t-\tau) Du(\tau)\, d\tau \\ \mathcal{A} \int_s^t C(t-\tau) Du(\tau)\, d\tau \end{bmatrix}, \tag{10.5.4.22}$$

where A generates a s.c. group e^{At} on Y, $t \in \mathbb{R}$, given by

$$e^{At} = \begin{bmatrix} C(t) & S(t) \\ -\mathcal{A}S(t) & C(t) \end{bmatrix}, \tag{10.5.4.23}$$

where $C(t)$ is even and $S(t)$ is odd.

The Case of $\mathcal{A}$ Self-Adjoint   Henceforth, until otherwise noted in Section 10.5.5, it will be expedient to first consider the case where all first-order terms vanish in (10.5.1.1): $b_j \equiv 0$, so that the operator $\mathcal{A}$ is self-adjoint on $L_2(\Omega)$: $\mathcal{A} = \mathcal{A}^*$. We shall, however, preserve the notation $\mathcal{A}^*$ to express intrinsic properties of relevant formulas. This case with $\mathcal{A}$ self-adjoint will result in particularly simplified formulas for B^* and A^*, and hence for e^{A^*t}. The general case with first-order terms b_j will be handled subsequently by perturbation methods via Theorem 7.5.1 of Chapter 7, in Section 10.5.5 below.

Thus, with $\mathcal{A}$ self-adjoint on $L_2(\Omega)$, we have that A^*, the adjoint of A with respect to $Y = L_2(\Omega) \times [\mathcal{D}(\mathcal{A}^{\frac{1}{2}})]'$, is given by

$$A^* = \begin{bmatrix} 0 & -I \\ \mathcal{A} & 0 \end{bmatrix} = -A, \quad \text{so that } A \text{ is skew-adjoint on } Y, \tag{10.5.4.24}$$

as one sees readily by (10.5.4.18), and then e^{At} is a unitary group on Y,

$$e^{A^*t} = e^{A(-t)} = \begin{bmatrix} C^*(t) & -S^*(t) \\ A^*S^*(t) & C^*(t) \end{bmatrix}, \quad t \in \mathbb{R}, \quad (10.5.4.25)$$

from (10.5.4.23), $C^*(t) = C(t)$, and $S^*(t) = S(t)$ in the present self-adjoint case.

Finally, with $B \in \mathcal{L}(U; [\mathcal{D}(A^*)]')$ and so $B^* \in \mathcal{L}(\mathcal{D}(A^*); U)$, after identifying $[\mathcal{D}(A^*)]''$ with $\mathcal{D}(A^*)$, we compute B^* with respect to $Y \equiv L_2(\Omega) \times [\mathcal{D}(A^{\frac{1}{2}})]'$: For $u \in U$ and $z = [z_1, z_2] \in Y$, via (10.5.4.19a) and (10.5.4.17) with $\mathcal{A}$ self-adjoint,

$$\begin{aligned} (Bu, z)_Y &= (\mathcal{A}Du, z_2)_{[\mathcal{D}(A^{\frac{1}{2}})]'} = (Du, z_2)_{L_2(\Omega)} \\ &= (u, D^*z_2)_{L_2(\Gamma)} = (u, B^*z)_U, \end{aligned} \quad (10.5.4.26)$$

so that we conclude that for $z = [z_1, z_2]$ (and $\mathcal{A}$ self-adjoint)

$$B^*z = B^* \begin{bmatrix} z_1 \\ z_2 \end{bmatrix} = D^*z_2 = \frac{\partial}{\partial \nu_A} \mathcal{A}^{*-1} z_2 : \mathcal{D}(A^*) \to U, \quad (10.5.4.27)$$

recalling (10.5.4.12). Specializing (10.5.4.27) to $z = e^{A^*t}x$ via (10.5.4.25), we obtain for $x = [x_1, x_2] \in Y$:

$$B^*e^{A^*t}x = D^*\mathcal{A}^*S^*(t)x_1 + D^*C^*(t)x_2 \quad (10.5.4.28)$$

$$= D^*\mathcal{A}^*[C^*(t)\mathcal{A}^{*-1}x_2 + S^*(t)x_1]$$

$$\text{(by (10.5.4.6))} \qquad = D^*\mathcal{A}^*\phi(t; \phi_0, \phi_1) \quad (10.5.4.29)$$

$$= \frac{\partial}{\partial \nu_A}\phi(t; \phi_0, \phi_1), \quad (10.5.4.30)$$

after recalling (10.5.4.6) with $f \equiv 0$ and (10.5.4.12), where $\phi(t; \phi_0, \phi_1)$ is the solution of problem (10.5.3.1) with $f \equiv 0$ and with initial conditions

$$\phi_0 = \mathcal{A}^{*-1}x_2, \quad \phi_1 = x_1, \quad (10.5.4.31)$$

so that, via (10.5.4.15), (10.5.4.16), and (10.5.4.30),

$$\|\phi_0\|_{H_0^1(\Omega)} \text{ equivalent to } \|\mathcal{A}^{*-1}x_2\|_{\mathcal{D}(A^{*\frac{1}{2}})} = \|\mathcal{A}^{*-\frac{1}{2}}x_2\|_{L_2(\Omega)}; \quad (10.5.4.32)$$

$$\|\phi_1\|_{L_2(\Omega)} = \|x_1\|_{L_2(\Omega)}; \quad (10.5.4.33)$$

$$x_1 \in L_2(\Omega); \quad x_2 \in [\mathcal{D}(A^{\frac{1}{2}})]'; \quad \text{or } x = [x_1, x_2] \in Y. \quad (10.5.4.34)$$

10.5.5 Verification of Assumption (H.1) = (10.1.6)

Self-Adjoint $\mathcal{A}$ At least when $\mathcal{A}$ is self-adjoint as in case 1, from (10.5.4.30)–(10.5.4.33), we see that (H.1) = (10.1.6) holds true:

$$\int_0^T \|B^*e^{A^*t}x\|_U^2 \, dt \le C_T \|x\|_Y^2, \quad x \in Y, \quad (10.5.5.1)$$

if and only if problem (10.5.3.1) with $f \equiv 0$ satisfies

$$\int_0^T \int_\Gamma \left(\frac{\partial \phi}{\partial \nu_A} \right)^2 d\Sigma \leq C_T \| \{ \phi_0, \phi_1 \} \|^2_{H_0^1(\Omega) \times L_2(\Omega)}, \qquad (10.5.5.2)$$

which is precisely the trace regularity result, guaranteed by Theorem 10.5.3.1, Eqn. (10.5.3.4). Then, according to Chapter 7, Theorem 7.2.1, estimate (10.5.5.1) [i.e., (10.5.5.2)] is, in turn, equivalent to the following property that

$$(L_s u)(t) = \int_s^t e^{A(t-\tau)} Bu(\tau) \, d\tau = \begin{bmatrix} w(t; 0, 0) \\ w_t(t; 0, 0) \end{bmatrix} \qquad (10.5.5.3a)$$

$$: \text{continuous } L_2(s, T; L_2(\Gamma)) \to C([s, T]; L_2(\Omega) \times H^{-1}(\Omega)),$$

$$(10.5.5.3b)$$

uniformly in s, where $w_0 = w_1 = 0$ in problem (10.5.1.3), which is precisely conclusion (10.5.3.8) of Theorem 10.5.3.2, the additional statement of uniformity in s being an immediate consequence of formula (10.5.5.3a) for L_s (via a change of variable). Moreover, recalling (10.5.4.28) and (10.5.4.21), we have, by duality on (10.5.5.3), that

$$(L_s^* v)(t) = B^* \int_t^T e^{A^*(\tau - t)} v(\tau) \, d\tau, \quad s \leq t \leq T \qquad (10.5.5.4)$$

$$\text{(by (10.5.4.28))} \quad = D^* A^* \int_t^T S^*(\tau - t) v_1(\tau) \, d\tau + D^* \int_t^T C^*(\tau - t) v_2(\tau) \, d\tau$$

$$(10.5.5.5)$$

$$: \text{continuous } L_1(s, T; L_2(\Omega) \times H^{-1}(\Omega)) \to L_2(s, T; L_2(\Gamma)),$$

$$(10.5.5.6)$$

uniformly in s. Notice that, if $\psi(t; f)$ is the solution of the problem

$$\begin{cases} \psi_{tt} = -\mathcal{A}^*(x, \partial)\psi + f & \text{in } Q; & (10.5.5.7a) \\ \psi(T, x) = 0, \quad \psi_t(T, x) = 0 & \text{in } \Omega; & (10.5.5.7b) \\ \psi|_\Sigma = 0 & \text{in } \Sigma, & (10.5.5.7c) \end{cases}$$

then the two terms in (10.5.5.5) can be rewritten as

$$v_1(\cdot) \to D^* A^* \int_t^T S^*(\tau - t) v_1(\tau) \, d\tau = D^* A^* \int_T^t S^*(t - \tau) v_1(\tau) \, d\tau \qquad (10.5.5.8)$$

$$\text{(by (10.5.4.12))} \quad = \frac{\partial}{\partial \nu_A} \psi(t; v_1) \qquad (10.5.5.9)$$

$$: \text{continuous } L_1(0, T; L_2(\Omega)) \to L_2(0, T; L_2(\Gamma)),$$

$$(10.5.5.10)$$

recalling (10.5.4.12), and similarly,

$$v_2(\cdot) \to -D^* \int_t^T C^*(\tau - t)v_2(\tau)\,d\tau = D^*\mathcal{A}^* \int_T^t C^*(t - \tau)\mathcal{A}^{*-1}v_2(\tau)\,d\tau$$

(10.5.5.11)

$$(\text{by } (10.5.4.12)) \qquad = \frac{\partial}{\partial v_A}\psi_t(t; \mathcal{A}^{*-1}v_2) \tag{10.5.5.12}$$

$$: \text{continuous } L_1(0, T; H^{-1}(\Omega)) \to L_2(0, T; L_2(\Gamma)). \tag{10.5.5.13}$$

Notice that regularity (10.5.5.10) of the normal trace (10.5.5.9) for the $\psi(t; v_1)$ solution of (10.5.5.7) with $f = v_1 \in L_1(0, T; L_2(\Omega))$ is precisely conclusion (10.5.3.4) of Theorem 10.5.3.1 for the time-reversed problem ϕ in (10.5.3.1) with initial data at $t = 0$, rather than at $t = T$ as for ψ, an inessential modification. The proof of Theorem 10.5.3.1 in Section 10.5.10 is by energy (PDE) methods. Instead, regularity (10.5.5.13) for the normal trace of the time derivative of the solution $\psi(t; f)$ of (10.5.5.7) with $f = \mathcal{A}^{*-1}v_2 \in L_1(0, T; H_0^1(\Omega))$ since $v_2 \in L_1(0, T; H^{-1}(\Omega))$ is obtained by duality via operator methods as in (10.5.5.4)–(10.5.5.6), but it does not appear to be provable by purely PDE methods. (Compare [Lasiecka et al., 1986, Theorem 10.2.2, p. 152] by purely PDE methods against [Lasiecka et al., 1986, Theorem 10.3.11, p. 182] by operator methods in the case of homogeneous boundary conditions, where the latter, unlike the former, requires an additional assumption of *time* regularity of the right-hand side, nonhomogeneous term.) Thus, assumption (H.1) is verified, at least in the self-adjoint case, where the first terms in (10.5.1.1) vanish: $b_j \equiv 0$.

General Case If $\mathcal{A}$ is not self-adjoint ($b_j \not\equiv 0$) and we choose $Y \equiv L_2(\Omega) \times [\mathcal{D}(\mathcal{A}^{*\frac{1}{2}})]'$ as topology on Y [thus exploiting (10.5.4.2) for $\theta = 1/2$], then the computations in (10.5.4.26) give now for $z = [z_1, z_2]$ the generalization of (10.5.4.27) as

$$B^*z = D^*\mathcal{A}^{*\frac{1}{2}}\mathcal{A}^{-\frac{1}{2}}z_2 = \frac{\partial}{\partial v_A}\mathcal{A}^{*-\frac{1}{2}}\mathcal{A}^{-\frac{1}{2}}z_2, \quad \mathcal{D}(A^*) \to U, \tag{10.5.5.14}$$

recalling (10.5.4.12) where $\mathcal{A}^{*\frac{1}{2}}\mathcal{A}^{-\frac{1}{2}}$ is an isomorphism on $L_2(\Omega)$ by (10.5.4.2) with $\theta = 1/2$. Moreover, now

$$A = \begin{bmatrix} 0 & I \\ \mathcal{A}_0 & 0 \end{bmatrix} + \begin{bmatrix} 0 & 0 \\ \mathcal{F} & 0 \end{bmatrix} = A_0 + \Pi, \tag{10.5.5.15}$$

where A_0 (corresponding to the second-order and the zero-order terms in (10.5.1.1)) is self-adjoint, whereas $\mathcal{F}$ corresponds to the first-order terms in (10.5.1.1)). Then, the $L_2(\Omega)$-adjoint $\mathcal{F}^*$ satisfies (for $b_j \in C^1(\bar{\Omega})$ in (10.5.3.2)):

$$\|\mathcal{F}^*f\|_{L_2(\Omega)} \le C\|\|\nabla f\|\|_{L_2(\Omega)}^2 = C\|\mathcal{A}^{*\frac{1}{2}}f\|_{L_2(\Omega)}, \quad f \in \mathcal{D}(\mathcal{A}^{\frac{1}{2}}), \tag{10.5.5.16}$$

or

$$\mathcal{F}^* \mathcal{A}^{*-\frac{1}{2}}, \quad \mathcal{A}^{-\frac{1}{2}} \mathcal{F} : \text{continuous } L_2(\Omega) \to L_2(\Omega); \qquad (10.5.5.17)$$

hence the perturbation operator Π in (10.5.5.15) satisfies

$$\Pi : \text{continuous } Y \to Y. \qquad (10.5.5.18)$$

It follows then, by (10.5.5.14), (10.5.5.15), (10.5.5.18), and the preceding self-adjoint case leading to the verification of (10.5.5.2), the inequality (10.5.5.1) holds true also in the general case, as a specialization of the perturbation theorem in Chapter 7, Section 7.5, Theorem 7.5.1. Assumption (H.1) is verified in general.

10.5.6 Selection of the Spaces U_θ and Y_θ in (10.1.10)

We select the spaces in (10.1.10) to be the following Sobolev spaces:

$$U_\theta \equiv H^\theta(\Gamma); \quad U_0 = U = L_2(\Gamma), \quad 0 \le \theta \le \frac{1}{2} + \delta, \quad \theta \ne \frac{1}{2}; \qquad (10.5.6.1)$$

$$Y_\theta \equiv H_0^\theta(\Omega) \times H^{\theta-1}(\Omega) = \mathcal{D}\left(\mathcal{A}^{\frac{\theta}{2}}\right) \times \mathcal{D}\left(\mathcal{A}^{\frac{\theta-1}{2}}\right),$$

$$0 \le \theta \le \frac{1}{2} + \delta, \quad \theta \ne \frac{1}{2}; \qquad (10.5.6.2)$$

$$H_0^\theta(\Omega) = H^\theta(\Omega), \quad \theta < \frac{1}{2}; \quad Y_0 = Y = L_2(\Omega) \times H^{-1}(\Omega),$$

via (10.5.4.2) and (10.5.4.3), in particular, the critical spaces for $\theta = 1/2 \pm \delta$ are

$$U_\delta^- \equiv U_{\frac{1}{2}-\delta} = H^{\frac{1}{2}-\delta}(\Gamma); \quad U_\delta^+ = U_{\frac{1}{2}+\delta} = H^{\frac{1}{2}+\delta}(\Gamma); \qquad (10.5.6.3)$$

$$Y_\delta^- = Y_{\frac{1}{2}-\delta} \equiv H^{\frac{1}{2}-\delta}(\Omega) \times H^{-\frac{1}{2}-\delta}(\Omega) \equiv \mathcal{D}\left(\mathcal{A}^{\frac{1}{4}-\frac{\delta}{2}}\right) \times \left[\mathcal{D}\left(\mathcal{A}^{\frac{1}{4}+\frac{\delta}{2}}\right)\right]', \qquad (10.5.6.4)$$

with equivalent norms, and duality with respect to the pivot space $L_2(\Omega)$,

$$Y_\delta^+ = Y_{\frac{1}{2}+\delta} \equiv H_0^{\frac{1}{2}+\delta}(\Omega) \times H^{-\frac{1}{2}+\delta}(\Omega) = \mathcal{D}\left(\mathcal{A}^{\frac{1}{4}+\frac{\delta}{2}}\right) \times \left[\mathcal{D}\left(\mathcal{A}^{\frac{1}{4}-\frac{\delta}{2}}\right)\right]', \qquad (10.5.6.5)$$

with equivalent norms, and duality with respect to the pivot space $L_2(\Omega)$. Next, in identifying $[Y_\delta^-]'$ and $[Y_\delta^+]'$, we need to use duality with respect to Y. We find

$$[Y_\delta^-]' = H^{-\frac{1}{2}+\delta}(\Omega) \times H^{-\frac{3}{2}+\delta}(\Omega); \qquad (10.5.6.6)$$

$$[Y_\delta^+]' = H^{-\frac{1}{2}-\delta}(\Omega) \times H^{-\frac{3}{2}-\delta}(\Omega), \qquad (10.5.6.7)$$

since the dual of $H^{-\frac{1}{2}\pm\delta}(\Omega)$ with respect to the pivot space $H^{-1}(\Omega)$ is $H^{-\frac{3}{2}\pm\delta}(\Omega)$. Thus, by (10.5.6.4) and (10.5.6.7) we verify the interpolation property

$$[Y_\delta^-, [Y_\delta^+]']_{\theta=\frac{1}{2}-\delta} = Y, \qquad (10.5.6.8)$$

as required in (10.1.10), since $(1/2 - \delta)(1 - \theta) - (1/2 + \delta)\theta = 0$ for the first component, and $(-\delta - 1/2)(1 - \theta) - (3/2 + \delta)\theta = 0$ for the second component, with $\theta = 1/2 - \delta$. Moreover, the injections $U_{\theta_1} \hookrightarrow U_{\theta_2}$, $Y_{\theta_1} \hookrightarrow Y_{\theta_2}$ are compact, $0 \leq \theta_2 < \theta_1 \leq 1/2 + \delta$, as required in (10.1.10), since Ω is a bounded domain. Thus, the spaces in (10.1.11) and (10.1.12) are in the present case as follows for $0 \leq \theta \leq 1/2 + \delta$, $\theta \neq 1/2$:

$$\mathcal{U}^\theta[s, T] \equiv H^{\theta,\theta}(\Sigma_{sT}) \equiv L_2(s, T; H^\theta(\Gamma)) \cap H^\theta(s, T; L_2(\Gamma)); \qquad (10.5.6.9)$$

$$\mathcal{Y}^\theta[s, T] \equiv L_2\big(s, T; H_0^\theta(\Omega) \times H^{\theta-1}(\Omega)\big) \cap H^\theta(s, T; L_2(\Omega) \times H^{-1}(\Omega)). \qquad (10.5.6.10)$$

10.5.7 Verification of Assumption (H.2) = (10.1.14)

The following regularity result, taken from Lasiecka and Triggiani [1986] and Lasiecka et al. [1986] is critical in verifying assumption (H.2) = (10.1.14).

Theorem 10.5.7.1 *With reference to the nonhomogeneous problem (10.5.1.3), assume*

$$\{w_0, w_1\} \in H^1(\Omega) \times L_2(\Omega), \quad \text{with the compatibility condition } w_0|_\Gamma = u(0); \qquad (10.5.7.1)$$

$$u \in C\big([0, T]; H^{\frac{1}{2}}(\Gamma)\big) \cap H^1(0, T; L_2(\Gamma)); \qquad (10.5.7.2)$$

[a fortiori guaranteed, if

$$u \in H^{1,1}(\Sigma) = L_2(0, T; H^1(\Gamma)) \cap H^1(0, T; L_2(\Gamma)) \qquad (10.5.7.3)$$

by [Lions, Magenes, 1972, Theorem 3.1, p. 19]].
  Then, the unique solution of problem (10.5.1.3) satisfies

$$\{w, w_t, w_{tt}\} \in C([0, T]; H^1(\Omega) \times L_2(\Omega) \times H^{-1}(\Omega)), \qquad (10.5.7.4)$$

a fortiori,

$$w \in H^{1,1}(Q) = L_2(0, T; H^1(\Omega)) \cap H^1(0, T; L_2(\Omega)), \qquad (10.5.7.5)$$

continuously.

Proof. The proof of Theorem 10.5.7.1 will be given in Section 10.5.12 below. □

Corollary 10.5.7.2 *With reference to the nonhomogeneous problem (10.5.1.3), assume*

$$w_0 = w_1 = 0; \quad u \in H^{\theta,\theta}(\Sigma) = \mathcal{U}^\theta[0, T] = L_2(0, T; H^\theta(\Gamma)) \cap H^\theta(0, T; L_2(\Gamma)),$$

$$0 \leq \theta < \frac{1}{2}. \qquad (10.5.7.6)$$

Then, the unique solution of problem (10.5.1.3) satisfies

$$\left.\begin{array}{l} w(\,\cdot\,; 0, 0) \in C([0, T]; H^\theta(\Omega)) \\ w_t(\,\cdot\,; 0, 0) \in C([0, T]; H^{\theta-1}(\Omega)) \end{array}\right\} \Rightarrow Lu \in \mathcal{Y}^\theta[0, T]; \quad (10.5.7.7)$$

$$w_{tt}(\,\cdot\,; 0, 0) \in C([0, T]; H^{\theta-2}(\Omega)); \quad\quad\quad\quad (10.5.7.8)$$

$$D_t^\theta w(\,\cdot\,; 0, 0) \in L_2(0, T; L_2(\Omega)). \quad\quad\quad\quad (10.5.7.9)$$

Proof of Corollary 10.5.7.2 For $\theta < 1/2$, the compatibility condition in (10.5.7.1), which now reads $u(0) = w_0|_\Gamma = 0$, does not interfere, and we then interpolate between (10.5.5.3b) or (10.5.3.7), (10.5.3.8) for $\theta = 0$ and (10.5.7.3), (10.5.7.4) in the form $u \in H^{1,1}(\Sigma)$ for $\theta = 1$, thereby obtaining (10.5.7.7) and (10.5.7.8), as desired.

Next, we apply the intermediate derivative theorem [Lions, Magenes, 1972, p. 19] and obtain

$$D_t^r w(\,\cdot\,; 0, 0) \in L_2(0, T; H^{\theta-r}(\Omega))$$

from which (10.5.7.9) follows. Hence, the implication in (10.5.7.7) is proved. $\square$

Corollary 10.5.7.2 plainly verifies assumption $(H.2) = (10.1.14)$, except for uniformity in s, which follows from (10.1.2b) for L_s, by a change of variable.

10.5.8 Verification of Assumption $(H.3) = (10.1.15)$

Verification of assumption $(H.3) = (10.1.15)$ is based upon the following key regularity result.

Theorem 10.5.8.1

(i) *With reference to the operator L^* defined by (10.5.5.4), we have for $0 \le r \le 1$:*

$$(L_s^* v)(t) = B^* \int_t^T e^{A^*(\tau-t)} v(\tau)\, d\tau, \quad s \le t \le T$$

: continuous $L_1(s, T; \mathcal{D}(A^{*r})) \to H^{r,r}(\Sigma_{sT}) = \mathcal{U}^r[s, T]$,

$$0 \le r \le 1, \quad (10.5.8.1)$$

uniformly in s, where

$$\mathcal{D}(A^{*r}) = \mathcal{D}\big(A^{*\frac{r}{2}}\big) \times \mathcal{D}\big(A^{*\frac{r-1}{2}}\big), \quad 0 \le r \le 1. \quad (10.5.8.2)$$

(ii) *Restricting r to $r \le 1/2 + \delta$, and recalling (10.5.4.3) and (10.5.6.2), we have, by (10.5.8.2), setting $\theta = r \le 1/2 + \delta$,*

$$\mathcal{D}(A^{*\theta}) = \mathcal{D}\big(A^{*\frac{\theta}{2}}\big) \times \mathcal{D}\big(A^{*\frac{\theta-1}{2}}\big) = H_0^\theta(\Omega) \times H^{\theta-1}(\Omega) \equiv Y_\theta,$$

$$0 \le \theta \le \frac{1}{2} + \delta, \quad \theta \ne \frac{1}{2}, \quad (10.5.8.3)$$

so that (10.5.8.1) specializes to

$$L_s^* : continuous \ L_2(0, T; Y_\theta) \to \mathcal{U}^\theta[0, T], \quad 0 \le \theta \le \frac{1}{2} + \delta, \quad \theta \ne \frac{1}{2},$$

$$(10.5.8.4)$$

uniformly in s.

Proof. The proof of Theorem 10.5.8.1 will be given in Section 10.5.13 below. Statement (10.5.8.4) verifies assumption (H.3) = (10.1.15). □

10.5.9 *Verification of Assumptions (H.4) = (10.1.18) through (H.7) = (10.1.21)*

Verification of Assumption (H.4) = (10.1.18) (Direct verification)   For

$$x = [x_1, x_2] \in Y_\delta^+ = \mathcal{D}\big(\mathcal{A}^{*\frac{1}{4} + \frac{\delta}{2}}\big) \times \big[\mathcal{D}\big(\mathcal{A}^{*\frac{1}{4} - \frac{\delta}{2}}\big)\big]'$$

by (10.5.6.5) and (10.5.4.2), we compute starting from (10.5.4.28), initially in the case of $\mathcal{A}$ self-adjoint:

$$\begin{aligned}
B^* e^{A^* t} x &= D^* A^* S^*(t) x_1 + D^* C^*(t) x_2 \\
&= D^* \mathcal{A}^{*\frac{1}{4} - \frac{\delta}{2}} \mathcal{A}^{*\frac{1}{2}} S^*(t) \mathcal{A}^{*\frac{1}{4} + \frac{\delta}{2}} x_1 \\
&\quad + D^* \mathcal{A}^{*\frac{1}{4} - \frac{\delta}{2}} C^*(t) \mathcal{A}^{*-\frac{1}{4} + \frac{\delta}{2}} x_2 \in C([0, T]; L_2(\Gamma)), \quad (10.5.9.1)
\end{aligned}$$

as required by (10.1.18), since by duality on (10.5.4.11) and by (10.5.4.8)

$$D^* \mathcal{A}^{*\frac{1}{4} - \frac{\delta}{2}} \in \mathcal{L}(L_2(\Omega); L_2(\Gamma));$$

$$t \to \mathcal{A}^{*\frac{1}{2}} S^*(t); \ C^*(t) \text{ strongly continuous on } L_2(\Omega); \quad (10.5.9.2)$$

$$\mathcal{A}^{*\frac{1}{4} + \frac{\delta}{2}} x_1 \in L_2(\Omega); \quad \mathcal{A}^{*-\frac{1}{2} + \frac{\delta}{2}} x_2 \in L_2(\Omega). \quad (10.5.9.3)$$

In the general case with $\mathcal{A}$ non–self-adjoint, we use the modification (10.5.5.14) for B^* over (10.5.4.27), where, however, the new term $\mathcal{A}^{*\frac{1}{2}} \mathcal{A}^{-\frac{1}{2}}$ is an isomorphism on $L_2(\Omega)$, as noted below (10.5.5.14), and thus conclusion (10.5.9.1) holds true in general.

Verification of Assumption (H.5) = (10.1.19)   With

$$Y_\delta^- \equiv \mathcal{D}\big(\mathcal{A}^{\frac{1}{4} - \frac{\delta}{2}}\big) \times \big[\mathcal{D}\big(\mathcal{A}^{\frac{1}{4} + \frac{\delta}{2}}\big)\big]'$$

by (10.5.6.4), $C(t)$ and $S(t)$ are likewise s.c. cosine/sine operators on any space $\mathcal{D}(\mathcal{A}^\theta)$ and its dual; hence e^{At} in (10.5.1.34) is an s.c. group on Y_δ^- as well. The space $\mathcal{D}(A)$ in (10.5.4.18) is clearly dense in the space Y_δ^- in (10.5.6.4).

Verification of Assumption (H.6) = (10.1.20)   For

$$x = [x_1, x_2] \in Y_\delta^- = \mathcal{D}\big(\mathcal{A}^{\frac{1}{4} - \frac{\delta}{2}}\big) \times \big[\mathcal{D}\big(\mathcal{A}^{\frac{1}{4} + \frac{\delta}{2}}\big)\big]',$$

we obtain, via (10.5.4.18),

$$Ax = \begin{bmatrix} 0 & I \\ -\mathcal{A} & 0 \end{bmatrix} \begin{bmatrix} x_1 \\ x_2 \end{bmatrix}$$

$$= \begin{bmatrix} x_2 \\ -\mathcal{A}x_1 \end{bmatrix} \in \left[\mathcal{D}(\mathcal{A}^{\frac{1}{4}+\frac{\delta}{2}}) \right]' \times \left[\mathcal{D}(\mathcal{A}^{\frac{3}{4}+\frac{\delta}{2}}) \right]' \qquad (10.5.9.4)$$

$$\subset \left[H_0^{\frac{1}{2}+\delta}(\Omega) \right]' \times \left[H_0^{\frac{3}{2}+\delta}(\Omega) \right]' \qquad (10.5.9.5)$$

$$\text{(by (10.5.6.7))} \qquad = H^{-\frac{1}{2}-\delta}(\Omega) \times H^{-\frac{3}{2}-\delta}(\Omega) = [Y_\delta^+]', \qquad (10.5.9.6)$$

where in going from (10.5.9.4) to (10.5.9.5) we have used (10.5.4.3) with $2\theta = 1/2 + \delta$ for the first component space, and

$$H_0^{\frac{3}{2}+\delta}(\Omega) = \left\{ f \in H^{\frac{3}{2}+\delta}(\Omega) : f|_\Gamma = 0, \left. \frac{\partial f}{\partial \nu} \right|_\Gamma = 0 \right\}$$

$$\subset \mathcal{D}(\mathcal{A}^{\frac{3}{4}+\frac{\delta}{2}}) = \left\{ f \in H^{\frac{3}{2}+\delta}(\Omega) : f|_\Gamma = 0 \right\}, \qquad (10.5.9.7)$$

via Grisvard [1967], for the second component space. Thus (10.5.9.6) proves (H.6)= (10.1.20).

Remark 10.5.9.1 Returning to (10.5.4.20), we see via (10.5.6.4) that, in the present case,

$$A^{-1}B : \text{continuous } U \rightarrow Y_\delta^-, \qquad (10.5.9.8)$$

which is property (10.1.32). Thus, as remarked below (10.1.32), property (10.5.9.8), along with (H.5) = (10.1.19) and (H.6) = (10.1.20), reprove (H.4) = (10.1.18).

Verification of Assumption (H.7) = (10.1.21)   Recalling (10.5.4.19a), (10.5.6.3), and (10.5.6.7), we show that

$$Bu = \begin{bmatrix} 0 \\ \mathcal{A}Du \end{bmatrix} : \text{continuous } U_\delta^- = H^{\frac{1}{2}-\delta}(\Gamma) \rightarrow [Y_\delta^+]' = H^{-\frac{1}{2}-\delta}(\Omega) \times H^{-\frac{3}{2}-\delta}(\Omega).$$

$$(10.5.9.9)$$

In fact, by (10.5.4.10), and (10.5.4.4) with $2\theta = 1/2 - \delta$:

$$u \in H^{\frac{1}{2}-\delta}(\Gamma) \Rightarrow Du \in H^{1-\delta}(\Omega) \subset H^{\frac{1}{2}-\delta}(\Omega) \equiv \mathcal{D}(\mathcal{A}^{\frac{1}{4}-\frac{\delta}{2}}), \quad (10.5.9.10)$$

so that, invoking (10.5.9.7) on (10.5.9.10),

$$\mathcal{A}Du \in \left[\mathcal{D}(\mathcal{A}^{*\frac{3}{4}+\frac{\delta}{2}}) \right]' \subset \left[H_0^{\frac{3}{2}+\delta}(\Omega) \right]' \equiv H^{-\frac{3}{2}-\delta}(\Omega). \qquad (10.5.9.11)$$

Then (10.5.9.11) proves (10.5.9.9), as desired.

10.5.10 Proof of Theorem 10.5.3.1

The interior regularity on $\{\phi, \phi_t, \phi_{tt}\}$ in (10.5.3.3) is standard [see paragraph below] whereas the key part of the theorem is the trace regularity (10.5.3.4) [which does *not* follow from the interior regularity of ϕ in (10.5.3.3) via trace theory]. Accordingly, we shall concentrate on (10.5.3.4).

Interior Regularity (10.5.3.3)   It is customary to prove (10.5.3.3) by energy methods: One multiplies Eqn. (10.5.3.1a) by ϕ_t (for smooth initial data) and integrates by parts using Green's first identity [Miklin, 1970, p. 211].

An alternative operator proof of (10.5.3.3) is as follows. We return to the explicit formula (10.5.4.6) for the solution ϕ, where, via (10.5.4.8) and $H_0^1(\Omega) = \mathcal{D}(\mathcal{A}^{*\frac{1}{2}})$ [see (10.5.4.3) for $\theta = 1/2$], we obtain

$$\{\phi_0, \phi_1\} \in \mathcal{D}\big(\mathcal{A}^{*\frac{1}{2}}\big) \times L_2(\Omega) \Rightarrow C^*(t)\phi_0 + S^*(t)\phi_1 \in C\big([0, T]; \mathcal{D}\big(A^{*\frac{1}{2}}\big)\big),$$

$$(10.5.10.1)$$

while, by convolution with $\mathcal{A}^{*\frac{1}{2}} S(t) : \mathcal{L}(L_2(\Omega); C([0, T]; L_2(\Omega))$ [see (10.5.4.8)] we have

$$f \in L_1(0, T; L_2(\Omega)) \Rightarrow \int_0^t S^*(t - \tau)f(\tau)\, d\tau \in C\big([0, T]; \mathcal{D}\big(\mathcal{A}^{*\frac{1}{2}}\big)\big).$$

$$(10.5.10.2)$$

Then, (10.5.10.1) and (10.5.10.2) prove (10.5.3.3) for ϕ. A similar argument performed on ϕ_t given by (10.5.4.7) yields (10.5.3.3) for ϕ_t.

Trace Regularity (10.5.3.4) in the Canonical Case $-\mathcal{A}^*(x, \partial) = \Delta$ We assume the boundary Γ to be of class C^1.

Step 1   For problem

$$\begin{cases} \phi_{tt} = \Delta\phi + f & \text{in } (0, T] \times \Omega \equiv Q; & (10.5.10.3\text{a}) \\ \phi(0, x) = \phi_0(x), \quad \phi_t(0, x) = \phi_1(x) & \text{in } \Omega; & (10.5.10.3\text{b}) \\ \phi|_\Sigma \equiv 0 & \text{in } (0, T] \times \Gamma \equiv \Sigma, & (10.5.10.3\text{c}) \end{cases}$$

we shall first show a key identity.

Lemma 10.5.10.1   *Let ϕ be a solution of Eqn. (10.5.10.3a) (with no boundary conditions imposed) for smooth data, say*

$$\{\phi_0, \phi_1, f\} \in \mathcal{D}(\mathcal{A}^*) \times \mathcal{D}\big(\mathcal{A}^{*\frac{1}{2}}\big) \times L_1\big(0, T; \mathcal{D}\big(\mathcal{A}^{*\frac{1}{2}}\big)\big). \quad (10.5.10.4)$$

Let $h(x) = [h_1(x), \ldots, h_n(x)] \in [C^1(\bar{\Omega})]^n$ be a given vector field. Then the following *identity holds true:*

$$\int_\Sigma \frac{\partial \phi}{\partial v}(h \cdot \nabla \phi)\, d\Sigma + \frac{1}{2}\int_\Sigma \phi_t^2 h \cdot v\, d\Sigma - \frac{1}{2}\int_\Sigma |\nabla \phi|^2 h \cdot v\, d\Sigma$$

$$= \int_Q H\nabla \phi \cdot \nabla \phi\, dQ + \frac{1}{2}\int_Q \left(\phi_t^2 - |\nabla \phi|^2\right) \operatorname{div} h\, dQ$$

$$+ \left[(\phi_t, h \cdot \nabla \phi)_{L_2(\Omega)}\right]_0^T - \int_Q f\, h \cdot \nabla \phi\, d\phi, \tag{10.5.10.5}$$

where

$$H(x) = \begin{bmatrix} \frac{\partial h_1}{\partial x_1}, & \cdots & \frac{\partial h_1}{\partial x_n} \\ & \cdots & \\ \frac{\partial h_n}{\partial x_1}, & \cdots & \frac{\partial h_n}{\partial x_n} \end{bmatrix}, \quad v(x) = \text{outward unit normal vector at } x \in \Gamma.$$

$$\tag{10.5.10.6}$$

[In fact, identity (10.5.10.5) may be extended to all initial data $\{\phi_0, \phi_1, f\} \in H_0^1(\Omega) \times L_2(\Omega) \times L_1(0, T; H_0^1(\Omega))$ by (10.5.3.3) or by (10.5.10.1) and (10.5.10.2).]

Proof of Lemma 10.5.10.1 In the proof, we shall frequently use the identity

$$\int_\Omega h \cdot \nabla \psi\, d\Omega = \int_\Gamma \psi h \cdot v\, d\Gamma - \int_\Omega \psi \operatorname{div} h\, d\Omega, \tag{10.5.10.7}$$

obtained by applying the divergence (Gauss) theorem on $\operatorname{div}(\psi h) = h \cdot \nabla \psi + \psi \operatorname{div} h$, $\psi \in H^1(\Omega)$.

To begin the proof, we take data as in (10.5.10.4). Then an operator argument, analogous to the one in (10.5.10.1) and (10.5.10.2), shows readily that

$$\{\phi, \phi_t\} \in C\left([0, T]; \mathcal{D}(\mathcal{A}^*) \times \mathcal{D}\left(\mathcal{A}^{*\frac{1}{2}}\right)\right), \tag{10.5.10.8}$$

so that, via (10.5.4.1), (10.5.4.2), and (10.5.4.15), the computations below are justified and all terms in (10.5.10.5) are well-defined.

We multiply Eqn. (5.10.3a) by $h \cdot \nabla \phi \in C([0, T]; H^1(\Omega))$ and integrate by parts. On the left-hand side of Eqn. (10.5.10.3a), we obtain integrating by parts in t:

$$\int_0^T \int_\Omega \phi_{tt} h \cdot \nabla \phi\, dt\, d\Omega = \int_\Omega [\phi_t h \cdot \nabla \phi]_0^T\, d\Omega - \int_0^T \int_\Omega \phi_t h \cdot \nabla \phi_t\, dt\, d\Omega$$

$$= \int_\Omega [\phi_t h \cdot \nabla \phi]_0^T\, d\Omega - \frac{1}{2}\int_0^T \int_\Omega h \cdot \nabla\left(\phi_t^2\right) dt\, d\Omega$$

$$\tag{10.5.10.9}$$

$$(\text{by } (10.5.10.7)) \quad = \int_\Omega [\phi_t h \cdot \nabla \phi]_0^T\, d\Omega - \frac{1}{2}\int_\Sigma \phi_t^2 h \cdot v\, d\Sigma$$

$$+ \frac{1}{2}\int_Q \phi_t^2 \operatorname{div} h\, dQ, \tag{10.5.10.10}$$

where in the last step, from (10.5.10.9) to (10.5.10.10), we have invoked identity (10.5.10.7) with $\psi = \phi_t^2$ on the last integral of (10.5.10.9).

On the right-hand side of Eqn. (10.5.10.3a), we obtain by Green's first identity, or (10.5.10.7),

$$\int_\Omega \Delta\phi \, h \cdot \nabla\phi \, d\Omega = \int_\Gamma \frac{\partial\phi}{\partial\nu} h \cdot \nabla\phi \, d\Gamma - \int_\Omega \nabla\phi \cdot \nabla(h \cdot \nabla\phi) \, d\Omega. \quad (10.5.10.11)$$

We next verify the following identity for $\psi \in H^2(\Omega)$ by direct computations:

$$\nabla\psi \cdot \nabla(h \cdot \nabla\psi) = H\nabla\psi \cdot \nabla\psi + \frac{1}{2} h \cdot \nabla(|\nabla\psi|^2), \quad (10.5.10.12)$$

where H is given by (10.5.10.6). Substituting (10.5.10.12) with $\psi = \phi$ into (10.5.10.11) yields

$$\int_\Omega \Delta\phi \, h \cdot \nabla\phi \, d\Omega = \int_\Gamma \frac{\partial\phi}{\partial\nu} h \cdot \nabla\phi \, d\Gamma - \int_\Omega H\nabla\phi \cdot \nabla\phi \, d\Omega$$

$$- \frac{1}{2} \int_\Omega h \cdot \nabla(|\nabla\phi|^2) \, d\Omega \quad (10.5.10.13)$$

$$\text{(by (10.5.10.7))} \quad = \int_\Gamma \frac{\partial\phi}{\partial\nu} h \cdot \nabla\phi \, d\Gamma - \int_\Omega H\nabla\phi \cdot \nabla\phi \, d\Omega$$

$$- \frac{1}{2} \int_\Gamma |\nabla\phi|^2 h \cdot \nu \, d\Gamma$$

$$+ \frac{1}{2} \int_\Omega |\nabla\phi|^2 \, \mathrm{div}\, h \, d\Omega, \quad (10.5.10.14)$$

where in the last step, from (10.5.10.13) to (10.5.10.14), we have invoked identity (10.5.10.7) with $\psi = |\nabla\phi|^2$. Finally, integrating (10.5.10.14) in t over $[0, T]$ yields the final identity for the right-hand side of Eqn. (10.5.10.3a):

$$\int_Q \Delta\phi \, h \cdot \nabla\phi \, dQ + \int_Q fh \cdot \nabla\phi \cdot dQ$$

$$= \int_\Sigma \frac{\partial\phi}{\partial\nu} h \cdot \nabla\phi \, d\Sigma - \int_Q H\nabla\phi \cdot \nabla\phi \, dQ - \frac{1}{2} \int_\Sigma |\nabla\phi|^2 h \cdot \nu \, d\Sigma$$

$$+ \frac{1}{2} \int_Q |\nabla\phi|^2 \, \mathrm{div}\, h \, dQ + \int_Q fh \cdot \nabla\phi \, dQ. \quad (10.5.10.15)$$

Equating (10.5.10.10) with (10.5.10.15), via (10.5.10.3a), yields the desired identity (10.5.10.5). □

Step 2   We next exploit also the boundary condition (10.5.3.1c), make a special choice of the vector field $h(x) \in [C^1(\bar\Omega)]^n$, and prove the desired trace regularity (10.5.3.4) as a consequence of Lemma 10.5.10.1 and of the interior regularity (10.5.3.3).

Theorem 10.5.10.2 *Let* Γ *be of class* C^1.

(i) *The solution of the mixed problem (10.5.10.3a,c) due to the data in (10.5.10.4)*
satisfies the identity

$$\frac{1}{2} \int_{\Sigma} \left(\frac{\partial \phi}{\partial \nu}\right)^2 d\Sigma = \int_Q H \nabla \phi \cdot \nabla \phi \, dQ + \frac{1}{2} \int_Q (\phi_t^2 - |\nabla \phi|^2) \, \text{div } h \, dQ$$

$$+ \left[(\phi, h \cdot \nabla \phi)_{L_2(\Omega)}\right]_0^T - \int_Q f h \cdot \nabla \phi \, dQ, \quad (10.5.10.16)$$

to which identity (10.5.10.5) reduces for any special choice of a vector field
$h(x) \in [C^1(\bar{\Omega})]^n$ *such that*

$$h(x)|_\Gamma = \nu(x) = \text{outward unit normal at } x \in \Gamma. \quad (10.5.10.17)$$

(ii) *Given any* $0 < T < \infty$, *there is a constant* $C_T > 0$ *such that*

$$\int_{\Sigma} \left(\frac{\partial \phi}{\partial \nu}\right)^2 d\Sigma \leq C_T \left[\|\{\phi_0, \phi_1\}\|^2_{H_0^1(\Omega) \times L_2(\Omega)} + \|f\|^2_{L_1(0,T;L_2(\Omega))}\right]. \quad (10.5.10.18)$$

Proof. (i) First, the BC (10.5.3.1c) yields

$$\phi_t \equiv 0 \text{ in } \Sigma; \quad \nabla \phi \perp \Gamma; \quad |\nabla \phi| \equiv \left|\frac{\partial \phi}{\partial \nu}\right|. \quad (10.5.10.19)$$

Moreover, if $\tau = \tau(x)$ denotes a unit tangent vector at $x \in \Gamma$,

$$h = (h \cdot \nu)\nu + (h \cdot \tau)\tau \text{ on } \Gamma; \quad \text{hence } h \cdot \nabla \phi = (h \cdot \nu)\frac{\partial \phi}{\partial \nu}, \quad (10.5.10.20)$$

since $\tau \cdot \nabla \phi \equiv 0$ on Γ. Inserting (10.5.10.19) and (10.5.10.20) on the left-hand side
(L.H.S.) of identity (10.5.10.5), we obtain

$$\text{L.H.S. of (10.5.10.5)} = \frac{1}{2} \int_{\Sigma} \left(\frac{\partial \phi}{\partial \nu}\right)^2 h \cdot \nu \, d\Sigma = \frac{1}{2} \int_{\Sigma} \left(\frac{\partial \phi}{\partial \nu}\right)^2 d\Sigma, \quad (10.5.10.21)$$

since $h \cdot \nu \equiv 1$ on Γ by (10.5.10.17). Then, (10.5.10.21) and the right-hand side of
(10.5.10.5) prove identity (10.5.10.16), as desired.

(ii) We next invoke the interior regularity (10.5.3.3) on the right-hand side of identity
(10.5.10.16), which can then be extended, and readily obtain inequality (10.5.10.18).
Theorem 10.5.10.2 is proved. □

Theorem 10.5.10.2(ii) establishes the trace regularity (10.5.3.4) of Theorem
10.5.3.1, at least in the canonical case $-\mathcal{A}^*(x, \partial) = \Delta$.

Trace Regularity (10.5.3.4) in the General Case of $-\mathcal{A}^*(x, \partial)$ **as in (10.5.3.2)**
The above proof in the canonical case can be directly extended via the general Green
identity to the general case as in (10.5.3.2), modulo additional technical details, which

are given in Lions [1983] and [Lasiecka et al., 1986, Section 4], where the coefficients are even allowed to be time dependent.

Remark 10.5.10.1   As a specialization of identity (10.5.10.5) in Lemma 10.5.10.1 to the one-dimensional case, say $\Omega = (0, 1)$, we obtain *trace regularity results for second-order hyperbolic equations with Neumann (mixed) BC:*

$$
\begin{cases}
\phi_{tt} = \phi_{xx} + f & \text{in } (0, T] \times \Omega; & (10.5.10.22a) \\
\phi(0, \cdot) = \phi_0, \quad \phi_t(0, \cdot) = \phi_1 & \text{in } \Omega; & (10.5.10.22b) \\
\phi|_{x=0} \equiv 0, \quad \phi_x|_{x=1} \equiv 0 & \text{in } (0, T] \times \Gamma; & (10.5.10.22c)
\end{cases}
$$

$$
\Gamma = \Gamma_0 \cup \Gamma_1, \qquad \Gamma_0 = \{x = 0\}, \qquad \Gamma_1 = \{x = 1\}.
$$

Theorem 10.5.10.3   *With reference to problem (10.5.10.22), the following trace regularity result holds true: The map*

$$
\begin{cases}
\{\phi_0, \phi_1, f\} \longrightarrow \{\phi_x|_{x=0}, \phi_t|_{x=1}\} \\
\text{is continuous} \\
H^1(\Omega) \times L_2(\Omega) \times L_1(0, T; L_2(\Omega)) \to L_2(0, T) \times L_2(0, T),
\end{cases} \quad (10.5.10.23)
$$

that is, there is a constant $C_T > 0$ such that

$$
\int_0^T [(\phi_x|_{x=0})^2 + (\phi_t|_{x=1})^2] \, dt \leq C_T \big[\|\{\phi_0, \phi_1\}\|^2_{H^1(\Omega) \times L_2(\Omega)} + \|f\|^2_{L_1(0,T;L_2(\Omega))} \big].
$$
$$(10.5.10.24)$$

Proof.   We return to identity (10.5.10.5). On its left-hand side (L.H.S.), we use

$$
\text{on } \Gamma_0, \, x = 0 : \phi_t \equiv 0, \quad \frac{\partial \phi}{\partial \nu} h \cdot \nabla \phi - \frac{1}{2} |\nabla \phi|^2 h \cdot \nu = \frac{1}{2} \phi_x^2 h \cdot \nu;
$$

$$
\text{on } \Gamma_1, \, x = 1 : \frac{\partial \phi}{\partial \nu} = \phi_x = \nabla \phi \equiv 0 \text{ (because of dim } \Omega = 1),
$$

so that

$$
\text{L.H.S. of (5.10.5)} = \int_\Sigma \frac{\partial \phi}{\partial \nu} h \cdot \nabla \phi \, d\Sigma + \frac{1}{2} \int_\Sigma \phi_t^2 h \cdot \nu \, d\Sigma - \frac{1}{2} \int_\Sigma |\nabla \phi|^2 h \cdot \nu \, d\Sigma
$$

$$
= \frac{1}{2} \int_0^T [(\phi_x|_{x=0})^2 + (\phi_t|_{x=1})^2] \, dt, \qquad (10.5.10.25)
$$

taking $h \cdot \nu = 1$. However, we have seen in part (ii) of the proof of Theorem 10.5.10.2, the right-hand side of (10.5.10.5) satisfies also in the Neumann case:

$$
\text{R.H.S. of (10.5.10.5)} = \mathcal{O}_T \big\{ \|\{\phi_0, \phi_1\}\|^2_{H^1(\Omega) \times L_2(\Omega)} + \|f\|^2_{L_1(0,T;L_2(\Omega))} \big\}.
$$
$$(10.5.10.26)$$

Combining (10.5.10.25) with (10.5.10.26) yields (10.5.10.24), as desired. $\square$

The above result (which holds true also with d^2/dx^2 replaced by a more general operator as in (10.5.1.1) and (10.5.1.2) in the one-dimensional case) includes (10.7.4.17b) of Theorem 10.7.4.1, as well as (10.8.2.34) of Proposition 10.8.2.2.

We recall that an elementary proof of dual regularity results, in the case dim $\Omega = 1$, was given in Chapter 9, Section 9.8.4.

10.5.11 Proof of Theorem 10.5.3.2

A proof that the trace regularity for the homogeneous ϕ-problem (10.5.3.1),

$$\left.\begin{array}{l} \{\phi_0, \phi_1\} \in H_0^1(\Omega) \times L_2(\Omega) \\ f \equiv 0 \end{array}\right\} \Rightarrow \frac{\partial \phi}{\partial \nu_A} \in L_2(0, T; L_2(\Gamma)) \equiv L_2(\Sigma), \quad (10.5.11.1)$$

as in Theorem 10.5.3.1, implies the interior regularity

$$\left.\begin{array}{l} u \in L_2(0, T; L_2(\Gamma)) \equiv L_2(\Sigma) \\ w_0 = w_1 = 0 \end{array}\right\} \Rightarrow$$

$$\begin{bmatrix} w(t) \\ w_t(t) \end{bmatrix} = \begin{bmatrix} A \int_0^t S(t-\tau) Du(\tau)\, d\tau \\ A \int_0^t C(t-\tau) Du(\tau)\, d\tau \end{bmatrix} \in C([0, T]; Y), \quad (10.5.11.2)$$

$$Y = L_2(\Omega) \times H^{-1}(\Omega) = L_2(\Omega) \times \left[\mathcal{D}\left(A^{\frac{1}{2}}\right)\right]' \quad (10.5.11.3)$$

(see (10.5.4.22)) for the nonhomogeneous w-problem (10.5.1.3) is, in effect, already contained in the preceding development. We shall briefly review it by singling out the main highlights.

Operator-Theoretic Proof of Theorem 10.5.3.2. Step 1   We have already seen in Section 10.5.4, Eqns. (10.5.4.28)–(10.5.4.31), that

$$\frac{\partial \phi}{\partial \nu_A}(t; \phi_0, \phi_1) = D^* A^* S^*(t) x_1 + D^* A^{*\frac{1}{2}} C^*(t) A^{*-\frac{1}{2}} x_2, \quad (10.5.11.4)$$

$$\phi_0 = A^{*-1} x_2, \quad \phi_1 = x_1, \quad (10.5.11.5)$$

and, at least for $A = A^*$ self-adjoint on $L_2(\Omega)$,

$$B^* e^{A^* t} x = \frac{\partial \phi}{\partial \nu_A}(t; \phi_0, \phi_1), \quad x = [x_1, x_2], \quad (10.5.11.6)$$

while in the general case analyzed at the end of Section 10.5 (from Eqn. (10.5.5.14) on)

$$\|B^* e^{A^*} \cdot x\|_{L_2(0,T;L_2(\Gamma))} \le C_T \left\| \frac{\partial \phi}{\partial \nu_A}(\cdot\,; \phi_0, \phi_1) \right\|_{L_2(0,T;L_2(\Gamma))}. \quad (10.5.11.7)$$

Thus, if we take

$$x_1 \in L_2(\Omega), \quad x_2 \in [\mathcal{D}(\mathcal{A}^{\frac{1}{2}})]' = H^{-1}(\Omega), \text{ i.e., } \mathcal{A}^{*-\frac{1}{2}}x_2 \in L_2(\Omega), \quad (10.5.11.8)$$

we see, via (10.5.11.4)–(10.5.11.8), that the implication (10.5.11.1) of Theorem 10.5.3.1 applies and yields the following operator-theoretic restatement, already noted in (10.5.5.1).

Theorem 10.5.11.1   *With reference to (10.5.11.4) and (10.5.11.6) we have with Y as in (10.5.11.3):*

$$B^* e^{A^* t} : continuous \ Y \to L_2(0, T; L_2(\Gamma)) \equiv L_2(\Sigma); \quad (10.5.11.9)$$

$$D^* \mathcal{A}^* S^*(t) : continuous \ L_2(\Omega) \to L_2(\Sigma); \quad (10.5.11.10)$$

$$D^* \mathcal{A}^{*\frac{1}{2}} C^*(t) : continuous \ L_2(\Omega) \to L_2(\Sigma). \quad (10.5.11.11)$$

Step 2   The following result then stems from Theorem 10.5.11.1, by an application of Chapter 7, Theorem 7.2.1 with $p = 2$: With reference to the (operator) explicit formula (10.5.4.21) for the solution $[w(t), w_t(t)]$ of the w-problem (10.5.1.3) with initial conditions $w_0 = w_1 = 0$ at $t = s = 0$, we have

$$\begin{bmatrix} w(t) \\ w_t(t) \end{bmatrix} = (Lu)(t) = \int_0^t e^{A(t-\tau)} Bu(\tau)\, d\tau = \begin{bmatrix} \mathcal{A} \displaystyle\int_0^t S(t-\tau)Du(\tau)\, d\tau \\ \mathcal{A} \displaystyle\int_0^t C(t-\tau)Du(\tau)\, d\tau \end{bmatrix}$$

$$: continuous \ L_2(\Sigma) \to C([0, T]; Y), \quad (10.5.11.12)$$

with Y as in (10.5.11.3). This establishes (10.5.11.2) and in turn yields the key part of Theorem 10.5.3.2 due to u. Indeed, using (10.5.1.3a), we obtain, as desired,

$$w_{tt} = -\mathcal{A}(x, \partial)w \in C([0, T]; H^{-2}(\Omega)), \quad (10.5.11.13)$$

from the established regularity of $w \in C([0, T]; L_2(\Omega))$, since $-\mathcal{A}(x, \partial)$: continuous $L_2(\Omega) \to H^{-2}(\Omega)$ [Lions, Magenes, 1972, p. 85]. Finally, the regularity due to the initial conditions

$$\{w_0, w_1\} \in L_2(\Omega) \times [\mathcal{D}(\mathcal{A}^{*\frac{1}{2}})]' \Rightarrow$$

$$\begin{cases} C(t)w_0 + S(t)w_1 \in C([0, T]; L_2(\Omega)), \\ \dfrac{d}{dt} [C(t)w_0 + S(t)w_1] \\ \quad = \mathcal{A}S(t)w_0 + C(t)w_1 \in C([0, T]; [\mathcal{D}(\mathcal{A}^{*\frac{1}{2}})]' = H^{-1}(\Omega)) \end{cases} \quad (10.5.11.14)$$

is clear (or else one uses the corresponding s.c. semigroup e^{At} on $L_2(\Omega) \times [\mathcal{D}(\mathcal{A}^{*\frac{1}{2}})]')$. Theorem 10.5.3.2 is fully proved. $\square$

PDE Proof in the Canonical Case $-\mathcal{A}(x, \partial) = \Delta$ A PDE version of the duality or transposition argument is as follows: Let w and ψ be solutions with regular data of the following two problems (recall (10.5.5.7)):

$$
\begin{cases} w_{tt} = \Delta w + F, \\ w(0, \cdot) = w_0, w_t(0, \cdot) \\ \qquad = w_1, \\ w|_{\Sigma} = u, \end{cases} \text{ and } \quad \begin{cases} \psi_{tt} = \Delta \psi + f & \text{in } Q; \quad (10.5.11.15a) \\ \psi(T, \cdot) = \psi_t(T, \cdot) & \qquad\qquad\quad (10.5.11.15b) \\ \qquad = 0 & \text{in } \Omega; \\ \psi|_{\Sigma} = 0 & \text{in } \Sigma. \quad (10.5.11.15c) \end{cases}
$$

We know in advance, via Theorem 10.5.3.1 applied to the ψ-problem (which is reversed in time over the ϕ-problem (10.5.3.1)), that

$$
f \in L_1(0, T; L_2(\Omega)) \Rightarrow \begin{cases} \{\psi, \psi_t\} \in C\big([0, T]; H_0^1(\Omega) \times L_2(\Omega)\big), \\ \text{in particular, } \psi(0) \in H_0^1(\Omega), \ \psi_t(0) \in L_2(\Omega); \\ \dfrac{\partial \psi}{\partial \nu}\bigg|_{\Sigma} \in L_2(0, T; L_2(\Gamma)) = L_2(\Sigma). \end{cases}
$$

$$(10.5.11.16)$$

We multiply the w-equation (10.5.11.15a) by ψ and integrate by parts.

On the left-hand side we obtain, after integrating by parts in t twice,

$$
\int_{\Omega} \int_0^T w_{tt} \psi \, dt \, d\Omega = \int_{\Omega} \left\{ [w_t \psi]_0^T - \int_0^T w_t \psi_t \, dt \right\} d\Omega
$$

$$
= \int_{\Omega} \left\{ [w_t \psi]_0^T - [w \psi_t]_0^T + \int_0^T w \psi_{tt} \, dt \right\} d\Omega
$$

$$
\text{(by (10.5.11.15))} \quad = -(w_1, \psi(0))_{L_2(\Omega)} + (w_0, \psi_t(0))_{L_2(\Omega)}
$$

$$
+ \int_0^T \int_{\Omega} w \Delta \psi \, d\Omega \, dt + \int_0^T \int_{\Omega} w f \, d\Omega \, dt \quad (10.5.11.17)
$$

$$
= \int_0^T \int_{\Omega} \Delta w \psi \, d\Omega \, dt + \int_0^T \int_{\Omega} F \psi \, d\Omega \, dt, \quad (10.5.11.18)
$$

after using (10.5.11.15a, b) for the ψ-problem in (10.5.11.17) and recalling the right-hand side of (10.5.11.15a) for the w-problem in (10.5.11.18). However, Green's second identity yields

$$
\int_{\Omega} \Delta w \psi \, d\Omega - \int_{\Omega} w \Delta \psi \, d\Omega = - \int_{\Gamma} u \frac{\partial \psi}{\partial \nu} \, d\Gamma, \quad (10.5.11.19)
$$

using the BC in (10.5.11.15c) for w and ψ. Inserting (10.5.11.19) in (10.5.11.18), we obtain the identity

$$
(f, w)_{L_2(Q)} + \left(u, \frac{\partial \psi}{\partial \nu} \right)_{L_2(\Sigma)} + (w_0, \psi_t(0))_{L_2(\Omega)}
$$

$$
= (F, \psi)_{L_2(Q)} + (w_1, \psi(0))_{L_2(\Omega)} \quad (10.5.11.20)
$$

for smooth solutions. Recalling the regularity results (10.5.11.16) for the ψ-problem, we see then that in identity (10.5.11.20) we can take by duality

$$\{w_0, w_1\} \in L_2(\Omega) \times H^{-1}(\Omega); \quad F \in L_1(0, T; H^{-1}(\Omega)); \quad u \in L_2(0, T; L_2(\Gamma)).$$
(10.5.11.21)

We thus obtain

$$w \in L_\infty(0, T; L_2(\Omega)),$$
(10.5.11.22)

a regularity result that can then be boosted to

$$w \in C([0, T]; L_2(\Omega))$$
(10.5.11.23)

by an approximation argument with smooth data in the w-problem. We have then shown that the data (10.5.11.21) imply the regularity (10.5.11.23) for w, hence specializing to $F \equiv 0$,

$$w_{tt} = \Delta w \in C([0, T]; H^{-2}(\Omega))$$
(10.5.11.24)

[Lions, Magenes, 1972, p. 85], and by interpolation between (10.5.11.23) and (10.5.11.24), we conclude that [Lions, Magenes, 1972, p. 95]

$$w_t \in C([0, T]; H^{-1}(\Omega)),$$
(10.5.11.25)

whereby Theorem 10.5.3.2 (where $F \equiv 0$) is then proved via (10.5.11.21)–(10.5.11.25), at least in the canonical case.

The above proof admits a direct extension to the general case of $-\mathcal{A}^*(x, \partial)$ as in (10.5.3.2).

Remark 10.5.11.1 An elementary proof of regularity (in the style of Chapter 9, Section 9.8.4. in the Neumann case) may be given in the case of the one-dimensional wave equation, defined on, say, $\Omega = (0, 1)$:

$$\begin{cases} w_{tt} = w_{xx} & \text{in } (0, T] \times \Omega; & (10.5.11.26a) \\ w(0, \cdot) \equiv 0, \quad w_t(0, \cdot) \equiv 0 & \text{in } \Omega; & (10.5.11.26b) \\ w|_{x=0} = u(t), \quad w|_{x=1} = 0 & \text{in } (0, T] \times \Gamma; & (10.5.11.26c) \end{cases}$$

$\Gamma = \Gamma_0 \cup \Gamma_1$, $\Gamma_0 = \{x = 0\}$, $\Gamma_1 = \{x = 1\}$. Then, the solution of problem (10.5.11.26) is given explicitly by

$$w(t, x) = u(t - x), \quad 0 \leq t < 1; \quad w(t, x) = u(t - x) - u(t - 2 + x), \quad 1 \leq t < 2,$$
(10.5.11.27a)

and generally,

$$w(t, x) = \sum_{\substack{k=0 \\ k \text{ even}}}^{K} u(t - k - x) - \sum_{\substack{k=1 \\ k \text{ odd}}}^{K} u(t - (k + 1) + x), \quad K \leq t < (K + 1),$$

(10.5.11.27b)

where $u(t)$ is extended by zero to vanish for negative argument $t < 0$. Formula (10.5.11.27) says that the boundary input applied at $x = 0$ travels with speed equal to one and is reflected at $x = 1$ in such a way as to satisfy the zero BC. From (10.5.11.27a) one obtains, say for $0 \leq t < 1$, that the map

$$t \to \|w(t, \cdot)\|_{L_2(\Omega)}^2 = \int_0^t u^2(t - x)\, dx = \int_0^t u^2(\sigma)\, d\sigma \leq \int_0^1 u^2(\sigma)\, d\sigma$$
$$(10.5.11.28)$$

is absolutely continuous in t, for $u \in L_2(0, T)$. Moreover, (10.5.11.28) yields $w(t, \cdot) \in L_\infty(0, 1; L_2(\Omega))$, and this can be boosted to $w(t, \cdot) \in C([0, 1]; L_2(\Omega))$, say by an approximating argument. If $u(\cdot)$ is continuous on $[0, 1]$, and hence uniformly continuous here, and, say, $0 \leq t_1 < t_2 < 1$, then

$$\|w(t_1, \cdot) - w(t_2, \cdot)\|_{L_2(\Omega)}^2 = \int_0^1 |u(t_1 - x) - u(t_2 - x)|^2\, dx$$
$$= \int_0^{t_1} |u(t_1 - x) - u(t_2 - x)|^2\, dx$$
$$+ \int_{t_1}^{t_2} |u(t_2 - x)|^2\, dx \to 0 \quad \text{as } [t_1 - t_2] \to 0,$$
$$(10.5.11.29)$$

and thus $w(t, \cdot) \in C([0, 1]; L_2(\Omega))$ in this case. Next, given $u \in L_2(0, 1)$, pick a sequence $u_n(\cdot)$ continuous on $[0, 1]$ such that $u_n \to u$ in $L_2(0, 1)$. Then, the solution $w_n(t, \cdot)$ corresponding to u_n satisfies $w_n(t, \cdot) \in C([0, 1]; L_2(\Omega)) \to w(t, \cdot)$ in $L_\infty(0, 1; L_2(\Omega))$ by (10.5.11.28); hence $w(t, \cdot) \in C([0, 1]; L_2(\Omega))$ as desired. The argument proceeds similarly for other intervals of t. Thus $w \in C([0, T]; L_2(\Omega))$, as desired.

10.5.12 Proof of Theorem 10.5.7.1

With reference to the nonhomogeneous w-problem (10.5.1.3), we assume

$$\{w_0, w_1\} \in H^1(\Omega) \times L_2(\Omega), \quad w_0|_\Gamma = u(0) \in H^{\frac{1}{2}}(\Gamma); \quad (10.5.12.1)$$

$$u \in C([0, T]; H^{\frac{1}{2}}(\Gamma)) \cap H^1(0, T; L_2(\Gamma)), \quad (10.5.12.2)$$

and we must show that

$$\{w, w_t, w_{tt}\} \in C([0, T]; H^1(\Omega) \times L_2(\Omega) \times H^{-1}(\Omega)). \quad (10.5.12.3)$$

Operator-Theoretic Proof We return to the explicit solution formula (10.5.4.21), (10.5.4.22), that is,

$$w(t) = C(t)w_0 + S(t)w_1 + \mathcal{A} \int_0^t S(t - \tau)Du(\tau)\, d\tau, \quad (10.5.12.4)$$

and integrate by parts the integral term with $u \in H^1(0, T; L_2(\Gamma))$, thus obtaining

$$w(t) = C(t)[w_0 - Du(0)] + S(t)w_1 + Du(t) - \int_0^t C(t - \tau)D\dot{u}(\tau)\,d\tau,$$

$$(10.5.12.5)$$

where, by the compatibility condition (10.5.12.1) and, respectively, $\dot{u} \in L_2(\Sigma)$, we have

$$w_0 = Du(0); \quad \int_0^t C(t - \tau)D\dot{u}(\tau)\,d\tau \in C\big([0, T]; \mathcal{D}\big(\mathcal{A}^{\frac{1}{2}}\big)\big), \qquad (10.5.12.6)$$

recalling the definition of D in (10.5.4.9) and, respectively, the regularity (10.5.11.2) (second component).

Likewise,

$$S(t)w_1 \in C\big([0, T]; \mathcal{D}\big(\mathcal{A}^{\frac{1}{2}}\big)\big), \quad \text{with } w_1 \in L_2(\Omega). \qquad (10.5.12.7)$$

We now use that $u \in C([0, T]; H^{\frac{1}{2}}(\Gamma))$ as well, from (10.5.12.2), so that, by elliptic theory [(10.5.4.10) with $s = 1/2$],

$$Du(t) \in C([0, T]; H^1(\Omega)). \qquad (10.5.12.8)$$

Thus, (10.5.12.6)–(10.5.12.8), used in (10.5.12.5), show (10.5.12.3) for w, via (10.5.4.15). As to w_t, we differentiate (10.5.12.5), thus obtaining

$$w_t(t) = C(t)w_1 + D\dot{u}(t) - D\dot{u}(t) + \mathcal{A}\int_0^t S(t - \tau)D\dot{u}(\tau)\,d\tau. \qquad (10.5.12.9)$$

Conclusion (10.5.12.3) for w_t now follows from (10.5.12.9) [where a cancellation of $D\dot{u}(t)$ occurs] via (10.5.11.2) (first component) with $\dot{u} \in L_2(\Sigma)$. Finally, conclusion (10.5.12.3) for w_{tt} follows then from the equation (10.5.1.3a), via the regularity of w just established and $-\mathcal{A}(x, \partial)$: continuous $H^1(\Omega) \to H^{-1}(\Omega)$ [Lions, Magenes, 1972, p. 85]. The proof of the implication (10.5.12.1), (10.5.12.2) $\Rightarrow$ (10.5.12.3), that is, of Theorem 10.5.7.1, is complete.

PDE Proof   Differentiating in t problem (10.5.1.3) with $u \in H^1(0, T; L_2(\Gamma))$ yields $(' = \frac{d}{dt})$:

$$\begin{cases} (w_t)'' = -\mathcal{A}(x, \partial)w_t & \text{in } Q; & (10.5.12.10a) \\ w_t(0, \cdot) = w_1, \quad w_t'(0, \cdot) = w_{tt}(0, \cdot) = -\mathcal{A}(x, \partial)w_0 & \text{in } \Omega; & (10.5.12.10b) \\ w_t|_\Sigma = u_t & \text{in } \Sigma, & (10.5.12.10c) \end{cases}$$

that is, the same problem (10.5.1.3), however in the variable w_t, with initial data $w_1 \in L_2(\Omega)$, $-\mathcal{A}(x, \partial)w_0 \in H^{-1}(\Omega)$, and $u_t \in L_2(\Sigma)$. Since estimates are first shown for regular data and then extended, we need the compatibility condition in (10.5.12.1) for regular data to invoke Theorem 10.5.3.2 and obtain

$$\{w_t, w_{tt}\} \in C([0, T]; L_2(\Omega) \times H^{-1}(\Omega)). \qquad (10.5.12.11)$$

As to w, we use elliptic theory on

$$\begin{cases} -\mathcal{A}(x, \partial)w = w_{tt} \in H^{-1}(\Omega), & (10.5.12.12) \\ \quad w|_\Gamma = u \in H^{\frac{1}{2}}(\Gamma) & (10.5.12.13) \end{cases}$$

at each t, recalling assumption (10.5.12.2), and conclude that

$$w \in C([0, T]; H^1(\Omega)), \qquad (10.5.12.14)$$

as desired. The proof of (10.5.12.3) is complete.   □

10.5.13 Proof of Theorem 10.5.8.1

At essentially no extra effort, we shall provide, in the next subsection, a more general result than what is strictly needed.

10.5.13.1 A Preliminary Trace Result

Theorem 10.5.13.1   *(i) With reference to the operators in (10.5.11.9)–(10.5.11.11), we have the following regularity properties for $0 \le r \le 1$, which generalize the case $r = 0$ of Theorem 10.5.10.2, Eqn. (10.5.10.18), and Theorem 10.5.3.1, Eqn. (10.5.3.4):*

$$D^* \mathcal{A}^* S^*(t) : continuous\ \mathcal{D}(\mathcal{A}^{*r}) \to H^{2r,2r}(\Sigma); \qquad (10.5.13.1)$$

$$D^* \mathcal{A}^{*\frac{1}{2}} C^*(t) : continuous\ \mathcal{D}(\mathcal{A}^{*r}) \to H^{2r,2r}(\Sigma); \qquad (10.5.13.2)$$

$$B^* e^{A^* t} : continuous\ \mathcal{D}(A^{*2r}) = \mathcal{D}(\mathcal{A}^{*r}) \times \mathcal{D}\big(\mathcal{A}^{*r-\frac{1}{2}}\big) \to H^{2r,2r}(\Sigma); \qquad (10.5.13.3)$$

$$H^{2r,2r}(\Sigma) \equiv L_2(0, T; H^{2r}(\Gamma)) \cap H^{2r}(0, T; L_2(\Gamma)). \qquad (10.5.13.4)$$

(ii) Equivalently, in PDE terms,

$$\{\phi_0, \phi_1\} \to \frac{\partial \phi}{\partial \nu_\mathcal{A}}(t; \phi_0, \phi_1) : continuous\ \mathcal{D}\big(\mathcal{A}^{*r+\frac{1}{2}}\big) \times \mathcal{D}(\mathcal{A}^{*r}) \to H^{2r,2r}(\Sigma), \qquad (10.5.13.5)$$

where $\phi(t; \phi_0, \phi_1)$ is the solution of problem (10.5.3.1) with $f \equiv 0$, and where we further recall (10.5.4.28)–(10.5.4.31) to justify the stated equivalence.

Note: In (10.5.13.3), $\mathcal{D}(\mathcal{A}^{*-\theta}) \equiv [\mathcal{D}(\mathcal{A}^\theta)]'$ for $\theta > 0$, duality with respect to $L_2(\Omega)$.

Proof.

Case $r = 0$ The case $r = 0$ is contained in Theorem 10.5.11.1 for (i) and in Eqn. (10.5.3.4) of Theorem 10.5.3.1 for (ii). The stated equivalence uses (10.5.4.28)–(10.5.4.31). Thus, it is sufficient to prove the case $r = 1$ and interpolate to establish Theorem 10.5.13.1.

Case r = 1 We first show the *time regularity*

$$D^* \mathcal{A}^* S^*(t)x, \quad D^* \mathcal{A}^{*\frac{1}{2}} C^*(t)x \in H^2(0, T; L_2(\Gamma)), \quad x \in \mathcal{D}(\mathcal{A}^*). \quad (10.5.13.6)$$

Indeed, with $x \in \mathcal{D}(\mathcal{A}^*)$, we compute

$$\frac{d^2}{dt^2} D^* \mathcal{A}^* S^*(t)x = \frac{d}{dt} D^* \mathcal{A}^* C^*(t)x = -D^* \mathcal{A}^* S^*(t) \mathcal{A}^* x \in L_2(\Sigma), \quad (10.5.13.7)$$

$$\frac{d^2}{dt^2} D^* \mathcal{A}^{*\frac{1}{2}} C^*(t)x = -\frac{d}{dt} D^* \mathcal{A}^{*\frac{1}{2}} S^*(t) \mathcal{A}^* x = -D^* \mathcal{A}^{*\frac{1}{2}} C^*(t) \mathcal{A}^* x \in L_2(\Sigma),$$
$$(10.5.13.8)$$

continuously in $x \in \mathcal{D}(\mathcal{A}^*)$, by (10.5.11.10) and (10.5.11.11). Thus (10.5.13.6) is proved by virtue of (10.5.13.7) and (10.5.13.8).

To show the space regularity

$$D^* \mathcal{A}^* S^*(t)x, \quad D^* \mathcal{A}^{*\frac{1}{2}} C^*(t)x \in L_2(0, T; H^2(\Gamma)), \quad x \in \mathcal{D}(\mathcal{A}^*), \quad (10.5.13.9)$$

we shall equivalently show by (10.5.4.29) and (10.5.4.30) that

$$\{\phi_0, \phi_1\} \to \frac{\partial \phi}{\partial \nu_A}(t; \phi_0, \phi_1) : \text{continuous } \mathcal{D}\left(\mathcal{A}^{*\frac{3}{2}}\right) \times \mathcal{D}(\mathcal{A}^*) \to L_2(0, T; H^2(\Gamma)).$$
$$(10.5.13.10)$$

To this end, we introduce a second-order operator

$$\mathcal{B} = \sum_{i,j} \beta_{ij} \frac{\partial^2}{\partial x_i \partial x_j}, \quad x \in \bar{\Omega}, \quad (10.5.13.11)$$

on $\bar{\Omega}$, *tangential* to Γ (i.e., without transversal derivatives to Γ, when expressed in local coordinates) and with smooth coefficients β_{ij} on $\bar{\Omega}$. Accordingly, we consider problem (10.5.3.1) with $f \equiv 0$:

$$\begin{cases} \phi_{tt} = -\mathcal{A}^*(x, \partial)\phi & \text{in } Q; & (10.5.13.12\text{a}) \\ \phi(0, \cdot) = \phi_0, \quad \phi_t(0, \cdot) = \phi_1 & \text{in } \Omega; & (10.5.13.12\text{b}) \\ \phi|_\Sigma \equiv 0 & \text{in } \Sigma; & (10.5.13.12\text{c}) \end{cases}$$

$$\{\phi_0, \phi_1\} \in \mathcal{D}\left(\mathcal{A}^{*\frac{3}{2}}\right) \times \mathcal{D}(\mathcal{A}^*) \subset H^3(\Omega) \times H^2(\Omega). \quad (10.5.13.13)$$

If we introduce a new variable

$$z = \mathcal{B}\phi, \quad (10.5.13.14)$$

then proving (10.5.13.10) is equivalent to proving that

$$\frac{\partial}{\partial \nu_A}(\mathcal{B}\phi)\bigg|_\Gamma = \frac{\partial z}{\partial \nu_A}\bigg|_\Gamma \in L_2(\Sigma). \quad (10.5.13.15)$$

The commutator

$$\mathcal{K} = -\mathcal{B}\mathcal{A}^*(x, \partial) + \mathcal{A}^*(x, \partial)\mathcal{B} \quad (10.5.13.16)$$

is an operator of order $2 + 2 - 1 = 3$ in x, with smooth coefficients in $\bar{\Omega}$. From

$(10.5.13.14)$ and $(10.5.13.12)$, we obtain the problem for z, recalling also $(10.5.13.13)$:

$$\begin{cases} z_{tt} = -\mathcal{A}^*(x, \partial)z + k, & (10.5.13.17\text{a}) \\ z(0, \cdot) = \mathcal{B}\phi_0 \in H_0^1(\Omega), \quad z_t(0, \cdot) = \mathcal{B}\phi_1 \in L_2(\Omega), & (10.5.13.17\text{b}) \\ z|_\Sigma \equiv 0, & (10.5.13.17\text{c}) \end{cases}$$

where we have set $k = \mathcal{K}\phi$. Thus, since

$$\phi(t) = C^*(t)\phi_0 + S^*(t)\phi_1 \in C\left([0, T]; \mathcal{D}\left(\mathcal{A}^{*\frac{3}{2}}\right)\right) \subset C([0, T]; H^3(\Omega)),$$
$$(10.5.13.18)$$

for $\{\phi_0, \phi_1\}$ as in $(10.5.13.13)$, it follows that

$$k = \mathcal{K}\phi \in C([0, T]; L_2(\Omega)) \subset L_2(Q). \qquad (10.5.13.19)$$

Thus, in view of the regularity of the data,

$$\{z(0, \cdot), z_t(0, \cdot), k\} \in H_0^1(\Omega) \times L_2(\Omega) \times L_2(0, T; L_2(\Omega)),$$

we can apply Theorem 10.5.3.1 to the z-problem $(10.5.13.17)$ and conclude, via Eqn. $(10.5.3.4)$, that $(10.5.13.15)$ holds true, as desired. The proof of Theorem 10.5.13.1 is complete. $\square$

10.5.13.2 Completion of the Proof of Theorem 10.5.8.1

Space Regularity   To show the space regularity

$$(L^*v)(t) = \int_t^T B^* e^{A^*(\tau - t)} v(\tau) \, d\tau \qquad (10.5.13.20)$$

$$: \text{continuous } L_1(0, T; \mathcal{D}(A^{*r})) \to L_2(0, T; H^r(\Gamma)),$$
$$0 \le r \le 1, \quad (10.5.13.21)$$

we simply invoke Theorem 7.2.1, Eqn. $(7.2.2)$ with $X^* = \mathcal{D}(A^{*r})$, $U^* = H^r(\Gamma)$, and $q = 2$, which is legal by virtue of the regularity $(10.5.13.3)$ of Theorem 10.5.13.1.

Time Regularity   It suffices to show the case $r = 1$, since the case $r = 0$ is contained in $(10.5.13.21)$ [or in $(10.5.5.6)$], and then interpolate. Thus, differentiating $(10.5.13.20)$ in t for

$$v \in L_2(0, T; \mathcal{D}(A^*)) \quad \text{or } A^*v \in L_2(0, T; Y) \qquad (10.5.13.22)$$

yields

$$\frac{d(L^*v)}{dt}(t) = -B^*v(t) - \int_t^T B^* e^{A^*(\tau - t)} A^*v(\tau) \, d\tau \in L_2(0, T; L_2(\Gamma)),$$
$$(10.5.13.23)$$

as desired, by $(10.5.5.6)$ [or $(10.5.13.21)$ with $r = 0$] using $(10.5.13.22)$, as well as $B^* A^{*-1} A^* v(t) \in L_2(0, T; L_2(\Gamma))$ by $B^* A^{*-1} \in \mathcal{L}(Y; U)$, via $(10.5.4.19)$, Y and U

in (10.5.4.14). Then (10.5.13.23) shows

$$L^* : \text{continuous } L_2(0, T; \mathcal{D}(A^*)) \rightarrow H^1(0, T; L_2(\Gamma)), \qquad (10.5.13.24)$$

as required. The proof of Theorem 10.5.8.1 is complete, except for noticing that uniformity in s is obtained by a change of variable on formula (10.5.13.20), as usual.

10.6 Application: Nonsymmetric, Nondissipative First-Order Hyperbolic Systems with Boundary Control. Regularity Theory

This section is the counterpart of Section 10.5, as applied this time to general first-order hyperbolic systems, which may be nonsymmetric and nondissipative, defined on a sufficiently smooth bounded domain of arbitrary dimension, and where the control function acts through the boundary conditions. As in the case of second-order hyperbolic equations treated in Section 10.5, we likewise show here that all abstract system's assumptions (H.1) = (10.1.6) through (H.7) = (10.1.21) of Section 10.1 are automatically satisfied in the natural mathematical setting. Accordingly, Theorems 10.2.1, 10.2.2, and 10.2.3 of Section 10.2 are then applicable to the present class, for any observation operator R with "minimal" smoothing as in (H.8) = (10.1.23). Basic references on the regularity theory are Kreiss [1970], Rauch [1971; 1972; 1973], and Ralston [1971].

10.6.1 Problem Formulation

The Dynamics   Let $\Omega \subset R^m$ be an open bounded domain with smooth boundary Γ. In Ω, we consider a differential operator of the form

$$A(x, \partial)y \equiv \sum_{j=1}^{m} A_j(x)\partial_j y + A_0(x)y, \qquad (10.6.1.1)$$

where $y(x)$ is a k-vector and $\partial_j = \partial/\partial x_j$. The coefficients A_j and A_0 are smooth $k \times k$ matrix-valued functions defined on the open bounded domain $\Omega \subset R^m$. We assume the following hypotheses throughout:

(h.1): $A(x, \partial)$ is strictly hyperbolic, that is, $\sum_{j=1}^{m} A_j(x)\xi_j$ has k distinct real eigenvalues for all $\xi \in R^m \setminus \{0\}$ and $x \in \bar{\Omega}$;

(h.2): The boundary Γ is noncharacteristic, that is, $\det A_\nu(x) \neq 0$ for $x \in \Gamma$, where $A_\nu(x) \equiv \sum_{j=1}^{m} A_j \nu_j(x)$, with $\nu = (\nu_1, \ldots, \nu_m)$ the inward unit normal.

It follows from (h.1) and (h.2) that, after a smooth change of coordinates, we may assume that A_ν be of the following form:

$$A_\nu = \begin{bmatrix} A_\nu^- & 0 \\ 0 & A_\nu^+ \end{bmatrix}, \quad A_\nu^- = \begin{bmatrix} a_1 & & & 0 \\ & a_2 & & \\ & & \ddots & \\ 0 & & & a_\ell \end{bmatrix} < 0, \quad A_\nu^+ = \begin{bmatrix} a_{\ell+1} & & 0 \\ & \ddots & \\ 0 & & a_k \end{bmatrix} > 0.$$

$$(10.6.1.2)$$

Accordingly, any vector $v \in R^k$ will be split consistently as $v = [v^-, v^+]$, with $v^- = [v_1, \ldots, v_\ell]$ and $v^+ = [v_{\ell+1}, \ldots, v_k]$.

Boundary conditions are imposed with the aid of a boundary operator $M(x)$, which is a smooth $\ell \times k$ matrix-valued function, where ℓ stands for the number of negative eigenvalues of A_ν. We assume further the following hypotheses:

(h.3): rank $M(x) = \ell, x \in \Gamma$.

(h.4): (Kreiss condition) The frozen (at the boundary point) mixed problem has no eigenvalues or generalized eigenvalues with nonnegative real parts.

This means that after making a local change of coordinates to map Ω into the half-space $\{x \in R^m; x_1 > 0\}$, the constant coefficient problem that arises by freezing $A_j, j = 1, \ldots, m$, and M at the boundary point and setting $A_0 = 0$, that is,

$$y_t - A_1 y_{x_1} - \sum_{j=2}^{m} A_j y_{x_j} = 0, \quad x_1 > 0; \tag{10.6.1.3a}$$

$$My = 0, \quad \text{at } x_1 = 0, \tag{10.6.1.3b}$$

has no eigenvalues or generalized eigenvalues with nonnegative real parts.

For the half-space problem (10.6.1.3), we have $A_\nu = A_1$; thus A_1 is invertible by (h.2). For a more detailed description of this condition we refer the reader to the fundamental papers [Kreiss, 1970; Rauch, 1971].

Convention   To streamline the notation, we shall write

$L_2(\Gamma)$ and $L_2(\Omega)$ to mean respectively, $L_2(\Gamma; R^\ell)$ and $L_2(\Omega; R^k)$, etc., and $L_2(\Sigma)$ and $L_2(Q)$ to mean, respectively, $L_2(0, T; L_2(\Gamma; R^\ell))$ and $L_2(0, T; L_2(\Omega; R^k))$, $\tag{10.6.1.4}$

without further mention, where $\Sigma = (0, T] \times \Gamma$; $Q = (0, T] \times \Omega$, for a fixed $0 < T < \infty$.

The mixed problem for the first-order hyperbolic system we consider is then

$$\begin{cases} y_t = A(x, \partial)y & \text{in } Q \equiv (0, T] \times \Omega, & (10.6.1.5a) \\ y(0, \cdot) = y_0(x) & \text{in } \Omega, & (10.6.1.5b) \\ [M(x)]y = u & \text{in } \Sigma \equiv (0, T] \times \Gamma, & (10.6.1.5c) \end{cases}$$

where the boundary control $u \in L_2(\Sigma) = L_2(0, T; L_2(\Gamma; R^\ell))$.

The Optimal Control Problem on $[s, T]$ Consistently with the (optimal) regularity theory for problem (10.6.1.5) presented in Theorem 10.6.3.1 below, we take the initial condition $y(s) = y_0 \in L_2(\Omega)$ at the initial time $0 \le s < T$ and the cost functional we seek to minimize over all $u \in L_2(\Sigma_{sT}) = L_2(s, T; L_2(\Gamma; R^\ell))$ to be

$$J(u, y) = \int_s^T \left\{ \|Ry(t)\|_{L_2(\Omega)}^2 + \|u(t)\|_{L_2(\Gamma)}^2 \right\} dt, \tag{10.6.1.6}$$

where the observation operator $R \in \mathcal{L}(L_2(\Omega))$ will be further specified below in (10.6.2.3).

10.6.2 Main Results

As a specialization to problem (10.6.1.5), (10.6.1.6), of the abstract theory presented in Theorems 10.2.1, 10.2.2, and 10.2.3, in the present section we establish the following result.

Theorem 10.6.2.1   *(a) With the observation operator R in (10.6.1.6) only assumed to satisfy*

$$R \in \mathcal{L}(L_2(\Omega)), \tag{10.6.2.1}$$

Theorem 9.2.1 of Chapter 9 applies to the optimal control problem (10.6.1.5), (10.6.1.6), with

$$U = L_2(\Gamma), \quad Y = L_2(\Omega) \tag{10.6.2.2}$$

[we are using the notational convention made in (10.6.1.4)] and produces a unique optimal pair $\{u^0(\,\cdot\,, s; y_0), y^0(\,\cdot\,, s; y_0)\}$, $y_0 \in Y$, satisfying in particular the pointwise feedback synthesis (9.2.17) of Chapter 9 and the optimal cost relation (9.2.18) of Chapter 9, as well as the other properties listed there.

(b) Assume, in addition to (10.6.2.1), that R satisfies the smoothing condition

$$R^*R : continuous \; H^{\frac{1}{2}-\delta}(\Omega) \to H_0^{\frac{1}{2}+\delta}(\Omega), \tag{10.6.2.3}$$

with R^* the adjoint of R in $Y = L_2(\Omega)$, for a fixed, arbitrarily small $\delta > 0$.

Then, Theorems 10.2.1, 10.2.2, and 10.2.3 of the present chapter hold true, with (in the convention made in (10.6.1.4))

$$\begin{aligned} Y_\delta^- &\equiv H^{\frac{1}{2}-\delta}(\Omega), \quad Y_\delta^+ \equiv H_0^{\frac{1}{2}+\delta}(\Omega), \\ U_\delta^- &\equiv H^{\frac{1}{2}-\delta}(\Gamma), \quad U_\delta^+ \equiv H^{\frac{1}{2}+\delta}(\Gamma) \end{aligned} \tag{10.6.2.4}$$

(see Section 10.6.6). In particular, explicitly:

(b_1) (Regularity of optimal pair) The optimal pair satisfies for $y_0 \in H^{\frac{1}{2}-\delta}(\Omega)$ and with $Q_{sT} = [s, T] \times \Omega$, $\Sigma_{sT} = [s, T] \times \Gamma$:

$$y^0(\,\cdot\,, s; y_0) \in C\big([s, T]; H^{\frac{1}{2}-\delta}(\Omega)\big) \cap H^{\frac{1}{2}-\delta}(s, T; L_2(\Omega)) \subset H^{\frac{1}{2}-\delta, \frac{1}{2}-\delta}(Q_{sT}), \tag{10.6.2.5}$$

$$u^0(\,\cdot\,, s; y_0) \in L_2\big(s, T; H^{\frac{1}{2}+\delta}(\Gamma)\big) \cap H^{\frac{1}{2}+\delta}(s, T; L_2(\Gamma)) \equiv H^{\frac{1}{2}+\delta, \frac{1}{2}+\delta}(\Sigma_{sT}), \tag{10.6.2.6}$$

a fortiori,

$$u^0(\,\cdot\,, s; y_0) \in C([s, T]; L_2(\Gamma)). \tag{10.6.2.7}$$

All of the above results hold uniformly in s, that is, with norms that may be made independent of s.

(b₂) (Gain operator) The gain operator $B^* P(t)$ *satisfies*

$$B^* P(t)x = -[A_\nu^-(\cdot)[P(t)x]^-]_\Gamma$$

$$: continuous\ H^{\frac{1}{2}-\delta}(\Omega) \to C([0, T]; L_2(\Gamma)), \qquad (10.6.2.8)$$

where $[P(t)x]^-$ *denotes the vector consisting of the first ℓ components of* $P(t)x$, *according to the decomposition in (10.6.1.2).*

(b₃) (Differential Riccati equation) With A the $L_2(\Omega)$-*realization of the differential operator* $A(x, \partial)$ *in (10.6.1.1) and given by (10.6.3.8) below, we have that* $P(t)$ *satisfies the DRE*

$$\begin{cases} \dfrac{d}{dt}(P(t)x, y)_{L_2(\Omega)} = -(Rx, Ry)_{L_2(\Omega)} - (P(t)x, Ay)_{L_2(\Omega)} - (P(t)Ax, y)_{L_2(\Omega)} \\ \qquad\qquad + ([A_\nu^-(\cdot)[P(t)x]^-]_\Gamma, [A_\nu^-(\cdot)[P(t)y]^-]_\Gamma)_{L_2(\Gamma)}, \\ \quad P(T) = 0 \end{cases}$$

$$(10.6.2.9)$$

$\forall\, x, y \in H^{\frac{1}{2}-\delta}(\Omega)$. *Moreover,* $P(t)$ *satisfies the corresponding integral Riccati equation, as in (10.2.7), for all* $x, y \in H^{\frac{1}{2}-\delta}(\Omega)$.

10.6.3 Regularity Theory for Problem (10.6.1.5) with $u \in L_2(\Sigma)$

A complete well-posedness theory for nonsymmetric, noncharacteristic first-order hyperbolic systems as in (10.6.1.5) has been provided by the fundamental paper Kreiss [1970], augmented by a note in Ralston [1971], and completed by Rauch [1972]; it provides justification for the selection of the cost functional J in (10.6.1.6).

Theorem 10.6.3.1 [Rauch, 1972, p. 272]   *Under the given hypotheses of Section 10.6.1, for any $T > 0$, assume*

$$y_0 \in L_2(\Omega), \quad u \in L_2(0, T; L_2(\Gamma)). \qquad (10.6.3.1)$$

(a) Then, the unique solution of problem (10.6.1.5) satisfies

$$y \in C([0, T]; L_2(\Omega)), \quad y|_\Gamma \in L_2(0, T; L_2(\Gamma)) \qquad (10.6.3.2)$$

continuously.

(b) Furthermore, assume

$$y_0 \in H_0^1(\Omega), \quad u \in H^1([0, T] \times \Gamma), \qquad (10.6.3.3)$$

along with the compatibility condition

$$u(0) = y_0|_\Gamma = 0. \qquad (10.6.3.4)$$

Then, the unique solution of problem (10.6.1.5) satisfies

$$y \in H^1([0, T] \times \Omega) \quad and \quad y|_\Gamma \in H^1([0, T] \times \Gamma) \qquad (10.6.3.5)$$

continuously.

Corollary 10.6.3.2   *Under the given hypotheses of Section 10.6.1, assume*

$$y_0 \in H^\theta(\Omega), \quad u \in H^\theta([0, T] \times \Gamma), \quad 0 < \theta < \frac{1}{2}. \qquad (10.6.3.6)$$

Then, the unique solution of problem (10.6.1.5) satisfies

$$y \in H^\theta([0, T] \times \Omega) \cap C([0, T]; H^\theta(\Omega)) \quad and \quad y|_\Gamma \in H^\theta([0, T] \times \Gamma) \qquad (10.6.3.7)$$

continuously.

Proof of Corollary 10.6.3.2.   We interpolate between the implication (10.6.3.1) $\Rightarrow$ (10.6.3.2) (case $\theta = 0$) and the implication (10.6.3.3), (10.6.3.4) $\Rightarrow$ (10.6.3.5) (case $\theta = 1$), where, however, for $\theta < 1/2$, the compatibility condition in (10.6.3.4) is irrelevant. $\square$

Next, we single out the result of the homogeneous case $u \equiv 0$ for problem (10.6.1.5) in a form that will be useful in the following sections. To this end, we introduce the operator A, by setting

$$Ah = A(x, \partial)h : L_2(\Omega) \supset \mathcal{D}(A) \to L_2(\Omega), \qquad (10.6.3.8a)$$

$$\mathcal{D}(A) = \{h \in L_2(\Omega) : A(x, \partial)h \in L_2(\Omega); Mh|_\Gamma = 0\}, \qquad (10.6.3.8b)$$

where $A(x, \partial)$ is the differential operator in (10.6.1.1).

Corollary 10.6.3.3   *Under the given hypotheses of Section 10.6.1, we have that the operator A in (10.6.3.8), corresponding to problem (10.6.1.5) with $u \equiv 0$, is the generator of a s.c. semigroup e^{At} on $L_2(\Omega), t \geq 0$, which, moreover, restricts/extends as a s.c. semigroup on all spaces $H_0^\theta(\Omega), 0 < \theta \leq 1, \theta \neq 1/2$, and, respectively, $H^{-\theta}(\Omega)$.*

Proof.   With $u \equiv 0$ in Theorem 10.6.3.1 and Corollary 10.6.3.2, we obtain the conclusion on $H_0^\theta(\Omega)$, and by duality on $H^{-\theta}(\Omega), 0 \leq \theta \leq 1, \theta \neq 1/2$. $\square$

10.6.4 Abstract Setting for Problem (10.6.1.5)

We follow Chang and Lasiecka [1986] and Lasiecka and Triggiani [1991(c)] closely. To put problem (10.6.1.5), (10.6.1.6) into the abstract model (10.1.1), (10.1.3), we need the following operators and spaces:

(i) The operator A defined by (10.6.3.8), which generates a s.c. semigroup e^{At} on the space

$$Y = L_2(\Omega). \qquad (10.6.4.1)$$

(ii) Next, we introduce the "Dirichlet" map (natural extension from the boundary Γ into the interior Ω, which uniquely solves (a suitable translation of) the corresponding static problem, defined by

$$D_\lambda g = v \iff \begin{cases} A(x, \partial)v - \lambda v = 0 & \text{in } \Omega, \\ Mv|_\Gamma = g & \text{in } \Gamma, \end{cases} \tag{10.6.4.2}$$

for a suitably large constant $\lambda \geq 0$, as justified by the following result.

Lemma 10.6.4.1   *With reference to problem (10.6.4.2), there exists a constant $\lambda \geq 0$, henceforth kept fixed, such that problem (10.6.4.2) admits a unique solution $v = D_\lambda g \in L_2(\Omega)$ for $g \in L_2(\Gamma)$. Moreover, the following estimate holds true: There is a constant $C_\lambda > 0$ depending on λ such that*

$$\|D_\lambda g\|_{L_2(\Omega)} + \|D_\lambda g|_\Gamma\|_{L_2(\Gamma)} \leq C_\lambda \|g\|_{L_2(\Gamma)}. \tag{10.6.4.3}$$

Thus,

$$D_\lambda : continuous\ L_2(\Gamma) \to L_2(\Omega), \tag{10.6.4.4}$$

$$D_\lambda^* : continuous\ L_2(\Omega) \to L_2(\Gamma), \tag{10.6.4.5}$$

where D_λ^* is the adjoint $(D_\lambda g, v)_{L_2(\Omega)} = (g, D_\lambda^* v)_{L_2(\Gamma)}$.

Proof.   The proof of Lemma 10.6.4.1, which is based on Kreiss's symmetrizer used for the dynamic problem (10.6.1.5) in Theorem 10.6.3.1, is given in Section 10.6.10. $\square$

(iii) We return to problem (10.6.1.5), and by virtue of the definition (10.6.4.2) of D_λ, λ henceforth as in Lemma 10.6.4.1, we rewrite it as

$$\begin{cases} y_t = (A(x, \partial) - \lambda)(y - D_\lambda u) + \lambda y & \text{in } (0, T] \times \Omega, & (10.6.4.6a) \\ y(0, x) = y_0(x) & \text{in } \Omega, & (10.6.4.6b) \\ M(y - D_\lambda u)|_\Gamma = 0 & \text{in } (0, T] \times \Gamma, & (10.6.4.6c) \end{cases}$$

or abstractly, by (10.6.3.9), as

$$y_t = (A - \lambda I)(y - D_\lambda u) + \lambda y \text{ in } L_2(\Omega), \quad y(0) = y_0 \in L_2(\Omega). \tag{10.6.4.7}$$

Moreover, extending the original operator A in (10.6.3.9) by $A : L_2(\Omega) \to [\mathcal{D}(A^*)]'$, that is, extending the original A in (10.6.3.9) to its double adjoint A^{**}, we obtain from (10.6.4.7)

$$y_t = Ay - (A - \lambda I)D_\lambda u \text{ in } [\mathcal{D}(A^*)]', \quad y(0) = y_0 \in L_2(\Omega), \tag{10.6.4.8}$$

which is precisely the abstract model (10.1.1), with A as in (10.6.3.9), and

$$B = -(A - \lambda I)D_\lambda : continuous\ U = L_2(\Gamma) \to [\mathcal{D}(A^* - \lambda I)]', \tag{10.6.4.9a}$$

or, equivalently,

$$(A - \lambda I)^{-1} B = -D_\lambda : \text{continuous } L_2(\Gamma) \to L_2(\Omega), \qquad (10.6.4.9b)$$

as guaranteed by (10.6.4.4).

Finally, with $B \in \mathcal{L}(U; [\mathcal{D}(A^* - \lambda I)]')$, and so $B^* \in \mathcal{L}(\mathcal{D}(A^*); U)$ after identifying $[\mathcal{D}(A^* - \lambda I)]''$ with $\mathcal{D}(A^*)$, we compute B^* as

$$B^* = -D_\lambda^*(A^* - \lambda I) : \text{continuous } \mathcal{D}(A^*) \to U. \qquad (10.6.4.10)$$

A more explicit representation of B^* is given by the next result.

Lemma 10.6.4.2   *With reference to (10.6.4.10), we have*

(i)

$$B^* y = -D_\lambda^*(A^* - \lambda I)y = [A_\nu^- y^-]_\Gamma, \quad y \in \mathcal{D}(A^*), \quad (10.6.4.11)$$

where A_ν^- is defined in (10.6.1.2) and the component y^- of y, consisting of the first ℓ coordinates, is likewise defined below (10.6.1.2).

(ii) Hence, by trace theory Lions and Magenes [1972] applied to (10.6.4.11):

$$B^* : \text{continuous } H^{\frac{1}{2} + \epsilon}(\Omega) \to L_2(\Gamma). \qquad (10.6.4.12)$$

Proof.   The proof of Lemma 10.6.4.2 is given in Section 10.6.11 below.

10.6.5   *Verification of Assumptions (H.1) = (10.1.6) through (H.3) = (10.1.15)*

Verification of Assumption (H.1) = (10.1.6)   The semigroup solution of the mixed problem (10.6.1.5), that is, of the corresponding abstract equation (10.5.4.8), with initial datum $y(s) = y_0 \in L_2(\Omega)$ given at the initial time, $0 \le s < T$, is

$$y(t, s; y_0) = e^{A(t-s)} y_0 + (L_s u)(t); \qquad (10.6.5.1)$$

$$(L_s u)(t) = -(A - \lambda I) \int_s^t e^{A(t-\tau)} D_\lambda u(\tau)\, d\tau \qquad (10.6.5.2a)$$

$$= \int_s^t e^{A(t-\tau)} B u(\tau)\, d\tau. \qquad (10.6.5.2b)$$

Theorem 10.6.3.1(a) yields for $y_0 = 0$:

$$L_s : u \to y(\cdot, s; y_0) \text{ continuous } L_2(s, T; L_2(\Gamma)) \to C([s, T]; L_2(\Omega)),$$
$$y_0 = 0, \qquad (10.6.5.3)$$

at least for s fixed, and indeed uniformly in s, as it follows from the integral formula of L_s in (10.6.5.2), via a change of variable.

Thus, assumption (H.1) = (10.1.6) is verified via (10.6.5.3) in its equivalent form (H.1*) = (10.1.7), with $U = L_2(\Gamma)$ and $Y = L_2(\Omega)$ as in (10.6.4.1).

Direct Trace Theory Verification   One can, of course, verify directly the abstract trace theory version (H.1) = (10.1.6) as well. To this end, we consider the homogeneous problem

$$
\begin{cases}
\zeta_t = -A^*(x, \partial)\zeta & \text{in } Q = (0, T] \times \Omega, & (10.6.5.4a) \\
\zeta(T, \cdot) = \zeta_0 & \text{in } \Omega, & (10.6.5.4b) \\
[M^*(x)]\zeta = 0 & \text{in } \Sigma = (0, T] \times \Gamma, & (10.6.5.4c)
\end{cases}
$$

where $A^*(x, \partial)$ is the formal adjoint of the differential operator $A(x, \partial)$ in (10.6.1.1) and is thus given by

$$
A^*(x, \partial) = -\sum_{j=1}^{m} A_j^*(x)\partial_j - \sum_{j=1}^{m} \partial_j A^*(x) + A_0^*(x), \qquad (10.6.5.5)
$$

so that the Green's identity

$$
(A(x, \partial)h, v)_{L_2(\Omega)} = (h, A^*(x, \partial)v)_{L_2(\Omega)} + (A_\nu h, v)_{L_2(\Gamma)} \quad \forall\, h, v \in C^1(\bar{\Omega})
$$
$$(10.6.5.6)$$

holds true. The form of M^* in (10.6.5.4c) will be given in (10.6.11.2), when needed. Problem (10.6.5.4) can be rewritten abstractly as

$$
\zeta_t = -A^*\zeta, \quad \zeta(T) = \zeta_0, \quad 0 \le t \le T, \qquad (10.6.5.7)
$$

where

$$
A^*h = A^*(x, \partial)h : L_2(\Omega) \supset \mathcal{D}(A^*) \to L_2(\Omega), \qquad (10.6.5.8a)
$$

$$
\mathcal{D}(A^*) = \{h \in L_2(\Omega) : A^*(x, \partial)h \in L_2(\Omega); M^*h|_\Gamma = 0\}. \quad (10.6.5.8b)
$$

Then A^* is the $L_2(\Omega)$-adjoint of the operator A in (10.6.3.9) and, moreover, via Corollary 10.6.3.3, A^* is the generator of a s.c. semigroup e^{A^*t} on $Y = L_2(\Omega), t \ge 0$. The unique solution of problem (10.6.5.4), that is, of its abstract version (10.6.5.7), is given by

$$
\zeta(t) = \zeta(t; \zeta_0) = e^{A^*(T-t)}\zeta_0, \qquad (10.6.5.9)
$$

whose regularity is discussed next. It is well known [Hersch, 1963, Lemma 4; Rauch, 1972, p. 270] that the pair $\{A(x, \partial), M(x)\}$ satisfies the Kreiss condition of assumption (h.4) in Section 10.6.1 for problem (10.6.1.5), if and only if so does the pair $\{A^*(x, \partial), M^*(x)\}$ for the problem

$$
\begin{cases}
\eta_t = -A^*(x, \partial) + f & \text{in } (0, T] \times \Omega \equiv Q, & (10.6.5.10a) \\
\eta(T, \cdot) = \eta_0 & \text{in } \Omega, & (10.6.5.10b) \\
[M^*(x)]\eta = g & \text{in } (0, T] \times \Gamma \equiv \Sigma, & (10.6.5.10c)
\end{cases}
$$

with time traveling backward. It follows that the regularity theory of problem (10.6.1.5) [with a right-hand side nonhomogeneous term, as well] has direct parallels for problem (10.6.5.10), hence for the special case of problem (10.6.5.4), presently of interest.

Recalling Remark 10.6.3.1, we therefore state the following version of Theorem 10.6.3.1, as applied to problem (10.6.5.4).

Theorem 10.6.5.1 *With reference to problem (10.6.5.4) [or its abstract version (10.6.5.7)], under the given hypotheses of Section 10.6.1, assume*

$$\zeta_0 \in H_0^\theta(\Omega), \quad 0 \le \theta \le 1, \ \theta \ne \frac{1}{2} \tag{10.6.5.11}$$

(so that the required compatibility condition is satisfied). Then, the solution (10.6.5.9) satisfies, for any $0 < T < \infty$,

$$\zeta(t) = e^{A^*(T-t)}\zeta_0 \in C([0, T]; H^\theta(\Omega)), \quad 0 \le \theta \le 1, \ \theta \ne \frac{1}{2}; \tag{10.6.5.12}$$

$$\zeta(t)|_\Gamma = \left[e^{A^*(T-t)}\zeta_0\right]\big|_\Gamma \in H^\theta([0, T] \times \Gamma), \quad 0 \le \theta \le 1, \ \theta \ne \frac{1}{2}, \tag{10.6.5.13}$$

continuously in ζ_0.

Then, as a corollary of (10.6.5.13) with $\theta = 0$, if $x \in L_2(\Omega)$, we then obtain, recalling (10.6.4.11),

$$B^* e^{A^*(T-t)} x = \left[A_\nu^-\left[e^{A^*(T-t)}x\right]^-\right]_\Gamma = [A_\nu^- \zeta^-(t)]_\Gamma \in L_2(0, T; L_2(\Gamma)), \tag{10.6.5.14}$$

continuously in $x \in L_2(\Omega)$, and (10.6.5.14) verifies directly (H.1) = (10.1.6), as desired.

Version of L_s^*   The adjoint L_s^* of the operator L_s in (10.6.5.2) in the L_2-sense:

$$(L_s u, f)_{L_2(s,T;L_2(\Omega))} = (u, L_s^* f)_{L_2(s,T;L_2(\Gamma))}$$

is given by

$$(L_s^* f)(t) = B^* \int_t^T e^{A^*(\tau-t)} f(\tau)\, d\tau, \quad s \le t \le T, \tag{10.6.5.15a}$$

$$\text{(by (10.6.4.10))} \quad = -D_\lambda^*(A^* - \lambda I) \int_t^T e^{A^*(\tau-t)} f(\tau)\, d\tau \tag{10.6.5.15b}$$

$$\text{(by (10.6.4.11))} \quad = [A_\nu^- \psi^-(\tau)]_\Gamma \tag{10.6.5.16}$$

$$: \text{continuous } L_1(s, T; L_2(\Omega)) \to L_2(s, T; L_2(\Gamma))$$

$$\text{uniformly with respect to } s, \tag{10.6.5.17}$$

where (recalling the splitting in (10.6.12) and below)

$$\psi(t) = [\psi^-(t), \psi^+(t)] = \int_t^T e^{A^*(\tau - t)} f(\tau) \, d\tau \qquad (10.6.5.18)$$

solves the problem

$$\begin{cases} \psi_t = -A^*(x, \partial)\psi + f & \text{in } (0, T] \times \Omega = Q, & (10.6.5.19a) \\ \psi(T, \cdot) = \zeta_0 & \text{in } \Omega, & (10.6.5.19b) \\ [M^*(x)]\psi = 0 & \text{in } (0, T] \times \Gamma = \Sigma, & (10.6.5.19c) \end{cases}$$

or, abstractly,

$$\psi_t = -A^*\psi + f, \quad \psi(T) = 0, \qquad (10.6.5.20)$$

whereby ψ is a specialization of η in (10.6.5.10) for $\eta_0 = 0$, $g \equiv 0$. The regularity (10.6.5.17), at least for s fixed, is a consequence of the property $f \to, \psi(t)|_\Gamma$: continuous $L_1(s, T; L_2(\Omega)) \to L_2(s, T; L_2(\Gamma))$ for problem (10.6.5.19), which is part of the regularity theory of first-order systems [Rauch, 1972]. Uniformity in s is, as usual, a consequence of formula (10.6.5.15a), via a change of variable.

10.6.6 Selection of the Spaces U_θ and Y_θ in (10.1.10)

We select the spaces in (10.1.10) to be the following Sobolev spaces:

$$U_\theta \equiv H^\theta(\Gamma), \quad U_0 = U = L_2(\Gamma), \quad 0 \le \theta \le \frac{1}{2} + \delta; \qquad (10.6.6.1)$$

$$Y_\theta \equiv H_0^\theta(\Omega), \quad Y_0 = Y = L_2(\Omega), \quad 0 \le \theta \le \frac{1}{2} + \delta, \quad \theta \ne \frac{1}{2} \qquad (10.6.6.2)$$

(so that $Y_\theta = H^\theta(\Omega), 0 \le \theta < 1/2$).

In particular, the critical spaces for $\theta = 1/2 \pm \delta$ are:

$$U_\delta^- \equiv U_{\frac{1}{2}-\delta} \equiv H^{\frac{1}{2}-\delta}(\Gamma), \quad U_\delta^+ \equiv U_{\frac{1}{2}+\delta} = H^{\frac{1}{2}+\delta}(\Gamma), \qquad (10.6.6.3)$$

$$Y_\delta^- \equiv Y_{\frac{1}{2}-\delta} \equiv H^{\frac{1}{2}-\delta}(\Omega), \quad Y_\delta^+ \equiv Y_{\frac{1}{2}+\delta} \equiv H_0^{\frac{1}{2}+\delta}(\Omega), \qquad (10.6.6.4)$$

so that the dual spaces with respect to the pivot space $Y = L_2(\Omega)$ are

$$[Y_\delta^-]' = H^{-\frac{1}{2}+\delta}(\Omega), \qquad [Y_\delta^+]' = H^{-\frac{1}{2}-\delta}(\Omega), \qquad (10.6.6.5)$$

and by (10.6.6.4) and (10.6.6.5) we verify the interpolation property

$$[Y_\delta^-, [Y_\delta^+]']_{\theta=\frac{1}{2}-\delta} = \left[H^{\frac{1}{2}-\delta}(\Omega), H^{-\frac{1}{2}-\delta}(\Omega) \right]_{\theta=\frac{1}{2}-\delta} = L_2(\Omega), \quad (10.6.6.6)$$

as required by (10.1.10), since $(1/2 - \delta)(1 - \theta) - (1/2 + \delta)\theta = 0$ for $\theta = 1/2 - \delta$. Compactness of the injections $U_{\theta_1} \hookrightarrow U_{\theta_2}, Y_{\theta_1} \hookrightarrow Y_{\theta_2}, 0 \le \theta_2 < \theta_1 \le 1/2 + \delta$, as required by (10.1.10), is true since Ω is bounded. The spaces in (10.1.11) and (10.1.12) are, in the present case, as follows:

$$\mathcal{U}^{\theta}[s, T] \equiv H^{\theta,\theta}(\Sigma_{sT}) \equiv L_2(s, T; H^{\theta}(\Gamma)) \cap H^{\theta}(s, T; L_2(\Gamma)); \tag{10.6.6.7}$$

$$\mathcal{Y}^{\theta}[s, T] \equiv L_2\left(s, T; H_0^{\theta}(\Omega)\right) \cap H^{\theta}(s, T; L_2(\Omega)), \quad 0 \le \theta \le \frac{1}{2} + \delta, \quad \theta \neq \frac{1}{2}. \tag{10.6.6.8}$$

10.6.7 Verification of Assumption (H.2) = (10.1.14)

With reference to problem (10.6.1.5), or its solution formula (10.6.5.1), Corollary 10.6.3.2 gives, in particular, via (10.6.6.2), (10.6.6.7), and (10.6.6.8), and for $0 \le \theta < 1/2$:

$$L_s : \begin{cases} u \to y(\cdot, s; y_0) \text{ continuous: } \mathcal{U}^{\theta}[s, T] = H^{\theta,\theta}(\Sigma_{sT}) = H^{\theta}([s, T] \times \Omega), \\ y_0 = 0 \end{cases}$$

$$\to \mathcal{Y}^{\theta}[s, T] \cap C([s, T]; Y_{\theta}) = H^{\theta}(s, T; L_2(\Omega)) \cap C([s, T]; H^{\theta}(\Omega)), \tag{10.6.7.1}$$

at least for fixed s, while uniformity with respect to s follows then by formula (10.6.5.2) via a change of variable, as usual. Thus, (10.6.7.1) verifies assumption (H.2) = (10.1.14), as desired.

10.6.8 Verification of Assumption (H.3) = (10.1.15)

We begin by noticing that the available regularity theory for first-order hyperbolic systems, say problem (10.6.5.10), assumes *symmetric hypotheses on the data, with equal regularity in time and space* and yields *symmetric solutions* with *equal regularity in time and space*. See [Rauch, 1972, Differentiability Theorem, p. 272] of which selected cases have been reported in Theorem 10.6.3.1 and Corollary 10.6.3.2 for problem (10.6.1.5), with equal results available for problem (10.6.5.10). In contrast, hypothesis (H.3) = (10.1.15) is of different type, in the sense that it assumes only space regularity in Y_{θ}, while only $L_2(s, T; \cdot)$-regularity in time, and asserts, however, equal regularity in time and space for the solution ψ of (10.6.5.19), which defines L_s^* via (10.6.5.16). Thus, we cannot simply quote the regularity theory for first-order hyperbolic systems from Rauch [1972], as done in Sections 10.6.5 and 10.6.7. We need to develop *nonsymmetric* regularity results. To this end, the operator representation for first-order systems, in particular, formula (10.6.5.15), (10.6.5.16) is very useful. [Similar consideration in contrasting purely PDE methods with operator methods were made in connection with second-order hyperbolic equations just below Eqn. (10.5.5.13)]. We begin with

Lemma 10.6.8.1 *With reference to the operator L^* (with $s = 0$) [see (10.6.5.15)] we have, under the hypotheses of Section 10.6.1,*

$$L^* : continuous \ L_2(0, T; \mathcal{D}(A^*)) \to H^1(0, T; L_2(\Gamma)). \qquad (10.6.8.1)$$

Proof. Without loss of generality for the present proof, we may take that the constant λ in (10.6.5.15b) [given by Lemma 10.6.4.1] is zero: $\lambda = 0$. Thus, let $f \in L_2(0, T; \mathcal{D}(A^*))$, or $A^* f \in L_2(0, T; L_2(\Omega))$, and differentiate (10.6.5.15b), that is,

$$(L^* f)(t) = -D_0^* \int_t^T e^{A^*(\tau - t)} A^* f(\tau) \, d\tau, \quad 0 \le t \le T, \qquad (10.6.8.2)$$

to obtain

$$\frac{d(L^* f)(t)}{dt} = D_0^* A^* f(t) + D_0^* A^* \int_t^T e^{A^*(\tau - t)} A^* f(\tau) \, d\tau$$

$$= D_0^* A^* f(t) + (L^* A^* f)(t) \in L_2(0, T; L_2(\Gamma)), \qquad (10.6.8.3)$$

where the regularity in (10.6.8.3) follows from $D_0^* \in \mathcal{L}(L_2(\Omega); L_2(\Gamma))$ [see (10.6.4.5)] and the regularity of L^* in (10.6.5.17) with $A^* f$ as assumed. Then, (10.6.8.3) proves (10.6.8.1), as desired. $\square$

Theorem 10.6.8.2 *Under the hypotheses of Section 10.6.1, and with reference to (10.6.5.15), say, for $s = 0$, or to (10.6.8.2), we have*

$$L^* : continuous \ L_2\big(0, T; H_0^1(\Omega)\big) \to H^{1,1}(\Sigma)$$

$$= L_2(0, T; H^1(\Gamma)) \cap H^1(0, T; L_2(\Gamma)). \qquad (10.6.8.4)$$

Proof. Since, recalling (10.6.5.5) and (10.6.5.8),

$$H_0^1(\Omega) \subset \mathcal{D}(A^*),$$

and thus Lemma 10.6.8.1 applies a fortiori, we see that all that is left to prove is half of the statement (10.6.8.4), more precisely,

$$f \in L_2\big(0, T; H_0^1(\Omega)\big) \Rightarrow L^* f \in L_2(0, T; H^1(\Gamma)) \text{ continuously} \qquad (10.6.8.5)$$

[a result that is *not* contained in Rauch [1972] explicitly (recall (10.6.5.16)) in connection with problem (10.6.5.19)].

To establish (10.6.8.5), we proceed as follows (we are arguing somewhat as in the proof of [Lasiecka et al., 1986, Theorem 3.11, p. 182]: The starting step is the homogeneous ζ-problem in (10.6.5.4) by PDE methods, followed by an operator analysis of (10.6.8.2).

Step 1 We apply Theorem 10.6.5.1 for $\theta = 1$: If $x \in H_0^1(\Omega)$, then a fortiori from (10.6.5.13) we obtain, for $\zeta(t; x) = e^{A^*(T - t)} x \in C([0, T]; H_0^1(\Omega))$, via (10.6.4.11), that

$$B^* e^{A^*(T-t)} x = [A_v^- \zeta^-(t; x)]_\Gamma = \left[A_v^- \left[e^{A^*(T-t)} x \right]^- \right]_\Gamma \in L_2(0, T; H^1(\Gamma)),$$
(10.6.8.6a)

continuously in x, that is,

$$B^* e^{A^* t} : \text{continuous } H_0^1(\Omega) \to L_2(0, T; H^1(\Gamma)). \qquad (10.6.8.6b)$$

Step 2 For $f \in L_2(0, T; H_0^1(\Omega))$ as assumed, we compute from (10.6.5.15a) for $s = 0$, via Schwarz's inequality,

$$\|L^* f\|_{L_2(0,T;H^1(\Gamma))}^2 = \int_0^T \left\| \int_t^T B^* e^{A^*(\tau-t)} f(\tau) \, d\tau \right\|_{H^1(\Gamma)}^2 dt$$

$$\leq T \int_0^T \int_t^T \left\| B^* e^{A^*(\tau-t)} f(\tau) \right\|_{H^1(\Gamma)}^2 d\tau \, dt$$

(after a change in the order of integration)

$$= T \int_0^T \int_0^\tau \left\| B^* e^{A^*(\tau-t)} f(\tau) \right\|_{H^1(\Gamma)}^2 dt \, d\tau$$

$$\leq T \int_0^T \int_0^T \left\| B^* e^{A^* \sigma} f(\tau) \right\|_{H^1(\Gamma)}^2 d\sigma \, d\tau$$

(by (10.6.8.6)) $\leq cT \int_0^T \|f(\tau)\|_{H_0^1(\Omega)}^2 d\tau = cT \|f\|_{L_2(0,T;H_0^1(\Omega))}^2,$ (10.6.8.7)

recalling (10.6.8.6) in the last step, and (10.6.8.7) proves (10.6.8.5), as desired. Theorem 10.6.8.2 is proved. □

Corollary 10.6.8.3 *Under the hypotheses of Section 10.6.1, and with reference to (10.6.5.15), we have*

$$L_s^* : \text{continuous } L_2(s, T; H_0^r(\Omega)) \to H^{r,r}(\Sigma_{sT}), \quad 0 \leq r \leq 1, \quad r \neq \frac{1}{2},$$

uniformly in s.
(10.6.8.8)

Restricting to $0 \leq \theta = r \leq 1/2 + \delta, \theta = r \neq 1/2$, then (10.6.8.8) specializes to

$$L_s^* : \text{continuous } L_2(s, T; Y_\theta) \to \mathcal{U}^\theta[s, T], \quad 0 \leq \theta \leq \frac{1}{2} + \delta, \quad \theta \neq \frac{1}{2},$$

uniformly in s,
(10.6.8.9)

by recalling (10.6.6.2) and (10.6.6.7).

Proof. We interpolate between (10.6.8.4) (for fixed s, in general) [case $r = 1$] and statement (10.6.5.17) in the weaker version L_s^* : continuous $L_2(s, T; L_2(\Omega)) \to L_2(s, T; L_2(\Gamma))$ [case $r = 0$], to obtain (10.6.8.8), as desired. [The space $H_{00}^{\frac{1}{2}}(\Omega) =$

$[H_0^1(\Omega), L_2(\Omega)]_{r=\frac{1}{2}}$ [Lions, Magenes, 1972, p. 66] applies for $r = 1/2$, but this is not needed for our purposes.] $\square$

Statement (10.6.8.9) of Corollary 10.6.8.3 verifies assumption (H.3) = (10.1.15).

10.6.9 Verification of Assumptions (H.4) = (10.1.18) through (H.7) = (10.1.21)

Verification of (H.4) = (10.1.18) (Direct verification)

Lemma 10.6.9.1 *Under the hypotheses of Section 10.6.1, we have*

$$B^* e^{A^* t} y_0 = [A_\nu^- [e^{A^* t} y_0]^-]_\Gamma : continuous \ Y_\delta^+ \equiv H_0^{\frac{1}{2}+\delta}(\Omega) \to C([0, T]; L_2(\Gamma)). \tag{10.6.9.1}$$

Proof. The expression for $B^* e^{A^* t} y_0$ in (10.6.9.1) was already obtained in (10.6.5.14), via (10.6.4.11). Thus, what we need to show is the regularity in (10.6.9.1). But this holds true. In fact, Eqn. (10.6.5.12), specialized to $\theta = 1/2 + \delta$, says that

$$y_0 \in H_0^{\frac{1}{2}+\delta}(\Omega) \Rightarrow e^{A^* t} y_0 \in C([0, T]; H^{\frac{1}{2}+\delta}(\Omega)), \tag{10.6.9.2}$$

while trace theory [Lions, Magenes, 1972] applied in (10.6.9.2) then yields

$$[A_\nu^- [e^{A^* t} y_0]^-]_\Gamma \in C([0, T]; L_2(\Gamma)), \tag{10.6.9.3}$$

continuously in y_0, and thus (10.6.9.1) is fully proved by (10.6.9.3). $\square$

Statement (10.6.9.1) verifies assumption (H.4) = (10.1.18).

Verification of Assumption (H.5) = (10.1.19) Corollary 10.6.3.3 gives that A in (10.6.3.8), hence A^* in (10.6.5.8), generate s.c. semigroups e^{At} and $e^{A^* t}$ on $Y_\theta \equiv H^\theta(\Omega), 0 \le \theta < 1/2$, in particular on $Y_\delta^- = H^{\frac{1}{2}-\delta}(\Omega)$. Moreover, $\mathcal{D}(A)$ given by (10.6.3.9b) is dense in Y_δ^-, as required.

Verification of Assumption (H.6) = (10.1.20) First, let $z_n \in \mathcal{D}(A)$ given by (10.6.3.8b) so that $Az_n = A(\cdot, \partial)z_n$, and then by applying [Lions, Magenes, 1972, p. 85] for a first-order differential operator, we obtain, recalling (10.6.6.4) and (10.6.6.5),

$$\|Az_n\|_{[Y_\delta^+]'} = \|Az_n\|_{H^{-\frac{1}{2}-\delta}(\Omega)} = \|A(\cdot, \partial)z_n\|_{H^{-\frac{1}{2}-\delta}(\Omega)}$$
$$\le C\|z_n\|_{H^{\frac{1}{2}-\delta}(\Omega)} = c\|z_n\|_{Y_\delta^-}, \quad z_n \in \mathcal{D}(A). \tag{10.6.9.4}$$

Next, we use the property (already observed in verifying (H.5) above) that $\mathcal{D}(A)$ is dense in Y_δ^- in the Y_δ^--topology, to extend (10.6.9.4) to all $z \in Y_\delta^-$:

$$\|Az\|_{[Y_\delta^+]'} \le c\|z\|_{Y_\delta^-}, \quad z \in Y_\delta^- \tag{10.6.9.5}$$

by closedness of A. Thus, (10.6.9.5) verifies assumption (H.6) = (10.1.20).

Verification of Assumption (H.7) $=$ (10.1.21)   Instead of verifying (H.7) as

$$B : \text{continuous } U_\delta^- \equiv H^{\frac{1}{2}-\delta}(\Gamma) \to [Y_\delta^+]' = H^{-\frac{1}{2}-\delta}(\Omega), \qquad (10.6.9.6)$$

we verify equivalently that

$$B^* : \text{continuous } H_0^{\frac{1}{2}+\delta}(\Omega) \to H^{-\frac{1}{2}+\delta}(\Gamma). \qquad (10.6.9.7)$$

Indeed, let $x \in H_0^{\frac{1}{2}+\delta}(\Omega)$ so that $x|_\Gamma = 0$ and then, recalling (10.6.4.11), we obtain $B^*x = [A_\nu^- x^-]_\Gamma = 0 \in H^{-\frac{1}{2}+\delta}(\Gamma)$, continuously in x, as desired.

Remark 10.6.9.1   We notice a contrast between the case of second-order hyperbolic equations treated in Section 10.5 and the case of first-order hyperbolic systems treated in this section: In the former case we verified the property $A^{-1}B \in \mathcal{L}(U; Y_\delta^-)$ [see Eqn. (10.5.9.8)], in the latter case, this property is not true, for all we have is $(A - \lambda I)^{-1}B = -D_\lambda \in \mathcal{L}(L_2(\Gamma); L_2(\Omega))$ [see (10.6.4.9)], with no additional smoothing available into $Y_\delta^- = H^{\frac{1}{2}-\delta}(\Omega)$.

10.6.10 Proof of Lemma 10.6.4.1: Existence and Regularity of the Map D_λ

Half-Space Problem   By partition of unity techniques, we are led to the corresponding half-space problem. Thus, without loss of generality, we may assume that Ω is the half-space

$$\Omega = \{x = [x_1, x^1]; \ x^1 = [x_2, \ldots, x_m]; \ x_1 > 0\}$$

$$\text{with boundary } \Gamma = \Omega|_{x_1=0} = \{x = [0, x^1]\}. \qquad (10.6.10.1)$$

If η is the dual variable corresponding to y, $\partial_j y \to i\eta_j$, then the original problem (10.6.4.2) is rewritten as

$$\begin{cases} A_1(x)\dfrac{dv}{dx_1} + i\mathcal{B}_\lambda(x; \eta)v = 0, & x_1 > 0; & (10.6.10.2a) \\[2mm] \qquad\qquad M(x^1)v = g, & x_1 = 0, & (10.6.10.2b) \end{cases}$$

where $\mathcal{B}_\lambda(x; \eta)$ is the pseudo-differential operator of first order corresponding to $\sum_{j=2}^m A_j(x)\partial_j - \lambda I$. Invoking assumption (h.2) in Section 10.6.1, we then obtain from (10.6.10.2)

$$\begin{cases} \dfrac{dv}{dx_1} = \mathcal{P}_\lambda(x; \eta)v, & x_1 > 0; & (10.6.10.3a) \\[2mm] M(x^1)v = g, & x_1 = 0, & (10.6.10.3b) \end{cases}$$

with $\mathcal{P}_\lambda(x; \eta) = -[A_1(x)]^{-1} i \mathcal{B}_\lambda(x; \eta)$.

Step 1 (Estimate (10.6.4.3)) We next obtain energy estimates for (10.6.10.3), using the Kreiss symmetrizer [Kreiss, 1970]. The Kreiss symmetrizer is a pseudo-differential operator of order zero, constructed locally, that is, in a canonical neighborhood of the point (x_1, η), with symbol $R(x; \eta; \lambda)$, which has an L_2-norm independent of λ, and which possesses the following properties:

(i) R is Hermitian.
(ii) R is homogeneous of order zero in $[\lambda, \eta]$, for $\lambda^2 + |\eta|^2 \geq 1$, and is a smooth function of all of its variables and of the coefficient matrices A_j and M.
(iii) The following estimate holds true: There exists constants $\delta > 0$ and $C > 0$ independent of λ such that

$$(u, Ru)_{L_2(\Gamma)} \geq \delta \|u\|^2_{L_2(\Gamma)} - C\|g\|^2_{L_2(\Gamma)} \qquad (10.6.10.4)$$

for all u such that $Mu = g$ on Γ.
(iv)

$$\mathrm{Re}(R\mathcal{P}_\lambda u, u)_{L_2(\Omega)} \geq \delta\lambda\|u\|^2_{L_2(\Omega)}. \qquad (10.6.10.5)$$

To obtain the *energy estimate* (10.6.4.3), we multiply both sides of Eqn. (10.6.10.3a) by R and take the $L_2(\Omega)$-inner product with v to obtain

$$\mathrm{Re}\left(R\frac{dv}{dx_1}, v\right)_{L_2(\Omega)} = \mathrm{Re}(R\mathcal{P}_\lambda v, v)_{L_2(\Omega)} \geq \delta\lambda\|v\|^2_{L_2(\Omega)}, \qquad (10.6.10.6)$$

recalling (iv) = (10.6.10.5). However, integration by parts in x_1, from 0 to ∞, on the left-hand side of (10.6.10.6) gives, recalling property (i),

$$\mathrm{Re}\left(R\frac{dv}{dx_1}, v\right)_{L_2(\Omega)} = \mathrm{Re}\left(\frac{dv}{dx_1}, Rv\right)_{L_2(\Omega)}$$

$$= \mathrm{Re}\left\{\left[(v, Rv)_{L_2(\Omega)}\right]^\infty_{x_1=0}\right.$$

$$\left. - \left(v, \left(\frac{dR}{dx_1}\right)v\right)_{L_2(\Omega)} - \left(v, R\frac{dv}{dx_1}\right)_{L_2(\Omega)}\right\}, \qquad (10.6.10.7)$$

or upon comparing the first and last term in (10.6.10.7),

$$\mathrm{Re}\left(R\frac{dv}{dx_1}, v\right)_{L_2(\Omega)} = \frac{1}{2}\left\{-(v, Rv)_{L_2(\Gamma)} - \mathrm{Re}\left(v, \left(\frac{dR}{dx_1}\right)v\right)_{L_2(\Omega)}\right\}. $$

$$(10.6.10.8)$$

Using estimate (iii) = (10.6.10.4), as well as the fact that dR/dx_1 is of zero order, we obtain from (10.6.10.8),

$$\mathrm{Re}\left(R\frac{dv}{dx_1}, v\right)_{L_2(\Omega)} \leq -\frac{1}{2}\delta\|v\|^2_{L_2(\Gamma)} + \frac{C}{2}\|g\|^2_{L_2(\Gamma)} + C_1\|v\|^2_{L_2(\Omega)}. \qquad (10.6.10.9)$$

The estimate from above (10.6.10.9), combined with the estimate from below (10.6.10.6), yields

$$(\delta\lambda - C_1)\|v\|^2_{L_2(\Omega)} + \frac{1}{2}\delta\|v\|^2_{L_2(\Gamma)} \le \frac{C}{2}\|g\|^2_{L_2(\Gamma)}, \qquad (10.6.10.10)$$

for all v satisfying the boundary condition (10.6.10.3b). Thus, selecting λ in (10.6.10.10) large enough leads to the desired estimate (10.6.4.3), in view of the definition (10.6.4.2).

Step 2 (Adjoint problem)   In a similar vein we obtain an estimate similar to (10.6.4.3) for the problem adjoint to (10.6.4.2), expressed in terms of the adjoints $A^*(x, \partial)$ and M^* (defined in Section 10.6.11).

Step 3   Estimate (10.6.10.10), that is, (10.6.4.3), yields *uniqueness* whereas Steps 1 and 2 on the original and adjoint problems yield *existence* (Fredholm alternative). The proof of Lemma 10.6.4.1 is complete.   □

Remark 10.6.10.1   In the special case when A_j are symmetric and block-diagonal, that is, of the form

$$A_j = \begin{bmatrix} A_j^- & 0 \\ 0 & A_j^+ \end{bmatrix},$$

construction of a symmetrizer is simple. Indeed, one takes, in this case,

$$R = \begin{bmatrix} -\epsilon I_1 & 0 \\ 0 & I_2 \end{bmatrix}; \quad \begin{array}{l} I_1 \text{ and } I_2 \text{ identity matrices of size } \ell \times \ell \text{ and} \\ (k - \ell) \times (k - \ell), \text{ respectively,} \end{array}$$

and $\epsilon > 0$ as a suitable small constant.

Remark 10.6.10.2   Lemma 10.6.4.1 was proved by Lax and Phillips [1980] in the case of symmetric problems with dissipative boundary conditions. It was later extended by Friedrichs and Lax [1965; 1968] to symmetrizable operators still with dissipative boundary conditions. The Kreiss symmetrizer allows one to solve the problem for nonsymmetric and nondissipative problems [Kreiss, 1970].

10.6.11 Proof of Lemma 10.6.4.2

To compute M^* corresponding to the problem adjoint of (10.6.1.5), we first note that in view of hypotheses (h.3) and (h.4) of Section 10.6.1, we can replace M by

$$M = [I_\ell, S]; \quad I_\ell : \ell \times \ell \text{ identity matrix}; \quad S : \ell \times (k - \ell) \text{ smooth valued matrix.}$$
$$(10.6.11.1)$$

From this, it follows [Rauch, 1972, p. 270] that the adjoint boundary operator M^* can be written as

$$M^* = [-(A_\nu^+)^{-1} S^{ct} A_\nu^-, \ I_{k-\ell}], \qquad (10.6.11.2)$$

where S^{ct} is the conjugate transpose (adjoint) of S. Next, let $A^*(x, \partial)$ be the formal adjoint of $A(x, \partial)$ defined by (10.6.1.1), so that the Green's formula yields with $g \in L_2(\Gamma)$, $y \in H^1(\Omega)$:

$$
\begin{aligned}
(A^*(x, \partial)y, D_\lambda g)_{L_2(\Omega)} &= (y, A(x, \partial)D_\lambda g)_{L_2(\Omega)} + (A_\nu y, D_\lambda g)_{L_2(\Gamma)} \\
\text{(by (10.6.4.2))} \quad &= (y, \lambda D_\lambda g)_{L_2(\Omega)} + (A_\nu y, D_\lambda g)_{L_2(\Gamma)} \\
&= (y, \lambda D_\lambda g)_{L_2(\Omega)} + (A_\nu^- y^-, (D_\lambda g)^-)_{L_2(\Gamma)} \\
&\quad + (A_\nu^+ y^+, (D_\lambda g)^+)_{L_2(\Gamma)}, \qquad (10.6.11.3)
\end{aligned}
$$

recalling (10.6.1.2) and splitting accordingly $y = [y^-, y^+]$, where $D_\lambda g = v$ solves (10.6.4.2). Next, we select y in (10.6.11.2) in $\mathcal{D}(A^*)$, defined by

$$\mathcal{D}(A^*) = \{u \in L_2(\Omega) : A^*(x, \partial)u \in L_2(\Omega) \text{ and } M^* u|_\Gamma = 0\}. \quad (10.6.11.4)$$

Thus, in particular, such y satisfies $M^* y|_\Gamma = 0$, or recalling (10.6.11.2),

$$-(A_\nu^+)^{-1} S^{ct} A_\nu^- y^- + y^+ = 0 \text{ on } \Gamma. \qquad (10.6.11.5)$$

However, by (10.6.4.2) and (10.6.11.1),

$$g = M D_\lambda g|_\Gamma = (D_\lambda g)^- + S(D_\lambda g)^+ \text{ on } \Gamma. \qquad (10.6.11.6)$$

Using (10.6.11.5) and (10.6.11.6) on the right-hand side of (10.6.11.3), we obtain with $y \in \mathcal{D}(A^*)$ and $g \in L_2(\Gamma)$:

$$
\begin{aligned}
(A^* y, D_\lambda g)_{L_2(\Omega)} &= (y, \lambda D_\lambda g)_{L_2(\Omega)} + (A_\nu^- y^-, g - S(D_\lambda g)^+)_{L_2(\Gamma)} \\
&\quad + (A_\nu^+ [(A_\nu^+)^{-1} S^{ct} A_\nu^- y^-], (D_\lambda g)^+)_{L_2(\Gamma)} \\
&= (y, \lambda D_\lambda g)_{L_2(\Omega)} + (A_\nu^- y^-, g)_{L_2(\Gamma)}, \qquad (10.6.11.7)
\end{aligned}
$$

moving S^{ct} on the left to S on the right of the inner-product and performing a cancellation. Thus, (10.6.11.7) gives, recalling D_λ^* below (10.6.4.5):

$$(D_\lambda^* A^* y - \lambda D_\lambda^* y, g)_{L_2(\Gamma)} = (A_\nu^- y, g)_{L_2(\Gamma)}, \qquad (10.6.11.8)$$

for all $g \in L_2(\Gamma)$ and $y \in \mathcal{D}(A^*)$. Equation (10.6.11.8) proves Lemma 10.6.4.2. □

Remark 10.6.11.1 Definition (10.6.11.4) yields that $H_0^1(\Omega) \subset \mathcal{D}(A^*)$.

10.7 Application: Kirchoff Equation with One Boundary Control. Regularity Theory

In this section we consider an optimal quadratic cost problem over a finite horizon for a Kirchoff equation, subject only to one control acting in the "moment" boundary condition. [The physical bending moment in the two-dimensional Kirchoff plate model is actually a modification of the boundary condition (10.7.1.1d) below.] The Kirchoff equation is hyperbolic with finite speed of propagation and displays a behavior similar to that of the wave equation in Section 10.5. As in Sections 10.5 and 10.6, we shall show likewise that all abstract system's assumptions $(H.1) = (10.1.6)$ through $(H.7) = (10.1.21)$ of Section 10.1 are automatically satisfied in a natural mathematical setting. Many can be chosen, and we shall select a particular interesting one where, as in the case of second-order hyperbolic equations of Section 10.5, the observation R^*R jumps across a boundary condition; see (10.7.2.3)–(10.7.2.5) below. Accordingly, Theorems 10.2.1, 10.2.2, and 10.2.3 of Section 10.2 are then applicable to the present class, for any observation operator R with "minimal" smoothing as in $(H.8) = (10.1.23)$.

10.7.1 Problem Formulation

The Dynamics   Let Ω be an open bounded domain in R^n with sufficiently smooth boundary Γ, say, of class C^2. The Kirchoff equation is given by

$$
\begin{cases}
w_{tt} - \rho\Delta w_{tt} + \Delta^2 w = 0 & \text{in } (0, T] \times \Omega \equiv Q; & (10.7.1.1a) \\
w(0, \cdot) = w_0, \quad w_t(0, \cdot) = w_1 & \text{in } \Omega; & (10.7.1.1b) \\
w|_\Sigma = 0 & \text{in } (0, T] \times \Gamma \equiv \Sigma, & (10.7.1.1c) \\
\Delta w|_\Sigma = u & \text{in } \Sigma, & (10.7.1.1d)
\end{cases}
$$

where $\rho > 0$ is a constant (proportional to the square of the thickness in the two-dimensional plate model), and where $u \in L_2(0, T; L_2(\Gamma)) \equiv L_2(\Sigma)$ is the control function acting in the "moment" BC (10.7.1.1d).

The Optimal Control Problem on $[s, T]$ Consistently with the (optimal) regularity theory for problem (10.7.1.1) presented in Theorem 10.7.3.2 below, the cost functional we seek to minimize over all $u \in L_2(s, T; L_2(\Gamma)) = L_2(\Sigma_{sT})$ is taken to be

$$
J(u, w) = \int_s^T \left\{ \left\| R \begin{bmatrix} w(t) \\ w_t(t) \end{bmatrix} \right\|^2_{H^2(\Omega) \times H^1(\Omega)} + \|u(t)\|^2_{L_2(\Gamma)} \right\} dt, \quad (10.7.1.2)
$$

with initial data $\{w_0, w_1\} \in [H^2(\Omega) \cap H_0^1(\Omega)] \times H_0^1(\Omega)$, where the observation operator $R \in \mathcal{L}([H^2(\Omega) \cap H_0^1(\Omega)] \times H_0^1(\Omega))$ will be further specified below in (10.7.2.3).

10.7.2 Main Results

As a specialization to problem (10.7.1.1), (10.7.1.2) of the abstract theory presented in Theorems 10.2.1, 10.2.2, and 10.2.3 of this chapter, in the present section we establish the following results.

Theorem 10.7.2.1   *(a) With the observation operator R in (10.7.1.2) only assumed to satisfy*

$$R \in \mathcal{L}\big(\big[H^2(\Omega) \cap H_0^1(\Omega)\big] \times H_0^1(\Omega)\big). \tag{10.7.2.1}$$

Theorem 9.2.1, part (a), of Chapter 9 applies to the optimal control problem (10.7.1.1), (10.7.1.2), with

$$y(t) = \begin{bmatrix} w(t) \\ w_t(t) \end{bmatrix}; \quad U = L_2(\Gamma); \quad Y \equiv \big[H^2(\Omega) \cap H_0^1(\Omega)\big] \times H_0^1(\Omega). \tag{10.7.2.2}$$

The problem yields a unique optimal pair $\{u^0(\cdot, s; y_0), y^0(\cdot, s; y_0)\}$, $y_0 = [w_0, w_1] \in Y$, satisfying, in particular, the pointwise feedback synthesis (9.2.17) and the optimal cost relation (9.2.18) of Chapter 9, as well as the other properties listed there.

(b) Assume, in addition to (10.7.2.1), that R satisfies the smoothing assumption

$$R^*R : continuous \ Y_\delta^- \to Y_\delta^+, \tag{10.7.2.3}$$

with R^* the adjoint of R in Y, where

$$Y_\delta^- \equiv \big\{h \in H^{\frac{5}{2}-\delta}(\Omega) : h|_\Gamma = 0\big\} \times \big\{h \in H^{\frac{3}{2}-\delta}(\Omega) : h|_\Gamma = 0\big\} \tag{10.7.2.4a}$$

$$= \big[H^{\frac{5}{2}-\delta}(\Omega) \cap H_0^1(\Omega)\big] \times H^{\frac{3}{2}-\delta}(\Omega) \cap H_0^1(\Omega), \tag{10.7.2.4b}$$

$$Y_\delta^+ \equiv \big\{h \in H^{\frac{5}{2}+\delta}(\Omega) : h|_\Gamma = \Delta h|_\Gamma = 0\big\} \times \big[H^{\frac{3}{2}+\delta}(\Omega) \cap H_0^1(\Omega)\big] \tag{10.7.2.5}$$

(see Section 10.7.6), and where $\delta > 0$ is an arbitrarily small constant, which is kept fixed throughout. [An additional characterization of Y_δ^- and Y_δ^+ will be given below in (10.7.6.4), (10.7.6.5).] Then Theorems 10.2.1, 10.2.2, and 10.2.3 of Section 10.2 hold true, with Y_δ^- and Y_δ^+ given by (10.7.2.4) and (10.7.2.5), and with

$$U_\delta^- = H^{\frac{1}{2}-\delta}(\Gamma), \quad U_\delta^+ = H^{\frac{1}{2}+\delta}(\Gamma) \tag{10.7.2.6}$$

(see Section 10.7.6). In particular, explicitly:

(b$_1$) (Regularity of the optimal pair) The optimal pair satisfies, for

$$y_0 = [w_0, w_1] \in \big[H^{\frac{5}{2}-\delta}(\Omega) \cap H_0^1(\Omega)\big] \times \big[H^{\frac{3}{2}-\delta}(\Omega) \cap H_0^1(\Omega)\big] \tag{10.7.2.7}$$

the following regularity properties:

(i)

$$\begin{bmatrix} w^0(\cdot, s; y_0) \\ w_t^0(\cdot, s; y_0) \end{bmatrix} = y^0(\cdot, s; y_0)$$

$$\in C\big([s, T]; \big[H^{\frac{5}{2}-\delta}(\Omega) \cap H_0^1(\Omega)\big] \times \big[H^{\frac{3}{2}-\delta}(\Omega) \cap H_0^1(\Omega)\big]\big)$$

$$\cap \big[H^{\frac{1}{2}-\delta}\big(s, T; \big[H^2(\Omega) \cap H_0^1(\Omega)\big] \times H_0^1(\Omega)\big); \tag{10.7.2.8}$$

(ii)

$$u^0(\,\cdot\,, s; y_0) \in L_2\big(s, T; H^{\frac{1}{2}+\delta}(\Gamma)\big) \cap H^{\frac{1}{2}+\delta}(s, T; L_2(\Gamma)) \equiv H^{\frac{1}{2}+\delta, \frac{1}{2}+\delta}(\Sigma_{sT}),$$

$$(10.7.2.9)$$

$\Sigma_{sT} = [s, T] \times \Gamma$, *a fortiori,*

$$u^0(\,\cdot\,, s; y_0) \in C([s, T]; H^\delta(\Gamma)). \tag{10.7.2.10}$$

All the above results hold true uniformly in s, that is, with norms that may be made independent of s.

(b_2) *(Gain operator) The gain operator $B^* P(t)$, with $P(t)$ defined by (9.2.14) of Chapter 9, satisfies the following regularity property (see (10.2.4)):*

$$B^* P(t)x = B^* \begin{bmatrix} P_{11}(t) & P_{12}(t) \\ P_{21}(t) & P_{22}(t) \end{bmatrix} \begin{bmatrix} x_1 \\ x_2 \end{bmatrix} \tag{10.7.2.11a}$$

$$= -\frac{\partial}{\partial\nu}[P_{21}(t)x_1 + P_{22}(t)x_2] \tag{10.7.2.11b}$$

$$: continuous \left[H^{\frac{5}{2}-\delta}(\Omega) \cap H^1_0(\Omega) \right] \times \left[H^{\frac{3}{2}-\delta}(\Omega) \cap H^1_0(\Omega) \right]$$

$$\to C([0, T]; L_2(\Gamma)), \tag{10.7.2.11c}$$

where the adjoint B^* is computed with respect to the space Y, topologized, however, as $Y \equiv \mathcal{D}(\mathcal{A}^{\frac{1}{2}}) \times \mathcal{D}(\mathcal{A}^{\frac{1}{4}}_\rho)$, see below in (10.7.4.22).

(b_3) *(Differential Riccati equation) With*

$$A = \begin{bmatrix} 0 & I \\ \mathbb{A} & 0 \end{bmatrix}, \quad \mathbb{A} = \big(I + \rho\mathcal{A}^{\frac{1}{2}}\big)^{-1}\mathcal{A}$$

as in (10.7.4.3) below, we have that $P(t)$ satisfies the following DRE:

$$\begin{cases} \dfrac{d}{dt}(P(t)x, y)_Y = -(Rx, Ry)_Y - (P(t)x, Ay)_Y - (P(t)Ax, y)_Y \\[2mm] \qquad\qquad + \left(\dfrac{\partial}{\partial\nu}[P_{21}(t)x_1 + P_{22}(t)x_2], \dfrac{\partial}{\partial\nu}[P_{21}(t)y_1 + P_{22}(t)y_2] \right)_{L_2(\Gamma)}, \\[2mm] P(T) = 0 \quad \forall\, x, y \in \left[H^{\frac{5}{2}-\delta}(\Omega) \cap H^1_0(\Omega) \right] \times \left[H^{\frac{3}{2}-\delta}(\Omega) \cap H^1_0(\Omega) \right]. \end{cases}$$

$$(10.7.2.12)$$

Furthermore, $P(t)$ satisfies the corresponding integral Riccati equation, as in (10.2.7), for all such x, y.

10.7.3 Regularity Theory for Problem (10.7.1.1) with $u \in L_2(\Sigma)$

The following well-posedness Theorem 10.7.3.2 provides the critical regularity result, which justifies as natural the selection of the cost functional J in (10.7.1.2), and which will permit us to verify assumption (H.1) = (10.1.6) below. To this end, as we shall see, it is expedient to associate with problem (10.7.1.1) the following boundary-homogeneous version (which would be the corresponding adjoint problem (10.7.5.7) below, if the initial conditions were given at $t = T$, an inessential modification since the problem is time reversible):

$$\begin{cases} \phi_{tt} - \rho \Delta \phi_{tt} + \Delta^2 \phi = f & \text{in } Q; & (10.7.3.1\text{a}) \\ \phi(0, \cdot) = \phi_0, \quad \phi_t(0, \cdot) = \phi_1 & \text{in } \Omega; & (10.7.3.1\text{b}) \\ \phi|_\Sigma \equiv 0, \quad \Delta \phi|_\Sigma \equiv 0 & \text{in } \Sigma. & (10.7.3.1\text{c}) \end{cases}$$

The following key result is taken from Lasiecka and Triggiani [1991(a)].

Theorem 10.7.3.1 *With reference to (10.7.3.1), assume Γ of class C^2, and*

$$\{\phi_0, \phi_1\} \in V \times \left[H^2(\Omega) \cap H_0^1(\Omega) \right], \quad f \in L_1(0, T; L_2(\Omega)), \quad (10.7.3.2)$$

$$V = \{h \in H^3(\Omega) : h|_\Gamma = \Delta h|_\Gamma = 0\}. \quad (10.7.3.3)$$

Then, the unique solution of problem (10.7.3.1) satisfies, continuously,

$$\{\phi, \phi_t\} \in C\left([0, T]; V \times \left[H^2(\Omega) \times H_0^1(\Omega) \right]\right), \quad (10.7.3.4\text{a})$$

$$\phi_{tt} \in L_1\left(0, T; H_0^1(\Omega)\right), \quad \text{or } \phi_{tt} \in C\left([0, T]; H_0^1(\Omega)\right) \quad \text{if } f \equiv 0, \quad (10.7.3.4\text{b})$$

and

$$\frac{\partial(\Delta \phi)}{\partial \nu} \in L_2(0, T; L_2(\Gamma)) \equiv L_2(\Sigma). \quad (10.7.3.5)$$

Proof. The proof of ((10.7.3.5) of) Theorem 10.7.3.1 is by PDE energy methods and will be given in Section 10.7.10 below. $\square$

Remark 10.7.3.1 The trace regularity (10.7.3.5) is, of course, the key result of Theorem 10.7.3.1. It does *not* follow from the optimal interior regularity (10.7.3.4): Indeed, (10.7.3.5) is "1/2 higher" in Sobolev space regularity on Ω over a formal application of trace theory to ϕ in (10.7.3.4).

Theorem 10.7.3.2 *With reference to the nonhomogeneous problem (10.7.1.1), assume Γ of class C^2 and*

$$\{w_0, w_1\} \in \left[H^2(\Omega) \cap H_0^1(\Omega) \right] \times H_0^1(\Omega), \quad u \in L_2(\Sigma). \quad (10.7.3.6)$$

Then, the unique solution of (10.7.1.1) satisfies

$$\{w, w_t\} \in C\left([0, T]; \left[H^2(\Omega) \cap H_0^1(\Omega) \right]\right) \times H_0^1(\Omega), \quad w_{tt} \in C([0, T]; L_2(\Omega))$$
$$(10.7.3.7)$$

continuously, that is, more precisely,

$$\|\{w, w_t, w_{tt}\}\|^2_{C([0,T];[H^2(\Omega) \cap H_0^1(\Omega)]) \times H_0^1(\Omega) \times L_2(\Omega))}$$
$$\leq C_T \left\{ \|u\|^2_{L_2(\Sigma)} + \|\{w_0, w_1\}\|^2_{[H^2(\Omega) \cap H_0^1(\Omega)] \times H_0^1(\Omega)} \right\}. \quad (10.7.3.8)$$

Proof. The proof of Theorem 10.7.3.2 is a consequence of Theorem 10.7.3.1 and will be given in Section 10.7.11 below. □

10.7.4 Abstract Setting for Problem (10.7.1.1)

We follow Lasiecka and Triggiani [1991a, c]. To put problems (10.7.1.1), (10.7.1.2) into the abstract model (10.1.1), (10.1.3), we introduce the following operators and spaces (recall Section 3.10 of Chapter 3):

(i)

$$\mathcal{A}h = \Delta^2 h; \quad \mathcal{A} : \mathcal{D}(\mathcal{A}) \to L_2(\Omega); \mathcal{D}(\mathcal{A}) = \{h \in H^4(\Omega) : h|_\Gamma = \Delta h|_\Gamma = 0\};$$

$$(10.7.4.1)$$

$$\mathcal{A}^{\frac{1}{2}}h = -\Delta h; \quad \mathcal{D}\big(\mathcal{A}^{\frac{1}{2}}\big) = H^2(\Omega) \cap H_0^1(\Omega); \qquad (10.7.4.2)$$

$$\mathbb{A} = \big(I + \rho \mathcal{A}^{\frac{1}{2}}\big)^{-1}\mathcal{A}; \quad \mathcal{D}(\mathbb{A}) = \mathcal{D}\big(\mathcal{A}^{\frac{1}{2}}\big). \qquad (10.7.4.3)$$

The operator $\mathcal{A}$ in (10.7.4.1) is positive self-adjoint on $L_2(\Omega)$. Furthermore, as in the case of Chapter 3, say Section 3.1, or as in Section 10.5.4 of the present chapter, we shall freely extend $\mathcal{A}$ originally defined in (10.7.4.1) [while maintaining the same symbol, with no fear of confusion] as $\mathcal{A} : L_2(\Omega) \to [\mathcal{D}(\mathcal{A}^*)]' = [\mathcal{D}(\mathcal{A})]'$, duality with respect to the pivot space $L_2(\Omega)$. The following space identifications are known (with equivalent norms) (see [Grisvard, 1967, Chapter 3, Appendix 3A]):

$$\mathcal{D}(\mathcal{A}^\theta) = \{h \in H^{4\theta}(\Omega) : h|_\Gamma = 0\}, \quad \frac{1}{8} < \theta < \frac{5}{8}; \qquad (10.7.4.4)$$

$$\mathcal{D}(\mathcal{A}^\theta) = \{h \in H^{4\theta}(\Omega) : h|_\Gamma = \Delta h|_\Gamma = 0\}, \quad \frac{5}{8} < \theta \le 1. \quad (10.7.4.5)$$

The following specializations thereof will be needed below:

$$\theta = \frac{1}{4} : \mathcal{D}\big(\mathcal{A}^{\frac{1}{4}}\big) = H_0^1(\Omega), \quad \text{and for } g \in H_0^1(\Omega); \qquad (10.7.4.6a)$$

$$\|g\|_{\mathcal{D}(\mathcal{A}^{\frac{1}{4}})} = \|\mathcal{A}^{\frac{1}{4}}g\|_{L_2(\Omega)} = \left\{\int_\Omega |\nabla g|^2 \, d\Omega\right\}^{\frac{1}{2}}, \quad \text{equivalent to the } \|g\|_{H_0^1(\Omega)}\text{-norm,}$$

$$(10.7.4.6b)$$

which is in turn equivalent to

$$\left\{\int_\Omega [g^2 + \rho|\nabla g|^2 \, d\Omega\right\}^{\frac{1}{2}} = \left\{\|g\|_{L_2(\Omega)}^2 + \rho\|\mathcal{A}^{\frac{1}{4}}g\|_{L_2(\Omega)}^2\right\}^{\frac{1}{2}}$$

$$\equiv \|g\|_{\mathcal{D}(\mathcal{A}_\rho^{\frac{1}{4}})} = \|g\|_{H_{0,\rho}^1(\Omega)}, \qquad (10.7.4.6c)$$

with the latter norm being denoted by $\mathcal{D}(A_\rho^{\frac{1}{4}})$-norm on $H_{0,\rho}^1(\Omega)$-norm;

$$\theta = \frac{1}{2} : \mathcal{D}(A^{\frac{1}{2}}) = \{h \in H^2(\Omega) : h|_\Gamma = 0\} = H^2(\Omega) \cap H_0^1(\Omega), \qquad (10.7.4.7a)$$

and for $g \in \mathcal{D}(A^{\frac{1}{2}})$:

$$\|g\|_{\mathcal{D}(A^{\frac{1}{2}})} = \left\|A^{\frac{1}{2}}g\right\|_{L_2(\Omega)} = \left\{\int_\Omega (\Delta g)^2 \, d\Omega\right\}^{\frac{1}{2}}, \qquad (10.7.4.7b)$$

which is equivalent to

$$\left\{\left\|A^{\frac{1}{4}}g\right\|_{L_2(\Omega)}^2 + \rho\left\|A^{\frac{1}{2}}g\right\|_{L_2(\Omega)}^2\right\}^{\frac{1}{2}} = \|g\|_{\mathcal{D}(A_\rho^{\frac{1}{2}})}, \qquad (10.7.4.7c)$$

with the latter norm being denoted as $\mathcal{D}(A_\rho^{\frac{1}{2}})$-norm;

$$\theta = \frac{3}{4}, \quad \mathcal{D}(A^{\frac{3}{4}}) = \{h \in H^3(\Omega) : h|_\Gamma = \Delta h|_\Gamma = 0\}, \qquad (10.7.4.8a)$$

and for $h \in \mathcal{D}(A^{\frac{3}{4}})$:

$$\|g\|_{\mathcal{D}(A^{\frac{3}{4}})} = \left\|A^{\frac{3}{4}}g\right\|_{L_2(\Omega)} = \left\|A^{\frac{1}{4}}\Delta g\right\|_{L_2(\Omega)} = \left\{\int_\Omega |\nabla(\Delta g)|^2 \, d\Omega\right\}^{\frac{1}{2}}, \qquad (10.7.4.8b)$$

by (10.7.4.2) and (10.7.4.6b).

Problem (10.7.3.1) can then be rewritten abstractly as

$$\left(I + \rho A^{\frac{1}{2}}\right)\phi_{tt} + A\phi = f, \quad \text{or } \phi_{tt} = -\mathbb{A}\phi + \left(I + \rho A^{\frac{1}{2}}\right)^{-1}f,$$
$$\phi(0) = \phi_0, \quad \phi_t(0) = \phi_1, \qquad (10.7.4.9)$$

recalling (10.7.4.1)–(10.7.4.3). The operator in (10.7.4.3), can be rewritten by using $R(\lambda, A^{\frac{1}{2}})A^{\frac{1}{2}} = -I + \lambda R(\lambda, A^{\frac{1}{2}})$ twice as

$$-\mathbb{A} = -\left(I + \rho A^{\frac{1}{2}}\right)^{-1}A^{\frac{1}{2}}A^{\frac{1}{2}} = -\left[\frac{1}{\rho}I - \frac{1}{\rho}\left(I + \rho A^{\frac{1}{2}}\right)^{-1}\right]A^{\frac{1}{2}}$$

$$= -\frac{A^{\frac{1}{2}}}{\rho} + \frac{1}{\rho^2}I - \frac{1}{\rho^2}\left(I + \rho A^{\frac{1}{2}}\right)^{-1}, \qquad (10.7.4.10a)$$

$$\mathbb{A} : \mathcal{D}(\mathbb{A}) = \mathcal{D}(A^{\frac{1}{2}}) \to L_2(\Omega), \qquad (10.7.4.10b)$$

which is a bounded perturbation of the negative self-adjoint operator $A^{\frac{1}{2}}$. It is therefore the generator of a s.c. cosine operator $\mathbb{C}(t)$ [Fattorini, 1985], with corresponding "sine" operator $\mathbb{S}(t) = \int_0^t \mathbb{C}(\tau) \, d\tau$, where the maps

$$t \to A^{\frac{1}{4}}\mathbb{S}(t), \mathbb{C}(t) \text{ are strongly continuous on } L_2(\Omega). \qquad (10.7.4.11)$$

Accordingly, the unique solution of problem (10.7.4.9), or (10.7.3.1), is given explicitly by:

$$\phi(t) = \mathbb{C}(t)\phi_0 + \mathbb{S}(t)\phi_1 + \int_0^t \mathbb{S}(t-\tau)\big(I + \rho\mathcal{A}^{\frac{1}{2}}\big)^{-1} f(\tau)\, d\tau, \qquad (10.7.4.12)$$

$$\phi_t(t) = -\mathbb{A}\mathbb{S}(t)\phi_0 + \mathbb{C}(t)\phi_1 + \int_0^t \mathbb{C}(t-\tau)\big(I + \rho\mathcal{A}^{\frac{1}{2}}\big)^{-1} f(\tau)\, d\tau, \qquad (10.7.4.13)$$

in appropriate function spaces, depending on $\{\phi_0, \phi_1, f\}$.

Moreover, returning to (10.7.4.3),

$\mathbb{A}$ is a positive self-adjoint operator on the space $\mathcal{D}\big(\mathcal{A}_\rho^{\frac{1}{4}}\big)$ defined

by (10.7.4.6c), with respect to the corresponding inner product;   (10.7.4.14a)

$$(x, y)_{\mathcal{D}(\mathcal{A}_\rho^{\frac{1}{4}})} = \big((I + \rho\mathcal{A}^{\frac{1}{2}})x, y\big)_{L_2(\Omega)}, \quad x, y \in H_0^1(\Omega). \qquad (10.7.4.14b)$$

(ii) As in Eqn. (3.6.3) of Chapter 3, we introduce the Green map G_2 by

$$y = G_2 v \iff \{\Delta^2 y = 0 \text{ in } \Omega;\ y|_\Gamma = 0;\ \Delta y|_\Gamma = v\}, \qquad (10.7.4.15)$$

and by elliptic theory [Lions, Magenes, 1972, Vol. I, p. 188]

$$G_2 : H^s(\Gamma) \to H^{s+\frac{5}{2}}(\Omega), \quad s \in \mathbb{R}. \qquad (10.7.4.16)$$

We have already shown in Eqn. (3.6.6) of Chapter 3 that

$$G_2 = -\mathcal{A}^{-\frac{1}{2}} D \text{ where } y = Dv \iff \{\Delta y = 0 \text{ in } \Omega;\ y|_\Gamma = v\}, \qquad (10.7.4.17)$$

where D is the Dirichlet map satisfying the regularity [Eqn. (3.6.7) of Chapter 3]

$$D : \text{continuous } H^s(\Gamma) \to H^{s+\frac{1}{2}}(\Omega); \text{ in particular,} \qquad (10.7.4.18a)$$

$$D : \text{continuous } L_2(\Gamma) \to H^{\frac{1}{2}}(\Omega) \subset H^{\frac{1}{2}-2\epsilon}(\Omega) = \mathcal{D}\big(\mathcal{A}^{\frac{1}{8}-\frac{\epsilon}{2}}\big), \quad \epsilon > 0; \qquad (10.7.4.18b)$$

$$\mathcal{A}^{\frac{1}{8}-\frac{\epsilon}{2}} D \in \mathcal{L}(L_2(\Gamma); L_2(\Omega)); \quad D^* \mathcal{A}^{\frac{1}{8}-\frac{\epsilon}{2}} \in \mathcal{L}(L_2(\Omega); L_2(\Gamma)), \qquad (10.7.4.18c)$$

with the property, by (3.1.15) of Chapter 3 as well as (10.7.4.2) and (10.7.4.17),

$$G_2^* \mathcal{A} h = -D^* \mathcal{A}^{\frac{1}{2}} h = -\frac{\partial}{\partial \nu} h, \quad h \in \mathcal{D}(\mathcal{A}). \qquad (10.7.4.19)$$

(iii) By (10.7.4.15) and (10.7.4.1), problem (10.7.1.1) can be written abstractly, first as

$$\big(I + \rho\mathcal{A}^{\frac{1}{2}}\big)w_{tt} + \mathcal{A}(w - G_2 u) = 0 \text{ on } L_2(\Omega), \text{ or } w_{tt} = -\mathbb{A}w + \mathbb{A}G_2 u \text{ on } [\mathcal{D}(\mathcal{A})]', \qquad (10.7.4.20)$$

recalling (10.7.4.3), and next as in (10.1.1), that is, as

$$y_t = Ay + Bu \text{ on } [\mathcal{D}(A^*)]', \quad y(0) = y_0 \in Y; \qquad (10.7.4.21)$$

$$Y \equiv \mathcal{D}(A^{\frac{1}{2}}) \times \mathcal{D}(A_\rho^{\frac{1}{4}}) = \left[H^2(\Omega) \cap H_0^1(\Omega) \right] \times H_0^1(\Omega); \quad U = L_2(\Gamma)$$
$$\text{(equivalent norms)}; \qquad (10.7.4.22)$$

$$y = \begin{bmatrix} w \\ w_t \end{bmatrix}; \quad A = \begin{bmatrix} 0 & I \\ -\mathbb{A} & 0 \end{bmatrix}, \quad \mathcal{D}(A) = \mathcal{D}(A^{\frac{3}{4}}) \times \mathcal{D}(A^{\frac{1}{2}}) \to Y; \qquad (10.7.4.23)$$

$$Bu = \begin{bmatrix} 0 \\ \mathbb{A}G_2 u \end{bmatrix} : \text{ continuous } U \to [\mathcal{D}(A^*)]' \text{ (duality with respect to } Y),$$
$$(10.7.4.24)$$

or, equivalently,

$$A^{-1}B : \text{ continuous } U \to Y, \qquad (10.7.4.25)$$

since, in fact, for $u \in L_2(\Gamma)$, recalling (10.7.4.23) and (10.7.4.17),

$$A^{-1}Bu = \begin{bmatrix} 0 & -\mathbb{A}^{-1} \\ I & 0 \end{bmatrix} \begin{bmatrix} 0 \\ \mathbb{A}G_2 u \end{bmatrix} = \begin{bmatrix} -G_2 u \\ 0 \end{bmatrix} = \begin{bmatrix} A^{-\frac{1}{2}}Du \\ 0 \end{bmatrix} \in Y,$$

$$A^{-\frac{1}{2}}Du \in \mathcal{D}(A^{\frac{5}{8}-\frac{\epsilon}{2}}). \qquad (10.7.4.26)$$

It is property (10.7.4.14) for $\mathbb{A}$ that makes the choice of $\mathcal{D}(A_\rho^{\frac{1}{4}})$ as the second component space of Y particularly convenient. In fact, with such a choice, we have that A in (10.7.4.23) is skew-adjoint on $Y = \mathcal{D}(A^{\frac{1}{2}}) \times \mathcal{D}(A_\rho^{\frac{1}{4}})$, that is, $A^* = -A$, and so it generates a s.c. unitary group e^{At} on Y.

Starting from (10.7.4.24) we compute B^* with respect to Y topologized as $\mathcal{D}(A^{\frac{1}{2}}) \times \mathcal{D}(A_\rho^{\frac{1}{4}})$, to obtain for $v = [v_1, v_2]$, recalling (10.7.4.14):

$$(Bu, v)_Y = (\mathbb{A}G_2 u, v_2)_{\mathcal{D}(A_\rho^{\frac{1}{4}})} = (\mathcal{A}G_2 u, v_2)_{L_2(\Omega)}$$

$$= (u, G_2^* \mathcal{A}v_2)_{L_2(\Gamma)}, \qquad (10.7.4.27)$$

that is, by virtue of (10.7.4.19):

$$B^* \begin{bmatrix} v_1 \\ v_2 \end{bmatrix} = G_2^* \mathcal{A}v_2 = -\frac{\partial v_2}{\partial \nu} : \mathcal{D}(A^*) = \mathcal{D}(A) \to U. \qquad (10.7.4.28)$$

As in the case of Section 10.5.4, Eqn. (10.5.4.22), the solution of problem (10.7.1.1), or (10.7.4.21), with initial condition $y(s) = y_0 = [w_0, w_1]$ is written abstractly as

$$\begin{bmatrix} w(t, s; y_0) \\ w_t(t, s; y_0) \end{bmatrix} = e^{A(t-s)} \begin{bmatrix} w_0 \\ w_1 \end{bmatrix} + (L_s u)(t); \qquad (10.7.4.29)$$

$$(L_s u)(t) = \int_s^t e^{A(t-\tau)} Bu(\tau)\,d\tau = \begin{bmatrix} \mathbb{A} \int_s^t \mathbb{S}(t-\tau) G_2 u(\tau)\,d\tau \\[2mm] \mathbb{A} \int_s^t \mathbb{C}(t-\tau) G_2 u(\tau)\,d\tau \end{bmatrix}$$

$$= \begin{bmatrix} -(I + \rho \mathcal{A}^{\frac12})^{-1} \mathcal{A}^{\frac12} \int_s^t \mathbb{S}(t-\tau) Du(\tau)\,d\tau \\[2mm] -(I + \rho \mathcal{A}^{\frac12})^{-1} \mathcal{A}^{\frac12} \mathbb{A} \int_s^t \mathbb{C}(t-\tau) Du(\tau)\,d\tau \end{bmatrix}, \qquad (10.7.4.30)$$

after recalling, in the last step, (10.7.4.3) and (10.7.4.17), where A generates a s.c. group e^{At} on Y, $t \in R$, which is given by

$$e^{At} = \begin{bmatrix} \mathbb{C}(t) & \mathbb{S}(t) \\ -\mathbb{A}\mathbb{S}(t) & \mathbb{C}(t) \end{bmatrix}. \qquad (10.7.4.31)$$

In (10.7.4.31), $\mathbb{C}(t)$ is even and $\mathbb{S}(t)$ is odd. By (10.7.4.28) and (10.7.4.31), A is skew-adjoint on Y, $A^* = -A$, and so $e^{A^* t} = e^{-At}$. We then compute with $x = [x_1, x_2] \in Y = \mathcal{D}(\mathcal{A}^{\frac12}) \times \mathcal{D}(\mathcal{A}_\rho^{\frac14})$:

$$B^* e^{A^* t} x = G_2^* A[\mathbb{A}\mathbb{S}(t) x_1 + \mathbb{C}(t) x_2] \qquad (10.7.4.32)$$

$$= G_2^* \mathcal{A} \mathcal{A}^{\frac12} \big[\mathbb{C}(t) \mathcal{A}^{-\frac12} x_2 + \mathbb{S}(t) (I + \rho \mathcal{A}^{\frac12})^{-1} \mathcal{A}^{\frac12} x_1 \big] \qquad (10.7.4.33)$$

$$\text{(by (10.7.4.28))} = G_2^* \mathcal{A} \mathcal{A}^{\frac12} \big[\mathbb{C}(t) \phi_0 + \mathbb{S}(t) \phi_1 \big] = -\frac{\partial}{\partial \nu} \Delta \phi(t; \phi_0, \phi_1), \qquad (10.7.4.34)$$

recalling (10.7.4.28), (10.7.4.2), and (10.7.4.12), where $\phi(t; \phi_0, \phi_1)$ solves problem (10.7.3.1) with $f \equiv 0$ and

$$\phi_0 = \mathcal{A}^{-\frac12} x_2 \in \mathcal{D}(\mathcal{A}^{\frac34}), \quad \text{for } x_2 \in \mathcal{D}(\mathcal{A}^{\frac14}) = \mathcal{D}(\mathcal{A}_\rho^{\frac14}); \qquad (10.7.4.35)$$

$$\phi_1 = (I + \rho \mathcal{A}^{\frac12})^{-1} \mathcal{A}^{\frac12} x_1 \in \mathcal{D}(\mathcal{A}^{\frac12}), \quad \text{for } x_1 \in \mathcal{D}(\mathcal{A}^{\frac12}). \qquad (10.7.4.36)$$

10.7.5 Verification of Assumption (H.1) = (10.1.6)

From (10.7.4.34)–(10.7.4.36), we see that (H.1) = (10.1.6) holds true,

$$\int_0^T \| B^* e^{A^* t} x \|_U^2 \, dt \le C_T \|x\|_Y^2, \quad x \in Y, \qquad (10.7.5.1)$$

if and only if problem (10.7.3.1) with $f \equiv 0$ satisfies

$$\int_0^T \int_\Gamma \left(\frac{\partial \Delta \phi}{\partial \nu} \right)^2 d\Sigma \le C_T \| \{\phi_0, \phi_1\} \|_{\mathcal{D}(\mathcal{A}^{\frac34}) \times \mathcal{D}(\mathcal{A}^{\frac12})}^2, \qquad (10.7.5.2)$$

which is precisely the trace regularity result, guaranteed by Theorem 10.7.3.1, Eqn. (10.7.3.5). Then, according to Chapter 7, Theorem 10.7.2.1, estimate (10.7.5.1), that is, (10.7.5.2), is, in turn, equivalent to the following property that

$$(L_s u)(t) = \int_s^t e^{A(t-\tau)} Bu(\tau)\, d\tau = \begin{bmatrix} w(t; 0, 0) \\ w_t(t; 0, 0) \end{bmatrix} \qquad (10.7.5.3a)$$

$$: \text{continuous } L_2(s, T; L_2(\Gamma)) \qquad (10.7.5.3b)$$

$$\to C\big([s, T]; \mathcal{D}(A^{\frac{1}{2}}) \times \mathcal{D}(A^{\frac{1}{4}}) \equiv \big[H^2(\Omega) \cap H_0^1(\Omega)\big] \times H_0^1(\Omega)\big) \qquad (10.7.5.3c)$$

uniformly in s, where $w_0 = w_1 = 0$ in problem (10.7.1.1). This is precisely conclusion (10.7.3.7) of Theorem 10.7.3.2; the additional statement of uniformity in s is an immediate consequence of formula (10.7.5.3a) for L_s (via a change of variable). Moreover, recalling (10.7.4.33), we have, by duality on (10.7.5.3) with $v = [v_1, v_2]$:

$$(L_s^* v)(t) = B^* \int_t^T e^{A^*(\tau-t)} v(\tau)\, d\tau, \quad s \le \tau \le T, \qquad (10.7.5.4)$$

$$(\text{by } (10.7.4.33)) \quad = G_2^* A A^{\frac{1}{2}} \bigg\{ \int_t^T \mathbb{S}(\tau - t)\big(I + \rho A^{\frac{1}{2}}\big)^{-1} A^{\frac{1}{2}} v_1(\tau)\, d\tau$$

$$+ \int_t^T \mathbb{C}(\tau - t) A^{-\frac{1}{2}} v_2(\tau)\, d\tau \bigg\} \qquad (10.7.5.5)$$

$$: \text{continuous } L_1\big(s, T; \mathcal{D}(A^{\frac{1}{2}}) \times \mathcal{D}(A^{\frac{1}{4}})\big) \to L_2(s, T; L_2(\Gamma)) \qquad (10.7.5.6)$$

uniformly in s. Now, let $\psi(t; h)$ be the solution of the (adjoint) problem

$$\begin{cases} \psi_{tt} - \rho \Delta \psi_{tt} + \Delta^2 \psi = f & \text{in } Q; & (10.7.5.7a) \\ \psi(T, \cdot) = 0, \quad \psi_t(T; \cdot) = 0 & \text{in } \Omega; & (10.7.5.7b) \\ \psi|_\Sigma \equiv \Delta\psi|_\Sigma \equiv 0 & \text{in } \Sigma, & (10.7.5.7c) \end{cases}$$

rewritten abstractly via (10.7.4.3) as

$$\psi_{tt} = -\mathbb{A}\psi + h; \quad h = \big(I + \rho A^{\frac{1}{2}}\big)^{-1} f; \quad \psi(T) = \psi_t(T) = 0, \qquad (10.7.5.8)$$

hence given explicitly by

$$\psi(t; h) = \int_T^t \mathbb{S}(t - \tau) h(\tau)\, d\tau; \quad \psi_t(t; h) = \int_T^t \mathbb{C}(t - \tau) h(\tau)\, d\tau. \qquad (10.7.5.9)$$

Then, recalling (10.7.4.19) and (10.7.4.2), we see via (10.7.5.9) that the two terms in

(10.7.5.5) can be rewritten, with $v = [v_1, v_2] \in L_1(0, T; \mathcal{D}(A^{\frac{1}{2}}) \times \mathcal{D}(A^{\frac{1}{4}}))$, as

$$(L_s^* v)(t) = \frac{\partial \Delta}{\partial v} \left\{ \int_T^t \mathbb{S}(t - \tau) h_1(\tau) \, d\tau - \int_T^t \mathbb{C}(t - \tau) h_2(\tau) \, d\tau \right\}$$

$$= \frac{\partial \Delta \psi(t; h_1)}{\partial v} - \frac{\partial \Delta \psi_t(t; h_2)}{\partial v} \qquad (10.7.5.10)$$

$$: \text{continuous } v = [v_1, v_2] \in L_1(0, T; \mathcal{D}(A^{\frac{1}{2}}) \times \mathcal{D}(A^{\frac{1}{4}}))$$

$$\to L_2(0, T; L_2(\Gamma)), \qquad (10.7.5.11)$$

where

$$h_1 = (I + \rho A^{\frac{1}{2}})^{-1} A^{\frac{1}{2}} v_1 \in L_1(0, T; \mathcal{D}(A^{\frac{1}{2}})), \qquad (10.7.5.12)$$

$$h_2 = A^{-\frac{1}{2}} v_2 \in L_1(0, T; \mathcal{D}(A^{\frac{3}{4}})). \qquad (10.7.5.13)$$

Notice that regularity (10.7.5.11) of the normal trace $\frac{\partial \Delta}{\partial v} \psi(t; h_1)$ for the $\psi(t; h_1)$ solution of (10.7.5.8) due to $h = h_1$, given by (10.7.5.12), is precisely conclusion (10.7.3.5) of Theorem 10.7.3.1 for the time-reversed problem ψ in (10.7.3.1) with initial data at $t = 0$, rather than $t = T$ as for ψ, an inessential modification. The proof of Theorem 10.7.3.1 in Section 10.7.10 is by energy (PDE)-methods. Instead, regularity (10.7.5.9) for the normal trace $\frac{\partial \Delta}{\partial v} \psi_t(t; h_2)$ for the time derivative ψ_t of the solution $\psi(t; h_2)$ of problem (10.7.5.8) due to $h = h_2$, given by (10.7.5.13), is obtained by duality via operator methods as in (10.7.5.6)–(10.7.5.11) [while it appears that purely PDE methods will require a *time* regularity assumption of the right-hand side, nonhomogeneous term]. Thus, assumption (H.1) = (10.1.6) is verified.

10.7.6 Selection of the Spaces U_θ and Y_θ in (10.1.10)

We select the spaces in (10.1.10) to be the following Sobolev spaces:

$$U_\theta = H^\theta(\Gamma), \quad U_0 = U = L_2(\Gamma), \quad 0 \le \theta \le \frac{1}{2} + \delta, \quad \theta \ne \frac{1}{2}; \qquad (10.7.6.1)$$

$$Y_\theta = \mathcal{D}(A^{\frac{1}{2} + \frac{\theta}{4}}) \times \mathcal{D}(A^{\frac{1}{4} + \frac{\theta}{4}}) \equiv \mathcal{D}(A^\theta), \quad Y_0 = Y = \mathcal{D}(A^{\frac{1}{2}}) \times \mathcal{D}(A^{\frac{1}{4}}), \qquad (10.7.6.2)$$

where A is as in (10.7.4.23), in particular, the critical spaces for $\theta = 1/2 \pm \delta$:

$$U_\delta^- = U_{\frac{1}{2} - \delta} = H^{\frac{1}{2} - \delta}(\Gamma), \quad U_\delta^+ = U_{\frac{1}{2} + \delta} = H^{\frac{1}{2} + \delta}(\Gamma); \qquad (10.7.6.3)$$

$$Y_\delta^- = Y_{\frac{1}{2} - \delta} = \mathcal{D}(A^{\frac{1}{2} - \delta}) = \mathcal{D}(A^{\frac{5}{8} - \frac{\delta}{4}}) \times \mathcal{D}(A^{\frac{3}{8} - \frac{\delta}{4}})$$

$$= [H^{\frac{5}{2} - \delta}(\Omega) \cap H_0^1(\Omega)] \times [H^{\frac{3}{2} - \delta}(\Omega) \cap H_0^1(\Omega)]; \qquad (10.7.6.4)$$

$$Y_\delta^+ = Y_{\frac{1}{2} + \delta} = \mathcal{D}(A^{\frac{1}{2} + \delta}) = \mathcal{D}(A^{\frac{5}{8} + \frac{\delta}{4}}) \times \mathcal{D}(A^{\frac{3}{8} + \frac{\delta}{4}})$$

$$= \{h \in H^{\frac{5}{2} + \delta}(\Omega) : h|_\Gamma = \Delta h|_\Gamma = 0\} \times [H^{\frac{3}{2} + \delta}(\Omega) \cap H_0^1(\Omega)], \qquad (10.7.6.5)$$

recalling (10.7.4.4) and, respectively, (10.7.4.5), with equivalent norms. The spaces $[Y_\delta^-]'$ and $[Y_\delta^+]'$, duality with respect to $Y = \mathcal{D}(A^{\frac{1}{2}}) \times \mathcal{D}(A^{\frac{1}{4}})$, are given by

$$[Y_\delta^-]' = \mathcal{D}\big(A^{\frac{3}{8}+\frac{\delta}{4}}\big) \times \mathcal{D}\big(A^{\frac{1}{8}+\frac{\delta}{4}}\big) = \big[H^{\frac{3}{2}+\delta}(\Omega) \cap H_0^1(\Omega)\big] \times H_0^{\frac{1}{2}+\delta}(\Omega), \quad (10.7.6.6)$$

$$[Y_\delta^+]' = \mathcal{D}\big(A^{\frac{3}{8}-\frac{\delta}{4}}\big) \times \mathcal{D}\big(A^{\frac{1}{8}-\frac{\delta}{4}}\big) = \big[H^{\frac{3}{2}-\delta}(\Omega) \cap H_0^1(\Omega)\big] \times H^{\frac{1}{2}-\delta}(\Omega). \quad (10.7.6.7)$$

Thus, by (10.7.6.4) and (10.7.6.7) we verify the interpolation property

$$[Y_\delta^-, [Y_\delta^+]']_{\theta=\frac{1}{2}-\delta} = Y = \mathcal{D}(A^{\frac{1}{2}}) \times \mathcal{D}(A^{\frac{1}{4}}), \quad (10.7.6.8)$$

as required in (10.1.10), since $(5/8 - \delta/4)(1-\theta) + (3/8 - \delta/4)\theta = 1/2$ for the first component, and $(3/8 - \delta/4)(1-\theta) + (1/8 - \delta/4)\theta = 1/4$ for $\theta = 1/2 - \delta$ for the second component space. Moreover, the injections $U_{\theta_1} \hookrightarrow U_{\theta_2}$, $Y_{\theta_1} \hookrightarrow Y_{\theta_2}$ are compact, $0 \le \theta_2 < \theta_1 \le 1/2 + \delta$, as required in (10.1.10), since Ω is a bounded domain. Thus, the spaces in (10.1.11) and (10.1.12) are in the present case as follows for $0 \le \theta \le 1/2 + \delta$, $\theta \ne 1/2$:

$$\mathcal{U}^\theta[s, T] = H^{\theta,\theta}(\Sigma_{sT}) = L_2(s, T; H^\theta(\Gamma)) \cap H^\theta(s, T; L_2(\Gamma)), \quad (10.7.6.9)$$

$$\mathcal{Y}^\theta[s, T] = L_2\big(s, T; \mathcal{D}(A^{\frac{1}{2}+\frac{\theta}{4}}) \times \mathcal{D}(A^{\frac{1}{4}+\frac{\theta}{4}})\big)$$

$$\cap H^\theta\big(s, T; \mathcal{D}(A^{\frac{1}{2}}) \times \mathcal{D}(A^{\frac{1}{4}})\big) \quad (10.7.6.10)$$

$$= L_2(s, T; \mathcal{D}(A^\theta)) \cap H^\theta(s, T; Y) \quad (10.7.6.11)$$

$$\mathcal{D}(A^\theta) = \mathcal{D}\big(A^{\frac{1}{2}+\frac{\theta}{4}}\big) \times \mathcal{D}\big(A^{\frac{1}{4}+\frac{\theta}{4}}\big). \quad (10.7.6.12)$$

10.7.7 Verification of Assumption (H.2) = (10.1.14)

The following regularity result is critical in verifying assumption (H.2) = (10.1.14).

Theorem 10.7.7.1 *With reference to the nonhomogeneous problem (10.7.1.1), assume*

$$\begin{cases} w_0 \in H^3(\Omega) \cap H_0^1(\Omega), \quad w_1 \in H^2(\Omega) \cap H_0^1(\Omega); \\ \text{with the compatibility relations} \\ w_0|_\Gamma = 0 \quad \text{and} \quad \Delta w_0|_\Gamma = u(0); \end{cases} \quad (10.7.7.1)$$

$$u \in C\big([0, T]; H^{\frac{1}{2}}(\Gamma)\big) \cap H^1(0, T; L_2(\Gamma)). \quad (10.7.7.2)$$

[Equation (10.7.7.2) is a fortiori guaranteed, if $u \in H^{1,1}(\Sigma)$, by [Lions, Magenes, 1972, I, Theorem 3.1, p. 19].]

Then, the unique solution to problem (10.7.1.1) satisfies

$$\{w, w_t, w_{tt}\} \in C\big([0, T]; \big[H^3(\Omega) \cap H_0^1(\Omega)\big] \times \big[H^2(\Omega) \cap H_0^1(\Omega)\big] \times H_0^1(\Omega)\big), \quad (10.7.7.3)$$

continuously.

Proof.   The proof of Theorem 10.7.7.1 will be given in Section 10.7.12 below.   □

Corollary 10.7.7.2   *With reference to the nonhomogeneous problem (10.7.1.1), assume $w_0 = w_1 = 0$, and for $0 \le \theta < 1/2$:*

$$u \in H^{\theta,\theta}(\Sigma) = \mathcal{U}^\theta[0, T] = L_2(0, T; H^\theta(\Gamma)) \cap H^\theta(0, T; L_2(\Gamma)). \qquad (10.7.7.4)$$

Then, the unique solution to problem (10.7.1.1) satisfies

$$\begin{cases} w(\,\cdot\,; 0, 0) \in C\big([0, T]; \mathcal{D}\big(A^{\frac{1}{2}+\frac{\theta}{4}}\big) = H^{2+\theta}(\Omega) \cap H_0^1(\Omega)\big); & (10.7.7.5) \\[4pt] w_t(\,\cdot\,; 0, 0) \in C\big([0, T]; \mathcal{D}\big(A^{\frac{1}{4}+\frac{\theta}{4}}\big) = H^{1+\theta}(\Omega) \cap H_0^1(\Omega)\big); & (10.7.7.6) \\[4pt] w_{tt}(\,\cdot\,; 0, 0) \in C\big([0, T]; \mathcal{D}\big(A^{\frac{\theta}{4}}\big) = H^\theta(\Omega)\big); & (10.7.7.7) \end{cases}$$

$$\begin{cases} D_t^r w(\,\cdot\,; 0, 0) \in L_2\big(0, T; \mathcal{D}\big(A^{\frac{1}{2}+\frac{\theta}{4}-\frac{r}{4}}\big)\big), & 0 \le r \le 1; & (10.7.7.8) \\[4pt] D_t^\theta w(\,\cdot\,; 0, 0) \in L_2\big(0, T; \mathcal{D}\big(A^{\frac{1}{2}}\big)\big); & (10.7.7.9) \end{cases}$$

$$\begin{cases} D_t^r w_t(\,\cdot\,; 0, 0) \in L_2\big(0, T; \mathcal{D}\big(A^{\frac{1}{4}+\frac{\theta}{4}-\frac{r}{4}}\big)\big), & 0 \le r \le 1; & (10.7.7.10) \\[4pt] D_t^\theta w_t(\,\cdot\,; 0, 0) \in L_2\big(0, T; \mathcal{D}\big(A^{\frac{1}{4}}\big)\big). & (10.7.7.11) \end{cases}$$

A fortiori,

$$Lu \in \mathcal{Y}^\theta[0, T] = L_2\big(0, T; \mathcal{D}\big(A^{\frac{1}{2}+\frac{\theta}{4}}\big) \times \mathcal{D}\big(A^{\frac{1}{4}+\frac{\theta}{2}}\big) = \mathcal{D}(A^\theta)\big)$$
$$\cap\, H^\theta\big(0, T; \mathcal{D}\big(A^{\frac{1}{2}}\big) \times \mathcal{D}\big(A^{\frac{1}{4}}\big) = Y\big). \qquad (10.7.7.12)$$

Proof of Corollary 10.7.7.2.   For $\theta < 1/2$, the compatibility relations in (10.7.7.1), which now read $u(0) = \Delta w_0|_\Gamma = 0$; $w_0|_\Gamma = 0$, do not interfere, and we then interpolate between (10.7.3.6) and (10.7.3.7), or (10.7.5.3b), for $\theta = 0$ and (10.7.7.2) and (10.7.7.3) for $\theta = 1$, thereby obtaining (10.7.7.5)–(10.7.7.7), as desired.

Next, application of the intermediate derivative theorem [Lions, Magenes, 1972, p. 15] to (10.7.7.5) and (10.7.7.6), as well as to (10.7.7.6) and (10.7.7.7), yields, respectively, (10.7.7.8) and (10.7.7.10), which then specialize to (10.7.7.9), and respectively, (10.7.7.11) for $r = \theta$. Thus, (10.7.7.12) is a consequence of (10.7.7.5), (10.7.7.6), and (10.7.7.9), (10.7.7.11).   □

Corollary 10.7.7.2 plainly verifies assumption (H.2) = (10.1.1.4).

10.7.8   Verification of Assumption (H.3) = (10.1.15)

Verification of assumption (H.3) = (10.1.15) is based upon the following regularity result.

Theorem 10.7.8.1 *(i) With reference to the operator L^* defined by (10.7.5.4), we have, for $0 \leq r \leq 1$,*

$$(L_s^* v)(t) = B^* \int_t^T e^{A^*(\tau - t)} v(\tau) \, d\tau, \quad s \leq t \leq T,$$

$$: \text{continuous } L_2\big(s, T; \mathcal{D}\big(\mathcal{A}^{\frac{1}{2} + \frac{r}{4}}\big) \times \mathcal{D}\big(\mathcal{A}^{\frac{1}{4} + \frac{r}{4}}\big) \equiv \mathcal{D}(A^r)\big)$$

$$\to H^{r,r}(\Sigma_{sT}) = \mathcal{U}^r[s, T], \quad (10.7.8.1)$$

uniformly in s.

Proof. The proof of Theorem 10.7.8.1 will be given in Section 10.7.13 below. □

Restricting (10.7.8.1) to $0 \leq \theta = r \leq 1/2 + \delta, \theta \neq 1/2$, we obtain verification of assumption (H.3) = (10.1.15).

10.7.9 Verification of Assumptions (H.4) = (10.1.18) through (H.7) = (10.1.21)

Verification of Assumption (H.4) = (10.1.18)   For $x = [x_1, x_2] \in Y_\delta^+ = \mathcal{D}(\mathcal{A}^{\frac{1}{2} + \delta}) = \mathcal{D}(\mathcal{A}^{\frac{5}{8} + \frac{\delta}{4}}) \times \mathcal{D}(\mathcal{A}^{\frac{3}{8} + \frac{\delta}{4}})$, by (10.7.6.5), we compute starting from (10.7.4.33), and recalling $G_2^* \mathcal{A} = -D^* \mathcal{A}^{-\frac{1}{2}}$ from (10.7.4.17):

$$B^* e^{A^* t} x = G_2^* \mathcal{A}\big[\mathbb{C}(t) x_2 + \mathbb{S}(t)\big(I + \rho \mathcal{A}^{\frac{1}{2}}\big)^{-1} \mathcal{A} x_1\big] \quad (10.7.9.1)$$

$$\text{(by (10.7.4.17))} \quad = -D^* \mathcal{A}^{\frac{1}{2}}\big[\mathbb{C}(t) x_2 + \mathbb{S}(t)\big(I + \rho \mathcal{A}^{\frac{1}{2}}\big)^{-1} \mathcal{A} x_1\big] \quad (10.7.9.2)$$

$$= -D^* \mathcal{A}^{\frac{1}{8} - \frac{\delta}{4}}\big[\mathbb{C}(t) \mathcal{A}^{\frac{3}{8} + \frac{\delta}{4}} x_2 + \mathcal{A}^{\frac{1}{4}} \mathbb{S}(t)\big(I + \rho \mathcal{A}^{\frac{1}{2}}\big)^{-1} \mathcal{A}^{\frac{9}{8} + \frac{\delta}{4}} x_1\big]$$
$$(10.7.9.3)$$

$$\in C([0, T]; L_2(\Gamma)), \quad (10.7.9.4)$$

where the desired regularity in (10.7.9.4) follows since, recalling (10.7.4.18c) and (10.7.4.11),

$$D^* \mathcal{A}^{\frac{1}{8} - \frac{\delta}{4}} \in \mathcal{L}(L_2(\Omega); L_2(\Gamma)); \quad t \to \mathcal{A}^{\frac{1}{4}} \mathbb{S}(t), \quad \mathbb{C}(t) \text{ strongly continuous on } L_2(\Omega),$$
$$(10.7.9.5)$$

as well as, via the assumptions on $[x_1, x_2]$,

$$\mathcal{A}^{\frac{3}{8} + \frac{\delta}{4}} x_2 \in L_2(\Omega); \quad \big(I + \rho \mathcal{A}^{\frac{1}{2}}\big)^{-1} \mathcal{A}^{\frac{9}{8} + \frac{\delta}{4}} x_1 \in L_2(\Omega). \quad (10.7.9.6)$$

Thus, (10.7.9.4) verifies assumption (H.4) = (10.1.18).

Verification of Assumption (H.5) = (10.1.19)   With $Y_\delta^- = \mathcal{D}(\mathcal{A}^{\frac{5}{8} - \frac{\delta}{4}}) \times \mathcal{D}(\mathcal{A}^{\frac{3}{8} - \frac{\delta}{4}})$ by (10.7.6.4), $\mathbb{C}(t)$ and $\mathbb{S}(t)$ are likewise s.c. cosine/sine operators on any space $\mathcal{D}(\mathcal{A}^\theta)$; hence e^{At} in (10.7.4.31) is a s.c. group on Y_δ^- as well. The space $\mathcal{D}(A)$ in (10.7.4.23) is clearly dense in Y_δ^-.

Verification of Assumption (H.6) = (10.1.20)   For $x = [x_1, x_2] \in Y_\delta^- = \mathcal{D}(A^{\frac{5}{8} - \frac{\delta}{4}}) \times \mathcal{D}(A^{\frac{3}{8} - \frac{\delta}{4}}) = \mathcal{D}(A^{\frac{1}{2} - \delta})$, we obtain, via (10.7.4.23) and (10.7.4.3),

$$
Ax = \begin{bmatrix} 0 & I \\ -(I + \rho A^{\frac{1}{2}})^{-1} A & 0 \end{bmatrix} \begin{bmatrix} x_1 \\ x_2 \end{bmatrix}
$$

$$
= \begin{bmatrix} x_2 \\ -(I + \rho A^{\frac{1}{2}})^{-1} A x_1 \end{bmatrix} \in \mathcal{D}(A^{\frac{3}{8} - \frac{\delta}{4}}) \times \mathcal{D}(A^{\frac{1}{8} - \frac{\delta}{4}}) \equiv [Y_\delta^+]', \quad (10.7.9.7)
$$

recalling in the last step (10.7.6.7) [and A is, in fact, an isomorphism from Y_δ^- onto $[Y_\delta^+]'$]. Equation (10.7.9.7) verifies assumption (H.6) = (10.1.20).

Remark 10.7.9.1   Returning to (10.7.4.26), we see via (10.7.6.4) that, in the present case,

$$
A^{-1} B : \text{continuous } U \to Y_\delta^-, \tag{10.7.9.8}
$$

which is property (10.1.32). Thus, as remarked below (10.1.32), property (10.7.9.8), along with (H.5) = (10.1.19) and (H.6) = (10.1.20) already verified, reproves (H.4) = (10.1.18).

Verification of Assumption (H.7) = (10.1.21)   Let

$$
u \in U_\delta^- = H^{\frac{1}{2} - \delta}(\Gamma),
$$

so that

$$
Du \in H^{1-\delta}(\Omega), \tag{10.7.9.9}
$$

by (10.7.4.18a). Thus, recalling (10.7.4.3), (10.7.4.24), and (10.7.4.17), we have

$$
Bu = \begin{bmatrix} 0 \\ \mathbb{A} G_2 u \end{bmatrix} = \begin{bmatrix} 0 \\ -(I + \rho A^{\frac{1}{2}})^{-1} A^{\frac{1}{2}} Du \end{bmatrix} \in [Y_\delta^+]'
$$

$$
= \left[H^{\frac{3}{2} - \delta}(\Omega) \cap H_0^1(\Omega) \right] \times H^{\frac{1}{2} - \delta}(\Omega), \tag{10.7.9.10}
$$

using, in the last step, (10.7.6.7) for $[Y_\delta^+]'$ and (10.7.9.9). Thus, (10.7.9.10) shows

$$
B : \text{continuous } U_\delta^- \to [Y_\delta^+]', \tag{10.7.9.11}
$$

as desired, and assumption (H.7) = (10.1.21) is verified.

10.7.10 Proof of Theorem 10.7.3.1

Interior Regularity (10.7.3.4) Rewriting the assumptions via (10.7.4.8) and (10.7.4.7), we must show that

$$
\begin{cases}
\{\phi_0, \phi_1\} \in \mathcal{D}(\mathcal{A}^{\frac{3}{4}}) \times \mathcal{D}(\mathcal{A}^{\frac{1}{2}}); \quad f \in L_1(0, T; L_2(\Omega)) \\
\Rightarrow \\
\{\phi, \phi_t\} \in C([0, T]; \mathcal{D}(\mathcal{A}^{\frac{3}{4}}) \times \mathcal{D}(\mathcal{A}^{\frac{1}{2}})); \quad \phi_{tt} \in L_1(0, T; \mathcal{D}(\mathcal{A}^{\frac{1}{4}})),
\end{cases}
\tag{10.7.10.1}
$$

for the solution ϕ of problem (10.7.3.1), that is, of (10.7.4.9), given by the explicit formula (10.7.4.12) and (10.7.4.13). It suffices to consider the effect of f, since the effect of $\{\phi_0, \phi_1\}$ plainly satisfies (10.7.10.1). Thus, for $\phi_0 = \phi_1 = 0$, we have by (10.7.4.12) that

$$
\mathcal{A}^{\frac{3}{4}}\phi(t) = \int_0^t \mathcal{A}^{\frac{1}{4}}\mathbb{S}(t-\tau)\mathcal{A}^{\frac{1}{2}}(I + \rho\mathcal{A}^{\frac{1}{2}})^{-1} f(\tau)\,d\tau \in C([0, T]; L_2(\Omega)),
$$

$$
\tag{10.7.10.2a}
$$

$$
\mathcal{A}^{\frac{1}{2}}\phi_t(t) = \int_0^t \mathbb{C}(t-\tau)\mathcal{A}^{\frac{1}{2}}(I + \rho\mathcal{A}^{\frac{1}{2}})^{-1} f(\tau)\,d\tau \in C([0, T]; L_2(\Omega)),
$$

$$
\tag{10.7.10.2b}
$$

for f as in (10.7.10.1), via properties (10.7.4.11) and by convolution, and Eqns. (10.7.10.2a,b) prove (10.7.10.1), as desired, at least for $\{\phi, \phi_t\}$. Either a further time differentiation or else return to (10.7.4.9) shows (10.7.10.1) for ϕ_{tt}.

Trace Regularity (10.7.3.5). Step 1 Key to this end is the following result:

Lemma 10.7.10.1 *Let ϕ be a solution of Eqn. (10.7.3.1a) (with no boundary conditions imposed) for smooth data, say*

$$
\{\phi_0, \phi_1, f\} \in \mathcal{D}(\mathcal{A}) \times \mathcal{D}(\mathcal{A}^{\frac{3}{4}}) \times L_1(0, T; \mathcal{D}(\mathcal{A}^{\frac{1}{4}})). \tag{10.7.10.3}
$$

Then, the following identity holds true:

$$
\int_\Sigma \frac{\partial(\Delta\phi)}{\partial\nu} h \cdot \nabla(\Delta\phi)\,d\Sigma + \int_\Sigma \frac{\partial\phi_t}{\partial\nu} h \cdot \nabla\phi_t\,d\Sigma
$$

$$
+ \frac{\rho}{2}\int_\Sigma (\Delta\phi_t)^2 h \cdot \nu\,d\Sigma - \frac{1}{2}\int_\Sigma |\nabla(\Delta\phi)|^2 h \cdot \nu\,d\Sigma
$$

$$
- \frac{1}{2}\int_\Sigma |\Delta\phi_t|^2 h \cdot \nu\,d\Sigma + \int_\Sigma \frac{\partial\phi_t}{\partial\nu}\phi_t\,\text{div}\,h\,d\Sigma - \int_\Sigma \phi_t\Delta\phi_t h \cdot \nu\,d\Sigma
$$

$$
= \int_Q H\nabla(\Delta\phi) \cdot \nabla(\Delta\phi)\,dQ + \int_Q H\nabla\phi_t \cdot \nabla\phi_t\,dQ
$$

$$+ \frac{1}{2} \int_Q \{|\nabla \phi_t|^2 + \rho(\Delta \phi_t)^2 - |\nabla(\Delta \phi)|^2\} \operatorname{div} h \, dQ$$

$$+ \int_Q \phi_t \nabla(\operatorname{div} h) \cdot \nabla \phi_t \, dQ + \int_Q f h \cdot \nabla(\Delta \phi) \, dQ$$

$$- \left[(\phi_t, h \cdot \nabla(\Delta \phi))_\Omega + \rho(\Delta \phi_t, h \cdot \nabla(\Delta \phi))_\Omega\right]_0^T, \tag{10.7.10.4}$$

where [as in (10.5.10.6)]

$$H(x) = \begin{bmatrix} \frac{\partial h_1}{\partial x_1}, & \cdots & \frac{\partial h_1}{\partial x_n} \\ & \cdots & \\ \frac{\partial h_n}{\partial x_1}, & \cdots & \frac{\partial h_n}{\partial x_n} \end{bmatrix}, \quad v(x) = \text{outward unit normal vector at } x \in \Gamma,$$

$$\tag{10.7.10.5}$$

and $h(x) = [h_1(x), h_2(x), \dots, h_n(x)] \in [C^2(\bar{\Omega})]^n$ *is a given vector field.*

Proof of Lemma 10.7.10.1   First, for data as in (10.7.10.3), it follows, by an argument similar to that below (10.7.10.1), that the following interior regularity holds true:

$$\{\phi, \phi_t\} \in C\left([0, T]; \mathcal{D}(\mathcal{A}) \times \mathcal{D}\left(\mathcal{A}^{\frac{3}{4}}\right)\right) \subset C([0, T]; H^4(\Omega) \times H^3(\Omega)), \tag{10.7.10.6}$$

so that, via (10.7.4.5) for $\theta = 1$ and $\theta = 3/4$, the computations below are justified and all terms in (10.7.10.4) are well-defined. We multiply Eqn. (10.7.3.1a) by $h \cdot \nabla(\Delta \phi)$ and integrate over Q. We shall use identity (10.5.10.7) as well as identity (10.5.10.4), rewritten here for convenience, after integration in time, as

$$\int_Q \Delta \psi (h \cdot \nabla \psi) \, dQ = \int_\Sigma \frac{\partial \psi}{\partial v}(h \cdot \nabla \psi) \, d\Sigma - \frac{1}{2} \int_\Sigma |\nabla \psi|^2 h \cdot v \, d\Sigma$$

$$- \int_Q H \nabla \psi \cdot \nabla \psi \, dQ + \frac{1}{2} \int_Q |\nabla \psi|^2 \operatorname{div} h \, dQ. \tag{10.7.10.7}$$

Term $\phi_{tt} h \cdot \nabla(\Delta \phi)$ Integrating at first by parts in t,

$$\int_\Omega \int_0^T \phi_{tt} h \cdot \nabla(\Delta \phi) \, dt \, d\Omega = \left[\int_\Omega \phi_t h \cdot \nabla(\Delta \phi) \, d\Omega\right]_0^T - \int_Q \phi_t h \cdot \nabla(\Delta \phi_t) \, dQ$$

[using (10.5.10.7) with h there replaced by $\phi_t h$ now, with $\psi = \Delta \phi_t$ and with $\operatorname{div}(\phi_t h) = \nabla \phi_t \cdot h + \phi_t \operatorname{div} h$]

$$= \left[\int_\Omega \phi_t h \cdot \nabla(\Delta \phi) \, d\Omega\right]_0^T - \int_\Sigma \phi_t \Delta \phi_t h \cdot v \, d\Sigma$$

$$= \left[\int_\Omega \phi_t h \cdot \nabla(\Delta \phi) \, d\Omega\right]_0^T - \int_\Sigma \phi_t \Delta \phi_t h \cdot v \, d\Sigma$$

$$+ \int_Q \Delta \phi_t h \cdot \nabla \phi_t \, dQ + \int_Q \Delta \phi_t \phi_t \operatorname{div} h \, dQ. \tag{10.7.10.8}$$

Using identity (10.7.10.7) with $\psi = \phi_t$ for the third integral on the right of (10.7.10.8) yields

$$\int_\Omega \int_0^T \phi_{tt} h \cdot \nabla(\Delta\phi) \, dt \, d\Omega$$

$$= \left[\int_\Omega \phi_t h \cdot \nabla(\Delta\phi) \, d\Omega \right]_0^T - \int_\Sigma \phi_t \Delta\phi_t h \cdot v \, d\Sigma$$

$$- \frac{1}{2} \int_\Sigma |\nabla\phi_t|^2 h \cdot v \, d\Sigma + \int_\Sigma \frac{\partial\phi_t}{\partial v} h \cdot \nabla\phi_t \, d\Sigma - \int_Q H \nabla\phi_t \cdot \nabla\phi_t \, dQ$$

$$+ \frac{1}{2} \int_Q |\nabla\phi_t|^2 \operatorname{div} h \, dQ + \int_Q \Delta\phi_t \phi_t \operatorname{div} h \, dQ. \qquad (10.7.10.9)$$

Applying Green's first theorem on the last integral at the right of (10.7.10.9) along with the identity

$$\nabla\phi_t \cdot \nabla(\phi_t \operatorname{div} h) = \phi_t \nabla(\operatorname{div} h) \cdot \nabla\phi_t + |\nabla\phi_t|^2 \operatorname{div} h, \quad (10.7.10.10)$$

we finally obtain from (10.7.10.9)

$$\int_Q \phi_{tt} h \cdot \nabla(\Delta\phi) \, dQ$$

$$= [(\phi_t, h \cdot \nabla(\Delta\phi))_\Omega]_0^T - \int_\Sigma \phi_t \Delta\phi_t h \cdot v \, d\Sigma - \frac{1}{2} \int_\Sigma |\nabla\phi_t|^2 h \cdot v \, d\Sigma$$

$$+ \int_\Sigma \frac{\partial\phi_t}{\partial v} h \cdot \nabla\phi_t \, d\Sigma + \int_\Sigma \frac{\partial\phi_t}{\partial v} \phi_t \operatorname{div} h \, d\Sigma - \int_Q H\nabla\phi_t \cdot \nabla\phi_t \, dQ$$

$$- \frac{1}{2} \int_Q |\nabla\phi_t|^2 \operatorname{div} h \, dQ - \int_Q \phi_t \nabla(\operatorname{div} h) \cdot \nabla\phi_t \, dQ. \qquad (10.7.10.11)$$

Term $\Delta^2\phi h \cdot \nabla(\Delta\phi)$ Using identity (10.7.10.7), this time with $\psi = \Delta\phi$, we obtain

$$\int_Q \Delta(\Delta\phi) h \cdot \nabla(\Delta\phi) \, dQ$$

$$= \int_\Sigma \frac{\partial(\Delta\phi)}{\partial v} h \cdot \nabla(\Delta\phi) \, d\Sigma - \frac{1}{2} \int_\Sigma |\nabla(\Delta\phi)|^2 h \cdot v \, d\Sigma$$

$$- \int_Q H\nabla(\Delta\phi) \cdot \nabla(\Delta\phi) \, dQ + \frac{1}{2} \int_Q |\nabla(\Delta\phi)^2 \operatorname{div} h \, dQ. \quad (10.7.10.12)$$

Term $\rho\Delta\phi_{tt} h \cdot \nabla(\Delta\phi)$ Using identity (10.5.10.7) with $\psi = \Delta\phi_t$ and h replaced by $\Delta\phi_t h$ we compute, by integration by parts on t,

$$\rho \int_Q \Delta\phi_{tt} h \cdot \nabla(\Delta\phi) \, dQ$$

$$= \rho[(\Delta\phi_t, h \cdot \nabla(\Delta\phi))_\Omega]_0^T$$

$$- \frac{\rho}{2} \int_\Sigma (\Delta\phi_t)^2 h \cdot v \, d\Sigma + \frac{\rho}{2} \int_Q (\Delta\phi_t)^2 \operatorname{div} h \, dQ. \quad (10.7.10.13)$$

Summing up (10.7.10.11), (10.7.10.12), and (10.7.10.13), and recalling Eqn. (10.7.3.1), results in identity (10.7.10.4), as desired, at least for data as in (10.7.10.3).

Step 2   We next exploit also the BCs (10.7.3.1c), make a special choice of the vector field $h \in C^2(\bar{\Omega})$, and prove the desired trace regularity (10.7.3.5).

Theorem 10.7.10.2   *Let the data $\{\phi_0, \phi_1, f\}$ be as in (10.7.10.1). Let $h \in [C^2(\bar{\Omega})]^n$ satisfy $h|_\Gamma = \nu$ on Γ. Then:*

(i) The solution ϕ of problem (10.7.3.1a,c) satisfies

$$\text{Left-Hand Side of (10.7.10.4)} = \frac{1}{2} \int_\Sigma \left\{ \left(\frac{\partial(\Delta\phi)}{\partial\nu} \right)^2 + \left(\frac{\partial\phi_t}{\partial\nu} \right)^2 \right\} d\Sigma,$$

$$(10.7.10.14)$$

and hence the identity

$$\frac{1}{2} \int_\Sigma \left\{ \left(\frac{\partial(\Delta\phi)}{\partial\nu} \right)^2 + \left(\frac{\partial\phi_t}{\partial\nu} \right)^2 \right\} d\Sigma$$

$$= \int_Q H\nabla(\Delta\phi) \cdot \nabla(\Delta\phi) \, dQ + \int_Q H\nabla\phi_t \cdot \nabla\phi_t \, dQ$$

$$+ \frac{1}{2} \int_Q \{ |\nabla\phi_t|^2 + \rho(\Delta\phi_t)^2 - |\nabla(\Delta\phi)|^2 \} \operatorname{div} h \, dQ$$

$$+ \int_Q \phi_t \nabla(\operatorname{div} h) \cdot \nabla\phi_t \, dQ + \int_Q fh \cdot \nabla(\Delta\phi) \, dQ$$

$$- \left[(\phi_t, h \cdot \nabla(\Delta\phi))_\Omega + \rho(\Delta\phi_t, h \cdot \nabla(\Delta\phi))_\Omega \right]_0^T. \quad (10.7.10.15)$$

(ii) Given any $0 < T < \infty$, there exists a constant $C_T > 0$ such that

$$\int_\Sigma \left[\left(\frac{\partial(\Delta\phi)}{\partial\nu} \right)^2 + \left(\frac{\partial\phi_t}{\partial\nu} \right)^2 \right] d\Sigma$$

$$\leq C_T \left\{ \| \{\phi_0, \phi_1\} \|^2_{\mathcal{D}(A^{\frac{3}{4}}) \times \mathcal{D}(A^{\frac{1}{2}}_\rho)} + \| f \|^2_{L_1(0,T; \mathcal{D}(A^{\frac{1}{2}}))} \right\}. \quad (10.7.10.16)$$

Proof.   Recalling the BCs (10.7.3.1c) for the smooth solution $\{\phi, \phi_t\}$ of (10.7.3.1a) as in (10.7.10.6) satisfying identity (10.7.10.4) of Lemma 10.7.10.1, we obtain

$$\begin{cases} \text{on } \Sigma : \phi_t \equiv 0; \quad \Delta\phi_t \equiv 0; \quad \nabla\phi_t \perp \Gamma; \quad \nabla(\Delta\phi) \perp \Gamma; \\[2mm] h \cdot \nabla\phi_t = \dfrac{\partial\phi_t}{\partial\nu} h \cdot \nu; \quad |\nabla\phi_t| = \left| \dfrac{\partial\phi_t}{\partial\nu} \right|; \\[3mm] h \cdot \nabla(\Delta\phi) = \dfrac{\partial(\Delta\phi)}{\partial\nu} h \cdot \nu; \quad |\nabla(\Delta\phi)| = \left| \dfrac{\partial(\Delta\phi)}{\partial\nu} \right|. \end{cases} \quad (10.7.10.17)$$

Using all of the identities in (10.7.10.17), along with $h \cdot \nu \equiv 1$ on Γ, on the left-hand side of (10.7.10.4) yields readily (10.7.10.14), from which (10.7.10.15) follows. We can now extend identity (10.7.10.15) to data $\{\phi_0, \phi_1, f\}$ as in (10.7.10.1), producing a solution $\{\phi, \phi_t\} \in C([0, T]; H^3(\Omega) \times H^2(\Omega))$, whereby the R.H.S. of (10.7.10.15) is well-defined, and then so is its L.H.S. Applying the implication (10.7.10.1) on the R.H.S. of identity (10.7.10.15), we obtain estimate (10.7.10.16), which then proves the boundary (trace) regularity property (10.7.3.5), as desired.   □

Remark 10.7.10.1   *A simpler proof may be given Lasiecka and Triggiani [1999] by setting $z = \Delta\phi$ and reducing to the wave equation problem for z, as given by Theorem 10.5.3.1.*

10.7.11 Proof of Theorem 10.7.3.2

We now provide the details, already contained in the preceding development, that the trace regularity for the homogeneous ϕ-problem (10.7.3.1),

$$\left.\begin{array}{l} \{\phi_0, \phi_1\} \in \mathcal{D}\left(\mathcal{A}^{\frac{3}{4}}\right) \times \mathcal{D}\left(\mathcal{A}^{\frac{1}{2}}\right) \\ f \equiv 0 \end{array}\right\} \Rightarrow \frac{\partial(\Delta\phi)}{\partial\nu} \in L_2(0, T; L_2(\Gamma)) \equiv L_2(\Sigma),$$

$$(10.7.11.1)$$

established in Theorem 10.7.3.1, implies by transposition the interior regularity

$$\left.\begin{array}{l} u \in L_2(0, T; L_2(\Gamma)) \equiv L_2(\Sigma) \\ w_0 = w_1 = 0 \end{array}\right\}$$

$$\Rightarrow \begin{bmatrix} w(t) \\ w_t(t) \end{bmatrix} = \begin{bmatrix} \mathbb{A}\displaystyle\int_0^t \mathbb{S}(t-\tau)G_2 u(\tau)\,d\tau \\ \mathbb{A}\displaystyle\int_0^t \mathbb{C}(t-\tau)G_2 u(\tau)\,d\tau \end{bmatrix} \in C([0, T]; Y); \quad (10.7.11.2)$$

$$Y = \mathcal{D}\left(\mathcal{A}^{\frac{1}{2}}\right) \times \mathcal{D}\left(\mathcal{A}_\rho^{\frac{1}{4}}\right) = \left[H^2(\Omega) \cap H_0^1(\Omega)\right] \times H_0^1(\Omega) \quad (10.7.11.3)$$

(see (10.7.4.30) and (10.7.4.22)) for the nonhomogeneous w-problem (10.7.1.1).

Operator-Theoretic Proof of Theorem 10.7.3.2. Step 1   We have already seen in Section 10.7.4, Eqns. (10.7.4.32)–(10.7.4.36), that

$$-\frac{\partial\Delta\phi}{\partial\nu}(t; \phi_0, \phi_1) = G_2^* \mathbb{A}\left[\mathbb{S}(t)\mathcal{A}x_1 + \mathbb{C}(t)\left(I + \rho\mathcal{A}^{\frac{1}{2}}\right)x_2\right] \quad (10.7.11.4)$$

$$= B^* e^{A^* t} x; \quad (10.7.11.5)$$

$$\phi_0 = \mathcal{A}^{-\frac{1}{2}} x_2; \quad \phi_1 = \left(I + \rho\mathcal{A}^{\frac{1}{2}}\right)^{-1}\mathcal{A}^{\frac{1}{2}} x_1. \quad (10.7.11.6)$$

Thus, if we take

$$x_1 \in \mathcal{D}\left(\mathcal{A}^{\frac{1}{2}}\right), \quad x_2 \in \mathcal{D}\left(\mathcal{A}^{\frac{1}{4}}\right) = \mathcal{D}\left(\mathcal{A}_\rho^{\frac{1}{4}}\right), \quad \text{hence } \{\phi_0, \phi_1\} \in \mathcal{D}\left(\mathcal{A}^{\frac{3}{4}}\right) \times \mathcal{D}\left(\mathcal{A}^{\frac{1}{2}}\right),$$

$$(10.7.11.7)$$

we then see, via (10.7.11.4) and (10.7.11.5), that implication (10.7.11.1) of Theorem 10.7.3.1 applies and yields the following operator-theoretic restatement, already noted in (10.7.5.1).

Theorem 10.7.10.1   *With reference to (10.7.11.4)–(10.7.11.6) and (10.7.11.1), we have with Y as in (10.7.11.3) and $\mathbb{A}$ as in (10.7.4.3):*

$$B^* e^{A^* t} : continuous \; Y \to L_2(0, T; L_2(\Gamma)) \equiv L_2(\Sigma); \qquad (10.7.11.8)$$

$$G_2^* \mathbb{A} \mathcal{A}^{\frac{1}{2}} \mathbb{S}(t), \; G_2^* \mathbb{A} \mathbb{S}(t) : continuous \; L_2(\Omega) \to L_2(\Sigma); \qquad (10.7.11.9)$$

$$G_2^* \mathbb{A} \mathcal{A}^{\frac{1}{4}} \mathbb{C}(t); \quad G_2^* \mathcal{A}^{\frac{3}{4}} \mathbb{C}(t) : continuous \; L_2(\Omega) \to L_2(\Sigma). \qquad (10.7.11.10)$$

Step 2   The following result then stems from Theorem 10.7.11.1, by an application of Theorem 7.2.1 of Chapter 7 with $p = 2$: With reference to the (operator) explicit formulas (10.7.4.29), (10.7.4.30) for the solution $\{w(t), w_t(t)\}$ of the w-problem (10.7.1.1) with initial conditions $w_0 = w_1 = 0$ at $t = s = 0$, we have

$$\begin{bmatrix} w(t) \\ w_t(t) \end{bmatrix} = (Lu)(t) = \int_0^t e^{A(t-\tau)} Bu(\tau) \, d\tau = \begin{bmatrix} \mathbb{A} \displaystyle\int_0^t \mathbb{S}(t - \tau) G_2 u(\tau) \, d\tau \\[2mm] \mathbb{A} \displaystyle\int_0^t \mathbb{C}(t - \tau) G_2 u(\tau) \, d\tau \end{bmatrix}$$

$$(10.7.11.11)$$

$$: continuous \; L_2(\Sigma) \to C([0, T]; Y), \qquad (10.7.11.12)$$

with Y as in (10.7.11.3). This establishes (10.7.11.2) and, in turn, yields the key part of Theorem 10.7.3.2 due to u. Then, recalling (10.7.4.20), (10.7.4.3), and (10.7.4.17), we obtain

$$w_{tt} = -\mathbb{A}w + \mathbb{A}G_2 u$$
$$= \left(I + \rho \mathcal{A}^{\frac{1}{2}}\right)^{-1} Aw - \left(I + \rho \mathcal{A}^{\frac{1}{2}}\right)^{-1} \mathcal{A}^{\frac{1}{2}} Du \in L_2(0, T; L_2(\Omega)), \quad (10.7.11.13)$$

where the indicated regularity stems from the established regularity $w \in C([0, T]; \mathcal{D}(\mathcal{A}^{\frac{1}{2}}))$ of w in (10.7.11.12) and from the regularity of D in (10.7.4.18b). To get $w_{tt} \in L_2(0, T; L_2(\Omega))$, we could also differentiate w_t in (10.7.11.1). Finally, we omit the details for the regularity due to the initial conditions $\{w_0, w_1\}$, using (10.7.4.11) [which are similar to the argument in (10.5.11.4) for the wave equation]. Theorem 10.7.3.2 is proved. $\square$

PDE Proof   A PDE version of the duality or transposition argument is as follows. Let w and ψ be solutions with regular data of the following two problems (recall (10.7.5.7) or (10.7.4.9)):

$$\begin{cases} w_{tt} - \rho \Delta w_{tt} + \Delta^2 w = F \\ w(0, \cdot) = w_0, w_t(0, \cdot) \\ \qquad\qquad = w_1 \\ w|_\Sigma \equiv 0 \\ \Delta w|_\Sigma = u \end{cases} \text{and} \begin{cases} \psi_{tt} - \rho \Delta \psi_{tt} + \Delta^2 \psi = f & \text{in } Q; \quad (10.7.11.14a) \\ \psi(T, \cdot) = 0; \; \psi_t(T, \cdot) & \text{in } \Omega; \quad (10.7.11.14b) \\ \qquad\qquad = 0 \\ \psi|_\Sigma \equiv 0 & \text{in } \Sigma; \quad (10.7.11.14c) \\ \Delta \psi|_\Sigma \equiv 0 & \text{in } \Sigma. \quad (10.7.11.14d) \end{cases}$$

We know in advance, via Theorem 10.7.3.1 applied to the ψ-problem (which is reversed in time over the ϕ-problem (10.7.3.1)), that

$$f \in L_1(0, T; L_2(\Omega)) \rightarrow \begin{cases} \{\psi, \psi_t\} \in C\big([0, T]; \mathcal{D}(A^{\frac{3}{4}}) \times \mathcal{D}(A^{\frac{1}{2}})\big), \\[2mm] \text{in particular: } \psi(0) \in \mathcal{D}(A^{\frac{3}{4}}), \quad \psi_t(0) \in \mathcal{D}(A^{\frac{1}{2}}); \\[2mm] \dfrac{\partial \Delta \psi}{\partial \nu} \in L_2(0, T; L_2(\Gamma)) = L_2(\Sigma). \end{cases}$$

$$(10.7.11.15)$$

We multiply the w-equation (10.7.11.14a) by $\Delta \psi$ and integrate by parts. We readily obtain:

$$\int_0^T \int_\Omega w_{tt} \Delta \psi \, dQ = \int_\Omega [w_t \Delta \psi]_0^T \, d\Omega - \int_\Omega [w \Delta \psi_t]_0^T \, d\Omega + \int_0^T \int_\Omega \Delta w \psi_{tt} \, dQ,$$

$$(10.7.11.16)$$

after integration by parts twice in t and subsequent application of Green's second theorem, where $w|_\Sigma = \psi_{tt}|_\Sigma = 0$ is used via (10.7.11.14c); likewise, integrating in t, we get

$$\int_0^T \int_\Omega w_{tt} \Delta \psi \, dQ = \int_\Omega [\Delta w_t \Delta \psi]_0^T \, d\Omega - \int_\Omega [\Delta w \Delta \psi_t]_0^T \, d\Omega$$

$$+ \int_0^T \int_\Omega \Delta w \Delta \psi_{tt} \, dQ; \qquad (10.7.11.17)$$

$$\int_0^T \int_\Omega \Delta^2 w \Delta \psi \, dQ = \int_0^T \int_\Omega \Delta w \Delta^2 \psi \, dQ - \int_0^T \int_\Gamma u \frac{\partial(\Delta \psi)}{\partial \nu} \, d\Sigma, \quad (10.7.11.18)$$

by application of Green's second theorem, where $\Delta \psi|_\Sigma \equiv 0$, and where $\Delta w|_\Sigma = u$ by (10.7.11.14d).

We multiply (10.7.11.17) by $-\rho$ and add the result to (10.7.11.16) and (10.7.11.18), thereby obtaining for smooth solutions

$$\int_0^T \int_\Omega [w_{tt} - \rho \Delta w_{tt} + \Delta^2 w] \Delta \psi \, dQ - \int_0^T \int_\Omega \Delta w [\psi_{tt} - \rho \Delta \psi_{tt} + \Delta^2 \psi] \, dQ$$

$$= \int_0^T \int_\Omega F \Delta \psi \, dQ - \int_0^T \int_\Omega (\Delta w) f \, dQ$$

$$= -\int_\Omega w_1 \Delta \psi(0) \, d\Omega + \int_\Omega w_0 \Delta \psi_t(0) \, d\Omega + \rho \int_\Omega \Delta w_1 \Delta \psi(0) \, d\Omega$$

$$- \rho \int_\Omega \Delta w_0 \Delta \psi_t(0) \, d\Omega - \int_0^T \int_\Gamma u \frac{\partial(\Delta \psi)}{\partial \nu} \, d\Sigma, \qquad (10.7.11.19)$$

after invoking (10.7.11.14a) and the initial conditions (10.7.11.14b) at $t = T$ for ψ. Recalling the regularity results (10.7.11.15) for the ψ-problem, we then see that in identity (10.7.11.19) we can take:

$$\Delta\psi = \mathcal{A}^{\frac{1}{2}}\psi \in C\big([0, T]; \mathcal{D}(\mathcal{A}^{\frac{1}{4}})\big); \quad \text{hence } F \in L_1\big(0, T; [\mathcal{D}(\mathcal{A}^{\frac{1}{4}})]' = H^{-1}(\Omega)\big);$$

$$(10.7.11.20)$$

$$\begin{cases} -\Delta w = \mathcal{A}^{\frac{1}{2}}w \in L_\infty(0, T; L_2(\Omega)); \text{ hence } w \in L_\infty\big(0, T; \mathcal{D}(\mathcal{A}^{\frac{1}{2}})\big), \\ \text{a regularity result that can then be boosted to} \\ w \in C\big([0, T]; \mathcal{D}(\mathcal{A}^{\frac{1}{2}}) = H^2(\Omega) \cap H_0^1(\Omega)\big) \end{cases} \quad (10.7.11.21)$$

by an approximating argument with smooth data in the w-problem;

$$\frac{\partial(\Delta\psi)}{\partial\nu} \in L_2(\Sigma); \quad \text{hence } u \in L_2(\Sigma); \quad\quad (10.7.11.22)$$

$$-\Delta\psi(0) = \mathcal{A}^{\frac{1}{2}}\psi(0) \in \mathcal{D}(\mathcal{A}^{\frac{1}{4}});$$

$$\text{hence } -\Delta w_1 = \mathcal{A}^{\frac{1}{2}}w_1 \in \big[\mathcal{D}(\mathcal{A}^{\frac{1}{4}})\big]' \quad \text{or } w_1 \in \mathcal{D}(\mathcal{A}^{\frac{1}{4}}); \quad (10.7.11.23)$$

$$-\Delta\psi_t(0) = \mathcal{A}^{\frac{1}{2}}\psi_t(0) \in L_2(\Omega);$$

$$\text{hence } -\Delta w_0 = \mathcal{A}^{\frac{1}{2}}w_0 \in L_2(\Omega), \text{ or } w_0 \in \mathcal{D}(\mathcal{A}^{\frac{1}{2}}). \quad (10.7.11.24)$$

Let now $F \equiv 0$. Finally, we establish the regularity of w_t. Equation (10.7.11.21) yields, via (10.7.3.1a),

$$(1 - \rho\Delta)w \in C\big([0, T]; L_2(\Omega)\big) \quad \text{and} \quad (1 - \rho\Delta)w_{tt} = -\Delta^2 w \in C\big([0, T]; H^{-2}(\Omega)\big).$$

$$(10.7.11.25)$$

Applying the intermediate derivative theorem [Lions, Magenes, 1972, p. 15] to (10.7.11.25) yields

$$(1 - \rho\Delta)w_t \in C\big([0, T]; H^{-1}(\Omega) = [\mathcal{D}(\mathcal{A}^{\frac{1}{4}})]'\big), \quad\quad (10.7.11.26)$$

or recalling the BC $w_t|_\Sigma = 0$, by (10.7.11.14c), we rewrite (10.7.11.26) as

$$\big(I + \rho\mathcal{A}^{\frac{1}{2}}\big)w_t = h \in C\big([0, T]; [\mathcal{D}(\mathcal{A}^{\frac{1}{4}})]'\big);$$

hence

$$w_t = \big(I + \rho\mathcal{A}^{\frac{1}{2}}\big)^{-1}h \in C\big([0, T]; \mathcal{D}(\mathcal{A}^{\frac{1}{4}}) = H_0^1(\Omega)\big). \quad (10.7.11.27)$$

Thus, Theorem 10.7.3.1 (where $F \equiv 0$) is then proved via (10.7.11.22) on u, (10.7.11.23) and (10.7.11.24) on $\{w_0, w_1\}$, and (10.7.11.21) and (10.7.11.27) on $\{w, w_t\}$. $\square$

10.7.12 Proof of Theorem 10.7.7.1

With reference to the nonhomogeneous w-problem (10.7.1.1), we assume

$$\{w_0, w_1\} \in \big[H^3(\Omega) \cap H_0^1(\Omega)\big] \times \big[H^2(\Omega) \cap H_0^1(\Omega)\big]; \quad w_0|_\Gamma = 0;$$

$$(10.7.12.1)$$

$$\Delta w_0|_\Gamma = u(0) \in H^{\frac{1}{2}}(\Gamma);$$

$$u \in C\big([0, T]; H^{\frac{1}{2}}(\Gamma)\big) \cap H^1(0, T; L_2(\Gamma)), \quad\quad (10.7.12.2)$$

and we must show that

$$\{w, w_t, w_{tt}\} \in C\big([0, T]; \big[H^3(\Omega) \cap H_0^1(\Omega)\big] \times \big[H^2(\Omega) \cap H_0^1(\Omega)\big] \times H_0^1(\Omega)\big).$$
(10.7.12.3)

Operator-Theoretic Proof We return to the explicit solution formula (10.7.4.30), that is,

$$w(t) = \mathbb{C}(t)w_0 + \mathbb{S}(t)w_1 + \mathbb{A}\int_0^t \mathbb{S}(t - \tau)G_2u(\tau)\,d\tau, \quad (10.7.12.4)$$

and integrate by parts the integral term with $u \in H^1(0, T; L_2(\Gamma))$, thus obtaining

$$w(t) = \mathbb{C}(t)[w_0 - \cancel{G_2u(0)}] + \mathbb{S}(t)w_1 + G_2u(t) - \int_0^t \mathbb{C}(t - \tau)G_2\dot{u}(\tau)\,d\tau.$$
(10.7.12.5)

Here, by the first compatibility condition in (10.7.12.1), and, respectively, $\dot{u} \in L_2(\Sigma)$, we have

$$w_0 = G_2u(0); \quad \int_0^t \mathbb{C}(t - \tau)G_2\dot{u}(\tau)\,d\tau \in C\big([0, T]; \mathcal{D}\big(\mathcal{A}^{\frac{3}{4}}\big)\big), \quad (10.7.12.6)$$

recalling the definition of G_2 in (10.7.4.15) and, respectively, the regularity (10.7.11.2) (second component) $[\mathbb{A}z \in \mathcal{D}(\mathcal{A}^{\frac{1}{4}}) \Longleftrightarrow z \in \mathcal{D}(\mathcal{A}^{\frac{3}{4}})]$. Likewise, by (10.7.4.11),

$$\mathbb{S}(t)w_1 \in C\big([0, T]; \mathcal{D}\big(\mathcal{A}^{\frac{3}{4}}\big)\big), \quad \text{with } w_1 \in \mathcal{D}\big(\mathcal{A}^{\frac{1}{2}}\big). \quad (10.7.12.7)$$

We now use that $u \in C([0, T]; H^{\frac{1}{2}}(\Gamma))$ as well, from (10.7.12.2), so that, by elliptic theory [(10.7.4.15), (10.7.4.16) with $s = 1/2$],

$$G_2u(t) \in C\big([0, T]; H^3(\Omega) \cap H_0^1(\Omega)\big), \quad (10.7.12.8)$$

since $G_2u(t)|_\Gamma = 0$ by definition (10.7.4.15). Thus, (10.7.12.6), (10.7.12.7), and (10.7.12.8) used in (10.7.12.5) show (10.7.12.3) for w, via (10.7.4.8a). As to w_t, we differentiate (10.7.12.5), thus obtaining

$$w_t(t) = \mathbb{C}(t)w_1 + G_2\dot{u}(t) - G_2\dot{u}(t) + \mathbb{A}\int_0^t \mathbb{S}(t - \tau)G_2\dot{u}(\tau)\,d\tau \in C\big([0, T]; \mathcal{D}\big(\mathcal{A}^{\frac{1}{2}}\big)\big).$$
(10.7.12.9)

Conclusion (10.7.12.3) for w_t now follows from (10.7.12.9) [where a cancellation of $G_2\dot{u}(t)$ occurs] via (10.7.11.2) [first component] and (10.7.11.3). Similarly, one differentiates (10.7.12.9) in t and obtains (10.7.12.3) for w_{tt} via (10.7.11.2) [second component] and (10.7.11.3).

PDE Proof   Differentiating in t problem (10.7.1.1) with $u \in H^1(0, T; L_2(\Gamma))$ yields

$$
\begin{cases}
(w_t)_{tt} - \rho \Delta (w_t)_{tt} + \Delta^2(w_t) = 0 & \text{in } Q; & (10.7.12.10a) \\
w_t|_{t=0} = w_1, \quad (w_t)_t|_{t=0} = -\bigl(I + \rho \mathcal{A}^{\frac{1}{2}}\bigr)^{-1}\Delta^2 w_0 & \text{in } \Omega; & (10.7.12.10b) \\
w_t|_\Sigma \equiv 0 & \text{in } \Sigma; & (10.7.12.10c) \\
\Delta w_t|_\Sigma = \dot u & \text{in } \Sigma, & (10.7.12.10d)
\end{cases}
$$

that is, the same problem (10.7.1.1), however, in the variable w_t, with initial data $w_t|_{t=0} \in \mathcal{D}(\mathcal{A}^{\frac{1}{2}})$ and $(w_t)_t|_{t=0} \in \mathcal{D}(\mathcal{A}^{\frac{1}{4}})$ [since $\Delta^2 w_0 \in H^{-1}(\Omega) = [\mathcal{D}(\mathcal{A}^{\frac{1}{4}})']$, and $\dot u \in L_2(\Sigma)$. Since estimates are first shown for regular data, and then extended, we need the compatibility conditions in (10.7.12.1) for regular data, to invoke Theorem 10.7.3.2 and obtain

$$
\{w_t, w_{tt}\} \in C\bigl([0, T]; Y = \mathcal{D}\bigl(\mathcal{A}^{\frac{1}{2}}\bigr) \times \mathcal{D}\bigl(\mathcal{A}^{\frac{1}{4}}\bigr)\bigr). \tag{10.7.12.11}
$$

As to w, we use elliptic theory on

$$
\begin{cases}
\Delta^2 w = -w_{tt} + \rho \Delta w_{tt} \in H^{-1}(\Omega) = \bigl[\mathcal{D}\bigl(\mathcal{A}^{\frac{1}{4}}\bigr)\bigr]', \\
w|_\Gamma = 0, \\
\Delta w|_\Gamma = u \in H^{\frac{1}{2}}(\Gamma),
\end{cases} \tag{10.7.12.12}
$$

at each t, recalling assumption (10.7.12.2), and conclude also via (10.7.4.16) that

$$
w \in C\bigl([0, T]; H^3(\Omega) \cap H^1_0(\Omega)\bigr), \tag{10.7.12.13}
$$

as desired. The proof of (10.7.12.3) is complete. $\square$

10.7.13 Proof of Theorem 10.7.8.1

10.7.13.1 A Preliminary Trace Result

Theorem 10.7.10.1   *(i) With reference to the operators in (10.7.11.8) and (10.7.11.9), we have the following regularity properties for $0 \le r \le 1$, which generalize the case $r = 0$ of Theorem 10.7.11.1:*

$$
G_2^* \mathbb{A} \mathcal{A}^{\frac{1}{2}} \mathbb{S}(t), \quad G_2^* \mathbb{A} \mathbb{S}(t) : continuous \; \mathcal{D}\bigl(\mathcal{A}^{\frac{r}{4}}\bigr) \to H^{r,r}(\Sigma), \tag{10.7.13.1}
$$

$$
G_2^* \mathbb{A} \mathcal{A}^{\frac{1}{4}} \mathbb{C}(t), \quad G_2^* \mathcal{A}^{\frac{3}{4}} \mathbb{C}(t) : continuous \; \mathcal{D}\bigl(\mathcal{A}^{\frac{r}{4}}\bigr) \to H^{r,r}(\Sigma), \tag{10.7.13.2}
$$

$$
B^* e^{A^* t} : continuous \; \mathcal{D}(A^r) \equiv \mathcal{D}\bigl(\mathcal{A}^{\frac{1}{2}+\frac{r}{4}}\bigr) \times \mathcal{D}\bigl(\mathcal{A}^{\frac{1}{4}+\frac{r}{4}}\bigr) \to H^{r,r}(\Sigma), \tag{10.7.13.3}
$$

$$
H^{r,r}(\Sigma) \equiv L_2(0, T; H^r(\Gamma)) \cap H^r(0, T; L_2(\Gamma)). \tag{10.7.13.4}
$$

(ii) Equivalently, in PDE terms (see (10.7.10.16) or (10.7.3.5) for $r = 0$)

$$
\{\phi_0, \phi_1\} \to \frac{\partial \Delta \phi(t; \phi_0, \phi_1)}{\partial \nu} : continuous, \tag{10.7.13.5}
$$

$$
\mathcal{D}(A^{1+r}) = \mathcal{D}\bigl(\mathcal{A}^{\frac{3}{4}+\frac{r}{4}}\bigr) \times \mathcal{D}\bigl(\mathcal{A}^{\frac{1}{2}+\frac{r}{4}}\bigr) \to H^{r,r}(\Sigma), \tag{10.7.13.6}
$$

where $\phi(t; \phi_0, \phi_1)$ is the solution of problem (10.7.3.1) with $f \equiv 0$, and where we *further recall (10.7.11.4)–(10.7.11.6) to justify the stated equivalence between parts (i) and (ii).*

Proof.

Case $r = 0$ The case $r = 0$ is contained in Theorem 10.7.11.1 for (i) and in Theorem 10.7.10.2, Eqn. (10.7.10.16), or Theorem 10.7.3.1, Eqn. (10.7.3.5), for (ii). The stated equivalence uses (10.7.11.4)–(10.7.11.6). Thus, it is sufficient to prove the case $r = 1$ and interpolate to establish Theorem 10.7.13.1.

Case $r = 1$ We first show the time regularity:

$$G_2^* \mathbb{A} \mathcal{A}^{\frac{1}{2}} \mathbb{S}(t)x, \quad G_2^* \mathbb{A} \mathcal{A}^{\frac{1}{4}} \mathbb{C}(t)x \in H^1(0, T; L_2(\Gamma)), \quad x \in \mathcal{D}(\mathcal{A}^{\frac{1}{4}}). \quad (10.7.13.7)$$

Indeed, with $x \in \mathcal{D}(\mathcal{A}^{\frac{1}{4}})$, we compute, recalling from (10.7.4.9) and the discussion following that $-\mathbb{A}$ is the infinitesimal generator of $\mathbb{C}(t)$:

$$\frac{d}{dt} G_2^* \mathbb{A} \mathcal{A}^{\frac{1}{2}} \mathbb{S}(t)x = G_2^* \mathbb{A} \mathcal{A}^{\frac{1}{4}} \mathbb{C}(t) \mathcal{A}^{\frac{1}{4}} x \in L_2(\Sigma), \quad (10.7.13.8)$$

$$\frac{d}{dt} G_2^* \mathbb{A} \mathcal{A}^{\frac{1}{4}} \mathbb{C}(t)x = -G_2^* \mathbb{A}^2 \mathbb{S}(t) \mathcal{A}^{\frac{1}{4}} x \in L_2(\Sigma), \quad (10.7.13.9)$$

where the regularity in (10.7.13.8) is a direct application of (10.7.11.10) (case $r = 0$), while the regularity of (10.7.13.9) is equivalent to (10.7.11.9) (case $r = 0$) by the definition of $\mathbb{A}$ in (10.7.4.3). Thus, (10.7.13.7) is proved. To show the space regularity,

$$G_2^* \mathbb{A} \mathcal{A}^{\frac{1}{2}} \mathbb{S}(t)x, \quad G_2^* \mathbb{A} \mathcal{A}^{\frac{1}{4}} \mathbb{C}(t)x \in L_2(0, T; H^1(\Gamma)), \quad x \in \mathcal{D}(\mathcal{A}^{\frac{1}{4}}), \quad (10.7.13.10)$$

we shall equivalently show, by (10.7.11.4)–(10.7.11.6), that

$$\{\phi_0, \phi_1\} \to \frac{\partial \Delta \phi(t; \phi_0, \phi_1)}{\partial \nu} : \text{continuous } \mathcal{D}(A^2) = \mathcal{D}(A) \times \mathcal{D}(A^{\frac{3}{4}})$$

$$\to L_2(0, T; H^1(\Gamma)) \quad (10.7.13.11)$$

with $\phi(t; \phi_0, \phi_1)$ solutions of problem (10.7.3.1) with $f \equiv 0$. To this end, we introduce

$$\begin{cases} \mathcal{B} = \sum_i b_i(x) \dfrac{\partial}{\partial x_i} = \text{first-order operator with} \\[2mm] \text{(time-independent) coefficients } b_i \text{ smooth in } \bar{\Omega} \text{ and such that} \\[2mm] \mathcal{B} \text{ is tangent to } \Gamma, \text{ that is, } \sum_i b_i \nu_i = 0 \text{ on } \Gamma. \end{cases} \quad (10.7.13.12)$$

Accordingly, we consider the problem

$$\begin{cases} \phi_{tt} - \rho \Delta \phi_{tt} + \Delta^2 \phi \equiv 0 & \text{in } Q; & (10.7.13.13\text{a}) \\ \phi(0, \cdot) = \phi_0, \quad \phi_t(0, \cdot) = \phi_1 & \text{in } \Omega, \quad \text{or } \phi_{tt} = -\mathbb{A}\phi; & (10.7.13.13\text{b}) \\ \phi|_\Sigma \equiv \Delta \phi|_\Sigma \equiv 0 & \text{in } \Sigma; & (10.7.13.13\text{c}) \end{cases}$$

$$\{\phi_0, \phi_1\} \in \mathcal{D}(A) \times \mathcal{D}(A^{\frac{3}{4}}) \subset H^4(\Omega) \times H^3(\Omega), \quad (10.7.13.14)$$

whose solution is

$$\phi(t) = \mathbb{C}(t)\phi_0 + \mathbb{S}(t)\phi_1 \in C([0, T]; \mathcal{D}(\mathcal{A})), \qquad (10.7.13.15)$$

$$\phi_t(t) = -\mathcal{A}\mathbb{S}(t)\phi_0 + \mathbb{C}(t)\phi_1 \in C([0, T]; \mathcal{D}(\mathcal{A}^{\frac{3}{4}})), \qquad (10.7.13.16)$$

$$\phi_{tt}(t) = -\mathcal{A}\mathbb{C}(t)\phi_0 - \mathcal{A}\mathbb{S}(t)\phi_1 \in C([0, T]; \mathcal{D}(\mathcal{A}^{\frac{1}{2}})). \qquad (10.7.13.17)$$

We then introduce a new variable

$$z = \mathcal{B}\phi, \qquad (10.7.13.18)$$

which, therefore, has a priori regularity from (10.7.13.12) and (10.7.13.15)–(10.7.13.17), given by

$$\begin{cases} z \in C\left([0, T]; H^3(\Omega) \cap H_0^1(\Omega)\right), & (10.7.13.19) \\[2mm] z_t \in C\left([0, T]; H^2(\Omega) \cap H_0^1(\Omega)\right), & (10.7.13.20) \\[2mm] z_{tt} \in C\left([0, T]; H_0^1(\Omega)\right). & (10.7.13.21) \end{cases}$$

Then proving (10.7.13.11) is equivalent to showing that

$$\frac{\partial(\Delta\mathcal{B}\phi)}{\partial\nu}\bigg|_{\Gamma} = \frac{\partial(\Delta z)}{\partial\nu}\bigg|_{\Gamma} \in L_2(\Sigma). \qquad (10.7.13.22)$$

The variable z satisfies the problem

$$\begin{cases} z_{tt} - \rho\Delta z_{tt} + \Delta^2 z = -[\mathcal{B}, \Delta^2]\phi + \rho[\mathcal{B}, \Delta]\phi_{tt} & \text{in } Q, & (10.7.13.23a) \\[2mm] z|_{\Sigma} \equiv 0 & \text{in } \Sigma, & (10.7.13.23b) \\[2mm] \Delta z|_{\Sigma} = -[\mathcal{B}, \Delta]\phi|_{\Gamma} & \text{in } \Sigma, & (10.7.13.23c) \end{cases}$$

as one readily sees by (10.7.3.18) and (10.7.3.13). Since the commutators

$$[\mathcal{B}, \Delta^2] = \text{operator of order } 1 + 4 - 1 = 4, \qquad (10.7.13.24a)$$

$$[\mathcal{B}, \Delta] = \text{operator of order } 1 + 2 - 1 = 2, \qquad (10.7.13.24b)$$

we see via the regularity (10.7.13.15) for ϕ and (10.7.13.17) for ϕ_{tt} that the right-hand side term k in (10.7.13.23a) satisfies

$$k \equiv [\mathcal{B}, \Delta^2]\phi + \rho[\mathcal{B}, \Delta]\phi_{tt} \in C([0, T]; L_2(\Omega)), \qquad (10.7.13.25)$$

recalling $\mathcal{D}(\mathcal{A}) \subset H^4(\Omega)$ and $\mathcal{D}(\mathcal{A}^{\frac{1}{2}}) \subset H^2(\Omega)$. Similarly, via (10.7.13.15), (10.7.13.16), and (10.7.13.24), as well as by using trace theory, we see that the boundary term g in (10.7.13.23c) satisfies

$$g = -[\mathcal{B}, \Delta]\phi|_{\Gamma} \in C([0, T]; H^{\frac{3}{2}}(\Gamma)), \qquad (10.7.13.26)$$

$$g_t = -[\mathcal{B}, \Delta]\phi_t|_{\Gamma} \in C([0, T]; H^{\frac{1}{2}}(\Gamma)). \qquad (10.7.13.27)$$

Thus, by (10.7.13.23), (10.7.13.25), (10.7.13.26), and (10.7.13.27), we see that the
z-problem becomes:

$$
\begin{cases}
z_{tt} - \rho \Delta z_{tt} + \Delta^2 z = k \in C([0, T]; L_2(\Omega)), & \text{(10.7.13.28a)} \\
z|_\Sigma \equiv 0, & \text{(10.7.13.28b)} \\
\Delta z|_\Sigma = g \in C\left([0, T]; H^{\frac{3}{2}}(\Gamma)\right) \cap C^1\left([0, T]; H^{\frac{1}{2}}(\Gamma)\right) & \text{(10.7.13.28c)}
\end{cases}
$$

with a priori interior regularity given by (10.7.13.19)–(10.7.13.21). We now return to
the basic identity (10.7.10.4). Because of the a priori interior regularity (10.7.13.19),
(10.7.13.20), for $\{z, z_t\}$ and that of k in (10.7.13.28a), the right-hand side (R.H.S.) of
identity (10.7.10.4) (with $\{\phi, f\}$ there replaced by $\{z, k\}$ now) is well defined. Thus,
the left-hand side of identity (10.7.10.4) is well defined. Taking the vector field h such
that $h|_\Gamma = \nu = $ outward unit normal vector on Γ, we have that

$$
\text{on } \Gamma : h \cdot \nabla(\Delta z) = \nabla(\Delta z) \cdot \nu = \frac{\partial \Delta z}{\partial \nu}, \quad h \cdot \nabla z_t = \frac{\partial \phi_t}{\partial \nu}, \quad \text{(10.7.13.29)}
$$

$$
(\Delta z_t)^2 h \cdot \nu = g_t^2, \quad \left|\frac{\partial z_t}{\partial \nu}\right| = |\nabla z_t|^2 \quad \text{by (10.7.13.28b);} \quad \text{(10.7.13.30)}
$$

$$
\nabla(\Delta z) = \frac{\partial(\Delta z)}{\partial \nu}\nu + \frac{\partial(\Delta z)}{\partial \nu}\tau, \quad \tau = \text{tangential unit vector on } \Gamma
$$

$$
= \frac{\partial(\Delta z)}{\partial \nu}\nu + \frac{\partial g}{\partial \tau}\tau; \quad \text{(10.7.13.31)}
$$

$$
|\nabla(\Delta z)|^2 = \left(\frac{\partial(\Delta z)}{\partial \nu}\right)^2 + \left(\frac{\partial g}{\partial \tau}\right)^2. \quad \text{(10.7.13.32)}
$$

Thus, the left-hand side (L.H.S.) of identity (10.7.10.4) can be rewritten, in the new
variable z as

$$
\text{L.H.S. of (10.7.10.4)} = \frac{1}{2}\int_\Sigma \left(\frac{\partial(\Delta z)}{\partial \nu}\right)^2 d\Sigma + \frac{1}{2}\int_\Sigma \left(\frac{\partial z_t}{\partial \nu}\right)^2 d\Sigma
$$

$$
+ \frac{\rho}{2}\int_\Sigma g_t^2 \, d\Sigma - \frac{1}{2}\int_\Sigma \left(\frac{\partial g}{\partial \tau}\right)^2 d\Sigma
$$

$$
= \text{well defined by R.H.S. of (10.7.10.4),} \quad \text{(10.7.13.33)}
$$

since the last two integral terms on the L.H.S. of (10.7.10.4) vanish due to the
BC (10.7.13.28b). The two boundary terms containing g in (10.7.13.33) are well-
defined by the regularity of g in (10.7.13.28c), whereas the boundary term containing $\partial z_t/\partial \nu$ is well-defined by (10.7.13.20) and trace theory. We conclude that the
remaining boundary term in (10.7.13.33) containing $\partial(\Delta z)/\partial \nu$ is well-defined, that
is, $\partial(\Delta z)/\partial \nu \in L_2(\Sigma)$, and thus (10.7.13.22) is established, as desired. The proof of
Theorem 10.7.13.1 is complete. $\quad \square$

Remark 10.7.10.1   In Theorem 10.7.10.2, the required interior regularity $\{\phi, \phi_t\} \in$
$C([0, T]; H^3(\Omega) \times H^2(\Omega))$ needed to guarantee that the right-hand side of identity
(10.7.10.15) [same as right-hand side of (10.7.10.4)] is well defined – is ensured by
the assumed regularity of the data $\{\phi_0, \phi_1, f\}$ as in (10.7.10.16), in particular $f \in$
$L_1(0, T; \mathcal{D}(\mathcal{A}^{\frac{1}{2}}))$, whereby then the (positive) left-hand side of identity (10.7.10.15)
establishes that $\partial(\Delta\phi)/\partial\nu \in L_2(\Sigma)$. By contrast, in the z-problem (10.7.13.28), the
right-hand side k is only in $C([0, T]; L_2(\Omega))$. However, the required regularity
$\{z, z_t\} \in C([0, T]; H^3(\Omega) \cap H^2(\Omega))$ for the right-hand side of identity (10.7.10.4)
is guaranteed by the a priori regularity (10.7.13.19), (10.7.13.20), which is a consequence of the regularity (10.7.13.15), (10.7.13.16) of $\{\phi, \phi_t\}$ via the change of
variable $z = \mathcal{B}\phi$ in (10.7.13.18). Thus, for the z-problem (10.7.13.28), k is only required to have the regularity that makes the term $\int_Q kh \cdot \nabla(\Delta z)\, dQ$ on the right-hand
side of (10.7.10.4) well defined, that is, say $k \in L_1(0, T; L_2(\Omega))$, and we still obtain
$\partial(\Delta z)/\partial\nu \in L_2(\Sigma)$. The above contrast between the ϕ-problem in Theorem 10.7.10.2
and the z-problem in (10.7.13.28) did not occur in the case of the wave equation of
Section 10.5.13 (see below 10.5.13.19)), although it is typical for the other illustrating
examples to follow: Euler–Bernoulli equation; Schrödinger equations, etc.

10.7.13.2 Completion of the Proof of Theorem 10.7.8.1
Space Regularity   To show space regularity,

$$(L^*v)(t) = \int_t^T B^* e^{A^*(\tau - t)} v(\tau)\, d\tau \qquad (10.7.13.34)$$

$$: \text{continuous } L_2\big(0, T; \mathcal{D}(A^r)\big) \equiv \mathcal{D}\big(\mathcal{A}^{\frac{1}{2}+\frac{r}{4}}\big) \times \mathcal{D}\big(\mathcal{A}^{\frac{1}{2}+\frac{r}{4}}\big)\big)$$

$$\rightarrow L_2(0, T; H^r(\Gamma)), \quad 0 \le r \le 1, \quad (10.7.13.35)$$

we simply invoke Chapter 7, Theorem 7.2.1, Eqn. (7.2.2) with $X^* = \mathcal{D}(A^r)$, $U^* =$
$H^r(\Gamma)$, and $q = 2$, which is legal by virtue of the regularity (10.7.13.3) of
Theorem 10.7.13.1.

Time Regularity   It suffices to show the case $r = 1$, since the case $r = 0$ is contained in (10.7.13.35), or in (10.7.5.6), and then interpolate. Thus, differentiating
(10.7.13.34) in t for

$$v \in L_2(0, T; \mathcal{D}(A)) \quad \text{or } A^*v \in L_2(0, T; Y), \qquad (10.7.13.36)$$

since A is skew-adjoint on Y (see below (10.7.4.26)), yields

$$\frac{d(L^*v)}{dt}(t) = -B^*v(t) - \int_t^T B^* e^{A^*(\tau - t)} A^*v(\tau)\, d\tau \in L_2(0, T; L_2(\Gamma)),$$

$$(10.7.13.37)$$

as desired, by (10.7.13.36) and $B^* A^{*-1} \in \mathcal{L}(Y; U)$ in (10.1.5a), with Y and U as in (10.7.4.22). Then (10.7.13.37) shows

$$L^* : \text{continuous } L_2(0, T; \mathcal{D}(A)) \to H^1(0, T; L_2(\Gamma)), \quad (10.7.13.38)$$

as required. The proof of Theorem 10.7.8.1 is complete except for noticing that uniformity in s is obtained by a change of variable on formula (10.7.13.34), as usual.

10.8 Application: Euler–Bernoulli Equation with One Boundary Control. Regularity Theory

In this section we consider the optimal quadratic cost problem over a finite horizon for the Euler–Bernoulli mixed problem, which is obtained from the Kirchoff equation of the preceding section by setting the constant $\rho = 0$ in Eqn. (10.7.1.1a). As a consequence, the problem ceases to be hyperbolic, has no finite speed of propagation, and the resulting function space setup is different: The smoothing effect of the operator $(I + \rho A^{\frac{1}{2}})^{-1}$ is now missing, and, moreover, the difference between the space of the displacement w and the space of the velocity w_t is now two Sobolev units, rather than one as in Section 10.7 (compare Y in (10.7.4.22) for the Kirchoff equation against Y in (10.8.2.2) for the Euler–Bernoulli equation). Our treatment will closely parallel that of Section 10.7, with notable technical differences, in verifying all the required abstract system's assumptions (H.1) = (10.1.6) through (H.7) = (10.1.21), in a natural mathematical setting. Again, infinitely many such settings are possible, and we shall select a particularly convenient one. Accordingly, Theorems 10.2.1, 10.2.2, and 10.2.3 of Section 10.2 are then applicable to the present class, for any observation operator R with "minimal" smoothing as in (H.8) = (10.1.23).

10.8.1 Problem Formulation

The Dynamics   Let Ω be an open bounded domain in R^n with sufficiently smooth boundary Γ, say of class C^2. The Euler–Bernoulli equation is given by

$$\begin{cases} w_{tt} + \Delta^2 w = 0 & \text{in } (0, T] \times \Omega = Q; & (10.8.1.1a) \\ w(0, \cdot) = w_0, \quad w_t(0, \cdot) = w_1 & \text{in } \Omega; & (10.8.1.1b) \\ w|_\Sigma \equiv 0 & \text{in } (0, T] \times \Gamma = \Sigma; & (10.8.1.1c) \\ \Delta w|_\Sigma = u & \text{in } \Sigma, & (10.8.1.1d) \end{cases}$$

with boundary control $u \in L_2(0, T; L_2(\Gamma)) \equiv L_2(\Sigma)$. Problem (10.8.1.1) results from setting $\rho = 0$ in the Kirchoff problem (10.7.1.1).

The Optimal Control Problem on $[s, T]$ Consistently with the (optimal) regularity theory for problem (10.8.1.1) presented in Theorem 10.8.3.2 below, the cost functional we seek to minimize over all $u \in L_2(s, T; L_2(\Gamma)) \equiv L_2(\Sigma_{sT})$ is taken to be

$$J(u, w) = \int_s^T \left\{ \left\| R \begin{bmatrix} w(t) \\ w_t(t) \end{bmatrix} \right\|^2_{H^1(\Omega) \times H^{-1}(\Omega)} + \|u(t)\|^2_{L_2(\Gamma)} \right\} dt, \qquad (10.8.1.2)$$

with initial data $\{w_0, w_1\} \in H_0^1(\Omega) \times H^{-1}(\Omega)$, where the observation operator $R \in \mathcal{L}(H_0^1(\Omega) \times H^{-1}(\Omega))$ will be further specified below in (10.8.2.3).

10.8.2 Main Results

As a specialization to problem (10.8.1.1), (10.8.1.2) of the abstract theory presented in Theorems 10.2.1, 10.2.2, and 10.2.3 of this chapter, in the present section we establish the following results.

Theorem 10.8.10.1 *(a) With the observation operator R in (10.8.1.2) only assumed to satisfy*

$$R \in \mathcal{L}\left(H_0^1(\Omega) \times H^{-1}(\Omega)\right), \qquad (10.8.2.1)$$

Theorem 9.2.1 of Chapter 9 applies to the optimal control problem (10.8.1.1), (10.8.1.2), with

$$y(t) = \begin{bmatrix} w(t) \\ w_t(t) \end{bmatrix}; \qquad U \equiv L_2(\Gamma); \qquad Y \equiv H_0^1(\Omega) \times H^{-1}(\Omega), \qquad (10.8.2.2)$$

and yields a unique optimal pair $\{u^0(\cdot, s; y_0), y^0(\cdot, s; y_0)\}$, $y_0 = [w_0, w_1] \in Y$, satisfying, in particular, the pointwise feedback synthesis (9.2.17) and the optimal cost relation (9.2.18) of Chapter 9, as well as the other properties listed there.

(b) Assume, in addition to (10.8.2.1), that R satisfies the smoothing assumption

$$R^*R : \text{continuous } Y_\delta^- \rightarrow Y_\delta^+, \qquad (10.8.2.3)$$

with R^* the adjoint of R in Y, where

$$Y_\delta^- = \{h \in H^{2-2\delta}(\Omega) : h|_\Gamma = 0\} \times H^{-2\delta}(\Omega)$$
$$= \left[H^{2-2\delta}(\Omega) \cap H_0^1(\Omega)\right] \times H^{-2\delta}(\Omega), \qquad (10.8.2.4)$$

$$Y_\delta^+ = \{h \in H^{2+2\delta}(\Omega) : h|_\Gamma = 0\} \times H^{2\delta}(\Omega)$$
$$= \left[H^{2+2\delta}(\Omega) \cap H_0^1(\Omega)\right] \times H^{2\delta}(\Omega) \qquad (10.8.2.5)$$

(see Section 10.8.6), and where $\delta > 0$ is an arbitrarily small constant, which is kept fixed throughout. [An additional characterization of Y_δ^- and Y_δ^+ will be given below in (10.8.6.4), (10.8.6.5).] Then, Theorems 10.2.1, 10.2.2, and 10.2.3 of Section 10.2 of the present chapter hold true, with Y_δ^- and Y_δ^+ given by (10.8.2.4) and (10.8.2.5), and with

$$U_\delta^- = H^{1-2\delta}(\Gamma); \qquad U_\delta^+ = H^{1+2\delta}(\Gamma) \qquad (10.8.2.6)$$

(see Section (10.8.6)). In particular, explicitly:

(b_1) *(Regularity of the optimal pair) For*

$$y_0 = \{w_0, w_1\} \in \left[H^{2-2\delta}(\Omega) \cap H_0^1(\Omega)\right] \times H^{-2\delta}(\Omega), \qquad (10.8.2.7)$$

the optimal pair satisfies the following regularity properties:

(i)

$$\begin{bmatrix} w^0(\,\cdot\,, s; y_0) \\ w_t^0(\,\cdot\,, s; y_0) \end{bmatrix} = y^0(\,\cdot\,, s; y_0) \in C\left([s, T]; \left[H^{2-2\delta}(\Omega) \cap H_0^1(\Omega)\right] \times H^{-2\delta}(\Omega)\right)$$

$$\cap \, H^{\frac{1}{2}-\delta}\left(s, T; H_0^1(\Omega) \times H^{-1}(\Omega)\right); \quad (10.8.2.8)$$

(ii)

$$u^0(\,\cdot\,, s; y_0) \in L_2(s, T; H^{1+2\delta}(\Gamma)) \cap H^{\frac{1}{2}+\delta}(s, T; L_2(\Gamma)) \equiv H^{1+2\delta, \frac{1}{2}+\delta}(\Sigma_{sT}), \tag{10.8.2.9}$$

$\Sigma_{sT} = [s, T] \times \Gamma$, *a fortiori [Lions, Magenes, Theorem 3.1, p. 19],*

$$u^0(\,\cdot\,, s; y_0) \in C([s, T]; H^{2\delta}(\Gamma)). \qquad (10.8.2.10)$$

All of the above results hold true uniformly in s, that is, with norms that may be made independent of s.

(b_2) *(Gain operator) The gain operator $B^*P(t)$, with $P(t)$ defined by (9.2.14) of Chapter 9, satisfies the following regularity property (see (10.2.4)):*

$$B^*P(t)x = B^* \begin{bmatrix} P_{11}(t) & P_{12}(t) \\ P_{21}(t) & P_{22}(t) \end{bmatrix} \begin{bmatrix} x_1 \\ x_2 \end{bmatrix} \qquad (10.8.2.11a)$$

$$= -\frac{\partial}{\partial \nu} \mathcal{A}^{-\frac{1}{2}} [P_{21}(t)x_1 + P_{22}(t)x_2] \qquad (10.8.2.11b)$$

$$: \text{continuous } \left[H^{2-2\delta}(\Omega) \cap H_0^1(\Omega)\right] \times H^{-2\delta}(\Omega)$$

$$\rightarrow C([0, T]; L_2(\Gamma)), \qquad (10.8.2.11c)$$

where the adjoint B^* is computed with respect to the space Y topologized, however, as $Y = [\mathcal{D}(\mathcal{A}^{\frac{1}{4}})] \times [\mathcal{D}(\mathcal{A}^{\frac{1}{4}})]'$ (see (10.8.4.7) and (10.8.4.14) below), and where $\mathcal{A}$ is the operator in (10.7.4.1).

(b_3) *(Differential Riccati equation) With*

$$A = \begin{bmatrix} 0 & I \\ -\mathcal{A} & 0 \end{bmatrix}$$

as in (10.8.4.8) below, we have that P(t) satisfies the following DRE:

$$\begin{cases} \dfrac{d}{dt}(P(t)x, y)_Y = -(Rx, Ry)_Y - (P(t)x, Ay)_Y - (P(t)Ax, y)_Y \\[2mm] \qquad + \left(\dfrac{\partial}{\partial v}\mathcal{A}^{-\frac{1}{2}}[P_{21}(t)x_1 + P_{22}(t)x_2], \right. \\[2mm] \qquad\qquad \left. \dfrac{\partial}{\partial v}\mathcal{A}^{-\frac{1}{2}}[P_{21}(t)y_1 + P_{22}(t)y_2] \right)_{L_2(\Gamma)}; \\[2mm] P(T) = 0; \quad \forall x, y \in \left[H^{2-2\delta}(\Omega) \times H_0^1(\Omega) \right] \times H^{-2\delta}(\Omega). \end{cases} \qquad (10.8.2.12)$$

Furthermore, P(t) satisfies the corresponding integral Riccati equation, as in (10.2.7), for all such x, y.

10.8.3 Regularity Theory for Problem (10.8.1.1) with $u \in L_2(\Sigma)$

The following well-posedness Theorem 10.8.3.2 provides the critical regularity result, which justifies as natural the selection of the cost functional J in (10.8.1.2), and which will permit us to verify assumption (H.1) = (10.1.6) below. To this end, we associate with problem (10.8.1.1) the following boundary-homogeneous version (which would be the corresponding adjoint problem (10.8.5.7) below, if the initial conditions were given at $t = T$, an inessential modification since the problem is time reversible):

$$\begin{cases} \phi_{tt} + \Delta^2\phi = f & \text{in } Q; & (10.8.3.1\text{a}) \\ \phi(0, \cdot) = \phi_0, \quad \phi_t(0, \cdot) = \phi_1 & \text{in } \Omega; & (10.8.3.1\text{b}) \\ \phi|_\Sigma \equiv 0, \quad \Delta\phi|_\Sigma \equiv 0 & \text{in } \Sigma. & (10.8.3.1\text{c}) \end{cases}$$

The following key result is taken from Lions [1983] and Lasiecka and Triggiani [1989(a)].

Theorem 10.8.10.1 *With reference to (10.8.3.1), assume Γ of class C^2 and*

$$\{\phi_0, \phi_1\} \in V \times H_0^1(\Omega), \quad f \in L_1\left(0, T; H_0^1(\Omega)\right); \qquad (10.8.3.2)$$

$$V \equiv \{h \in H^3(\Omega) : h|_\Gamma = \Delta h|_\Gamma = 0\}. \qquad (10.8.3.3)$$

Then, the unique solution of problem (10.8.3.1) satisfies, continuously,

$$\{\phi, \phi_t\} \in C\left([0, T]; V \times H_0^1(\Omega)\right), \quad \phi_{tt} \in L_1(0, T; H^{-1}(\Omega)),$$

$$\phi_{tt} \in C([0, T]; H^{-1}(\Omega)) \quad \text{if } f \equiv 0; \qquad (10.8.3.4)$$

and

$$\frac{\partial(\Delta\phi)}{\partial v}, \frac{\partial\phi_t}{\partial v} \in L_2(0, T; L_2(\Gamma)) \equiv L_2(\Sigma). \qquad (10.8.3.5)$$

Proof. The interior regularity (10.8.3.4) for $\{\phi, \phi_t\}$ follows, for example, from the explicit solution formula (10.8.4.3), (10.8.4.4) below. The key boundary (trace) regularity results (10.8.3.5) can be shown by specializing the proof by energy methods in Section 10.7.10 to the present case $\rho = 0$. Details are given in Section 10.8.10.

$\square$

Remark 10.8.10.1   The trace regularity results for $\Delta\phi$ and ϕ_t in (10.8.3.5) do *not* follow from the optimal interior regularity (10.8.3.4): Indeed both trace regularity results in (10.8.3.5) are "1/2 higher" in Sobolev space regularity over a formal application of trace theory to ϕ and ϕ_t in (10.8.3.4).

Theorem 10.8.10.1   *With reference to the nonhomogeneous problem (10.8.1.1), assume*

$$\{w_0, w_1\} \in H_0^1(\Omega) \times H^{-1}(\Omega); \quad u \in L_2(\Sigma). \tag{10.8.3.6}$$

Then, the unique solution of (10.8.1.1) satisfies

$$\{w, w_t, w_{tt}\} \in C\left([0, T]; H_0^1(\Omega) \times H^{-1}(\Omega) \times H^{-3}(\Omega)\right), \tag{10.8.3.7}$$

continuously, that is, more precisely, there exists $C_T > 0$ such that

$$\|\{w, w_t, w_{tt}\}\|^2_{C([0,T]; H_0^1(\Omega) \times H^{-1}(\Omega) \times H^{-3}(\Omega))}$$
$$\leq C_T \left\{ \|u\|^2_{L_2(\Sigma)} + \|\{w_0, w_1\}\|^2_{H_0^1(\Omega) \times H^{-1}(\Omega)} \right\}. \tag{10.8.3.8}$$

Proof.   The proof of Theorem 10.8.3.2 is a consequence of Theorem 10.8.3.1 and will be given in Section 10.8.11 below. $\square$

10.8.4 Abstract Setting for Problem (10.8.1.1)

We follow Lasiecka and Triggiani [1989(a)] and Lasiecka and Triggiani [1991(c)]. We set $\rho = 0$ in the treatment of Section 10.7.4 for the Kirchoff equation.

(i) We let $\mathcal{A}, \mathcal{A}^{\frac{1}{2}}$ be the operators in (10.7.4.1), (10.7.4.2), for which relations (10.7.4.4), (10.7.4.5) on domains of fractional powers hold true, in particular (10.7.4.6a,b) [but not now (10.7.4.6c)], (10.7.4.7a,b) [but not now (10.7.4.7c)], as well as (10.7.4.8a,b). Instead of (10.7.4.9), we now have

$$\phi_{tt} + \mathcal{A}\phi = f \tag{10.8.4.1}$$

as abstract equation for problem (10.8.3.1). The self-adjoint operator $-\mathcal{A}$ generates a s.c. (self-adjoint) cosine operator $C(t)$ and corresponding sine operator $S(t) = \int_0^t C(\tau)\,d\tau$, where now the maps

$$t \to \mathcal{A}^{\frac{1}{2}} S(t), C(t) \text{ are strongly continuous on } L_2(\Omega) \tag{10.8.4.2}$$

[note the difference over (10.7.4.11)]. The unique solution of problem (10.8.4.1), or (10.8.3.1), is given explicitly by

$$\phi(t) = C(t)\phi_0 + S(t)\phi_1 + \int_0^t S(t-\tau)f(\tau)\,d\tau, \tag{10.8.4.3}$$

$$\phi_t(t) = -\mathcal{A}S(t)\phi_0 + C(t)\phi_1 + \int_0^t C(t-\tau)f(\tau)\,d\tau, \tag{10.8.4.4}$$

in appropriate function spaces, depending on $\{\phi_0, \phi_1, f\}$.

(ii) Let G_2 be the Green map defined by (10.7.4.15), satisfying properties (10.7.4.16)–(10.7.4.19).

(iii) Accordingly, problem (10.8.1.1) can be written abstractly first as the second-order equation

$$w_{tt} + \mathcal{A}(w - G_2 u) = 0 \quad \text{on } L_2(\Omega) \tag{10.8.4.5}$$

[compare with (10.7.4.20)], and next as in (10.1.1), that is, as

$$y_t = Ay + Bu \text{ on } [\mathcal{D}(A^*)]', \quad y(0) = y_0 \in Y; \tag{10.8.4.6}$$

$$Y = \mathcal{D}(\mathcal{A}^{\frac{1}{4}}) \times [\mathcal{D}(\mathcal{A}^{\frac{1}{4}})]' \equiv H_0^1(\Omega) \times H^{-1}(\Omega) \text{ (equivalent norms); } U = L_2(\Gamma); \tag{10.8.4.7}$$

$$y = \begin{bmatrix} w \\ w_t \end{bmatrix}; \quad A = \begin{bmatrix} 0 & I \\ -\mathcal{A} & 0 \end{bmatrix}; \quad \mathcal{D}(A) = \mathcal{D}(\mathcal{A}^{\frac{3}{4}}) \times \mathcal{D}(\mathcal{A}^{\frac{1}{4}}) \to Y; \tag{10.8.4.8}$$

$$Bu = \begin{bmatrix} 0 \\ \mathcal{A}G_2 u \end{bmatrix} : \text{continuous } U \to [\mathcal{D}(A^*)]' \text{ (duality with respect to } Y\text{)}, \tag{10.8.4.9}$$

or, equivalently,

$$A^{-1}B : \text{continuous } U \to Y, \tag{10.8.4.10}$$

since for $u \in L_2(\Gamma)$, recalling (10.8.4.8), (10.8.4.9), and (10.7.4.17), we have

$$A^{-1}Bu = \begin{bmatrix} 0 & -\mathcal{A}^{-1} \\ I & 0 \end{bmatrix} \begin{bmatrix} 0 \\ \mathcal{A}G_2 u \end{bmatrix} = \begin{bmatrix} -G_2 u \\ 0 \end{bmatrix} = \begin{bmatrix} \mathcal{A}^{-\frac{1}{2}}Du \\ 0 \end{bmatrix} \in Y;$$

$$\mathcal{A}^{-\frac{1}{2}}Du \in \mathcal{D}(\mathcal{A}^{\frac{5}{8} - \frac{\epsilon}{2}}), \tag{10.8.4.11}$$

since $Du \in \mathcal{D}(\mathcal{A}^{\frac{1}{8} - \frac{\epsilon}{2}})$ by (10.7.4.18b), the counterpart of (10.7.4.26). The operator A is skew-adjoint, $A^* = -A$, on $Y \equiv \mathcal{D}(\mathcal{A}^{\frac{1}{4}}) \times [\mathcal{D}(\mathcal{A}^{\frac{1}{4}})]'$ and generates a s.c. unitary group on it, given by

$$
e^{At} = \begin{bmatrix} C(t) & S(t) \\ -AS(t) & C(t) \end{bmatrix}; \quad e^{A^*t} = e^{-At} = \begin{bmatrix} C(t) & -S(t) \\ AS(t) & C(t) \end{bmatrix}. \tag{10.8.4.12}
$$

Starting from (10.8.4.9), we compute B^* with respect to Y topologized as $\mathcal{D}(A^{\frac{1}{4}}) \times [\mathcal{D}(A^{\frac{1}{4}})]'$, to obtain, for $v = [v_1, v_2]$,

$$
\begin{aligned}
(Bu, v)_Y &= (AG_2 u, v_2)_{[\mathcal{D}(A^{\frac{1}{4}})]'} = \left(A^{\frac{1}{2}} G_2 u, v_2\right)_{L_2(\Omega)} \\
&= \left(u, G_2^* A^{\frac{1}{2}} v_2\right)_{L_2(\Gamma)} = (u, B^* v)_{L_2(\Gamma)=U},
\end{aligned} \tag{10.8.4.13}
$$

that is, by virtue of (10.7.4.19),

$$
B^* \begin{bmatrix} v_1 \\ v_2 \end{bmatrix} = G_2^* A^{\frac{1}{2}} v_2 = G_2^* A A^{-\frac{1}{2}} v_2 = -\frac{\partial}{\partial \nu} A^{-\frac{1}{2}} v_2. \tag{10.8.4.14}
$$

Thus, from (10.8.4.12) and (10.8.4.14) we compute with $x = [x_1, x_2] \in Y$:

$$
\begin{aligned}
B^* e^{A^*t} x &= G_2^* A^{\frac{1}{2}} [AS(t)x_1 + C(t)x_2] \\
&= G_2^* A A^{\frac{1}{2}} [S(t)x_1 + C(t)A^{-1}x_2] \tag{10.8.4.15}
\end{aligned}
$$

$$
\text{[(by 10.7.4.19)]} \qquad = -\frac{\partial}{\partial \nu} A^{\frac{1}{2}} [C(t)\phi_0 + S(t)\phi_1] = -\frac{\partial \Delta \phi(t; \phi_0, \phi_1)}{\partial \nu}, \tag{10.8.4.16}
$$

recalling (10.7.4.19) and (10.7.4.2), where $\phi(t; \phi_0, \phi_1)$ solves problem (10.8.3.1), or (10.8.4.1), with $f \equiv 0$ [see (10.8.4.3)] and

$$
\phi_0 = A^{-1} x_2 \in \mathcal{D}(A^{\frac{3}{4}}); \quad \phi_1 = x_1 \in \mathcal{D}(A^{\frac{1}{4}}). \tag{10.8.4.17}
$$

As in the case of Section 10.5.4, Eqn. (10.5.4.22), or Section 10.7.4, Eqns. (10.7.4.29) and (10.7.4.30), the solution of problem (10.8.1.1), or (10.8.4.6), with initial condition $y(s) = y_0 = [w_0, w_1] \in Y$, is written abstractly as

$$
\begin{bmatrix} w(t, s; y_0) \\ w_t(t, s; y_0) \end{bmatrix} = e^{A(t-s)} \begin{bmatrix} w_0 \\ w_1 \end{bmatrix} + (L_s u)(t); \tag{10.8.4.18}
$$

$$
(L_s u)(t) = \int_s^t e^{A(t-\tau)} Bu(\tau) \, d\tau = \begin{bmatrix} A \displaystyle\int_s^t S(t-\tau) G_2 u(\tau) \, d\tau \\ A \displaystyle\int_s^t C(t-\tau) G_2 u(\tau) \, d\tau \end{bmatrix}. \tag{10.8.4.19}
$$

10.8.5 Verification of Assumption (H.1) = (10.1.6)

From (10.8.4.16) and (10.8.4.17), we see that (H.1) = (10.1.6) holds true:

$$\int_0^T \| B^* e^{A^* t} x \|_U^2 \, dt \leq C_T \| x \|_Y^2, \quad x \in Y, \tag{10.8.5.1}$$

if and only if problem (10.8.3.1) with $f \equiv 0$ satisfies

$$\int_0^T \int_\Gamma \left(\frac{\partial(\Delta\phi)}{\partial\nu} \right)^2 d\Sigma \leq C_T \| \{\phi_0, \phi_1\} \|_{\mathcal{D}(A^{\frac{3}{4}}) \times \mathcal{D}(A^{\frac{1}{4}})}^2, \tag{10.8.5.2}$$

which is precisely the trace regularity result guaranteed by Theorem 10.8.3.1, Eqn. (10.8.3.5). [Compare (10.8.5.2) for the Euler–Bernoulli problem (10.8.3.1) with (10.7.5.2) for the corresponding Kirchoff problem (10.7.3.1)]. Then, according to Theorem 7.2.1 of Chapter 7, estimate (10.8.5.1), that is, (10.8.5.2), is, in turn, equivalent to the following property that, with Y as in (10.8.4.7):

$$(L_s u)(t) = \int_s^t e^{A(t-\tau)} Bu(\tau) \, d\tau = \begin{bmatrix} w(t; 0, 0) \\ w_t(t; 0, 0) \end{bmatrix} \tag{10.8.5.3a}$$

$$: \text{continuous } L_2(s, T; L_2(\Gamma))$$
$$\rightarrow C\big([s, T]; \mathcal{D}(A^{\frac{1}{4}}) \times [\mathcal{D}(A^{\frac{1}{4}})]'\big) = H_0^1(\Omega) \times H^{-1}(\Omega)), \tag{10.8.5.3b}$$

uniformly in s, where $w_0 = w_1 = 0$ in problem (10.8.1.1). This result is precisely contained in conclusion (10.8.3.7) of Theorem 10.8.3.2, the additional statement of uniformity in s being an immediate consequence of formula (10.8.5.3a) for L_s (via a change of variable). Moreover, recalling (10.8.4.15), we have by duality on (10.8.5.3) with $v = [v_1, v_2]$:

$$(L_s^* v)(t) = B^* \int_t^T e^{A^*(\tau - t)} v(\tau) \, d\tau, \quad s \leq t \leq T$$

$$[\text{by } (10.8.4.15)] = G_2^* \mathcal{A} A^{\frac{1}{2}} \left\{ \int_t^T S(\tau - t) v_1(\tau) \, d\tau + \int_t^T C(\tau - t) \mathcal{A}^{-1} v_2(\tau) \, d\tau \right\} \tag{10.8.5.4}$$

$$: \text{continuous } L_1\big(s, T; \mathcal{D}(A^{\frac{1}{4}}) \times [\mathcal{D}(A^{\frac{1}{4}})]'\big)$$
$$\rightarrow L_2(s, T; L_2(\Gamma)), \tag{10.8.5.5}$$

uniformly in s. Now, let $\psi(t; f)$ be the solution of the (adjoint) problem

$$\begin{cases} \psi_{tt} + \Delta^2 \psi = f & \text{in } Q; & (10.8.5.6a) \\ \psi(T, \cdot) = 0, \quad \psi_t(T, \cdot) = 0 & \text{in } \Omega; & (10.8.5.6b) \\ \psi|_\Sigma \equiv \Delta\psi|_\Sigma \equiv 0 & \text{in } \Sigma, & (10.8.5.6c) \end{cases}$$

rewritten abstractly as

$$\psi_{tt} = -\mathcal{A}\psi + f, \quad \psi(T) = \psi_t(T) = 0, \tag{10.8.5.7}$$

and hence given explicitly by

$$\psi(t; f) = \int_T^t S(t - \tau) f(\tau) \, d\tau, \quad \psi_t(t; f) = \int_T^t C(t - \tau) f(\tau) \, d\tau. \tag{10.8.5.8}$$

Then, recalling (10.7.4.19) and (10.7.4.2), we see via (10.8.5.9) that the two terms in the definition of L_s^* in (10.8.5.5) can be rewritten, with $v = [v_1, v_2] \in L_1(0, T; \mathcal{D}(\mathcal{A}^{\frac{1}{4}}) \times [\mathcal{D}(\mathcal{A}^{\frac{1}{4}})]')$, as

$$(L_s^* u)(t) = \frac{\partial \Delta}{\partial v} \left\{ \int_T^t S(t - \tau) v_1(\tau) \, d\tau - \int_T^t C(t - \tau) \mathcal{A}^{-1} v_2(\tau) \, d\tau \right\} \tag{10.8.5.9}$$

$$= \frac{\partial \Delta \psi(t; f_1)}{\partial v} - \frac{\partial \Delta \psi_t(t; f_2)}{\partial v} \tag{10.8.5.10}$$

$$: \text{continuous } v = [v_1, v_2] \in L_1\left(0, T; \mathcal{D}(\mathcal{A}^{\frac{1}{4}}) \times [\mathcal{D}(\mathcal{A}^{\frac{1}{4}})]'\right)$$
$$\to L_2(0, T; L_2(\Gamma)); \tag{10.8.5.11}$$

$$f_1 = v_1 \in L_1\left(0, T; \mathcal{D}(\mathcal{A}^{\frac{1}{4}})\right); \quad f_2 = \mathcal{A}^{-1} v_2 \in L_1\left(0, T; \mathcal{D}(\mathcal{A}^{\frac{3}{4}})\right). \tag{10.8.5.12}$$

We notice that the regularity (10.8.5.12) of the normal trace $\partial \Delta \psi(t; f_1)/\partial v$ for the $\psi(t; f_1)$ solution of (10.8.5.7) due to $f = f_1$, as given by (10.8.5.13), is precisely the conclusion (10.8.3.5) of Theorem 10.8.3.1 for the time-reversed problem ϕ in (10.8.3.1) with initial data at $t = 0$ rather than at $t = T$ as for ψ, an inessential modification. Instead, the regularity (10.8.5.12) for the normal trace $\partial \Delta \psi_t(t; f_2)/\partial v$ of the time derivative ψ_t of the solution $\psi(t; f_2)$ of problem (10.8.5.7), due to $f = f_2$ as given by (10.8.5.13), is obtained by duality via operator methods as in (10.8.5.4)–(10.8.5.12) [whereas it appears that purely PDE methods will require a *time* regularity assumption of the right-hand side, nonhomogeneous term]. Thus, assumption (H.1) = (10.1.6) is verified.

10.8.6 Selection of the Spaces U_θ and Y_θ in (10.1.10)

We select the spaces in (10.1.10) to be the following Sobolev spaces:

$$U_\theta = H^{2\theta}(\Gamma), \quad U_0 = U = L_2(\Gamma), \quad 0 \le \theta \le \frac{1}{2} + \delta, \quad \theta \ne \frac{1}{2}; \tag{10.8.6.1}$$

$$Y_\theta \equiv \mathcal{D}\left(\mathcal{A}^{\frac{1}{4} + \frac{\theta}{2}}\right) \times \mathcal{D}\left(\mathcal{A}^{-\frac{1}{4} + \frac{\theta}{2}}\right) \equiv \mathcal{D}(A^\theta) = \mathcal{D}(A^{*\theta}),$$
$$Y_0 = Y = \mathcal{D}\left(\mathcal{A}^{\frac{1}{4}}\right) \times \left[\mathcal{D}\left(\mathcal{A}^{\frac{1}{4}}\right)\right]', \tag{10.8.6.2}$$

where A is given by (10.8.4.8), using henceforth the convention $\mathcal{D}(A^{-s}) = [\mathcal{D}(A^s)]'$ for $s > 0$, duality with respect to $L_2(\Omega)$. In particular, the critical spaces for $\theta = 1/2 \pm \delta$ are

$$U_\delta^- \equiv U_{\theta=\frac{1}{2}-\delta} = H^{1-2\delta}(\Gamma), \quad U_\delta^+ \equiv U_{\theta=\frac{1}{2}+\delta} = H^{1+2\delta}(\Gamma); \qquad (10.8.6.3)$$

$$Y_\delta^- \equiv Y_{\theta=\frac{1}{2}-\delta} = \mathcal{D}\big(A^{\frac{1}{2}-\frac{\delta}{2}}\big) \times \mathcal{D}\big(A^{-\frac{\delta}{2}}\big) = \mathcal{D}\big(A^{\frac{1}{2}-\delta}\big) = \mathcal{D}\big(A^{*\frac{1}{2}-\delta}\big)$$

$$\text{(by (10.7.4.4))} \qquad = \big[H^{2-2\delta}(\Omega) \cap H_0^1(\Omega)\big] \times H^{-2\delta}(\Omega); \qquad (10.8.6.4)$$

$$Y_\delta^+ \equiv Y_{\theta=\frac{1}{2}+\delta} = \mathcal{D}\big(A^{\frac{1}{2}+\frac{\delta}{2}}\big) \times \mathcal{D}\big(A^{\frac{\delta}{2}}\big) = \mathcal{D}\big(A^{\frac{1}{2}+\delta}\big) = \mathcal{D}\big(A^{*\frac{1}{2}+\delta}\big)$$

$$\text{(by (10.7.4.4))} \qquad = \big[H^{2+2\delta}(\Omega) \cap H_0^1(\Omega)\big] \times H^{2\delta}(\Omega). \qquad (10.8.6.5)$$

From (10.7.4.4) we have

$$\mathcal{D}\big(A^{\frac{1}{2}\pm\frac{\delta}{2}}\big) = \{h \in H^{2\pm2\delta}(\Omega) : h|_\Gamma = 0\} = H^{2\pm2\delta}(\Omega) \cap H_0^1(\Omega). \qquad (10.8.6.6)$$

Unlike in the preceding sections, the choice of Y_θ as in (10.8.6.2) does not imply that the passage from Y_δ^- to Y_δ^+ represents a jump across boundary conditions. This could be done, however; but it would require additional higher-order regularity results over Theorem 10.8.7.1 and Theorem 10.8.13.1 below. To streamline the present illustration, we have opted for the case in (10.8.6.2). The dual space $[Y_\delta^+]'$, duality with respect to $Y = \mathcal{D}(A^{\frac{1}{4}}) \times [\mathcal{D}(A^{\frac{1}{4}})]'$, is given by

$$[Y_\delta^+]' = \big[Y_{\frac{1}{2}+\delta}\big]' = \big[\mathcal{D}\big(A^{\frac{\delta}{2}}\big)\big]' \times \big[\mathcal{D}\big(A^{\frac{1}{2}+\frac{\delta}{2}}\big)\big]' \qquad (10.8.6.7)$$

$$= \mathcal{D}\big(A^{-\frac{\delta}{2}}\big) \times \mathcal{D}\big(A^{-\frac{1}{2}-\frac{\delta}{2}}\big), \qquad (10.8.6.8)$$

in the convention noted below (10.8.6.2), where the duality on the right-hand side of (10.8.6.7) is intended instead with respect to $L_2(\Omega)$. Thus, by (10.8.6.4) and (10.8.6.7), we verify the interpolation property

$$[Y_\delta^-, [Y_\delta^+]']_{\theta=\frac{1}{2}-\delta} = Y = \mathcal{D}\big(A^{\frac{1}{4}}\big) \times \big[\mathcal{D}\big(A^{\frac{1}{4}}\big)\big]', \qquad (10.8.6.9)$$

as required in (10.1.10), since $(1/2 - \delta/2)(1 - \theta) - \delta\theta/2 = 1/4$ for $\theta = 1/2 - \delta$ for the first component space; and $-\delta/2(1-\theta) - (1/2+\delta/2)\theta = -1/4$ for $\theta = 1/2 - \delta$ for the second component space. Moreover, the injections $U_{\theta_1} \hookrightarrow U_{\theta_2}$, $Y_{\theta_1} \hookrightarrow Y_{\theta_2}$ are compact, $0 \le \theta_2 < \theta_1 \le 1/2 + \delta$, as required in (10.1.10), since Ω is a bounded domain. Thus, by (10.8.6.1) and (10.8.6.2), the spaces in (10.1.11) and (10.1.12) are in the present case as follows, for $0 \le \theta \le 1/2 + \delta, \theta \ne 1/2$:

$$\mathcal{U}^\theta[s, T] = H^{2\theta,\theta}(\Sigma_{sT}) = L_2(s, T; H^{2\theta}(\Gamma)) \cap H^\theta(s, T; L_2(\Gamma)), \qquad (10.8.6.10)$$

$$\mathcal{Y}^\theta[s, T] = L_2\big(s, T; \mathcal{D}(A^\theta) = \mathcal{D}\big(A^{\frac{1}{4}+\frac{\theta}{2}}\big) \times \mathcal{D}\big(A^{-\frac{1}{4}+\frac{\theta}{2}}\big)\big)$$

$$\cap H^\theta\big(s, T; Y = \mathcal{D}\big(A^{\frac{1}{4}}\big) \times \big[\mathcal{D}\big(A^{\frac{1}{4}}\big)\big]'\big). \qquad (10.8.6.11)$$

10.8.7 Verification of Assumptions (H.2) = (10.1.14)

The following regularity result is critical in verifying assumptions (H.2) = (10.1.14).

Theorem 10.8.10.1   *With reference to the nonhomogeneous problem (10.8.1.1), assume:*

$$\{w_0, w_1\} \in \left[H^3(\Omega) \cap H_0^1(\Omega)\right] \times H_0^1(\Omega), \quad \text{with the compatibility relations}$$

$$w_0|_\Gamma = 0 \quad \text{and} \quad \Delta w_0|_\Gamma = u(0) \in H^{\frac{1}{2}}(\Gamma); \tag{10.8.7.1}$$

$$u \in C\left([0, T]; H^{\frac{1}{2}}(\Gamma)\right) \cap H^1(0, T; L_2(\Gamma)) \tag{10.8.7.2}$$

[(10.8.7.2) is a fortiori guaranteed if $u \in H^{1,1}(\Sigma)$, by [Lions, Magenes, 1972, I, Theorem 3.1, p. 19]. Then, the unique solution to problem (10.8.1.1) satisfies

$$\{w, w_t, w_{tt}\} \in C\left([0, T]; \left[H^3(\Omega) \cap H_0^1(\Omega)\right] \times H_0^1(\Omega) \times H^{-1}(\Omega)\right) \tag{10.8.7.3}$$

continuously.

Proof.   The proof of Theorem 10.8.7.1 will be given in Section 10.8.12 below.   □

Corollary 10.8.10.1   *With reference to the nonhomogeneous problem (10.8.1.1), assume $w_0 = w_1 = 0$ and*

$$u \in H^{\theta,\theta}(\Sigma) = L_2(0, T; H^\theta(\Gamma)) \cap H^\theta(0, T; L_2(\Gamma)), \quad 0 \le \theta < \frac{1}{2}, \tag{10.8.7.4}$$

in particular,

$$u \in \mathcal{U}^\theta[0, T] = L_2(0, T; H^{2\theta}(\Gamma)) \cap H^\theta(0, T; L_2(\Gamma)) \subset H^{\theta,\theta}(\Sigma). \tag{10.8.7.5}$$

Then, the unique solution of problem (10.8.1.1) satisfies for $0 \le \theta < 1/2$:

$$\begin{cases} w(\,\cdot\,; 0, 0) \in C\left([0, T]; H^{1+2\theta}(\Omega) \cap H_0^1(\Omega) = \mathcal{D}\big(\mathcal{A}^{\frac{1}{4}+\frac{\theta}{2}}\big)\right), & (10.8.7.6) \\[2mm] w_t(\,\cdot\,; 0, 0) \in C\left([0, T]; \left[H_0^{1-2\theta}(\Omega)\right]' = \left[\mathcal{D}\big(\mathcal{A}^{\frac{1}{4}-\frac{\theta}{2}}\big)\right]'\right), & (10.8.7.7) \\[2mm] w_{tt}(\,\cdot\,; 0, 0) \in C\left([0, T]; H^{-3+2\theta}(\Omega)\right), \quad \theta \ne \frac{1}{4}. & (10.8.7.8) \end{cases}$$

Thus, for $0 \le r \le 1$,

$$\begin{cases} D_t^r w(\,\cdot\,; 0, 0) \in L_2\big(0, T; \mathcal{D}\big(\mathcal{A}^{\frac{1}{4}+\frac{\theta-r}{2}}\big)\big), \text{ in particular,} & (10.8.7.9) \\[2mm] D_t^\theta w(\,\cdot\,; 0, 0) \in L_2\big(0, T; \mathcal{D}\big(\mathcal{A}^{\frac{1}{4}}\big)\big); & (10.8.7.10) \end{cases}$$

$$\begin{cases} D_t^r w_t(\,\cdot\,; 0, 0) \in L_2\big(0, T; H^{-1+2\theta-2r}(\Omega)\big), \quad \theta \ne \frac{1}{4}, \text{ in particular,} & (10.8.7.11) \\[2mm] D_t^\theta w_t(\,\cdot\,; 0, 0) \in L_2\big(0, T; H^{-1}(\Omega) = \left[\mathcal{D}\big(\mathcal{A}^{\frac{1}{4}}\big)\right]'\big), \quad \theta \ne \frac{1}{4}. & (10.8.7.12) \end{cases}$$

A fortiori, for $0 \le \theta < 1/2, \theta \ne 1/4$, via (10.8.6.11) and (10.8.6.2),

$$Lu = \{w(\,\cdot\,; 0, 0), w_t(\,\cdot\,; 0, 0)\} \in \mathcal{Y}^\theta[0, T] \cap C([0, T]; Y_\theta). \tag{10.8.7.13}$$

Proof of Corollary 10.8.7.2.   For $\theta < 1/2$, the compatibility relations in (10.8.7.1), which now read: $w_0|_\Gamma = 0$; $\Delta w_0|_\Gamma = u(0) = 0$, do not interfere. We then interpolate between (10.8.3.6), (10.8.3.7), or (10.8.5.3b) for $\theta = 0$ in the one hand; and (10.8.7.2) [read as $u \in H^{1,1}(\Sigma)$], (10.8.7.3) for $\theta = 1$ in the other hand. We thereby obtain (10.8.7.6), (10.8.7.7), and (10.8.7.8), as desired (see [Lions, Magenes, 1972, Theorem 9.6, p. 43; Theorem 12.2, p. 71; Theorem 12.3, p. 72]).

Next, applying the intermediate derivative theorem [Lions, Magenes, 1972, p. 15] between (10.8.7.6) (case $r = 0$) and (10.8.7.7) (case $r = 1$), and between (10.8.7.7) (case $r = 0$) and (10.8.7.8) (case $r = 1$), yields (10.8.7.9), and, respectively, (10.8.7.11), from which (10.8.7.10) and, respectively, (10.8.7.12) are obtained by specializing to $r = \theta$. Then, (10.8.7.6) and (10.8.7.7) yield $Lu \in C([0, T]; Y_\theta)$ by (10.8.6.2), whereas (10.8.7.10) and (10.8.7.12) yield $Lu \in H^\theta(0, T; Y)$, and then (10.8.7.13) follows via (10.8.6.11). The proof of Corollary 10.8.7.2 is complete. $\square$

Thus, $u \in \mathcal{U}^\theta[0, T] \subset H^{\theta,\theta}(\Sigma)$ produces Lu in (10.8.7.13) and assumption (H.2) = (10.1.14) is verified, except for the statement of uniformity in s, which is an immediate consequence, via a change of variable, of the formula (10.8.4.19) for L_s.

10.8.8 Verification of Assumption (H.3) = (10.1.15)

Verification of assumption (H.3) = (10.1.15) is based upon the following regularity result.

Theorem 10.8.10.1   *With reference to the operator L_s^* defined by (10.8.5.4), we have for $0 \le r \le 1$:*

$$(L_s^* v)(t) = B^* \int_t^T e^{A^*(\tau - t)} v(\tau)\, d\tau, \quad s \le t \le T$$

$$: continuous\; L_2\big(s, T; \mathcal{D}\big(A^{\frac{1}{4}+\frac{r}{2}}\big) \times \mathcal{D}\big(A^{-\frac{1}{4}+\frac{r}{2}}\big) = \mathcal{D}(A^{*r})\big)$$

$$\to H^{2r,r}(\Sigma_{sT}) \equiv \mathcal{U}^r[s, T], \tag{10.8.8.1}$$

uniformly in s (with the convention made below (10.8.6.2)).

Proof.   The proof of Theorem 10.8.8.1 will be given in Section 10.8.13 below. $\square$

Restricting (10.8.8.1) to $0 \le \theta = r < 1/2 + \delta, \theta \neq 1/2$, we obtain verification of assumption (H.3) = (10.1.15) by recalling (10.8.6.2).

10.8.9 Verification of Assumption (H.4) = (10.1.18) through (H.7) = (10.1.21)

Verification of Assumption (H.4) = (10.1.18)   For $x = [x_1, x_2] \in Y_\delta^+ = \mathcal{D}(A^{\frac{1}{2}+\frac{\delta}{2}}) \times \mathcal{D}(A^{\frac{\delta}{2}}) = \mathcal{D}(A^{*\frac{1}{2}+\delta})$ by (10.8.6.5), we compute starting from (10.8.4.15), and recalling $G_2^* A = -D^* A^{\frac{1}{2}}$ from (10.7.4.17):

$$B^* e^{A^* t} x = G_2^* A^{\frac{1}{2}} [AS(t)x_1 + C(t)x_2]$$

$$= G_2^* A \big[A^{\frac{1}{2}} S(t)x_1 + C(t) A^{-\frac{1}{2}} x_2\big] \tag{10.8.9.1}$$

$$\text{(by (10.7.4.17))} \qquad = -D^* \mathcal{A}^{\frac{1}{2}} \big[\mathcal{A}^{\frac{1}{2}} S(t) x_1 + C(t) \mathcal{A}^{-\frac{1}{2}} x_2 \big] \qquad (10.8.9.2)$$

$$= -D^* \mathcal{A}^{\frac{1}{8} - \frac{\delta}{4}} \big[\mathcal{A}^{\frac{1}{2}} S(t) \mathcal{A}^{\frac{3}{8} + \frac{\delta}{4}} x_1 + C(t) \mathcal{A}^{-\frac{1}{8} + \frac{\delta}{4}} x_2 \big] \quad (10.8.9.3)$$

$$\in C([0, T]; L_2(\Gamma)), \qquad (10.8.9.4)$$

where the desired regularity in (10.8.9.4) follows, since, recalling (10.7.4.18c) [or (10.7.9.5)] and (10.8.4.2),

$$\begin{cases} D^* \mathcal{A}^{\frac{1}{8} - \frac{\delta}{4}} \in \mathcal{L}(L_2(\Omega); L_2(\Gamma)); \\[4pt] t \to \mathcal{A}^{\frac{1}{2}} S(t), \, C(t) \text{ strongly continuous on } L_2(\Omega); & (10.8.9.5) \\[4pt] \mathcal{A}^{\frac{3}{8} + \frac{\delta}{4}} x_1 = \mathcal{A}^{-\frac{1}{8} - \frac{\delta}{4}} \mathcal{A}^{\frac{1}{2} + \frac{\delta}{2}} x_1 \in L_2(\Omega); \\[4pt] \mathcal{A}^{-\frac{1}{8} + \frac{\delta}{4}} x_2 = \mathcal{A}^{-\frac{1}{8} - \frac{\delta}{4}} \mathcal{A}^{\frac{\delta}{2}} x_2 \in L_2(\Omega), & (10.8.9.6) \end{cases}$$

a fortiori from the choice of x in Y_δ^+. Thus, (10.8.9.4) verifies assumption (H.4) = (10.1.18).

Verification of Assumption (H.5) = (10.1.19) With $Y_\delta^- = \mathcal{D}(\mathcal{A}^{\frac{1}{2} - \frac{\delta}{2}}) \times \mathcal{D}(\mathcal{A}^{-\frac{\delta}{2}})$ by (10.8.6.4) = $\mathcal{D}(\mathcal{A}^{\frac{1}{2} - \delta})$, we have that e^{At} in (10.8.4.12) is a s.c. group on Y_δ^- as well. The space $\mathcal{D}(A)$ in (10.8.4.8) is clearly dense in Y_δ^-.

Verification of Assumption (H.6) = (10.1.20) For $x = [x_1, x_2] \in Y_\delta^- = \mathcal{D}(A^{\frac{1}{2} - \delta}) = \mathcal{D}(\mathcal{A}^{\frac{1}{2} - \frac{\delta}{2}}) \times \mathcal{D}(\mathcal{A}^{-\frac{\delta}{2}})$, we obtain, via (10.8.4.8),

$$Ax = \begin{bmatrix} 0 & I \\ -\mathcal{A} & 0 \end{bmatrix} \begin{bmatrix} x_1 \\ x_2 \end{bmatrix} = \begin{bmatrix} x_2 \\ -\mathcal{A}x_1 \end{bmatrix} \in \mathcal{D}(\mathcal{A}^{-\frac{\delta}{2}}) \times \mathcal{D}(\mathcal{A}^{-\frac{1}{2} - \frac{\delta}{2}}) = [Y_\delta^+]',$$

$$(10.8.9.7)$$

recalling (10.8.6.8) [and A is actually an isomorphism from Y_δ^- onto $[Y_\delta^+]'$]. Equation (10.8.9.7) verifies assumption (H.6) = (10.1.20).

Remark 10.8.10.1 Returning to (10.8.4.11) we see via (10.8.6.4) that, in the present case,

$$A^{-1} B : \text{continuous } U \to Y_\delta^- \qquad (10.8.9.8)$$

[since $\mathcal{D}(\mathcal{A}^{\frac{5}{8} - \frac{\epsilon}{2}}) \subset \mathcal{D}(\mathcal{A}^{\frac{1}{2} - \frac{\delta}{2}})$], which is property (10.1.32). Thus, as remarked below (10.1.32), property (10.8.9.8), along with (H.5) = (10.1.19) and (H.6) = (10.1.20), which were already verified, reprove (H.4) = (10.1.18).

Verification of Assumption (H.7) = (10.1.21) Let $u \in U_\delta^- = H^{1 - 2\delta}(\Omega)$ by (10.8.6.3). Then via (10.7.4.18) and (10.7.4.4) we have

$$Du \in H^{\frac{3}{2} - 2\delta}(\Omega) \subset H^{\frac{1}{2} - 4\epsilon}(\Omega) = \mathcal{D}(\mathcal{A}^{\frac{1}{8} - \epsilon}), \qquad (10.8.9.9)$$

so that

$$\mathcal{A}^{\frac{1}{2}} Du \in \mathcal{D}(\mathcal{A}^{-\frac{3}{8} - \epsilon}) = \big[\mathcal{D}(\mathcal{A}^{\frac{3}{8} + \epsilon}) \big]' \subset \big[\mathcal{D}(\mathcal{A}^{\frac{1}{2} + \frac{\delta}{2}}) \big]', \qquad (10.8.9.10)$$

and thus by (10.8.4.9) and (10.7.4.17),

$$Bu = \begin{bmatrix} 0 \\ \mathcal{A}G_2 u \end{bmatrix} = \begin{bmatrix} 0 \\ -\mathcal{A}^{\frac{1}{2}}Du \end{bmatrix} \in [Y_\delta^+]', \qquad (10.8.9.11)$$

using in the last step inclusion (10.8.9.10) and the characterization (10.8.6.7) for $[Y_\delta^+]'$. Thus, (10.8.9.11) shows

$$B : \text{continuous } U_\delta^- \to [Y_\delta^+]', \qquad (10.8.9.12)$$

as desired, and assumption (H.7) = (10.1.21) is verified. [Property (10.1.22) may be likewise verified.]

10.8.10 Proof of Theorem 10.8.3.1

Interior Regularity Rewriting the assumptions via (10.7.4.8) and (10.7.4.6), we must show that

$$
\begin{cases}
\{\phi_0, \phi_1\} \in \mathcal{D}\big(\mathcal{A}^{\frac{3}{4}}\big) \times \mathcal{D}\big(\mathcal{A}^{\frac{1}{4}}\big), \quad f \in L_1\big(0, T; \mathcal{D}\big(\mathcal{A}^{\frac{1}{4}}\big)\big) \\
\Rightarrow \\
\{\phi_0, \phi_1\} \in C\big([0, T]; \mathcal{D}\big(\mathcal{A}^{\frac{3}{4}}\big) \times \mathcal{D}\big(\mathcal{A}^{\frac{1}{4}}\big)\big); \\
\phi_{tt} \begin{cases} \in C\big([0, T]; [\mathcal{D}(\mathcal{A}^{\frac{1}{4}})]'\big) & \text{if } f \equiv 0; \\ \in L_1\big(0, T; [\mathcal{D}(\mathcal{A}^{\frac{1}{4}})]'\big) & \text{if } f \not\equiv 0, \end{cases}
\end{cases}
\qquad (10.8.10.1)
$$

for the solution ϕ of problem (10.8.3.1) [i.e., of (10.8.4.1)] given explicitly by the formulas (10.8.4.3) and (10.8.4.4). Indeed, these formulas yield

$$\mathcal{A}^{\frac{3}{4}}\phi(t) = C(t)\mathcal{A}^{\frac{3}{4}}\phi_0 + \mathcal{A}^{\frac{1}{2}}S(t)\mathcal{A}^{\frac{1}{4}}\phi_1 + \int_0^t \mathcal{A}^{\frac{1}{2}}S(t-\tau)\mathcal{A}^{\frac{1}{4}}f(\tau)\,d\tau, \quad (10.8.10.2)$$

$$\mathcal{A}^{\frac{1}{4}}\phi_t(t) = -\mathcal{A}^{\frac{1}{2}}S(t)\mathcal{A}^{\frac{3}{4}}\phi_0 + C(t)\mathcal{A}^{\frac{1}{4}}\phi_1 + \int_0^t C(t-\tau)\mathcal{A}^{\frac{1}{4}}f(\tau)\,d\tau, \quad (10.8.10.3)$$

which show at once the implication (10.8.10.1) on $\{\phi, \phi_t\}$ via convolution, by use of (10.8.4.2).

Trace Regularity (10.8.3.5) We shall rely on Lemma 10.7.10.1, and Theorem 10.7.10.2 on the Kirchoff equation by specializing those results to the case $\rho = 0$ in Eqn. (10.7.1.1a) needed to obtain the Euler–Bernoulli equation (10.8.3.1).

Step 1. Lemma 10.8.10.1   *Let ϕ be a solution of Eqn. (10.8.3.1a) (with no boundary conditions imposed) for smooth data, say*

$$\{\phi_0, \phi_1, f\} \in \mathcal{D}(\mathcal{A}) \times \mathcal{D}\big(\mathcal{A}^{\frac{1}{2}}\big) \times L_1\big(0, T; \mathcal{D}\big(\mathcal{A}^{\frac{1}{2}}\big)\big). \qquad (10.8.10.4)$$

Then, the following identity holds true:

$$\int_\Sigma \frac{\partial(\Delta\phi)}{\partial\nu} h \cdot \nabla(\Delta\phi) \, d\Sigma + \int_\Sigma \frac{\partial\phi_t}{\partial\nu} h \cdot \nabla\phi_t \, d\Sigma + \int_\Sigma \frac{\partial\phi_t}{\partial\nu} \phi_t \, \mathrm{div}\, h \, d\Sigma$$

$$- \frac{1}{2} \int_\Sigma |\nabla\phi_t|^2 h \cdot \nu \, d\Sigma - \frac{1}{2} \int_\Sigma |\nabla(\Delta\phi)|^2 h \cdot \nu \, d\Sigma - \int_\Sigma \phi_t \Delta\phi_t h \cdot \nu \, d\Sigma$$

$$= \int_Q H\nabla(\Delta\phi) \cdot \nabla(\Delta\phi) \, dQ + \int_Q H\nabla\phi_t \cdot \nabla\phi_t \, dQ$$

$$+ \frac{1}{2} \int_Q \{|\nabla\phi_t|^2 - |\nabla(\Delta\phi)|^2\} \, \mathrm{div}\, h \, dQ + \int_Q \phi_t \nabla(\mathrm{div}\, h) \cdot \nabla\phi_t \, dQ$$

$$- [(\phi_t, h \cdot \nabla(\Delta\phi))_\Omega]_0^T + \int_Q f h \cdot \nabla(\Delta\phi) \, dQ, \tag{10.8.10.5}$$

where $H(x)$, $h(x)$, and $\nu(x)$ are the quantities defined in (10.7.10.5).

Proof. Set $\rho = 0$ in identity (10.7.10.4). The regularity (10.8.10.4) on the data guarantees the following interior regularity

$$\{\phi, \phi_t\} \in C([0, T]; \mathcal{D}(\mathcal{A}) \times \mathcal{D}(\mathcal{A}^{\frac{1}{2}})) \subset C([0, T]; H^4(\Omega) \times H^2(\Omega)) \tag{10.8.10.6}$$

by an argument similar to that used to establish (10.8.10.1) [compare with (10.7.10.6)]. Thus, the computations in the proof of Lemma 10.7.10.1 are justified. $\square$

Step 2 We next exploit also the BC (10.8.3.1c), make a special choice of the vector field $h \in C^2(\bar\Omega)$ as $h|_\Gamma = \nu$ on Γ, and obtain the counterpart of Theorem 10.7.10.2, yielding the desired trace regularity (10.8.3.5).

Theorem 10.8.10.1 *Let the data $\{\phi_0, \phi_1, f\}$ be as in (10.8.10.1). Let $h \in [C^2(\bar\Omega)]^n$ satisfy $h|_\Gamma = \nu$ on Γ. Then:*

(i) The solution ϕ of problem (10.8.3.1a,c) satisfies

$$\text{L.H.S. of } (10.8.10.5) = \frac{1}{2} \int_\Sigma \left\{ \left(\frac{\partial(\Delta\phi)}{\partial\nu}\right)^2 + \left(\frac{\partial\phi_t}{\partial\nu}\right)^2 \right\} d\Sigma, \tag{10.8.10.7}$$

and hence the identity

$$\frac{1}{2} \int_\Sigma \left\{ \left(\frac{\partial(\Delta\phi)}{\partial\nu}\right)^2 + \left(\frac{\partial\phi_t}{\partial\nu}\right)^2 \right\} d\Sigma$$

$$= \int_Q H\nabla(\Delta\phi) \cdot \nabla(\Delta\phi) \, dQ + \int_Q H\nabla\phi_t \cdot \nabla\phi_t \, dQ$$

$$+ \frac{1}{2} \int_Q \{|\nabla\phi_t|^2 - |\nabla(\Delta\phi)|^2\} \, \mathrm{div}\, h \, dQ + \int_Q \phi_t \nabla(\mathrm{div}\, h) \cdot \nabla\phi_t \, dQ$$

$$- [(\phi_t, h \cdot \nabla(\Delta\phi))_\Omega]_0^T + \int_Q f h \cdot \nabla(\Delta\phi) \, dQ. \tag{10.8.10.8}$$

(ii) Given any $0 < T < \infty$, there exists a constant $C_T > 0$, such that

$$\int_{\Sigma} \left[\left(\frac{\partial(\Delta\phi)}{\partial\nu} \right)^2 + \left(\frac{\partial\phi_t}{\partial\nu} \right)^2 \right] d\Sigma$$

$$\leq C_T \left\{ \|\{\phi_0, \phi_1\}\|^2_{\mathcal{D}(A^{\frac{3}{4}})\times\mathcal{D}(A^{\frac{1}{4}})} + \|f\|^2_{L_1(0,T;\mathcal{D}(A^{\frac{1}{4}}))} \right\}. \quad (10.8.10.9)$$

Proof.   The proof proceeds as in that of Theorem 10.7.10.2.   □

The proof of Theorem 10.8.3.1 is complete.

10.8.11 Proof of Theorem 10.8.3.2

We now provide the details, already contained in the preceding development, that the trace regularity for the homogeneous ϕ-problem (10.8.3.1),

$$\left. \begin{array}{l} \{\phi_0, \phi_1\} \in \mathcal{D}(A^{\frac{3}{4}}) \times \mathcal{D}(A^{\frac{1}{4}}) \\ f \equiv 0 \end{array} \right\} \Rightarrow \frac{\partial(\Delta\phi)}{\partial\nu} \in L_2(0, T; L_2(\Gamma)) \equiv L_2(\Sigma),$$

$$(10.8.11.1)$$

established in Theorem 10.8.3.1, implies by transposition the interior regularity

$$\left. \begin{array}{l} u \in L_2(0, T; L_2(\Gamma)) \equiv L_2(\Sigma) \\ w_0 = w_1 = 0 \end{array} \right\}$$

$$\Rightarrow \begin{bmatrix} w(t) \\ w_t(t) \end{bmatrix} = (Lu)(t) = \begin{bmatrix} A \int_0^t S(t-\tau)G_2u(\tau)\,d\tau \\ A \int_0^t C(t-\tau)G_2u(\tau)\,d\tau \end{bmatrix} \in C([0, T]; Y);$$

$$(10.8.11.2)$$

$$Y = \mathcal{D}(A^{\frac{1}{4}}) \times [\mathcal{D}(A^{\frac{1}{4}})]' = H_0^1(\Omega) \times H^{-1}(\Omega) \text{ (norm equivalence)} \quad (10.8.11.3)$$

(see (10.8.4.18) and (10.8.4.19)) for the nonhomogeneous w-problem (10.8.1.1).

Operator-Theoretic Proof of Theorem 10.8.3.2. Step 1   We have already seen in Section 10.8.4, Eqns. (10.8.4.16) and (10.8.4.17), that

$$-\frac{\partial\Delta\phi(t; \phi_0, \phi_1)}{\partial\nu} = G_2^* A^{\frac{3}{4}} \left[A^{\frac{1}{2}} S(t) A^{\frac{1}{4}} x_1 + C(t) A^{-\frac{1}{4}} x_2 \right] \quad (10.8.11.4)$$

$$= B^* e^{A^* t} x; \quad (10.8.11.5)$$

$$\phi_0 = A^{-1} x_2, \quad \phi_1 = x_1. \quad (10.8.11.6)$$

Thus, if we take

$$\{x_1, x_2\} \in \mathcal{D}(A^{\frac{1}{4}}) \times [\mathcal{D}(A^{\frac{1}{4}})]', \quad \text{hence } \{\phi_0, \phi_1\} \in \mathcal{D}(A^{\frac{3}{4}}) \times \mathcal{D}(A^{\frac{1}{4}}), \quad (10.8.11.7)$$

we then see, via (10.8.11.4) and (10.8.11.5), that implication (10.8.11.1) of Theorem 10.8.3.1 applies and yields the following operator-theoretic restatement, already noted in (10.8.5.1).

Theorem 10.8.10.1 *With reference to (10.8.11.4)–(10.8.11.6) and (10.8.11.1), we have with Y as in (10.8.11.3):*

$$B^* e^{A^* t} : continuous\ Y \to L_2(0, T; L_2(\Gamma)) \equiv L_2(\Sigma); \quad (10.8.11.8)$$

$$G_2^* A^{1+\frac{1}{4}} S(t),\ G_2^* A^{\frac{3}{4}} C(t) : continuous\ L_2(\Omega) \to L_2(\Sigma). \quad (10.8.11.9)$$

Step 2 The following result then stems from (10.8.11.8) of Theorem 10.8.11.1, by an application of Chapter 7, Theorem 7.2.1 with $p = 2$: With reference to the (operator) explicit formulas (10.8.4.19) for the solution $\{w(t), w_t(t)\}$ of the w-problem (10.8.1.1) with initial conditions $w_0 = w_1 = 0$ at $t = s = 0$, we have

$$
\begin{bmatrix} w(t) \\ w_t(t) \end{bmatrix} = (Lu)(t) = \int_0^t e^{A(t-\tau)} B u(\tau)\, d\tau
$$

$$
= \begin{bmatrix} A \int_0^t S(t-\tau) G_2 u(\tau)\, d\tau \\ A \int_0^t C(t-\tau) G_2 u(\tau)\, d\tau \end{bmatrix} \quad (10.8.11.10)
$$

$$: continuous\ L_2(\Sigma) \to C([0, T]; Y),$$

with Y as in (10.8.11.3). This establishes (10.8.11.2) and, in turn, yields the key part of Theorem 10.8.3.2 due to u. Then, from (10.8.1.1a),

$$w_{tt} = -\Delta^2 w \in C([0, T]; H^{-3}(\Omega)), \quad (10.8.11.11)$$

using the already established regularity $w \in C([0, T]; H_0^1(\Omega))$ and [Lions, Magenes, 1972, p. 85]. The regularity due to the initial conditions is straightforward using (10.8.4.12) and (10.8.4.2).

PDE Proof. A PDE version of the duality or transposition argument is as follows. Let w and ψ be solutions with regular data of the following two problems (recall (10.8.5.7)):

$$
\begin{cases}
w_{tt} + \Delta^2 w = F \\
w(0, \cdot) = w_0,\ w_t(0, \cdot) = w_1 \\
w|_\Sigma \equiv 0 \\
\Delta w|_\Sigma = u
\end{cases}
\qquad
\begin{cases}
\psi_{tt} + \Delta^2 \psi = f & \text{in } Q; \quad (10.8.11.12\text{a}) \\
\psi(T, \cdot) = \psi_t(T, \cdot) = 0 & \text{in } \Omega; \quad (10.8.11.12\text{b}) \\
\psi|_\Sigma \equiv 0 & \text{in } \Sigma; \quad (10.8.11.12\text{c}) \\
\Delta \psi|_\Sigma \equiv 0 & \text{in } \Sigma. \quad (10.8.11.12\text{d})
\end{cases}
$$

We know in advance, via Theorem 10.8.3.1 applied to the ψ-problem (which is reversed in time over the ϕ-problem (10.8.3.1)), that

$$f \in L_1(0, T; H_0^1(\Omega)) \Rightarrow \begin{cases} \{\psi, \psi_t\} \in C([0, T]; \mathcal{D}(\mathcal{A}^{\frac{3}{4}}) \times H_0^1(\Omega)), \text{ in particular,} \\ \psi(0) \in \mathcal{D}(\mathcal{A}^{\frac{3}{4}}), \quad \psi_t(0) \in H_0^1(\Omega); \\ \dfrac{\partial \Delta \psi}{\partial \nu} \in L_2(0, T; L_2(\Gamma)) \equiv L_2(\Sigma). \end{cases}$$

(10.8.11.13)

The argument below (10.7.11.14) for the Kirchoff equation, specialized to the present situation of the Euler–Bernoulli equation where $\rho = 0$, yields the counterpart of identity (10.7.11.19) in the form

$$\int_0^T \int_\Omega F \Delta \psi \, dQ - \int_0^T \int_\Omega (\Delta w) f \, dQ$$

$$= - \int_\Omega w_1 \Delta \psi(0) \, d\Omega + \int_\Omega w_0 \Delta \psi_t(0) \, d\Omega - \int_0^T \int_\Gamma u \frac{\partial \Delta \psi}{\partial \nu} \, d\Sigma. \quad (10.8.11.14)$$

Recalling the regularity results (10.8.11.13) for the ψ-problem, as well as (10.7.4.2), we see that in identity (10.8.11.14) we can take

$$-\Delta \psi = \mathcal{A}^{\frac{1}{2}} \psi \in C([0, T]; H_0^1(\Omega) = \mathcal{D}(\mathcal{A}^{\frac{1}{4}})), \quad \text{hence } F \in L_1(0, T; H^{-1}(\Omega));$$

(10.8.11.15)

$$\begin{cases} -\Delta w = \mathcal{A}^{\frac{1}{2}} w \in L_\infty(0, T; H^{-1}(\Omega)) = [\mathcal{D}(\mathcal{A}^{\frac{1}{4}})]', \\ \text{hence } w \in L_\infty(0, T; H_0^1(\Omega) = \mathcal{D}(\mathcal{A}^{\frac{1}{4}})), \\ \text{a regularity result that can then be boosted to } w \in C([0, T]; H_0^1(\Omega)) \end{cases}$$

(10.8.11.16)

by an approximating argument with smooth data in the w-problem;

$$\frac{\partial (\Delta \psi)}{\partial \nu} \in L_2(\Sigma), \quad \text{hence } u \in L_2(\Sigma); \quad (10.8.11.17)$$

$$-\Delta \psi(0) = \mathcal{A}^{\frac{1}{2}} \psi(0) \in \mathcal{D}(\mathcal{A}^{\frac{1}{4}}) = H_0^1(\Omega), \quad \text{hence } w_1 \in H^{-1}(\Omega); \quad (10.8.11.18)$$

$$-\Delta \psi_t(0) = \mathcal{A}^{\frac{1}{2}} \psi_t(0) \in H^{-1}(\Omega), \quad \text{hence } w_0 \in H_0^1(\Omega). \quad (10.8.11.19)$$

Having the regularity $w \in C([0, T]; H_0^1(\Omega))$ from (10.8.11.16), we obtain $w_{tt} \in C([0, T]; H^{-3}(\Omega))$ as in (10.8.11.11), and by interpolation we obtain $w_t \in C([0, T]; H^{-1}(\Omega))$ (see [Lions, Magenes, 1972, I, Theorem 12.3, p. 72]), as desired. □

10.8.12   Proof of Theorem 10.8.7.1

With reference to the nonhomogeneous w-problem (10.8.1.1), we assume

$$\{w_0, w_1\} \in [H^3(\Omega) \cap H_0^1(\Omega)] \times H_0^1(\Omega), \quad w_0|_\Gamma = 0; \quad \Delta w_0|_\Gamma = u(0) \in H^{\frac{1}{2}}(\Gamma);$$

(10.8.12.1)

$$u \in C([0, T]; H^{\frac{1}{2}}(\Gamma)) \cap H^1(0, T; L_2(\Gamma)), \quad (10.8.12.2)$$

and we must show that

$$\{w, w_t, w_{tt}\} \in C\big([0, T]; \big[H^3(\Omega) \cap H_0^1(\Omega)\big] \times H_0^1(\Omega) \times H^{-1}(\Omega)\big). \quad (10.8.12.3)$$

Operator-Theoretic Proof This is very similar to the case of Section 10.7.12. We return to the explicit solution formula (10.8.4.18), (10.8.4.19), via (10.8.4.12), that is,

$$w(t) = C(t)w_0 + S(t)w_1 + \mathcal{A} \int_0^t S(t - \tau)G_2 u(\tau)\,d\tau, \quad (10.8.12.4)$$

and integrate by parts the integral term with $u \in H^1(0, T; L_2(\Gamma))$, thus obtaining

$$w(t) = C(t)[w_0 - G_2 u(0)] + S(t)w_1 + G_2 u(t) - \int_0^t C(t - \tau)G_2 \dot{u}(\tau)\,d\tau.$$

$$(10.8.12.5)$$

As in (10.7.12.6), by the first compatibility condition in (10.8.12.1), and, respectively, $\dot{u} \in L_2(\Sigma)$, we have

$$w_0 = G_2 u(0), \quad \int_0^t C(t - \tau)G_2 \dot{u}(\tau)\,d\tau \in C\big([0, T]; \mathcal{D}(\mathcal{A}^{\frac{3}{4}})\big), \quad (10.8.12.6)$$

recalling the definition of G_2 in (10.7.4.15) and, respectively, the regularity (10.8.11.2) for w_t (second component). Moreover, with $w_1 \in \mathcal{D}(\mathcal{A}^{\frac{1}{4}})$, and $u \in C([0, T]; H^{\frac{1}{2}}(\Gamma))$, we obtain

$$S(t)w_1 \in C\big([0, T]; \mathcal{D}(\mathcal{A}^{\frac{3}{4}})\big), \quad G_2 u(t) \in C\big([0, T]; H^3(\Omega) \cap H_0^1(\Omega)\big),$$

$$(10.8.12.7)$$

by (10.8.4.2) and, respectively, as in (10.7.12.8), via (10.7.4.16). Thus, (10.8.12.6) and (10.8.12.7), used in (10.8.12.5) yield the desired regularity of w in (10.8.12.3). As to w_t, we differentiate (10.8.12.5), thus obtaining

$$w_t(t) = C(t)w_1 + G_2 \dot{u}(t) - G_2 \dot{u}(t) + \mathcal{A} \int_0^t S(t - \tau)G_2 \dot{u}(\tau)\,d\tau$$

$$\in C\big([0, T]; \mathcal{D}(\mathcal{A}^{\frac{1}{4}}) = H_0^1(\Omega)\big), \quad (10.8.12.8)$$

where the regularity of w_t in (10.8.12.8) – in which a cancellation of $G_2 \dot{u}(t)$ occurs – stems from (10.8.11.2) for the first component w and $\dot{u} \in L_2(\Sigma)$. The regularity for w_{tt} in (10.8.12.3) stems then from (10.8.1.1a).

PDE Proof. The proof is identical to that of Section 10.7.12, for problem (10.7.12.10), specialized with $\rho = 0$. □

10.8.13 Proof of Theorem 10.8.8.1

10.8.13.1 A Preliminary Trace Result

Theorem 10.8.10.1

(i) *With reference to the operators in (10.8.11.8) and (10.8.11.9), we have the fol-lowing regularity properties for $0 \le r \le 1$, which generalize the case $r = 0$ of*

Theorem 10.8.11.1:

$$G_2^* \mathcal{A}^{1+\frac{1}{4}} S(t), \quad G_2^* \mathcal{A}^{\frac{3}{4}} C(t) : continuous \ \mathcal{D}\big(\mathcal{A}^{\frac{r}{2}}\big) \to H^{2r,r}(\Sigma), \quad (10.8.13.1)$$

$$B^* e^{A^* t} : continuous \ \mathcal{D}(A^{*r}) = \mathcal{D}\big(\mathcal{A}^{\frac{1}{4}+\frac{r}{2}}\big) \times \mathcal{D}\big(\mathcal{A}^{-\frac{1}{4}+\frac{r}{2}}\big) \to H^{2r,r}(\Sigma),$$
$$(10.8.13.2)$$

$$H^{2r,r}(\Sigma) = L_2(0, T; H^{2r}(\Gamma)) \cap H^r(0, T; L_2(\Gamma)). \quad (10.8.13.3)$$

(ii) Equivalently, in PDE terms (see (10.8.10.9) or (10.8.3.5) for $r = 0$), we have

$$\{\phi_0, \phi_1\} \to \frac{\partial \Delta \phi(t; \phi_0, \phi_1)}{\partial \nu} : continuous,$$

$$\mathcal{D}(A^{1+r}) = \mathcal{D}\big(\mathcal{A}^{\frac{3}{4}+\frac{r}{2}}\big) \times \mathcal{D}\big(\mathcal{A}^{\frac{1}{4}+\frac{r}{2}}\big) \to H^{2r,r}(\Sigma), \quad (10.8.13.4)$$

where $\phi(t; \phi_0, \phi_1)$ is the solution of problem (10.8.3.1) with $f \equiv 0$, and where we further recall (10.8.11.4)–(10.8.11.6), to justify the stated equivalence between parts (i) and (ii).

Proof.

Case $r = 0$ The case $r = 0$ is contained in Theorem 10.8.11.1 for (i), and in Theorem 10.8.10.2, Eqn. (10.8.10.9) or Theorem 10.8.3.1, Eqn. (10.8.3.5) for (ii). The stated equivalence uses (10.8.11.4)–(10.8.11.6), as already noted. Thus, it is sufficient to prove the case $r = 1$ and interpolate to establish Theorem 10.8.13.1.

Case $r = 1$ We first show the time regularity:

$$G_2^* \mathcal{A}^{1+\frac{1}{4}} S(t)x, \quad G_2^* \mathcal{A}^{\frac{3}{4}} C(t)x \in H^1(0, T; L_2(\Gamma)) \quad \text{for } x \in \mathcal{D}\big(\mathcal{A}^{\frac{1}{2}}\big). \quad (10.8.13.5)$$

Indeed, with $x \in \mathcal{D}(\mathcal{A}^{\frac{1}{2}})$, we compute recalling from (10.8.4.1) that $-\mathcal{A}$ is the infinitesimal generator of $C(t)$:

$$\frac{d}{dt} G_2^* \mathcal{A}^{1+\frac{1}{4}} S(t)x = G_2^* \mathcal{A}^{\frac{3}{4}} C(t) \mathcal{A}^{\frac{1}{2}} x \in L_2(\Sigma), \quad (10.8.13.6)$$

$$\frac{d}{dt} G_2^* \mathcal{A}^{\frac{3}{4}} C(t)x = -G_2^* \mathcal{A}^{1+\frac{1}{4}} S(t) \mathcal{A}^{\frac{1}{2}} x \in L_2(\Sigma), \quad (10.8.13.7)$$

where the regularity in (10.8.13.6) and (10.8.13.7) is a direct application of (10.8.11.9) (case $r = 0$). Thus, (10.8.13.5) is proved. To show the space regularity,

$$G_2^* \mathcal{A}^{1+\frac{1}{4}} S(t)x, \quad G_2^* \mathcal{A}^{\frac{3}{4}} C(t)x \in L_2(0, T; H^2(\Gamma)), \quad x \in \mathcal{D}\big(\mathcal{A}^{\frac{1}{2}}\big), \quad (10.8.13.8)$$

we shall equivalently show, by (10.8.11.4)–(10.8.11.6), part (ii), that

$$\{\phi_0, \phi_1\} \to \frac{\partial \Delta \phi(t; \phi_0, \phi_1)}{\partial \nu}$$

$$: \ continuous \ \mathcal{D}(A^2) = \mathcal{D}\big(\mathcal{A}^{1+\frac{1}{4}}\big) \times \mathcal{D}\big(\mathcal{A}^{\frac{3}{4}}\big) \to L_2(0, T; H^2(\Gamma)),$$
$$(10.8.13.9)$$

with $\phi(t; \phi_0, \phi_1)$ a solution of problem (10.8.3.1) with $f \equiv 0$.

To this end, we introduce

$$
\begin{cases}
\mathcal{B} = \displaystyle\sum_{i,j}^{n} \beta_{i,j} \frac{\partial^2}{\partial x_i \partial x_j} = \text{second-order operator with (time independent)} \\[2mm]
\text{coefficients } \beta_{ij} \text{ smooth in } \bar{\Omega}, \text{ and such that } \mathcal{B} \text{ is } tangential \text{ to } \Gamma, \text{ that is,} \\[1mm]
\text{without transversal derivatives to } \Gamma \text{ when expressed in local coordinates.}
\end{cases}
$$
(10.8.13.10)

Accordingly, we consider the problem

$$
\begin{cases}
\phi_{tt} + \Delta^2 \phi = 0 & \text{in } Q; & (10.8.13.11a) \\
\phi(0, \cdot) = \phi_0, \quad \phi_t(0, \cdot) = \phi_1 & \text{in } \Omega; & (10.8.13.11b) \\
\phi|_\Sigma = \Delta\phi|_\Sigma = 0 & \text{in } \Sigma; & (10.8.13.11c)
\end{cases}
$$

$$
\{\phi_0, \phi_1\} \in \mathcal{D}\big(A^{1+\frac{1}{4}}\big) \times \mathcal{D}\big(A^{\frac{3}{4}}\big) \subset H^5(\Omega) \times H^3(\Omega),
$$
(10.8.13.12)

whose solution is

$$
\begin{cases}
\phi(t) = C(t)\phi_0 + S(t)\phi_1 \in C\big([0, T]; \mathcal{D}\big(A^{1+\frac{1}{4}}\big)\big), & (10.8.13.13) \\[2mm]
\phi_t(t) = -AS(t)\phi_0 + C(t)\phi_1 \in C\big([0, T]; \mathcal{D}\big(A^{\frac{3}{4}}\big)\big), & (10.8.13.14) \\[2mm]
\phi_{tt}(t) = -AC(t)\phi_0 - AS(t)\phi_1 \in C\big([0, T]; \mathcal{D}\big(A^{\frac{1}{4}}\big)\big). & (10.8.13.15)
\end{cases}
$$

We then introduce a new variable

$$
z = \mathcal{B}\phi,
$$
(10.8.13.16)

which, therefore has a priori regularity from (10.8.13.10) and (10.8.13.12)–(10.8.13.15), given by

$$
\begin{cases}
z \in C\big([0, T]; H^3(\Omega) \cap H_0^1(\Omega)\big), & (10.8.13.17) \\[2mm]
z_t \in C\big([0, T]; H_0^1(\Omega)\big), & (10.8.13.18) \\[2mm]
z_{tt} \in C([0, T]; H^{-1}(\Omega)). & (10.8.13.19)
\end{cases}
$$

Then, proving (10.8.13.9) is equivalent to showing that

$$
\frac{\partial(\Delta\mathcal{B}\phi)}{\partial\nu}\bigg|_\Gamma = \frac{\partial(\Delta z)}{\partial\nu}\bigg|_\Gamma \in L_2(\Sigma).
$$
(10.8.13.20)

The variable z satisfies the problem

$$
\begin{cases}
z_{tt} + \Delta^2 z = -[\mathcal{B}, \Delta^2]\phi, & (10.8.13.21a) \\
z|_\Sigma \equiv 0, & (10.8.13.21b) \\
\Delta z|_\Sigma = -[\mathcal{B}, \Delta]\phi|_\Gamma, & (10.8.13.21c)
\end{cases}
$$

as one readily sees from (10.8.13.10), (10.8.13.11), and (10.8.13.16). Compare with

(10.7.13.23). Since the commutators

$$[\mathcal{B}, \Delta^2] = \text{operator of order } 2 + 4 - 1 = 5, \qquad (10.8.13.22)$$

$$[\mathcal{B}, \Delta] = \text{operator of order } 2 + 2 - 1 = 3, \qquad (10.8.13.23)$$

we see via (10.8.13.22) and the regularity (10.8.13.13) for $\phi \in C([0, T]; H^5(\Omega))$ that the right-hand side term k in (10.8.13.21a) satisfies

$$k \equiv -[\mathcal{B}, \Delta^2]\phi \in C([0, T]; L_2(\Omega)), \qquad (10.8.13.24)$$

since $\mathcal{D}(\mathcal{A}^{1+\frac{1}{4}}) \subset H^5(\Omega)$. Similarly, via (10.8.13.23), (10.8.13.13), and trace theory, we see that the boundary term g in (10.8.13.21c) satisfies

$$g \equiv -[\mathcal{B}, \Delta]\phi|_\Gamma \in C([0, T]; H^{\frac{3}{2}}(\Gamma)). \qquad (10.8.13.25)$$

Thus, by (10.8.13.21), (10.8.13.24), and (10.8.13.25), we see that the z-problem becomes (compare with (10.7.13.27))

$$\begin{cases} z_{tt} + \Delta^2 z = k \in C([0, T]; L_2(\Omega)), & (10.8.13.26a) \\ z|_\Sigma = 0, & (10.8.13.26b) \\ \Delta z|_\Sigma = g \in C([0, T]; H^{\frac{3}{2}}(\Gamma)), & (10.8.13.26c) \end{cases}$$

with a priori interior regularity given by (10.8.13.17)–(10.8.13.19). We now return to the basic identity (10.8.10.5). Because of the a priori interior regularity (10.8.13.17), (10.8.13.18) for $\{z, z_t\}$ and that of k in (10.8.13.24), the right-hand side (R.H.S.) of identity (10.8.10.5) [with ϕ, f there replaced by z, k now, respectively] is well-defined. Thus, the left-hand side of identity (10.8.10.5) is well-defined. Proceeding as in (10.7.13.28) through (10.7.13.32) in the Kirchoff equation case [which is now specialized to the Euler–Bernoulli case $\rho = 0$], we obtain that the left-hand side (L.H.S.) of identity (10.8.10.5) can be rewritten in the new variable z, once the BC (10.8.13.26b,c) are taken into account, as

L.H.S. of (10.8.10.5)

$$= \frac{1}{2} \int_\Sigma \left(\frac{\partial(\Delta z)}{\partial \nu} \right)^2 d\Sigma + \frac{1}{2} \int_\Sigma \left(\frac{\partial z_t}{\partial \nu} \right)^2 d\Sigma - \frac{1}{2} \int_\Sigma \left(\frac{\partial g}{\partial \tau} \right)^2 d\Sigma$$

$$= \text{well defined by R.H.S. of (10.8.10.5)}, \qquad (10.8.13.27)$$

that is, Eqn. (10.7.13.32) with $\rho = 0$. But the tangential derivative $\partial g / \partial \tau \in C([0, T];$ $H^{\frac{1}{2}}(\Gamma))$ by (10.8.13.25), so that the last integral term in (10.8.13.27) is well-defined. We conclude from (10.8.13.27) that

$$\int_\Sigma \left[\left(\frac{\partial(\Delta z)}{\partial \nu} \right)^2 + \left(\frac{\partial z_t}{\partial \nu} \right)^2 \right] d\Sigma \leq C_T \| \{\phi_0, \phi_1\} \|^2_{\mathcal{D}(\mathcal{A}^{1+\frac{1}{4}}) \times \mathcal{D}(\mathcal{A}^{\frac{3}{4}})}, \qquad (10.8.13.28)$$

and thus (10.8.13.20) is established, as desired. The proof of Theorem 10.8.13.1 is complete. $\quad \square$

Remark 10.8.13.1   A remark similar to Remark 10.7.13.1 applies in the present Euler–Bernoulli equation: It is the a priori interior regularity (10.8.13.17), (10.8.13.18) for $\{z, z_t\}$ that permits one to obtain the desired trace regularity (10.8.13.28) from identity (10.8.10.5), even though k in (10.8.13.26a) is only in $L_2(\Omega)$ by (10.8.13.24) (compare with f in $H_0^1(\Omega)$ in (10.8.3.1), (10.8.3.2)).

10.8.13.2 Completion of the Proof of Theorem 10.8.8.1

Space Regularity   To show the space regularity

$$(L^*v)(t) = \int_t^T B^* e^{A^*(\tau - t)} v(\tau)\, d\tau \tag{10.8.13.29}$$

$$: \text{ continuous } L_2\big(0, T; \mathcal{D}(A^{*r}) = \mathcal{D}\big(\mathcal{A}^{\frac{1}{4}+\frac{r}{2}}\big) \times \mathcal{D}\big(\mathcal{A}^{-\frac{1}{4}+\frac{r}{2}}\big)\big)$$

$$\rightarrow L_2(0, T; H^{2r}(\Gamma)), \quad 0 \le r \le 1, \tag{10.8.13.30}$$

we simply invoke Theorem 7.2.1, Eqn. (7.2.2) with $X^* = \mathcal{D}(A^{*r})$, $U^* = H^{2r}(\Gamma)$, and $q = 2$, which is legal by virtue of the regularity (10.8.13.2) of Theorem 10.8.13.1.

Time Regularity   It suffices to show the case $r = 1$, since the case $r = 0$ is contained in (10.8.13.30), or in (10.8.5.6), and then interpolate. Thus, differentiating (10.8.13.29) in t for

$$v \in L_2(0, T; \mathcal{D}(A^*)), \quad \text{or } A^*v \in L_2(0, T; Y) \tag{10.8.13.31}$$

[recall that A is skew-adjoint, $A^* = -A$, on $Y = \mathcal{D}(\mathcal{A}^{\frac{1}{4}}) \times [\mathcal{D}(\mathcal{A}^{\frac{1}{4}})]']$], yields, as in (10.7.13.36),

$$\frac{d(L^*v)}{dt}(t) = -B^*v(t) - \int_t^T B^* e^{A^*(\tau - t)} A^* v(\tau)\, d\tau \in L_2(0, T; L_2(\Gamma)), \tag{10.8.13.32}$$

as desired, by (10.8.13.31) and (10.8.13.30) with $r = 0$ and $B^* A^{*-1} \in \mathcal{L}(Y; U)$ in (10.1.15a), with U and Y as in (10.8.4.7). Then (10.8.13.32) shows

$$L^* : \text{continuous } L_2(0, T; \mathcal{D}(A^*)) \rightarrow H^1(0, T; L_2(\Gamma)), \tag{10.8.13.33}$$

as required. The proof of Theorem 10.8.8.1 is complete, except for noticing that uniformity in s is obtained by a change of variable on formula (10.8.13.29), as usual.

Final Remark   It is surely possible to provide additional illustrations of the abstract theory comprising Theorems 10.2.1, 10.2.2, and 10.2.3 of the present chapter also in the case of Kirchoff equations (10.7.1.1), or Euler–Bernoulli equations (10.8.1.1), under different boundary conditions.

10.9 Application: Schrödinger Equations with Dirichlet Boundary Control.
Regularity Theory

In the present section we show that the abstract theory of Theorems 10.2.1, 10.2.2, and 10.2.3 is applicable to the class of Schrödinger equations defined on a smooth bounded domain of arbitrary dimension, with control function acting in the Dirichlet boundary conditions. The observation operator satisfies a "minimal" smoothing property as in $(H.8) = (10.1.23)$.

10.9.1 Problem Formulation

The Dynamics   Let Ω be an open bounded domain in R^n with sufficiently smooth boundary Γ, say of class C^1. On Ω we consider the Schrödinger equation

$$\begin{cases} y_t = -i\,\Delta y + F(y) & \text{in } (0, T] \times \Omega \equiv Q, & (10.9.1.1a) \\ y(0, \cdot) = y_0 & \text{in } \Omega, & (10.9.1.1b) \\ y|_\Sigma = u & \text{in } (0, T] \times \Gamma \equiv \Sigma, & (10.9.1.1c) \end{cases}$$

with control function $u \in L_2(0, T; L_2(\Gamma)) \equiv L_2(\Sigma)$ acting in the Dirichlet boundary conditions, where $F(y)$ is a first-order differential operator in the space variables $x_1, x_2, \ldots, x_n$ with $L_\infty(\bar{Q})$-coefficients (as in (10.9.3.1d) below).

The Optimal Control Problem on $[s, T]$ Consistently with the (optimal) regularity theory for problem (10.9.1.1) presented in Theorem 10.9.3.2 below, the cost functional we seek to minimize over all $u \in L_2(s, T; L_2(\Gamma)) = L_2(\Sigma_{sT})$ is taken to be

$$J(u, y) = \int_s^T \left\{ \|Ry(t)\|^2_{H^{-1}(\Omega)} + \|u(t)\|^2_{L_2(\Gamma)} \right\} dt, \qquad (10.9.1.2)$$

with initial datum $y_0 \in H^{-1}(\Omega)$ at the initial time $s \geq 0$, where the observation operator $R \in \mathcal{L}(H^{-1}(\Omega))$ will be further specified below in (10.9.2.3).

10.9.2 Main Results

As a specialization to problem (10.9.1.1), (10.9.1.2) of the abstract theory presented in Theorems 10.2.1, 10.2.2, and 10.2.3 of this chapter, in the present section we establish the following results.

Theorem 10.9.10.1   *(a) With the observation operator R in (10.9.1.2) only assumed to satisfy*

$$R \in \mathcal{L}(H^{-1}(\Omega)), \qquad (10.9.2.1)$$

Theorem 9.2.1 of Chapter 9 applies to the optimal control problem (10.9.1.1), (10.9.1.2) with

$$U = L_2(\Gamma), \quad Y = H^{-1}(\Omega), \qquad (10.9.2.2)$$

and yields a unique optimal pair $\{u^0(\,\cdot\,, s; y_0), y^0(\,\cdot\,, s; y_0)\}$, $y_0 \in Y$, *satisfying, in particular, the pointwise feedback synthesis (9.2.17) and the optimal cost relation (9.2.18) of Chapter 9, as well as the other properties listed there.*

(b) Assume, in addition to (10.9.2.1), that R satisfies the smoothing assumption:

$$R^*R : continuous \ H^{-2\delta}(\Omega) \to H^{2\delta}(\Omega), \tag{10.9.2.3}$$

with R^* the adjoint of R in Y, for a fixed arbitrarily small $\delta > 0$. Then, Theorems 10.2.1, 10.2.2, and 10.2.3 of the present chapter hold true, with

$$Y_\delta^- \equiv H^{-2\delta}(\Omega), \quad Y_\delta^+ = H^{2\delta}(\Omega), \quad U_\delta^- \equiv H^{1-2\delta}(\Gamma), \quad U_\delta^+ = H^{1+2\delta}(\Gamma) \tag{10.9.2.4}$$

(see Section 10.9.6). In particular, explicitly:

(b_1) (Regularity of optimal pair). The optimal pair satisfies the following regularity properties: For $y_0 \in H^{-2\delta}(\Omega) = Y_\delta^-$, we have

$$y^0(\,\cdot\,, s; y_0) \in C([s, T]; H^{-2\delta}(\Omega)) \cap H^{\frac{1}{2}-\delta}(s, T; H^{-1}(\Omega)), \quad (10.9.2.5)$$

$$u^0(\,\cdot\,, s; y_0) \in L_2(s, T; H^{1+2\delta}(\Gamma)) \cap H^{\frac{1}{2}+\delta}(s, T; L_2(\Gamma)) \equiv H^{1+2\delta,\frac{1}{2}+\delta}(\Sigma_{sT}), \tag{10.9.2.6}$$

$\Sigma_{sT} = [s, T] \times \Gamma$, *a fortiori [Lions, Magenes, Theorem 3.1, pp. 19 and 23],*

$$u^0(\,\cdot\,, s; y_0) \in C([s, T]; H^{2\delta}(\Gamma)). \tag{10.9.2.7}$$

All of the above results hold true uniformly in s, that is, with norms that may be made independent of s.

(b_s) (Gain operator) The gain operator $B^*P(t)$, with $P(t)$ defined by (9.2.14) of Chapter 9, satisfies the following regularity property (see (10.2.4)):

$$B^*P(t)x = -i\frac{\partial}{\partial \nu}(\mathcal{A}^{-1}P(t)x) \tag{10.9.2.8a}$$

$$: \ continuous \ H^{-2\delta}(\Omega) \to C([0, T]; L_2(\Gamma)), \tag{10.9.2.8b}$$

where $\mathcal{A}$ is the realization of $-\Delta$ with Dirichlet boundary condition defined in (10.9.4.2) and (10.9.4.8) [see (10.9.4.13) for B^*].

(b_3) (Differential Riccati equation) We have that $P(t)$ satisfies the following DRE:

$$\begin{cases} \dfrac{d}{dt}(P(t)x, y)_Y = -(Rx, Ry)_Y - (P(t)x, i\mathcal{A}y)_Y - (iP(t)\mathcal{A}x, y)_Y \\ \qquad\qquad + \left(i\dfrac{\partial}{\partial \nu}\mathcal{A}^{-1}P(t)x, i\dfrac{\partial}{\partial \nu}\mathcal{A}^{-1}P(t)y\right)_{L_2(\Gamma)}, \\ P(T) = 0, \end{cases} \tag{10.9.2.9}$$

with $Y = H^{-1}(\Omega)$, $\forall x, y \in Y_\delta^- = H^{-2\delta}(\Omega)$.

10.9.3 Regularity Theory for Problem (10.9.1.1) with $u \in L_2(\Sigma)$

The following well-posedness Theorem 10.9.3.2 provides the critical regularity result that justifies as natural the selection of the cost functional J in (10.9.1.2) and will permit us to verify assumption (H.1) = (10.1.6) below. As in preceding examples, we associate with problem (10.9.1.1) the following boundary-homogeneous version (which would be the corresponding adjoint problem (10.9.5.8), if the initial conditions were given at $t = T$, an inessential modification since the problem is time reversible):

$$\begin{cases} i\phi_t = \Delta\phi + F(\phi) + f & \text{in } Q, & (10.9.3.1a) \\ \phi(0, \cdot) = \phi_0 & \text{in } \Omega, & (10.9.3.1b) \\ \phi|_\Sigma \equiv 0 & \text{in } \Sigma, & (10.9.3.1c) \end{cases}$$

where $F(\phi)$ is a (linear) first-order differential operator in the space variables $x_1, \ldots, x_n$, with $L_\infty(\bar{Q})$-coefficients, thus satisfying the pointwise bound

$$|F(\phi)| \leq C_T[|\nabla\phi|^2 + |\phi|^2], \quad \forall, x, t \in \bar{Q}. \tag{10.9.3.1d}$$

The following key results are taken from Lasiecka and Triggiani [1991(b)].

Theorem 10.9.10.1   *With reference to (10.9.3.1), assume*

$$\phi_0 \in H_0^1(\Omega), \quad f \in L_1(0, T; H_0^1(\Omega)). \tag{10.9.3.2}$$

Then, the unique solution of (10.9.3.1) satisfies continuously:

$$\phi \in C([0, T]; H_0^1(\Omega)); \quad \phi_t \in L_1(0, T; H^{-1}(\Omega)),$$

$$\text{or } \phi_t \in C([0, T]; H^{-1}(\Omega)) \text{ if } f \equiv 0; \tag{10.9.3.3}$$

and

$$\frac{\partial\phi}{\partial\nu} \in L_2(0, T; L_2(\Gamma)) \equiv L_2(\Sigma). \tag{10.9.3.4}$$

Proof.   The proof of [(10.9.3.4) of] Theorem 10.9.3.1 is by PDE energy methods and will be given in Section 10.9.10 below. $\square$

Remark 10.9.10.1   The trace regularity (10.9.3.4) is, of course, the key result of Theorem 10.9.3.1. It does not follow from the optimal interior regularity (10.9.3.3); indeed, (10.9.3.4) is "1/2 higher" in Sobolev space regularity on Ω over *a formal* application of trace theory to ϕ in (10.9.3.3).

Theorem 10.9.10.1   *With reference to the nonhomogeneous problem (10.9.1.1), assume*

$$y_0 \in H^{-1}(\Omega), \quad u \in L_2(\Sigma). \tag{10.9.3.5}$$

Then, the unique solution of (10.9.1.1) satisfies

$$y \in C([0, T]; H^{-1}(\Omega)), \quad y_t \in C([0, T]; H^{-3}(\Omega)), \tag{10.9.3.6a}$$

continuously; that is, more precisely, there exists $C_T > 0$ such that

$$\|y\|^2_{C([0,T];\,H^{-1}(\Omega))} + \|y_t\|^2_{C([0,T];\,H^{-3}(\Omega))} \le C_T\{\|y_0\|^2_{H^{-1}(\Omega)} + \|u\|^2_{L_2(\Sigma)}\}. \quad (10.9.3.6b)$$

Proof. The proof of Theorem 10.9.3.2 is a consequence of Theorem 10.9.3.1 (by transposition) and will be given in Section 10.9.11 below. □

10.9.4 Abstract Setting for Problem (10.9.1.1)

We follow Lasiecka and Triggiani [1991b, c]. To put problem (10.9.1.1), (10.9.1.2) into the abstract model (10.1.1), (10.1.3), we introduce the following operators and spaces and compare with Chapter 3, Section 3.1, or with Section 10.5.4:

(i) Problem (10.9.1.1) can be written abstractly as in (10.1.1), that is, as

$$\dot{y} = Ay + Bu, \quad \text{on } [\mathcal{D}(A^*)]' = [\mathcal{D}(\mathcal{A})]', \quad y(0) = y_0 \in Y; \quad (10.9.4.1)$$

$$A = i\mathcal{A}; \quad \mathcal{A}h = -\Delta h; \quad \mathcal{D}(A) = \mathcal{D}(\mathcal{A}) = H^2(\Omega) \cap H^1_0(\Omega), \quad (10.9.4.2)$$

where $\mathcal{A}$ and A will be freely extended below, preserving the same symbols, as operators: $L_2(\Omega) \to [\mathcal{D}(\mathcal{A})]'$, where $[\mathcal{D}(\mathcal{A})]'$ denotes duality with respect to the pivot space $L_2(\Omega)$;

$$Y \equiv \left[\mathcal{D}\left(\mathcal{A}^{\frac{1}{2}}\right)\right]' = H^{-1}(\Omega); \quad (x, y)_Y = (\mathcal{A}^{-1}x, y)_{L_2(\Omega)}; \quad U \equiv L_2(\Gamma), \quad (10.9.4.3)$$

since

$$\mathcal{D}\left(\mathcal{A}^{\frac{1}{2}}\right) = H^1_0(\Omega); \quad (x, y)_{\mathcal{D}(\mathcal{A}^{\frac{1}{2}})} = (\mathcal{A}x, y)_{L_2(\Omega)}, \quad (10.9.4.4)$$

and more generally [recall, say, (10.5.4.3) and (10.5.4.4) of Section 10.5.4]

$$\mathcal{D}(\mathcal{A}^\theta) = \begin{cases} H^{2\theta}_0(\Omega), & 2\theta < \dfrac{3}{2}, \ \theta \neq \dfrac{1}{4}, & (10.9.4.5) \\[2ex] H^{2\theta}(\Omega), & 2\theta < \dfrac{1}{2}, & (10.9.4.6) \end{cases}$$

with equivalent norms;

$$B = -i\mathcal{A}D, \quad h = Dv \Longleftrightarrow \{\Delta h = 0 \text{ in } \Omega; \ h|_\Gamma = v\}, \quad (10.9.4.7)$$

where D is the Dirichlet map satisfying the regularity properties as in (10.5.4.10) and (10.5.4.11), as well as (10.5.4.12):

$$D : \text{continuous } H^s(\Gamma) \to H^{s+\frac{1}{2}}(\Omega); \text{ in particular} \quad (10.9.4.8)$$

$$: \text{continuous } L_2(\Gamma) \to H^{\frac{1}{2}}(\Omega) \subset H^{\frac{1}{2}-2\epsilon}(\Omega) \equiv \mathcal{D}\left(\mathcal{A}^{\frac{1}{4}-\epsilon}\right), \quad \forall \epsilon > 0; \quad (10.9.4.9)$$

$$D^*\mathcal{A}h = -\frac{\partial h}{\partial \nu}, \quad h \in \mathcal{D}(\mathcal{A}); \quad (Du, h)_{L_2(\Omega)} = (u, D^*h)_{L_2(\Gamma)}. \quad (10.9.4.10)$$

From (10.9.4.2) and (10.9.4.7), we obtain a fortiori, as required,

$$A^{-1}B = -D : \text{continuous } U \equiv L_2(\Gamma) \to \mathcal{D}(A^{\frac{1}{4}-\epsilon}) \subset Y; \quad (10.9.4.11\text{a})$$

$$B \in \mathcal{L}(U; [\mathcal{D}(A^*)]'), \quad A^* \text{ being the } Y\text{-adjoint of } A, \quad (10.9.4.11\text{b})$$

where the duality in (10.9.4.11b) is with respect to Y as a pivot space, and

$$Y \equiv [\mathcal{D}(A^{\frac{1}{2}})]' \supset \mathcal{D}(A) = \mathcal{D}(A^{\frac{1}{2}}) \to Y, \quad (10.9.4.11\text{c})$$

where the duality in (10.9.4.11c) is instead with respect to $L_2(\Omega)$ as a pivot space. The adjoint $B^* \in \mathcal{L}(\mathcal{D}(A^*); U)$ is computed via (10.9.4.7), (10.9.4.3), and (10.9.4.10) as

$$
\begin{aligned}
(Bu, v)_{Y \equiv [\mathcal{D}(A^{\frac{1}{2}})]'} &= -i(A^{-1}ADu, v)_{L_2(\Omega)} \\
&= (u, iD^*v)_{L_2(\Gamma)} = (u, B^*v)_{U=L_2(\Gamma)}, \quad (10.9.4.12)
\end{aligned}
$$

$$B^*v = iD^*v = iD^*AA^{-1}v = -i\frac{\partial(A^{-1}v)}{\partial v}. \quad (10.9.4.13)$$

(ii) The operator A is skew-adjoint [on $L_2(\Omega)$, but, more importantly, also] on the space Y defined by (10.9.4.3), as stated in (10.9.4.11b): $A^* = -A = -iA$, where $*$ refers either to the $L_2(\Omega)$-norm or to the Y-norm. Thus, A generates a s.c. unitary group e^{At} on Y. By (10.9.4.2), problem (10.9.3.1) can then be written abstractly as [here $*$ refers to $L_2(\Omega)$]:

$$i\phi_t = A\phi + f, \quad \text{or } \phi_t = A^*\phi - if; \quad \phi(0) = \phi_0. \quad (10.9.4.14)$$

The solution of (10.9.4.14), that is, of (10.9.3.1), is given explicitly by

$$\phi(t) = e^{A^*t}\phi_0 - i\int_0^t e^{A^*(t-\tau)}f(\tau)\,d\tau \quad (10.9.4.15)$$

in appropriate functions spaces, depending on $\{\phi_0, f\}$. By (10.9.4.13), we obtain with $y \in Y$:

$$B^*e^{A^*t}y = -i\frac{\partial}{\partial v}(A^{-1}e^{-iAt}y) = -i\frac{\partial}{\partial v}(e^{A^*t}A^{-1}y) = -i\frac{\partial}{\partial v}\phi(t; \phi_0),$$
$$\quad (10.9.4.16)$$

where $\phi(t; \phi_0)$ is the solution of problem (10.9.4.14), that is, of problem (10.9.3.1), with $f \equiv 0$ and

$$\phi_0 = A^{-1}y \in \mathcal{D}(A^{\frac{1}{2}}) = H_0^1(\Omega), \quad \text{for } y \in Y = [\mathcal{D}(A^{\frac{1}{2}})]'. \quad (10.9.4.17)$$

The solution to problem (10.9.1.1), that is, to Eqn. (10.9.4.1) is, by (10.9.4.2) and (10.9.4.7):

$$y(t) = e^{At}y_0 + \int_0^t e^{A(t-\tau)}Bu(\tau)\,d\tau \quad (10.9.4.18)$$

$$= e^{At}y_0 - iA\int_0^t e^{iA(t-\tau)}Du(\tau)\,d\tau. \quad (10.9.4.19)$$

10.9.5 Verification of Assumption (H.1) = (10.1.6)

By (10.9.4.16) and (10.9.4.17), we see that the required assumption (H.1) = (10.1.6),

$$\int_0^T \| B^* e^{A^* t} y \|_U^2 \, dt \le C_T \| y \|_Y^2, \quad y \in Y, \tag{10.9.5.1}$$

is satisfied if and only if the solution ϕ of problem (10.9.3.1) with $f \equiv 0$ satisfies

$$\int_0^T \int_\Gamma \left(\frac{\partial \phi}{\partial \nu} \right)^2 d\Sigma \le C_T \| \phi_0 \|_{H_0^1(\Omega)}^2, \tag{10.9.5.2}$$

which is precisely the trace regularity result guaranteed by Theorem 10.9.3.1, Eqn. (10.9.3.4). Then, according to Chapter 7, Theorem 7.2.1, estimate (10.9.5.1) that is, (10.9.5.2) is, in turn, equivalent to the following property that, with $y_0 = 0$ in (10.9.1.1b):

$$(L_s u)(t) = \int_s^t e^{A(t-\tau)} Bu(\tau) \, d\tau = y(t; 0) \tag{10.9.5.3a}$$

$$\text{: continuous } L_2(s, T; L_2(\Gamma)) \to C([s, T]; H^{-1}(\Omega)), \tag{10.9.5.3b}$$

uniformly in s, via a change of variable. Moreover, recalling (10.9.4.16), or (10.9.4.13), or (10.9.4.2), we obtain from (10.9.5.3):

$$(L_s^* v)(t) = B^* \int_t^T e^{A^*(\tau - t)} v(\tau) \, d\tau, \quad s \le t \le T \tag{10.9.5.4}$$

$$= -B^* \int_T^t e^{-A^*(t-\tau)} v(\tau) \, d\tau$$

$$\text{(by (10.9.4.13))} \qquad = -i D^* \int_T^t e^{i \mathcal{A}(t-\tau)} v(\tau) \, d\tau$$

$$= i(-D^* \mathcal{A}) \int_T^t e^{i \mathcal{A}(t-\tau)} \mathcal{A}^{-1} v(\tau) \, d\tau \tag{10.9.5.5}$$

$$\text{(by (10.9.4.10))} \qquad = i \frac{\partial}{\partial \nu} \psi(t; f) \tag{10.9.5.6}$$

$$\text{: continuous } v \in L_1(0, T; H^{-1}(\Omega)) \to L_2(s, T; L_2(\Gamma)),$$

$$\tag{10.9.5.7}$$

where $\psi(t; f)$ is the solution of the (dual) problem

$$\begin{cases} \psi_t = -i \Delta \psi + f & \text{in } Q; & (10.9.5.8a) \\ \psi(T, \cdot) = 0 & \text{in } \Omega, \text{ or } \psi_t = i \mathcal{A} \psi + f, & (10.9.5.8b) \\ \psi|_\Sigma \equiv 0 & \text{in } \Sigma, & (10.9.5.8c) \end{cases}$$

given explicitly by

$$\psi(t) = \int_T^t e^{i \mathcal{A}(t-\tau)} f(\tau) \, d\tau; \quad f(t) = \mathcal{A}^{-1} v(t) \in L_1\big(0, T; H_0^1(\Omega)\big). \tag{10.9.5.9}$$

Notice that the regularity (10.9.5.7) for the normal trace of $\psi(t; f)$, where f has the regularity as in (10.9.5.9), and ψ solves (10.9.5.8), is precisely conclusion (10.9.3.4) for the time-reversed problem ψ in (10.9.3.1) with initial data at $t = 0$, rather than at $t = T$ as for ψ. Assumption (H.1) = (10.1.6) is verified.

10.9.6 Selection of the Spaces U_θ and Y_θ in (10.1.10)

We choose among infinitely many possibilities:

$$U_\theta \equiv H^{2\theta}(\Gamma); \quad U_0 = U = L_2(\Gamma), \quad 0 \leq \theta \leq \frac{1}{2} + \delta; \quad \theta \neq \frac{1}{2}; \quad (10.9.6.1)$$

$$Y_\theta \equiv \mathcal{D}\big(\mathcal{A}^{-\frac{1}{2}+\theta}\big) = H^{-1+2\theta}(\Omega), \quad 0 \leq \theta \leq \frac{1}{2} + \delta; \quad (10.9.6.2)$$

$$Y_0 = Y = \big[\mathcal{D}\big(\mathcal{A}^{\frac{1}{2}}\big)\big]' = H^{-1}(\Omega), \quad (10.9.6.3)$$

using in (10.9.6.2) the convention $\mathcal{D}(\mathcal{A}^{-s}) = [\mathcal{D}(\mathcal{A}^s)]'$ for $s > 0$, duality with respect to $L_2(\Omega)$. In particular, the critical spaces for $\theta = 1/2 \pm \delta$ are

$$U_\delta^- \equiv U_{\frac{1}{2}-\delta} = H^{1-2\delta}(\Gamma), \quad U_\delta^+ \equiv U_{\frac{1}{2}+\delta} = H^{1+2\delta}(\Omega), \quad (10.9.6.4)$$

$$Y_\delta^- \equiv Y_{\theta=\frac{1}{2}-\delta} \equiv \mathcal{D}(\mathcal{A}^{-\delta}) = [\mathcal{D}(\mathcal{A}^\delta)]' = H^{-2\delta}(\Omega), \quad (10.9.6.5)$$

$$Y_\delta^+ \equiv Y_{\theta=\frac{1}{2}+\delta} = \mathcal{D}(\mathcal{A}^\delta) = H^{2\delta}(\Omega), \quad (10.9.6.6)$$

with equivalent norms. Next, we see that $[Y_\delta^+]'$, the dual of $\mathcal{D}(\mathcal{A}^\delta)$ with respect to $Y = [\mathcal{D}(\mathcal{A}^{\frac{1}{2}})]'$, is given by

$$[Y_\delta^+]' = [\mathcal{D}(\mathcal{A}^{1+\delta})]' = \mathcal{D}(\mathcal{A}^{-1-\delta}), \quad (10.9.6.7)$$

duality on the right-hand side being instead with respect to $L_2(\Omega)$, and we verify the interpolation property that

$$[Y_\delta^-, [Y_\delta^+]']_{\theta=\frac{1}{2}-\delta} = [[\mathcal{D}(\mathcal{A}^\delta)]', [\mathcal{D}(\mathcal{A}^{1+\delta})]']_{\theta=\frac{1}{2}-\delta} = \big[\mathcal{D}(\mathcal{A}^{\frac{1}{2}})\big]' \equiv Y, \quad (10.9.6.8)$$

as required in (10.1.10), since $-\delta(1-\theta) - (1+\delta)\theta = -1/2$ for $\theta = 1/2 - \delta$. Moreover, the injections $U_{\theta_1} \hookrightarrow U_{\theta_2}, Y_{\theta_1} \hookrightarrow Y_{\theta_2}$ are compact, as required in (10.1.10), since Ω is a bounded domain. Thus, in the present case, the spaces in (10.1.11) and (10.1.12) are as follows for $0 \leq \theta \leq 1/2 + \delta, \theta \neq 1/2$:

$$\mathcal{U}^\theta[s, T] = H^{2\theta,\theta}(\Sigma_{sT}) = L_2(s, T; H^{2\theta}(\Gamma)) \cap H^\theta(s, T; L_2(\Gamma)); \quad (10.9.6.9)$$

$$\mathcal{Y}^\theta[s, T] = L_2\big(s, T; \mathcal{D}\big(\mathcal{A}^{-\frac{1}{2}+\theta}\big)\big) \cap H^\theta\big(s, T; [\mathcal{D}(\mathcal{A}^{\frac{1}{2}})]'\big), \quad (10.9.6.10)$$

recalling (10.9.6.1)–(10.9.6.3) and the convention below (10.9.6.3).

10.9.7 Verification of Assumptions (H.2) = (10.1.14)

The following regularity result is critical in verifying assumptions (H.2) = (10.1.14).

Theorem 10.9.7.1 *With reference to the nonhomogeneous problem (10.9.1.1), assume*

$$y_0 \in H^1(\Omega) \text{ with the compatibility relation } y_0|_\Gamma = u(0) \in H^{\frac{1}{2}}(\Gamma); \quad (10.9.7.1)$$

$$u \in C\left([0, T]; H^{\frac{1}{2}}(\Gamma)\right) \cap H^1(0, T; L_2(\Gamma)) \quad (10.9.7.2)$$

[(10.9.7.2) is a fortiori guaranteed if

$$u \in H^{1,1}(\Sigma) = L_2(0, T; H^1(\Gamma)) \cap H^1(0, T; L_2(\Gamma))$$

by [Lions, Magenes, 1972, I, Theorem 3.1, p. 19]. Then, the unique solution of problem (10.9.1.1) satisfies

$$\{y, y_t\} \in C([0, T]; H^1(\Omega) \times H^{-1}(\Omega)) \quad (10.9.7.3)$$

a fortiori,

$$y \in H^{1,\frac{1}{2}}(Q) \equiv L_2(0, T; H^1(\Omega)) \cap H^{\frac{1}{2}}(0, T; L_2(\Omega)) \quad (10.9.7.4)$$

continuously.

Proof. The proof of Theorem 10.9.7.1 will be given in Section 10.9.12 below. □

Corollary 10.9.7.1 *With reference to the nonhomogeneous problem (10.9.1.1), assume, for $0 \le \theta < 1/2$:*

$$y_0 = 0, \quad u \in H^{\theta,\theta}(\Sigma) = L_2(0, T; H^\theta(\Gamma)) \cap H^\theta(0, T; L_2(\Gamma)), \quad (10.9.7.5)$$

in particular,

$$u \in \mathcal{U}^\theta[0, T] = L_2(0, T; H^{2\theta}(\Gamma)) \cap H^\theta(0, T; L_2(\Gamma)) \subset H^{\theta,\theta}(\Sigma). \quad (10.9.7.6)$$

Then, the unique solution of problem (10.9.1.1) satisfies for $0 \le \theta < 1/2$:

$$y(\,\cdot\,; 0) \in C([0, T]; H^{-1+2\theta}(\Omega)), \quad \theta \ne \frac{1}{4}; \quad (10.9.7.7)$$

$$y_t(\,\cdot\,; 0) \in C([0, T]; H^{-3+2\theta}(\Omega)), \quad \theta \ne \frac{1}{4}; \quad (10.9.7.8)$$

$$D_t^r y(\,\cdot\,; 0) \in L_2(0, T; H^{-1+2\theta-2r}(\Omega)), \quad 0 \le r \le 1, \quad \theta \ne \frac{1}{4}; \quad (10.9.7.9a)$$

$$D_t^\theta y(\,\cdot\,; 0) \in L_2(0, T; H^{-1}(\Omega)) = Y, \quad (10.9.7.9b)$$

where $D_t^r = (d/dt)^r$ and $D_t^\theta = (d/dt)^\theta$ are fractional time derivatives.

Proof. With $y_0 = 0$, we interpolate between the case $\theta = 0$ in (10.9.3.5), (10.9.3.6) and the case $\theta = 1$ in (10.9.7.2), (10.9.7.3) in the form $u \in H^{1,1}(\Sigma)$ in the range

$0 \le \theta < 1/2$, so that the compatibility relation in (10.9.7.1) is irrelevant. We use [Lions, Magenes, 1972, Theorem 12.4, p. 73, and Theorem 12.2, p. 71] and obtain (10.9.7.7) and (10.9.7.8). To get (10.9.7.9a), we then apply the intermediate derivative theorem [Lions, Magenes, 1972, pp. 15 and 23], from which (10.9.7.9b) follows by setting $r = \theta$. $\square$

The implications $u \in \mathcal{U}^\theta[0, T] \Rightarrow D_t^\theta y(\cdot; 0) \in L_2(0, T; Y)$ from (10.9.7.6) to (10.9.7.9b) as well as $\Rightarrow y \in C([0, T]; H^{-1+2\theta}(\Omega))$ in (10.9.7.7) provide the desired result that $Lu = y \in \mathcal{Y}^\theta[0, T]$, $\theta < \frac{1}{2}$, via (10.9.6.10) and (10.9.6.3). Assumption (H.2) = (10.1.14) is thus verified.

10.9.8 Verification of Assumption (H.3) = (10.1.15)

Theorem 10.9.8.1   (a) With reference to the operator $B^* e^{A^* t}$ in (10.9.4.16), we have the following regularity properties for $0 \le r \le 1$, which generalize the case $r = 0$ of Theorem 10.9.3.1, Eqn. (10.9.3.4), or of Eqn. (10.9.5.1):

(i)

$$B^* e^{A^* t} : \text{continuous } \mathcal{D}\big(\mathcal{A}^{-\frac{1}{2}+r}\big) \to H^{2r,r}(\Sigma), \qquad (10.9.8.1)$$

$$H^{2r,r}(\Sigma) = L_2(0, T; H^{2r}(\Gamma)) \cap H^r(0, T; L_2(\Gamma)), \qquad (10.9.8.2)$$

where we continue to use the convention $\mathcal{D}(\mathcal{A}^{-s}) = [\mathcal{D}(\mathcal{A}^s)]'$ for $s > 0$, as below (10.9.6.3);

(ii) equivalently, via (10.9.4.16) and (10.9.4.17), we have in PDE terms

$$\phi_0 \to \frac{\partial \phi}{\partial \nu}(\cdot; \phi_0) : \text{continuous } \mathcal{D}\big(\mathcal{A}^{\frac{1}{2}+r}\big) \to H^{2r,r}(\Sigma), \qquad (10.9.8.3)$$

where $\phi(t; \phi_0)$ is the unique solution of problem (10.9.3.1) with $f \equiv 0$.

(b) With reference to the operator L_s^* in (10.9.5.4), we have, uniformly in s:

$$L_s^* : \text{continuous } L_2\big(s, T; \mathcal{D}\big(\mathcal{A}^{-\frac{1}{2}+r}\big)\big) \to H^{2r,r}(\Sigma_{sT}). \qquad (10.9.8.4)$$

Proof.   The proof of Theorem 10.9.8.1 will be given in Section 10.9.13 below. $\square$

Restricting r in (10.9.8.4) to $0 \le r = \theta < 1/2 + \delta$, $\theta \ne 1/2$, and recalling (10.9.6.1), (10.9.6.2), and (10.9.6.9), we obtain

$$L_s^* : L_2\big(s, T; Y_\theta \equiv \mathcal{D}\big(\mathcal{A}^{-\frac{1}{2}+\theta}\big)\big)$$
$$\to L_2(s, T; U_\theta \equiv H^{2\theta}(\Gamma)) \cap H^\theta(0, T; L_2(\Gamma)) \equiv \mathcal{U}^\theta[0, T], \qquad (10.9.8.5)$$

and assumption (H.3) = (10.1.15) is a fortiori verified, as the condition of uniformity in s follows as usual by a change of variable.

10.9.9 Verification of Assumptions (H.4) = (10.1.18) through (H.7) = (10.1.21)

Verification of Assumption (H.4) = (10.1.18)   Due to the choice of a smooth space Y_δ^+ in (10.9.6.6), this assumption holds true trivially. In fact, let $z \in Y_\delta^+ = \mathcal{D}(\mathcal{A}^\delta)$; then recalling (10.9.4.13) and (10.9.4.2), we obtain

$$B^* e^{A^* t} z = i D^* e^{-iAt} z = i D^* \mathcal{A}^{\frac{1}{4}-\epsilon} e^{-i\mathcal{A}t} \mathcal{A}^{-\frac{1}{4}+\epsilon} z \in C([0, T]; L_2(\Gamma)),$$

since $D^* \mathcal{A}^{\frac{1}{4}-\epsilon} \in \mathcal{L}(L_2(\Omega))$ by (10.9.4.9), and $\mathcal{A}^{-\frac{1}{4}+\epsilon} z \in L_2(\Omega)$ surely.

Verification of Assumption (H.5) = (10.1.19)   Surely, $e^{At} = e^{i\mathcal{A}t}$ is a s.c. semigroup also in $Y_\delta^- = [\mathcal{D}(\mathcal{A}^\delta)]'$ with $\mathcal{D}(A)$ dense in Y_δ^-; see (10.9.6.5).

Verification of Assumption (H.6) = (10.1.20)   We have, as desired,

$$A = i\mathcal{A} : \text{continuous } Y_\delta^- = \mathcal{D}(\mathcal{A}^{-\delta}) \text{ onto } \mathcal{D}(\mathcal{A}^{-1-\delta}) = [Y_\delta^+]', \qquad (10.9.9.1)$$

recalling (10.9.6.5) and (10.9.6.7).

Verification of Assumption (H.7) = (10.1.21)   We have for $u \in U_\delta^- = H^{1-2\delta}(\Gamma)$, hence $Du \in H^{\frac{1}{2}-2\epsilon}(\Omega) = \mathcal{D}(\mathcal{A}^{\frac{1}{4}-\epsilon})$ a fortiori, that

$$Bu = -i\mathcal{A}Du \in \mathcal{D}(\mathcal{A}^{-\frac{3}{4}-\epsilon}) \subset \mathcal{D}(\mathcal{A}^{-1-\delta}) = [Y_\delta^+]',$$

recalling (10.9.6.7) and (10.9.4.9).

10.9.10 Proof of Theorem 10.9.3.1

Interior Regularity   With reference to problem (10.9.3.1a,c) with, say $F \equiv 0$, the required implication (by (10.9.4.5) with $\theta = 1/2$):

$$
\begin{cases}
\phi_0 \in \mathcal{D}(\mathcal{A}^{\frac{1}{2}}) = \mathcal{D}(A^{\frac{1}{2}}) = H_0^1(\Omega); \quad f \in L_1(0, T; H_0^1(\Omega)) & (10.9.10.1a) \\[2mm]
\Rightarrow \ \phi(t) = e^{At}\phi_0 + \int_0^t e^{A(t-\tau)} f(\tau)\, d\tau \in C([0, T]; H_0^1(\Omega)); & (10.9.10.1b) \\[2mm]
\phi_t(t) = A\phi(t) + f(t) \in L_1(0, T; H^{-1}(\Omega)), & (10.9.10.1c)
\end{cases}
$$

is immediate, by convolution and semigroup properties. This regularity is not altered by the presence of the first-order operator F as in (10.9.3.1d).

Boundary Regularity (10.9.3.4). Step 1   Key to this end is the following result.

Lemma 10.9.10.1 *[Lasiecka, Triggiani, 1991(b)] Let ϕ be a solution of Eqn. (10.9.3.1a) (with no boundary conditions imposed) for smooth data, say as in (10.9.10.1a), hence with regularity as in (10.9.10.1b,c). Then, the following identity* *holds true for any $0 < T < \infty$:*

$$\mathrm{Re}\left(\int_\Sigma \frac{\partial \phi}{\partial v} h \cdot \nabla \bar{\phi}\, d\Sigma\right) - \frac{1}{2}\int_\Sigma |\nabla \phi|^2 h \cdot v\, d\Sigma - \frac{i}{2}\int_\Sigma \phi_t \bar{\phi} h \cdot v\, d\Sigma$$

$$= \mathrm{Re}\left(\int_Q H\nabla\phi \cdot \nabla\bar{\phi}\, dQ\right) - \frac{1}{2}\int_Q |\nabla\phi|^2 \,\mathrm{div}\, h\, dQ - \frac{i}{2}\int_Q \phi_t\bar{\phi}\,\mathrm{div}\, h\, dQ$$

$$- \frac{i}{2}\left[\int_\Omega \bar{\phi}h \cdot \nabla\phi\, d\Omega\right]_0^T - \mathrm{Re}\left(\int_Q [F(\phi) + f]\right) h \cdot \nabla\bar{\phi}\, dQ, \quad (10.9.10.2)$$

where $h(x) = [h_1(x), \ldots, h_n(x)]$ is any real vector field in $[C^1(\bar{\Omega})]^n$ and, where [as in (10.5.10.6), (10.7.10.5), ...]

$$H(x) = \begin{bmatrix} \frac{\partial h_1}{\partial x_1} & \cdots & \frac{\partial h_1}{\partial x_n} \\ \cdots & \cdots & \cdots \\ \frac{\partial h_n}{\partial x_1} & \cdots & \frac{\partial h_n}{\partial x_n} \end{bmatrix}, \quad v(x) = \text{outward unit normal vector at } x \in \Gamma.$$

$$(10.9.10.3)$$

Proof. With ϕ having the regularity given by (10.9.10.1b), we multiply both sides of Eqn. (10.9.3.1a) by $h \cdot \nabla\bar{\phi}$ and integrate by parts (similar computations were performed in the proof of Lemma 10.5.10.1). On the left-hand side (L.H.S.) of (10.9.3.1a), we obtain, after using the divergence theorem identity (10.5.10.7), with h there replaced by $\phi_t h$ now, and ψ there replaced by $\bar{\phi}$ now:

$$\text{L.H.S.} = ia \equiv i\int_0^T \int_\Omega \phi_t h \cdot \nabla\bar{\phi}\, d\Omega\, dt \quad (10.9.10.4)$$

$$= i\int_0^T \left\{ \int_\Gamma \phi_t\bar{\phi}h \cdot v\, d\Gamma - \int_\Omega \bar{\phi}h \cdot \nabla\phi_t\, d\Omega - \int_\Omega \phi_t\bar{\phi}\,\mathrm{div}\, h\, d\Omega \right\} dt$$

$$= i\int_\Sigma \phi_t\bar{\phi}h \cdot v\, d\Sigma - i\left[\int_\Omega \bar{\phi}h \cdot \nabla\phi\, d\Omega\right]_0^T$$

$$+ i\bar{a} - i\int_Q \phi_t\bar{\phi}\,\mathrm{div}\, h\, dQ. \quad (10.9.10.5)$$

From (10.9.10.5) we readily obtain

$$ia - i\bar{a} = -2\,\mathrm{Im}\, a$$

$$= i\int_\Sigma \phi_t\bar{\phi}h \cdot v\, d\Sigma - i\left[\int_\Omega \bar{\phi}h \cdot \nabla\phi\, d\Omega\right]_0^T - i\int_Q \phi_t\bar{\phi}\,\mathrm{div}\, h\, dQ.$$

$$(10.9.10.6)$$

As to the right-hand side (R.H.S.) of (10.9.3.1a), we first note that

$$\int_0^T \int_\Omega \Delta\phi h \cdot \nabla\bar\phi \, d\Omega \, dt = \int_0^T \int_\Gamma \frac{\partial\phi}{\partial\nu} h \cdot \nabla\bar\phi \, d\Gamma \, dt - \int_0^T \int_\Omega \nabla\phi \cdot \nabla(h \cdot \nabla\bar\phi) \, d\Omega \, dt,$$
(10.9.10.7)

and then, recalling $\nabla\phi \cdot \nabla(h \cdot \nabla\bar\phi) = H\nabla\phi \cdot \nabla\bar\phi + \frac{1}{2} h \cdot \nabla(|\nabla\phi|^2)$ from, for example, (10.5.10.12), and then again identity (10.5.10.7), this time with $\psi = |\nabla\phi|^2$, we obtain from (10.9.10.4) and (10.9.10.7):

$$\text{L.H.S.} = ia = -\text{Im} \, a + i \, \text{Re} \, a$$

$$= \text{R.H.S.} = \int_0^T \int_\Omega \Delta\phi h \cdot \nabla\bar\phi \, d\Omega \, dt + \int_Q [F(\phi) + f] h \cdot \nabla\bar\phi \, dQ$$

$$= \int_\Sigma \frac{\partial\phi}{\partial\nu} h \cdot \nabla\bar\phi \, d\Sigma - \int_Q H\nabla\phi \cdot \nabla\bar\phi \, dQ - \frac{1}{2} \int_\Sigma |\nabla\phi|^2 h \cdot \nu \, d\Sigma$$

$$+ \frac{1}{2} \int_Q |\nabla\phi|^2 \, \text{div} \, h \, dQ + \int_Q [F(\phi) + f] h \cdot \nabla\bar\phi \, dQ.$$
(10.9.10.8)

Equating the real part on the left-hand side of (10.9.10.8), that is, $[-\text{Im} \, a]$ as given by (10.9.10.6), with the real part on the right-hand side of (10.9.10.8) yields (10.9.10.2), as desired.   □

Step 2   As in Section 10.5.10, we now exploit also the boundary condition (10.9.3.1c), make a special choice of the vector field $h(x) \in [C^1(\bar\Omega)]^n$, and prove the desired trace regularity (10.9.3.4) as a consequence of Lemma 10.9.10.1 and of the interior regularity (10.9.10.1).

Theorem 10.9.10.2   *Let the data $\{\phi_0, f\}$ be as in (10.9.10.1a). Let $h \in [C^1(\bar\Omega)]^n$ satisfy $h|_\Gamma = \nu$ on Γ. Then:*

(i) The solution ϕ of problem (10.9.3.1a–c) satisfies

$$\text{L.H.S. of } (10.9.10.2) = \frac{1}{2} \int_\Sigma \left| \frac{\partial\phi}{\partial\nu} \right|^2 d\Sigma,$$
(10.9.10.9)

and hence the identity

$$\frac{1}{2} \int_\Sigma \left| \frac{\partial\phi}{\partial\nu} \right|^2 d\Sigma = \text{Re} \left(\int_Q H\nabla\phi \cdot \nabla\bar\phi \, dQ \right) - \frac{1}{2} \int_Q |\nabla\phi|^2 \, \text{div} \, h \, dQ$$

$$- \frac{i}{2} \int_Q \phi_t \bar\phi \, \text{div} \, h \, dQ$$

$$- \frac{i}{2} \left[\int_\Omega \bar\phi h \cdot \nabla\phi \, d\Omega \right]_0^T - \text{Re} \left(\int_Q [F(\phi) + f] h \cdot \nabla\bar\phi \, dQ \right).$$
(10.9.10.10)

(ii) Given any $0 < T < \infty$, there exists a constant $C_T > 0$, such that

$$\int_\Sigma \left| \frac{\partial \phi}{\partial \nu} \right|^2 d\Sigma \leq C_T \left\{ \| \phi_0 \|^2_{H^1_0(\Omega)} + \| f \|^2_{L_1(0,T;H^1_0(\Omega))} \right\}. \qquad (10.9.10.11)$$

Proof.   (i) Upon using the BC (10.9.3.1c), we have on Σ:

$$h \cdot \nabla \bar\phi = \frac{\partial \bar\phi}{\partial \nu} h \cdot \nu, \quad |\nabla \phi|^2 = \left| \frac{\partial \phi}{\partial \nu} \right|^2, \qquad (10.9.10.12)$$

and with $h|_\Gamma = \nu$ on Γ, we then readily obtain (10.9.10.9) from (10.9.10.2) and (10.9.10.12), and hence (10.9.10.10).

(ii) Applying the implication in (10.9.10.1) on the right-hand side of (10.9.10.10) readily yields (10.9.10.11).   □

Remark 10.9.10.1   Multiplying Eqn. (10.9.3.1a) by $\bar\phi \operatorname{div} h$, $h \in [C^2(\bar\Omega)]^2$, and integrating over Q yields by the first Green theorem

$$i \int_Q \phi_t \bar\phi \operatorname{div} h \, dQ = \int_\Sigma \frac{\partial \phi}{\partial \nu} \bar\phi \operatorname{div} h \, d\Sigma - \int_Q \nabla \phi \cdot \nabla(\bar\phi \operatorname{div} h) \, dQ$$

$$+ \int_Q [F(\phi) + f] \bar\phi \operatorname{div} h \, dQ. \qquad (10.9.10.13)$$

Substituting (10.9.10.13) for the third integral term on the right-hand side of identity (10.9.10.2) yields the following identity for smooth solutions ϕ of problem (10.9.3.1) after a cancellation:

$$\operatorname{Re}\left(\int_\Sigma \frac{\partial \phi}{\partial \nu} h \cdot \nabla \bar\phi \, d\Sigma \right) - \frac{1}{2} \int_\Sigma |\nabla \phi|^2 h \cdot \nu \, d\Sigma$$

$$- \frac{i}{2} \int_\Sigma \phi_t \bar\phi h \cdot \nu \, d\Sigma + \frac{1}{2} \int_\Sigma \frac{\partial \phi}{\partial \nu} \bar\phi \operatorname{div} h \, d\Sigma$$

$$= \operatorname{Re}\left(\int_Q H \nabla \phi \cdot \nabla \bar\phi \, dQ \right) + \frac{1}{2} \int_Q \bar\phi \nabla \phi \cdot \nabla(\operatorname{div} h) \, dQ$$

$$- \frac{i}{2} \left[\int_\Omega \bar\phi h \cdot \nabla \phi \, d\Omega \right]_0^T - \operatorname{Re}\left(\int_Q [F(\phi) + f] h \cdot \nabla \bar\phi \, dQ \right)$$

$$- \frac{1}{2} \left(\int_Q [F(\phi) + f] \bar\phi \operatorname{div} h \, dQ \right). \qquad (10.9.10.14)$$

10.9.11   Proof of Theorem 10.9.3.2

We now provide the details, already contained in the preceding development, that the trace regularity for the homogeneous ϕ-problem (10.9.3.1),

$$f \equiv 0, \quad \phi_0 \Rightarrow \frac{\partial \phi}{\partial \nu} : \text{continuous } H^1_0(\Omega) \to L_2(0, T; L_2(\Gamma)), \qquad (10.9.11.1)$$

established in Theorem 10.9.3.1, Eqn. (10.9.3.4), or (10.9.10.9), implies by transposition the interior regularity

$$\left.\begin{cases} u \in L_2(0, T; L_2(\Gamma)) \equiv L_2(\Sigma) \\ y_0 = 0 \end{cases}\right\} \Rightarrow$$

$$y(t) = (Lu)(t) = \int_0^t e^{A(t-\tau)} Bu(\tau)\, d\tau \in C([0, T]; Y), \qquad (10.9.11.2)$$

$$Y = \left[\mathcal{D}(A^{\frac{1}{2}})\right]' = H^{-1}(\Omega), \qquad (10.9.11.3)$$

see (10.9.4.18) and (10.9.3.6) for the nonhomogeneous y-problem (10.9.1.1).

Operator-Theoretic Proof We return to the trace result (10.9.5.1), equivalent to (10.9.5.2), and apply Chapter 7, Theorem 7.2.1 with $p = 2$, to obtain the regularity of y in (10.9.11.2) at once. With such regularity, we return to Eqn. (10.9.1.1a) and obtain

$$y_t = -i\,\Delta y \in C([0, T]; H^{-3}(\Omega)) \qquad (10.9.11.4)$$

by using [Lions, Magenes, 1972, p. 85]. Moreover, as to the initial condition

$$y_0 \in H^{-1}(\Omega) = \left[\mathcal{D}(A^{\frac{1}{2}})\right]'$$

$$\Rightarrow \begin{cases} e^{At} y_0 = e^{iAt} y_0 \in C([0, T]; H^{-1}(\Omega)), & (10.9.11.5) \\ A e^{At} y_0 \in C\left([0, T]; \left[\mathcal{D}(A^{\frac{3}{2}})\right]'\right) \subset C([0, T]; H^{-3}(\Omega)), & (10.9.11.6) \end{cases}$$

since $H_0^3(\Omega) \subset \mathcal{D}(A^{\frac{3}{2}})$. The proof is complete.

PDE Proof. A PDE version of the duality or transposition argument is as follows: Let y and ψ be solutions with regular data of the following two problems (recall (10.9.5.8), and (10.9.3.1)):

$$\begin{cases} y_t = -i\,\Delta y + F_1, \\ y(0, \cdot) = y_0, \\ y|_\Sigma = u, \end{cases} \quad \text{and} \quad \begin{cases} \psi_t = -i\,\Delta\psi + f & \text{in } Q; & (10.9.11.7a) \\ \psi(T, \cdot) = 0 & \text{in } \Omega; & (10.9.11.7b) \\ \psi|_\Sigma \equiv 0 & \text{in } \Sigma. & (10.9.11.7c) \end{cases}$$

We know in advance, via Theorem 10.9.3.1 applied on the ψ-problem, which is reversed in time over the ψ-problem (10.9.3.1), that

$$f \in L_1\left(0, T; H_0^1(\Omega)\right) \Rightarrow \begin{cases} \psi \in C\left([0, T]; H_0^1(\Omega)\right), \\ \text{in particular, } \psi(0) \in H_0^1(\Omega); & (10.9.11.8) \\ \dfrac{\partial\psi}{\partial\nu} \in L_2(\Sigma). \end{cases}$$

We multiply the y-equation (10.9.11.7a) by $\bar{\psi}$ and integrate by parts (Green's second theorem) using initial and boundary conditions (10.9.11.7b,c). We obtain

$$
\int_\Omega \int_0^T [y_t + i\Delta y]\bar{\psi}\, dQ = -\int_\Omega \int_0^T y[\bar{\psi}_t - i\Delta\bar{\psi}]\, dQ - i\int_\Sigma u\frac{\partial\bar{\psi}}{\partial\nu}\, d\Sigma
$$

$$
- \int_\Omega y_0\bar{\psi}(0)\, d\Omega, \qquad (10.9.11.9)
$$

$$
\int_Q F_1\bar{\psi}\, dQ + \int_Q y\,\bar{f}\, dQ + i\int_\Sigma u\frac{\partial\bar{\psi}}{\partial\nu}\, d\Sigma + \int_\Omega y_0\bar{\psi}(0)\, d\Omega = 0, \quad (10.9.11.10)
$$

for smooth solutions. Recalling the regularity results (10.9.11.8) for the ψ-problem, we see that in identity (10.9.11.10) we can take by duality

$$
y_0 \in H^{-1}(\Omega), \quad F_1 \in L_1(0, T; H^{-1}(\Omega)), \qquad (10.9.11.11)
$$

and thus obtain

$$
y \in L_\infty(0, T; H^{-1}(\Omega)), \qquad (10.9.11.12)
$$

a regularity result that can then be boosted to

$$
y \in C([0, T]; H^{-1}(\Omega)), \qquad (10.9.11.13)
$$

by an approximating argument with smooth data in the y-problem. The remaining argument is as in the preceding proof, from (10.9.11.4) to (10.9.11.6). $\square$

10.9.12 Proof of Theorem 10.9.7.1

With reference to the nonhomogeneous y-problem (10.9.1.1), we assume

$$
y_0 \in H^1(\Omega), \quad y_0|_\Gamma = u(0) \in H^{\frac{1}{2}}(\Gamma); \qquad (10.9.12.1)
$$

$$
u \in C\big([0, T]; H^{\frac{1}{2}}(\Gamma)\big) \cap H^1(0, T; L_2(\Gamma)), \qquad (10.9.12.2)
$$

and we must show that

$$
\{y, y_t\} \in C([0, T]; H^1(\Omega) \times H^{-1}(\Omega)). \qquad (10.9.12.3)
$$

Operator-Theoretic Proof Integration by parts on the input-solution formula (10.9.4.18) with $\dot{u} \in L_2(\Sigma)$ and with $A^{-1}B = -D$ by (10.9.4.11) yields

$$
y(t) = e^{At}y_0 + \int_0^t Ae^{A(t-\tau)}A^{-1}Bu(\tau)\, d\tau
$$

$$
= e^{At}[y_0 - Du(0)] + Du(t) + A^{-1}\int_0^t e^{A(t-\tau)}B\dot{u}(\tau)\, d\tau, \quad (10.9.12.4)
$$

where, by (10.9.11.2) and (10.9.11.3), with u replaced by $\dot{u} \in L_2(\Sigma)$, (10.9.4.2), and respectively, (10.9.4.9) on D and $u \in C([0, T]; H^{\frac{1}{2}}(\Gamma))$, we obtain

$$(A^{-1}L\dot{u})(t) = A^{-1}\int_0^t e^{A(t-\tau)}B\dot{u}(\tau)\,d\tau \in C\big([0, T]; H_0^1(\Omega) = \mathcal{D}\big(\mathcal{A}^{\frac{1}{2}}\big)\big),$$

$$Du(t) \in C([0, T]; H^1(\Omega)). \tag{10.9.12.5}$$

Thus, (10.9.12.5) used in (10.9.12.4), along with $y_0 - Du(0) = 0$ by the compatibility relation (10.9.12.1), yields the desired regularity for y in (10.9.12.3). Then, differentiating (10.9.12.4) in t yields since $A^{-1}B = -D$:

$$y_t = D\dot{u}(t) - D\dot{u}(t) + \int_0^t e^{A(t-\tau)}B\dot{u}(\tau)\,d\tau \in C([0, T]; H^{-1}(\Omega)), \tag{10.9.12.6}$$

after a cancellation of $D\dot{u}(t)$, by (10.9.11.2) and (10.9.11.3) used on $\dot{u} \in L_2(\Sigma)$.

PDE Proof   Differentiating in t problem (10.9.1.1) with $u \in H^1(0, T; L_2(\Gamma))$ yields

$$\begin{cases} (y_t)_t = -i\,\Delta y_t & \text{in } Q, & (10.9.12.7a) \\ y_t(0, \cdot) = -i\,\Delta y_0 & \text{in } \Omega, & (10.9.12.7b) \\ y_t|_\Sigma = u_t & \text{in } \Sigma, & (10.9.12.7c) \end{cases}$$

that is, the same problem (10.9.1.1), however, in the variable y_t, with initial datum $-i\,\Delta y_0 \in H^{-1}(\Omega)$ by (10.9.12.1) on y_0. Then, Theorem 10.9.3.2 applies and yields conclusion (10.9.12.3) for y_t. Next, recalling (10.9.12.2) we look at the elliptic problem

$$\begin{cases} \Delta y = i y_t \in C([0, T]; H^{-1}(\Omega)), \\ y|_\Sigma = u \in C\big([0, T]; H^{\frac{1}{2}}(\Gamma)\big), \end{cases}$$

at each t and then obtain conclusion (10.9.12.3) for y. Since estimates are first shown for regular data, and then extended, we need the compatibility relation (10.9.12.1). The proof of Theorem 10.9.7.1 is complete.

10.9.13 Proof of Theorem 10.9.8.1

Part (a). Case $r = 0$ The case $r = 0$ is contained in Theorem 10.9.3.1, Eqn. (10.9.3.4), or Eqn. (10.9.5.1), with the stated equivalence between (10.9.7.10) and (10.9.7.12) being established via (10.9.4.16) and (10.9.4.17). Thus, it suffices to prove the case $r = 1$ and interpolate to establish part (a) of Theorem 10.9.7.3.

Part (a). Case $r = 1$ The required time regularity in (10.9.8.1) [with $A^* = -i\mathcal{A}$ by (10.9.4.2)],

$$D_t^1(B^* e^{A^* t} x) = B^* e^{A^* t} A^* x \in L_2(\Sigma), \quad x \in \mathcal{D}\big(\mathcal{A}^{\frac{1}{2}}\big), \tag{10.9.13.1}$$

is an immediate consequence of (10.9.5.1), that is, of

$$B^* e^{A^* t} : Y = \left[\mathcal{D}\!\left(A^{\frac{1}{2}}\right)\right]' \to L_2(\Sigma). \tag{10.9.13.2}$$

To show the space regularity for $r = 1$ in (10.9.8.1) [equivalently in (10.9.8.3)],

$$\phi_0 \in \mathcal{D}\!\left(A^{\frac{3}{2}}\right) \to \frac{\partial \phi}{\partial \nu}(t, \phi_0) \in L_2(0, T; H^2(\Gamma)), \tag{10.9.13.3}$$

we introduce the operator

$$\begin{cases} \mathcal{B} = \Sigma \beta_{ij} \dfrac{\partial^2}{\partial x_i \partial x_j} = \text{ second-order operator with (time-independent)} \\ \text{coefficients } b_{ij} \text{ smooth on } \bar{\Omega} \text{ and such that } \mathcal{B} \text{ is tangential to } \Gamma. \end{cases} \tag{10.9.13.4}$$

Accordingly, we consider problem (10.9.3.1) with $f \equiv 0$:

$$\begin{cases} i\phi_t = \Delta\phi & \text{in } Q, & (10.9.13.5a) \\ \phi(0, \,\cdot\,) = \phi_0 \in \mathcal{D}\!\left(A^{\frac{3}{2}}\right) & \text{in } \Omega, & (10.9.13.5b) \\ \phi|_\Sigma \equiv 0 & \text{in } \Sigma, & (10.9.13.5c) \end{cases}$$

whose solution enjoys therefore the following regularity:

$$\{\phi, \phi_t\} \in C\!\left([0, T]; \mathcal{D}\!\left(A^{\frac{3}{2}}\right) \times \mathcal{D}\!\left(A^{\frac{1}{2}}\right)\right) \subset C\!\left([0, T]; H^3(\Omega) \times H_0^1(\Omega)\right). \tag{10.9.13.6}$$

We then introduce a new variable

$$z = \mathcal{B}\phi, \tag{10.9.13.7}$$

which, therefore, by virtue of (10.9.13.4) and (10.9.13.6), has the following a priori regularity:

$$\{z, z_t\} \in C\!\left([0, T]; H_0^1(\Omega) \times H^{-1}(\Omega)\right). \tag{10.9.13.8}$$

Then, proving (10.9.13.3) is equivalent to proving

$$\frac{\partial(\mathcal{B}\phi)}{\partial \nu} = \frac{\partial z}{\partial \nu} \in L_2(0, T; L_2(\Gamma)) \equiv L_2(\Sigma). \tag{10.9.13.9}$$

The new variable z satisfies the following problem:

$$\begin{cases} iz_t = \Delta z + [\mathcal{B}, \Delta]\phi, & (10.9.13.10a) \\ z|_\Sigma = 0, & (10.9.13.10b) \end{cases}$$

as one readily sees by (10.9.13.5), (10.9.13.7), and (10.9.13.4). Since the commutator

$$[\mathcal{B}, \Delta] = \text{operator of order } 2 + 2 - 1 = 3, \tag{10.9.13.11}$$

we see via (10.9.13.6) on ϕ that

$$f \equiv [\mathcal{B}, \Delta]\phi \in C\!\left([0, T]; L_2(\Omega)\right). \tag{10.9.13.12}$$

We now return to the basic identity (10.9.10.8) for problem (10.9.13.10), which satisfies the homogeneous BC (10.9.13.10b) as required by (10.9.10.8), with ϕ there replaced by z now. Because of the a priori interior regularity (10.9.13.8) for $\{z, z_t\}$ and the a priori regularity (10.9.13.12) of f, we then see that the right-hand side (R.H.S.) of identity (10.9.10.8) is well-defined. Thus, the left-hand side of (10.9.10.8) is well-defined and yields (10.9.13.9), as desired. The proof of part (a), case $r = 1$, of Theorem 10.9.8.1 is complete.

Part (c)   Part (c), Eqn. (10.9.8.4), of Theorem 10.9.8.1 follows from Part (a), via Chapter 7, Theorem 7.2.1, with $p = 2$. $\square$

Remark 10.9.13.1   A remark similar to Remark 10.7.13.1 applies also in the present case of the Schrödinger equation: It is the a priori interior regularity (10.9.13.8) for $\{z, z_t\}$ that permits one to obtain the desired trace regularity of $\partial z/\partial \nu$, even though f in (10.9.13.12) is only in $L_2(\Omega)$, and not in $H_0^1(\Omega)$ as in (10.9.10.1).

Notes on Chapter 10

Sections 10.1–10.4

Both the abstract framework and the treatment of the optimal control problem in Sections 10.1 through 10.4 (with a slightly smoothing observation operator R) are contributions of the present chapter, following Triggiani [1996]. This framework represents a unifying abstract setting, which builds upon the only two specific cases available in the literature: (i) the case of second-order hyperbolic equations with Dirichlet control in Lasiecka and Triggiani [1986] and, subsequently, (ii) the case of first-order hyperbolic systems with boundary control in Chang and Lasiecka [1986] [and in Chang's Ph.D. thesis at the University of Florida]. The abstract setting of Triggiani [1996] extracts and singles out essential and intrinsic dynamical properties – assumptions (H.1) through (H.7) – common to the two aforementioned classes, as well as to the other hyperbolic and Petrowski-type problems of this chapter. Additional candidates include the system of elasticity, Maxwell equations, etc. The abstract proofs of Sections 10.3 and 10.4 are modeled after the treatment of Lasiecka and Triggiani [1986], except that here they are given in a semigroup framework, while in Lasiecka and Triggiani [1986] they were given within the framework of cosine operator theory, since – for simplicity of exposition (not mathematics) – only the position was penalized in the cost functional, not the velocity. An integral, essential part of the present abstract treatment is the optimal regularity theory of mixed PDE problems of hyperbolic or Petrowsky type, of which Sections 10.5 through 10.9 give an enlightening, but nonexhaustive, sample; it is this regularity theory, mostly of recent origin (see Triggiani [1999] for a most recent contribution) that motivates and ultimately justifies the abstract setting. Indeed, all abstract hypotheses, (H.1) through

(H.7), are *instrinsic dynamical properties* of the relevant PDE mixed problems. No artificial assumptions are imposed on the pair $\{A, B\}$ in (10.1.1).

Section 10.5. Regularity Theory

In the critical case $\dim \Omega \geq 2$ (the case $\dim \Omega = 1$ is elementary; see Remark 10.5.10.1), the basic interior regularity result (10.5.3.7)–(10.5.3.9) with boundary nonhomogeneous Dirichlet term $u \in L_2(\Sigma)$ was first obtained in Lasiecka and Triggiani [1981; 1983], in a two-step procedure. First, Lasiecka and Triggiani [1981] obtained $\{w, w_t\} \in L_2(0, T; L_2(\Omega) \times H^{-1}(\Omega))$: In the case of special geometries (parallelopipeds and sphere), by a sharp Fourier analysis (or Laplace analysis) carried out on the abstract operator model (10.5.4.22), once expressed in terms of the appropriate eigenfunction expansions; in the general (time-independent) case, by pseudo-differential energy methods. The original $L_2(0, T; L_2(\Omega) \times H^{-1}(\Omega))$ result (or, L_p, $2 \leq p < \infty$) of Lasiecka and Triggiani [1981] was subsequently boosted in time in Lasiecka and Triggiani [1983] to $C([0, T]; L_2(\Omega) \times H^{-1}(\Omega))$ as in (10.5.3.8), while preserving the space regularity, by means of an operator-theoretic proof such as the one in Chapter 7, Theorem 7.3.1. In the process, the equivalent (dual) trace regularity result (10.5.3.4) of Theorem 10.5.3.1 for the (boundary) homogeneous problem (10.5.3.1) was obtained in [Lasiecka, Triggiani, 1983, Theorem 2.1 and Remark 2.1], by virtue of a duality analysis such as that of Chapter 7, Theorem 7.2.1 with $p = 2$], except in the cosine-operator framework, rather than in the semigroup model (refer to Eqn. (10.5.4.28)). Instead, in the work of J. L. Lions [1983] (and references therein), the reverse approach was taken: first a proof by multiplier methods of the trace regularity (10.5.3.4) of the homogeneous problem (10.5.3.1), and then, by duality, a proof of the interior regularity (10.5.3.8) of the mixed problem (10.5.1.3). A general treatment for problem (10.5.1.3) that provides also "higher level" and "lower level" regularity results such as Theorem 10.5.7.1, Theorem 10.5.8.1 (or Theorem 10.5.13.1) was subsequently given in Lasiecka et al. [1986] and in [Lasiecka, Triggiani, 1986, Section 10.3]. Prior literature on the mixed problem (10.5.1.3) includes: [Lions, Magenes, 1972, Vols. I and II], who gave the (weaker) most updated account available in the late 1960s, as well as Sakamoto [1970], who studied the regularity of the hyperbolic mixed problem (10.5.1.3) only for $u \in H^k(\Sigma)$, $k \geq 1$, on the half-space, by means of pseudo-differential energy methods. The approach that seeks first the appropriate trace regularity of the (boundary) homogeneous problem, and then, by duality or transposition, the interior regularity of the corresponding mixed problem, has since been recognized as being simpler and more convenient than the other way around, also in other mixed problems, such as those in Sections 10.7 through 10.9. The proofs of regularity given in these sections reflect this philosophy. The key multiplier $h \cdot \nabla \phi$ used in Section 10.5.10 after Lions [1983] and Lasiecka et al. [1986] goes back, apparently, to Rellich, for elliptic problems. This technique has served also as a starting point for obtaining the reversed (continuous observability) inequalities or the dissipation inequalities for uniform stabilization problems at least for canonical models; see Notes at the end of Chapter 12.

Operator Models

The abstract operator formulation for second-order hyperbolic equations of Sections 10.5.4 and 10.5.5 was introduced by Triggiani [1978] and pursued further in Lasiecka and Triggiani [1981]. Since then, this abstract setting has been successfully used by these authors in dealing with a variety of problems, including: regularity in the Dirichlet case Lasiecka and Triggiani [1981; 1983] and regularity in the Neumann case (papers by Lasiecka and Triggiani quoted in the reference list of Chapter 8); strong and uniform stabilization; eigenvalue and Riesz basis assignment by boundary feedback; exact controllability; and quadratic cost problems and Riccati equations. In particular, formulae such as (10.5.4.28) and (10.5.4.30) were first given in Lasiecka and Triggiani [1983]. They motivated, since then the "abstract trace condition" (H.1) = (10.1.6), by theorems such as Theorem 10.5.3.1. These operator formulations were soon introduced by Lasiecka and Triggiani also in the study of the first-order systems, Kirchoff equations, Euler–Bernoulli equations, Schrödinger equations, etc. (see below). Under most boundary conditions, assumption (H.1) = (10.1.6) continues to hold true for Y being an explicitly identified space of optimal regularity. Many examples are given in Lasiecka and Triggiani [1991(c)] and will be included in Chapter 13 of Volume 3, along with new ones (e.g., spherical shells). The introduction of the second-order models of this section, in both differential and integral form, benefited from the original contribution of Balakrishnan [1976], who pushed further an idea of Fattorini [1968] (for both first- and second-order equations), by a simple yet critical integration by parts. This was noted also in the Notes at the end of Chapter 3. Indeed, in Balakrishnan [1976] the abstract model in integral form $A \int_0^t \exp[A(t - \tau)] Du(\tau) \, d\tau$ is introduced for a parabolic problem with u in the Dirichlet BC and where the operator $A = \Delta$ with zero BC generates an analytic semigroup $\exp(At)$. It was then proved in Balakrishnan [1976] in the special case of Ω being a square (via eigenfunctions expansion) and subsequently in Washburn [1979] in the general analytic semigroup case by using the theory of intermediate spaces as in Butzer and Berens [1967] that the bound $(\#) : |Ae^{At}D| = \mathcal{O}(t^{-\frac{3}{4}-\epsilon})$ holds for $0 < t \leq 1$, $\forall \epsilon > 0$, in the $L_2(\Omega)$-uniform norm. That this important bound $(\#)$ can be directly and more easily proved by using (10.5.4.11) was observed by Triggiani [1979; 1980a, b]. Note that (10.5.4.11) is based on elliptic regularity plus identification of domains of appropriate fractional powers with Sobolev spaces. In addition to reproving $(\#)$, this introduction of fractional powers was proven to be a useful technical device in the analysis. We refer to the Notes of Chapter 3.

At the end of these Notes, we shall point out the drastic difference in the role played by abstract models in hyperbolic/plate-like mixed problems, second-order in time, or Schrödinger equations, on the one hand, and parabolic mixed problems (first-order in time) on the other, regarding the preliminary basic issue of regularity theory.

Section 10.6

The basic Theorem 10.6.3.1, part (a), Eqn. (10.6.3.2), giving interior and boundary regularity of the mixed problem (10.6.1.5) with boundary term $u \in L_2(0, T; L_2(\Gamma))$ is due to Kreiss [1970], at least in the case of zero initial datum $y_0 = 0$, augmented by Ralston's note [Ralston, 1971]. The proof is by pseudo-differential methods and introduces the so-called Kreiss's symmetrizer, defined in Step 1 of Section 10.6.10. The general case $y_0 \in L_2(\Omega)$ was completed by Rauch [1971; 1972], who also proved corresponding higher regularity results (the Differentiability Theorem of [Rauch, 1972, p. 272]), such as Theorem 10.6.3.1, part (b), Eqn. (10.6.3.5). This settled the outstanding problem on the well-posedness of nonsymmetric, nondissipative, first-order hyperbolic systems with noncharacteristic boundary. Treatments of this theory in book form include those of Chazarian and Piriou [1977; 1982]. Prior literature includes Friedrichs and Lax [1965; 1968] and Lax and Phillips [1980], etc.

Sections 10.6.4 through 10.6.11 (including the operator model) follow closely the work of Chang and Lasiecka [1986], who present a parallel treatment of the case of second-order hyperbolic equations in Lasiecka and Triggiani [1986]. The control problem for a *symmetric* hyperbolic system in *one* space dimension was studied previously by Russell [1973] by means of the method of characteristics (which is known to fail in the case of several space variables).

Section 10.7

The basic trace regularity result (10.7.3.5) for the (boundary) homogeneous problem (10.7.3.1) in Theorem 10.7.3.1 as well as the dual interior regularity result (10.7.3.7) for the mixed problem (10.7.1.1) were given in Lasiecka and Triggiani [1991(a)]. The respective proofs in Section 10.7.10 (by multipliers or energy methods) and in Section 10.7.11 (by duality or transposition) follow closely this reference, which also contains the abstract operator formulation of Sections 10.7.4 and 10.7.5 (this is reproduced also in Lasiecka and Triggiani [1991(c)]). Besides providing regularity results for problem (10.7.1.1), Lasiecka and Triggiani [1991(a)] give an exact controllability result and, more demandingly, a uniform stabilization result via a boundary feedback. The remaining Sections 10.7.6–10.7.9, as well as 10.7.12–10.7.13, contain new material, after Triggiani [1996]: in particular, the regularity results of Theorem 10.7.7.1 and Theorem 10.7.8.1. Regarding the latter, Remark 10.7.13.1 points out new technical phenomena over, and some basic differences with, the case of second-order hyperbolic equations of Section 10.5 in establishing higher level trace regularity results in the space variable in both cases.

Section 10.8

The basic trace regularity result (10.8.3.5) for the (boundary) homogeneous problem (10.8.3.1) in Theorem 10.8.3.1 as well as the dual interior regularity result (10.8.3.7) for the mixed problem (10.8.1.1) were given, independently, in Lions [1988] and in Lasiecka and Triggiani [1989(a)]. The proofs in Section 10.8.10 (by multipliers or energy methods) and in Section 10.8.11 (by duality or transposition) follow closely Lasiecka and Triggiani [1989(a)], which contains also the abstract operator formulation of Sections 10.8.4 and 10.8.5 (this is reproduced also in Lasiecka and Triggiani [1991(c)]). The remaining Sections 10.8.6–10.8.9, as well as 10.8.12–10.8.13, contain new material, after Triggiani [1999]: in particular, the regularity results of Theorem 10.8.7.1 and Theorem 10.8.8.1. Regarding the latter, Remark 10.8.13.1 points out the new technical phenomena and the basic differences that arise over the case of second-order hyperbolic equations of Section 10.5.

Finally, we point out that Euler–Bernoulli equations (10.8.1.1) with different boundary conditions such as those in [Lasiecka, Triggiani, 1991(c), Section 7.2 and Section 7.6] can also be treated in a similar manner Camurdan and Triggiani [1995]. They include, for example, the following cases:

$$w|_\Sigma \equiv 0 \quad \text{and} \quad \left.\frac{\partial w}{\partial \nu}\right|_\Sigma = u; \ w|_\Sigma = u \quad \text{and} \quad \left.\frac{\partial w}{\partial \nu}\right|_\Sigma \equiv 0;$$

$$w|_\Sigma \equiv u \quad \text{and} \quad \Delta w|_\Sigma = 0; \ w|_\Sigma = 0 \quad \text{and} \quad [\Delta w + (1-\mu)B_1 w]_\Sigma = u,$$

the last one being two-dimensional, where B_1 is a boundary operator such that u then acts as a (physical) bending moment. Relevant references to deal with these cases in a parallel manner are Lasiecka and Triggiani [1991(c)] and references quoted therein.

Section 10.9

The basic trace regularity result (10.9.3.4) for the (boundary) homogeneous problem (10.9.3.1) in Theorem 10.9.3.1 as well as the dual interior regularity (10.9.3.6) for the mixed problem (10.9.1.1) were given in Lasiecka and Triggiani [1991(b)]. The respective proofs in Section 10.9.10 (by multipliers or energy methods) and in Section 10.9.11 (by duality or transposition) follow closely this reference, which contains also the abstract operator formulation of Sections 10.9.4 and 10.9.5 (this is reproduced also in Lasiecka and Triggiani [1991(c)]). Besides providing regularity results for problem (10.9.1.1), Lasiecka and Triggiani [1991(b)] give also an exact controllability result, and, more demandingly, a uniform stabilization result via a boundary feedback. The remaining Sections 10.9.6–10.9.9, as well as 10.9.12–10.9.13, contain new material, after Triggiani [1999]: in particular, the regularity results of Theorem 10.9.7.1 and

Theorem 10.9.8.1. Regarding the latter, Remark 10.9.13.1 points out the new technical phenomena and the basic differences that arise over the case of second-order hyperbolic equations of Section 10.5, in establishing higher-level trace regularity results in the space variable.

Finally, in conclusion, we remark again that given a mixed problem such as the illustrative examples of the present chapter and those of [Lasiecka, Triggiani, 1991(c), Section 10.7] there are many, in fact infinitely many, choices possible for the selection of the spaces U_θ and Y_θ. For example, other choices are studied in Camurdan and Triggiani [1995].

Abstract Models and Regularity Theory. Parabolic Versus Hyperbolic/Petrowsky Case

Parabolic Case

The regularity theory of (linear) parabolic mixed problems has been in essentially optimal shape for quite some time, for example, Lions and Magenes [1972] within the Hilbert space framework. Later, a purely operator approach was given, yielding an explicit representation formula as in Balakrishnan [1976] in the general case where the free problem generates an analytic semigroup. This representation – when merged with elliptic theory and Sobolev theory as pointed out in Triggiani [1979; 1980a, b] – permits one to reobtain the preceding theory (with some improvements) and allows one the flexibility to work outside the L_2-framework, as in the work of Lasiecka [1980]. This approach has been used in Chapters 1 through 6.

Hyperbolic/Platelike Case

The situation is drastically different in this case. Here, the first crucial step or building block of a regularity theory comes from purely partial differential equation methods (energy or multiplier methods, either in differential form or else in pseudo-differential form), which were brought to bear on these problems only rather recently, beginning with second-order hyperbolic equations with Dirichlet control (Lasiecka and Triggiani [1981; 1983]; Lions [1962]; Lasiecka et al. [1986]; Sakamoto [1970]). (The case dim $\Omega = 1$ is elementary, see Remark 10.5.10.1 in the Dirichlet case and Chapter 9, Section 9.9.4 in the Neumann case.) In the case of hyperbolic/plate/Schrödinger mixed problems, abstract operator methods provide useful tools only at a subsequent level (for higher/lower data, duality or transposition, etc.), after a key preliminary regularity result (typically a trace regularity property of the corresponding homogeneous problem) has been obtained from energy methods. These "trace regularity" properties can then be abstracted and unified in the "abstract trace property" (H.1) = (1.6), first introduced in Lasiecka and Triggiani [1983] for second-order hyperbolic equations, for example, Eqns. (10.5.4.28) and (10.5.4.30).

Glossary of Selected Symbols for Chapter 10

A, B, R, J 276–277
L_s, L_s^* 277–278
$U_\theta, Y_\theta, Y_s^\pm, U_s^\pm$ 278, 279, 309, 337, 356, 383, 384, 404
$\mathcal{U}^\theta[s, T], \mathcal{Y}^\theta[s, T]$ 278

References and Bibliography

P. M. Anselone, *Collectively Compact Operator Approximation Theory,* Prentice-Hall, 1971.

A. V. Balakrishnan, *Applied Functional Analysis*, Springer-Verlag, 1976.

P. L. Butzer and H. Berens, *Semigroups of Operators and Approximation*, Springer-Verlag, 1967.

M. Camurdan and R. Triggiani, Higher level regularity of Petrowsky systems with application to Differential Riccati Equations Report, University of Virginia, August 1995.

S. Chang and I. Lasiecka, Riccati equations for nonsymmetric and nondissipative hyperbolic systems with L_2-boundary controls, *J. Math. Anal. Appl.* **116** (1986), 378–414.

J. Chazarian and A. Piriou, Problemes mixtes hyperboliques, in *Hyperbolicity*, Liguori Editore, 1977, C.I.M.E., 1976.

J. Chazarian and A. Piriou, *Introduction to the Theory of Linear Partial Differential Equations*, North-Holland, Studies in Mathematics and Its Applications, 1982.

H. O. Fattorini, Ordinary differential equations in linear topological spaces I, II, *J. Diff. Eqn.* **5** (1968), 72–105; and **6** (1969), 50–70.

H. O. Fattorini, Boundary control systems, *SIAM J. Control* **6** (1968), 349–385.

H. O. Fattorini, *Second Order Linear Differential Equations in Banach Spaces*, North-Holland, 1985.

K. O. Friedrichs and P. D. Lax, Boundary value problems for first order operators, *Comm. Pure Appl. Math.* **18** (1965), 355–388.

K. O. Friedrichs and P. D. Lax, On symmetrizable differential operators, *Proc. Symp. Pure Math.* **10**, American Mathematics Society, (1968), 128–137.

D. Fujiwara, Concrete characterizations of domains of fractional powers of strongly elliptic differential operators of the second order, *Proc. Japan Acad.* **43** (1967), 82–86.

L. Gärding, Solution directe du problème de Cauchy pour les hyperboliques, *Coll. Int. C.N.R.S.*, Nancy (1956), 71–90.

P. Grisvard, Caracterization de qualques espaces d'interpolation, *Arch. Rational Mech. Anal.* **25** (1967), 40–63.

R. Hersh, Mixed problems in several variables, *J. Math. Mech.* **12** (1963), 317–334.

H. O. Kreiss, Initial boundary value problems for hyperbolic systems, *Comm. Pure Appl. Math.* **13** (1970), 277–298.

I. Lasiecka, Unified theory for abstract parabolic boundary problems – A semigroup approach, *Appl. Math. Optim.* **6** (1980), 287–333.

I. Lasiecka, J. L. Lions, and R. Triggiani, Nonhomogeneous boundary value problems for second order hyperbolic operators, *J. Math. Pures Appl.* **65** (1986), 149–192.

I. Lasiecka and R. Triggiani, A cosine operator approach to modelling $L_2(0, T; L_2(\Omega))$ boundary input hyperbolic equations, *Appl. Math. and Optim.* **7** (1981), 35–83.

I. Lasiecka and R. Triggiani, Regularity of hyperbolic equations under $L_2(0, T; L_2(\Gamma))$-Dirichlet boundary terms, *Appl. Math. Optim.* **10** (1983), 275–286.

I. Lasiecka and R. Triggiani, Riccati equations for hyperbolic partial differential equations with $L_2(\Sigma)$-Dirichlet boundary terms, *SIAM J. Contr. Optim.* **24** (1986), 884–926. (A preliminary version, entitled "An L_2-theory for the quadratic optimal cost problem of hyperbolic equations with control in the Dirichlet BC," in Springer-Verlag *Lecture Notes* *in Control and Information Sciences* **54**, 138–152, Proceedings of the Conference on Control Theory for Distributed Parameter Systems, held at Vorau, Austria, July 1982.)

I. Lasiecka and R. Triggiani, Exact controllability of the Euler–Bernoulli equation with $L_2(\Sigma)$-control only in the Dirichlet boundary condition, *Atti Accademia Nazionale dei Lincei*, Anno CCCLXXXIV Rendiconti Classe di Scienze fisiche, mathematiche e naturali, Roma (1988), 35–42.

I. Lasiecka and R. Triggiani, Regularity theory for a class of nonhomogeneous Euler–Bernoulli equations: A cosine operator approach, *Bollettino Unione Matematica Italiana* **7**, (3-B) (1989)(a), 199–228. (Also, invited paper for special volume. *Topics in Mathematical Analysis*, volume dedicated to A. L. Cauchy; T. M. Rassias, ed., World Scientific, 1989, pp. 623–657.)

I. Lasiecka and R. Triggiani, Exact controllability of the Euler–Bernoulli equation with controls in the Dirichlet and Neumann boundary conditions: A non-conservative case, *SIAM J. Control Optim.* **27** (1989)(b), 330–373. (Also, preliminary version in *Semigroups and Applications*, Marcel Dekker Lecture Notes **116** (1988), 241–261.)

I. Lasiecka and R. Triggiani, Exact controllability of the Euler–Bernoulli equation with boundary controls for displacement and moment, *J. Math. Anal. Appl.* **146** (1990), 1–33. (Also preliminary version as an invited paper for special volume, *Topics in Mathematical Analysis*, volume dedicated to A. L. Cauchy; T. M. Rassias, ed., World Scientific 1989, pp. 577–622.)

I. Lasiecka and R. Triggiani, Exact controllability and uniform stabilization of Kirchoff plates with boundary controls only in $\Delta w|_\Sigma$, *J. Diff. Eqn.* **93** (1991)(a), 62–101. Preliminary version in *Semigroup and Evolution Equations*, Marcel Dekker Lecture Notes in Pure and Applied Mathematics **135**, 267–295.

I. Lasiecka and R. Triggiani, Optimal regularity, exact controllability and uniform stabilization of the Schrödinger equation, *Diff. Int. Eqn.* **5** (1991)(b), 521–535.

I. Lasiecka and R. Triggiani, *Differential and Algebraic Riccati Equations with Applications to Boundary/Point Control Problems: Continuous Theory and Approximation Theory*, **164**, Springer-Verlag Lecture Notes in Information and Control 1991(c), 160 pp.

I. Lasiecka and R. Triggiani, Recent advances in regularity theory of second-order hyperbolic mixed problems, in book series *Dynamics Reported*, **3** New Series (1994), 104–162, C. K. R. T. Jones, U. Kirchgraber, and H. O. Walther, managing eds.

I. Lasiecka and R. Triggiani, A sharp trace result on a thermo-elastic plate equation with copuled hinged/Neumann Boundary conditions, *Discrete and Continuous DYNAMICAL SYSTEMS*, **5** (1999), 585–598.

P. D. Lax and R. S. Phillips, Local boundary conditions for dissipative symmetric linear differential operators, *Comm. Pure Appl. Math.* **13** (1980), 427–455.

J. L. Lions, Espaces d'interpolation et domains de puissances fractionnaires d'operateurs, *J. Math. Soc. Japan* **14** (1962), 233–241.

J. L. Lions, *Control of Singular Distributed Systems*, Gauthier Villars, 1983.

J. L. Lions, Exact controllability, stabilization and perturbations for distributed systems, *SIAM Review* **30** (1988), 1–68.

J. L. Lions, *Controlabilite Exacte et Stabilisation de Systemes Distribues*, vol. 1, Masson, 1988.

J. L. Lions and J. Magenes, *Non-Homogeneous Boundary Value Problems and Applications*, Vol. 1, Springer-Verlag, 1972.

S. G. Miklin, *Mathematical Physics, an Advanced Course*, North Holland Series in Applied Mathematics and Mechanics, 1970.

J. Ralston, Note on a paper of Kreiss, *Comm. Pure Appl. Math.* **24** (1971), 759–762.

J. Rauch, Energy inequalities for hyperbolic initial boundary value problems, thesis, New York University, February 1971.

J. Rauch, L_2 is a continuable initial condition for Kreiss' mixed problems, *Comm. Pure Appl. Math.* **25** (1972), 263–285.

J. Rauch, General theory of hyperbolic mixed problems, Proceedings of Symposium in Pure Mathematics, vol. 23, Am. Math. Soc., 1973.

D. L. Russell, Quadratic performance criteria in boundary control of linear symmetric hyperbolic systems, *SIAM J. Control* **11** (1973), 475–509.

R. Sakamoto, Mixed problems for hyperbolic equations, I and II, *J. Math. Kyoto Univ.* **10** (1970), 349–373; 403–417.

M. Sova, *Cosine Operator Functions*, Rozprawy Matematyczne, **XLIX**, 1966.

C. C. Travis and G. Webb, Second order differential equations in Banach space, in *Nonlinear Equations in Abstract Spaces*, V. Lakshmikantham, ed., Academic Press, 1978, 331–361.

R. Triggiani, A cosine operator approach to modeling boundary input problems for hyperbolic systems, Proceedings of 8th IFIP Conference on Optimization Techniques, University of Wurzburg, West Germany, September 1977, Springer-Verlag Lectures Notes on Control Sciences, no. 6 (1978), 380–390.

R. Triggiani, On Nambu's boundary stabilizability problem for diffusion processes, *J. Diff Eqn.* **33** (1979), 189–200. (Preliminary versions in Proceedings International Conference on Recent Advances in Differential Equations, Miramare-Trieste (Italy), August 1978, Academic Press, and Proceedings International Conference on Systems Analysis, I.R.I.A., Paris, 1978.)

R. Triggiani, Well-posedness and regularity of boundary feedback parabolic systems, *J. Diff. Eqn.* **36** (1980)(a), 347–362.

R. Triggiani, Boundary feedback stabilizability of parabolic equations, *Appl. Math. Optim.* **6** (1980)(b), 201–220.

R. Triggiani, An abstract framework for differential Riccati equations arising in the quadratic optimal control of hyperbolic/Petrowsky-type equations, with boundary control, *Abstract Appl. Anal.* **1** (1996), 435–484.

R. Triggiani, Higher level interior and boundary regularity results of the Euler–Bernoulli equation with application to differential Riccati equations in optimal control, presented at International Conference on Applied Analysis in the Aegean Archipelagos, Samos, Greece, July 1996; *Numerical Functional Analysis & Optimization*, Vol. 20, n. 3–4, 1999, 367–386.

D. C. Washburn, A semigroup theoretic approach to modeling of input problems, *SIAM J. Control* **17** (1979), 652–671.

Index

Index

Index